The Ecology and Management of Wood in World Rivers

The Ecology and Management of Wood in World Rivers

Edited by

Stan V. Gregory

Department of Fisheries and Wildlife, Oregon State University
Corvallis, Oregon 97331, USA

Kathryn L. Boyer

USDA Natural Resources Conservation Service
Wildlife Habitat Management Institute, Department of Fisheries and Wildlife
Oregon State University, Corvallis, Oregon 97331, USA

Angela M. Gurnell

Department of Geography, King's College London
Strand, London WC2R 2LS, UK

American Fisheries Society Symposium 37

International Conference on Wood in World Rivers
held at Oregon State University, Corvallis, Oregon
23–27 October 2000

American Fisheries Society
Bethesda, Maryland
2003

The American Fisheries Society Symposium series is a registered serial. Suggested citation formats follow.

Entire book

Gregory, S. V., K. L. Boyer, and A. M. Gurnell, editors. 2003. The ecology and management of wood in world rivers. American Fisheries Society, Symposium 37, Bethesda, Maryland.

Chapter within the book

Abbe, T. B., A. P. Brooks, and D. R. Montgomery. 2003. Wood in river rehabilitation and management. Pages 367–389 *in* S. V. Gregory, K. L. Boyer, and A. M. Gurnell, editors. The ecology and management of wood in world rivers. American Fisheries Society, Symposium 37, Bethesda, Maryland.

Printed in the United States of America on acid-free paper.

Library of Congress Control Number 2003112769
ISBN 1-888569-56-5
ISSN 0892-2284

American Fisheries Society website address: www.fisheries.org

American Fisheries Society
5410 Grosvenor Lane, Suite 110
Bethesda, Maryland 20814-2199
USA

Contents

Preface

Toward a Synthesis of Knowledge of Wood in World Rivers

Streamside forests provide inputs of large wood, an important element of conservation of biological diversity at landscape scales. Historically, large wood has been an important component in streams and rivers throughout the world. Research results have been applied by natural resource managers to protect and restore biodiversity, fisheries, and aquatic ecosystem functions. Wood is placed in streams as part of restoration efforts to increase channel complexity and associated ecological functions, and streamside forests are actively managed to supply large wood to river networks. Transferring research results from one region of the world to others has raised both technical and cultural questions.

To address these questions, researchers at Oregon State University, in cooperation with the USDA Forest Service and Natural Resources Conservation Service, the University of Washington, and the University of Birmingham, organized an international conference of 30 invited experts from around the world and more than 400 participants. The "First International Conference on Wood in World Rivers" was held 23–27 October 2000 on the campus of Oregon State University, in Corvallis, Oregon. The purpose of this conference was to

(1) synthesize the status of knowledge of the physical dynamics and ecological interactions of large wood in streams and rivers in different geographical regions,
(2) create a framework for interpreting and applying the results of research in different geographical regions and management systems,
(3) identify different management systems for large wood in rivers,
(4) assess physical and biological responses of large wood in stream restoration,
(5) explore links between primary information of the physical and ecological dynamics of large wood, resource management systems, and the communities and cultures in which they are applied.

During the conference, 30 invited scientists and more than 100 contributing scientists presented current knowledge of the role of wood in aquatic ecosystems and implications for landscape planning and management. Most of the invited scientists have contributed a chapter to this book, creating a broad overview of what is known about wood in rivers and the context for applying research results from different regions, management systems, and cultures. Plenary speakers David Montgomery and Herve Piégay also co-edited a special issue of Geomorphology (Volume 51, Issues 1–3) entitled *Interactions between wood and channel forms and processes.* This volume presented another 12 research papers of scientists who contributed papers at the Wood Conference.

This book synthesizes worldwide research on the ecology and management of wood in world rivers. Chapter 1 provides an overview of the direction wood research has taken over time and suggestions for future research trajectories that focus on long-term river conservation needs. Chapters 2 through 6 focus on the geomorphic aspects of wood in rivers. Biological implications of river wood are treated in Chapters 7 through 13. Chapters 14 through 17 consider larger scale perspectives related to the dynamic nature of wood in rivers, including modeling. Resource management applications resulting from wood research are covered in chapters 18–22. And finally, Chapter 23 offers a perspective on the human dimensions of wood in rivers and provides insights on how to tailor restoration and management of large wood to different landscapes, rivers, forests, regions, and countries of the world.

One of the first challenges of the conference was the plethora of names and acronyms used for wood by resource professionals. In his plenary address to the conference, Dr. Ken Gregory provided an insightful account of the numerous terms used for wood in streams and rivers and their applications in different disciplines. The term "debris" was first used to refer to the wood slash and debris left on the land and in the streams after timber harvest. Unfortunately, the term "woody debris" negatively connotes garbage or trash to the general public. One of the editors, Stan Gregory, thus discouraged further use of the terms *large woody debris, coarse woody debris, large organic debris,* etc. by conference participants. In the interest of consistency and effective communication, the editors and authors in this book use the word "wood" and encourage resource professionals to adopt this simpler and more accurate term.

Several speakers and members of the audience called for efforts to standardize definitions and methods for measuring wood in streams. Throughout this book, the authors reported the operational definitions of wood and the units of measures in different studies. These definitions and units of measure differ for different sizes of streams, ecological or physical applications, and regional conventions. Abundance of wood can be quantified as numbers, volume, mass, or surface area of wood and expressed as amount per area, volume, or length of stream. All of these measures are appropriate and offer different insights about the potential functions of wood in streams and rivers. For example, a study of fish habitat may select wood numbers per length of river as an appropriate measure of wood, a nutrient budget study may express wood abundance as mass per area of stream, and a study of microbial ecology may express the amount of wood as surface area of wood available for colonization per area of stream. All of these are appropriate when determined by the objectives of the study, characteristics of wood, and sizes of the stream and river systems. We encourage resource professionals to carefully select the most accurate and informative measures of wood, maintain consistency where appropriate, but resist mindless uniformity without regard to management applications and research objectives.

One of the most challenging and exciting outcomes of the conference was the recognition of the diversity of scientific and cultural perspectives of wood in streams and rivers around the world. This richness of world views was noted by many of the authors and explored more thoroughly in the two chapters by Ken Gregory and Geoff Petts and Robin Welcomme. The international nature of the conference and its participants is apparent in this book with authors from nine countries and four continents. Their diverse backgrounds, cultures, and native languages offer the reader a global perspective and alternative world views about river wood not previously available in one volume. Their universal recognition of the importance of large wood to the ecology and management of the world's rivers is clear and will hopefully compel scientists, managers, and citizens to consider wood, and the forests from which it comes, an integral part of conservation and restoration at global scales.

Stan Gregory, Oregon State University
Kathryn Boyer, Natural Resources Conservation Service
Angela Gurnell, King's College London

Acknowledgments

The idea for the "First International Conference on Wood in World Rivers," held in October 2000, originated in animated discussions between Jim Sedell, Geoff Petts, Angela Gurnell, and Stan Gregory. These conspirators quickly enlisted the guidance of Glen Contreras, Kathryn Boyer, Pete Bisson, Mel Warren, Andy Dolloff, Susan Bolton, and Bob Bilby as conference coordinators. In addition to the experience and advice of the coordinators, the extraordinary organizational skills of Kelly Wildman contributed to the overall success of the conference. And throughout the development of the conference and book, from inspiration to completion, Jim Sedell provided advice and encouragement.

Many of the authors of this book individually acknowledge those colleagues who assisted with specific chapter development and peer reviews. We thank all of the authors who both contributed to this book and to the success of the conference that precipitated its publication. Our colleagues who speak French, German, Spanish, Polish, and Japanese humble us with their mastery of English and their ability to write articulate, fluent chapters in English. Their respective usage of some words and phrases may seem curious to American readers, yet maintains the international essence and perspective that made the Wood Conference so successful.

Most of the authors also graciously reviewed chapters at the request of the editors. In addition, we thank the following scientists for their reviews of chapters: Linda Ashkenas, Norm Anderson, Robert Anthony, Nick Aumen, Fred Benfield, Robert Beschta, John Bolte Christian Braudrick, Alan Covich, Kurt Fausch, Craig Fischenich, Ian Fleming, Jerry Franklin, Steve Golladay, Dixon Landers, Tom Nickelson, Sherri Johnson, Edward Keller, Mike Maki, Michael Manga, Richard Marston, Christine May, Brenda McComb, Art McKee, Robert Naiman, William Pearcy, Doug Shields, Courtland Smith, Eric Tabacchi, and Jack Webster,

Martha Brookes provided expert editorial assistance with chapter texts. Kelly Wildman contributed her outstanding technical acuity in enhancing the visual quality of graphics. Paulo Petry offered expertise and insights on Amazonian fishes and habitats not yet available in published literature.

The editorial staff at the American Fisheries Society, especially Debby Lehman and Aaron Lerner, provided patience and encouragement during chapter development and excellent guidance during the editorial stages of the book's production.

Funding for the "International Conference on the Ecology and Management of Wood in World Rivers" and production of this book was generously provided by the U.S. Forest Service and Natural Resources Conservation Service. Additional funds for the conference were provided by the Western Division of the American Fisheries Society, U.S. Bureau of Land Management, U.S. Geological Survey, Oregon Watershed Enhancement Board, Boise Cascade Corporation, and Oregon State University. We would like to thank Linda Yung, Larry Schmidt, Pete Heard, Ed Shepard, Michael Mac, Ken Bierly, Bob Danehy, Pete Bisson, and Erik Fritzell for their supportive roles in acquiring financial support for this effort.

Lastly, we wish to acknowledge the scientists and managers that participated in the conference, especially those that presented their research findings to the more than 400 participants from around the world. This book is dedicated to the many natural resource professionals in countries around the world who have explored the role of wood in streams and rivers.

The editors

Symbols and Abbreviations

The following symbols and abbreviations may be found in this book without definition. Also undefined are standard mathematical and statistical symbols given in most dictionaries.

A	ampere
AC	alternating current
Bq	becquerel
C	coulomb
°C	degrees Celsius
cal	calorie
cd	candela
cm	centimeter
Co.	Company
Corp.	Corporation
cov	covariance
DC	direct current; District of Columbia
D	dextro (as a prefix)
d	day
d	dextrorotatory
df	degrees of freedom
dL	deciliter
E	east
E	expected value
e	base of natural logarithm (2.71828…)
e.g.	(exempli gratia) for example
eq	equivalent
et al.	(et alii) and others
etc.	et cetera
eV	electron volt
F	filial generation; Farad
°F	degrees Fahrenheit
fc	footcandle (0.0929 lx)
ft	foot (30.5 cm)
ft^3/s	cubic feet per second ($0.0283\ m^3/s$)
g	gram
G	giga (10^9, as a prefix)
gal	gallon (3.79 L)
Gy	gray
h	hour
ha	hectare (2.47 acres)
hp	horsepower (746 W)
Hz	hertz
in	inch (2.54 cm)
Inc.	Incorporated
i.e.	(id est) that is
IU	international unit
J	joule
K	Kelvin (degrees above absolute zero)
k	kilo (10^3, as a prefix)
kg	kilogram
km	kilometer
l	levorotatory
L	levo (as a prefix)
L	liter (0.264 gal, 1.06 qt)
lb	pound (0.454 kg, 454g)
lm	lumen
log	logarithm
Ltd.	Limited
M	mega (10^6, as a prefix); molar (as a suffix or by itself)
m	meter (as a suffix or by itself); milli (10^{23}, as a prefix)
mi	mile (1.61 km)
min	minute
mol	mole
N	normal (for chemistry); north (for geography); newton
N	sample size
NS	not significant
n	ploidy; nanno (10^{29}, as a prefix)
o	ortho (as a chemical prefix)
oz	ounce (28.4 g)
P	probability
p	para (as a chemical prefix)
p	pico (10^{212}, as a prefix)
Pa	pascal

pH negative log of hydrogen ion activity
ppm parts per million
qt quart (0.946 L)
R multiple correlation or regression coefficient
r simple correlation or regression coefficient
rad radian
S siemens (for electrical conductance); south (for geography)
SD standard deviation
SE standard error
s second
T tesla
tris tris(hydroxymethyl)-aminomethane (a buffer)
UK United Kingdom
U.S. United States (adjective)
USA United States of America (noun)
V volt
V, Var variance (population)
var variance (sample)
W watt (for power); west (for geography)
Wb weber
yd yard (0.914 m, 91.4 cm)
α probability of type I error (false rejection of null hypothesis)
β probability of type II error (false acceptance of null hypothesis)
Ω ohm
μ micro (10^{26}, as a prefix)
′ minute (angular)
″ second (angular)
° degree (temperature as a prefix, angular as a suffix)
% per cent (per hundred)
‰ per mille (per thousand)

American Fisheries Society Symposium 37:1–19, 2003

The Limits of Wood in World Rivers: Present, Past, and Future

KEN J. GREGORY

Department of Geography, University of Southampton, Southampton S09 5NH, UK

Abstract.—A context is provided by focusing on research on wood in rivers in the present, the past and the future. Recent, **present** research has grown rapidly since 1967, terminology has expanded so that standardization of terms is now desirable, and six themes focus upon channel morphology, extent and significance of wood, process investigations, distribution and spatial pattern, dynamics, and channel management. Study of wood in **past** river systems, including paleohydrology, can show how past situations provide information additional to that from the period of continuous monitoring and furnish information about flow data, sensitive reaches, the temporal sequence and also indicate what is "natural" for restoration. In the **future**, knowledge of wood in rivers present and past should be used for guiding management especially where restoration is practiced. Community views must be obtained on the basis of meaningful explanations, a catchment-based holistic approach should be adopted wherever possible, and implications to be considered are suggested.

Introduction

It is a great honor to be asked to give the keynote address at the start of this world conference, but it is also a daunting prospect because subsequent papers will cover many aspects of wood in world rivers in more detail. However, I want to introduce major issues to provide a context, highlighting three aspects of the limits of research investigations of wood related to the range of disciplines concerned with the present, the past, and the future. I am not attempting to mirror Tom Stoppard's *Arcadia,* but I do want to show how the links between these three time frameworks can provide a context for this volume. These three aspects have been used before, for example, in the order of past, present, and future, in "Large woody debris in forested streams in the Pacific Northwest: past, present and future" (Bisson et al. 1987), but I want to use them in a rather different way, to focus on limits of investigation in science, while appreciating that, inevitably, my background in geomorphology may be showing.

The Present

Deciding when the "present" began is not easy. Early references to wood in rivers are available, such as Lobeck's description (Lobeck 1939) of the great raft of the Red River of NW Louisiana (see Keller and Macdonald 1995), allusion in Russian papers to woody alluvial accumulation (Chemekov 1951, 1959), and recognition of the damming effects of wood during flood events such as the Lynmouth flood (Dobbie and Wolf 1953). What is surprising is the lack of reference to wood in rivers in many early books and in the research literature. I naturally think of papers of geomorphological significance like the one by Zimmerman et al. (1967), which was extremely influential because their approach focused on the influence of vegetation, including in-channel wood, and on-channel morphology. It recognized two thresholds—at drainage areas of 2 and 10 km^2—, and it demonstrated spatial variations in relation to controls on morphology. This paper may have been influenced (F. J. Swanson, personal communication) by analysis of vegetation significance in geomorphology (Hack and Goodlett 1960). Subsequent influential papers (such as Swanston and Swanson 1976; Keller and Swanson 1979) utilized field data collected in the Pacific Northwest. Key ecological references (for example, Gregory et al. 1991) and the synthesis by 13 authors of the "Ecology of coarse woody debris in temperate ecosystems" (Harmon et al. 1986) were also extremely influential. Which books referred to these early papers and to the subject of wood in world rivers is of interest: whereas some mention the topic (such as Gregory and Walling 1973; Richards 1982), others (such as Morisawa 1968, 1985; Schumm 1977; Knighton 1984, 1998) do not. It has taken time for research on the subject to appear in books, so that even in 2000, only one indexed reference to woody debris appears in *Global Perspectives on River Conservation* (Boon et al. 2000). The lack of allusion to wood in world rivers is surprising in view of the demonstration that, although it may occupy only less than 2% of channel length, it can be responsible for half the total flow resistance (Manga and Kirchner 2000), account for 4% of the vertical drop in a channel long profile (Thompson 1995) and for 70% of the sediment stored in the channel. Wood dams can be densely distributed, with average spacing as little as every 2.8 m (Heede 1981).

Although research on wood in world rivers could have been more extensively referenced, in the three decades since 1967, interest in and publications on wood in world rivers have grown significantly. The impressive bibliography of *World Literature on Wood in Streams, Rivers, Estuaries and Riparian Areas* (Gregory et al. 2000), which includes 1,172 references, was used to compile Figure 1. This figure shows how the related research literature has grown substantially since the paper by Zimmerman et al. (1967). During this growth, some early papers avoided proposing terms, but "large organic material" was one used by Keller and Swanson (1979), and at least 15 other terms have been used subsequently (Table 1), some obviously related, and examples of their usage and definitions are cited (Table 1). A culmination of the growth of research (Figure 1) up to the millennium suggests that now could be the time to standardize terminology. An implication arising from this conference could be to standardize the use of terms (Table 1) with the benefit of 30 years' hindsight, as implied in Table 8 (implication 1). The diversity of terms could be one reason why tracing references to wood in world rivers in the research literature is so difficult; standardization could facilitate finding references in indexes and bibliographies, helping to raise awareness of the significance of the research. Growth of research (Figure 1) has been a great success story, and the significance of wood in world rivers is now better understood than some three decades ago when it

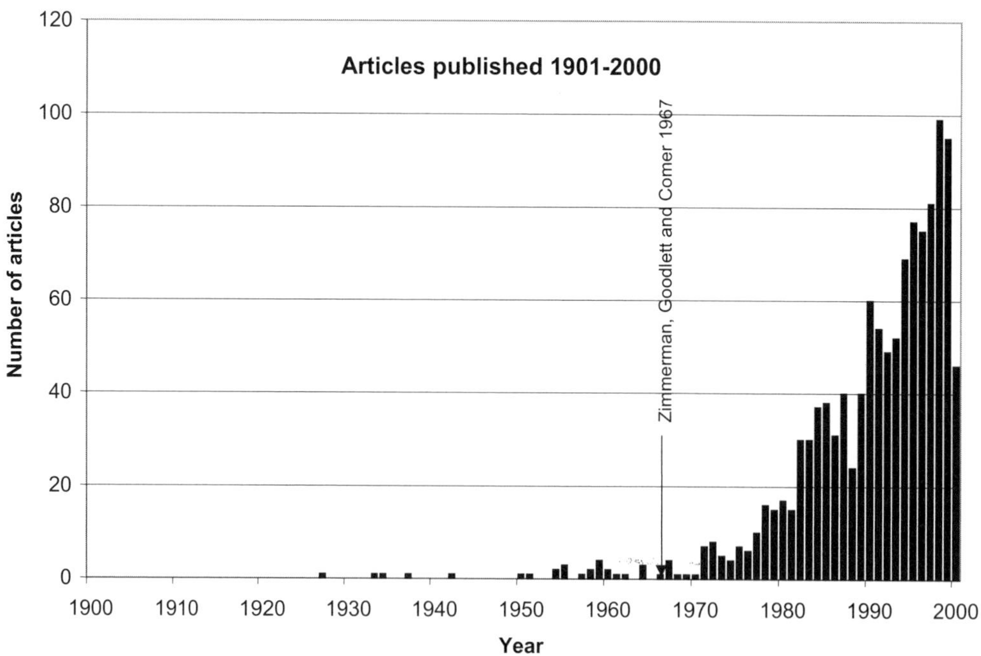

FIGURE 1. Articles from Bibliography (Gregory et al. 2000) of *World Literature on Wood in Streams, Rivers, Estuaries and Riparian Areas.*

TABLE 1. Examples of terms used in titles of research papers.

Term used	Related to	Source
Large organic material	Channel form and fluvial processes	Keller and Swanson 1979
Organic debris	Channel morphology and bedload transport	Mosley 1981
	Logging treatments and channel morphology	Towes and Moore 1982
	Cutthroat trout	Lestelle and Cederholm 1984
Organic debris dams	Role in streams	Likens and Bilby 1982
Large organic debris	Channel morphology	Swanson et al. 1976
		Swanson and Lienkaemper 1978
		Keller and Tally 1979
		Macdonald et al. 1982
		Beschta 1983
		Hogan 1987
Obstructions	Sediment storage	Megahan 1982
Organic debris dams	Function of stream ecosystems	Bilby and Likens 1980
	Development, maintenance and role	Likens and Bilby 1982
	Effect of deforestation	Hedin et al. 1988
Organic matter budgets	Stream ecosystems	Cummins et al. 1983
Debris dams	River channel processes	Gregory et al. 1985
	Stream structure and functioning	Smock et al. 1989
Streamside obstructions	Gravel channel processes	Lisle 1986
Woody debris	Salmonid nursery streams	Bryant 1983
	Fish habitat	Angermeier and Karr 1984
	Stream channel stability	Bilby 1984
	Pool formation	Andrus et al. 1988
	Source of fine particulate organic matter	Murphy and Koski 1989
	Fisheries and streamside management	Ward and Aumen 1986
	Macroinvertebrates	Carlson et al. 1990
Wood debris	Channel morphology and riparian areas	Triska 1984
Coarse woody debris	Ecology in temperate ecosystems	Harmon et al. 1986
	Ecological aspects	Spies et al. 1988
	Channel morphology	Robison and Beschta 1990
Logging debris	Dolly Varden population and macrobenthos	Elliott 1986
Large woody debris	Stream channel response	Macdonald and Keller 1987
	Forestry and fishery interactions	Bisson et al. 1987
	Dynamics in streams	Lienkaemper and Swanson 1987
Log steps	Geomorphic significance in forest streams	Marston 1982
Organic matter storage	Spatial and temporal variation in headwater streams	Smock 1990

was relatively ignored. A second implication that we need to further proselytize, however, is the view that—in forested areas—wood in channels is natural and has significant effects, with dynamics that need to be managed (see Table 8).

At the end of the 20th century (Figure 1), six themes (Table 2) have been prominent in the study of wood in rivers over more than three decades, and these six themes have to some extent developed sequentially. The **channel morphology** theme (the first in Table 2) was evident in headwater streams in the Sleepers River basin,

Table 2. Developing themes in research on wood in world rivers.

Major theme	Types of investigation and examples of aspects investigated
1. Channel morphology	• Effects on headwater streams in relation to thresholds
2. Extent and significance	• Detailed studies of specific reaches • Classification of types of wood • Effects on bank stability, channel roughness, pool-riffle sequence
3. Processes	• Hydrological and hydraulic: effects on discharge, during major floods; travel times, energy losses • Ecological: food source, sinks for nutrients, nutrient dynamics, diversity of stream habitats, cover for fish, magnitude of fish populations, diversity of macroinvertebrates, invertebrate communities associated with habitats associated with wood dams, biodiversity • Geomorphological: effects on channel banks, sediment transport, water quality, sediment storage, channel and floodplain sedimentation, growth of channel bars
4. Distribution and spatial pattern	• Spacing of dams • Volume of wood in spatial distribution • Sediment storage
5. Dynamics	• Delivery rates of wood • Residence times, decomposition, wood stability, dating of dams • How much change in individual dams and spatial patterns • Impact of storm events
6. Management implications	• Implications of logging • Significance of wood removal- channel instability, scour, downcutting, reduction in ecosystem processing of organic matter, sediment release effects of spawning and rearing grounds, reduction in quality of fish habitats, fish population size and densities reduced, significant reduction in diversity and density of invertebrates • Restoration, adaptive restoration, stream reconstruction • Consequences of conflicts with recreation, navigation

Vermont, analyzed in relation to thresholds (Zimmerman et al. 1967), developed in relation to the mass movement implications of timber harvesting (Swanston and Swanson 1976), and studied more specifically by showing the effects of large organic material on channel form and fluvial processes (Keller and Swanson 1979). Whereas initial studies concentrated in particular areas, especially the Pacific Northwest, subsequent research began to focus on the **extent and significance of wood** (2, in Table 2) achieved by detailed survey and investigation of specific reaches (for example, Keller and Swanson 1979; Keller and Tally 1979), which allowed comparison of detailed surveys at different dates (Gregory 1997), by classifications of types of accumulation, such as active, complete, and partial (Gregory et al. 1985), and for incised channels in underflow jam, debris jam, deflector jam, and parallel bar jam types (Wallerstein et al. 1997). After these studies, the effects on bank stability, channel roughness, and the pool-riffle sequence could be demonstrated. **Process investigations** (3, in Table 2) subsequently became a major theme; they were investigated in a range of ways. Processes controlling the formation, structure, and stability of natural wood jams are fundamental to the dynamics of forested ecosystems and provide insights into the design of both habitat restoration structures and ecosystem-based watershed management (Abbe and Montgomery 1996). Effects on discharge, particularly during major floods, influenced hydrological and hydraulic processes and the changes in travel times, and energy losses could be scrutinized. Ecological processes provided a range of research topics, including wood as a food source, sinks for nutrients, diversity of stream habitats, cover for fish, magnitude of fish populations, diversity of macroinvertebrates, invertebrate communities associated with habitats produced by, or as a result of wood accumulations, and thence implications for biodiversity. Such ecological investigations were some of the earliest bases for research on wood in world rivers, initially independent of geomorphological research, but subsequently, more comprehensive as studies of wood budget and large wood balance (for example, Fetherston et al. 1995) became feasible. A range of geomorphic process investigations included studies of sediment transport, sediment storage (such as Nakamura and Swanson 1993), channel and floodplain sedimentation, meander cutoffs (Piégay et al. 1998), growth of channel bars, effects on channel banks, and on-channel widening. Along an undisturbed reach of the Tolt River, the channel widened only where logjams diverted flow into the banks (Montgomery et al. 1995), and a new type of associated mass movement was described on the banks of the Highland Water (Davis and Gregory 1993). Ecological and geomorphic process investigations demonstrated the diversity of ways in which fluvial processes were influenced by wood in world rivers and were significant to stream ecology and geomorphology.

On this basis, a fourth theme could emerge (4, in Table 2), the **distribution and spatial pattern** of wood from studies of the spacing of dams and their frequency, the volume of wood spatially distributed throughout a basin (Gregory et al. 1993), and the amount of sediment storage occasioned by the wood accumulations in rivers. Such studies showed how spatial patterns were important, paving the way for investigations of **dynamics** (5, in Table 2) in studies of delivery rates of wood, residence times, rates of decomposition, dating of wood dams, and how much individual dams and spatial patterns changed as a result of treefall and storms. That patterns of wood distribution along networks were not as stable and as persistent as first thought rapidly became clear. The effects of storm events were demonstrated in several basins such as the Highland Water (Gregory 1992), where the amount of change for 100-m reaches was demonstrated (Gurnell and Gregory 1995). Further research on short-term changes was aided by environmental hydraulic experiments (such as Gippel 1995) and flume studies (Young 1991). Physically based models could be used to identify the conditions under which logs move in rivers (Braudrick and Grant 2000) using a first-order approach to evaluate either the stability of naturally derived wood or of material emplaced. On the basis of these five kinds of investigation, research was directed (6, in Table 2) towards **channel management**. This had been an early subject for research stimulated by the implications of logging (for example, by Swanston and Swanson 1976), complemented by studies of the significance of wood removal, which embraced a range of topics, such as (Table 2) channel instability, scour, downcutting; reduction in ecosystem processing of organic matter; sediment-release effects on spawning and rearing grounds; reduction in quality of fish habitats, fish population size and densities; and significant reduction in diversity and density of invertebrate populations. Thus, Shields and Gippel (1995) showed how artificially placed wood increased stability along incised channels and removing wood increased degradation. In the White Mountains, Heede (1981) showed that, when all log steps were removed, gravel bars replaced 74% of them in 5 years through increased bedload movement. Wood should be removed conservatively when streams and aquatic habitats are managed (Keller et al. 1995). Wood in world rivers is now managed in a context of approaches to restoration (see NRC 1992), adaptive restoration, and stream reconstruction.

Research developments in the disciplines related to wood in river channels have to be seen against the context of change both in research in the basin and in hydrological research. A series of new themes emerged in the three decades after 1967, and some examples are given in Table 3, suggesting how these new themes focused on the dynamics of the drainage basin, on increasingly holistic or integrated basin approaches, on the components of the basin, and on the ways in which water, sediment, and biotic materials are transferred. The drainage basin has become perceived as increasingly dynamic, with greater emphasis placed on transfers of water, sediment, and energy. A third implication (Table 8), therefore, could be to undertake analysis of wood in channels in the holistic basin context (such as Harper et al. 1999). New fields of research have been established, with the creation of new subdisciplines and new journals. Examples are provided in Table 4 exemplifying the more energetic approach used and the types of subdisciplines advocated. Such developments are pertinent to the study of wood in world rivers; for example, ecohydrology has been suggested as one of the most exciting frontiers of the future (Rodriguez-Iturbe 2000; Zalewski et al. 2003, this volume), and another implication could be that we need to further develop interdisciplinary as well as multidisciplinary approaches, particularly for wood in rivers. Montgomery (1997) indicates the need for a corps of river professionals whose

Table 3. Some examples of the inclusion of new terminology.

Subject or theme	Example of new terminology
Basin theme	Spatial and temporal scales
	Contributing areas
	Channel and network dynamics
	Sediment and solute dynamics and budgets
	Integrated basin management
	Holistic approach
Components of basin	Patches
	River corridor
	Riparian zone
	Buffer zones
Transfers of water, sediment, energy	Flood pulse concept
	Sediment slugs, sediment waves

expertise spans geomorphology, ecology, and public policy.

A tremendous amount of understanding has been achieved of wood in present world rivers, and it owes much to the researchers present at this conference. We are certainly now aware of the range of themes in wood research (Table 2); our future research can benefit from an appreciation of developments both within the basin (Table 3) and in the context of new disciplinary developments (Table 4). A better scientific understanding of processes can lead to more enlightened management practices (Harmon et al. 1986), but we need to be more aware of differences in dynamics from area to area so that a further implication (Table 8) could be that we should seek to understand the present differences between contrasting world environments.

The Past

How far have research results been applied to interpreting past environments? Certain aspects of woody debris have been researched in relation to a short time scale associated with effects of storm events, dating of wood, influence of human activity, and management changes; some understanding has been gained from the effects of experimental removal of wood (Smith et al. 1993). But longer time scales are, as yet, insufficiently explored, and an enhanced understanding of past changes could also be relevant to the future. In central Europe, the great phase of deforestation (Darby 1956) for 200 years after AD 1050 must have had a significant influence on the supply of wood and on the dynamic influence it had on riv-

Table 4. The discipline context: examples developing new disciplinary areas.

Type of development	Example	Illustrative references
Energetic approach	Geosystem	Rumney 1970
	Bioenergetics	Broda 1975
	Synergetics	Haken 1985
Subdisciplines	Landscape ecology	
	Biogeomorphology	Viles 1988
		Hupp et al. 1995
	Eco-hydraulics	New international symposia series launched 1994
	Hydrosystems	Petts and Amoros 1996
	Eco-hydrology	Wassen and Grootjans 1996
	Ecohydrology	Baird and Wilby 1999
		Zalewski et al. 1997
		Rodriguez-Iturbe 2000
		Gurnell 2000
New journals	Regulated Rivers	Initiated by G. E. Petts, 1984

ers. Prehistoric clearance of wood from the United Kingdom's landscape brought about a major change in its hydrology (Robinson et al. 2000), although no exploration has been made of the role of wood. Riparian woodland persisted despite deforestation, although more than 80% of riparian river corridors in North America and Europe are estimated to have disappeared in the last 200 years (Naiman et al. 1993; cited in Hupp 1999). Although studies of wood in world rivers have been limited to particular time scales, earlier phases could be investigated. Episodes of extensive woodland clearance in Northeast England were followed by cultivation and grazing (Macklin et al. 1992), and this sequence was repeated in many other basins (for example, Knox 1972, 1977) affecting catchment runoff, flood flashiness, slope erosion, and delivery of fine sediment to valley floors. Although wood in rivers is not mentioned in such research, it must nonetheless have been significant. Some investigations of temporal change relate to the influence of beavers (Naiman et al. 1988; Pollock et al. 2003, this volume) whose dams were more important in some areas in the past (for example, Gurnell 1998), such as in Scotland where beavers became extinct about 300 years ago. Evidence of beavers in England and Wales indicates they felled trees and were probably building dams; they became extinct in England after the 12th century (Coles 1992).

Analysis of past environments needs to be undertaken in a framework of timescales associated with spatial scales (for example, Knighton 1984, 1998; Newson 1992; Lawler 1993) up to the basin scale. Considering longer time scales requires palaeohydrology, a multidisciplinary field that developed, rather like research on wood in world rivers, after the 1970s, and is defined as

> the science of the waters of the earth, their composition, distribution, and movement on ancient landscapes from the beginning of the first rainfall to the beginning of continuous hydrological records [Schumm 1977; Gregory 1983, 1996].

Paleohydrological research has progressed by including the effects of human activity (Gregory 1995a), relating to other environmental disciplines (Baker 1996), and becoming relevant to global change (Gregory 1997; Gregory and Benito 2003). It is now undertaken at a world scale (Gregory 1995b) under an INQUA Commission on Global Continental Palaeohydrology (GLOCOPH). Reconstructing past processes has been undertaken with insufficient attention to wood in river channels, despite the fact that woodland was more extensive in the late Quaternary (Brown 1995a). Where woodland was more extensive in the past, vegetation influences, including wood in channels, affected longer sections of rivers, especially in the headwaters of drainage basins, until channels reached a width at least equal to tree height. Wood could have a significant influence and the characteristics affected are suggested in Table 5, with asterisks to indicate those that were probably most affected and could be detected from records.

The paucity of attention given to wood in rivers in the past means that we do not yet have a comprehensive understanding of the implications in past environments. Several implications, especially those associated with human activity, have been explored however. Paleohydrological research has increasingly benefited from a retrospective approach by using enhanced knowledge of contemporary environments to facilitate understanding of earlier ones. Thus, in the Siberian Taiga, tree dams and debris dams determine alluvium accumulations upstream (Yamskikh et al. 1999); on the historical time scale, major clearance is reflected in remains of trees in floodplain sediments as in the Labe floodplain in central Europe (Ruzickova and Zeman 1994); and prehistorically, the Langford Quarry downstream of Newark on Trent enabled reconstruction of lowland environmental conditions (Howard et al. 1999) to show a logjam in a quarry section, the origin of which being uncertain but possibly caused by deliberate impounding of the river. Dating of the sediment associated with the logjam between 4300 and 3980 BP led to the conclusion that reliance should not be placed on a single site. Early fishing baskets and wood accumulations have been recorded of at least Neolithic age. For Quaternary reconstructions, large wood has been used to provide a dendrochronological framework for floodplain evolution of the Main and the Danube (Becker and Schirmer 1977), and in interpreting the evolution of floodplain channels under 20 m wide, prone to damming by wood by beaver activity, creating ponds and by-pass channels (Brown 1995b, 1995c). The upper Brue mid-Holocene valley of the Somerset levels was choked with tree trunks and small, almost imperceptible channels (A. G. Brown, personal communication). Analysis of the Soar and Nene in the East Midlands (Brown et al. 1994) provides evidence of an early form of engineering use of groynes to deflect flow into one

TABLE 5. Wood in rivers in the past: where woodland was more extensive in the past, vegetation influences, including wood in rivers, affected longer sections of channels in headwater basins.

System characteristics	Aspects affected
Channel morphology	• Width, channel widening adjacent to dams*
	• Pools and riffles*
	• Plunge pools below dams
	• Bank stability*
	• Channel roughness and resistance
	• Channel pattern*
	• Cut offs*
Channel process	• Flow pulses
	• Flood peak travel times
	• Sediment transport
	• Sediment storage*
	• Erosion distribution*
	• Potential energy dissipation
	• Sediment/channel erosion*
	• Flood plain sedimentation*
Channel ecology	• Organic matter storage*
	• Fish populations
	• Fish/aquatic habitats
	• Habitat diversity

* Potentially significant features of the paleohydrologic record in sediments or morphology

channel, encouraging siltation in secondary channels and aided by wood and beaver activity, especially during the Flandrian when the channel changed from a stable anastomosing to a stable, single, sinuous one. The basic contrast is between gravel bed, high-energy interactions, and mobile channel systems, such as the Tagliamento (for example, Gurnell et al. 2000), that were braided and meandering and included large tree trunks, as opposed to meandering and anastomosing channels that had mixed loads, beds with silt/clay banks and channels, including classic wood dams, in which channels changed through backwater effects and avulsion.

Such specific examples demonstrate that, although we know something of the potential implications of wood in rivers in the past, probably much more is to be learned (for example, Brown 2002; Montgomery et al. 2003, this volume). The question of why it matters has at least two answers. First, because wood in channels can affect the evidence, whether synchronic or diachronic, as used to interpret past environments, so that a single site may be atypical. Second, it affects what is perceived to be natural, which may, in turn, be influential when channels are managed or restored. Brown (1998, 2002) has suggested that multiple channel systems characterized Northwest European floodplains before deforestation and channelization (Brown 1997), leading to an argument for restoring multiple channel systems, including regular flooding of parts of the floodplain, with secondary channels allowed to exist and dead wood being left in the system even when obstructing channel flow. Such an approach would significantly alter the way channels are restored and, as knowledge of past river systems is increasingly relied on to inform decisions about future ones, a further implication could be to communicate the consequences and significance of long-term change (see Table 8).

Paleohydrological studies can be valuable (Gregory 2003a) by furnishing flow data for periods prior to instrumental records, indicating the location of sensitive channel reaches, providing a temporal sequence of hydrological and sediment history within which the channel can be managed, and illuminating what is "natural" for a particular area. Conclusions from research suggest (Gregory 2003b) that each river has a history, reflected in landscape memory; that late Quaternary environmental changes provide the "initial" conditions for present day processes, their activity rates, and the resulting morphology; that the effects of climate change still affect contemporary river systems, so that longer term studies are required to explain river processes, with long-term evolution providing understanding of the proximity of individual reaches to threshold conditions. Considering wood in rivers on a longer time scale should allow strategies at the reach and watershed scale to be based on both historical and contemporary assessment of geomorphological dynamics, thus relaxing the dependence of engineers on inflexible design periods, when varying time scales should be considered. Recommendations made to guide river channel management have given insufficient attention to inputs from paleohydrological research, so that existing protocols could be augmented (Gregory 2003b).

The Future

Knowledge of past and present channel environments can influence how we approach managing future environments, with an increasing number of papers about managing the amount of wood in the future (for example, Gregory and Davis 1992) and whether wood is subtracted or added, with the sequence of management options available for different areas (Gurnell et al. 1995), and with management guidelines recognizing that different strategies may be required for different types of reach (Piégay and Gurnell 1997). Concern about future river channels arises from experience of snagging in relation to channelization. Implications of channelization are appreciated from the mapped extent in England and Wales (Brookes et al. 1983) from the particular implications in urban areas (Keller 1976) and from the idea that snagging implicit in many channelization schemes can be beneficial if undertaken selectively (Shields and Nunnally 1984).

From experience of the effects of channelization (Brookes 1988) came a new approach, and stream restoration was seen as the

> process of altering urban streams so that their behavior corresponds as closely as possible with that of natural streams while providing some measure of flood prevention [Keller and Hoffman 1977].

Many projects were like Briar Creek, Mecklenburg Co., North Carolina, where

> prior to the project, all sorts of urban debris, brush, large trees, and several logjams choked the stream channel....The design philosophy was to produce a meandering stream that appeared as a country stream wandering through an urban area [Keller and Hoffman 1977].

In Ohio, the Sugar Creek Protection Society was formed in 1973 to provide an alternative to channelization, by removing obstructions (Magsig 1990). Stream restoration emerged as a new concept that could maintain the integrity of the stream system and provide a more esthetically pleasing environment, while helping to reduce the flood hazard. Restoration, visualized as a general approach, could simply mean making the environment look more natural, but with a danger (Budiansky 1995):

> Restoration has always been viewed with a certain amount of aloof disdain by ecologists. At best it's landscaping; at worst it's nothing but glorified civil engineering.

A potential danger is that the hard engineering approach to river channel management could be succeeded by a softer approach that creates more natural-looking channels but not necessarily those appropriate for the area concerned, so, as Riley (1998) contended:

> the first problem the restorationist needs to address is what historical and indigenous (nature to the location) conditions to restore to. In some circumstances it may be the most practical to restore a waterway to its condition during a particular period of history....

In the move towards restoring rivers and streams based on definitions collated in Table 6, three broad approaches are possible: a general one, often called "restoration"; an attempt to make the channel more natural, including enhancement, rehabilitation, creation, naturalization, and mitigation; and full restoration, which includes recovery, re-establishment, reinstatement, and restoration in the strictest sense.

When deciding which of these three groups of approaches to follow, advocates must remember that the environment is of increasing interest to a range of disciplines, extending to the social sciences, philosophy, and ethics. Several environmental developments (Table 7) can be grouped into those that have emerged from the "environmental disciplines," those that have arisen from other disciplines, including social sciences and philosophy, and those that can be categorized as national and international developments. In restor-

TABLE 6. Terms used for river restoration.

Approach	Specific term	Definition
General	Restoration	*The act of restoring (a river) to a former or original condition* "the complete structural and functional return of a bio-physical system to a predisturbance state" (NRC 1992)
More natural condition	Re-establishment	*To make (a river) secure in a former condition*
	Enhancement	Any improvement of a structural or functional attribute (NRC 1992) Any improvement in environmental quality (Brookes and Shields 1996)
	Rehabilitation	*To help (a river) adapt to a new environment* "a partial structural or functional return to the predisturbance state" (Cairns 1991; NRC 1992) Partial return to a pre-disturbance structure or function. (Brookes and Shields 1996)
	Creation	Development of a resource that did not previously exist at the site. Includes the term 'naturalization,' which determines morphological and ecological configuration with contemporary magnitudes and rates of fluvial processes (Brookes and Shields 1996).
	Naturalization	Recognizes that the concept of "natural" is defined by the community relative to the modified state of the system and that the goal of naturalization is to drive the system as a whole toward a state of increasing morphologic, hydraulic, and ecological diversity, but to do so in a manner that is acceptable to the local community and sustainable by natural processes, including human intervention (Rhoads et al. 1999)
	Mitigation	Action taken to avoid, reduce or compensate for the effects of environmental damage (Holmes 1998)
Full restoration	Recovery	*The act of restoration (of a river) to an improved/former condition*
	Full restoration	The complete structural and functional return to a pre-disturbance state (Brookes and Shields 1996)
	Reinstatement	*To restore (a river) to a former condition*

Developed from Sear (1994) and Gregory (2000)

ing nature, or as one philosopher expressed it "Faking Nature"(Elliott 1997), the choice is between restoring the nature that actually existed before original modification or the one that would have evolved if unaffected by human actions. Suggesting design principles has often been possible for creating and restoring wetlands (for example, Mitsch and Gosselink 1993), but are those guidelines feasible for wood in rivers? Guides have been developed for desnagging and resnagging, and people generally agree that hard engineering should be avoided. Thus the salient "restoration" issues may be listed as

- Hard engineering should not be used except where absolutely essential
- Examples of unchanged landscape should be maintained for future generations to see, including Sites of Special Scientific Interest, heritage sites, wilderness areas, museums
- Restoration in a general sense should be attempted wherever possible to restore landscapes and environments to more "natural" conditions
- Interdisciplinary effort should focus on determining what "natural" means (see Table 7)
- Consideration should be given to how the variety of approaches envisioned (Table 6) apply to wood in rivers
- Restoration of process is as important or more important than restoration of structure

TABLE 7. The environmental context: examples of developments related to concern for the environment.

Origin	Example	Definition and source
Environmental disciplines	Geomorphic engineering	"The geomorphic engineer is interested in maintaining (and working towards the accomplishment of) the maximum integrity and balance of the total land-water ecosystem as it relates to landforms, surface materials and processes" (Coates 1976)
	Ecological engineering	"[T]he design of human society with its natural environment for the benefit of both" (Mitsch and Jorgensen 1989)
	Shallow and deep ecology	Shallow ecology considers the values of nature instrumental to humans and is strongly anthropocentric, whereas deep ecology maintains that all species have an intrinsic right to exist in the natural environment (Lemons 1987, 1999)
	Biological habitat reconstruction	"Attitudes towards habitat reconstruction rapidly polarize into a position taken by the restorative pragmatist or 'engineer' on the one hand, and the nature conservationist on the other" (Buckley 1989)
	Restoration ecology	"[R]estore badly damaged ecosystems to pre-disturbance condition" (Cairns 1989)
Other disciplines	Environmental ethics	"[C]oncerned with the moral relations that hold between humans and the natural world" (Taylor 1986)
National and international developments	Sustainability Agenda 21 Community participation	
	Strategies for America's Watersheds	"Successful watershed management strives for a better balance between ecosystem and watershed integrity and provision of human and social goals" (NRC 1999)
	Blue Revolution Watersheds Blue revolution	"The blue revolution, although supported by technological advances, is more a philosophical revolution in the way we respect the world's environment and one of its most precious assets, water" (Calder 1999)
	Quality of Life	UK government includes rivers in its new "Quality of Life" Index

- To determine restoration policy for a particular area, community views should be included, with attention given to how community values are obtained and accommodated in determining restoration policies
- Practices that diminish the role of wood in streams should be reduced or eliminated prior to restoration
- How can risk and uncertainty be included in restoration strategies
- A catchment-based, holistic, approach should be developed

Many implications arise from these issues, and one approach is preserving wood in river channels at designated sites. The issues need to be addressed, and two particular ones merit further comment. *First,* the status of community views needs to be considered. Although this is implicit in some of the approaches in Table 6, a hierarchy exists in public involvement: Arnstein (1969) outlined eight levels ranging from nonparticipation through to full community control in a ladder consisting of manipulating, providing therapy, informing, consulting, placating, partnering, delegating power, and citizen control. MacKay (1998) suggests a range from systematic information gathering to information dissemination to interested parties, consultation, and participation. In addition, public perception of riverscape esthetics may not always be well informed. Thus, House and Sangster (1991) showed that public perception of water and river corridor quality, according to two on-site questionnaire surveys of several river user groups, showed that signs of bad water quality were most influential, with an overwhelming desire for trees and an equally strong preference for mature, sinuous rivers with natural channels and banks. A study in Hampshire showed that three groups questioned preferred "pretty" not natural (Gregory and Davis 1993), and a study in New Zealand (Mosley 1989) demonstrated that the river's environment was more important than the river itself and that rivers in strongly modified landscapes are more highly regarded than those in wilderness settings.

Second, can a catchment-based, holistic, restoring approach be developed to succeed a more site-based, local approach? Greater emphasis to the drainage basin as the unit of water management was advocated generally in Action 15.5 of *Caring for the Earth* (IUCN, UNEP, and WWF 1991), and a catchment framework was recommended as a basis for the physical restoration of lowland UK rivers (Harper et al. 1999). Such suggestions could be helpful in restoring channels with wood, in that a catchment framework should provide the most sustainable basis for restoration. Three further implications can be added to Table 8. We should use results from research on wood in rivers to communicate what is possible for "restoration," so that, if floodplain forests are to be retained or recreated, then a hydrogeomorphologically active channel and floodplain environment is essential (Gurnell 1997; Brown 2002). We should present alternatives to the community in a meaningful way and devise guidelines for "restoration" that could be generally applicable case-by-case, as outlined by Montgomery (1997), certainly embracing a wide range of alternatives, with selective removal included because of the benefits for fish habitats (Bisson and Sedell 1984).

Conclusion

In addressing the limits of wood in world rivers, with particular reference to the limits of disciplinary research, not only do we have further scope for research on the past and on the future—both of which need to develop from our understanding of the present—but also the present, past, and future are tightly interrelated. Indeed, Messerli et al. (2000) quote F. Broudel, a French historian (1902–1985), cited in Geoparks (1999), that "The present without a past has no future"—a sentiment appropriate to wood in world rivers. A review of the present, past, and future raises the implications collectively summarized in Table 8, together with the goals set for the conference. In considering both the implications and the goals, we should remember the broadening interpreta-

TABLE 8. Implications arising from this paper as a basis for the goals set for this conference

Section of paper	Implications suggested
Present	• Standardize the use of terms (with the benefit of hindsight) • Proselytize the view that in wooded areas wood in channels is natural and its dynamics need to be managed • Analyze wood in channels in the basin context • Develop interdisciplinary as well as multidisciplinary approaches • Seek to understand the differences between world environments
Past	• Communicate the consequences and significance of long-term change
Future	• Use research results to communicate what is possible in "restoration" • Present meaningful alternatives to the community • Devise principles and guidelines for "restoration"

The goals for the International Conference on Wood in World Rivers are to:

- Synthesize what is known around the world about large wood in streams and rivers for physical and ecological processes and stream restoration.
- Present status of knowledge of the physical dynamics and ecological interactions of large wood in streams and rivers in different geographical regions.
- Create a framework for interpreting and potentially applying the results of research in different geographical regions and management systems.
- Identify different management systems for large wood in rivers.
- Assess physical and biological responses of large wood in stream restoration.
- Explore links between primary information of the physical and ecological dynamics of large wood, resource management systems, and the communities and cultures in which they are applied.

tion of environment. Thus, Attfield (1999) draws attention to rival views of nature: as a living organism, as a mechanism, Gaia as a self-restoring organism, as a bottomless mine of resources, or as a sanctuary or temple like Mount Ranier or the Grand Canyon. While remembering such diversity of views, perhaps we should adopt the Rene Dubos environmentalist slogan "Think globally, act locally." Much has been achieved in the last 30 years of research growth (Figure 1) that has diffused to hydrology and to the earth and environmental sciences. This conference is extremely timely and has a strategic opportunity to influence thinking in the future and to disseminate the results of contemporary research as widely as possible.

Acknowledgment

The author is grateful to the Leverhulme Trust for a Leverhulme Emeritus Fellowship 1998–2001 during the tenure of which this paper was written.

References

Abbe, T. B., and D. R. Montgomery. 1996. Large woody debris jams, channel hydraulics and habitat formation in large rivers. Regulated Rivers Research and Management 12:201–221.

Andrus, C. W., B. A. Long, and N. A. Froehlich. 1988. Woody debris and its contribution to pool formation in a coastal stream 50 years after logging. Canadian Journal of Fisheries and Aquatic Sciences 45:2080–2086.

Angermeier, P. L., and J. R. Karr. 1984. Relationships between woody debris and fish habitat in a small warmwater stream. Transactions of the American Fisheries Society 113:716–726.

Arnstein, S. R. 1969. A ladder of citizen participation. American Institute of Planners Journal 35:216–224.

Attfield, R. 1999. The ethics of the global environment. Edinburgh University Press, Edinburgh, Scotland.

Baird, A. J., and R. L. Wilby, editors 1999. Ecohydrology. Routledge, London.

Baker, V. R. 1996. Discovering earth's future in its past: palaeohydrology and global environmental change. Pages 73–88 *in* J. Branson, A. G. Brown, and K. J. Gregory, editors. Global continental changes: the context of palaeohydrology. Geological Society Publication 115, London.

Becker, B., and W. Schirmer. 1977. Palaeoecological study on the Holocene valley development of the River Main, Southern Germany. Boreas 6:303–321.

Beschta, R. L. 1983. The effects of large organic debris upon channel morphology: a flume study. Pages 8.63–8.78 *in* Proceedings of the Symposium on Erosion and Sediments. Simons, Li and Assoc., Fort Collins, Colorado.

Bilby, R. E. 1984. Removal of woody debris may affect stream channel stability. Journal of Forestry 82:609–613.

Bilby, R. E., and G. E. Likens. 1980. Importance of organic debris in the structure and function of stream ecosystems. Ecology 61:1107–1113.

Bisson, P. A., and J. R. Sedell. 1984. Salmonid populations in stream in clearcut vs old growth forests of Western Washington. Pages 121–129 *in* W. R. Meehan, T. R. Merrell, Jr., and T. A. Hanley, editors. Proceedings of a symposium: fish and wildlife relationships in old-growth forests. April 12–15, Juneau, Alaska, USA. American Institute Fish Research Biology, Juneau, Alaska.

Bisson, P. A., R. E. Bilby, M. D. Bryant, C. A. Dolloff, G. B. Grette, R. A. House, M. L. Murphy, K. V. Koski, and J. R. Sedell, 1987. Large woody debris in forested streams in the Pacific Northwest: past, present and future. Pages 143–190 *in* E. O Salo and T. W. Cundy, editors. Streamside management: forestry and fishery implications. University of Washington, Seattle.

Boon, P. J., B. R. Davies, and G. E. Petts. 2000. Global perspectives on river conservation. John Wiley, Chichester, UK.

Brookes, A. 1988. Channelized rivers: perspectives for environmental management. John Wiley, Chichester, UK.

Brookes, A., K. J. Gregory, and F. H. Dawson. 1983. An assessment of river channelization in England and Wales. Science of the Total Environment 27:97–112.

Brookes, A., and F. D. Shields, Jr. 1996. River channel restoration: guiding principles for sustainable projects. John Wiley, Chichester, UK.

Braudrick, C. A., and G. E. Grant. 2000. When do logs move in rivers? Water Resources Research 36:571–583.

Broda, E. 1975. The evolution of the bioenergetic processes. Pergamon, Oxford, UK.

Brown, A. G. 1995a. Vegetation and lake level change. Pages 131–150 *in* K. J. Gregory, L. Starkel, and V. R. Baker, editors. Global continental palaeohydrology. Wiley, Chichester, UK.

Brown, A. G. 1995b. The biogeomorphology of a wooded anastomosing river: the Gearagh on the river Lee in County Cork, Ireland. Occasional Paper 32, Department of Geography, University of Leicester, Leicester, UK.

Brown, A. G. 1995c. Lateglacial - Holocene sedimentation in lowland temperate environments: floodplain metamorphosis and multiple channel systems. Pages 21–35 *in* B. Frenzel, editor. European river activity and climatic change during the Lateglacial and early Holocene. Palaeoklimforschung Palaeoclimate Research, volume 14, Special Issue: ESF Project European Palaeoclimate and Man. 9. Gustav Fischer Verlag, Stuttgart, Germany.

Brown, A. G. 1997. Biogeomorphology and diversity in multiple - channel river systems. Global Ecology and Biogeography Letters 6:179–185.

Brown, A. G. 1998. The maintenance of diversity in multiple channel floodplains. Pages 83–92 *in* R. G. Bailey, P. V. Jose, and B. R. Sherwood, editors. United Kingdom floodplains. Otley, Westbury, UK.

Brown, A. G. 2002. Learning from the past: palaeohydrology and palaeoecology. Freshwater Biology 47:817–829.

Brown, A. G., M. K. Keough, and R. J. Rice. 1994. Floodplain evolution in the East Midlands, United Kingdom: the Lateglacial and Flandrian alluvial record from the Soar and Nene valleys. Philosophical Transactions of the Royal Society London A 348:261–293.

Bryant, M. D. 1983. The role and management of woody debris in west coast salmonid nursery streams. North American Journal of Fisheries Management 3:322–330.

Buckley, G. P., editor. 1989. Biological habitat reconstruction. Belhaven, London.

Budiansky, S. 1995. Nature's keepers: the new science of nature management. Weidenfield and Nicolson, London.

Cairns, J. 1989. Restoring damaged ecosystems: is pre-disturbance condition a viable option? The Environmental Professional 11:152–159.

Cairns, J. 1991. The status of the theoretical and applied science of restoration ecology. The Environmental Professional 13:186–194.

Calder, I. R. 1999. The Blue Revolution. Land use and integrated water resources management. Earthscan, London.

Carlson, J. Y., C. W. Andrus, and N. A Froehlich. 1990. Woody debris, channel features, and macroinvertebrates of streams with logged and undisturbed riparian timber in Northeastern Oregon, U.S.A. Canadian Journal of Fisheries and Aquatic Sciences 47:1103–1181.

Chemekov, I. F. 1951. Drevesnye alliuvialnye otlozheniia [Woody alluvial deposits]. Priroda 11.

Chemekov, Iu. F. 1959. Zalomy, ikh obrazo-vanie i razvitie [Debris jams; formation and development]. Izvestiia Vsesoiuznogo Geograficheskogo Obshchestva 87(2):134–146.

Chemekov, I. F. 1959. Zalomy, ikh obrazovanie i razvitie [Debris jams; formation and development]. Izvestiia Vsesoiuznogo Geograficheskogo Obshchestva 87(2):134 -146.

Coates, D. R. 1976. Geomorphic engineering. Pages 3–21 *in* D. R. Coates, editor. Geomorphology and engineering. Dowden, Hutchinson and Ross, Stroudsburg, Pennsylvania.

Coles, B. 1992. Further thoughts on the impact of beaver on temperate landscapes. Pages 93–99 *in* S. Needham and M. G. Macklin, editors. Alluvial archaeology in Britain. Oxbow Books, Oxford, UK.

Cummins, K. W., J. R. Sedell, F. J. Swanson, G. W. Minshall, S. G. Fisher, C. E. Cushing, R. C Peterson, and R. L. Vannote. 1983. Organic matter budgets for stream ecosystems. Problems in the evaluation. Pages 299–353 *in* J. R. Barnes and G. W. Minshall, editors. Stream ecology, application and testing of general ecological theory. Plenum, New York.

Darby, H. C. 1956. The clearing of the woodland in Europe. Pages 183–216 *in* W. L. Thomas, editor. Man's role in changing the face of the earth. University of Chicago Press, Chicago.

Davis, R. J., and K. J. Gregory, 1993. A new distinct mechanism of river bank erosion in a forested catchment. Journal of Hydrology 157:1–11.

Dobbie, C. H., and P. O. Wolf. 1953. The Lynmouth flood of August 1952. Proceedings of the Institution of Civil Engineers 2:522–588.

Elliott, S. T. 1986. Reduction of a Dolly varden population and macrobenthos after removal of logging debris. Transactions of the American Fisheries Society 115:392–400.

Elliott, R. 1997. Faking nature. the ethics of environmental restoration. Routledge, London.

Fetherston, K. L., R. J Naiman, and R. E. Bilby. 1995. Large woody debris, physical process, and riparian development in montane river networks of the Pacific Northwest. Geomorphology 13:133–144.

Geoparks, 1999. UNESCO network of geoparks. UNESCO, Division of Earth Sciences, Paris, France.

Gippel, C. J. 1995. Environmental hydraulics of large woody debris in streams and rivers. Journal of Hydraulic Engineering, American Society of Civil Engineers 121(5):388–395.

Gregory, K. J., editor. 1983. Background to Palaeohydrology. John Wiley, Chichester, UK.

Gregory, K. J. 1992. Vegetation and river channel process interactions. Pages 255–269 *in* P. J. Boon, P. Calow, and G. E. Petts, editors. River conservation and management. John Wiley, Chichester, UK.

Gregory, K. J. 1995a. Human activity and palaeohydrology. Pages 151–172 *in* K. J. Gregory, L. Starkel, and V. R. Baker, editors. Global continental palaeohydrology. John Wiley, Chichester, UK.

Gregory, K. J. 1995b. The increased significance of vegetation on river channel dynamics. Quaestiones Geographicae Special Issue 4:117–120.

Gregory, K. J. 1996. Introduction. Pages 1–8 *in* J. Branson, A. G. Brown, and K. J. Gregory, editors. Global continental changes: the context of palaeohydrology. The Geological Society, London.

Gregory, K. J. 1997. Highland Water, Hampshire. Pages 260–265 *in* K. J. Gregory, editor. Fluvial geomorphology of Great Britain. Joint Nature Conservation Committee. Chapman and Hall, London.

Gregory, K. J. 2003a. Palaeohydrology, environmental change and river channel management. Pages 357–378 *in* K. J. Gregory and G. Benito, editors. Palaeohydrology: understanding global change. Wiley, Chichester, UK.

Gregory, K. J. 2003b. Palaeohydrology and river channel management. Geological Society of India. In press.

Gregory, K. J., and G. Benito, editors. 2003. Palaeohydrology: understanding global change. Wiley, Chichester, UK.

Gregory, K. J., and R. J. Davis, 1992. Coarse woody debris in stream channels in relation to river channel management in woodland areas. Regulated Rivers Research and Management 7:117–136.

Gregory, K. J., and R. J. Davis, 1993. The perception of riverscape aesthetics: an example from two Hampshire rivers. Journal of Environmental Management 39:171–185.

Gregory, K. J., R. J. Davis, and S. Tooth. 1993. Spatial distribution of coarse woody debris dams in the Lymington basin, Hampshire, UK. Geomorphology 6:207–224.

Gregory, K. J., A. M. Gurnell, and C. T. Hill. 1985. The permanence of debris dams related to river channel processes. Journal of Hydrological Sciences 30:371–381.

Gregory, K. J., and D. E. Walling, 1973. Drainage basin form and process. Arnold, London.

Gregory, S. V., F. J. Swanson, W. A. McKee, and K. W. Cummins. 1991. An ecosystem perspective of riparian zones. BioScience 41:540–551.

Gregory, S. V., A. M. Gurnell, K. J. Gregory, S. Bolton, L. A. Medvedeva, A. Semenchenco, A. N. Mahkinov, D. Sobota, J. Baurer, and K. Staley. 2000. Bibliography: world literature on wood in streams, rivers, estuaries and riparian areas. From International Conference on Wood in World Rivers. Version 1. Oregon State University, Corvallis.

Gurnell, A. M. 1997. The hydrological and geomorphological significance of forested floodplains. Global Ecology and Biogeography Letters, Floodplain Forests Special Issue 6:219–229.

Gurnell, A. M. 1998. The hydrogeomorphological effects of beaver dam-building activity. Progress in Physical Geography 22:167–189.

Gurnell, A. M., C. R. Hupp, and S. V. Gregory, editors. 2000. Linking hydrology and ecology. Special Issue of Hydrological Processes 14:2813–3179.

Gurnell, A. M., and K. J. Gregory. 1995. Drainage basin perspectives on the interaction between seminatural vegetation and hydrogeomorphological processes. Geomorphology 13:49–69.

Gurnell, A. M., K. J. Gregory, and G. E. Petts. 1995. The role of a coarse woody debris in forest aquatic habitats: implications for management. Aquatic Conservation: Marine and Freshwater Ecosystems 5:143–166.

Gurnell, A. M., G. E. Petts, D. M Hannah, B. P. G. Smith, P. J. Edwards, J. Kollmann, J. V. Ward, and K. Tockner. 2000. Wood storage within the active zone of a large European gravel-bed river. Geomorphology 34:55–72.

Hack, J. T., and J. C. Goodlett. 1960. Geomorphology and forest ecology of a mountain region in the central Appalachians. U.S. Geological Survey Professional Paper 347, GPO, Washington, D.C.

Haken, H. 1985. Synergetics - an interdisciplinary approach to phenomena of self organisation. Geoforum 16:205–211.

Harmon, M. E., J. F. Franklin, F. J. Swanson, P. Sollins, S. V. Gregory, J. D. Lattin, N. H. Anderson, S. P. Cline, N. G. Aumen, J. R. Sedell, G. W. Lienkaemper, K. Cromack Jr., and K. W. Cummins. 1986. Ecology of coarse woody debris in temperate ecosystems. Advances in Ecological Research 15:133–302.

Harper, D. M., M. Ebrahimnezhad, E. Taylor, S. Dickinson, O. Decamp, G. Verniers, and T. Balbi. 1999. A catchment-scale approach to the physical restoration of lowland UK rivers. Aquatic Conservation: Marine and Freshwater Ecosystems 9:141–157.

Hedin, L. O., M. S. Mayer, and G. E. Likens. 1988. The effect of deforestation on organic debris dams. Verhandlungen Internationale Vereinigung fur

Theoretische und Angewandte Limnologie 23:1135–1141.

Heede, B. H. 1981. Dynamics of selected mountain streams in the western United States of America. Zeitschrift fur Geomorphologie 25:17–32.

Heede, B. H. 1985. Channel adjustments to the removal of log steps: an experiment in a mountain stream. Environmental Management 9:427–432.

Hogan, D. L. 1987. The influence of large organic debris on channel recovery in the Queen Charlotte Islands, British Columbia, Canada. International Hydrological Sciences Publication 165:343–353.

Holmes, N. T. H. 1998. Floodplain restoration. Pages 331–348 *in* R. G. Bailey, P. V. Jose, and B. R. Sherwood, editors. United Kingdom floodplains. Westbury, Otley, UK.

House, M. A., and E. K. Sangster. 1991. Public perception of river-corridor management. Journal Institution of Water Engineers and Managers 5:312–317.

Howard, A. J., D. N. Smith, D. Garton, J. Hillam, and M. Pearce. 1999. Middle to late Holocene environments in the middle to lower Trent Valley. Pages 165–178 *in* A. G. Brown and T. Quine, editors. Fluvial processes and environmental change. John Wiley, Chichester, UK.

Hupp, C. R. 1999. Relations among riparian vegetation, channel incision processes and forms, and large woody debris. Pages 219–245 *in* S. E. Darby and A. Simon, editors. Incised river channels. John Wiley, Chichester, UK.

Hupp, C. R., W. R. Osterkamp, and A. D. Howard, editors. 1995. Biogeomorphology, terrestrial and freshwater systems. Proceedings of the 26th Binghamton Symposium in Geomorphology. Elsevier, Amsterdam, Netherlands.

IUCN, UNEP, and WWF (International Union for the Conservation of Nature, United Nations Environment Programme, and World Wildlife Fund). 1991. Caring for the Earth: a strategy for sustainable living. Earthscan, London.

Keller, E. A. 1976. Channelization: environmental, geomorphic and engineering aspects. Pages 115–140 *in* D. R. Coates, editor. Geomorphology and engineering. George Allen and Unwin, London.

Keller, E. A., and A. Macdonald. 1995. River channel change: the role of large woody debris. Pages 217–235 *in* A. Gurnell and G. Petts, editors. Changing river channels. John Wiley, Chichester, UK.

Keller, E. A., and E. K. Hoffman. 1977. Urban streams: sensual blight or amenity. Journal of Soil and Water Conservation 32:237–242.

Keller, E. A., and F. J. Swanson. 1979. Effects of large organic material on channel form and fluvial process. Earth Surface Processes 4:361–380.

Keller, E. A., and T. Tally. 1979. Effects of large organic debris on channel form and fluvial processes in the coastal redwood environment. Pages 169–179 *in* D. D. Rhodes and G. P. Williams. Adjustments to the fluvial system. Kendall-Hunt, Dubuque, Iowa.

Keller, E. A., A. MacDonald, T. Tally, and N. J. Merrit. 1995. Effects of large organic debris on channel morphology and sediment storage in selected tributaries of Redwood Creek, northwestern California. U.S. Geological Survey Professional Paper 1454-P, Menlo Park, California.

Knighton, A. D. 1984. Fluvial forms and processes. Arnold, London.

Knighton, A. D. 1998. Fluvial forms and processes. A new perspective. Arnold, London.

Knox, J. C. 1972. Valley alluviation in southwestern Wisconsin. Annals of the Association of American Geographers 62:401–410.

Knox, J. C. 1977. Human impacts on Wisconsin stream channels. Annals of the Association of American Geographers 67:325–342.

Lawler, D. 1993. The measurement of river bank erosion and lateral channel change: a review. Earth Surface Processes and Landforms 18:777–821.

Lemons, J., editor. 1987. Special focus on environmental ethics. Environment Professional 9:277–368.

Lemons, J. 1999. Environmental ethics. Pages 204–206 *in* D. E. Alexander and R. W. Fairbridge, editors. Encyclopedia of environmental science. Kluwer Academic Publishers, Dordrecht, Netherlands.

Lestelle, L. C., and C. J. Cederholm. 1984. Short- term effects of organic debris removal on resident cutthroat trout. Pages 131–140 *in* W. R. Meehan, T. R. Merrell Jr., and T. A. Hanley, editors. Proceedings of a symposium on fish and wildlife relationships in old-growth forests, April 12–15, 1982, Juneau, Alaska. American Institute of Fisheries Research Biology, Juneau, Alaska.

Lienkaemper, G. W., and F. J. Swanson. 1987. Dynamics of large woody debris in streams in old-growth Douglas-fir forests. Canadian Journal of Forest Research 17:150–156.

Likens, G. E., and R. E. Bilby. 1982. Development, maintenance, and role of organic debris dams in New England streams. Pages 122–128 *in* F. J. Swanson, R. J. Janda, T. Dunne and D. N. Swanston, editors. Sediment budgets and routing in forested drainage basins. Pacific Northwest Research Station, USDA Forest Service Research Paper PNW-141, Portland, Oregon.

Lisle, T. E. 1986. Effects of woody debris on anadromous salmonid habitat, Prince of Wales Island, southeast Alaska. North American Journal of Fisheries Management 6:538–550.

Lobeck, A. K. 1939. Geomorphology. McGraw Hill, New York.

MacDonald, A., and E. A. Keller. 1987. Stream channel response to the removal of large woody debris, Larry Damm Creek, northwestern California. Pages 405–406 *in* R. L. Beschta, T. Blinn, G. E. Grant, F. J. Swanson, and G. G. Ice, editors. Erosion and sedimentation in the Pacific Rim.

International Association of Hydrological Sciences Publication 165.

MacDonald, A., E. A. Keller, and T. Tally. 1982. The role of large organic debris on stream channels draining redwood forests northwestern California. Pages 226–245 *in* D. K. Darden, D. C. Marren, and A. MacDonald, editors. Friends of the Pleistocene 1982, Pacific Cell fieldtrip guidebook, late Cenozoic history and forest geomorphology of Humboldt Co., California.

MacKay, A. 1998. Concepts and process of public participation: conceptual briefing note. Pages 3–13 *in* Public participation in electric power projects. Bangkok: United Nations Economic and Social Commission for Asia and the Pacific, Bangkok, Thailand.

Macklin, M. G., D. G. Passmore, and B. T. Rumsby. 1992. Climatic and cultural signals in Holocene alluvial sequences: the Tyne basin, northern England. Pages 123–139 *in* S. Needham, and M. G. Macklin, editors. Alluvial archaeology in Britain. Oxbow Books, Oxford, UK.

Magsig, J. 1990. Volunteer stream restoration in an agricultural watershed in northwest Ohio. Pages 228–233 *in* J. J. Berger, editor. Environmental restoration: science and strategies for restoring the Earth. Island Press, Washington D.C.

Manga, M., and J. W. Kirchner. 2000. Stress partitioning in streams by large woody debris. Water Resources Research 36:2373–2379.

Marston, R. A. 1982. The geomorphic significance of log steps in forest streams. Annals Association of American Geographers 72:99–108.

Megahan, W. F. 1982. Channel sediment storage behind obstructions in forest drainage basins draining the granitic bedrock of the Idaho batholith. Pages 114–121 *in* F. J. Swanson, R. J. Janda, T. Dunne, and D. N. Swanston, editors. Sediment budgets and routing in forested drainage basins. Pacific Northwest Research Station, U.S. Department of Agriculture Forest Service General Technical Report PNW-141, Portland, Oregon.

Messerli, B., M. Grosjean, T. Hofer, L. Nunez, and C. Pfister. 2000. From nature dominated to human-dominated environmental changes. Quaternary Science Reviews 19:459–479.

Mitsch, W. J., and J. G. Gosselink. 1993. Wetlands. Van Nostrand Reinhold, New York.

Mitsch, W. J., and S. E. Jorgensen, editors. 1989. Ecological engineering: an introduction to ecotechnology. Wiley, New York.

Montgomery, D. R. 1997. River management: what's best on the banks? Nature 388:328–329.

Montgomery, D. R., B. D. Collins, J. M. Buffington, and T. M. Abbe. 2003. Geomorphic effects of wood in rivers. Pages 21–47 *in* S. V. Gregory, K. L. Boyer, and A. M. Gurnell, editors. The ecology and management of wood in world rivers. American Fisheries Society, Symposium 37, Bethesda, Maryland.

Montgomery, D. R., J. M. Buffington, R. D. Smith, K. M. Schmidt, and G. Pess. 1995. Pool spacing in forest channels. Water Resources Research 31:1097–1105.

Morisawa, M. A. 1968. Streams. Their dynamics and morphology. McGraw Hill, New York.

Morisawa, M. A. 1985. Rivers. Form and process. Longman, London.

Mosley, M. P. 1981. The influence of organic debris on channel morphology and bedload transport in a New Zealand forest stream. Earth Surface Processes 6:571–579.

Mosley, M. P. 1989. Perceptions of New Zealand river scenery. New Zealand Geographer 45:2–13.

Murphy, M. L., and K. V. Koski. 1989. Input and depletion of woody debris in Alaska streams and implications for streamside management. North American Journal of Fisheries Management 9:427–436.

Naiman, R. J., C. A. Johnston, and J. C. Kelley. 1988. Alteration of North American streams by beaver. Bioscience 38:753–762.

Naiman, R. J., H. Decamps, and M. Pollock. 1993. The role of riparian corridors in maintaining regional diversity. Ecological Applications 3:209–212.

Nakamura, F., and F. J. Swanson. 1993. Effects of coarse woody debris on morphology and sediment storage of a mountain stream system in western Oregon. Earth Surface Processes and Landforms 18:43–61.

NRC (National Research Council). 1992. Restoration of aquatic ecosystems. National Academy Press, Washington, D.C.

Newson, M. D. 1992. Land, water and development: river basin systems and their sustainable management. Routledge, London.

Petts, G. E., and C. Amoros, editors. 1996. Fluvial hydrosystems. Chapman and Hall, London.

Piégay, H., and A. M. Gurnell. 1997. Large woody debris and river geomorphological pattern: examples from S. E. France and S. England. Geomorphology 19:99–116.

Piégay, H., A. Citterio, and L. Astrade. 1998. Interactions between large woody debris and meander cut-offs (example of the Mollon reach on the Ain River, France). Zeitschrift fur Geomorphologie 42:187–208.

Pollock, M. M., M. Heim, and R. J. Naiman. 2003. Hydrologic and geomorphic effects of beaver dams and their influence on fishes. Pages 213–233 *in* S. V. Gregory, K. L. Boyer, and A. M. Gurnell, editors. The ecology and management of wood in world rivers. American Fisheries Society, Symposium 37, Bethesda, Maryland.

Rhoads, B. L., D. Wilson, M. Urban, and E. Herricks. 1999. Interaction between scientists and nonscientists in community-based watershed management: emergence of the concept of stream naturalization. Environmental Management 24:297–308.

Richards, K. S. 1982. Rivers. Form and process in alluvial channels. Methuen, London.

Riley, A. L. 1998. Restoring streams in cities. A guide for planners, policy makers and citizens. Island Press, Washington, D.C.

Robinson, M. J., Boardman, R. Evans, K. Heppell, J. Packman, and G. Leeks. 2000. Land use change. Pages 30–54 *in* M. Acreman, editor. The hydrology of the UK. Routledge, London.

Robison, E. G., and R. L. Beschta. 1990. Coarse woody debris and channel morphology interactions for undisturbed streams in southeast Alaska, U.S.A. Earth Surface Processes and Landforms 15:149–156.

Rodriguez-Iturbe, I. 2000. Ecohydrology: a hydrologic perspective of climate-soil-vegetation dynamics. Water Resources Research 36:3–9.

Rumney, G. R. 1970. The geosystem: dynamic integration of land, sea and air. Wm. C. Brown Company, Dubuque, Iowa.

Ruzickova, E., and A. Zeman. 1994. Paleogeographic development of the Labe River flood plain during the Holocene. Geological Institute of the Academy of Sciences, Prague:104–113.

Schumm, S. A. 1977. The fluvial system. John Wiley, New York.

Shields, F. D., and C. J. Gippel. 1995. Prediction of effects of woody debris removal on flow resistance. Journal of Hydraulic Engineering 121:341–354.

Shields, F. D., and N. R. Nunnally. 1984. Environmental aspects of clearing and snagging. Journal of Environmental Engineering, American Society of Civil Engineers 110:152–165.

Smith, R. D., R. C. Sidle, P. E. Porter, and J. R. Noel. 1993. Effects of experimental removal of woody debris on the channel morphology of a forest, gravel bed stream. Journal of Hydrology 152:153–178.

Smock, L. A. 1990. Spatial and temporal variation in organic matter storage in low-gradient headwater streams. Archiv fur Hydrobiologie 118:169–184.

Smock, L. A., G. M. Metzler, and J. E. Gladden. 1989. Role of debris dams in the structure and functioning of low gradient headwater streams. Ecology 70:764–775.

Spies, T. A., J. F. Franklin, and T. B. Thomas. 1988. Coarse woody debris in Douglas fir forests of western Oregon and Washington. Ecology 69:1689–1702.

Swanson, F. J., and G. W. Lienkaemper. 1978. Physical consequences of large organic debris in Pacific Northwest Streams. Pacific Northwest Research Station, USDA Forest Service General Technical Report PNW-166, Portland, Oregon.

Swanson, F. J., G. W. Lienkaemper, and J. R. Sedell. 1976. History, physical effects and management implications of large organic debris in western Oregon streams. USDA Forest Service General Technical Report PNW-56, Portland, Oregon.

Swanston, D. H., and F. J. Swanson. 1976. Timber harvesting, mass erosion and steepland forest geomorphology in the Pacific Northwest. Pages 199–221 *in* D. R. Coates, editor. Geomorphology and engineering. Dowden, Hutchinson and Ross, Stroudsburg, Pennsylvania.

Taylor, P. 1986. Respect for nature. Princeton University Press, Princeton, New Jersey.

Thompson, D. M. 1995. The effect of large organic debris on sediment processes and stream morphology in Vermont. Geomorphology 11:235–244.

Towes, D. A. A., and M. K. Moore. 1982. The effects of three streamside logging treatments on organic debris and channel morphology of Carnation Creek. Pages 129–153 *in* G. F. Hartman, editor. Proceedings of the Carnatian Creek Workshop: A Ten Year Review. Nanaimo, BC.

Triska, F. J. 1984. Role of wood debris in modifying channel geomorphology and riparian areas of a large lowland river under pristine conditions: a historical case study. Verhanlungen Internationale Vereinigung fur Theoretische und Angewandte Limnologie 22:1876–1892.

Viles, H. A., editor. 1988. Biogeomorphology. John Wiley, Chichester, UK.

Wallerstein, N., C. R. Thorne, and M. W. Doyle. 1997. Spatial distribution and impact of large woody debris in northern Mississippi. Pages 145–150 *in* S. S. Y. Wang, E. J. Langendoen, and F. D. Shields, editors. Management of landscapes disturbed by channel incision. University of Mississippi, Oxford.

Ward, G. M., and N. G. Aumen. 1986. Woody debris as a source of fine particulate organic matter in coniferous forest stream ecosystems. Canadian Journal of Fisheries and Aquatic Sciences 43:1635–1642.

Wassen, M. J., and A. P. Grootjans. 1996. Ecohydrology: an interdisciplinary approach for wetland management and restoration. Vegetatio 126:1–4.

Yamskikh, A. F., A. A. Yamskikh, and A. G. Brown. 1999. Siberian - type Quaternary floodplain sedimentation: the example of the Yenisei River. Pages 241–252 *in* A. G. Brown and T. Quine, editors. Fluvial processes and environmental change. John Wiley, Chichester, UK.

Young, W. J. 1991. Flume study of the hydraulic effects of large woody debris in lowland rivers. Regulated Rivers Research and Management 6:203–211.

Zalewski, M., M. Lapinska, and P. B. Bayley. 2003. Fish relationships with wood in large rivers. Pages 195–211 *in* S. V. Gregory, K. L. Boyer, and A. M. Gurnell, editors. The ecology and management of wood in world rivers. American Fisheries Society, Symposium 37, Bethesda, Maryland.

Zalewski, M., G. A. Janauer, and G. Jolankai. 1997. Ecohydrology. A new paradigm for the sustainable use of aquatic resources. International Hydrological Programme - V Projects 2.3/2.4. UNESCO, Paris, France.

Zimmerman, R. C., J. C. Goodlett and G. H. Comer. 1967. The influence of vegetation on channel form of small streams. Symposium on River Morphology, International Association of Scientific Hydrology 75:255–275.

American Fisheries Society Symposium 37:21–47, 2003

Geomorphic Effects of Wood in Rivers

David R. Montgomery

Department of Earth and Space Sciences
University of Washington, Seattle, Washington 98195, USA

Brian D. Collins

Department of Earth and Space Sciences
University of Washington, Seattle, Washington 98195, USA

John M. Buffington

Department of Civil Engineering, University of Idaho
800 Park Boulevard, Suite 200, Boise, Idaho 83712, USA

Timothy B. Abbe

Herrera Environmental Consultants, Inc.
2200 Sixth Avenue, Suite 601, Seattle, Washington 98121, USA

Abstract.—Wood has been falling into rivers for millions of years, resulting in both local effects on channel processes and integrated influences on channel form and dynamics over a wide range of spatial and temporal scales. Effects of stable pieces of wood on local channel hydraulics and sediment transport can influence rates of bank erosion, create pools, or initiate sediment deposition and bar formation. At larger spatial scales, changes in the supply of large wood can trigger changes in both river-reach morphology and the interaction between a river and its floodplain. Over long time scales, wood-rich rivers may retain more sediment and have lower sediment transport rates and steeper slopes than comparable wood-poor channels. Most geomorphic effects of wood in rivers arise from large, stable logs that catalyze changes in the routing and storage of both smaller wood and sediment. The size of a log relative to the channel provides a reasonable gauge of the potential stability of in-channel wood. Channels with a high supply of large, potentially stable wood may experience substantial vertical variability in bed elevation independent from external forcing (e.g., climate variability, temporal variations in sediment supply, or tectonic activity). In some river systems, changes in the wood regime, as described by the size and amount of wood supplied to a river, can result in effects as great as those arising from changes in the sediment supply or the discharge regimes. Consequently, an understanding of the geomorphic effects of wood is crucial for assessing the condition and potential response of forest channels.

Introduction

Over the past several decades, researchers have recognized that large wood has a significant effect on many channel processes. Recent studies show that wood affects channel processes at scales from channel roughness (Shields and Gippel 1995; Buffington and Montgomery 1999b; Manga and Kirchner 2000) to bed-surface grain size (Lisle 1995; Buffington and Montgomery 1999b), pool formation (Keller and Tally 1979; Lisle 1995; Montgomery et al. 1995; Abbe and Montgomery 1996), channel-reach morphology (Keller and Swanson 1979; Lisle 1986; Nakamura and Swanson 1993; Montgomery et al. 1996; Montgomery and Buffington 1997, 1998; Piégay and Gurnell 1997), and the formation of valley-bottom landforms (Abbe and Montgomery 1996; Montgomery et al. 1996; Gurnell et al. 2001). This chapter addresses the geological context of wood in fluvial environments, the extent of historical changes in wood loading, and the geomorphic effects of wood across this range of scales. We also propose a framework for using the relative size of large wood for assessing its geomorphic effectiveness in rivers. Our intent is not to simply review the recent literature on the geomorphic effects of wood in rivers, but to focus on developing a framework for interpreting both general and region-specific geomorphic effects of large wood across a range of scales. We begin with overviews of the geological and historical contexts in which to consider the changing roles of wood as a geomorphic agent in river systems.

Geological Context

Modern forests cover almost a third of Earth's land surface (Atjay et al. 1979), and wood has been entering streams and rivers for more than 400 million years, although its relative influence on freshwater ecosystems has varied substantially and becomes progressively more speculative back through geologic time (Figure 1). Indirect evidence suggests that primitive plants first colonized the land surface in the Middle to Late Ordovician (470–438 m.y.a.), but fossils document the existence of terrestrial plants by the Late Silurian (423–408 m.y.a.; DiMichele and Hook 1992). The evolution of land plants coincides with the first appearance of meandering stream deposits in the geologic record, before which only braided channel deposits appear in fluvial deposits (Schumm 1968; Cotter 1978). Development of a meandering channel pattern requires bank cohesion (Schumm 1963), which, with the advent of terrestrial vegetation, was presumably provided by riparian root strength. By the Late Devonian (384–360 m.y.a.), forested areas dominated by small trees had become common, although trees with diameters up to 1 m were present in low abundances (DiMichele and Hook 1992). Vegetation covered much of the world's land mass during the Carboniferous (360–286 m.y.a.), and logjams probably were common in many Carboniferous rivers, such as those whose deposits have been mined for coal in recent history. Presumably, the geomorphic role of wood in rivers increased as trees grew larger and more abundant later in the Paleozoic, at least until the Permian-Triassic (P/T) extinction event (Figure 2), when meandering channels worldwide reverted to the

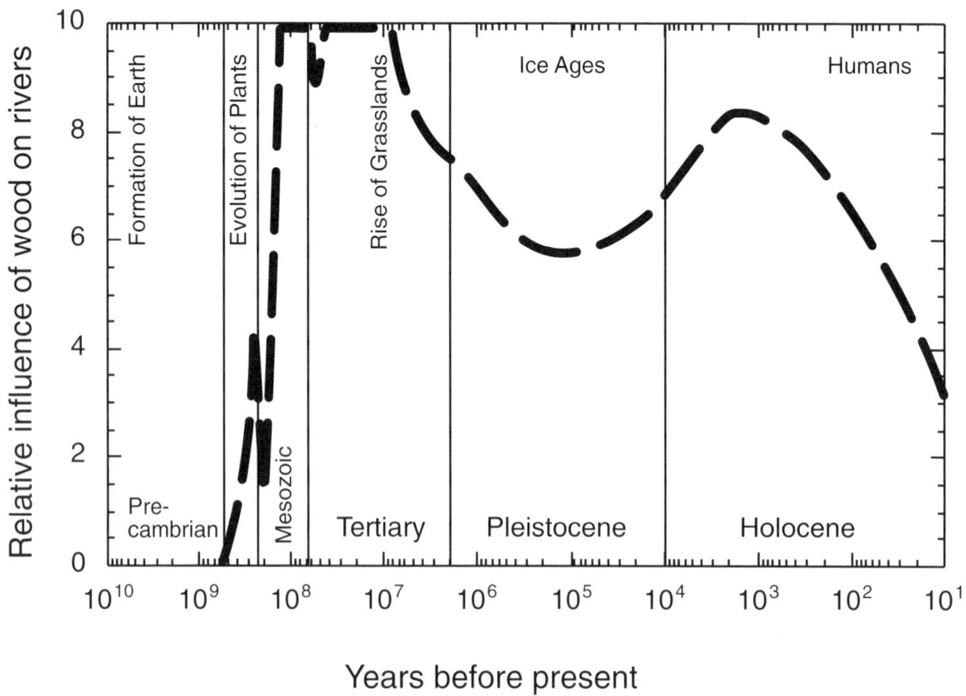

FIGURE 1. Hypothesized relative influence of wood on the world's rivers throughout geologic time.

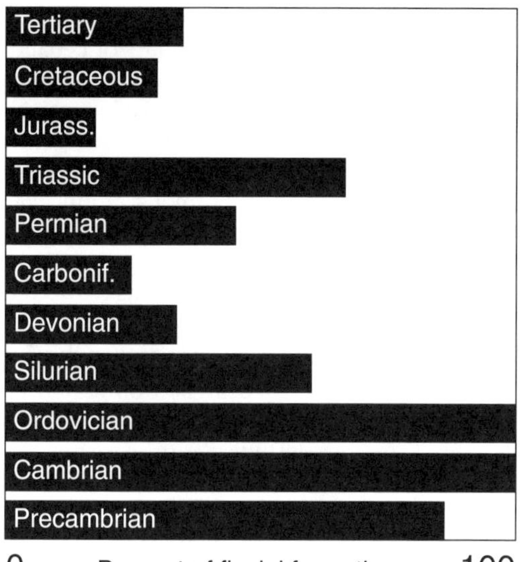

FIGURE 2. Plot of the percentage of fluvial deposits interpreted as deposited by braided channel systems for periods of geologic time from the Precambrian to the Tertiary (data from Cotter 1978). Note the original dominance of braided channel deposits, their gradual decline in relative abundance as they were replaced by meandering channel deposits, and the increase in the relative abundance of braided channels at the Permian/Triassic boundary (Ward et al. 2000).

braided morphology typical of Precambrian fluvial deposits (Ward et al. 2000). The apparent global die-off of terrestrial vegetation at the P/T event would have dramatically reduced the role of wood in world rivers.

After the P/T event, trees generally increased in size during the Mesozoic (248–65 m.y.a.), hence we hypothesize that so did the influence of wood in rivers. Temperatures were warmer than at present, especially at high latitudes, and the equatorial regions were more arid than today. Primitive conifers were the largest trees in Triassic forests, and some Late Triassic (228–208 m.y.a.) floodplain forests supported dense stands of trees up to 60 m tall (Wing and Sues 1992). In the warm Jurassic climate (208–144 m.y.a.), vegetation assemblages that included numerous conifer species extended to high latitudes (Gee 1989). Conifers were still the dominant large trees, and Late Jurassic (175–148 m.y.a.) forests of northwestern China supported conifers up to at least 2.5 m in diameter (McKnight et al. 1990). Spicer and Parrish (1987) report fossil conifer logs up to 50 cm in diameter in Late Cretaceous fluvial deposits as far north as 80° to 85° latitude. Moreover, they show evidence that logs were not transported far from their site of growth, which implies that they were relatively stable in the streams in which they were deposited. The diversity of smaller angiosperms increased during the Cretaceous to the point that they accounted for most of the species in typical fossil assemblages and particularly those from riparian corridors. The asteroid impact at the Cretaceous/Tertiary (K/T) boundary 65 m.y.a. caused devastating ecological disruption in western North America but less severe effects on terrestrial biota in the Southern Hemisphere (Wing and Sues 1992). After the K/T event in the Early Tertiary, floral diversity recovered gradually, but preferential extinction of conifer species at the K/T boundary is hypothesized to have enriched Northern Hemisphere floras in deciduous taxa (Wolfe 1987). All told, the influence of wood in rivers is likely to have been substantial throughout the Mesozoic.

Cenozoic changes in global vegetation patterns seem likely to have imparted significant variability to the role of wood in world rivers. In the Early Oligocene, broad-leaved forests expanded into areas previously covered by conifers (Wolfe 1985) and ultimately came to cover large regions of the Northern Hemisphere (Potts and Behrensmeyer 1992). The global climate began cooling in the Oligocene (37–23 m.y.a.), a trend that accelerated in the latter half of the Miocene (14–5 m.y.a.). Extensive savannas and grasslands became established by the Pliocene, and closed canopy forests were increasingly replaced by more open woodlands (Potts and Behrensmeyer 1992). With the onset of Pleistocene glaciations 2 m.y.a., the extent of forests oscillated in the temperate and polar latitudes, with forest cover expanding during warm interglacial periods and contracting during glacial periods (Porter 1983; Velichko 1984; Huntley and Webb 1988). During glacial maxima, only a few European rivers in small isolated areas of forest refugia would have had significant wood loading. In contrast, parts of eastern Australia have been continuously forested for more than 100 million years. The recession of glaciers in regions such as northern Europe was followed by rapid forest expansion early in the Holocene, about 10,000 years ago (Wright 1983; Velichko 1984). European forests reached their maximum Holocene extent before the onset of deforestation that fueled the rise of human civilization (Perlin 1989). Humans have reduced global forest cover to about half its maximum Holocene extent and eliminated all but a fraction of the world's frontier forest (Figure 3). The influence of wood on world rivers during most of the Holocene was likely intermediate between the more extensive effects hypothesized for the warm Mesozoic climate and the more limited influence during major glaciations.

Historical Context

In most of the industrialized nations, the primeval character of rivers is lost or unknown because ancient deforestation and river clearing to improve navigation changed the input of wood to rivers long before recorded history. The United States and Australia are unusual in that they were technologically advanced bureaucracies at the time that many of their rivers were first modified and their forests first cleared. Consequently, historical records of instream wood removal and clearing of riparian forests can now be used to reconstruct and evaluate how the role of wood in some river systems has changed in the last two centuries. We must rely on more indirect methods for regions cleared in earlier times.

The effect of forest clearing on soil erosion in the ancient Mediterranean had dramatic long-term effects on the region and its inhabitants (Dale and Carter 1955). The famous cedars of Lebanon once covered nearly 2,000 square miles, but by Roman times the forest had been so extensively cutover that Emperor Hadrian protected groves for the use of the Roman fleet (Lowdermilk 1950). The Lebanese cedars were enormous; around 2600 B.C., a shipment from Phoenicia to Egypt arrived with logs 100 cubits long (just over 170 ft; Perlin 1989). By 1950, this ancient forest was reduced to several tiny groves (Lowdermilk 1950). The similar, but more recent, forest destruction in the Chinese Province of Shansi caused extensive soil erosion and substantial adverse effects to regional river systems (Lowdermilk 1926).

Human alteration of forests in Europe has been significant for at least 6,000 years (Williams 2000), and river clearing and engineering date to the Roman era (Herget 2000). Forests in southern Europe were already confined to mountainous areas by classical times, and the clearing of central and western European forests that began in the Neolithic expanded dramatically in the Middle Ages (Darby 1956). Even with such early forest clearing, logs and logjams were depicted in early European books on rivers (Figure 4). By the 18th century, the fortunes of European naval powers were influenced by access to distant sources of timber for building ships, particularly trees large

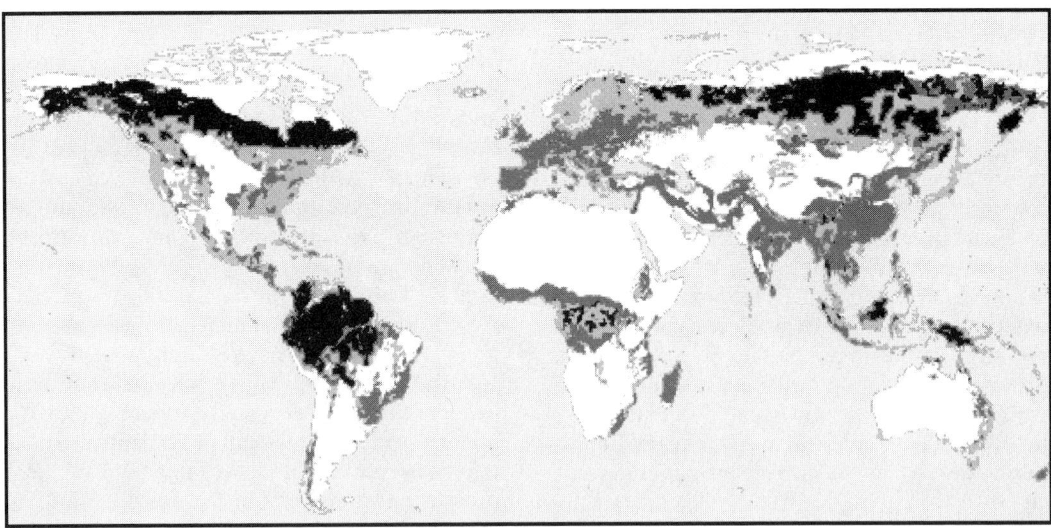

FIGURE 3. Global extent of frontier forest today (darkest shade), nonvirgin forest (medium shade), and forest cleared during the Holocene (lightest shade), and nonforested (white) showing that about half of the original Holocene forest cover is gone and about one-fifth remains in large tracts of relatively undisturbed forest. Modified from World Resources Institute (http://www.wri.org/wri/ffi/maps/).

FIGURE 4. Portions of woodcuts showing logs and log jams in 17th century European rivers (Herbinio 1678).

enough to be suitable for masts (Albion 1926; Bamford 1956). The English navy, for example, depended on large, old-growth white pine from the American colonies to provide masts for their ships (Albion 1926).

The forests of eastern North America were vast and dense; John Bartram writing in 1751 characterized the tree tops as "so close to one another for many miles together, that there [was] no seeing which way the clouds drive, nor which way the wind sets: and it seem[ed] almost as if the sun had never shown on the ground, since the creation" (quoted in Whitney 1996). Remnant stands of eastern old-growth mixed conifer-hardwood or hemlock-white pine stands have a biomass of 560–820 t/ha (Whitney 1996), nearly as much as that of the dense, highly productive coniferous forests of the Pacific Northwest (Franklin and Dyrness 1973). White pine often grew to diameters of 2–3 m and heights of 45–60 m, while sycamore attained diameters greater than 4 m, tulip tree diameters of almost 2 m, and cottonwood and oak reached diameters well in excess of 2 m (Whitney 1996).

Although the European forests were cleared over centuries and millennia, the vast forests of North America vanished much more rapidly, often over the course of decades (Whitney 1996). The earliest known clearings by indigenous populations began about 1,000 years ago (Williams 2000). European settlers began more widespread clearings for farming and fuel in the 18th and early 19th centuries. Forest stands near rivers were usually among the first logged because early transport was primarily by log rafts (beginning about 1600) and log drives (from about 1800) on rivers. Industrial timbering greatly expanded the rate of clearing in the late 19th century, due in part to the development of canals and railroads for timber transport (Whitney 1996). Harvest of the eastern forests was complete and thorough; by 1920, the Northeast and Midwest had lost 96% of their old-growth timber. By late in the 20th century, estimates of original old-growth forest remaining in individual states ranged from less than 0.01% to 0.36% (Table 8-4 in Whitney 1996).

Trees that had accumulated in the rivers draining these massive forests were cleared throughout the 19th and 20th centuries. The large rivers of the eastern and Midwestern United States were perceived as "national highways and arms of the sea," and their clearing was a matter of great commercial and military importance (Hill 1957). In 1824, Congress made its first appropriation to remove snags from the Mississippi and Ohio rivers, and the first snagboat designed for that purpose was built by 1829. More than 800,000 snags were pulled in a 50-year period along the lower Mississippi alone; these cottonwood and sycamore snags averaged 1.5 m in diameter at the base and 0.6 m at the top and had an average length of 37 m (Sedell et al. 1982). Triska (1984) reported that the largest logs in the Red River, Louisiana were up to 36 m long and 1.75 m in diameter. In the following decades, snagging extended to rivers throughout the Southeast and Midwest.

Later in the 19th century, snagging extended to rivers in the West Coast region (Sedell and Froggatt 1984; Collins et al. 2002). Rivers were sometimes snagged and massive log rafts dismantled, even before the surrounding forests were logged, because rivers were the primary means for settlers and loggers to access upstream lands. In the Puget Sound region, river clearing began around 1880 and was followed immediately by clearing of valley-bottom forests. By 1900, essentially all virgin low-elevation riparian forests were gone (Plummer et al. 1902). Snagging was part of a general effort for channelization that also included plugging and disconnecting secondary channels and floodplain sloughs (Sedell and Froggatt 1984). Snagging records attest to the giant sizes that trees could attain in the Pacific Northwest; snags removed from rivers in western Washington were as large as 5.3 m in diameter (Collins and Montgomery 2001). Snagging records also suggest wood loading in large Pacific Northwest rivers 100 times greater than now (Sedell and Froggatt 1984). A similar difference was found in several Puget Sound rivers by comparing present-day wood loading in a protected reach of the lower Nisqually River to cleared reaches of the Stillaguamish and Snohomish rivers (Collins et al. 2002).

While large rivers were being cleaned of snags and jams, smaller tributary streams were catastrophically cleared through the ubiquitous practice of splash damming, in which a dam-break flood was induced to transport trees to the larger rivers from where logs were then rafted to market. Splash damming was common in the Northwest (Sedell and Duvall 1985), Midwest, and Northeast (Sedell et al. 1982). These torrents scoured sediment and wood from streambeds and banks and reduced roughness and obstructions to flow, leaving some channels scoured down to bedrock.

Few places remain where the effects of wood can be studied in rivers relatively unmodified by human actions. The Queets River on the Olympic Peninsula, Washington is one such river where extensive field work has documented that logjams trap sediment and deflect flow and thereby affect channel morphology and processes throughout a mountain channel network (Abbe and Montgomery 1996, 2003). We have observed wood accumulations identical to types found in the Queets system in gravel-bed rivers that drain old-growth rainforest in the coast range of northern New South Wales, Australia and in the southern Alps of New Zealand. The Rio Beni, Bolivia, a large tributary to the Amazon, has huge snags that create navigation hazards in some reaches and that armor riverbanks or form large logjams in other reaches. Types and patterns of in-channel wood seen by the senior author during a reconnaissance expedition down the Rio Beni resemble those portrayed in Karl Bodmer's paintings of the Mississippi River in the 1840s (Bodmer et al. 1984), suggesting that rivers of the upper Amazon may offer reasonable analogs for the prehistoric role of wood in other large rivers. Areas in far eastern Siberia and central Africa may offer additional sites to study the dynamics and effects of wood in rivers that flow through relatively pristine forests, but these regions are being deforested rapidly. Although the role of wood has been studied in some relatively undisturbed forest streams, most of the research to date on wood in large rivers has been from relatively altered systems.

Geomorphic Effects of Wood

Three themes dominate any discussion of the geomorphic effects of wood on rivers: changes in sediment routing and storage, channel dynamics and processes, and channel morphology. These different effects are produced through direct and indirect influences of wood across a wide range of scales. Our discussion will proceed from small length scales to large because the local direct effects of wood are easiest to see. Indirect effects generally become apparent over larger spatial and temporal scales, but they have been more widely affected and obscured by historic changes and are generally less well recognized.

When a tree falls into a river, it may remain intact or break into smaller, more mobile pieces. Depending on the size of the tree and the size of the channel it fell into, the tree may remain stable at or near where it entered the channel or it may be transported downstream to lodge against the bank or in a logjam, become stranded on a bar top or floodplain during falling flow, lodge in the riverbed as a snag, or transit the river system and leave the basin. Many of the geomorphic effects of wood in rivers arise from the influence of large stable wood as obstructions to flow and sediment transport. The influence of "key" pieces of large wood on stabilizing other pieces in logjams has been recognized for many years (Habersham 1881; Deane 1888; Russell 1909; Keller and Tally 1979; Nakamura and Swanson 1993; Abbe and Montgomery 1996). The distribution of instream wood and the accumulation of logs into logjams have been studied extensively in Europe and North America (Gregory et al. 1985, 1993; Robison and Beschta 1990; Gregory and Davis 1992; Piégay 1993; Abbe and Montgomery 1996; Gurnell and Sweet 1998; Piégay and Marston 1998; Downs and Simon 2001). A number of workers have noted how the organization of wood, the changes catalyzed by its introduction into channels, and styles of wood transport vary with position in the channel network (Keller and Swanson 1979; Swanson et al. 1982; Abbe and Montgomery 1996, 2003; Wallerstein et al. 1997; Gurnell et al. 2001). Many of the commonly recognized effects of wood on aquatic habitat are manifest at the scale of individual channel units (Bisson et al. 1982), but these local influences can generate emergent properties at larger spatial scales of channel reaches, valley bottoms, and even at the landscape scale (Figure 5).

Channel unit

At the channel-unit scale, wood can dramatically affect the size and type of pools, bars, and steps in coarse-grained channels (Lisle 1986; Keller et al. 1995; Montgomery et al. 1995; Woodsmith and Buffington 1996; Hogan et al. 1998).

Pool scour.—Wood is an effective flow obstruction that alters channel hydraulics and enhances scour of pools (Figure 6). Depending on its orientation and position above the bed, wood can form four basic obstruction types (vertical, pitched, horizontal, and step), each of which corresponds with common channel unit types (Figure 7).

Vertical obstructions behave like bridge abutments and piers, creating horseshoe vortices and turbulent eddies that scour the channel bed (Melville 1992, 1997), with scour depth increas-

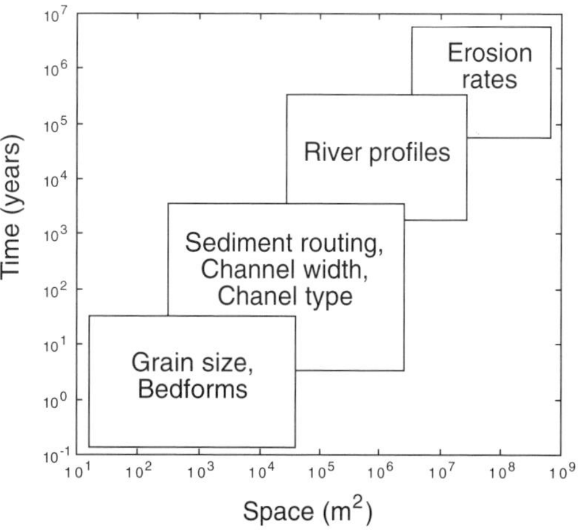

FIGURE 5. Scales of geomorphic influences of wood in the world's rivers.

FIGURE 6. Plunge pool formed by log step in headwater channel.

ing as the obstructed width increases relative to flow depth (Buffington et al. 2002 b). Lateral vertical obstructions (abutments) can be formed by wood jams sutured to one or both banks: isolated vertical obstructions (piers) can be created by in-channel rootwads and debris piles (Buffington et al. 2002b). In the channel-unit vernacular, vertical obstructions form scour pools, eddy pools, or dammed pools (Robison and Beschta 1990).

Logs located above the bed and subparallel to the water surface form horizontal obstructions that advect flow downward toward the bed. Pitched obstructions are logs oriented obliquely to the water surface. The flow structure and scouring mechanisms of pitched logs are similar to pier-type vertical obstructions, but they are complicated by flow accelerations over, under, and around the debris. Pool scour for both horizontal and pitched obstructions depends on log diameter, angle of attack, flow velocity, height above the bed, and pitch from horizontal (Beschta 1983; Cherry and Beschta 1989). Pitched logs create scour pools and eddy pools, and horizontal logs form underscour pools (Robison and Beschta 1990; Woodsmith and Buffington 1996).

Wood steps create downstream turbulent jets that scour plunge pools (Heede 1972; Marston 1982; Chin 1989; Wohl et al. 1997). The depth to which plunge pools are scoured depends on jet energy, flow depth, and the degree of turbulence (Mason and Arumugam 1985; Bormann and Julien 1991). Pools scoured by flow around large debris jams typically have larger and more variable depths than other types of pools (Abbe and Montgomery 1996; Collins et al. 2002). In addition to the orientation and size of wood, scour depth is influenced by channel geometry, bed-surface grain size, particle form drag (relative roughness), supply and caliber of bed load, and channel gradient (Buffington et al. 2002b).

Bar deposition and sediment storage.—Wood can force the formation of bars by creating organic dams that physically block sediment transport or by forcing local flow divergence and consequent sediment deposition. Where flow deflection by wood scours a pool, a complimentary bar will develop, partially defining the boundaries of the associated pool. Sediment deposition forced by wood can be significant. In some systems sediment storage associated with wood exceeds the annual sediment yield by more than 10-fold, thereby regulating the transport of sediment through the channel system (Megahan and Nowlin 1976; Swanson et al. 1976; Mosley 1981; Hogan 1986; Bilby and Ward 1989; Nakamura and Swanson 1993; Keller et al. 1995; Pitlick 1995). In a channel system with a high load of wood, sediment storage associated with logs and logjams can act as a sediment capacitor and significantly damp variability in sediment transport rates (Massong and Montgomery 2000; Lancaster

Obstruction *Pool*

Vertical → Dammed, Scour, Eddy

Pitched → Scour, Eddy

Horizontal → Underscour

Step → Plunge

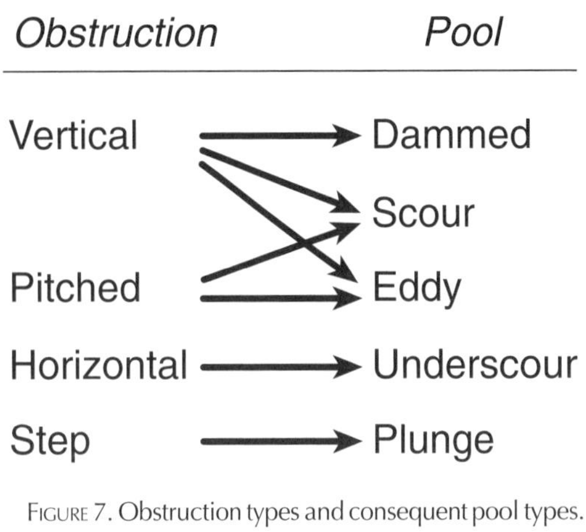

FIGURE 7. Obstruction types and consequent pool types.

et al. 2001). Conversely, the destruction of debris dams can result in large decreases in storage and related increases in sediment transport (Beschta 1979; Bilby 1981; Megahan 1982; Heede 1985; Smith et al. 1993a). Whether or not the failure of an in-channel debris dam causes a debris flow depends, in part, on the volume and stability of wood in a channel (Swanston and Swanson 1976).

Channel width.—Wood influences channel width by either armoring channel banks and maintaining relatively narrow sections or by locally directing flow into the banks, causing localized erosion and channel widening. Consequently, channel width can vary considerably within a single reach of a forest channel. For example, reach-scale variations in the bank-full width along portions of the Tolt River, Washington, flowing through old-growth forest, show that logjams can superimpose dramatic local variability on the hydraulic geometry of river systems (Figure 8). Zimmerman et al. (1967) also found that local channel widening due to flow deflection by vegetation dominated the variability of channel width at drainage areas less than about 1 km^2. Similar observations have been made in forest channels in northern California (Keller et al. 1995), western Oregon (Nakamura and Swanson 1993), coastal British Columbia (Hogan 1986), and southeastern Alaska (Robison and Beschta 1990; Buffington and Montgomery 1999b). Independent studies from Wisconsin (Trimble 1997) and New Zealand (Davies-Colley 1997) have shown that forested channel reaches are both wider and have greater variance in width than do channels flowing through grasslands. Murgatroyd and Ternan (1983) found that reforestation increased bank erosion and channel width of Narrator Brook, England. In contrast, Stott (1997) reported that erosion rates of forested streambanks in Scotland were less than in moorland channels because a forest canopy inhibited the formation of frost that triggered bank failure. In a series of field experiments, Smith (1976) found that riverbank erosion rate was inversely related to the percent of roots in channel banks, indicating that vegetation can substantially retard bank erosion. Triska (1984) reported that the huge raft jams on the Red River dramatically narrowed the channel; above and below the reach influenced by the raft the channel was 180–230 m wide,

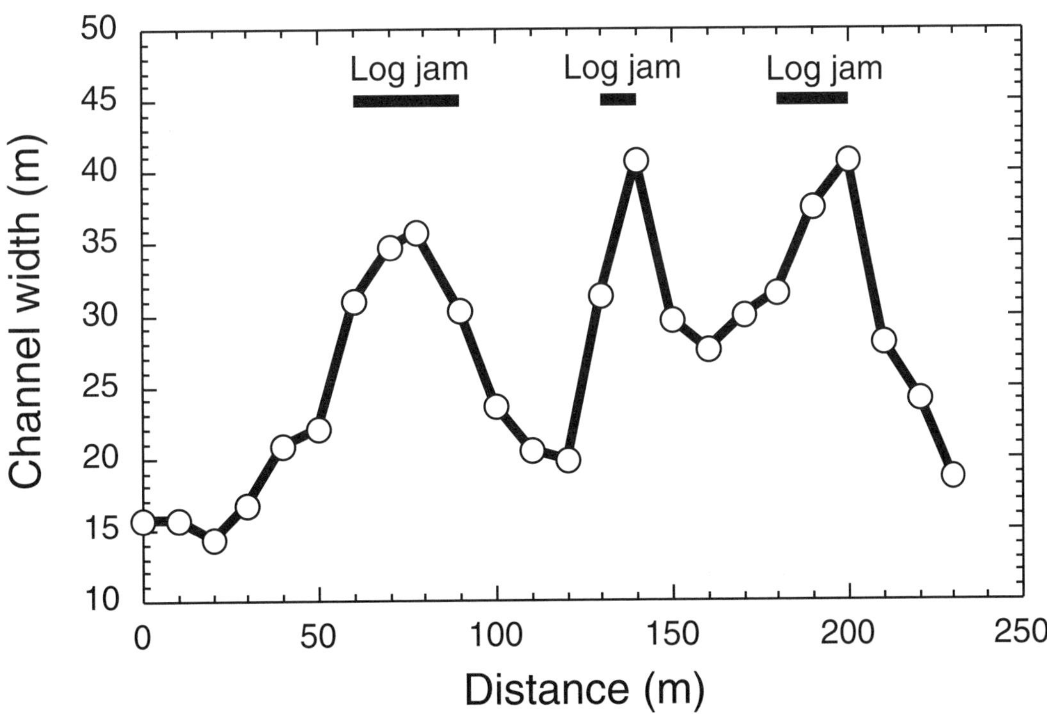

FIGURE 8. Channel width measured every 10 m along a reach of the Tolt River, Washington on 3 August 1993.

whereas the channel width in the area of the raft was only 27–46 m. After removal of the raft, the constricted portion of the channel widened dramatically. Flow splitting by stable logs or logjams can reduce channel widths by increasing the number of channels and splitting a single-thread channel into a system of smaller anastomosing channels (Harwood and Brown 1993). Hence, in-channel wood and riparian vegetation can either increase or decrease both the local and average width of channel reaches depending upon the geomorphic context of the reach in question (Thorne 1990).

Channel reach

Pool spacing.—At the reach scale, wood can control the frequency of pools and bars in gravel-bed rivers. Previous studies have shown that, in forest rivers, mean pool spacing is inversely related to wood frequency (Montgomery et al. 1995; Beechie and Sibley 1997). However, pool spacing can be quite variable for a given wood frequency (Figure 9), reflecting both regional and site-specific differences in channel and wood characteristics. Regional differences in mean pool spacing

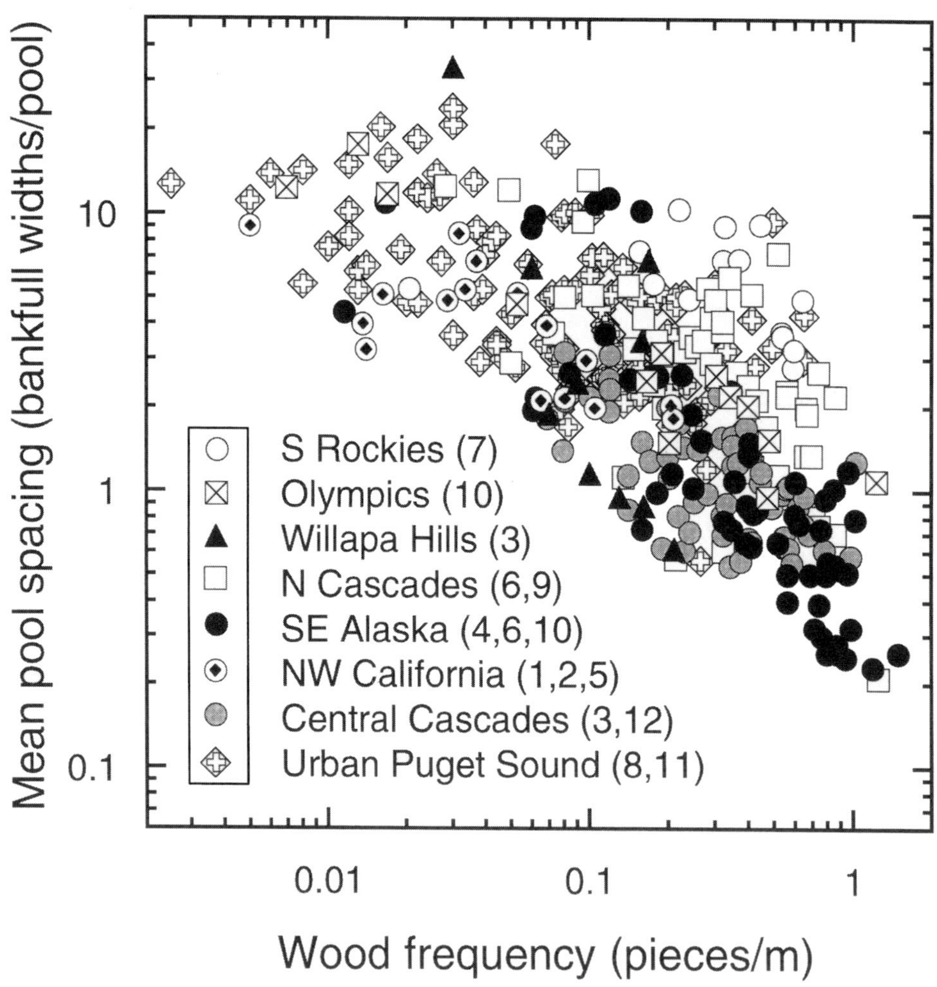

FIGURE 9. Mean pool spacing versus wood frequency. Numbers in parentheses show data sources: (1) Keller and Tally (1979); (2) Florsheim (1985); (3) Bilby and Ward (1989, 1991); (4) Buffington and Montgomery (unpublished data); (5) Keller et al. (1995); (6) Montgomery et al. (1995); (7) Richmond (1994); (8) May (1996); (9) Beechie and Sibley (1997); (10) Buffington and Montgomery (1999b); (11) Larson (1999); (12) Turaski (2000). Modified from Buffington et al. (2003a).

also reflect differences in lithology, climate, tree species, wood size, and land management.

Hydraulic roughness, channel competence, and bed-surface grain size.—Wood can create significant hydraulic roughness, which influences flow velocity, discharge, and shear stress. In a study of gravel-bed channels in old-growth forests of southeastern Alaska, Buffington (2001) found that nearly 60% of the total bank-full shear stress was spent on form drag caused by wood. Manga and Kirchner (2000) report a similar value for a spring-fed channel in the Oregon Cascades. Assani and Petit (1995) found that removing logjams in a small gravel-bed channel in Belgium caused a more than 50% increase in near-bed shear stress and commensurately reduced bedload transport rates. Shields and Gippel (1995) report that historical wood removal from rivers in Australia, the United States, and the United Kingdom reduced values of Manning's *n* by 10% to more than 90% in those channels. Loss of large wood in alluvial channels, however, may be partially compensated for by development of larger amplitude bedforms or coarser bed-surface grain sizes after wood removal (MacDonald and Keller 1987; Smith et al. 1993b). Smith et al. (1993b) found that removing wood from a small pool-riffle channel reorganized pool and bar topography and increased bed-load transport rates, both from release of sediments previously stored behind wood dams and from increased boundary shear stress and greater transport capacity.

In addition to creating its own hydraulic resistance, wood can force spatial variations in shear stress and stream power that alter bed and bank topography. Specifically, wood increases the number of pools and bars and forces spatial variations in channel width. Consequently, wood creates more irregular bed and bank topography, generating additional form drag that further increases channel resistance (Buffington and Montgomery 1999b).

The combined effects of roughness caused by wood, the bed, and banks can significantly reduce bed shear stress and channel competence, diminishing reach-average surface grain sizes (MacDonald and Keller 1987; Assani and Petit 1995; Lisle 1995; Buffington and Montgomery 1999b; Manga and Kirchner 2000). Buffington and Montgomery (1999b) found that diminished channel competence resulting from bar, bank, and wood roughness can cause up to a 90% reduction in median surface grain size relative to that predicted for a low-roughness, wide, planar channel (Figure 10). At subreach scales, wood can force strong spatial variations in shear stress and sediment transport, resulting in spatially variable textural patches (grain-size facies; Figure 11). The effect of wood on both absolute grain size and spatial variability of particle sizes may influence both the availability and diversity of aquatic habitats in forest channels. For example, recent theoretical models indicate that bar and wood roughness can lead to the deposition of spawning gravels in steep mountain drainage basins that otherwise would only have large bed material inhospitable to salmonids because of high shear stresses (Buffington 1998).

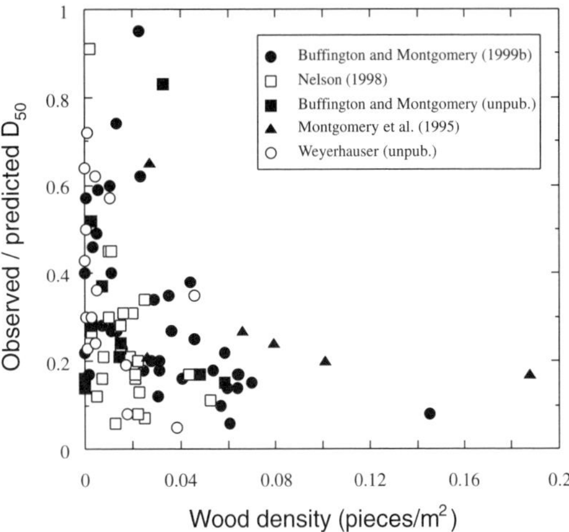

FIGURE 10. Ratio of observed to predicted median surface grain size (D_{50}) versus wood pieces/m^2. The predicted D_{50} is the competent value for the total bank-full shear stress (that of a wide, planar channel): it is determined from the Shields equation for a dimensionless critical shear stress of 0.03. Bar and bank roughness as well as sediment supply limitations cause the D_{50} ratio to be less than 1 at zero wood loadings.

Channel type.—The effect of wood on channel hydraulics and sediment transport provides a strong control on reach-scale channel morphology in forested basins. For example, wood may create steps that dissipate energy otherwise available for sediment transport (Keller and Swanson 1979; Heede 1981; Marston 1982). Such organic steps may trap sediment and maintain an alluvial bed in channels with an otherwise bedrock bed due to either rapid downstream fining or lack of coarse sediment. Removing organic steps can transform a forced step-pool channel into either a step-pool, cascade, or bedrock channel depending on slope, discharge, and sediment load. Similarly, wood may force a pool-riffle morphology in otherwise plane-

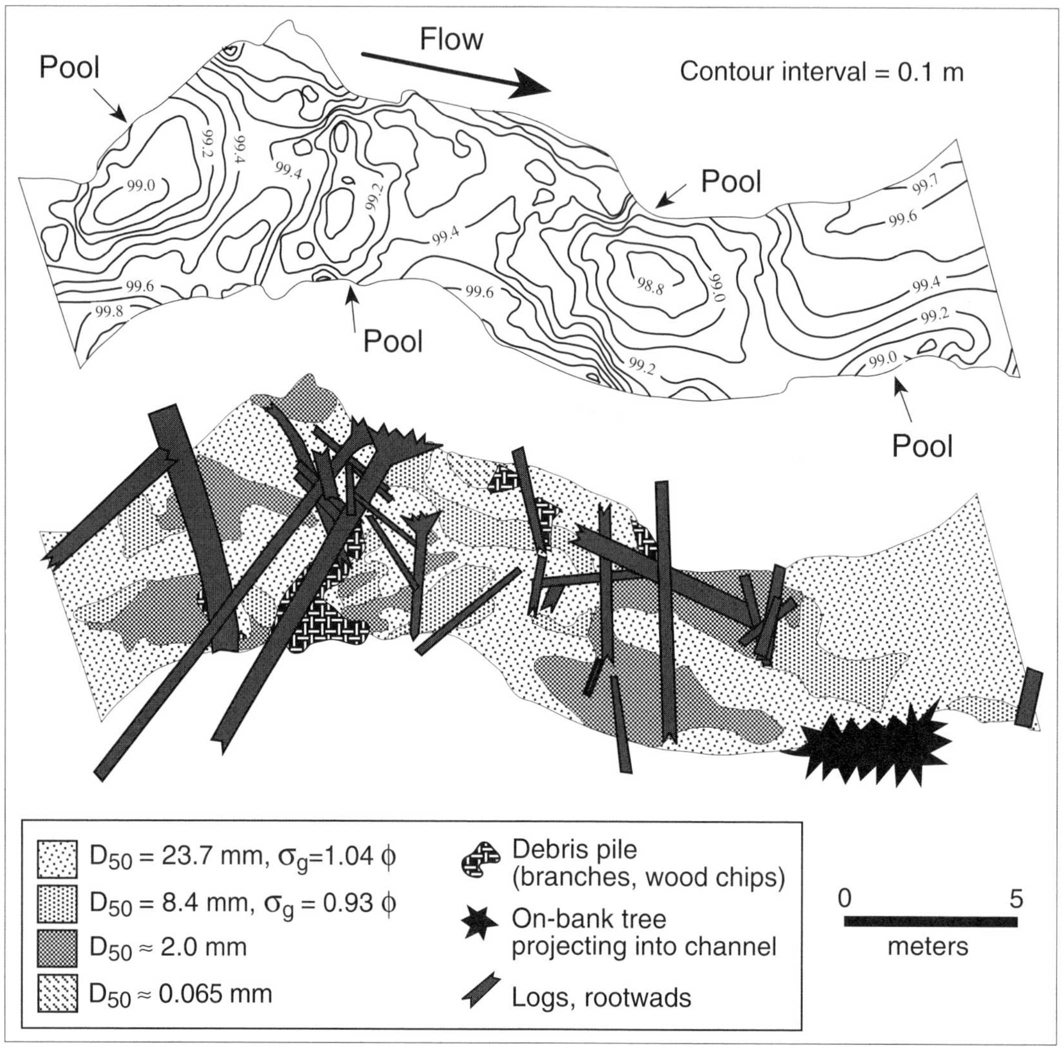

FIGURE 11. Channel topography, wood loading, and bed-surface textures in Mill Creek, western Washington (modified from Buffington and Montgomery 1999a). σ_g is the graphic standard deviation of particle sizes (Folk 1974).

bed or bedrock reaches (Swanson et al. 1976; Montgomery et al. 1995, 1996).

Because of the major influence of wood on channel morphology, finding free-formed pools and bars can be difficult in moderate-gradient (that is, >0.01), cobble- and gravel-bed forest rivers. Many studies indicate that most pools in such channels are either formed by or strongly influenced by wood (Zimmerman et al. 1967; Keller and Swanson 1979; Lisle 1986; Andrus et al. 1988; Robison and Beschta 1990; Keller et al. 1995; Montgomery et al. 1995; Abbe and Montgomery 1996; Woodsmith and Buffington 1996).

Valley bottom

In his revolutionary *Principles of Geology*, English geologist Charles Lyell commented on the geological significance of vast accumulations of logs in North American rivers (Lyell 1837). Wood rafts that completely blocked large lowland rivers significantly influenced the channel and valley-bottom morphology, gradient, and sediment transport capacity (Veatch 1906; Guardia 1933; Triska 1984; Harvey et al. 1988). These effects could be both extensive and persistent. Spanish explorers reported a large, channel-spanning wood raft on the lower Colorado River, Texas, in 1690 (Clay 1949). In the three decades after the raft was removed by the U.S. Army in 1927, sediment eroded from the channel created a delta extending all the way across Matagorda Bay (Hartopo 1991). Early European explorers also recorded massive wood rafts on the Red River, Louisiana and reported that Native Americans could not recall a time when jams did not block the river (Lowrey 1968). Veatch (1906) estimated that the "Great Raft" of the Red River, which in 1875 affected 390–480 km of the river, began accumulating in the late 1400s. Veatch also reported incision of 1–5 m over a 24-km reach of the Red River in the two decades after the last logjam was removed. In addition to incision following raft removal, Triska (1984) reported how channel blockage by logjams on the Red River flooded large tracts of riparian lands creating a series of lakes where tributaries entered the main river. Consistent with this observed entrenchment, Harvey et al. (1988) estimated that removing the Red River jams increased the river's transport capacity by sixfold. A raft jam that blocked more than 1 km of the Skagit River, Washington was observed to cause "the river to overflow its banks annually, flooding 150 square miles, more or less, of rich bottom lands" (Habersham 1881, p. 2606). After the jam was removed, even the highest spring flood in the memory of local settlers did not overtop the riverbanks, though before jam removal, the floodplain was often inundated during "snow floods" (Habersham 1881). These large raft jams strongly influenced interaction between the channel and the valley bottom.

When accumulated in jams, wood can influence channel pattern and floodplain processes in large forest channels (Figure 12) and especially the formation of side channels and channel avulsions (Bryant 1980; Sedell and Froggatt 1984; Harwood and Brown 1993; Nakamura and Swanson 1993; Abbe and Montgomery 1996; Collins et al. 2002). Harwood and Brown (1993) found that stable logs and logjams split flow into multiple channels and maintained an anastomosing channel form in a wooded, seminatural channel in Ireland. Although anastomosing channels are now rare in northwest Europe, they were more common earlier in the Holocene (Brown and Keough 1992). Studies of historical characteristics of valley bottoms in the Puget Sound area show that logjams can maintain floodplain sloughs, trigger avulsions, and thereby both create and maintain an anastomosing channel pattern (Collins and Montgomery 2001, 2002). Sedell and Froggatt (1984) showed how the

FIGURE 12. Large meander jam on Queets River, Washington.

Willamette River was a complex of anastomosing channels in 1854 that was progressively confined to a single-thread channel as debris jams were removed, the channel dredged, the side channels closed off, and the valley-bottom forest cleared. In many areas, anastomosing channel patterns appear to have been more common in natural forest channels than they are today.

The naturalist John Muir also recognized how wood on valley bottoms affected small mountain streams. In describing an early visit to the sequoia groves of the southern Sierra Nevada, he noted that

> a single trunk falling across a stream often forms a dam 200 ft long and ten to thirty feet high, giving rise to a pond which kills the trees within its reach; these dead trees fall in turn, thus clearing the ground; while sediments gradually accumulate, changing the pond into a bog or drier meadow... In some instances a chain of small bogs rise above one another on a hill-side, which are gradually merged into one another, forming sloping bogs and meadows. [Muir 1878, p. 827.]

The changes Muir described parallel those observed in the different context of confined headwater channels on the Olympic Peninsula, where sediment impounded by logjams can form terraces that exceed 10 m in height and store more than 10,000 m^3 of sediment (Abbe 2000). Even in unconfined reaches of large channels, sediment deposited in the lee of individual logjams can become integrated into a patchwork mosaic of individual surfaces that coalesce to form a multilevel floodplain with substantial local relief (Abbe and Montgomery 1996). Brooks and Brierley (2000) showed that removing instream wood and riparian forest in a river in southeastern Australia resulted in a 10-fold increase in channel size and sediment transport capacity. The formation of valley-bottom landforms—from pieces of a floodplain to entire floodplains and even terraces—can be catalyzed by a single log, logjam, or the integrated effects of many logjams.

Deposition associated with stable logs and logjams adds an intrinsic vertical dynamic to forest channels in which the channel can incise or aggrade without changes in external boundary conditions such as climate or tectonics. The effect of such vertical variation in bed surface elevation on valley-bottom topography is shown by cross-

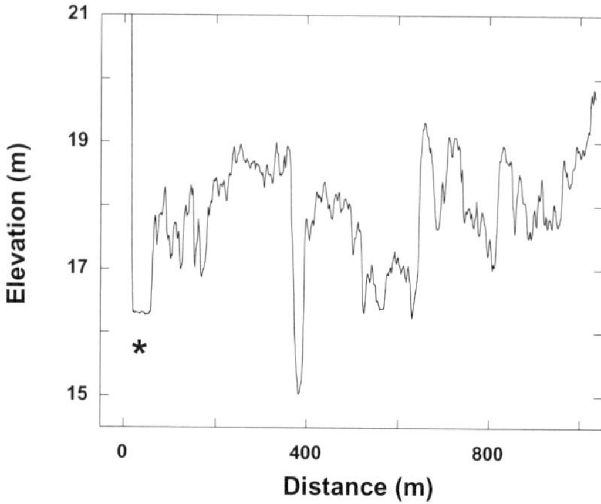

FIGURE 13. Topographic cross-sections across the floodplain of the Nisqually River, Washington, showing the multi-elevation nature of the active floodplain surface due to flow splitting and deposition associated with logjams.

sections from detailed elevation transects across active river floodplains in western Washington (Figure 13). In the Nisqually river system where logjams maintain an anastomosing channel morphology, the floodplain surface is a patchwork of surfaces with a net relief of 2–4 m. This vertical dynamic complicates predicting channel migration because past meander migration rates do not provide ready guidance for projecting future channel migration. Local aggradation of the channel bed in response to sediment deposition associated with a logjam can affect the potential for the channel to avulse into a side channel or cause the river to simply carve a new channel across the floodplain. Due to such influences, changes in the age structure of riparian forests, and thereby the supply of large trees available to a river, can dramatically influence the dynamics of channels and ultimately lateral channel migration.

A channel unconfined by valley walls can aggrade to at least the level of two stacked logs before avulsing across the floodplain surface. Hence, a minimum estimate of the long-term portion of a valley bottom potentially occupied by forest channels can be defined as the area with elevations lower than twice the diameter of the site-potential tree diameter above the bank-full elevation. Vertical changes in riverbed elevation due to deposition around or behind logjams are higher in confined channels, where there is little leeway for such channels to move laterally around the jams. In addition, the potential for wood-mediated vertical changes in bed-surface elevation complicates the assessment of channel conditions, as a wood-rich channel can incise after logjam removal (for example, Stover and Montgomery 2001). Hence, interpretation of the causes of entrenchment in forest channels requires consideration of how wood loading has changed. Evaluations of where forest channels are likely to move in the future need to consider the role of local aggradation associated with logjams in affecting the potential for channel avulsion.

Landscape

Over time, the integrated effects of wood on erosion and sediment transport processes can translate into geomorphic effects at the landscape scale. For example, log steps provide for dissipation of potential energy otherwise available to transport sediment, and studies of the total elevation drop attributable to flow over log steps report a range of from 6% to 80% of the total elevation drop in stream systems (Keller and Tally 1979; Keller and Swanson 1979; Marston 1982; Abbe 2000). In small streams, wood can also trap and store more sediment than the average annual rate of bedload transport (Marston 1982), and log removal can trigger increased transport of formerly stored material (Bilby 1984). Lancaster and others (2001) showed that sediment storage associated with logjams could transform a stochastic input of sediment into a relatively steady sediment output in mountain channel networks. In steep channel systems developed on weak or unconsolidated substrate, such as in the glacial sediments of the Puget Lowland, loss of the roughness attributable to wood can lead to catastrophic incision, especially when combined with increased discharges due to urbanization (Booth 1990, 1991). Hence, wood can influence watershed-scale patterns of erosion and sediment transport and can greatly influence channel response to disturbance.

Recognition of the potential for large-scale effects of forests to influence landscape evolution is not new. At the dawn of the 20th century, Toumey (1905, p. 94) argued that a forest cover decreased the ability of mountain streams to incise bedrock by moderating both streamflow and the supply of "grinding material carried by the moving water." He concluded that the effects of a forest cover on erosion rates were such that, over geologic time, a forest cover influences physiographic form. More recent work on the processes controlling the distribution of bedrock and alluvial channels offers support to Toumey's conclusion. Montgomery et al. (1996) found that, in mountain channel networks, logjams can convert bedrock reaches to alluvial reaches by trapping bedload sediments. Rates of bedrock river incision will be lower if a streambed has a thick alluvial cover because the bedrock surface beneath the stream will be protected from abrasion (Sklar and Dietrich 1998). The effect of such influences of wood on the steepness of river systems can be addressed by formalizing the controls on the form of river profiles.

The longitudinal profile of mountain channel systems, and thereby landscape relief, is controlled by both the erodibility of the bedrock surface and the ability of the channel to cut into rock. The equilibrium stream profile will be that for which the bedrock erosion rate equals the rock uplift rate set by tectonic processes. An analytical solution for the form of river profiles can be based on a general expression for the rate of river incision (E) as a function of drainage area (A) and local slope (S):

$$E = KA^mS^n \, , \tag{1}$$

where K is a constant that incorporates climatic factors and erodibility and m and n are thought to vary with different erosional processes. For steady-state topography, where E equals the rock uplift rate U, equation (1) can be recast as

$$S = (U/K)^{(1/n)} A^{-m/n} \, . \tag{2}$$

Equation (2) shows that a less erodible (or "harder") streambed with a lower erodibility coefficient (a lower value of K) will lead to a steeper equilibrium river profile. Hence, a channel network with extensive, stable, and persistent deposits of wood that trap sediment and shield the bedrock surface will have a lower K and maintain steeper profiles than comparable channels lacking wood. In addition to altering stream gradients, forests can also influence the surrounding topography. The maximum stable angle predicted by limit-equilibrium models of slope stability increases with increased soil strength due to the apparent cohesion of an interlocking root network. Hence, vegetative cover with substantial root strength, such as a forest (for example, Schmidt et al. 2001), can allow soil-mantled hillslopes to stand at steeper angles than would be predicted based on soil strength alone. In these ways, a forest cover can allow steeper topography, and therefore greater relief, to develop and persist than would occur without a forest cover.

Controls on Wood Stability and Flux

The species of wood and its size relative to the channel in which it lies are primary controls on the stability of wood. Although dry wood generally has a specific gravity less than water and therefore readily floats, saturated wood has an effective density greater than water and can therefore be stable in some circumstances, even if submerged. Saturated wood can have an effective density greater than water, and some species, such as teak and some varieties of eucalyptus, have dry densities exceeding unity and sink without being saturated. The geometry of wood also influences its stability. A rootwad raises the center of mass of a wood piece to well over that for a simple log with the same bole diameter. The presence of a rootwad is therefore a fundamental control on log stability (Abbe 2000; Braudrick and Grant 2000).

The size of stable wood changes as a river widens and deepens downstream, and the style in which large wood is organized in river systems changes commensurately (Bisson et al. 1987; Abbe and Montgomery 1996, 2003). In headwater channels, even relatively small logs can remain where they fall, and wood is therefore arrayed in random orientations; in larger channels, wood is more mobile and tends to form both snags and become reorganized into different types of discrete logjams. Some jams have an exquisite architecture of pieces laid in one after another as if woven together. Other accumulations, such as many debris-flow-deposited jams, are chaotic jumbles of logs oriented at all angles to one another. Based on field observations in the Pacific Northwest, New Zealand, Australia, and South America, we recognize a generalized downstream change from randomly oriented wood, log steps, and debris flow jams in steep headwater channels to progressively larger, more organized jams in mainstem channels.

Several workers have identified different types of wood accumulations. Abbe and Montgomery (1996) described the development of bar top (BTJ), bar apex (BAJ), and meander (MJ) jams on the Queets River, a more than 1,000-km^2 old-growth watershed on the Olympic Peninsula, Washington. They noted that chaotically organized BTJs were relatively unstable and had little impact on channel morphology, whereas the more organized and stable BAJs and MJs created deep pools, bars, and patches of floodplain. Abbe and Montgomery (2003) later described 10 distinct types of logjams based on the orientation of key, racked, and loose wood mapped in field surveys along the Queets River (Figure 14), where stable wood accumulations dominate alluvial morphology from the scale of pool formation to channel anabranching and floodplain topography. Wallerstein et al. (1997) developed a wood jam classification based on the functional relation of the jam to the channel, with underflow jams causing local scour but limited deposition, dam jams trapping a wedge of sediment and forming log steps in the stream profile, deflector jams causing local pool scour and bar formation due to flow deflection, and parallel/bar head jams, which encompass Abbe and Montgomery's (1996) bar apex and meander jams. The relative abundance of these different wood jam types changes systematically downstream through a channel network (Abbe and Montgomery 1996, 2003).

The geomorphic effects of wood jams can change through time. Hogan et al. (1998) described patterns in the temporal evolution of logjams deposited by debris flows in small gravel-bed streams of coastal British Columbia. They observed that channel morphology changes from morphologically diverse prior to jam formation to simplified morphology following emplacement of a logjam by a debris-flow, followed by a progressive change back to a complex channel morphology as a jam decays (Hogan et al. 1998). In contrast, Abbe (2000) documented increased morphologic complexity of channels influenced by the development of valley jam complexes in the headwaters of the Queets River, Washington. Hence, the temporal evolution of wood jams, and their attendant geomorphic effects, varies according to processes governing jam formation, preservation, decay, and destruction.

Key pieces appear crucial for altering channel morphology and processes in large rivers. River systems filled with stable wood can trap and retain smaller, mobile pieces. A comparison of wood loading in Puget Sound rivers showed strong contrasts in the influence of wood between relatively undisturbed rivers and those that had been both cleared of debris and no longer have a forested floodplain (Collins et al. 2002). The Nisqually River, which retains a source of key-member-size trees in its floodplain, has numerous logjams and a wood loading (pieces of large wood per unit river length) one to two orders of magnitude greater than the Snohomish and Snoqualmie rivers, which lack key-member-size

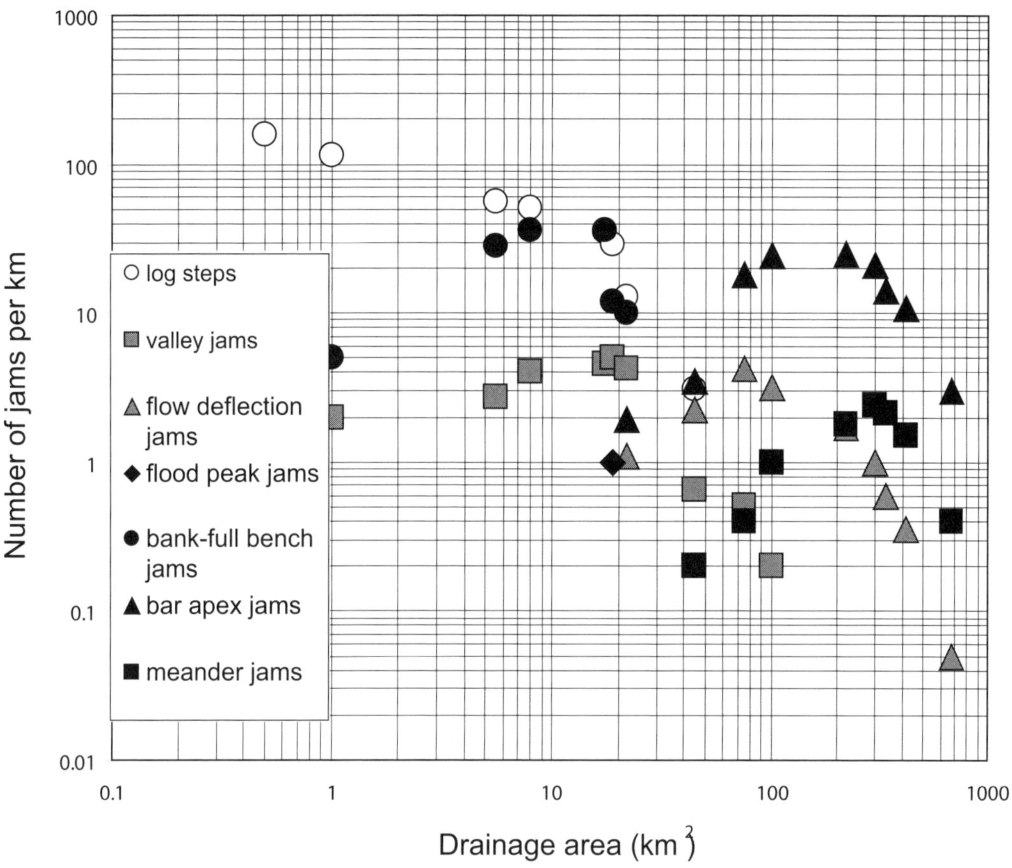

FIGURE 14. Downstream variations in the style of log accumulations in the Queets River, Washington based on field surveys from 1993 to 1996. Modified from Abbe and Montgomery (in press): open symbols represent in situ jams that remain close to where they were introduced to the channel; gray symbols represent *combination* jams composed of transported wood racked onto in situ key pieces; and black symbols represent *transport* jams in which key pieces of wood were transported within the channel.

debris because of snagging, prevention of channel migration (and thus wood recruitment) due to levee construction, and floodplain forest clearing. If even a small portion of the wood load supplied to a river is large enough to prove stable, these pieces can catalyze retaining additional mobile wood that racks up onto the stable pieces. An example of such an effect is the trapping of more than 400 pieces of wood by a single "engineered logjam" built on the North Fork Stillaguamish River in the first year after the structure was built (Abbe et al. 2000).

Shape is another primary influence on the stability of wood. Based on observations in areas with large deciduous trees, we suggest that widely spreading or multiple-stemmed hardwoods are more prone to forming snags than accumulating as racked members in large logjams because their geometry makes them extend laterally to well beyond their bole diameter. In contrast, conifers tend to create longer, more cylindrical pieces more readily transported and routed through river systems, resulting in enhanced development of logjams. Although both snags and jams form in many forested regions, regional forest composition influences the relative tendency for large wood to form either individual snags or large logjams in river systems.

The effects of large wood on channel morphology depend on wood stability, which is, in turn, a function of wood dimensions and wood size relative to channel size (Bisson et al. 1987). The typical definition of large wood based on uniform size criteria (for example, >0.1 m in diameter and 1 m in length) does not account for wood stability relative to channel size and stream power. A 0.3-m-diameter, 5-m-long log may have significant geomorphic effects in a small headwater channel but be ineffective flotsam in a large river. Abbe and Montgomery (2003) showed that in the drainage basin of the Queets River, the relative size of key piece sizes varies with wood dimensions relative to channel size (Figure 15). They found that, for wood longer than about half the bank-full width, those pieces with a diameter larger than about half the bank-full depth tend to form key pieces. Progressively larger relative diameters are required to form stable pieces of shorter wood. Abbe and Montgomery (2003) also found that wood size relative to channel size differentiates metastable pieces that tend to form racked members of logjams from unstable wood that provides little, if any, structural integrity to wood accumulations. Controls on wood stability, and thus its geomorphic effects, depend on both regional factors (such as the size and shape of wood delivered to the channel) and location in the channel network.

Wood regime

The supply and size of wood delivered to a channel system defines a wood regime, analogous to the sediment or discharge regimes, respectively

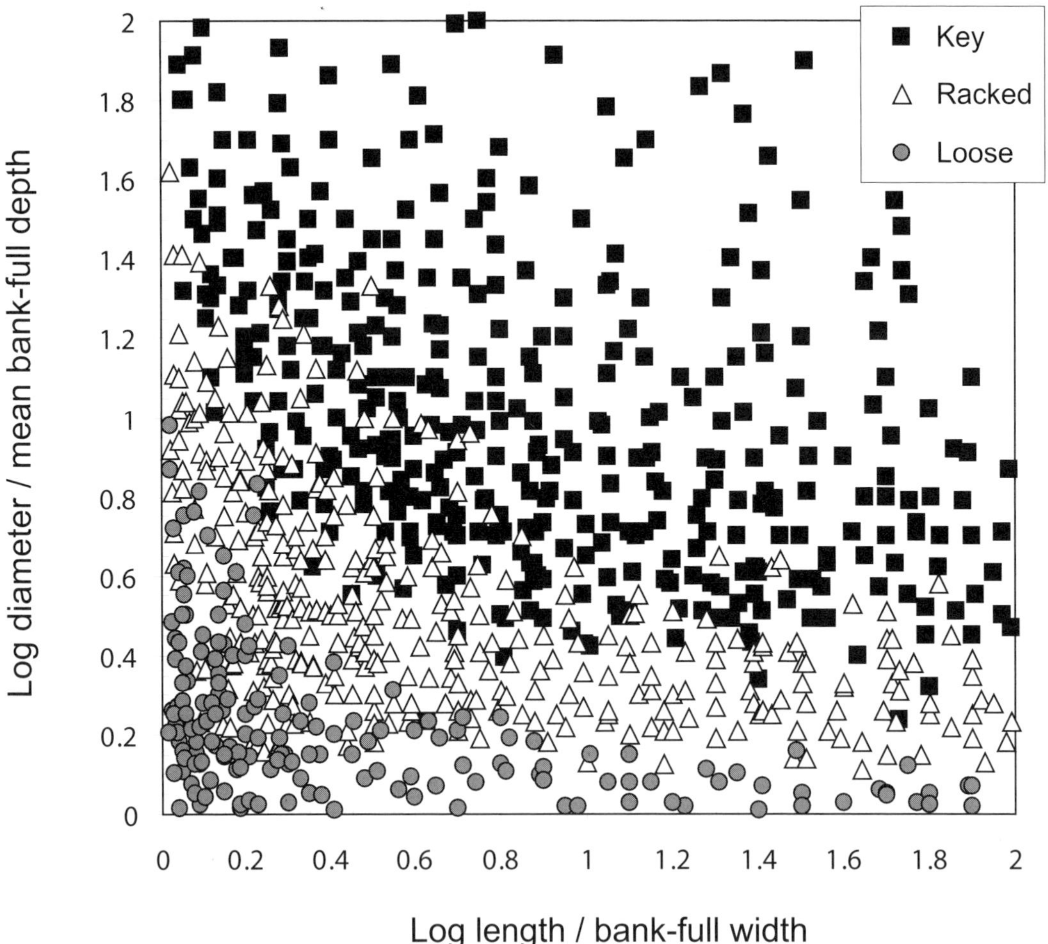

FIGURE 15. Dimensionless plot of the ratio of log diameter to bank-full channel depth versus the ratio of log length to channel width, showing distinct fields defined by loose, racked, and stable key members of logjams from throughout the Queets River system, Washington. Modified from Abbe and Montgomery (in press).

determined by the caliber (size) and volume (amount) of sediment delivered to the channel and the discharge magnitude and variability. In forested landscapes, the wood regime interacts with the sediment and water regimes to set channel morphology and control channel dynamics. Channels with a large supply of stable logs would be expected to have different characteristics, dynamics, and sensitivity to changes in land use than channels supplied only with small wood, or little wood at all.

The quantity and characteristics of wood delivered to a channel depend on the nature of the forest supplying the wood and the processes that introduce wood into the channel. Tree fall can directly deliver trees to channels from streamside forests (McDade et al. 1990). Slope instability and particularly debris flows or large earthflows can deliver substantial amounts of wood to headwater channels from hillside forests (Grant and Swanson 1995; Johnson et al. 2000). Bank erosion can introduce both standing trees and wood stored in floodplain sediments into channels (Murphy and Koski 1989; Piégay et al. 1999). The relative importance of processes that directly or indirectly deliver wood to rivers and streams vary by region but, in general, include biological processes (such as insect outbreaks and disease), fire, floods, landsliding, wind storms, and snow avalanches. Different processes deliver wood to different portions of a river system, with a typical pattern of landsliding and tree fall dominating wood inputs in steep headwater channels and bank erosion dominating wood inputs to larger floodplain rivers. Riparian stand growth models have been used to predict in-channel wood loading based on recruitment of nearstream trees under a range of land use and disturbance histories (McDade et al. 1990; Beechie et al. 2000; Bragg 2000). These processes affect the age, amount, diameter, and species of trees recruited to channels, as well as where wood enters the channel system. Natural or anthropogenic changes to the recruitment, transport and storage of wood can affect a stream's physical and biological condition.

As discussed above, the size of wood is the primary control on its stability and transport in a river. The strong influence of large "key" pieces of wood debris on stabilizing other debris in logjams is well known (Keller and Tally 1979; Nakamura and Swanson 1993; Abbe and Montgomery 1996). The role of key pieces of wood appears crucial for triggering geomorphological effects of wood in large rivers, as river systems filled with stable wood debris can trap and retain smaller, mobile debris (Abbe and Montgomery 1996; Collins et al. 2002). In addition to wood size and channel size, the species of wood also influences the stability of instream wood through both its density and shape. The controls on wood stability and transport—wood size relative to the channel, shape, species, and recruitment mechanism—vary both with position in a channel network and regionally.

The amount of wood stored in a river system varies greatly between catchments and can exhibit substantial variability over time. Harmon and et al. (1986) compiled reported values of wood biomass stored in natural (unmanaged) channel systems in temperate forests and found substantial differences in wood loading between redwood forest (>1,000 m^3/ha), other coniferous forest (200–1,000 m^3/ha), and deciduous forest (<200 m^3/ha). The frequency of logjams also varies widely both within and among river systems. Wood dam frequencies as high as 400/km have been reported in small channels, and the frequency of organic obstructions typically decreases with increasing channel size, or basin area (Bilby and Likens 1980; Gregory et al. 1993), although the size of stable jams likely increases with increasing channel size. Deforestation and debris clearing have reduced wood jam frequency by 3-fold to more than 10-fold in different regions (Hedin et al. 1988; Webster et al. 1992; Gregory et al. 1993; Collins et al. 2002). In addition to storage in the active channel, substantial amounts of wood can be buried in floodplain sediments where wood can persist for thousands of years under anaerobic conditions (Becker and Schirmer 1977; Nanson et al. 1995).

The transport and decay of wood debris also varies greatly among river systems, and much of this variability is due to the influence of both wood size and channel size in controlling the stability and retention of wood delivered to channels. In small channels flowing through old-growth forests in the Pacific Northwest, key pieces of wood can remain stable for hundreds of years (Keller and Tally 1979; Murphy and Koski 1989). Studies on the Queets River, Washington, have shown that decay-resistant species can last centuries to more than 1,000 years in even large rivers (Abbe 2000; Hyatt and Naiman 2001). In contrast, wood storage on the Drôme River, France is only several times the annual wood input, and therefore, the residence time of wood accumulations is on the order of just several years (Piégay et al. 1999), most likely due to the small size of forest trees recruited to the channel. The transport and decay of wood also depends on the environment; hardwoods generally decay faster than coniferous species; small wood is more mobile than larger wood; and wood decay rates are high in tropical forests due to high temperatures and rainfall.

Just as for sediment, the wood regime can be assessed using a mass balance approach to characterize the wood budget for a channel (Swanson et al. 1982; Cummins et al. 1983; Gregory 1992; Piégay et al. 1999). A wood budget is a straightforward quantitative statement of the inputs, outputs, and changes in storage of wood in a river system. A wood budget can be constructed for a particular reach of river or for a whole river system and used to characterize or explore differences in the recruitment, storage, or export of wood. The relative mobility and potential for wood transport and storage can be modeled using analyses of buoyant and drag effects (Abbe 2000; Braudrick and Grant 2000). The transport of logs by debris flows also has been incorporated into simulation models (Lancaster et al. 2001). A key consideration for interpreting wood budgets is the potential for wood storage to depend on the presence of key pieces of wood, which strongly influences retention of organic matter in a river system (Figure 16).

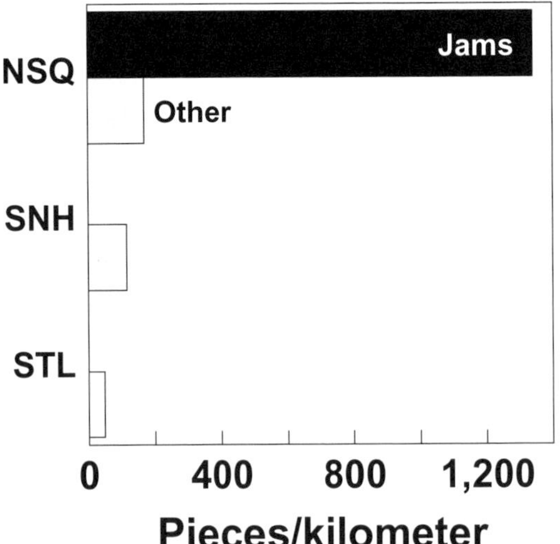

FIGURE 16. Comparison of wood loading in the Nisqually (NSQ), Stillaguamish (STL), and Snohomish (SNH) rivers (from Collins et al. 2002).

Regional patterns in the influence of wood on rivers

Regional variation in the size and shape of trees and the characteristics of their wood imparts global patterns to the geomorphic influence of wood. Rivers flowing through boreal forests are likely to be less influenced by wood because the trees are generally small. The effect of small-wood jams in tropical forest channels may be more short lived and ephemeral than in temperate latitudes because of the rapid breakdown and decay of debris in equatorial regions, although equatorial regions also have many decay-resistant hardwoods. Forest stand density will also strongly influence wood recruitment, with channels in open woodlands less influenced by wood than channels flowing through dense closed-canopy forests. The greatest effects of wood in rivers may be in the old-growth temperate rainforests dominated by massive, decay-resistant wood such as cedar in the Pacific Northwest of North America and eucalyptus in Australia. Regional biogeographic patterns are therefore primary determinants of the geomorphic effects of wood in rivers.

Different zones of wood influence should shift up and down river systems based on the absolute size of the available wood. In this sense, variations in the geomorphic effects of wood in different regions may correspond to variations in different environments and positions in a single watershed through the common currency of relative wood size. Hence, the geomorphic effectiveness of wood in rivers has two independent dimensions: tree size (and shape) and river size. Tree size sets regional limits on how much of the channel network could be influenced and in what way. Regional geography controls drainage-basin size and therefore the size of the rivers. Coastal mountains have small rivers; thus, the available wood should have a relatively large influence on fluvial processes and morphology. Huge continental rivers such as the Amazon are less influenced by wood, even if their tributaries are (or were) strongly influenced by wood. But the potential for even large rivers to be influenced by logjams emphasizes the need to incorporate the effects of wood into geomorphic models and thinking across all scales in forest river systems.

Wood, History, and the Landscape

The role of wood on fine-scale geomorphological features such as pools has been widely recognized for decades. Such effects can be observed directly in the field without specialized training or consideration of geologic time. But the recently recognized importance of wood in influencing channel geometry, channel patterns, and floodplain topography has been masked by decades to centuries of river cleaning and forest clearing in Europe and North America. Geomorphologists tend to downplay the potential effects of vegetation on landform evolution, but the stabilizing effects of forests and wood on sediment transport and storage processes could require increased river slopes to counteract a given rate of rock uplift. Such large-scale effects remain somewhat speculative, but vegetation and its effects on earth surface processes may be a greater influence on landscape evolution than typically imagined. Although the geomorphic effects of wood on rivers has been obscured by a legacy of anthropogenic changes to river systems, wood has been entering and affecting fluvial systems for more than 100 million years. Fluvial geomorphology is only now starting to link wood to channel properties and processes at spatial scales larger than individual habitat units (for example, single pools and bars). Just as the sediment and water regimes are widely acknowledged to influence channel systems across a wide range of scales, the wood regime is an important component of fluvial systems.

Acknowledgments

Our discussion of the effects of wood on bar deposition and sediment storage is based on a literature summary developed by R.D. Woodsmith.

References

Abbe, T. B. 2000. Patterns, mechanics and geomorphic effects of large wood accumulations in a forest river system. Doctoral dissertation. University of Washington, Seattle.

Abbe, T. B., and D. R. Montgomery. 1996. Interaction of large woody debris, channel hydraulics and habitat formation in large rivers. Regulated Rivers Research & Management 12:201–221.

Abbe, T. B., and D. R. Montgomery. 2003. Patterns and process of wood debris accumulation in the Queets River basin, Washington. Geomorphology 51:81–107.

Abbe, T. B., G. R. Pess, D. R. Montgomery, M. L. White, and B. Smith. 2000. Preliminary performance of engineered log jam technology in protecting infrastructure and rehabilitating fluvial ecosystems. EOS, Transactions of the American Geophysical Union 81(19):S264–265.

Albion, R. G. 1926. Forests and sea power: the timber problem of the Royal Navy, 1652–1862. Harvard University Press, Cambridge, Massachusetts.

Andrus, C. W., B. A. Long, and F. H. Froehlich. 1988. Woody debris and its contribution to pool formation in a coastal stream 50 years after logging. Canadian Journal of Fisheries and Aquatic Sciences 45:2080–2086.

Assani, A. A., and F. Petit. 1995. Log-jam effects on bed-load mobility from experiments conducted in a small gravel-bed forest ditch. Catena 25:117–126.

Atjay, G. L., P. Ketner, and P. Duvigneaud. 1979. Terrestrial primary production and phytomass. Pages 129–182 *in* B. Bolin, E. T. Degens, S. Kempe, and P. Ketner, editors. The global carbon cycle. Wiley, New York.

Bamford, P. W. 1956. Forests and French sea power 1660–1789. University of Toronto Press, Toronto.

Becker, B., and W. Schirmer. 1977. Palaeoecological study on the Holocene valley development of the River Main, southern Germany. Boreas 4:303–321.

Beechie, T. J., and T. H. Sibley. 1997. Relationship between channel characteristics, woody debris, and fish habitat in northwestern Washington streams. Transactions of the American Fisheries Society 126:217–229.

Beechie, T. J., G. Pess, P. Kennard, R. E. Bilby, and S. Bolton. 2000. Modeling recovery rates and pathways for woody debris recruitment in northwestern Washington streams. North American Journal of Fisheries Management 20:436–452.

Beschta, R. L. 1979. Debris removal and its effects on sedimentation in an Oregon Coast Range stream. Northwest Science 53:71–77.

Beschta, R. L. 1983. The effects of large organic debris upon channel morphology: a flume study. Pages 63–78 *in* D. B. Simons, editor. Symposium on erosion and sedimentation. Simons, Li & Associates, Fort Collins, Colorado.

Bilby, R. E. 1981. Role of organic debris dams in regulating the export of dissolved and particulate matter from a forested watershed. Ecology 62:1234–1243.

Bilby, R. E. 1984. Removal of woody debris may affect stream channel stability. Journal of Forestry 82:609–613.

Bilby, R. E., and G. E. Likens. 1980. Importance of organic debris dams in the structure and function of stream ecosystems. Ecology 61:1107–1113.

Bilby, R. E., and J. W. Ward. 1989. Changes in characteristics and function of woody debris with increasing size of streams in western Washington. Transactions of the American Fisheries Society 118:368–378.

Bilby, R. E., and J. W. Ward. 1991. Characteristics and function of large woody debris in streams draining old-growth, clear-cut, and second-growth forests in southwestern Washington. Canadian Journal of Fisheries and Aquatic Sciences 48:2499–2508.

Bisson, P. A., R. E. Bilby, M. D. Bryant, C. A. Dolloff, G. B. Grette, R. A. House, M. L. Murphy, K. V. Koski, and J. R. Sedell. 1987. Large wood debris in forested streams in the Pacific Northwest: past, present, and future. Pages 143–190 *in* E. O. Salo and T. W. Cundy, editors. Streamside management: forestry and fishery interactions. Institute of Forest Resources, no. 57, University of Washington, Seattle.

Bisson, P. A., J. L. Nielson, R. A. Palmason, and L. E. Gore. 1982. A system of naming habitat types in small streams, with examples of habitat utilization by salmonids during low stream flow. Pages 62–73 *in* N. B. Armantrout, editor. Acquisition and utilization of aquatic habitat information. American Fisheries Society, Western Division, Portland, Oregon.

Bodmer, K., D. C. Hunt, and M. V. Gallegher. 1984. Karl Bodmer's America. University of Nebraska Press, Lincoln.

Booth, D. B. 1990. Stream-channel incision following drainage-basin urbanization. Water Resources Bulletin 26:407–417.

Booth, D. B. 1991. Urbanization and the natural drainage system — impacts, solution, and prognoses. The Northwest Environmental Journal 7:93–118.

Bormann, N. E., and P. Y. Julien. 1991. Scour downstream of grade-control structures. Journal of Hydraulic Engineering 117:579–594.

Bragg, D. C. 2000. Simulating catastrophic and individualistic large woody debris recruitment for a small riparian system. Ecology 81:1383–1394.

Braudrick, C. A., and G. E. Grant. 2000. When do logs move in rivers? Water Resources Research 36:571–583.

Brooks, A. P., and G. J. Brierley. 2000. The role of European disturbance in the metamorphosis of the Lower Bega River. Pages 221–246 *in* S. Brizga and B. Finlayson, editors. River management: the Australasian experience. Wiley, Chichester, UK.

Brown, A. G., and M. Keough. 1992. Palaeochannels, palaeoland-surfaces and the three-dimensional reconstruction of floodplain environmental change. Pages 185–202 *in* P. A. Carling and G. E. Petts, editors. Lowland floodplain rivers: geomorphological perspectives. Wiley, Chichester, UK.

Bryant, M. D. 1980. Evolution of large, organic debris after timber harvest: Maybeso Creek, 1949 to 1978. Pacific Northwest Forest and Range Experiment Station, U.S.D.A. Forest Service General Technical Report PNW-101, Portland, Oregon.

Buffington, J. M. 1998. The use of streambed texture to interpret physical and biological conditions at watershed, reach, and subreach scales. Doctoral dissertation. University of Washington, Seattle.

Buffington, J. M. 2001. Hydraulic roughness and shear stress partitioning in forest pool-riffle channels. In T. J. Nolan, and C. R. Thorne, editors. Gravel-Bed Rivers 2000 CD ROM. Special Publication of the New Zealand Hydrological Society.

Buffington, J. M., and D. R. Montgomery. 1999a. A procedure for classifying textural facies in gravel-bed rivers. Water Resources Research 35:1903–1914.

Buffington, J. M., and D. R. Montgomery. 1999b. Effects of hydraulic roughness on surface textures of gravel-bed rivers. Water Resources Research 35:3507–3522.

Buffington, J. M., R. D. Woodsmith, D. B. Booth, D. B., and D. R. Montgomery. 2002 a. Fluvial processes in the Pacific Northwest and Puget Sound rivers. Pages 46–78 *in* D. R. Montgomery, D. B. Booth, S. B. Bolton, and L. Wall, editors. Restoration of Puget Sound rivers. University of Washington Press.

Buffington, J. M., T. E. Lisle, R. D. Woodsmith, and S. Hilton. 2002 b. Controls on the size and occurrence of pools in coarse-grained forest rivers. River Research and Applications 18:507–531.

Cherry, J., and R. L. Beschta. 1989. Coarse woody debris and channel morphology: a flume study. Water Resources Bulletin 25:1031–1036.

Chin, A. 1989. Step pools in stream channels. Progress in Physical Geography 13:391–407.

Clay, C. 1949. The Colorado River raft. The Southwestern Historical Quarterly 102:400–426.

Collins, B. D., and D. R. Montgomery. 2001. Importance of archival and process studies to characterizing pre-settlement riverine geomorphic processes and habitat in the Puget Lowland. Pages 227–243 *in* J. B. Dorava, D. R. Montgomery, B. Palcsak, and F. Fitzpatrick, editors. Geomorphic processes and riverine habitat. American Geophysical Union, Washington, D.C.

Collins, B. D., and D. R. Montgomery. 2002. Forest development, wood jams, and restoration of floodplain rivers in the Puget Lowland, Washington. Restoration Ecology 10: 237–247.

Collins, B. D., D. R. Montgomery, and A. Haas. 2002. Historical changes in the distribution and functions of large wood in Puget Lowland rivers. Canadian Journal of Fisheries and Aquatic Sciences 59:66–76.

Cotter, E. 1978. The evolution of fluvial style, with special reference to the central Appalachian Paleozoic. Pages 361–383 *in* A. D. Miall, editor. Fluvial sedimentology. Canadian Society of Petroleum Geologists Memoir 5, Calgary, AB.

Cummins, K W., J. R. Sedell, F. J. Swanson, G. W.

Minshall, S. G. Fisher, C. E. Cushing, R. C. Petersen, and R. L. Vannote. 1983. Organic matter budgets for stream ecosystems: problems in their evaluation. Pages 299–353 *in* G. W. Minshall and J. R. Barnes, editors. Stream ecology: application and testing of general ecological theory. Plenum, New York.

Dale, T., and V. G. Carter. 1955. Topsoil and civilization. University of Oklahoma Press, Norman.

Darby, H. C. 1956. The clearing of the woodland of Europe. Pages 183–216 *in* W. L. Thomas, Jr., C. O. Sauer, M. Bates, and L. Mumford, editors. Man's role in changing the face of the earth. University of Chicago Press, Chicago.

Davies-Colley, R. J. 1997. Stream channels are narrower in pasture than in forest. New Zealand Journal of Marine and Freshwater Research 31:599–608.

Deane, W. 1888. A New Hampshire log-jam. New England Magazine 30:97–103.

DiMichele, W. A., and R. W. Hook. 1992. Paleozoic terrestrial ecosystems. Pages 205–325 *in* A. K. Behrensmeyer, J. D. Damuth, W. A. DiMichele, R. Potts, H-D. Sues, and S. L. Wing, editors. Terrestrial ecosystems through time. The University of Chicago Press, Chicago.

Downs, P. W., and A. Simon. 2001. Fluvial geomorphological analysis of the recruitment of large woody debris in the Yalobusha River network, Central Mississippi, USA. Geomorphology 37:65–91.

Florsheim, J. L. 1985. Fluvial requirements for gravel bar formation in northwestern California. Master's thesis. Humboldt State University, Arcata, California.

Folk, R. L. 1974. Petrology of sedimentary rock. Hemphill Publishing Co., Austin, Texas.

Franklin, J. F., and C. T. Dyrness. 1973. Natural vegetation of Oregon and Washington. USDA Forest Service, Pacific Northwest Forest and Range Experiment Station, General Technical Report PNW-8, Portland, Oregon.

Gee, C. T. 1989. Revision of the Late Jurassic/Early Cretaceous flora from Hope Bay, Antarctica. Palaeontographica, B 213:149–214.

Grant, G. E., and F. J. Swanson. 1995. Morphology and processes of valley floors in mountain streams, western Cascades, Oregon. Pages 83–101 *in* J. E. Costa, A. J. Miller, K. W. Potter, and P. R. Wilcock, editors. Natural and anthropogenic influences in fluvial geomorphology. American Geophysical Union, Geophysical Monograph 89, Washington, D.C.

Gregory, K. J. 1992. Vegetation and river channel process interactions. Pages 255–269 *in* P. J. Boon, P. Calow, and G. E. Petts, editors. River conservation and management. Wiley, Chichester.

Gregory, K. J., and R. J. Davis. 1992. Coarse woody debris in stream channels in relation to river channel management in woodland areas. Regulated Rivers Research & Management 7:117–136.

Gregory, K. J., R. J. Davis, and S. Tooth. 1993. Spatial distribution of coarse woody debris dams in the Lymington Basin, Hampshire, UK. Geomorphology 6:207–224.

Gregory, K. J., A. M. Gurnell, and C. T. Hill. 1985. The permanence of debris dams related to river channel processes. Hydrological Sciences Journal 30:371–381.

Guardia, J. E. 1933. Some results of the log jams in the Red River. The Bulletin of the Geographical Society of Philadelphia 31:103–114.

Gurnell, A. M., and R. Sweet. 1998. The distribution of large woody debris accumulations and pools in relation to woodland stream management in a small, low-gradient stream. Earth Surface Processes and Landforms 23:1101–1121.

Gurnell, A. M., G. E. Petts, D. M. Hannah, B. P. G. Smith, P. J. Edwards, J. Kollmann, J. V. Ward, and K. Tockner. 2001. Riparian vegetation and island formation along the gravel-bed Riume Tagliamento, Italy. Earth Surface Processes and Landforms 26:31–62.

Habersham, R. A. 1881. Report of Mr. Robt. A. Habersham, Assistant Engineer, in Report of the Chief of Engineers, U.S. Army, Appendix OO, p. 2605–2607.

Harmon, M. E., J. F. Franklin, F. J. Swanson, P. Sollins, S. V. Gregory, J. D. Lattin, N. H. Anderson, S. P. Cline, N. G. Aumen, J. R. Sedell, G. W. Lienkaemper, K. Cromack, and K. W. Cummins. 1986. Ecology of coarse woody debris in temperate ecosystems. Advances in Ecological Research 15:133–302.

Hartopo. 1991. The effect of raft removal and dam construction on the lower Colorado River, Texas. Master's thesis. Texas A&M University, College Station.

Harvey, M. D., D. S. Biedenharn, and P. Combs. 1988. Adjustments of Red River following removal of the Great Raft in 1873. EOS, Transactions of the American Geophysical Union 69(18):567.

Harwood, K., and A. G. Brown. 1993. Fluvial processes in a forested anastomosing river: flood partitioning and changing flow patterns. Earth Surface Processes and Landforms 18:741–748.

Hedin, L. O., M. S. Mayer, and G. E. Likens. 1988. The effect of deforestation on organic debris dams. Verhandlungen-Internationale Vereinigung für Theoretifche und Angewandte Limnologie 23:1135–1141.

Heede, B. H. 1972. Influences of a forest on the hydraulic geometry of two mountain streams. Water Resources Bulletin 8:523–530.

Heede, B. H. 1981. Dynamics of selected mountain streams in the western United States of America. Zeitschrift für Geomorphologie 25:17–32.

Heede, B. H. 1985. Channel adjustment to the removal of log steps: an experiment in a mountain stream. Environmental Management 9:427–432.

Herbinio, J. M. 1678. Dissertationes de admirandis

mundi cataractis supra & subterraneis. Janssonio-Waesbergios, Amstelodami.

Herget, J. 2000. Holocene development of the River Lippe valley, Germany: a case study of anthropogenic influence. Earth Surface Processes and Landforms 25:293–305.

Hill, F. G. 1957. Roads, rails & waterways, the army engineers and early transportation. University of Oklahoma Press, Norman.

Hogan, D. L. 1986. Channel morphology of unlogged, logged, and debris torrented streams in the Queen Charlotte Islands. British Columbia Ministry of Forests and Lands, Land Management Report 49, Victoria, BC.

Hogan, D. L., S. A. Bird, and M. A. Hassan. 1998. Spatial and temporal evolution of small coastal gravel-bed streams: influence of forest management on channel morphology and fish habitats. Pages 365–392 *in* P. C. Klingeman, R. L. Beschta, P. D. Komar, and J. B. Bradley, editors. Gravel-bed rivers in the environment. Water Resources Publications, Englewood, New Jersey.

Huntley, B., and T. Webb III, editors. 1988. Vegetation history. Kluwer Academic Publishers, Dordrecht, Netherlands.

Hyatt, T. L., and R. J. Naiman. 2001. The residence time of large woody debris in the Queets River, Washington, USA. Ecological Applications 11:191–202.

Johnson, A. C., D. N. Swanston, and K. E. McGee. 2000. Landslide initiation, runout, and deposition within clearcuts and old-growth forests of Alaska. Journal of the American Water Resources Association 36:17–30.

Keller, E. A., and F. J. Swanson. 1979. Effects of large organic material on channel form and fluvial processes. Earth Surface Processes 4:361–380.

Keller, E. A., and T. Tally. 1979. Effects of large organic debris on channel form and fluvial processes in the coastal redwood environment. Pages 169–197 *in* D. D. Rhodes and G. P. Williams, editors. Adjustments of the fluvial system. Kendal-Hunt, Dubuque, Iowa.

Keller, E. A., A. MacDonald, T. Tally, and N. J. Merritt. 1995. Effects of large organic debris on channel morphology and sediment storage in selected tributaries of Redwood Creek. U.S. Geological Survey Professional Paper 1454-P, Menlo Park, California.

Lancaster, S. T., S. K. Hayes, and G. E. Grant. 2001. Modeling sediment and wood storage and dynamics in small mountainous watersheds. Pages 85–102 *in* J. B. Dorava, D. R. Montgomery, B. Palcsak, and F. Fitzpatrick, editors. Geomorphic processes and riverine habitat. American Geophysical Union, Washington, D.C.

Larson, M. G. 1999. Effectiveness of large woody debris in stream rehabilitation projects in urban basins. Master's thesis. University of Washington, Seattle.

Lisle, T. E. 1986. Stabilization of a gravel channel by large streamside obstructions and bedrock bends, Jacoby Creek, northwestern California. Water Resources Research 97:999–1011.

Lisle, T. E. 1995. Effects of coarse woody debris and its removal on a channel affected by the 1980 eruption of Mount St. Helens, Washington. Water Resources Research 31:1797–1808.

Lowdermilk, W. C. 1926. Forest destruction and slope denudation in the Province of Shansi. The China Journal of Science & Arts 4(3):127–135.

Lowdermilk, W. C. 1950. Conquest of the land through seven thousand years. U.S. Department of Agriculture Soil Conservation Service, S. C. S. MP-32, Washington, D.C.

Lowrey, W. M. 1968. The red. Pages 53–73 *in* E. A. Davis, editor. The rivers and bayous of Louisiana. Louisiana Education Research Association, Baton Rouge, Louisiana.

Lyell, C. 1837. Principles of geology: being an inquiry how far the former changes of the Earth's surface are referable to causes now in operation. volume I. James Kay Jun. & Brother, Philadelphia.

MacDonald, A., and E. A. Keller. 1987. Stream channel response to the removal of large woody debris, Larry Damm Creek, northwestern California. Pages 405–406 *in* Proceedings of the Symposium on Erosion and Sedimentation in the Pacific Rim. International Association of Hydrological Sciences Publication 165.

Manga, M., and J. W. Kirchner. 2000. Stress partitioning in streams by large woody debris. Water Resources Research 36:2373–2379.

Marston, R. A. 1982. The geomorphic significance of log steps in forest streams. Annals of the American Association of Geographers 72:99–108.

Mason, P. J., and K. Arumugam. 1985. Free jet scour below dams and flip buckets. Journal of Hydraulic Engineering 111:220–235.

Massong, T. M., and D. R. Montgomery. 2000. Influence of lithology, sediment supply, and wood debris on the distribution of bedrock and alluvial channels. Geological Society of America Bulletin 112:591–599.

May, C. W. 1996. Assessment of cumulative effects of urbanization on small streams in the Puget Sound lowland ecoregion: implications for salmonid resource management. Doctoral dissertation. University of Washington, Seattle.

McDade, M. H., F. J. Swanson, W. A. McKee, J. F. Franklin, and J. Van Sickle. 1990. Source distances for coarse woody debris entering small streams in Oregon and Washington. Canadian Journal of Forest Research 20:326–330.

McKnight, C. L., S. A. Graham, A. R. Carroll, Q. Gan, D. L. Dilcher, M. Zhao, and Y. H. Liang. 1990. Fluvial sedimentology of an Upper Jurassic petrified forest assemblage, Shishu Formation, Junggar Basin, Xinjiang, China. Palaeogeography, Palaeoclimatology, Palaeoecology 79:1–9.

Megahan, W. F. 1982. Channel sediment storage behind obstructions in forested drainage basins draining the granitic bedrock of the Idaho Batholith. Pages 114–121 *in* F. J. Swanson, R. J. Janda, T. Dunne, and D. N. Swanston, editors. Sediment budgets and routing in forested drainage basins. U.S. Department of Agriculture, Forest Service, General Technical Report PNW-141, Portland, Oregon.

Megahan, W. F. and R. A. Nowlin. 1976. Sediment storage in channels draining small forested watersheds. Pages 115–126 *in* Proceedings of the Third Federal Interagency Sedimentation Conference. Water Resources Council, Washington, D.C.

Melville, B. W. 1992. Local scour at bridge abutments. Journal of Hydraulic Engineering 118:615–631.

Melville, B. W. 1997. Pier and abutment scour: integrated approach. Journal of Hydraulic Engineering 123:125–136.

Montgomery, D. R., and J. M. Buffington. 1997. Channel reach morphology in mountain drainage basins. Geological Society of America Bulletin 109:596–611.

Montgomery, D. R., and J. M. Buffington. 1998. Channel processes, classification, and response potential. Pages 13–42 *in* R. J. Naiman and R. E. Bilby, editors. River ecology and management. Springer-Verlag Inc., New York.

Montgomery, D. R., J. M. Buffington, R. Smith, K. Schmidt, and G. Pess. 1995. Pool spacing in forest channels. Water Resources Research 31:1097–1105.

Montgomery, D. R., T. B. Abbe, N. P. Peterson, J. M. Buffington, K. Schmidt, and J. D. Stock. 1996. Distribution of bedrock and alluvial channels in forested mountain drainage basins. Nature (London) 381:587–589.

Mosley, M. P. 1981. The influence of organic debris on channel morphology and bedload transport in a New Zealand forest stream. Earth Surface Processes and Landforms 6:571–579.

Muir, J. 1878. The new sequoia forests of California. Harper's New Monthly Magazine 57:813–827.

Murgatroyd, A. L., and J. L. Ternan. 1983. The impact of afforestation on stream bank erosion and channel form. Earth Surface Processes and Landforms 8:357–369.

Murphy, M. L., and K. V. Koski. 1989. Input and depletion of woody debris in Alaska streams and implications for streamside management. North American Journal of Fisheries Management 9:427–436.

Nakamura, F., and F. J. Swanson. 1993. Effects of coarse woody debris on morphology and sediment storage of a mountain stream system in western Oregon. Earth Surface Processes and Landforms 18:43–61.

Nanson, G. C., M. Barbetti, and G. Taylor. 1995. River stabilization due to changing climate and vegetation during the late Quaternary in western Tasmania, Australia. Geomorphology 13:145–158.

Nelson, K. 1998. The influence of sediment supply and large woody debris on pool characteristics and habitat diversity. Master's thesis. University of Washington, Seattle.

Perlin, J. 1989. A forest journey: the role of wood in the development of civilization. Harvard University Press, Cambridge, Massachusetts.

Piégay, H. 1993. Nature, mass and preferential sites of coarse woody debris deposits in the lower Ain Valley (Mollon Reach), France. Regulated Rivers Research & Management 8:359–372.

Piégay, H., and A. M. Gurnell. 1997. Large woody debris and river geomorphological pattern: examples from S. E. France and S. England. Geomorphology 19:99–116.

Piégay, H., and R. A. Marston. 1998. Distribution of large woody debris along the outer bend of meanders in the Ain River, France. Physical Geography 19:318–340.

Piégay, H., A. Thévenet, and A. Citterio. 1999. Input, storage and distribution of large woody debris along a mountain river continuum, the Drôme River, France. Catena 35:19–39.

Pitlick, J. 1995. Sediment routing in tributaries of the Redwood Creek basin, northwestern California. U.S. Geological Survey Professional Paper 1454-K, GPO, Washington, D.C.

Plummer, G. H., F. G. Plummer, and J. H. Rankine. 1902. Map of Washington showing classification of lands. Plate 1 *in* H. Gannet, editor. The forests of Washington, a revision of estimates. U.S. Geological Survey Professional Paper 5, Series H, Forestry, 2.

Porter, S. C., editor. 1983. Late-Quaternary environments of the United States, volume 1: the late Pleistocene. University of Minnesota Press, Minneapolis.

Potts, R., and A. K. Behrensmeyer. 1992. Late Cenozoic terrestrial ecosystems. Pages 419–541 *in* A. K. Behrensmeyer, J. D. Damuth, W. A. DiMichele, R. Potts, H-D. Sues, and S. L. Wing, editors. Terrestrial ecosystems through time. The University of Chicago Press, Chicago.

Richmond, A. D. 1994. Characteristics and function of large woody debris in mountain streams on northern Colorado. Master's thesis. Colorado State University, Fort Collins.

Robison, E. G., and R. L. Beschta. 1990. Coarse woody debris and channel morphology interactions for undisturbed streams in southeast Alaska, USA. Earth Surface Processes and Landforms 15:149–156.

Russell, I. C. 1909. Rivers of North America. G. P. Putnam's Sons, New York.

Schumm, S. A. 1963. Sinuosity of alluvial rivers on the Great Plains. Bulletin of the Geological Society of America 74:1089–1100.

Schumm, S. A. 1968. Speculations concerning

paleohydrologic controls of terrestrial sedimentation. Geological Society of America Bulletin 79:1573–1588.

Schmidt, K. M., J. R. Roering, J. D. Stock, W. E. Dietrich, D. R. Montgomery, and T. Schaub. 2001. Root cohesion variability and shallow landslide susceptibility in the Oregon Coast Range. Canadian Geotechnical Journal 38:995–1024.

Sedell, J. R., F. H. Everest, and F. J. Swanson. 1982. Fish habitat and streamside management: past and present. Pages 244–255 *in* Proceedings of the 1981 Convention of the Society of American Foresters, September 27–30, 1981. Society of American Foresters, Publication 82–01, Bethesda, Maryland.

Sedell, J. R. and W. S. Duvall. 1985. Water transportation and storage of logs. USDA Forest Service, Pacific Northwest Research Station, General Technical Report PNW-186, Portland, Oregon.

Sedell, J. R., and J. L. Froggatt. 1984. Importance of streamside forests to large rivers: the isolation of the Willamette River, Oregon, U.S.A., from its floodplain by snagging and streamside forest removal. Verhandlungen-Internationale Vereinigung für Theoretifche und Angewandte Limnologie 22:1828–1834.

Shields, F. D., and C. J. Gippel. 1995. Prediction of effects of woody debris removal on flow resistance. American Society of Civil Engineers, Journal of Hydraulic Engineering 121:341–354.

Sklar, L., and W. E. Dietrich. 1998. River longitudinal profiles and bedrock incision models: stream power and the influence of sediment supply. Pages 237–260 *in* K. J. Tinkler and E. E. Wohl, editors. Rivers over rock: fluvial processes in bedrock channels. American Geophysical Union Monograph 107, Washington, D.C.

Smith, D. G. 1976. Effect of vegetation on lateral migration of anastomosed channels of a glacier meltwater river. Geological Society of America Bulletin 87:857–860.

Smith, R. D., R. C. Sidle, and P. E. Porter. 1993a. Effects on bedload transport of experimental removal of woody debris from a forest gravel-bed stream. Earth Surface Processes and Landforms 18:455–468.

Smith, R. D., R. C. Sidle, P. E. Porter, and J. R. Noel. 1993b. Effects of experimental removal of woody debris on channel morphology of a forest, gravel-bed stream. Journal of Hydrology 152:153–178.

Spicer, R. A., and J. T. Parrish. 1987. Plant megafossils, vertebrate remains, and paleoclimate of the Kogosukruk tongue (Late Cretaceous), North Slope Alaska. Pages 47–48 *in* T. D. Hamilton and J. P. Galloway, editors. Geologic studies in Alaska by the U.S. Geological Survey during 1986. U.S. Geological Survey Circular 998.

Stott, T. 1997. A comparison of stream bank erosion processes on forested and moorland streams in the Balquhidder catchments, central Scotland. Earth Surface Processes and Landforms 22:383–399.

Stover, S. C., and D. R. Montgomery. 2001. Channel change and flooding, Skokomish River, Washington. Journal of Hydrology 243:272–286.

Swanson, F. J., G. W. Lienkaemper, and J. R. Sedell. 1976. History, physical effects, and management implications of large organic debris in western Oregon streams. USDA Forest Service, Pacific Northwest Forest and Range Experiment Station, General Technical Report GTR-PNW-56, Portland, Oregon.

Swanson, F. J. S. V. Gregory, J. R. Sedell, and A. G. Campbell. 1982. Land-water interactions: the riparian zone. Pages 267–291 *in* R. L. Edmonds, editor. Analysis of coniferous forest ecosystems in the western United States. Hutchinson Ross, Stroudsburg, Pennsylvania.

Swanston, D. N. and F. J. Swanson. 1976. Timber harvesting, mass erosion, and steepland forest geomorphology in the Pacific Northwest. Pages 199–221 *in* D. R. Coates, editor. Geomorphology and engineering. Dowden, Hutchinson, and Ross. Inc., Stroudsburg, Pennsylvania.

Thorne, C. R. 1990. Effects of vegetation on riverbank erosion and stability. Pages 125–144 *in* J. B. Thornes, editor. Vegetation and erosion. Wiley, Chichester.

Toumey, J. W. 1905. Forests as a factor in shaping the physiographic form of mountains. Pages 93–96 *in* American Forest Congress Proceedings. American Forestry Association, Washington, D.C.

Trimble, S. W. 1997. Stream channel erosion and change resulting from riparian forests. Geology 25:467–469.

Triska, F. J. 1984. Role of large wood in modifying channel morphology and riparian areas of a large lowland river under pristine conditions: a historical case study. Verhandlungen-Internationale Vereinigung für Theorelifche und Angewandte Limnologie 22:1876–1892.

Turaski, M. R. 2000. Temporal and spatial patterns of stream channel response to watershed restoration, Cedar Creek, southwest Oregon. Master's thesis. University of Wisconsin, Madison.

Veatch, A. C. 1906. Geology and underground water resources of northern Louisiana and southern Arkansas. U.S. Geological Survey, Professional Paper 46.

Velichko, A. A., editor. 1984. Late-Quaternary environments of the Soviet Union. University of Minnesota Press, Minneapolis.

Wallerstein, N., C. R. Thorne, and M. W. Doyle. 1997. Spatial distribution and impact of large woody debris in northern Mississippi. Pages 145–150 *in* C. C. Wang, E. J. Langendoen, and F. D. Shields, editors. Proceedings of the Conference on Management of Landscapes Disturbed by

Channel Incision. University of Mississippi, Oxford.

Ward, P., D. R. Montgomery, and R. Smith. 2000. Altered river morphology in South Africa associated with the Permian-Triassic mass extinction. Science 289:1740–1743.

Webster, J. R., S. W. Golladay, E. F. Benfield, J. L. Meyer, W. T. Swank, and J. B. Wallace. 1992. Catchment disturbance and stream response: an overview of stream research at Coweeta Hydrologic Laboratory. Pages 231–253 *in* P. J. Boon, P. J. Calow, and G. E. Petts, editors. River conservation and management. Wiley, New York.

Whitney, G. G. 1996. From coastal wilderness to fruited plain: a history of environmental change in temperate North America from 1500 to the present. Cambridge University Press, Cambridge, UK.

Williams, M. 2000. Dark ages and dark areas: global deforestation in the deep past. Journal of Historical Geography 26(1):28–46.

Wing, S. L., and H.-D. Sues. 1992. Mesozoic and early Cenozoic terrestrial ecosystems. Pages 205–325 *in* A. K. Behrensmeyer, J. D. Damuth, W. A. DiMichele, R. Potts, H-D. Sues, and S. L. Wing, editors. Terrestrial ecosystems through time. The University of Chicago Press, Chicago.

Wohl, E., S. Madsen, and L. MacDonald. 1997. Characteristics of log and clast bed-steps in step-pool streams of northwestern Montana. Geomorphology 20:1–10.

Wolfe, J. A. 1985. Distribution of major vegetational types during the Tertiary. Pages 357–375 *in* E. T. Sundquist and W. S. Broecker, editors. The carbon cycle and atmospheric CO_2. American Geophysical Union, Monograph 32, Washington, D.C.

Wolfe, J. A. 1987. Late Cretaceous—Cenozoic history of deciduousness and the terminal Cretaceous event. Paleobiology 13:215–226.

Woodsmith, R. D., and J. M. Buffington. 1996. Multivariate geomorphic analysis of forest streams: implications for assessment of land use impact on channel condition. Earth Surface Processes and Landforms 21:377–393.

Wright, H. E. editors. 1983. Late-Quaternary environments of the United States, volume 2: the Holocene. University of Minnesota Press, Minneapolis.

Zimmerman, R. C., J. C. Goodlet, and G. H. Comer. 1967. The influence of vegetation on channel form of small streams. Pages 255–275 *in* Symposium on River Morphology. International Association of Hydrological Sciences Publication 75.

American Fisheries Society Symposium 37:49–73, 2003

Wood Recruitment Processes and Wood Budgeting

LEE BENDA AND DANIEL MILLER

Earth Systems Institute, 3040 NW 57th Street, Seattle, Washington 98107, USA

JOAN SIAS

Hydrologic Research and Consulting, 6532 42nd Avenue, NE, Seattle, Washington 98115, USA

DOUGLAS MARTIN

2103 N. 62nd Street, Seattle, Washington 98103, USA

ROBERT BILBY

Weyerhaeuser Co., WTC 1A5, P.O. Box 9777, Federal Way, Washington 98063–9777, USA

CURT VELDHUISEN

Skagit System Cooperative, 25944 Community Plaza Way, Sedro Wooley, Washington 98284, USA

THOMAS DUNNE

Donald Bren School of Environmental Science & Management, 4670 Physical Sciences North University of California, Santa Barbara, California 93106-5131, USA

Abstract.—Wood is recruited to rivers by a diversity of processes, including chronic mortality, windstorms, wildfires, bank erosion, landslides, and ice storms. Recruitment, storage, and transport of large wood in streams can be understood in terms of a mass balance, or quantitative wood budget, similar to the study of other material fluxes in watersheds. A wood budgeting framework is presented that includes numerical expressions for punctuated forest mortality by fire, chronic mortality and tree fall, bank erosion, mass wasting, decay, and stream transport. When used with appropriate parameter values derived for specific conditions or regions, the wood budget equations can be used to make predictions on the importance of various landscape processes on wood abundance in streams in any locale. For example, wood budgets can be used to predict how variations in climate (wet – dry), topography (steep – gentle), basin size (small – large), and land management could affect abundance and distribution of large wood in streams. Wood budgets also can be integrated into numerical simulation models for estimating the natural range of variability, specifically temporal fluctuations of wood supply driven by large storms, floods, fires, and mass wasting, and spatial variability driven by topographic heterogeneity and variations in wood transport. Field studies of wood in streams may be enhanced by the use of a wood budget framework. This includes specifying what measurements are required over what length of stream for estimating recruitment rates of all relevant inputs processes, wood loss by decay, and stream transport of wood. Finally, wood budgets can be used to estimate rates of bank erosion, forest mortality, and landsliding, given appropriate field measurements of wood in streams and riparian conditions.

Introduction

More than 25 years of research have created a foundation for the development of a theory and quantitative framework for evaluating the mass balance, or budget, of large wood in rivers and streams. A wood budget is used to estimate the relative importance of different climatic, vegetative, and geomorphic processes on wood abundance in streams, including mortality, bank erosion, and landsliding across a range of spatial and temporal scales. In addition, wood budgets can also be used to predict the importance of instream wood supply from large regional disturbances, such as wildfires, floods, hurricane-force windstorms, and widespread mass wasting. This information could be helpful in quantifying the range of variability in wood supply and storage and to make predictions about how differences in landscape attributes (climate, topography, etc.) and land management lead to differences in instream wood abundance.

From a resource management perspective, interest in defining the necessary amount of instream wood is increasing. Most existing approaches and models consider input from mortality only (Bragg et al. 2000; Welty et al. 2002), and none of the current regulatory approaches considers recruitment from other processes (that is, bank erosion, landsliding, etc.). Wood budgeting applied at the scale of whole watersheds can provide a useful tool for establishing realistic goals for wood management that consider spatial and temporal variability in recruitment processes. Fisheries biologists and foresters can apply that information to develop forest management prescriptions to ensure adequate wood supply to streams. Wood budgeting can also be used to estimate rates of forest mortality, bank erosion, and landsliding, information useful to foresters, ecologists, and geomorphologists.

In this chapter, the diversity of wood recruitment processes documented in the world's rivers is presented. To help understand the relative importance of different wood recruitment agents, we outline a new technology referred to as "wood budgeting." We begin with the quantitative framework for constructing wood budgets for any landscape in the world, addressing effects of fires, chronic tree mortality (suppression, disease, insects, and sporadic blowdown), bank erosion, and landsliding. Other less well-known processes, such as ice storms and ice-breakage in rivers, could be added to tailor the approach to different landscapes. Next, the numerical expressions are used to identify the data and field methods needed to estimate wood recruitment rates, source distance curves, wood transport, and other components of a wood budget. This is illustrated using data from several regions in the Pacific Northwest because, to the authors' knowledge, they represent the only studies conducted in the context of the wood budget methodology presented in this chapter. We also show how field data can be used to calculate rates of forest mortality and bank erosion. And lastly, we couple a wood budget to a landscape simulation model to predict the natural range of variability in wood abundance over centuries and to examine the role of rare and episodic processes.

Many studies have defined elements of wood budgets, and collectively they comprise the foundation for this chapter; only a partial list can be presented here. Keller and Swanson (1979) developed a conceptual wood budget for streams in the western Cascade Range by identifying the major inputs, outputs, and storage reservoirs. Likens and Bilby (1982) proposed a temporal relation among forest age, wood inputs, and the formation of wood jams in New England. Field measurements of in-channel wood in southeast Alaska by Murphy and Koski (1989) were used to define the relative contribution from stand mortality, bank erosion, and landsliding at the stream reach scale. From these data, they also estimated a wood depletion rate. Measurement of the diameters and lengths of wood in streams in the Oregon Cascades, southwest Washington, and southeast Alaska characterized the dimensions of pieces susceptible to fluvial transport (Lienkaemper and Swanson 1987; Bilby and Ward 1989; Martin and Benda 2001). Van Sickle and Gregory (1990) developed a wood recruitment model based on random tree fall. Field studies by McDade et al. (1990) and Robison and Beschta (1990) identified the source locations of recruited wood to streams. The importance of mass wasting on wood recruitment was identified by Swanson and Lienkaemper (1978), Everest and Meehan (1981), and Hogan et al. (1998). Recruitment of wood by hurricanes along coastal areas has been studied by Greenberg and McNab (1998). The importance of bank erosion as a tree recruitment agent in larger rivers was identified by Sedell and Froggatt (1984), Palik et al. (1998), and Piégay et al. (1999). Finally, simulation models have been developed to predict wood recruitment (Beechie et al. 2000; Bragg et al. 2000; USDA Forest Service 2002; Welty et al. 2002; Meleason, in press).

Wood Recruitment Processes

We begin this chapter by reviewing the rich diversity of wood recruitment processes that have been documented worldwide. Much emphasis has been placed on wood recruitment by chronic mortality from the adjacent riparian forest, particularly in the Pacific Northwest region of North America. However, other processes of wood recruitment include hurricanes, floods, wildfires, bank erosion, landslides, and ice storms. Wood recruitment by different mechanisms reflects regional gradients of climatic, hydrologic, and geomorphic processes. For example, hurricane-force winds are more likely to occur near coastal areas, although massive blowdown has been documented in the middle of continents. Landslides that recruit large trees to streams are often concentrated in wet and steep coastal areas, such as along the Pacific Rim. Wood recruitment by bank erosion is more ubiquitous, although variation within watersheds occurs because bank erosion processes and rates vary downstream or, more locally, due to tributary confluences and other topographic knick points. Channel avulsion in floodplains is a major source of wood in large rivers. Wildfires occur wherever large forests exist, perhaps with the exception of very humid coastal environments and tropical areas. This section briefly discusses each of the major wood recruitment processes in turn, describing some of the governing climatic and geomorphic conditions.

Forest mortality refers to a suite of tree killing processes, including blowdown (but distinguished from widespread, catastrophic blowdown; see below), insects, pathogens, and water logging; chronic mortality during early seral stages of forest growth is also referred to as "suppression mortality" or "stem exclusion" (Figure 1). Rates of forest mortality vary over time in any forest (Bormann and Likens 1979), and mortality

FIGURE 1. Wood recruitment to streams and rivers occurs by a diversity of processes. Shown here are forest mortality, bank erosion, landsliding, and postfire toppling.

rates also vary across regional climatic gradients (Benda et al. 2002). In early seral stages, instream wood is often associated with previous disturbances, such as fires, because wood recruitment in young forests is minimal (Hedman et al. 1996; Figure 1).

Catastrophic blowdown refers to widespread toppling of trees during a single event, such as during hurricanes (Greenberg and McNab 1998) or other downbursts (Wesley et al. 1998). As such, catastrophic blowdown occurs episodically, has recurrence interval of several centuries, and may dominate wood recruitment for decades.

Wildfires, particularly stand replacing events, can cause widespread tree death, including in riparian forests (Figure 1). Trees not killed outright by fire may later succumb to insect outbreaks or disease. In general, tree boles survive fire, although most branches, particularly the finer ones, can be consumed in the blaze (Agee 1993). Following fires, dead trees topple over after one to two decades, as their rooting systems decay or their weakened boles collapse in wind storms (Agee and Huff 1987). The importance of fires in tree recruitment depends on the frequency and severity of fires and their spatial extent, characteristics of fire regimes that vary over climatic gradients (Harmon et al. 1986; USDA Forest Service 2002; Benda and Sias 2003).

Bank erosion is an effective process that recruits trees to streams and rivers, in part because trees that are undercut tend to fall towards channels (Murphy and Koski 1989; Palik et al. 1998; Piégay et al. 1999; Martin and Benda 2001; Acker et al. 2003). Although bank erosion generally increases downstream (Hooke 1980), it also occurs nonuniformly and may even peak in areas associated with logjams, tributary confluences, and other fluvial topographic knick points (Figure 1). In large rivers, extensive sections of floodplains may be eroded during major floods, delivering large volumes of wood from floodplain forests (Piégay et al. 1999). Recruitment of wood to streams by bank erosion depends on the frequency and magnitude of floods, erodibility of stream banks, and the nature of streamside forests (Benda and Sias 2003). Bank erosion may not be differentiated from other recruitment processes in stand-level measures of chronic mortality in some studies.

Wood recruitment by landsliding is yet another important agent of wood recruitment, although its role in watershed-scale wood budgets is only recently being documented (Hogan et al. 1998; Benda et al. 2002; Reeves et al. 2003). Wood recruitment occurs by a diversity of mass wasting processes, including small, streamside landslides and larger, deep-seated failures that transfer wood from hill slopes to channels (Figure 1). In contrast, debris flows scour the long-accumulated wood in first- and second-order channels and deposit jams downstream in larger, often fish-bearing streams. Although debris flows fall into the domain of mass wasting, they are considered primarily an agent of wood redistribution at the channel network scale. Conditions necessary for wood recruitment by mass wasting include steep slopes, narrow valley floors, and intense precipitation. Therefore, streams and rivers in mountain regions are more likely to have significant contributions of wood from mass wasting.

There are a number of less well known wood recruitment processes that may be regionally important. Ice storms can kill trees outright, although, in many cases, ice coating, combined with wind, is more effective at breaking off limbs. Ice storms have increased wood loading to first- through third-order channels in the northeastern United States and Canada (Kraft et al. 2002). Another process that may be locally important is ice break and rafting in rivers. Ice dams may form in rivers during spring thaws, and floating ice can scour riverbanks, creating a form of bank erosion. Yet another process is dam-break floods. Landslide dams that breach often send a flood wave downstream that can be highly erosive and scour streambanks (Costa 1988).

The diversity of wood recruitment processes presents a challenge to researchers, resource managers, and regulators. Most studies of instream wood have not differentiated among various wood recruitment agents. Nevertheless, it is increasingly necessary to evaluate the relative importance of different recruitment processes, in particular how they vary across watersheds or regions. This information can be used to help design wood recruitment strategies (for example, riparian buffer strips), to understand the role of rare and episodic processes on long-term wood recruitment (such as fires, windstorms, landslides), and to begin to understand natural variability of recruitment that may have consequences for designing river restoration and monitoring programs. The theory, technology, and modeling of wood budgeting, presented in the remaining parts of the chapter, can help address the challenge of understanding the relative importance of the diversity of wood recruitment processes worldwide.

Quantitative Framework

Mass budget

Environmental systems with definable inputs, outputs, and residence or storage times lend themselves to an accounting of material fluxes over time and space in the form of a mass balance or budget. Techniques for evaluating mass budgets for other watershed processes have been developed, including erosion and sediment supply (Dietrich and Dunne 1978; Reid and Dunne 1996) and the hydrologic cycle (Dunne and Leopold 1978). Similarly, a wood budget is concerned with the differences among input, output, and decay of wood, a relationship that can be expressed as

$$\Delta S = [I\,\Delta x - L\,\Delta x + (Q_i - Q_o) - D]\Delta t\,, \quad (1)$$

where ΔS is a change in wood storage in a reach of some length Δx over the time interval Δt (Benda and Sias 2003). Change in wood storage is a consequence of wood recruitment (I), loss of wood from over-bank deposition in floods and abandonment of jams (L), fluvial transport of wood into (Q_i) and out of (Q_o) the segment, and in situ decay (D). The terms I and L have units of volume per unit reach-length per time, and the remaining terms (Q_i, Q_o, and D) have units of volume per time (Table 1). The values of these terms will vary depending on position in the channel network. Figures 2A and 3 provide a flowchart and a schematic illustrating the components of a wood budget.

Wood is delivered to channels from a variety of sources. Total input can be summarized as

$$I = I_m + I_{be} + I_s + I_e \quad (2)$$

Inputs include tree mortality from disease, suppression, and sporadic blowdown (I_m); toppling of trees after stand-replacing fires and during windstorms (I_f); inputs from flood-induced bank erosion (I_{be}); wood delivered by landslides, debris flows, and snow avalanches (I_s); and exhumation of wood buried in the bed or bank or recapture of wood previously deposited on the banks (I_e). Mortality refers to the death and toppling of trees, and, though these processes may be offset in time, they are represented by a single rate (that is, long-term chronic mortality is equivalent to long-term toppling). Other processes could be added as needed, for example, ice breakage in rivers.

Forest mortality and growth

Wood delivery to streams from forest death can be viewed as the product of either chronic input of relatively small volumes of wood or rare, episodic events that can add massive quantities of wood over a short time (hours to years). Chronic inputs are caused by competition-induced suppression, insects, and disease. Episodic inputs of large quantities of wood can include wildfires (Agee 1993) and windstorms, processes that often cause widespread tree death and initiation of new forests. Blowdown is also an important process in managed forests (Grizzel and Wolff 1998).

The rate of recruitment from chronic mortality (I_m in equation (2)) can be expressed as

$$I_m = [B_L * M * H * P_m] * N\,, \quad (3)$$

where I_m is the average flux of wood per unit channel length per unit time; B_L is volume of standing live and dead trees per unit area; M is the rate of forest mortality; H is average stand height; P_m [dimensionless] refers to the stand-average fraction of stem volume or length that becomes in-channel wood when trees fall by mortality; and N is 1 or 2, depending on whether one or both sides of the channel are forested (Table 1; Benda and Sias 2003). The term P_m is described later in the chapter. All parameters are functions of time and position, and over any given channel length and time, all exhibit a distribution of values that may be characterized by a mean and some measure of variability. For simplicity, the effect of time is not explicitly included in equation (3) or in subsequent equations in this chapter, and steady state assumptions may be acceptable over short periods (years to a few decades) for most field studies. Over longer periods, however, the effect of time and stochastic processes on the parameters of all the mass balance equations may need to be considered.

The recruitment of fire- or wind-derived wood (I_f in equation (2)) is calculated similarly:

$$I_f = [B_f * T_f * H_f * P_m] * N\,, \quad (4)$$

where I_f is the average annual flux of fire- or wind-killed trees (I_f is zero during all other times), B_f is the volume of standing trees just prior to the fire or windstorm, T_f is the annual proportion of the volume toppled during a specific period during or after the event, and H_f is the average height of trees (Table 1). The frequency of fires or wind-

Table 1. Notation, variable descriptions, and variable dimensions in wood budgeting.

Notation	Variable description	Dimensions	Notation	Variable description	Dimensions
ΔS	Wood storage	m^3	T_f	Toppling period	year
I	Wood input	$m^3\ m^{-1}\ year^{-1}$	E	Bank erosion	m/year
x	Measurement length	m	P_{be}	Probability of tree fall (bank erosion)	%
L	Wood loss	$m^3\ m^{-1}\ year^{-1}$	S_s	Wood storage in landslide zone	m^3/m^2
Q_i, Q_o	Wood transport	m^3/year	A_s	Landslide area	m^2
D	Decay	m^3/year	N_s	Number of landslide per channel length	#/m
I_m	Mortality recruitment	$m^3\ m^{-1}\ year^{-1}$	T_s	Frequency of landslides	per year
I_f	Fire recruitment	$m^3\ m^{-1}\ year^{-1}$	R_c	Landslide delivery ratio	%
I_s	Landslide recruitment	$m^3\ m^{-1}\ year^{-1}$	k_d	Decay constant	#/year
I_e	Exhumation recruitment	$m^3\ m^{-1}\ year^{-1}$	ϕ	Proportion of mobile pieces	%
I_{be}	Bank erosion recruitment	$m^3\ m^{-1}\ year^{-1}$	ζ	Lifetime travel distance	M
B	Forest volume per unit area	m^3/m^2	L_j	Interjam distance	M
M	Mortality rate	%/year	T_p	Lifetime of wood	year
H	Tree height	m	T_j	Lifetime of jam	year
P_m	Probability of tree fall (mortality)	%	L_p	Piece length	m
N	Number of banks	#	β	L_p/channel width	%

storms will govern the relative importance of episodic tree recruitment processes compared to chronic forms of mortality.

Stream bank erosion

Rates of tree recruitment from bank erosion during floods depend on erodibility of banks, flow energy, flood frequency and magnitude, and stand density. The resistance of stream banks to erosion is influenced by composition of the bank material and reinforcement by roots (Hooke 1980). Bank erosion is often greatest in lower, actively migrating portions of channel networks, although it may also peak in the mid-regions of river networks (O'Connor et al., in press). Banks also erode when flow is diverted around debris jams and other obstructions. An expression for mean wood recruitment from bank erosion depends on standing forest volume, rate of bank retreat, and the fraction of tree length that can intersect a channel, or

$$I_{be} = [B_L * E * P_{be}] * N\ , \tag{5}$$

where I_{be} is annual wood supply to streams, E is mean bank erosion rate (lateral distance eroded per year), and P_{be} is the fraction of stem length of fallen trees that is deposited into the channel ($0 < P_{be} \leq 1.0$) (Benda and Sias 2003). P_{be} is analogous to P_m in equations (3) and (4) but generally has a larger value, since all trees recruited by bank erosion are immediately adjacent to the channel, and trees undercut by bank erosion tend to fall toward the channel (Murphy and Koski 1989). Equation (6) predicts annual wood recruitment for a given value of B_L and could be used to predict episodic wood influx by treating E as a stochastic variable.

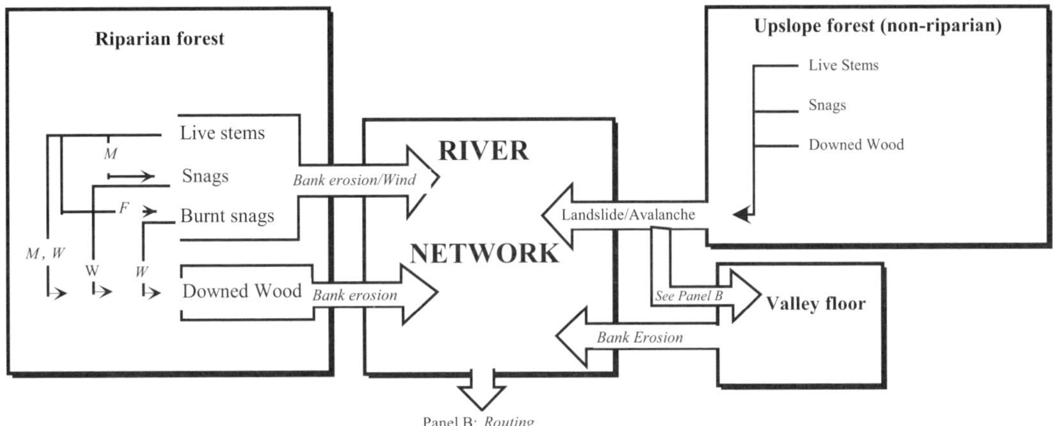

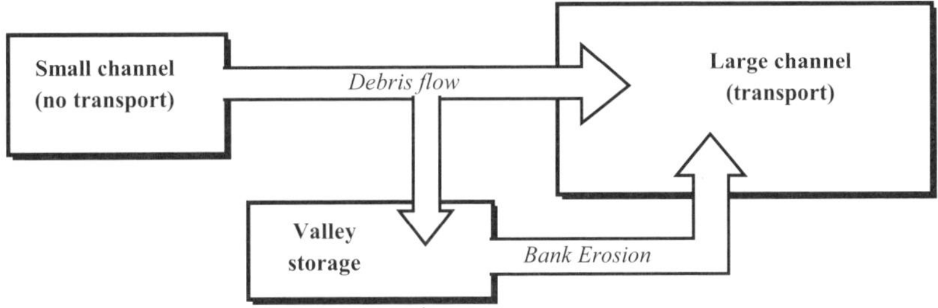

FIGURE 2. Flowchart indicating the major components of a wood budget (from Benda and Sias 2003). Panel A: Fire (F), wind (W), and mortality (M) transfer woody debris to streams and forest floors. In riparian forests, wind and bank erosion transfers wood to rivers. Landslides and snow avalanches recruit live and dead trees to streams, a portion of which may be deposited on valley floors. Panel B: fluvial transport, including debris flows in small, headwater channels.

Mass wasting and snow avalanches

Shallow and deep-seated landslides, debris flows, and snow avalanches recruit wood to channels and valley floors (Swanson and Lienkaemper 1978; Fetherston et al. 1995; Hogan et al. 1998). The importance of wood recruitment by mass wasting depends on the type and area of the landslide or debris flow, sizes of trees recruited, number of landslide or debris flow source areas intersecting a channel segment of a given length, temporal frequency of landsliding or debris flows, and fraction of wood entrained by the event. Landslides and avalanches may deposit partially on fans and terraces at the base of hill slopes, thereby reducing the amount of wood delivered to a channel. The influx of wood from landslides, therefore, can be expressed as

$$I_s = [S_s * A_s * N_s * T_s^{-1}] * R_c , \qquad (6)$$

where I_s is the wood recruitment by mass wasting or by avalanche; S_s is the storage of live and dead wood in the areas entrained; A_s is landslide, debris flow, or avalanche path area; N_s is the number of landslide sites or debris flow tributaries that intersect the downstream (receiving) channel (number per channel length); T_s is the average landslide or debris flow recurrence interval (that is, 1/year); and R_c is the delivery ratio (the proportion of trees that enter the channel) (Table 1). Although equation (7) predicts an average annual flux, mass wasting and avalanches occur as stochastic events, and the episodic nature of wood recruitment by mass wasting can be simulated by stochastic models (see below).

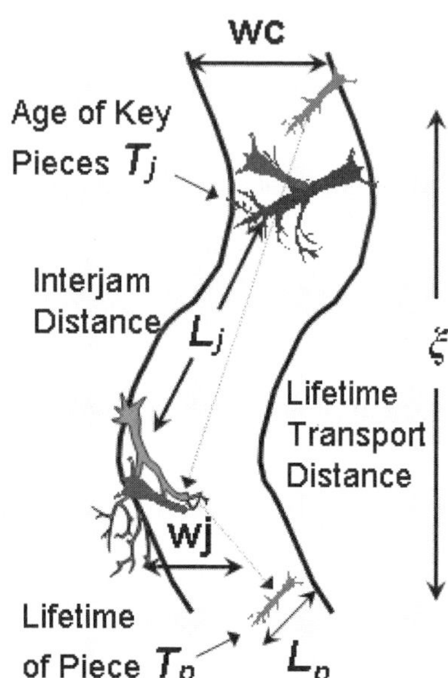

FIGURE 3. (a) A schematic illustrating the major inputs and outputs of a wood budget, including parameters for wood transport (b) (from Martin and Benda 2001).

Wood decay

Wood decay (D in equation (1)) limits the longevity of wood that falls on forest floors or into streams, and it is governed by numerous physical and biological factors. Field studies have shown that annual decay of conifer wood in forest-floor environments commonly ranges from 2% to 7% per year (Harmon et al. 1986; Spies et al. 1988). Streams also exert hydraulic forces that abrade wood or breakup decayed and mechanically weakened wood into smaller transportable pieces. Estimates of annual decay rates for submerged wood ranged from 2% to 3% per year, depending on tree species found in the Pacific Northwest region of North America (Bilby et al. 1999). Estimates of wood loss in unmanaged streams (including decay, abrasion, and transport) have ranged between 1%/year in southeast Alaska (Murphy and Koski 1989) and 3%/year in the Olympic Peninsula (Hyatt and Naiman 2001).

Decay can be expressed as an exponential process:

$$D(x,t) = k_d\ S\ , \tag{7}$$

where k_d is decay loss per unit time and S is storage volume (Harmon et al. 1986). Integrating equation (7) with time yields an exponential loss of wood volume. Wood decays primarily in equation (7) due to a loss of mass (that is, decreasing wood density) (Hartley 1958). Loss of mass, however, should equate with loss of strength and, therefore, wood decay in fluvial environments is assumed to occur by breakup of wood into very small pieces that cannot be effectively captured by jams (or other obstructions) and that exit the system as floatable wood pieces. Transport of wood is covered below, and abrasion of wood during transport is not included.

Stream transport of wood

Understanding how wood moves in a channel network may be an important component of a wood budget. For example, wood transport can alter the distribution of wood, increase jam size, and export wood to estuaries and marine environments. Wood transport may also be of interest when managing the supply of wood to streams (Martin and Benda 2001). Field studies have shown that wood transport depends on several factors. Transported pieces tend to be shorter than bank-full width because larger pieces become lodged between banks (Lienkaemper and Swanson 1987; Nakamura and Swanson 1993; Martin and Benda 2001). In addition, transport distances are limited by obstructions such as debris jams (Likens and Bilby 1982). Hence, because channel width generally increases downstream, an increasing proportion of all wood becomes mobile if the distribution of recruited piece sizes remains constant (Bilby and Ward 1989; Martin and Benda 2001). Wood transport is also affected by stream power (slope and stream cross-sectional area) and flow depth (Haga et al. 2002). Other complexities include the diameter of logs (Bilby and Ward 1989), piece orientation and the presence of root boles (Abbe and Montgomery 1996; Braudrick and Grant 2000), and wood density (Piégay and Gurnell 1997).

Here, we present a wood-transport equation based on the following assumptions. First, wood transport is dependent on the proportion of pieces that are mobile, defined as pieces shorter than channel width at bank-full stage. Second, the transport distance of wood during the lifetime of a piece is dependent upon the lifetime of wood, the distance between transport-impeding jams, the longevity of jams, and the proportion of channel width spanned by jams. The transport equations are more suitable for examining large-scale patterns of wood redistribution and the jam frequencies and sizes that would arise throughout watersheds over decades. They are less suitable for predicting wood movement at the reach scale over a few years because of the complexities that were omitted. Fluvial transport of wood is defined here as

$$Q_w = I \phi \xi , \qquad (8)$$

where Q_w is the volumetric wood transport or flux rate at a cross section (equivalent to Q_i or Q_o in equation (1), I is the average volumetric rate of lateral recruitment, ϕ is the long-term average proportion of all recruited wood with piece lengths (L_p) less than the channel width, and ξ is transport distance over the lifetime of a piece (Benda and Sias 2003). In equation (8) the relative proportions of mobile to nonmobile wood remain constant over time (although they may vary spatially in a network) because of continuous tree recruitment (this assumption may not hold during episodes of very high or very low recruitment). The transport distance (ξ) over the lifetime of wood is predicted by

$$\xi = L_j(T_p / T_j)\beta^{-1} \quad \text{for } T_p >= T_j , \qquad (9)$$

where L_j is the average distance between transport-impeding obstructions, T_p is the lifetime of wood in fluvial environments, T_j is jam longevity, and b is the proportion of channel spanned by a jam (Figure 3b). Equation (9) expresses a hypothesis that transport of wood can exceed inter-jam spacing when wood longevity exceeds jam longevity, and/or when less than 100% of jams are channel-spanning ($\beta < 1.0$). Location and time indices are omitted in equation (9), although our expectation is that all dependent variables (and therefore also the independent variable ξ) will be a function of network position and of time. The main influence of time is stand-age dependence of size and longevity of jam-forming pieces and mobile wood.

Given that β cannot exceed unity, the constraint $T_p = T_j$ ensures that ξ cannot be less than L_j. This fulfills an assumption that wood travel time from location of recruitment to the next downstream jam is much shorter than jam longevity (that is, mobile wood is quickly transported downstream until its migration is impeded by a partial and channel-spanning jam). Accordingly, wood will tend to accumulate at jams, rather than being distributed along channel margins throughout the inter-jam space. This model does not require any consideration of flood frequency and how it changes, for example, with drainage area and climate.

Equations (8) and (9) apply only to streams and rivers where transport is limited by jams; they do not address transport in larger rivers with other forms of wood storage, such as on floodplain and in off-channel areas.

Estimating the proportion of wood falling into streams

The stand-average proportion of wood volume or length that becomes in-channel pieces from all trees in a streamside forest is referred to as P_m and P_{be} in equations (3)–(5). These dimensionless parameters take into account variable fall angle (not all trees will fall directly toward the channel) and variable source distance (any stem within a distance H from the streambank has the potential to contribute wood to the channel). Van Sickle and Gregory's (1990) geometric fall model is used to calculate P_m for all possible combinations of source distances and fall angles (piece breakage can also be incorporated, see Benda and Sias 2003; Sobota 2003). P_{be} is estimated in the same manner as P_m, except that source distance is limited to one meter and trees are constrained to falling within an 180° arc circumscribed by the adjacent bank. Further, our calculation of P assumes that trees are cylinders to avoid the complexity of how the bole's taper varies with species, height, and tree age; taper could be added to the estimation of P when information is available. At any specific time, the random nature of tree fall will cause the value of P, appropriate for a given reach, to fall within some range of values. For any given channel reach, P will vary according to mean tree height (or taper), distance of trees from a channel, and channel width (Figure 4). P is independent of tree mortality rates and simply reflects the cumulative proportion of all tree lengths in a riparian forest that would intersect a stream.

Using this approach, average P_m is about 0.10 for a 15-m-wide channel and an average 50-m tree height (that is, 10% of the cumulative length of all trees intersect the channel and becomes instream wood; Figure 4A). In contrast, P_m is 0.05 in 5-m-wide channels with the same tree height. The term P_{be} values for bank erosion are significantly higher, assuming a 100% fall probability towards the channel when trees are undercut (Figure 4B). P-values decrease dramatically with distance from stream, and higher values are associated with smaller tree heights (Figure 4C). Field measurements should be used to define P in terrains where random fall assumption may not apply or where studies occur over relatively short reach lengths. A recent study in Oregon, Washington, Idaho, and Montana found that tree fall angle was significantly directional toward the stream channel and variance in tree fall angle decreased with increasing hill slope gradient (Sobota 2003).

Field Methods

The quantitative framework provided by wood dynamics models, as illustrated in equations (1)–(9), dictates the type of field measurements necessary to define a wood budget (Figures 2 and 3). In general, wood storage should be tabulated in terms of volume, rather than pieces, because pieces do not discriminate between very small and very large wood. However, wood storage defined as pieces may have more ecological significance. Constructing a field-based wood budget requires making quantitative estimates of wood recruitment (volume/length/time) by fires, chronic mortality, bank erosion, landsliding, and snow avalanches. Field based wood budgets will also require determining the time of fall of individual trees; (Murphy and Koski 1989; Hyatt and Naiman 2001; Martin and Benda 2001; Benda et al. 2002). In general, trees and shrubs that originate by the falling tree (that is, either dependent samplings growing on boles or rootwads or vegetation established by disruption of pre-existing groundcover) is used to date timing of tree falls.

The time over which pieces of wood are recruited to streams can be estimated by

$$\Delta T = (\sum_{i=1}^{n} a_i p_i) \ , \tag{10}$$

where a_i is the mean age of wood in decay class I, and p_i is the proportion of wood in decay class i in any segment (Harmon et al. 1986; Murphy and Koski 1989; Hennon et al. in press); The term ΔT over short time periods is sensitive to the sequence of recruited trees of various sizes (that is, ΔT would be significantly different if a large tree fell in year 10 versus year 1 during a 10-year period). Hence, the proportion of wood in each decay class is based on number of trees, rather than on volume, to reduce the variability in ΔT that can arise due to variation in the temporal sequence of recruitment. In addition, equation (10) gives more weight to trees that have been recruited longer ago to account for the assumed increasing loss of trees (and hence their undercount) with increasing time since recruitment (Murphy and Koski 1989).

Other pertinent field measurements may include forest age, forest volume per unit area, tree height, wood decay, jam spacing and size, jam longevity (age), and wood storage on floodplains, terraces, and fans. Measurements of other watershed attributes may also be necessary, including landslide history and slide-prone topography. Although full wood budgets may be useful for certain purposes, individual components of a wood budget may focus more narrowly on certain aspects, including defining recruitment processes, size distribution of organic debris, source distances, and wood transport. Data required for these more focused questions may not be as extensive as those needed for a complete wood budget.

Determining length of study reaches

Those developing wood budgets are confronted with two important sampling questions: how long

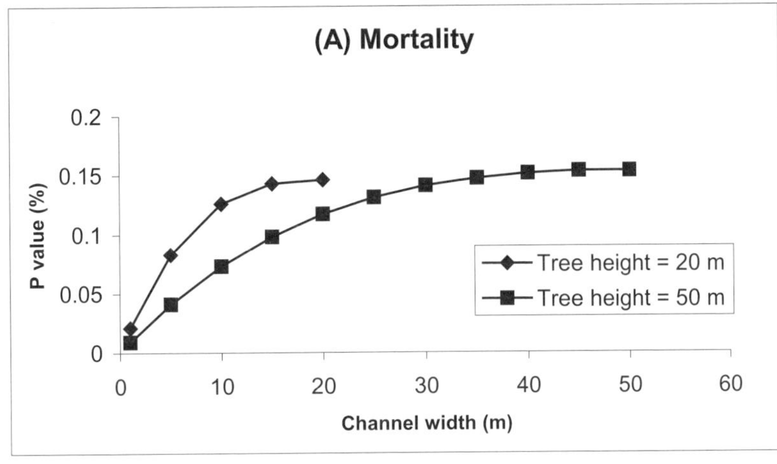

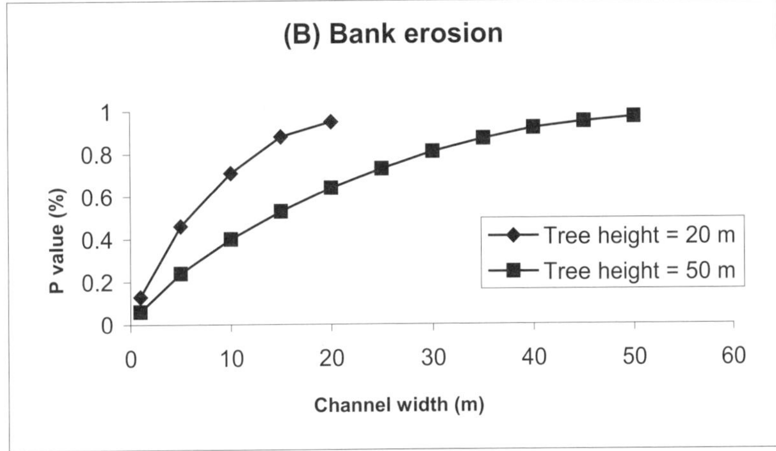

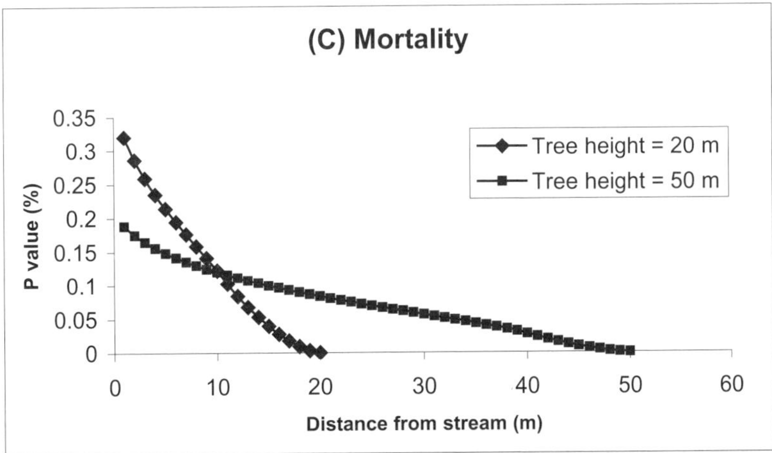

FIGURE 4. (A) Stand average proportion of forest biomass that becomes instream wood depending on tree height and channel width (assumes random fall over 360°). (B) Stand average proportion of forest biomass that becomes instream wood due to bank erosion according the tree height and channel width (assumes 100% fall probability within a streamside 180° arc that intersects the channel). (C) Stand average proportion of forest biomass that becomes instream wood according to distance away from channel edge for two different tree heights (10-m-wide channel).

should study reaches be, and over what periods should studies be conducted? Generally, instream wood is measured in a single year, but in some rare instances, studies have repeatedly measured wood at a site to determine wood longevity or transport rate (for example, Gregory et al. 2000). Although the duration of studies may be restricted, generally more flexibility exists in designating the length of sample reaches. To obtain a good estimate of the relative importance of different recruitment processes, the length of the study reach will depend on the rate at which wood is recruited (and possibly the rate at which wood is lost). A short study reach (~hundreds of meters) may be acceptable in areas of high recruitment, but a longer reach may be necessary in regions of low wood recruitment to accurately characterize input rates. Another confounding aspect is the stochastic behavior of wood recruitment, in which a single large storm, flood, or fire delivers (or removes) large volumes of wood in streams.

The wood recruitment equations in this chapter can be used to estimate lengths of study reaches that might be suitable. To illustrate, we estimate the reach length necessary to measure wood recruitment in areas of different bank erosion rates (a similar technique can be applied to mortality or landslides). The analysis assumes a constant rate of tree recruitment; more sophisticated analyses (such as Monte Carlo simulation) could incorporate the stochastic nature of bank erosion and of other recruitment processes. Bank erosion rates can frequently range from 0.01 m/year to more than 1 m/year (Hooke 1980). To estimate a survey distance, first define the amount of wood to measure (that amount accumulating over a particular time). In this example, our target is a minimum of three trees that entered a channel over a period of 10 years. Begin by estimating the volume of in-channel wood contained in three trees in a 10-m-wide channel. If an average diameter of 1 m is used, the required volume to measure is about 94 m^3 (applying the geometry of a cylinder). Next, the standing forest volume is estimated; here, we use a B_L of 0.25 m^3/m^2. We can ignore P because measured instream wood already accounts for the proportion of wood intersecting a channel. Solving for distance in equation (5) requires a survey of about 3 km of stream to measure three trees with a bank erosion rate of 0.01 m/year (for both sides of the stream) and a survey of 0.03 km for an erosion rate of 1 m/year. Temporal variability in mortality rates and in P will cause variation in the amount of wood actually encountered along 6 km of stream, and survey distances may be longer or shorter than those predicted.

Estimating sources and rates of wood recruitment

Most wood studies have not estimated recruitment rates, partly because of the absence of a wood-budgeting technology. We present results from two recent studies that have estimated recruitment rates: southeast Alaska (Martin and Benda 2001) and redwood forests of northern California (Benda et al. 2002). Game Creek (132 km^2), on Chichagof Island in southeast Alaska, is forested by old-growth western hemlock *Tsuga heterophylla* and Sitka spruce *Picea sitchensis*. The study sites in old-growth redwoods *Sequoia sempervirens* are located in Redwood State Park (Prairie Creek, 57 km^2), northern California. The southeast Alaska and northern California wood budgets estimated recruitment rates for chronic mortality, bank erosion, and landsliding over 40 years and 20 years, respectively (reflecting the time over which wood entered channels. For example, equation (10)).

Both wood budget studies revealed a high degree of spatial variability driven by stream differences in recruitment processes and wood transport. For example, in Prairie Creek, instream wood volumes varied by a factor of 30 (maximum) at the scale of 100-m reaches (Figure 5). Some of the variability is linked directly to variation in recruitment processes.

To estimate recruitment rates over relatively short periods (<2 decades), we can omit stream transport (that is, Q_i and Q_o are assumed be equivalent), loss of wood from over-bank deposition in floods and abandonment of jams (L), and in situ decay (D).

Consequently, equation (1) reduces to

$$\Delta S/(\Delta T \Delta X) = (I_m + I_{be} + I_s)\ , \qquad (11)$$

where ΔS is the change in recruited wood storage (m^3/m), ΔX is length of study segments over some elapsed time period ΔT (that is, equation (10)). The estimated recruitment rate is high in second-growth redwoods compared to old-growth redwoods (4 versus 2.5 $m^3/km/year$; Figure 6), a difference driven by a low mortality rate in old growth (see below).

In both regions, recruitment from bank erosion, landsliding, or both dominated the wood

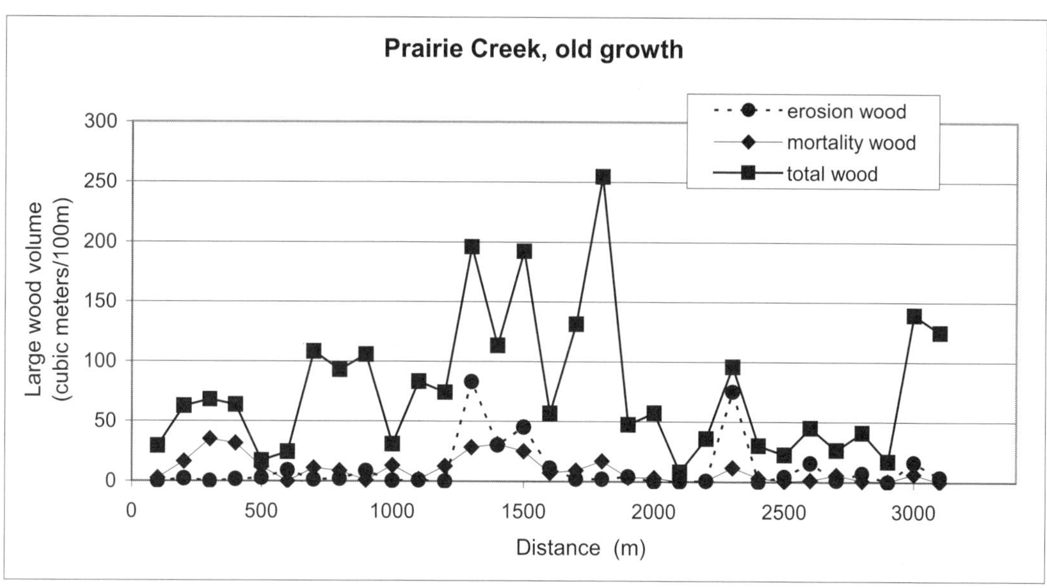

FIGURE 5. Wood storage and recruitment processes in old-growth redwood forests reveal a high degree of spatial variability.

budget (Figure 6). The southeast Alaska budget exhibited a trend of increasing bank-recruited wood with increasing drainage area (Martin and Benda 2001), a finding consistent with increasing bank erosion with increasing basin size. Theoretically, a crossover point in a channel network should be reached where bank erosion recruitment exceeds mortality recruitment (estimated at a bank erosion rate of 0.05 m/year (one side of channel) in mature Douglas-fir forests if an average mortality rate of 0.5%/year is used (Benda and Sias 2003). In the Game Creek watershed, the average mortality recruitment rate of about 4 m^3/km/year (corresponding to an average mortality rate of 1.5%/year) was exceeded by bank erosion recruitment at a drainage area of about 20 km^2 (equivalent to a bank erosion rate of 0.07 m/year).

Estimating source-distance curves

Defining the distances to wood sources in a riparian zone is important in designing forest management and applying regulatory policies. The proportion of wood (either in length or volume) that enters a channel declines with increasing distance from the channel edge. This relation has been demonstrated both empirically and through model simulations (McDade et al. 1990; Robison and Beschta 1990; Meleason et al., in press). The cumulative distribution plot that indicates how the proportion of wood input declines with distance from the channel is referred to as a "source-distance curve." The source-distance curve of wood volume (or length) is sensitive to both tree height and channel width. The proportion of wood volume decreases continuously with distance from a stream because a decreasing proportion of random-fall trajectories intersect the channel (for example, Figure 4C), and the diameter of the bole decreases. To estimate source distance curves during field studies, the distance from the channel edge to the source of wood is measured for each piece where the source can be determined.

Source distance curves are sensitive to different recruitment processes. A theoretical prediction of the source distance curve for mortality recruitment only (assuming a 360° random fall probability) for two different tree heights in a 10-m-wide channel is shown in Figure 7. For comparison, two empirically derived source-distance curves are also plotted, but they include bank erosion and landsliding, recruitment processes that cause a greater proportion of wood to enter closer to the channel. Landslides entering streams not initiated in the streamside zone, especially those that propagate as debris flows, can cause a greater proportion of wood to enter channels from distances further away (May 2001).

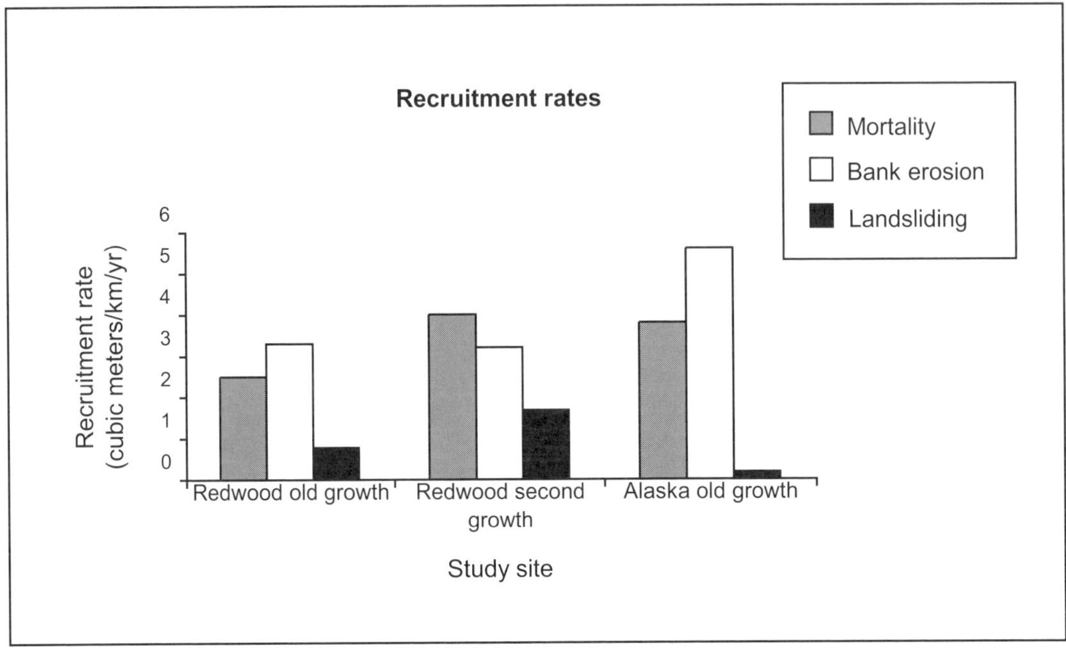

FIGURE 6. Nonmortality sources dominate wood recruitment rates in unmanaged forests in southeast Alaska and in northern California.

Estimating wood recruitment from mass wasting

Numerous field studies have observed that landslides and debris flows deliver large amounts of wood to streams in the Pacific Northwest ecoregion (Swanson and Lienkaemper 1978; Murphy and Koski 1989; Hogan et al. 1998; May 2001). Our experience in the Pacific Northwest indicates that wood delivered to streams by landslides can be measured in two ways. The first method requires conducting long, continuous surveys (~kilometers) to identify the number of pieces of wood recruited by mass wasting. Either the proximity of pieces to landslide debris or the piece condition (landslides and debris flows often leave large scars) can often be used for identification. The second method, which does not require associating pieces with recruitment, evaluates all wood as to distance from mass-wasting source areas, such as debris flow deposits at headwater tributary junctions. This second method is a statistical analysis of relationships between wood accumulations and potential sources of mass wasting and identifies potential delivery from mass wasting rather than actual delivery. Both types of survey procedures are plotted in Figure 8.

Mass wasting, particularly debris flows, may create a clumped distribution of wood in both unmanaged and managed basins (Figure 8). Between debris-flow deposits in our field example in an unmanaged basin in the Oregon Coast Range (Figure 8A), little wood is found, in part because of low forest mortality (in 150-year stands) and the prevalence of small deciduous trees in riparian forests (Nierenberg and Hibbs 2000). In the Oregon Coast Range study, mass wasting was responsible for 80% of instream wood. In second-growth forests in the Olympic Peninsula, Washington, there was a statistically significant correlation (p = 0.1) between in-channel wood storage (across 6 km of third- and fourth-order channels) and proximity to debris flow deposits at low-order confluences; the largest volumes of wood were located 25–50 m from low-order confluences (Benda et al., in press; Figure 8B). Other studies in the Coast Range have observed that approximately half of the wood was derived from mass wasting (May 2001; Reeves et al. 2003). The concentration of wood, and also boulders from debris flows, may lead to clumping or wave-like distribution of aquatic habitat features (Everest and Meehan 1981; Reeves et al. 1995; Benda 1990; Benda et al., in press).

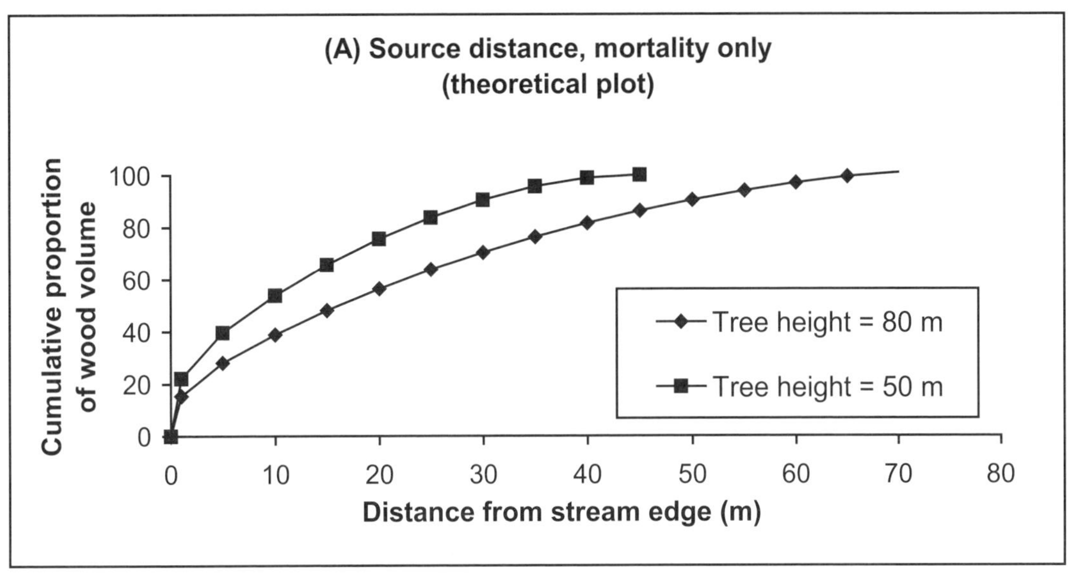

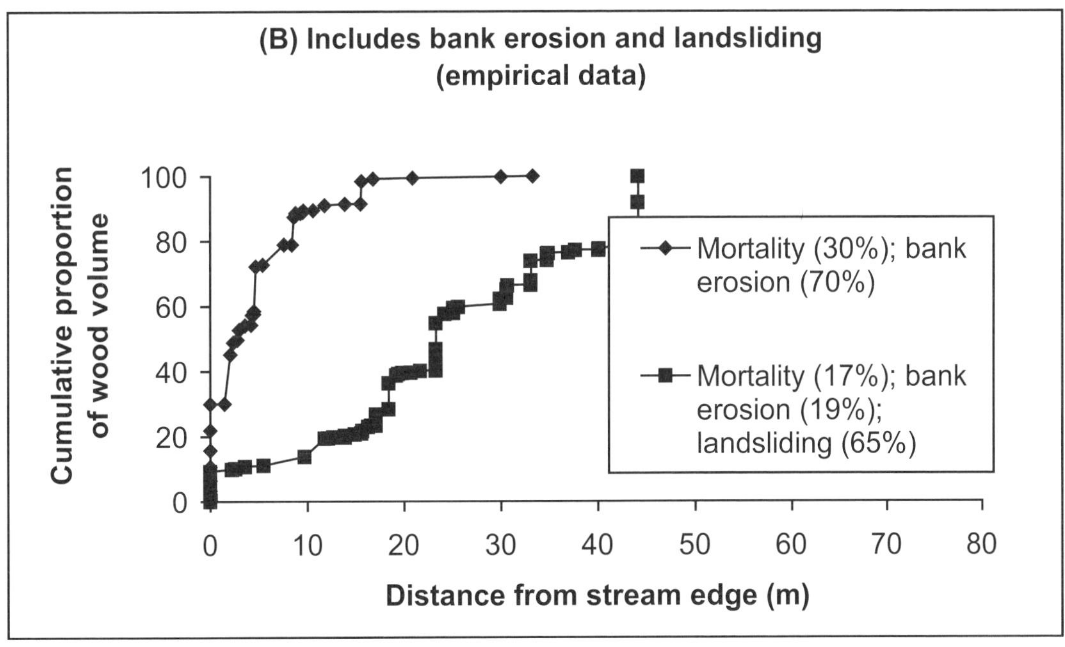

FIGURE 7. (A) Theoretical predictions of source distances are shown for chronic mortality for two different tree heights. (B) Field data reveal differences in source distances due to recruitment by bank erosion and landsliding.

Calculating rates of forest mortality

Estimates of forest mortality are necessary for predicting recruitment of wood to streams and rivers (Beechie et al. 2000; Welty et al. 2002), and they may be useful to foresters and ecologists for other reasons. Forest mortality in upslope stands has been estimated by repeated surveys of stands over long periods (multiple decades). Comparable information is often not available for riparian stands, and estimating mortality rates from current stand conditions is often difficult because of problems in estimating the age of standing dead and downed trees. Mortality rates were measured in seven stands in upland forest, mid-order riparian forests, and low-order riparian

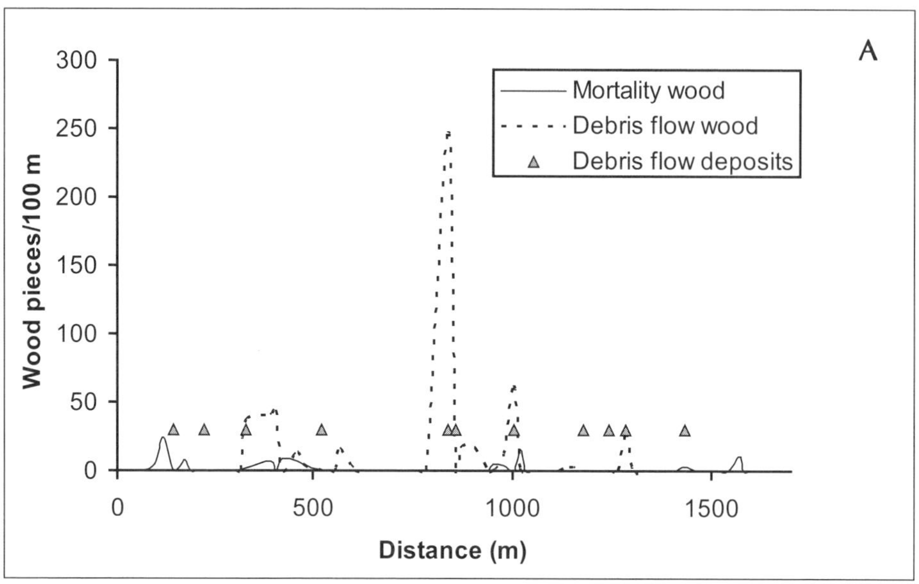

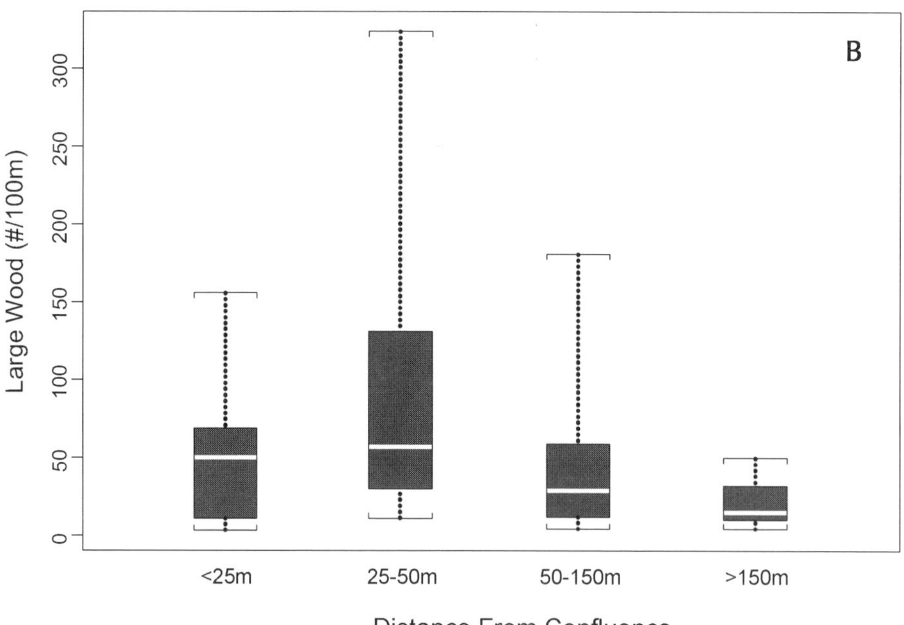

FIGURE 8. (A) Spatial distribution of wood along an unmanaged Oregon Coast Range stream (~150-years-old forest) showing clumps of wood in association with debris flow deposits. 80% of wood was originated from debris flows. (B) Wood densities along 6 km of third- and fourth-order channels in managed forests (Olympic Peninsula, Washington) increase with increasing proximity to low-order confluences prone to debris flow deposition ($P = 0.1$) (Benda et al., in press).

forests in the western Cascades (Acker et al. 2003). Period of record was 15 years for four sites, 17 years for one site, 17 years for one site, and 7 years for one site. Average annual mortality rates for the entire period of record ranged from a low of 0.4%/year to a high of 4.4%/year. Five of the seven sites exhibited mortality rates between 1.0% and 1.6% per year. The high mortality rate came from the unconstrained reach in Lookout Creek, and most of the mortality occurred as a result of trees being knocked over or swept away in the 1996 floods.

A wood budget can be used to estimate forest mortality rates in riparian forests. Solving for mortality in equation (3) requires data on wood recruitment, standing forest volume, tree height, and the proportion of tree length that intersects the channel (P). Generally, the temporal variability of the variables can be ignored when estimating mortality over short periods (years to a few decades). Mortality recruitment (I_m) is obtained from field surveys. Forest inventory surveys can be used to estimate B_L and H. For example, standing forest biomass for Alaskan mixed spruce–hemlock is estimated at 625 m^3/ha; average tree height is 20 m. In contrast, forest biomass in old-growth and second-growth redwoods can be 10,000 m^3/ha (Westman and Whittaker 1975) and 500 m^3/ha, respectively; average tree heights are 80 and 30 m. The P-values are selected from Figure 4.

Using those values in equation (3), average mortality rates in Alaska, redwood old-growth, and redwood second-growth conifer forests varied from 1.6%/year, 0.01%/year, and 1%/year (Table 2). The very low mortality rate in old-growth redwood forests is similar to one estimated by using a tree-replacement-rate estimated by Viers (1978) of two to three redwood trees per ha every 50 years (equivalent to 0.01–0.03%/year). For comparison, a forest mortality rate of 0.5%/year was estimated for mature Douglas fir forests in western Washington and Oregon using other methods (Franklin 1979). Higher mortality rates have been measured in riparian forests (Acker et al. 2003). From the data, a latitudinal control on forest mortality, as well as tree size, is apparent. For instance, mortality is highest in the forests with the smallest (spruce–hemlock) trees in southeast Alaska. Mortality is intermediate in the mid-sized Douglas fir forests in Washington and Oregon, and it is least in the largest (old-growth redwood) trees of the northern California redwoods. Mortality rates can also be estimated for inclusions of stands of deciduous trees within predominantly coniferous forests; rates of 0.02%/year and 0.6%/year for deciduous stands in old-growth and second-growth redwood forests have been documented (Benda et al. 2002). Forest mortality will also vary with forest age, a process not addressed in this example.

Calculating rates of bank erosion and soil creep

Observed rates of wood input from the undercutting of banks can also be used to calculate bank-erosion or soil creep rates, though the time scale represented is constrained by equation (10). Knowledge of bank-erosion rates can aid in developing sediment budgets and in analyses of fluvial geomorphology. Estimating these rates, however, is often difficult because of the paucity of long-term field measurements or the complexity of mortality and undercutting of trees on stream-banks. Solving for bank erosion in equation (5) in southeast Alaska and in the redwood sites (Table 2) required data on wood recruitment, forest volume per unit area, tree height, and P. When values described previously for old-growth redwood forests were used, bank erosion was low (0.01–0.006 m/year), in part because large trees grow on a 3-m-high terrace underlain by erosion-resistant sedimentary rock. Calculated bank erosion rates in southeast Alaska were higher (0.005–0.25 m/year) and increased downstream (Martin and Benda 2001). Soil creep rates can also be estimated using a similar approach (Benda et al. 2002).

Predicting wood recruitment in different climatic regions

We now apply the estimated forest mortality rates to examine how wood supply should vary with distance from stream edge for three different unmanaged forest zones along the Pacific Coast, specifically southeast Alaska spruce–hemlock forests, Washington Douglas-fir forests, and northern California redwoods. A 10-m-wide channel is used to estimate P for all three cases (Figure 4). For Washington's mature Douglas-fir forests, an average forest volume of 0.15 m^3/m^2 and a tree height of 60 m is used (McArdle et al. 1961). The data in Table 2 are used for Alaska and California. Using equation (3), significant differences

TABLE 2. Calculated rates of forest mortality and bank erosion in southeast Alaska (Martin and Benda 2001) and in northern California (Benda et al. 2002).

Alaska	Site 1	Site 2	Site 3	Site 4
Channel width (m)/drainage area (km^2)	7/3.6	11/18	30/79	5/2.5
Forest biomass (m^3/m^2)/tree height (m)	0.0625/20	0.0625/20	0.0625/20	0.0625/20
Mortality/bank erosion recruitment (m^3/km/year)	4.41.87	4.711.8	3.70.3	4.63.2
P: Mortality/bank erosion	0.10/0.57	0.13/0.75	0.15/1.0	0.08/0.62
Forest mortality (%/year)	1.7	1.4	0.9	2.3
Bank erosion (m/year)	0.05	0.25	0.005	0.08
California	**Site 1**	**Site 2**	**Site 3**	**Site 4**
Channel width (m)/drainage area (km^2)	14/7.4	14/7.4	17/24	17/24
Forest biomass (m^3/m^2)/tree height (m)	1.0/80	1.0/80	1.0/80	1.0/80
Mortality/bank erosion recruitment (m^3/km/year)	2.15.9	01.9	1.21.1	4.22.7
P: Mortality/bank erosion	0.08/0.35	0.08/0.35	0.09/0.41	0.09/0.41
Conifer mortality (%/year)	0.02	0	0.013	0.01
Bank erosion (m/year)	0.006	0.01	0.003	0.01

appear in wood recruitment from mortality across all three regions with Washington Douglas-fir forests having the highest rates and redwoods the lowest (Figure 9), a result driven primarily by large differences in forest mortality rates. The analysis indicates how different climatic and vegetation zones can affect wood loading and storage, patterns that could be used to inform management and regulatory programs.

Predicting wood transport

Field data from southeast Alaska are used in equation (9) to predict the transport distance of wood during its expected lifetime. Variables in the transport equation that need defining include L_j (inter-jam spacing), T_j (jam lifetime), T_p (lifetime of wood in fluvial environments), and β (proportion of a channel spanned by a jam). In Game Creek, Alaska, the distance between jams increased with increasing channel size or drainage area (L_j =

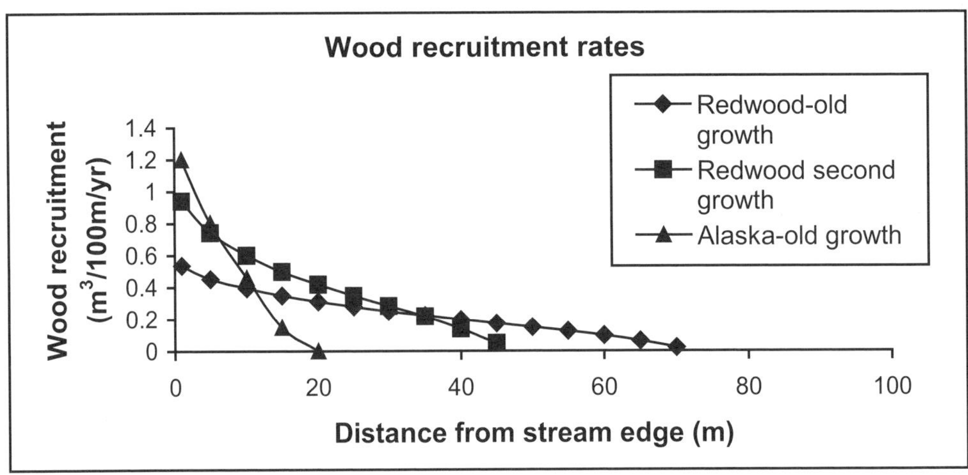

FIGURE 9. Wood recruitment rates according to distance from stream edge for three different unmanaged forest zones are predicted by using equation (3) and parameter values in Table 2.

3.28(A); r^2 = 0.56; Martin and Benda 2001). This relation is anticipated if the piece size distribution of wood input throughout a network remains roughly constant with increasing channel width (that is, more pieces are mobile with increasing stream width). Coinciding with this pattern is decreasing jam longevity with increasing channel size or drainage area (15–30 years in small to large channels). These patterns will, by themselves, lead to systematic increases in transport of wood with increasing drainage area.

Predicted transport distances of mobile wood (piece size < channel width) were calculated using the Alaska data described above, a T_p of 100 years (based on equation (9) and using a 3%/year annual decay rate), and an average β of 0.76. Average transport distances over the lifetime of wood ranged from 100 to 300 m in the smallest streams (drainage areas < 5 km^2, channel width < 5 m) to 800–1,400 m in the largest channels (40–80 km^2 and 20–25 m wide). The predicted wood transport should impose spatial patterns on wood distribution in a watershed. For example, because lateral recruitment (I) depends on stream length (inter-jam distance), jam size (volume or pieces) should increase with increasing transport distance (that is, Q in equation (1)) will increase downstream). A pattern of increasing jam size with increasing drainage areas was observed in the Alaska field study and elsewhere (Likens and Bilby 1982; Bilby and Ward 1989).

Predictive Modeling

Developing testable hypotheses

Equations (1)–(9) can be used to develop hypotheses on the relative importance of different climatic or erosional regimes in the long-term (century) wood budget. To illustrate the approach here, we examine the role of two different stand-replacing fire regimes on the long-term wood budget: (1) an average fire recurrence interval of 500 years for a coastal rainforest regime, and (2) a recurrence interval of 150 years, applicable to drier landscapes. Rough approximations for the parameters in equations (3) and (4) were used in their solution, including (1) fire-killed trees topple over several decades (Agee and Huff 1987) (that is, T_f in equation (4) is 0.025 per year for $11 \leq t_f \leq 50$, where t_f is time, in years, since most recent fire); (2) although hardwoods often dominate the riparian forest in the first century of growth after a stand-eliminating fire, their contribution to the total long-term wood budget is small (Harmon et al. 1986) and therefore is neglected; (3) western coniferous forests accumulate live biomass at a linear rate until about year 500, a rate that may remain stable or decline slightly thereafter (Spies et al. 1988); (4) significant mortality and therefore production of wood from large conifer trees does not begin until about a century after stand initiation (Spies et al. 1988); (5) by the first century, the majority of site potential tree height is attained (McArdle et al. 1961); (6) mortality in mature conifer forests is estimated to be 0.5%/year (Franklin 1979). The term P is defined for a 10-m-wide channel, and equation (7) is used with an average annual decay rate of 3%/year.

Using this approach, large differences in the wood budget between wetter and drier forests are predicted (Benda and Sias 2003). The largest recruitment in both regions occurs immediately post fire as burnt snags topple within several decades after forest death (Figure 10). Because of the longer growth interval between disturbances, the rainforest produces a considerably larger volume of wood than the drier forest from postfire toppling of burnt snags. Moreover, the magnitude of wood recruitment associated with chronic stand mortality is significantly higher in the 500-year cycle because the constant rate of stand mortality is applied against the larger standing volume of older forests (Figure 10). Because the average time between fires in the 150-year cycle is similar to the time when significant mortality of conifers begins (100 years in our solution), the proportion of the total conifer wood supply from postfire toppling of trees in the 150-year cycle is about 50%, compared to 15% for the 500-year cycle. Finally, the range of values of wood recruitment likely to be observed is much greater in forest environments with the 500-year fire cycle compared to the 150-year fire cycle, although finding lower values of wood are more likely in the drier forest.

Model simulation: analysis of landscape dynamics and natural variability

Field surveys of short durations may be insufficient to define natural variability in wood recruitment and storage, in part due to the difficulty of measuring the role of rare and episodic processes in the long-term wood budget, including wildfires, windstorms, landslides, and major floods.

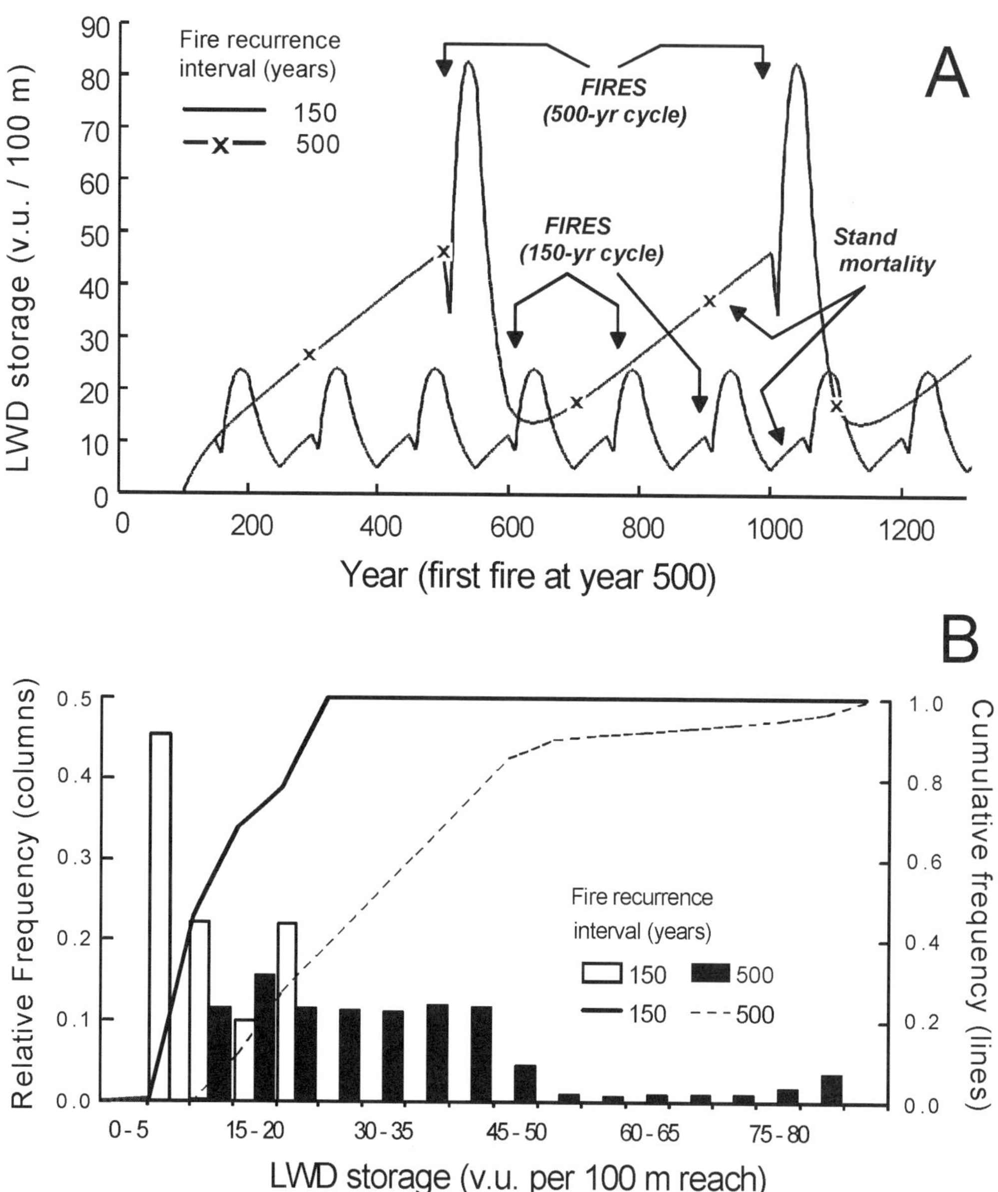

FIGURE 10. Theoretical predictions of the storage of wood for two different fire cycles. The terms B_L and B_f in equations (3) and (4) are expressed in arbitrary volume units (v.u.) in this exercise to avoid specifying a particular growing condition (Benda and Sias 2003).

Simulation models can be used to circumvent that limitation. To illustrate this approach, a stochastic simulation model that includes fires, storms, debris flows, and bank erosion (Benda and Dunne 1997a, 1997b) is used to solve equations (1)–(7) over a period of 4,000 years in a 200 km^2 watershed located in southwest Washington (USDA Forest Service 2002).

The model illustrates how disturbances (fires and storms) and forest succession can lead to marked temporal variability in wood storage (Figure 11). During periods of low disturbance (old-growth forest, no fires or large storms), wood volumes throughout most of the network are relatively low (Figure 11), with the exception of a few persistent landslide and debris flow areas. At

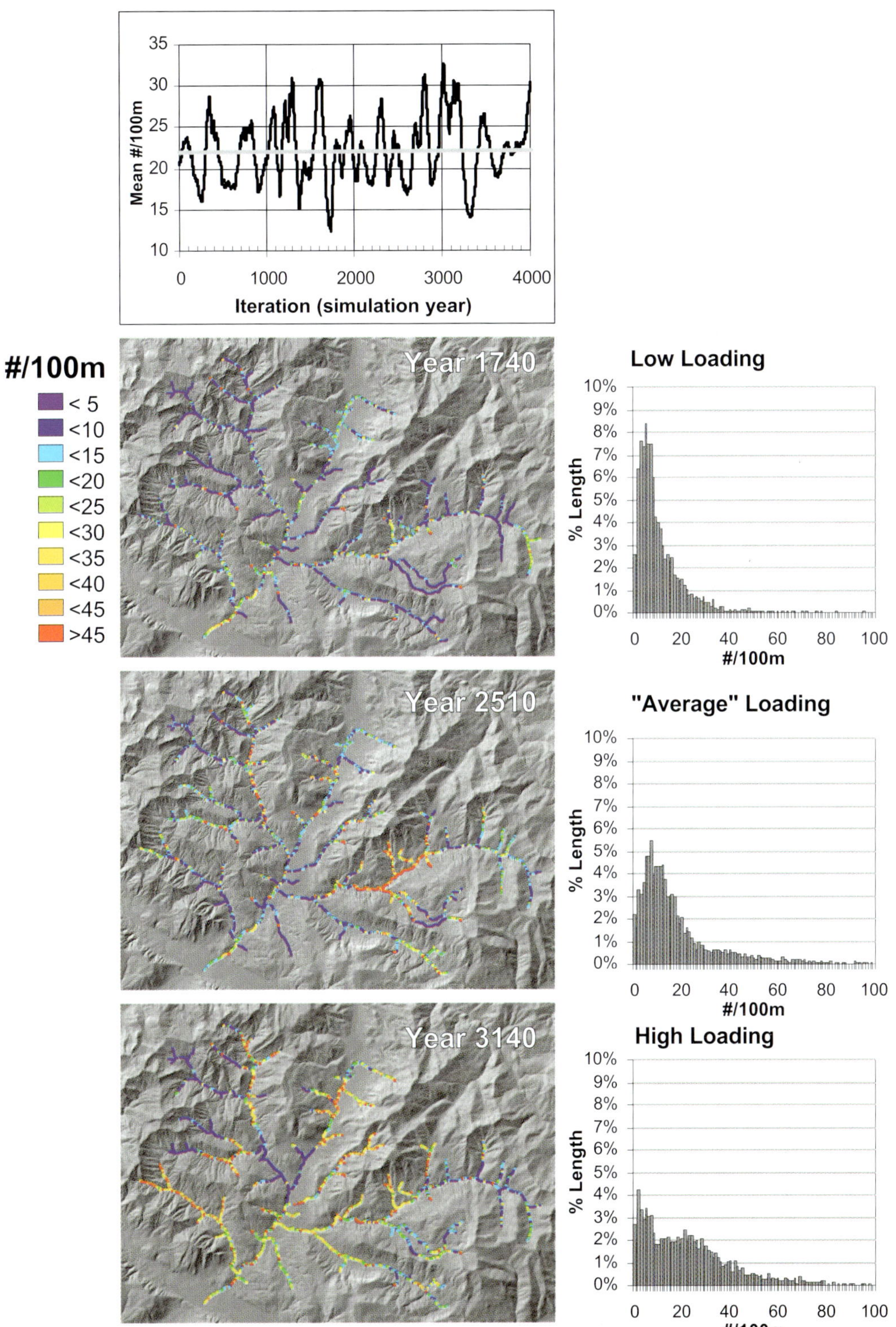

FIGURE 11. Stochastic simulation modeling of wood over 4,000 years in a 200 km² basin in southwest Washington indicates periods of high, average, and low wood storage (USDA Forest Service 2002).

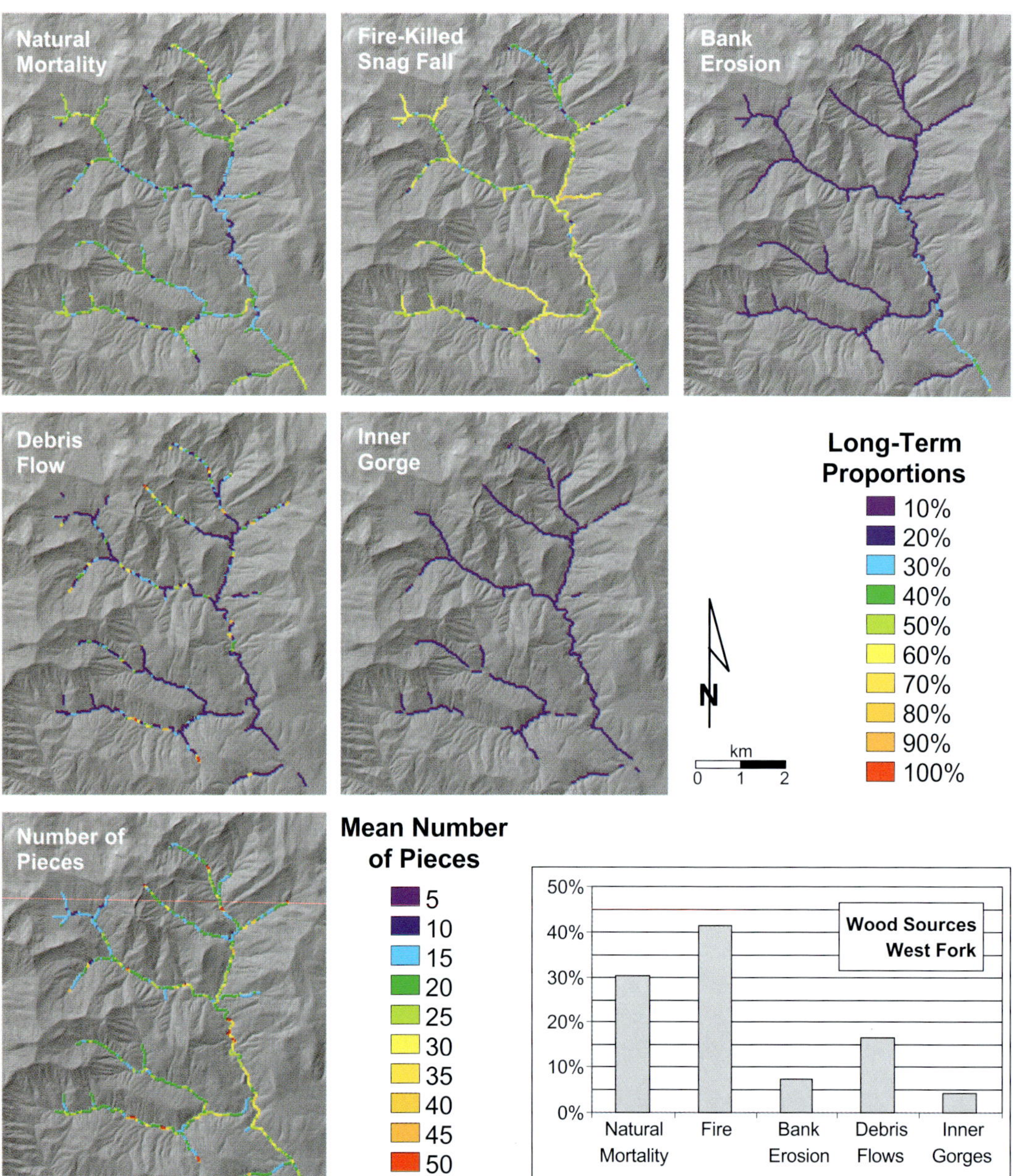

FIGURE 12. Stochastic simulation modeling of wood over 4,000 years in a 200 km^2 basin in southwest Washington illustrating the relative importance of five different recruitment processes and their spatial distribution (USDA Forest Service 2002).

other times, wood storage is predicted to be considerably higher. Hence, the model indicates that measures of wood storage taken at a single time reveal little about the dynamic nature of wood recruitment and storage.

Model predictions are also useful for illustrating how variation in topography (steep versus gentle hill slopes) and basin size (small versus large bank erosion rates) can create both random and systematic spatial variability in wood storage at the scale of a watershed (Figure 12). Debris flows and inner-gorge landslides create localized areas of persistently high wood loading. The model also illustrates how the proportion of wood supplied from the five recruitment processes varies spatially throughout the network (Figure 12). In some areas, fire-killed snag fall dominated, but in others, bank erosion or landsliding dominates. These types of model predictions can inform strategies that pertain to managing, restoring, regulating, and monitoring wood in streams and rivers.

Conclusions

The predictive and testable quantitative relations among landscape process rates, their spatial variance in watersheds or across landscapes, and long-term patterns of wood abundance and distribution described in this chapter comprise a general theoretical framework for the study of wood input processes to streams. The equations can be used to construct hypotheses about wood loading across gradients in climate, basin size, topography, and land management. Anticipated shifts in wood recruitment and storage along environmental gradients can also provide keys to understanding natural variability. When applying the quantitative relations, some places may lack one or more of the processes identified here and perhaps other, less well-known processes may need to be added. Nevertheless, the general principles developed here can aid in constructing field-based wood budgets, designing simulation models, estimating the range of variability, and generating testable hypotheses on future trends of wood in rivers.

Acknowledgments

Development of wood budgeting technology was supported by the Bureau of Land Management (Portland, Oregon), Washington Forest Practices Association, Sealaska Corporation, and Earth Systems Institute. We thank Fred Swanson and an anonymous reviewer for their critique of an earlier version of this chapter.

References

Abbe, T. A., and D. R. Montgomery. 1996. Large woody debris jams, channel hydraulics, and habitat formation in large rivers. Regulated Rivers: Research and Management 12:201–221.

Acker, S. A., S. V. Gregory, G. Lienkaemper, W. A. McKee, F. J. Swanson, and S. D. Miller. 2003. Composition, complexity, and tree mortality in riparian forests in the central western Cascades of Oregon. Forest Ecology and Management 173:293–308.

Agee, J. K., and M. F. Huff. 1987. Fuel succession in a western hemlock/Douglas-fir forest. Canadian Journal of Forest Research 17:697–704.

Agee, J. K. 1993. Fire ecology of Pacific Northwest forests. Island Press, Washington, D.C.

Beechie, T. J., G. Pess, P. Kennard, R. E. Bilby, and S. Bolton. 2000. Modeling recovery rates and pathways for woody debris recruitment in northwestern Washington streams. North American Journal of Fisheries Management 20:436–452.

Benda, L. 1990. The influence of debris flows on channels and valley floors in the Oregon Coast Range, U.S.A. Earth Surface Processes and Landforms 15:457–466.

Benda, L., and T. Dunne. 1997a. Stochastic forcing of sediment supply to the channel network from landsliding and debris flow. Water Resources Research 33(12):2849–2863.

Benda, L., and T. Dunne. 1997b. Stochastic forcing of sediment routing and storage in channel networks. Water Resources Research 33(12):2865–2880.

Benda, L., P. Bigalow, and T. Worsley. 2002. Processes and rates of in-stream wood recruitment in old growth and second-growth redwood forests, northern California. Canadian Journal of Forest Research 32:1460–1477.

Benda, L., and J. Sias. 2003. A quantitative framework for evaluating the wood budget. Forest Ecology and Management 172:1–16.

Benda, L., C. Veldhuisen, and J. Black. In press. Debris flows as agents of morphological heterogeneity at low-order confluences, Olympic Mountains, Washington. Geological Society of America Bulletin.

Bilby, R. E., and J. W. Ward. 1989. Changes in characteristics and function of woody debris with increasing size of streams in western Washington. Transactions of the American Fisheries Society 118:368–378.

Bilby, R., J. Heffner, B. Fransen, J. Ward, and P. Bisson. 1999. Effects of submergence on deterioration of wood from five species of trees used for habitat enhancement projects. North American Journal Fish Management 19:687–695.

Bormann, F. H., and G. E. Likens. 1979. Pattern and process in a forested ecosystem. Springer-Verlag, New York.

Bragg, D. C., J. L. Kershner, and D. W. Roberts. 2000. Modeling large woody debris recruitment for small streams of the central Rocky Mountains. U.S. Department of Agriculture, Rocky Mountain Research Station, General Technical Report RMRS-GTR-55, Fort Collins, Colorado.

Braudrick, C. A., and G. E. Grant. 2000. When do

logs move in rivers? Water Resources Research 36:571–583.

Costa, J. E. 1988. Floods from dam failures. Pages 439–464 *in* V. R. Baker, R. C. Kochel and P. C. Patton, editors. Flood geomorphology. John Wiley, New York.

Dietrich, W. E., and T. Dunne. 1978. Sediment budget for a small catchment in mountainous terrain: Zietschrift fur Geomorphologie, Suppl. Bd 29:191–206.

Dunne, T., and L. B. Leopold. 1978. Water in environmental planning. Freeman, New York.

Everest, F. H., and W. R. Meehan. 1981. Forest management and anadromous fish habitat productivity. Pages 521–530 *in* Transactions of the 46th North American Wildlife and Natural Resources Conference. Wildlife Management Institute, Washington, D.C.

Fetherston, K., J. R. Naiman, and R. E. Bilby. 1995. Geomorphology, terrestrial and freshwater systems. Geomorphology 13(1–4):133–144.

Franklin, J. F. 1979. Vegetation of the Douglas fir region. Pages 93–112 *in* P. E. Heilman, H. W. Anderson, and D. M. Baumgartner, editors. Forest soils of the Douglas-fir region. Washington State University Cooperative Extension Service, Pullman, Washington.

Greenberg, C. H., and W. H. McNab. 1998. Forest disturbance in hurricane-related downbursts in the Appalachian mountains of North Carolina. Forest Ecology and Management 104:179–191.

Gregory, S. V.; L. R. Ashkenas, R. C. Wildman, M. A. Meleason, and G. A. Lienkaemper. 2000. Long-term dynamics of large wood in a third-order Cascade mountain stream. Abstract *in* International Conference on Wood in World Rivers, 2000 October 23–27, Corvallis, Oregon. Oregon State University, Corvallis.

Grizzel, J. D., and N. Wolff. 1998. Occurrence of windthrow in forest buffer strips and its effect on small streams in northwest Washington. Northwest Science 72(3):214–223.

Haga, H., T. Kumagai, K. Otsuki, and S. Ogawa. 2002. Transport and retention of coarse woody debris in mountain streams: an in situ field experiment of log transport and a field survey of coarse woody debris distribution. Water Resources Research 38(8)1–16.

Harmon, M. E., J. F. Franklin, F. J. Swanson, P. Sollins, S. V. Gregory, J. D. Lattin, N. H. Anderson, S. P. Cline, N. G. Aumen, J. R. Sedell, G. W. Lienkaemper, K. Cromack, Jr., and K. W. Cummins. 1986. Ecology of coarse woody debris in temperate ecosystems. Advances in Ecological Research 15:133–302.

Hartley, C. 1958. Evaluation of wood decay in experimental work. USDA Forest Service, Forest Products Laboratory, Report No. 2119, Madison, Wisconsin.

Hedman, C. W. D., H. Van Lear, and W. T. Swank. 1996. Instream large woody debris loading and riparian forest seral stage associations in the southern Appalachian Mountains. Canadian Journal of Forest Research 26:1218–1227.

Hennon, P. E., M. McClellan, and P. Palkovic. In press. Comparing deterioration and ecosystem function of decay-resistant and decay-susceptible species of dead trees. In P. Shea, editor. Symposium on the Ecology and Management of Dead Wood in Western Forests, Reno, Nevada, November 2–4, 1999. The Wildlife Society, Berkeley, California.

Hogan, D. L., S. A. Bird, and M. A. Hassan. 1998. Spatial and temporal evolution of small coastal gravel-bed streams: the influence of forest management on channel morphology and fish habitats. Pages 365–392 *in* P. C. Klingeman, R. L. Beschta, P. D. Komar, and J. B. Bradley, editors. Gravel-bed rivers in the environment. Gravel-bed rivers IV. Water Resources Publications, LLC, Colorado.

Hooke, J. M. 1980. Magnitude and distribution of rates of river bank erosion. Earth Surface Processes and Landforms 5:143–157.

Hyatt, T. L., and R. Naiman. 2001. The residence time of large woody debris in the Queets River, Washington, USA. Ecological Applications 11(1):191–202.

Keller, E. A., and F. J. Swanson. 1979. Effects of large organic material on channel form and fluvial processes. Earth Surface Processes 4:361–380.

Kraft, C. E., R. L. Schneider, and D. R. Warren. 2002. Ice storm impacts on woody debris and debris dam formation in northeastern U.S. streams. Canadian Journal of Fish and Aquatic Sciences 59:1677–1684.

Lienkaemper, G. W., and F. J. Swanson. 1987. Dynamics of large woody debris in streams in old growth Douglas-fir forests. Canadian Journal of Forest Research 17:150–156.

Likens, G. E., and R. E. Bilby. 1982. Development, maintenance, and role of organic-debris dams in New England streams. Pages 122–128 *in* F. J. Swanson, R. J. Janda, T. Dunne and D. N. Swanston, editors. Workshop on Sediment Budgets and Routing in Forested Drainage Basins. PNW – 141. USDA, Forest Service, Portland, Oregon. Products Laboratory, Report No. 2119, Madison, Wisconsin.

Martin, D., and L. Benda. 2001. Patterns of in-stream wood recruitment and transport at the watershed scale. Transactions of the American Fisheries Society 130:940–958.

May, C. L. 2001. Spatial and temporal dynamics of sediment and wood in headwater streams of the Oregon Coast Range. Doctoral dissertation. Oregon State University, Corvallis.

McArdle, R. E., W. H. Meyer, et al. 1961. The yield of Douglas fir in the Pacific Northwest. USDA Forest Service Technical Bulletin No. 201, Washington, D.C.

McDade, M. H., F. J. Swanson, W. A. McKee, J. F. Franklin, and J. Van Sickle. 1990. Source distances for coarse woody debris entering small streams in western Oregon and Washington. Canadian Journal of Forest Research 20(3):326–330.

Meleason, M. A., S. V. Gregory, and J. Bolte. In press. Implications of riparian management strategies on wood in streams of the Pacific Northwest. Ecological Applications.

Murphy, M. L., and K. V. Koski. 1989. Input and depletion of woody debris in Alaska streams and implementation for streamside management. North American Journal of Fisheries Management 9:427–436.

Nakamura, F., and F. J. Swanson. 1993. Effects of coarse woody debris on morphology and sediment storage of a mountain stream in western Oregon. Earth Surface Processes and Landforms 18:43–61.

Nierenberg, T. R., and D. E. Hibbs. 2000. A characterization of unmanaged riparian overstories in the central Oregon Coast Range. Forest Ecology and Management 129:195–206.

O'Connor, J. E., M. A. Jones, T. L. Haluska, T. Marie, and D. J. Polette. In press. Floodplain and channel dynamics of the Quinault and Queets rivers. Geomorphology.

Palik, B., S. W. Golladay, P. C. Goebel, and B. W. Taylor. 1998. Geomorphic variation in riparian tree mortality and stream coarse woody debris recruitment from record flooding in a coastal plain stream. Ecoscience 5:551–560.

Piégay, H., and A. M. Gurnell. 1997. Large woody debris and river geomorphological pattern: examples from S. E. France and S. England. Geomorphology 19(1–2):99–116.

Piégay, H., A. Thevenet, and A. Citterio. 1999. Input, storage and distribution of large woody debris along a mountain river continuum, the Drome River, France. Catena 35(1):19–39.

Reeves, G., L. Benda, K. Burnett, P. Bisson, and J. Sedell. 1995. A disturbance-based ecosystem approach to maintaining and restoring freshwater habitats of evolutionarily significant units of anadromous salmonids in the Pacific Northwest. Pages 334–349 *in* J. L. Nielson, editor. Evolution and the aquatic system: defining unique units in population conservation. American Fisheries Society, Symposium 17, Bethesda, Maryland.

Reeves, G. H., K. M. Burnett, and E. V. McGarry. 2003. Sources of large wood in the mainstem of a fourth order watershed in coastal Oregon. Canadian Journal of Forest Research 33(8):1363-1370.

Reid, L. M., and T. Dunne. 1996. Rapid construction of sediment budgets for drainage basins. Catena-Verlag, Cremlingen, Germany.

Robison, G. E., and R. L. Beschta. 1990. Characteristics of coarse woody debris for several coastal streams of southeast Alaska, USA. Canadian Journal of Aquatic Sciences 47:1684–1693.

Sedell, J. R., and J. L. Froggatt. 1984. Importance of streamside forests to large rivers: the isolation of the Willamette River, Oregon, U.S.A., from its floodplain by snagging and streamside forest removal. Internationale Vereinigung fur theoretische und angewandte Limnologie Verhandlungen 22:1828–1834.

Sobota, D. J. 2003. Fall directions and breakage of riparian trees along streams in the Pacific Northwest. Master's thesis. Oregon State University, Corvallis.

Spies, T. A., J. F. Franklin, and T. B. Thomas. 1988. Coarse woody debris in Douglas-fir forests of Western Oregon and Washington. Ecology 696:1689–1702.

Swanson, F. J., and G. W. Lienkaemper. 1978. Physical consequences of large organic debris in Pacific Northwest streams. USDA Forest Service, Pacific Northwest Forest and Range Experiment Station, General Technical Report PNW-69, Portland, Oregon.

USDA Forest Service. 2002. Landscape dynamics and forest management. USDA Forest Service, Rocky Mountain Research Station, General Technical Report RMRS-GTR-101-CD, Fort Collins, Colorado.

Van Sickle, J., and S. V. Gregory. 1990. Modeling inputs of large woody debris to streams from falling trees. Canadian Journal of Forest Research 20:1593–1601.

Viers, S. D. 1978. Redwood vegetation dynamics. Bulletin Ecological Society of America 56:34–35.

Welty, J. W., T. Beechie, K. Sullivan, D. M. Hyink, R. E. Bilby, C. Andrus, and G. Pess. 2002. Riparian aquatic interaction simulator (RAIS): a model of riparian forest dynamics for the generation of large woody debris and shade. Forest Ecology and Management 162/2:299–318.

Wesley, D. A., G. S. Poulos, M. P. Meyers, J. S. Snook, and A. Judson. 1998. Observations and forcing mechanisms during the October 1997 Front Range blizzard and forest destruction. Preprints, Eighth Conference on Mountain Meteorology, 3–7 August 1998, Flagstaff, Arizona, American Meteorological Society, 25–30.

Westman, W. E., and R. H. Whittaker. 1975. The pygmy forest region of northern California: studies on biomass and primary production. Ecological Monographs 63:453–520.

American Fisheries Society Symposium 37:75–91, 2003

Wood Storage and Mobility

ANGELA M. GURNELL

Department of Geography, King's College London, Strand, London WC2R 2LS, UK

Abstract.—This paper considers wood storage and mobility in relation to four themes illustrated by both literature review and case studies.

- The first theme presents estimates of wood storage in relatively unmanaged river systems by using information derived from the open literature. More than 150 sets of observations of wood storage are synthesized. In this international data set, major contrasts are found according to tree species, stand age, and geographic setting. However, the identification of trends in wood storage characteristics from the data set is confounded by interdependency in potential explanatory variables.
- The above analyses provide a context for the second theme, an examination of some of the factors that influence wood storage within river reaches. A number of influences are discussed, including the characteristics of the wood; the size and morphology of the river channel; a range of climatic, hydrologic, and geomorphic processes; and the nature of wood management.
- The aggregate effect of these influences is not only variability in the amount of wood stored in river reaches but also variability in its mode of storage, which forms the third theme of this paper.
- The amount and mode of wood storage are a product of the interaction between wood supply, transport, and retention characteristics of river reaches. The fourth theme of the paper is the presentation of two case studies that illustrate wood anchorage and mobility within rivers of different size.

Introduction

Although a number of researchers have quantified wood storage in river systems, comparison between studies has proved difficult. While many researchers have adopted the definition of large wood as wood pieces greater than 1 m in length and 10 cm in diameter, this is not universally the case. The contextual information that is provided also varies widely from one study to another, making comparisons and generalizations concerning wood storage difficult. Furthermore, very few studies have explicitly considered wood mobility, and so this aspect of the present review is developed from a particularly limited information base.

As a result of these limitations, this paper starts by evaluating a large sample of studies in an inevitably simplistic way, and then builds understanding by concentrating on information drawn from a smaller number of studies. The review contains four sections. First, a simple analysis of information from more than 150 studies provides a broad assessment of wood storage within an international context. This is followed by two sections that consider (1) the complexities of some of the key controls on wood storage and (2) variations in the mode of wood storage within different systems, particularly within rivers of different size. Finally, two case studies illustrate some more detailed properties of wood anchorage and mobility.

Wood Storage in River Channels: An International Perspective

To provide a broad context for this review, information was assembled from the open literature on wood storage in river channels draining unmanaged or only lightly managed river corridors. Interpretation of what was lightly managed from published sources was inevitably subjective, but the intention was to include sites where cutting of trees along the river margins and removal of wood from the river margins and channel was minimal. As a result of the limited quantity and nonstandard presentation of data describing riparian zone and floodplain wood storage, analysis was confined to wood storage in active river channels. Wherever possible, information was drawn from studies specifying not only the amount of stored wood but also the width and slope of the river channel and the dominant tree species. In addition to the excellent compilation provided in Harmon et al. (1986), data were drawn from many other sources, including Heede (1981), MacDonald et al. (1982), Swanson et al. (1984), Wallace and Benke (1984), Hogan (1986), Murphy et al. (1986), Smock et al. (1989), Carlson et al. (1990), Potts and Anderson (1990), Robison and Beschta (1990), Fausch and North-cote (1992), O'Connor (1992), Evans et al. (1993), Diez (1999), Piégay et al. (1999), Treadwell et al. (1999), Gurnell et al. (2000a and unpublished), and Rutherfurd et al. (2000). These sources provided data for 152 river reaches, although information on the full range of variables was not available for all reaches.

The amalgamated data set had several limitations. The definition of "large wood" and the method used to estimate the amount of wood varied among studies and often was not clearly specified. The amalgamated data set was also geographically biased towards North America, particularly the northwest portion of the United States. However, sufficient information was available from other areas to provide a thought-provoking if not comprehensive analysis. The data were standardized in a variety of ways to permit amalgamation. To avoid problems generated by differences in wood density between tree species, the load of wood was expressed by volume in cubic meters per hectare of channel area. Channel width was expressed in meters. Two characteristics of the riparian woodland were recorded: age and type. Typical woodland stand age was coded according to three classes (<100 years, 100–200 years, >200 years). While a greater resolution in stand age would have been very informative, it was not possible to achieve this reliably from the published information. Five types of woodland were identified according to the dominant tree species: redwoods (n = 14), other conifers (n = 90), mixed (usually conifers with deciduous hardwoods, n = 6), deciduous hardwoods (n = 20), evergreen hardwoods (eucalyptus, n = 11), and deciduous softwoods (Salicaceae, n = 11).

The continuous variables wood volume, channel width, and slope had highly skewed distributions. Each of these variables was log-transformed to support box and whisker plots and scatter plots that were easier to interpret and also to conform as closely as possible to the variance homogeneity assumptions of analysis of variance (ANOVA; equal variance of samples) and regression analysis (homogeneity of residual variance). Where, after transformation, the data still did not conform to the equal variance assumption, non-parametric (Kruskal-Wallace) ANOVA was used to analyze the data.

The stored wood volume data set is described using box and whisker plots in Figures 1 and 2. These figures describe the distribution in the log-transformed wood storage volume estimates according to three different subdivisions of the data set (woodland type, stand age, and geographical region).

Differences in wood storage according to woodland type are illustrated in Figure 1A. Wood volume storage associated with redwoods, other

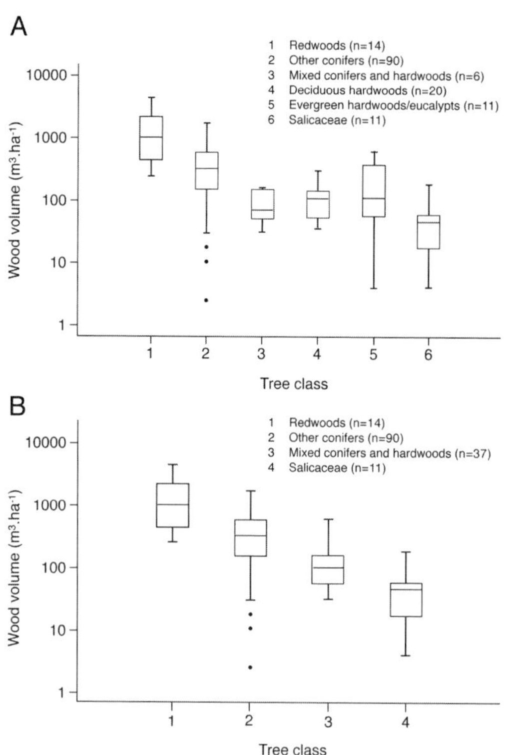

FIGURE 1. Box and whisker plots of the variation in the in-channel volume of stored wood in relation to type of woodland bordering the river.

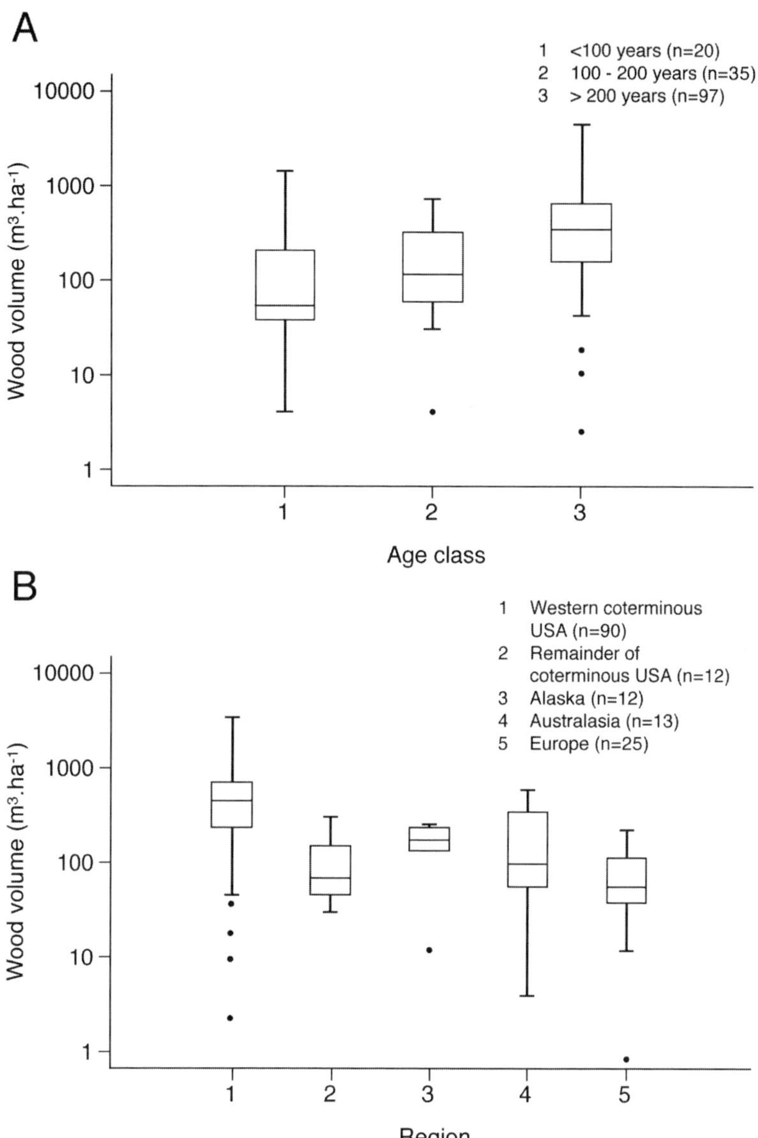

FIGURE 2. Box and whisker plots of the variation in the in-channel volume of stored wood in relation to (A) woodland age; (B) geographical region.

conifers, and the Salicaceae were all significantly different from one another ($P < 0.05$) and from wood storage within the remaining three groups (mixed, deciduous hardwoods, and evergreen hardwoods). However, there was no significant difference in wood storage between the mixed, deciduous hardwood, and evergreen hardwood categories. Figure 1B combines the latter three groups and illustrates four woodland type categories that possessed significantly different in-channel wood storage volumes ($P < 0.05$).

Grouping wood storage by stand age (Figure 2A) identified a significantly higher wood volume stored in streams draining woodland stands that were more than 200 years old ($P < 0.05$), but no significant difference between the remaining two age categories (<100 years and 100–200 years). However, there were confounding factors that may influence these results. The oldest woodland sites were predominantly under redwood and coniferous woodland, which have already been shown to have the highest in-channel wood storage. The lack of discrimination between the two younger age-classes reflects the wide variation in wood storage within the youngest class. Unfortunately, the information on woodland age presented in the published sources was insufficiently precise to allow a more detailed discrimination of age on which to develop analysis.

The volume of stored wood in relation to the geographical region from which the data were drawn is illustrated in Figure 2B. There was a significantly higher volume of stored wood observed in the western coterminous United States ($P < 0.05$) than in the other four areas for which data were available (eastern coterminous United States, Alaska, Australia, and Europe), but this was largely a reflection of the dominance of old, coniferous woodland sites, including all of the redwood sites, within this geographic region. In conclusion, the differences identified in the volume of wood stored in river channels draining woodland of differing type need to be interpreted with care because the patterns are confounded by the inter-correlation between stand age and woodland type and the geographic region within which the study sites were located.

The log-transformed scatter plot of the volume of in-channel wood storage against channel width (Figure 3A) shows a significant, although weak, negative relationship between the two variables ($R^2 = 0.115$, t statistic for slope coefficient = –4.17 [$P < 0.001$]; $F = 17.42$ [$P < 0.001$]; $n = 135$).

$$\text{Stored wood volume} = 385(\text{width})^{-0.340}$$

The negative relationship is hardly surprising since stored wood volume is expressed in cubic meters per hectare, and thus, other factors being equal, a decrease in storage per unit area would be expected as the channel widens. More significant are the distinct patches within the scatter plot in Figure 3A, reflecting the four categories of woodland type presented in Figure 1B (and thus stand age and geographic area as well as tree species). The young Salicaceae cover (willows, poplar) along the margins of some European rivers is clearly associated with wider channels than for the other woodland types. Although this may represent an artifact of the data set, this is the type of disturbed marginal woodland cover that would be expected along large, dynamic, temperate river systems. The interaction between $\log_{10}$(stored wood volume), $\log_{10}$(channel width), and the type of woodland was explored by incorporating dummy variables into the analysis. A multiple regression model was estimated using seven independent variables: log-transformed channel width, three dummy variables (value 1 or 0 for presence or absence) for woodland types other than Redwoods, which were used as the base woodland type, and interaction terms between the three individual dummy variables and $\log_{10}$(channel width). All of the estimated slope coefficients for the four variables incorporating channel width (that is, $\log_{10}$[channel width], [conifer dummy].$\log_{10}$[channel width]; [mixed + hardwood dummy].$\log_{10}$[channel width], [Salicaceae dummy].$\log_{10}$[channel width]) were not significantly different from zero, and so the multiple regression relationship was re-estimated excluding these variables. The weak, negative relationship between wood volume and channel width was no longer statistically significant, and the dominant role of woodland type on stored wood volume per unit channel area, regardless of channel width, was expressed by the following relationship (adjusted $R^2 = 0.332$, t statistics for slope coefficients = –4.20, –6.68, –7.54 [all $P < 0.001$]; $F = 26.00$ [$P < 0.001$]; $n = 151$ – the increased sample size incorporates sites for which no channel width was available).

$$\begin{aligned}\text{Log}_{10}(\text{Volume}) = 2.999 &- 0.590(\text{coniferdummy}) \\ &- 1.024(\text{mixed+hardwood dummy}) \\ &- 1.485(\text{Salicaceae dummy})\end{aligned}$$

This equation can be expressed as a set of four constant wood volume estimates according to woodland type:

Redwoods	Volume = 998 m^3/ha
Conifers	Volume = 256 m^3/ha
Mixed + hardwoods	Volume = 94.4 m^3/ha
Salicaceae	Volume = 32.7 m^3/ha

No significant simple linear relationship was found between log-transformed stored wood volume and channel slope, although, once again, the woodland types defined distinct zones within the scatter plot between these two variables (Figure 3B). Multiple linear regression analysis using dummy variables for woodland type and interaction terms between woodland type and $\log_{10}$(slope) defined the following significant relationship (adjusted $R^2 = 0.672$, t statistics for slope coefficients = 2.45, –3.33, –4.41, –5.53, –2.28, –2.32, –3.58 [P = 0.017, 0.001, 0.000, 0.000, 0.026, 0.023, 0.001]; $F = 23.88$ [$P < 0.001$]; $n = 78$).

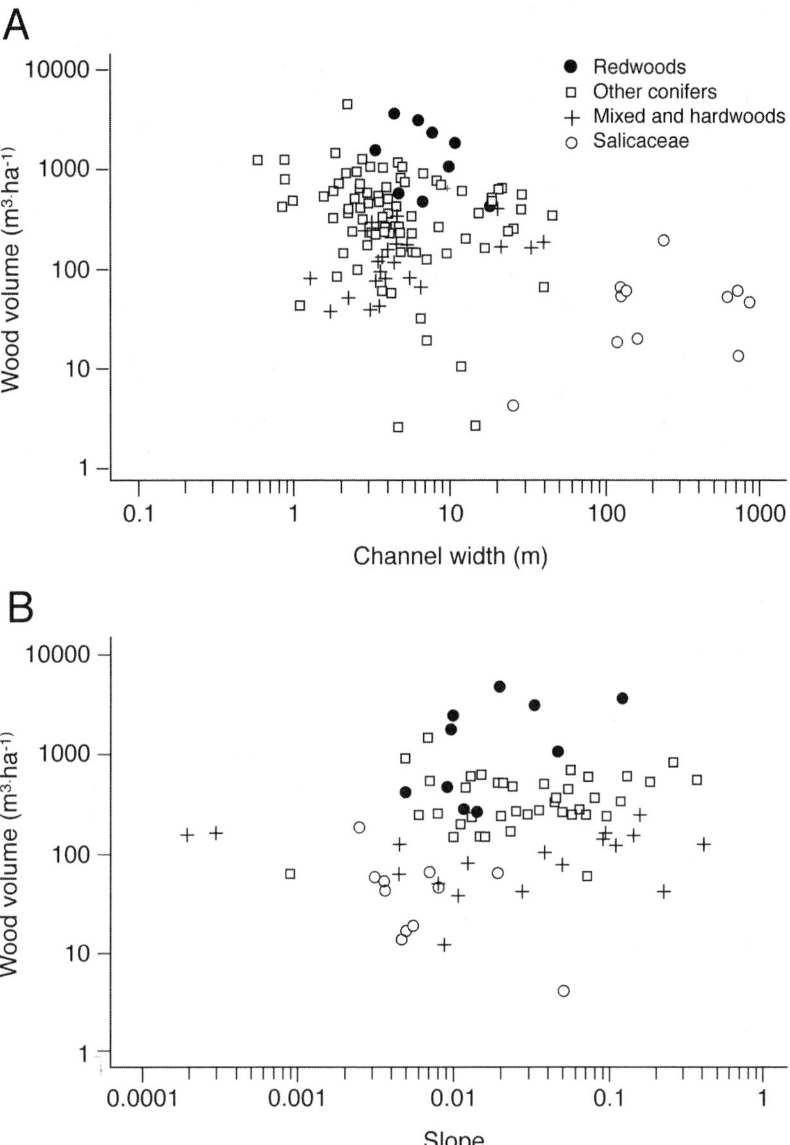

FIGURE 3. (A) Scatter plot of the in-channel volume of stored wood in relation to channel width. (B) Scatter plot of the in-channel volume of stored wood in relation to channel slope.

$$
\begin{aligned}
\text{Log}_{10}(\text{Volume}) = {} & 4.127 + 0.631 \cdot \log_{10}(\text{slope}) \\
& - 1.644(\text{conifer dummy}) \\
& - 2.146(\text{mixed + hardwood dummy}) \\
& - 4.109(\text{Salicaceae dummy}) \\
& - 0.637 \cdot \log_{10}(\text{slope}) \cdot (\text{conifer dummy}) \\
& - 0.628 \cdot \log_{10}(\text{slope}) \cdot (\text{mixed + hardwood dummy}) \\
& - 1.311 \cdot \log_{10}(\text{slope}) \cdot (\text{Salicaceae})
\end{aligned}
$$

This equation can be expressed as the following four relationships for each woodland type (note that for conifers and mixed + hardwoods, the exponent is very close to zero, indicating that stored wood volume is essentially constant, regardless of channel slope).

Redwoods	Volume = $13{,}396(\text{slope})^{0.631}$ m³/ha
Conifers	Volume = $304(\text{slope})^{-0.006}$ m³/ha
Mixed + hardwoods	Volume = $95.7(\text{slope})^{0.003}$ m³/ha
Salicaceae	Volume = $1.04(\text{slope})^{-0.680}$ m³/ha

The above data and analyses provide baseline estimates of typical wood volumes stored within river channels where both the wood and marginal woodland are unmanaged or lightly managed. While the detailed form of the relationships between wood volume and channel slope or width in relation to each of the woodland types is likely to be heavily influenced by the character and limited size of the data set, the analyses illustrate major differences between the woodland type groups, which are likely to be a genuine reflection of the type of woodland. However, these differences may also reflect some influence of woodland age; climatic, hydrological, and sediment transport regimes; catchment topography; and level and type of woodland and river management (even though sites were selected where the marginal woodland and wood were unmanaged or only lightly managed).

Influences on Wood Storage in River Channels

Wood density, decay, and buoyancy

The analyses in the previous section were deliberately applied to estimates of stored wood volumes rather than loads. This approach avoided distortion of the analyses by an important but often neglected factor influencing the amount of wood stored in rivers: its density. According to Harmon et al. (1986, p. 190),

> In the case of aquatic studies, density has rarely been measured, and a density of 0.5 Mg/m^3 has often been applied. While this value probably does represent the average for undecayed wood of all species, it overestimates the density of undecayed coniferous wood by as much as 20–40% and underestimates the density of undecayed hardwood by the same amount.

Although the wide range in undecayed wood density of 0.3–0.7 Mg/m^3 implied by this statement is probably appropriate for many environments, important exceptions exist. For example, Rutherfurd et al. (2000) quote densities ranging from 0.8 to 1.3 Mg/m^3 for 15 Australian tree species, and 11 of these species have densities in excess of 1.0 Mg/m^3. These high densities imply that the wood from many Australian tree species will not float in freshwater. They explain why wood is often found lining the bed of Australian river channels (Rutherfurd et al. 2000).

Wood density is also affected by the degree of waterlogging and decay of the wood. Laboratory experiments by Thévenet et al. (1998) showed an increase in density of more than 140% as a result of submerging oven dried cubes of wood in water for 10 d. They observed greatest water absorption by older wood, which they suggested might enhance the rate of wood decay.

These very significant differences in wood density attributable to species, age, and degree of waterlogging affect the buoyancy of wood pieces and so are likely to strongly affect wood mobility, particularly where the wood is not firmly anchored. Indeed, where the wood is denser than water, it may act like very coarse bed material, moving infrequently and providing a large, long-residence wood store that armors the riverbed and provides an important control on the mobilization of mineral bed material.

River channel size

Another important factor affecting wood retention is the size of the river channel. This is partly because wood pieces can become jammed in small channels but also because the discharge regime in small channels may be insufficient to transport large pieces of wood. The association of enhanced wood storage with small river channels in England and Wales (Figure 4) was identified by Linstead and Gurnell (1999). The figure presents cumulative frequency distributions of the widths of channels where wood debris is extensive (>33% channel area), present, or absent (Figure 4A) and where debris dams are present or absent (Figure 4B), based on information from River Habitat Surveys of more than 4,000 river reaches in England and Wales undertaken for the UK Environment Agency (Raven et al. 1997). Linstead and Gurnell (1999) showed how the abundance of large wood and wood jams was strongly associated with river margin land use. The presence and extent of woodland, particularly broadleaf and mixed woodland, was found to be important. Moreover, the absence of wood from many channels may reflect wood clearance, particularly in larger channels, where wood accumulations are often seen as a flood defense problem. Nevertheless, the enhanced retention of wood and dams in channels narrower than 15 m is also likely to reflect the accumulation and jamming or wedging of large wood pieces within these narrow channels.

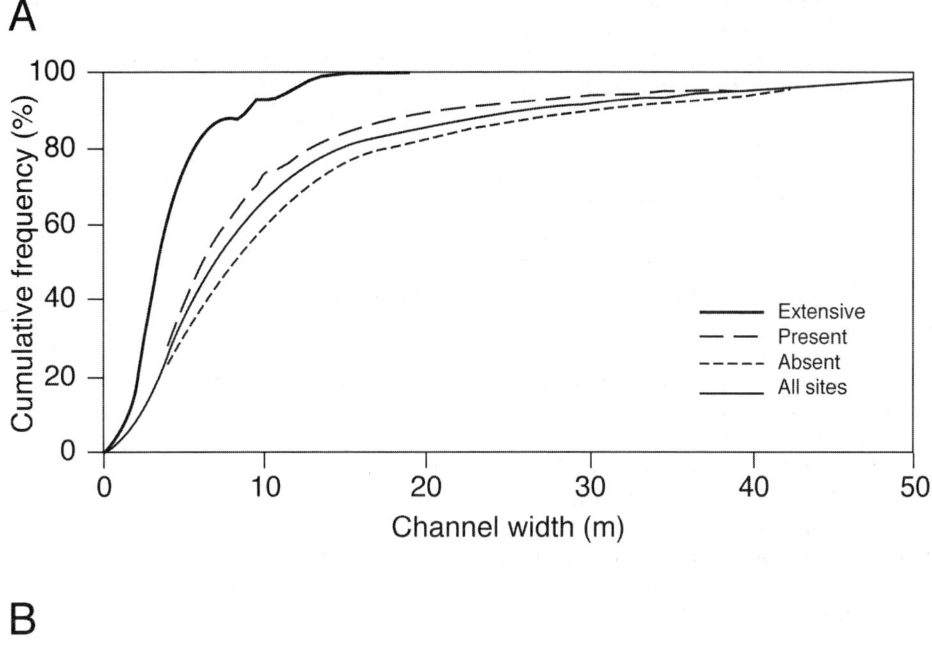

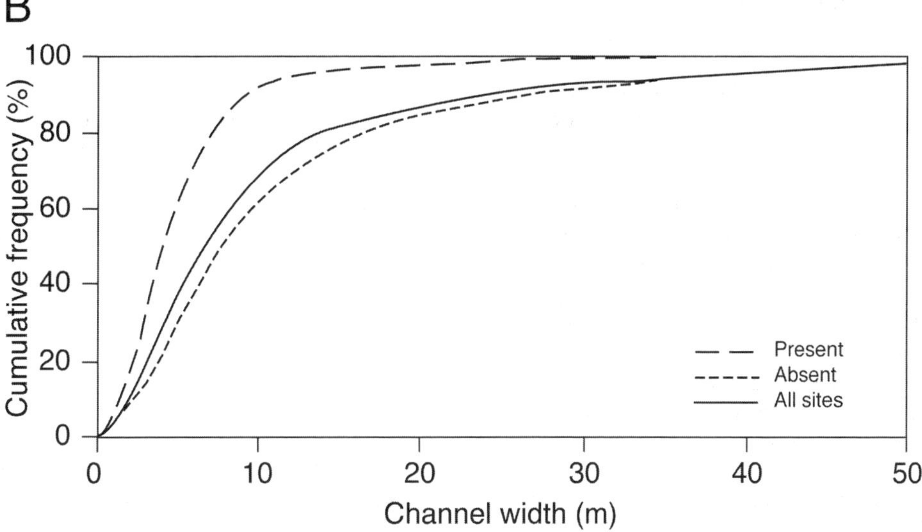

FIGURE 4. Cumulative frequency distributions of the widths of channels where (A) wood debris is extensive, present, or absent and where (B) debris dams are present or absent. (Based upon information from River Habitat Surveys of more than 4,000 river reaches in England and Wales undertaken for the UK Environment Agency, after Linstead and Gurnell 1999).

Church (1992) defined three sizes of channel and criteria for distinguishing them in his discussion of channel morphology and typology: small, intermediate, and large. He defined these three sizes according to channel dimensions scaled by bed-material size. The potential for river channels to retain wood can be placed in a similar context by scaling channel size, particularly active channel width, to the piece sizes of wood delivered to the channel. Gurnell et al. (2002) drew analogies between wood and sediment dynamics and retention within rivers, assessing the degree to which the science of mineral sedimentary features and processes (sedimentology) might be instructive in considering wood features and processes. One simple aspect of this is to propose different sizes of river channel in relation to their potential to retain wood. *Small* channels are narrower than the locally typical wood-piece length (for example, they might be defined as those channels whose width is less than the median wood piece length). Thus, in small channels, the majority of the wood pieces are long enough to be easily jammed within the channel width or to overlap the banks and span the channel. *Medium* channels are slightly narrower than the longer wood pieces present (for example they might be defined as those channels whose width is less than the upper quartile wood piece length), where a relatively small proportion of the wood pieces are sufficiently long to have the potential to become jammed within the channel width. *Large* channels are wider than the length of all of the wood pieces delivered to them. These size limitations reflect an important aspect of the ability of channels to retain stable, key wood pieces around which smaller pieces can accumulate and so also be retained.

Thus, in relation to wood storage and mobility, channel size can be viewed as a function of tree size, which in turn reflects tree age and species because these factors are the main natural controls on wood piece sizes and on the form of wood entering a river. For example, some species, such as the crack willow *Salix fragilis*, are particularly susceptible to breakage, shedding individual branches freely. In contrast, other species (for example, many conifers) are more susceptible to toppling, so that entire trees enter the river. The age of trees is also important because older trees of most species shed diseased or decaying branches. The growth habit of the tree may also be an important influence on the potential of wood pieces to become jammed within channels and thus on wood retention. Irregularly shaped wood pieces have a greater hydraulic roughness and are more likely to become trapped or snagged in river channels than more streamlined wood pieces of a similar size.

Management

The above factors are important influences on wood storage and mobility in undisturbed river reaches, as is the overall reach wood budget (that is, the relative importance of wood input rates and processes, output, and in situ decay of wood; Benda et al. 2003, this volume) and its influence on the degree to which wood moves in a congested or uncongested manner (Braudrick et al. 1997). Management activities also profoundly affect the amount, size characteristics, and spatial pattern of wood storage as well as the retention time and mobility of the wood.

Riparian woodland management has short- and long-term effects on wood storage and mobility. In the short term, tree thinning and felling can contribute many small pieces of wood, which are readily moved by the river. Tree harvest also reduces the input of the largest pieces of wood that would normally form the key pieces in wood accumulations. In the long term, woodland management may directly or indirectly lead to changes in the species composition and age structure of riparian trees, with further consequences for the amount, size distribution, and decay rates of wood that goes into the river. However, the most important long-term impact of river margin and upstream woodland management in comparison with the unmanaged situation is that the quantity and size of wood input to the river is reduced, leading to a decrease in wood storage and an increase in wood mobility.

Riparian woodland management may also have far-reaching effects on the resistance of riverbanks to erosion and, thus, on their rates of erosion and susceptibility to failure, which in turn may impact on the distribution, frequency, and rates of local sediment and wood delivery to the river, on the spatial patterns and rates of sediment scour and deposition, and on vegetation colonization of river margins and the age structure of river margin woodlands.

Hydrologic management induces changes in the river flow regime with direct effects on the ability of the river to mobilize and transport wood, changing the frequency and transport distance of wood pieces of different sizes. It also influences sediment transport, affecting the degree to which partial or complete burial may help to retain wood. In particular, the reduction of peak flows can greatly reduce wood mobilization and transport.

River channel management, which often aims to increase flow conveyance of river channels, reduces channel and riparian wood retention capacity by the removal or reduction of the geomorphologic or vegetation structures that retain wood. Management also frequently involves the active removal of wood obstructions from the river channel, thus directly reducing wood storage, removing the most stable wood pieces, and so reducing the wood piece sizes that remain in channel storage. In extreme cases, river channel management involves bank reinforcement and the construction of levees, which prevent bank erosion and disconnect the river from its floodplain, so preventing wood supply to the river.

To summarize, the impact of woodland and river channel management is to reduce the quantity and size of wood pieces delivered to and stored within river channels and to disturb much stable wood within the channel. This has the effect of changing the apparent size of a small or medium channel, according to the above definitions based on wood piece size. As a result, the transitions between small, medium, and large channels move upstream. Since wood retention is generally higher, the smaller the river, management is likely to increase wood transfer rates and decrease wood retention and thus local decomposition throughout the river system. In addition, more severe attempts to engineer channels to increase their conveyance not only reduce the size and quantity of wood pieces input to the river but also reduce the wood retention capacity of the channel so that the mobility of wood increases regardless of its size. Finally, flow regulation that involves the reduction of high flows to some extent counteracts the trends induced by other forms of management because it decreases the frequency of events that can mobilize and transport large wood.

The time taken for wood storage and dynamics to recover from management is likely to vary greatly among different systems, and thus, sensitivity to management is an important issue to consider in attempting restoration. For example, recovery from channel wood clearance may be rapid in systems where wood delivery and turnover by decomposition and redistribution is rapid. Many European rivers are bordered by species that mature within a century and that deliver wood that decays within decades. Here recovery from wood clearance, particularly where the marginal woodland remains relatively intact, can occur within 10–20 years. However, the very large, slowly recruited wood that is typical of streams in the Pacific Northwest of the United States can retain the imprint of wood clearance for centuries.

Variations in the Mode of Wood Storage

The mode of wood storage is heavily influenced by how the controlling factors vary in their relative importance along the river continuum. The controls on retaining and storing large wood fall into four main groups: the character of the riparian woodland and wood debris delivered to the river (tree sizes, species, density, and age; wood-piece sizes, morphology, and buoyancy); geomorphology (river corridor width, slope, and form; river channel bank and bed stability; river channel size, pattern, and dynamics; the relative roughness presented by bed material and other obstructions in the channel); physical, particularly hydrologic processes (lateral delivery processes, river discharge, and sediment transport regimes); and management, as it affects these three groups of factors.

In less-managed systems, where river channels are morphologically intact and bordered by riparian forest throughout their length, the relative importance of forest character, hydrologic processes and geomorphology changes in a downstream direction (Gurnell et al. 1995) and the predominant modes of wood storage also change (Piégay and Gurnell 1997; Gurnell et al. 2002).

In very small headwater streams, the character of the riparian woodland coupled with the type and frequency of wood-input processes are of overriding importance. Because the river channels are narrow, many wood pieces are large enough to span the channel, sometimes being supported above the channel by the valley sides in very narrow river corridors. Once they have fallen into the river channel, large wood pieces are relatively immobile because of their high relative roughness compared with the channel depth, and because stream discharges are not sufficient to move them, although smaller pieces may be redistributed. The result is an apparently random distribution of wood pieces within and across the channel, governed largely by the sites of wood input and the rate of wood decay. Input mechanisms, such as blow down, avalanches, riverbank and slope failures, and debris torrents (Keller and Swanson 1979), dictate the distribution of large wood in the stream system.

As streams increase in size and depth in a downstream direction, wood pieces are less likely to span or jam across the channel, have a lower relative roughness, and are more likely to be mobilized by the increasing stream discharges. Although stream discharges may not be able to move the largest pieces of wood, intermediate-sized pieces can be retained only if structures are available to trap them or brace them against the flow. Such retention structures include the largest wood pieces, which may span the channel and/or which cannot easily be floated or moved by river flows; other wood pieces, which are stable as a result of burial, bracing, or snagging; riparian vegetation (particularly the trunks and exposed roots of riparian trees); other large roughness elements in the channel such as boulders, channel morphological constrictions (for example, channel narrowing, major bars, or benches); and planform irregularities (for example, sharp bends). The result is the development of accumulations or dams of wood supported by larger pieces of wood, but which trap smaller, mobile pieces. This process results in a trend where the average size of wood pieces retained and the spacing between accumulations increases with the width of the stream (Bilby and Ward 1989). The dominant control category here is the hydrologic regime because it drives the periodic transport of wood pieces during high flows, controls the size of wood pieces transported, and influences the breakup of wood accumulations. In high energy systems, time and space transitions between small, medium, and large channels may be abrupt in terms of their wood dynamics and storage. For example, debris torrents may result in complete removal of wood stored in steep streams or, in less-confined situations, flood events may force sediment and wood onto the adjacent narrow floodplain (for example, Johnson et al. 2000).

Once the river channel becomes so wide that wood pieces no longer span the channel, and the discharge regime is sufficient to transport most of the available wood during higher flows, then the river's geomorphic structure becomes the most important control on wood retention. In particular, the geomorphic style or form of river channel (such as meandering, braided, island braided) dictates the availability of sites for large wood retention (Piégay and Gurnell 1997). The geomorphic sites within which wood may be retained also reflect the size, form, and buoyancy of the wood pieces. Sites observed to retain wood within large rivers include

- In side streams and distributary channels, where wood becomes jammed (Piégay 1993);
- In floodplain woodland, where wood pieces may accumulate in ribbons parallel to the channel margin, in patches trapped on the upstream side of individual trees (Piégay and Gurnell 1997), or as plugs in the entrance to avulsion channels (Gurnell et al. 2000b);
- In association with vegetated islands, where large wood can become braced against upstream margins, accumulate in sheltered areas along the sides and in the lee of islands, and accumulate in sheets, patches, or plugs on island surfaces (Hickin 1984; Gurnell et al. 2000b); and
- In association with other morphologic features in the active zone of the channel where wood can be
    - Braced at the apex of bars (Abbe and Montgomery 1996) and on the outer up stream-facing margins of channel bends (Piégay and Marston 1998); or
    - Deposited in backwater areas as, for ex ample, in downstream facing, concave bank benches (Hickin 1984); or
    - Stranded during flood recessions along channel margins and on bar surfaces (for example Malanson and Butler 1990; Gurnell et al. 2000a).

Wood Anchorage and Mobility

The mobility of wood in river systems depends on three main factors: its size, density, and the degree to which it is anchored in position by factors other than its weight. The dominant methods of anchorage are likely to change with the size of both the channel and the wood pieces; the character of the river channel and the flow regime. Research on two European river systems can be used to illustrate some contrasts between small, medium, and large rivers.

The Highland Water, New Forest, UK

The Highland Water drains a wooded corridor. The results presented here are based on observations of more than 300 in-channel accumulations of wood in sectors that have not been channelized nor subject to wood clearance for at least 10 years. Indeed, through most of the studied area, there has been no significant channel modification or debris clearance for considerable longer (records are limited, but at least 50 years is estimated). The marginal tree species are typically deciduous hardwoods, the majority of which are more than 100 years old. Wood supply from these aging trees is quite high, and the wood entering the river is often decaying, branch-sized pieces that have broken from the parent tree, although undercutting and windthrow of mature trees are also common events. In relation to the definitions previously given, the study reaches of the Highland Water cover the range from small to medium channels.

The channel sectors studied are grouped into three classes, according to their size: small headwater channels (first-order streams), larger headwater channels (second-order streams) and the main channel (fourth-order in the reaches surveyed). Table 1 illustrates how increasing channel width and depth is accompanied by an increase in the spacing and size of wood accumulations and a decrease in the ratio of the length of the largest wood piece in each accumulation to the width of the channel. In relation to the definitions given above, the smaller headwater streams can be classified as *small*, the larger headwater streams as transitional between *small* and *medium*, and the main channel as *medium*. Figures 5, 6, and 7 show variations in the horizontal orientation of the largest piece of wood in each accumulation in relation to the channel center line, the nature of the anchorage of the wood accumulations, and the nature of the key piece of wood that supports each accumulation. The frequency of key piece orientation in 15-degree classes (Figure 5, centered on 0, 15, 30, 45, 60, 75, and 90 degrees) shows that, in the smallest streams, the largest wood pieces are predominantly oriented across the channel at between 45 and 90 degrees. In the larger headwater streams, an increasing proportion of the largest wood pieces are oriented at less than 45 degrees to the channel center line. This is because these pieces become snagged and wedged across the channel. In the main channel, the most commonly occurring category is parallel to the stream center line. This suggests increasing orientation of wood pieces parallel to the flow as they are able to move more freely in the largest channels. Figure 6 indicates the importance of living tree roots, exposed by bank erosion, in snagging and anchoring wood in small channels. In larger channels, bracing of the wood against living tree trunks and overlapping of the channel bank are more important anchorage mechanisms. Furthermore, in the smallest channels, tree roots can span the channel (growing across the channel from one bank to the other) to form the key pieces in dams, and pieces of large wood form the most common type of key piece (Figure 7). In contrast, in the largest channels, entire fallen trees provide the most commonly-occurring key pieces, although large wood pieces are also important (Figure 7).

The variation in wood accumulation size (Table 1) and the orientation of the largest wood piece in each accumulation (Figure 5) are two indications of wood mobility. Research on this river has also illustrated how the location and type of wood accumulation varies rapidly as flood events disrupt and move individual wood pieces (for example, Gurnell and Gregory 1995; Piégay and Gurnell 1997). Changes in the location and type of dams within a first-order and a fourth-order reach of the river between 1983 and 1996 are summarized in

TABLE 1. Average channel dimensions and wood dam characteristics in first-, second-, and fourth-order reaches of the Highland Water.

	All dams and streams surveyed	Channel type: Small headwater (first order)	Larger headwater (second order)	Main channel (fourth order)
Dams surveyed (no.)	303	100	105	98
Channel width (m)	1.9	1.2	1.8	4.9
Channel depth (m)	0.45	0.27	0.45	1.11
Dam spacing (m)	25	12	21	42
In-channel dam volume (m^3)	4.9	0.5	1.6	12.9
Length of longest wood piece (m)/ channel width (m) in individual dams	1.9	2.5	2.0	1.4
Number of large wood pieces in individual dams[a] (>1m long and >0.1m diameter)	3	1	2	5

[a] Twenty is the maximum number of large wood pieces recorded in a single dam.

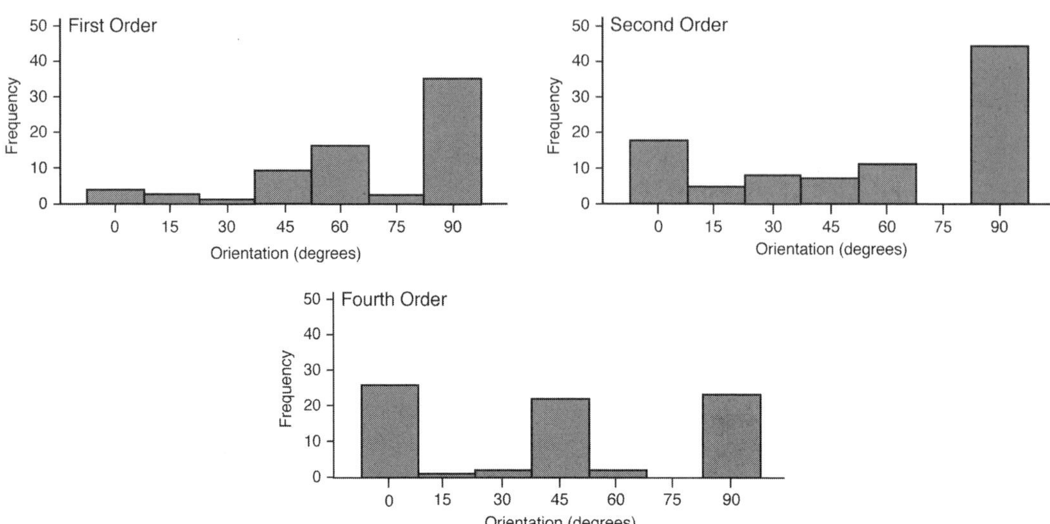

FIGURE 5. Variations in the horizontal orientation of the largest wood pieces within 303 wood accumulations in first-, second-, and fourth-order channels of the Highland Water, New Forest, UK (Deviation from the channel center line in 15-degree classes. Surveys undertaken 1996–1999).

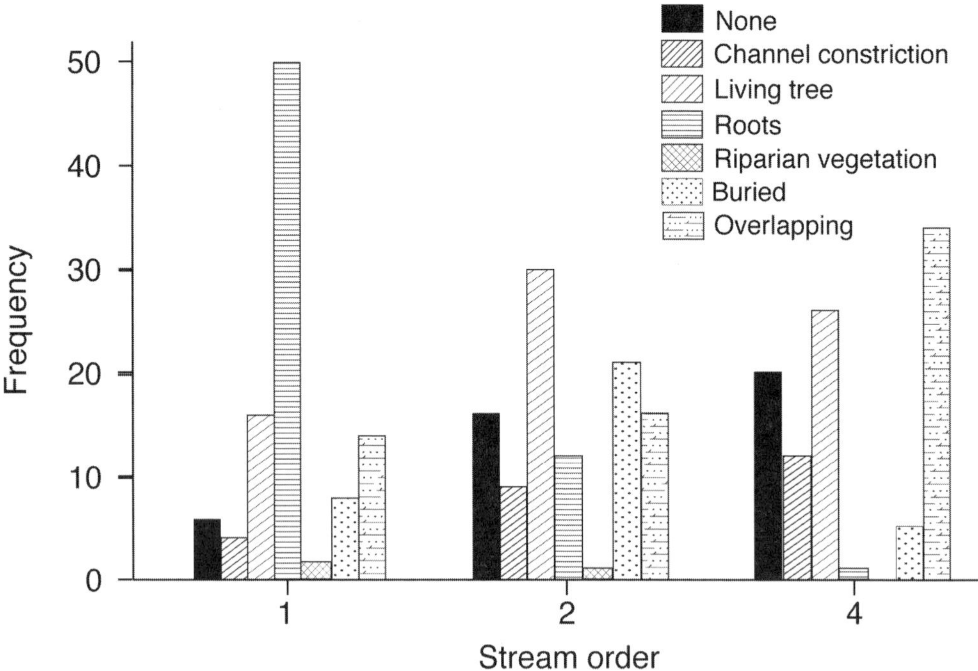

FIGURE 6. Variations in the frequency of different means of anchorage of 303 wood accumulations in first-, second-, and fourth-order channels of the Highland Water, New Forest, UK (Classes are: no apparent means of anchorage; wedged by a channel constriction; braced against a living tree or trees; caught in exposed tree roots; caught in riparian vegetation; part-buried in banks and/or bed; overlapping the bank(s). Surveys undertaken 1996–1999).

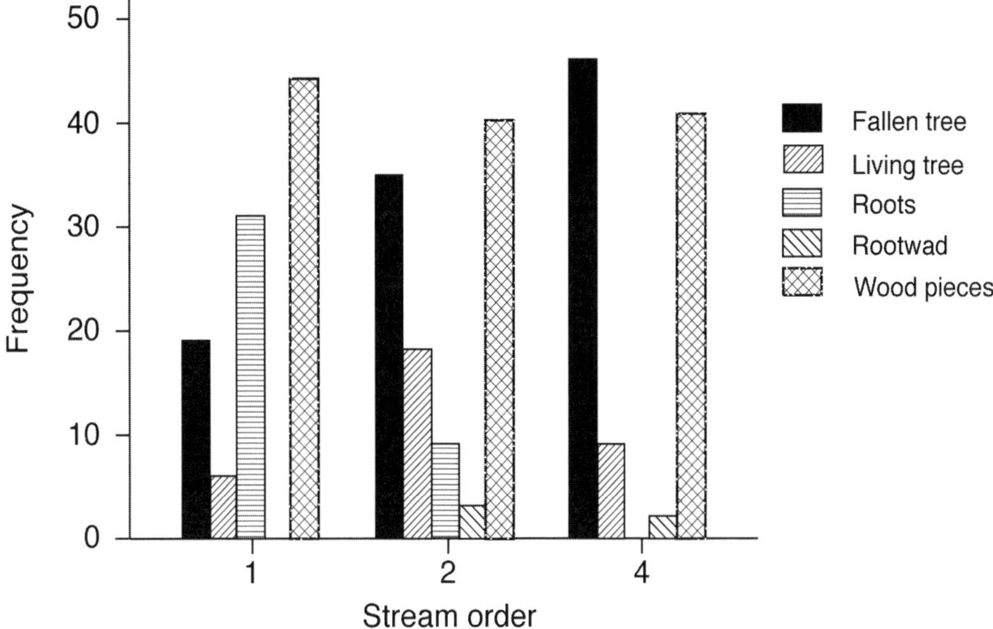

FIGURE 7. Variations in the frequency of different controlling wood piece types for 303 wood accumulations in first-, second- and fourth-order channels of the Highland Water, New Forest, UK (Classes indicate the nature of the controlling piece(s) as: a fallen tree; a living tree that is growing across the channel; tree roots extending across the channel and sometimes growing out of one bank and into the opposite bank; a rootwad deposited in the channel; or one or more large unattached wood pieces. Surveys undertaken 1996–1999).

Table 2. Dams are categorized into three types: active dams (which completely cross the channel and cause a step in the water surface profile even at low flows), complete dams (which completely cross the channel but are sufficiently open in structure that they do not cause a step in the water surface profile at low flows), and partial dams (which cross a part of the channel, but extend to at least 25% channel width). The table shows the contrast between the small first-order channel in which active dams dominate, whereas in the larger fourth-order channel, most of the dams are only partial. The table also illustrates that although the total number of wood accumulations of each type in each of the reaches between 1983 and 1996 changed little, the position of dams and, where dams have persisted in the same location, the type of dam has changed dramatically. Indeed, virtually every dam mapped in 1996 was either of a different type from the one mapped in 1983 or accumulated after 1983. In many cases, change in dam type reflects little change in the key wood pieces, but rather changes in the quantity and structure of the other wood pieces within the dam. In other cases, break up of key pieces has resulted in dam failure followed by the accumulation of new wood pieces at the same site.

The Fiume Tagliamento, Italy

The Fiume Tagliamento is a large gravel-bed river characterized by island-braided reaches bordered by riparian woodland. The woodland is dominated by *Alnus incana, Populus nigra, Salix elaeagnos, S. alba, S. purpurea*, and *S. daphnoides*. The low-density wood in this system floats freely and is retained by three main mechanisms: bracing against roughness elements, such as living riparian trees and the upstream margins of islands; deposition on high areas of the active zone, particularly the crests of bars during the falling stages of flood events; and the rapid sprouting of roots to anchor pieces of living wood, including whole trees, that have been transported and deposited by the river within its active zone. This last mechanism is particularly important in the middle and lower reaches of the river, enhancing wood retention and changing wood piece size and shape. The mechanism has not been previ-

TABLE 2. Dynamics of wood accumulations of differing type within reaches of first- and fourth-order channels of the Highland Water.

Dam type	Partial	Complete	Active	Total
First-order reach				
1983 total	9	8	37	54
Additions 1983–1996				
Change in class	5	13	3	21
Dams created	2	1	0	3
Total additions	7	14	3	24
Losses 1983–1996				
Change in class	2	1	18	21
Lost dams	6	2	5	13
Total losses	8	3	23	34
Net change 1983–1996	–1	11	–20	–10
1996 total	8	19	16	43
Fourth-order reach				
1983 total	25	10	7	42
Additions 1983–1996				
Change in class	6	3	1	10
Dams created	13	8	1	22
Total additions	19	11	2	32
Losses 1983–1996				
Change in class	2	3	2	7
Lost dams	15	5	5	25
Total losses	17	8	7	3
Net change 1983–1996	2	3	–5	0
1996 total	27	13	2	42

ously reported in relation to river wood dynamics. The very rapid growth of the Salicaceae, particularly *Populus nigra,* from uprooted trees and broken wood pieces allows small, pioneer islands to develop in less than 12 months. These rooted wood pieces and the trees sprouting from them can evolve into established islands by coalescing and trapping additional living wood, dead wood, and other vegetation propagules. The survival of these new patches of vegetation depends on their growth rate in comparison with the frequency of major, destructive flood events. The interaction between active-zone morphology, vegetation, and the aggradation and degradation of the active zone by flood events dictates the quantity and location of dead and living wood stored in the system. Gurnell and Petts (2002) explore these interactions and highlight the benefit of both dead and living wood in sustaining different vegetation recruitment and growth trajectories that may fundamentally influence the character of island-braided rivers.

A conceptual model of wood and island dynamics has been developed for the Fiume Tagliamento (Figure 8). The model and its components are fully described in Gurnell et al. (2000a, 2000b) and are further developed in Gurnell and Petts (2002). In the present context, the model serves to illustrate the important interactions between sediment, wood, and trees, which govern wood storage and wood dynamics on large rivers. As in all large rivers, wood can accumulate only on discrete retention sites whose position and frequency reflect the patterns and forms produced by hydrogeomorphic processes and vegetation growth. In the Tagliamento, the role of living wood is to increase rates of vegetation establishment so that the balance between the different types of wood retention site changes in both time (rapid recovery of retentive vegetation after destructive floods) and space (a high frequency of vegetation-related retention sites).

Summary

The storage and mobility of wood in river systems is undoubtedly highly variable from site to site and responds to many controlling factors. However, certain common patterns or trends can be identified.

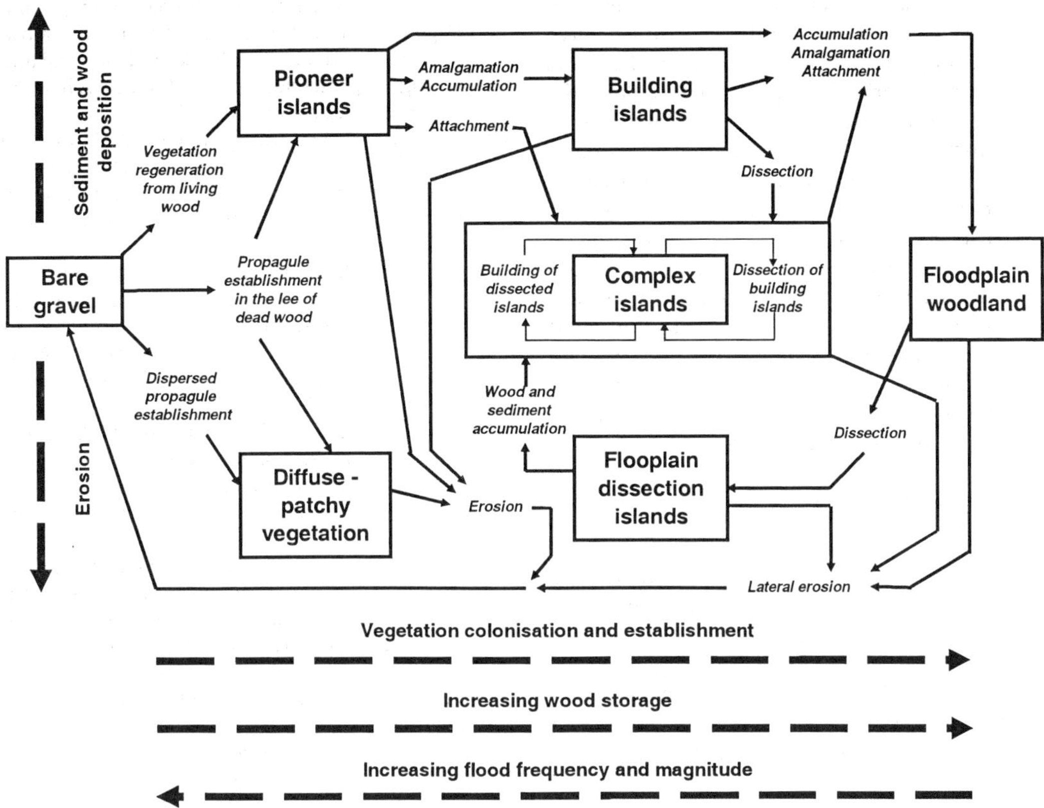

FIGURE 8. A conceptual model of island formation on the Fiume Tagliamento, Italy.

- Tree type (the broad composition of the woodland bordering the river) is a key control on the amount of wood stored in relatively unmanaged river systems.
- The species composition of the woodland may also be extremely important in defining the dynamics and in-channel geomorphic influence of wood through its effects on wood buoyancy. The bouyancy of wood input to the river system can vary greatly according to species and degree of decay. Waterlogging can cause further buoyancy changes in the river channel.
- Wood storage and dynamics can be usefully described according to channel size. Broad changes in storage and dynamics may be associated with small (for example, channel width < median wood-piece length), medium (for example, channel width < upper quartile wood piece length), and large (for example, channel width >> wood piece length) river channels. This description implies that the size of the channel in relation to wood dynamics can vary in time and space as wood-piece size varies and may reflect not only changes in tree species but also changes in the management of riparian trees and large wood. By scaling river channels in relation to the size of the stored wood, generalizations can be drawn that apply to rivers and riparian woodland, both managed and unmanaged.
- Variations in the size of river channel can be related to changes in the mode of wood storage. This relation reflects gross changes in the relative importance of groups of controlling factors: riparian woodland characteristics dominate in small rivers; hydrologic events and hydraulic characteristics dominate in medium rivers; and geomorphological characteristics provide the key controls in large rivers.
- Although most research has concentrated on the storage and dynamics of dead wood, an

additional property of wood that requires further investigation is its ability to stay alive during transport and deposition. In rivers where riparian tree species (particularly the Salicaceae) adopt vigorous vegetative reproduction strategies, the dynamics of living as well as dead wood may support extremely rapid vegetation establishment on areas of exposed sediment. This rapid establishment can significantly change the balance between destructive and constructive processes, which can yield a major increase in wood retention and a change in the predominant modes of wood storage, compared to geomorphically similar rivers that only receive inputs of dead wood.

Acknowledgments

The ideas and case-study material presented in this paper were developed largely as the result of research sponsored by the UK Natural Environment Research Council under grants GR9/03249, NER/B/S/2000/00298, and GR3/CO036. Three anonymous referees are also thanked for their very helpful comments, which have resulted in significant improvements to this manuscript.

References

Abbe, T. B., and D. R. Montgomery. 1996. Large woody debris jams, channel hydraulics and habitat formation in large rivers. Regulated Rivers Research and Management 12:201–221.

Benda, L., D. Miller, J. Sias, D. Martin, R. Bilby, C. Veldhuisen, and T. Dunne. 2003. Wood recruitment processes and wood budgeting. Pages 49–73 *in* S. V. Gregory, K. Boyer, and A. M. Gurnell, editors. The ecology and management of wood in world rivers. American Fisheries Society, Symposia 37, Bethesda, Maryland.

Bilby, R. E., and J. W. Ward. 1989. Changes in characteristics and function of woody debris with increasing size of streams in western Washington. Transactions of the American Fisheries Society 118:368–378.

Braudrick, C. A., G. E. Grant, Y. Ishikawa, and H. Ikeda. 1997. Dynamics of wood transport in streams: a flume experiment. Earth Surface Processes and Landforms 22:669–683.

Carlson, J. Y., C. W. Andrus, and H. A. Froehlich. 1990. Woody debris, channel features and macroinvertebrates of streams with logged and undisturbed riparian timber in Northeastern Oregon, USA. Canadian Journal of Fisheries and Aquatic Sciences 47:1103–1111.

Church, M. 1992. Channel morphology and typology. Pages 185 to 202 *in* G. E. Petts and P. Calow, editors. The rivers handbook: hydrological and ecological principles, volume 1. Blackwell Scientific Publications, Oxford, UK.

Diez, J. R. 1999. Dinámica y función de la madera en el sistema fluvial del Agüera. Relaciones con la vegetatión de la cuenca. Doctoral dissertation. University of the Basque Country, Bilbao, Spain.

Evans, B. F., C. R. Townsend, and T. A. Crowl. 1993. Distribution and abundance of coarse woody debris in some southern New Zealand streams from contrasting forest catchments, New Zealand. Journal of Marine and Freshwater Research 27:227–239.

Fausch, K. D., and T. G. Northcote. 1992. Large woody debris and salmonid habitat in a small coastal British Columbia stream. Canadian Journal of Fisheries and Aquatic Sciences 49:682–693.

Gurnell, A. M., and K. J. Gregory. 1995. Drainage basin perspectives on the interaction between seminatural vegetation and hydrogeomorphological processes. Geomorphology 13:49–69.

Gurnell, A. M., K. J. Gregory, and G. E. Petts. 1995. The role of coarse woody debris in forest aquatic habitats: implications for management. Aquatic Conservation 5:143–166.

Gurnell, A. M., and G. E. Petts. 2002. Island-dominated landscapes of large floodplain rivers, a European perspective. Freshwater Ecology 47:581–600.

Gurnell, A. M., G. E. Petts, D. M. Hannah, B. P. G. Smith, P. J. Edwards, J. Kollmann, J. V. Ward, and K. Tockner. 2000a. Wood storage within the active zone of a large European gravel-bed river. Geomorphology 34:55–72.

Gurnell, A. M., G. E. Petts, D. M. Hannah, B. P. G. Smith, P. J. Edwards, J. Kollmann, J. V. Ward, and K. Tockner. 2000b. Riparian vegetation and island formation along the gravel-bed Fiume Tagliamento, Italy. Earth Surface Processes and Landforms 26:31–62.

Gurnell, A. M., H. Piégay, F. J. Swanson, and S. V. Gregory. 2002. Large wood and fluvial processes. Freshwater Biology 47:601–619.

Harmon, M. E., J. F. Franklin, F. J. Swanson, P. Sollins, S. V. Gregory, J. D. Lattin, N. H. Anderson, S. P. Cline, N. G. Aumen, J. R. Sedell, G. W. Lienkaemper, K. Cromack, and K. W. Cummins. 1986. Ecology of coarse woody debris in temperate ecosystems. Advances in Ecological Research 15:133–302.

Heede, B. H. 1981. Dynamics of selected mountain streams in western United States of America. Zeitschrift fur Geomorphologie N. F 25:17–32.

Hickin, E. J. 1984. Vegetation and river channel dynamics. Canadian Geographer 28:111–126.

Hogan, D. L. 1986. Channel morphology of unlogged,

logged and debris torrented streams in the Queen Charlotte Islands. British Columbia Ministry of Forests and Lands, Land Management Report No. 49, Victoria, BC.

Johnson, S. L., F. J. Swanson, G. E. Grant, and S. M. Wondzell. 2000. Riparian forest disturbance by a mountain flood – the influence of floated wood. Hydrological Processes 14:3031–3050.

Keller, E. A., and F. J. Swanson. 1979. Effects of large organic debris on channel form and fluvial process. Earth Surface Processes 4:361–380.

Linstead, C., and A. M. Gurnell. 1999. Large woody debris in British Headwater rivers (main report). Environment Agency R&D Technical Report W181.

MacDonald, A., E. A. Keller, and T. Tally. 1982. The role of large organic debris on stream channels draining redwood forests northwestern California. Pages 226–245 *in* D. K. Harden, D. C. Marran, and A. MacDonald, editors. Late Cenozoic history and forest geomorphology of Humboldt Co., California, Friends of the Pleistocene 1982, Pacific cell fieldtrip guidebook.

Malanson, G. P., and D. R. Butler. 1990. Woody debris, sediment, and riparian vegetation of a subalpine river, Montana, USA. Arctic and Alpine Research 22:183–194.

Murphy, M. L., J. Heifetz, S. W. Johnson, K. V. Koski, and J. F. Thedinga. 1986. Effects of clear-cut logging with and without buffer strips on juvenile salmonids in Alaskan streams. Canadian Journal of Fisheries and Aquatic Sciences 43:1521–1533.

O'Connor, N. A. 1992. Quantification of submerged wood in a lowland Australian stream system. Freshwater Biology 27:387–395.

Piégay, H. 1993. Nature, mass and preferential sites of coarse woody debris deposits in the lower Ain Valley (Mollon Reach), France. Regulated Rivers Research and Management 8:359–372.

Piégay, H., and A. M. Gurnell. 1997. Large woody debris and river geomorphological pattern: examples from S. E. France and S. England. Geomorphology 19:99–116.

Piégay, H., and R. A. Marston. 1998. Distribution of large woody debris along the outer bend of meanders in the Ain River, France. Physical Geography 19:318–340.

Piégay, H., A. Thévenet, and A. Citterio. 1999. Input, storage and distribution of large woody debris along a mountain river continuum. The Drôme River, France. Catena 35:19–39.

Potts, D. F., and B. K. M. Anderson. 1990. Organic debris and management of small stream channels. Western Journal of Applied Forestry 5:25–38.

Raven, P. J., P. Fox, M. Everard, N. T. H. Holmes, and F. H. Dawson. 1997. River habitat survey: a new system for classifying rivers according to their habitat quality. Pages 215–234 *in* P. J. Boon and D. L. Howell, editors. Freshwater quality: defining the indefinable? Stationary Office, Edinburgh, UK.

Robison, E. G., and R. L. Beschta. 1990. Characteristics of coarse woody debris for several coastal streams of southeast Alaska, USA. Canadian Journal of Aquatic Science 47:1684–1693.

Rutherfurd, I., K. White, N. Marsh, and K. Jerie. 2000. Some observations on the amount and distribution of large woody debris in Australian streams. Riparian Land's Management Newsletter 16:10–16, Land and Water Resources Research and Development Corporation, Canberra ACT 2601, Australia.

Smock, L. A., G. M. Metzler, and J. E. Gladden. 1989. Role of debris dams in the structure and functioning of low gradient headwater streams. Ecology 70:764–775.

Swanson, F. J., M. D. Bryant, G. W. Lienkaemper, and J. R. Sedell. 1984. Organic debris in small streams, Prince of Wales Island, southeast Alaska. USDA Forest Service, Pacific Northwest Forest and Range Experiment Station, General Technical Report PNW-166.

Thévenet, A., A. Citterio, and H. Piégay. 1998. A new methodology for the assessment of large woody debris accumulations on highly modified rivers. Regulated Rivers Research and Management 14:467–483.

Treadwell, S., J. Koehn, and S. Bunn. 1999. Large woody debris and other aquatic habitats. Pages A79–A96 *in* Riparian land management guidelines, volume 1. Land and Water Resources Research and Development Corporation, Canberra ACT 2601, Australia.

Wallace, J. B., and A. C. Benke. 1984. Quantification of wood habitat in subtropical coastal plain streams. Canadian Journal of Fisheries and Aquatic Sciences 41:1643–1652.

American Fisheries Society Symposium 37:93–107, 2003

Hydraulic Effects of Wood in Streams and Rivers

MICHAEL MUTZ

Research Station Bad Saarow, Seestr. 45, D-15526, Bad Saarow, Germany

Abstract.—Although submerged wood obviously influences the flow, little information exists on its various hydraulic effects in streams and rivers. This chapter gives a brief overview of the current knowledge about hydraulic effects of circular cylinders and simple tree shaped models and summarizes the few field data on wood induced hydraulics in streams and rivers. The focus is on the flow pattern and other effects of importance for instream ecology. The principal cross-flow field of a singular log perpendicular to flow is determined by the Reynolds number related to the log's diameter. For the range of Reynolds numbers of logs and branches in streams and rivers ($1 \cdot 10^2$ to $1 \cdot 10^6$), the cross flow pattern is symmetrical, vortex streets shed, and a wake with reduced mean velocity develops behind the log. In the vertical confined flow of streams and rivers, the hydraulic effects depend on the blockage caused by the log, its distances to the water surface, and its distance to the streambed. The blockage determines the resistance to flow, the upstream afflux, the local flow acceleration, and the intensity of flow deflection. For logs within distances of 2 diameters to the water surface, the relative submergence and the Froude Number determine the highly variable local cross-flow field. For logs near the streambed, the form and roughness of the bed and the size of the gap to the bed control the hydraulics. Submerged jet-like flows, which cause local scour, are reported, but detailed information on the hydraulics of logs close to a natural streambed is missing. For logs in close contact to or partly embedded into the bed, the principal flow pattern of recirculating vortices attached to the bed develop in front and behind the logs. The extent of these vortices and the extent of the wake behind the logs appear to be larger in sand-bed streams than in flumes with smooth and level beds. Complex dense wooden objects and wood accumulations are comparable to solid structures. Their flow field is determined by the size of the bluff surfaces and the shedding from edges obtuse to flow. Wood spread out at the streambed causes skin roughness, and models based on technical roughness approximate the resulting near-bed flow regime. The general validity of most findings in streams and rivers is still vague since they are supported by only few data. Further flow data from the field and from flume experiments that simulate the complexity of the natural environment are needed.

Introduction

River flow and the associated forces are responsible for the most important environmental effects in running waters (Statzner et al. 1988; Allan 1995). Although their importance seems obvious and has long been recognized (Ambühl 1959), many aspects of the flow in natural streams and rivers are still poorly understood. Most knowledge about flow in running waters is based on semi-empirical techniques and flume studies with simplified conditions—for example, steady, uniform flow, and smooth and level boundaries.

A set of methods now available ranges from empirical uniform-flow equations to sophisticated mathematical modeling. These tools seem to work well for most applications in engineering. From the ecological point of view, however, much knowledge about the hydraulics in natural streams and rivers is still lacking. As the flow is not steady and interacts with complex morphological features, such as wood, in many ways, field hydraulics is highly variable. How well current laboratory-based models express the natural environment is unknown.

In natural streams, wood is an abundant substrate greatly influencing the current. Wood obstructs and diverts the flow of the water and induces complex interactions between the flow and bed sediments. The interdependency of large wood and stream morphology is well established (Montgomery et al. 2003, this volume). Harmon et al. (1986) have shown that the wood can control the streambed morphology in headwaters. With the exception of wood completely buried in the sediments, however, the primary effect of wood in running waters is on their hydraulics. Therefore, the hydrodynamic and hydraulic effects of wood in streams and rivers are fundamental to understanding river morphology and ecology. How little information on these subjects appears in the literature is surprising.

This chapter is an overview of current knowledge about the hydraulic effects produced by wood in streams and rivers. I bring together findings from the few field studies on hydraulic effects of wood in the natural environment and the current knowledge about hydraulic effects of circular cylinders or hemi-cylinders under controlled flow in flumes. I assess the transferability of laboratory-based findings to wood in streams and rivers. Beginning with the basic hydraulic features of circular cylinders, I go on to complex hydraulic patterns induced by wood in the natural environment.

As wood influences flow on a wide range of spatial scales, being aware of the scale at which the hydrodynamics of wood is addressed is important. Logs in a river may divert and accelerate the adjacent flow, but—at the reach scale—the wood may increase flow resistance and thereby decrease mean flow velocity (Lisle 1995; Manga and Kirchner 2000). I focus on the hydraulic effects near individual logs or debris accumulations. The hydraulic patterns generated at this scale have implications for the microhabitats of benthic organisms and fish, and they influence, for example, the retention and turnover of organic matter. I have had to choose from the many hydrodynamic and hydraulic factors that could be addressed; my selection covers a variety of factors related to the water column as well as to the near-bed flow associated with wood. The significance of wood for stream reach parameters such as flow resistance, discharge capacity, and dispersion of the flow are not covered.

Hydraulics of Logs

As it is important to engineering, sound knowledge is available about drag, pressure distribution, and the flow field for flow around circular cylinders in flumes. Such cylinders are a reasonable approximation of isolated logs or cylindrical parts of more complex wood structures. The next section briefly reviews the principal findings from engineering literature. For further details, the reader is referred to textbooks on fluid mechanics. The limits of transferring findings from flume experiments to the natural environment are also discussed.

Logs in free flow

A symmetrical cross-flow pattern develops around a cylinder oriented perpendicular to the flow. At the upstream center of the cylinder, there is a point of zero velocity called the stagnation point. From it, a boundary layer develops at the cylinder surface as the water loses energy to friction or turbulence when it passes around the cylinder in both directions. At a certain distance along the cylinder surface, the boundary layer can separate from the cylinder. At the point of separation, vortices shed downstream (Figure 1). Separation and vortex shedding happen in the Reynolds number (Re) range of $1 \cdot 10^2$ to $1 \cdot 10^6$, with

$$Re = U \cdot D / v ,$$

where U is the free stream velocity, D is the diameter of the circular cylinder, and ν is the kinematic viscosity of the fluid. This range of Reynolds numbers spans the range of Reynolds numbers of wood in most streams and rivers (Figure 2) and also spans the effect of varying velocity with change of discharge. In the region behind the cylinder, the mean stream velocity is reduced; turbulence and energy dissipation intensify as the vortices extend downstream. This region is called the wake. Directly behind the cylinder, the flow can even recirculate at reduced speed, an effect called a near-wake flow regime, sometimes called "dead water" in ecological papers. The distance between the stagnation point and separation, the properties of the boundary layer at the cylinder, and the character of the wake depend largely on the Reynolds number (Munson et al. 1998; Figure 1). At low Reynolds numbers ($Re \approx 10^1$), the wake shows two symmetrical recirculations. If the Reynolds number is slightly increased ($Re \approx 10^2$), the recirculations change to alternate shedding of vortices, and the so-called "Karman vortex street" is induced. Sometimes a Karman vortex street can be seen at the water surface behind bridge-piers in very calm flow. Higher Reynolds numbers ($Re \approx 10^4$) induce regular shedding and two separate mixing layers along the shedding vortices, which bound or 'fence' the wake. The mixing takes place between the undisturbed flow and the flow at reduced speed in the wake. The boundary layer along the cylinder surface is still laminar through-

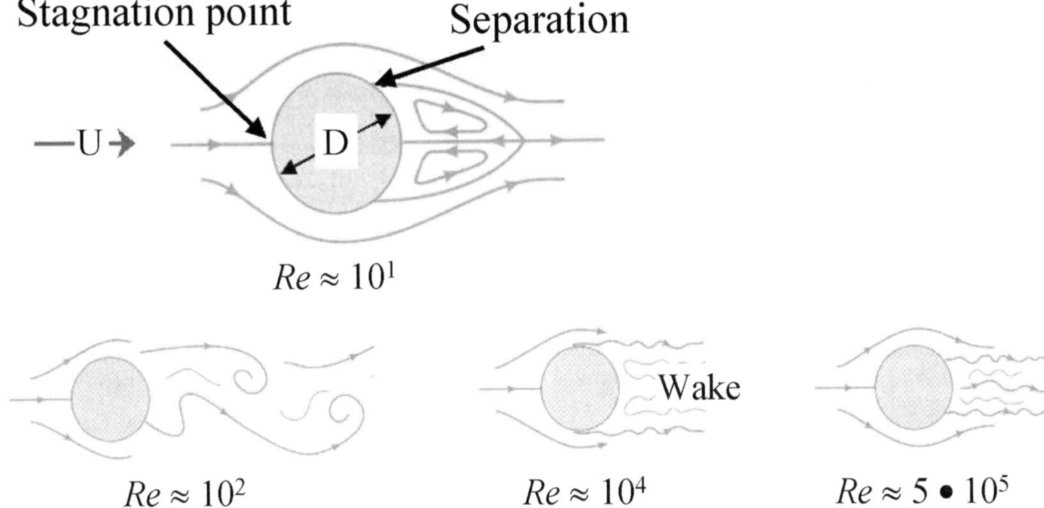

FIGURE 1. Cross-flow field produced by circular cylinders at different Reynolds numbers (Re): D = diameter of cylinder, U = free stream velocity. Arrows on streamlines indicate direction of flow. Adapted from Munson et al. (1998).

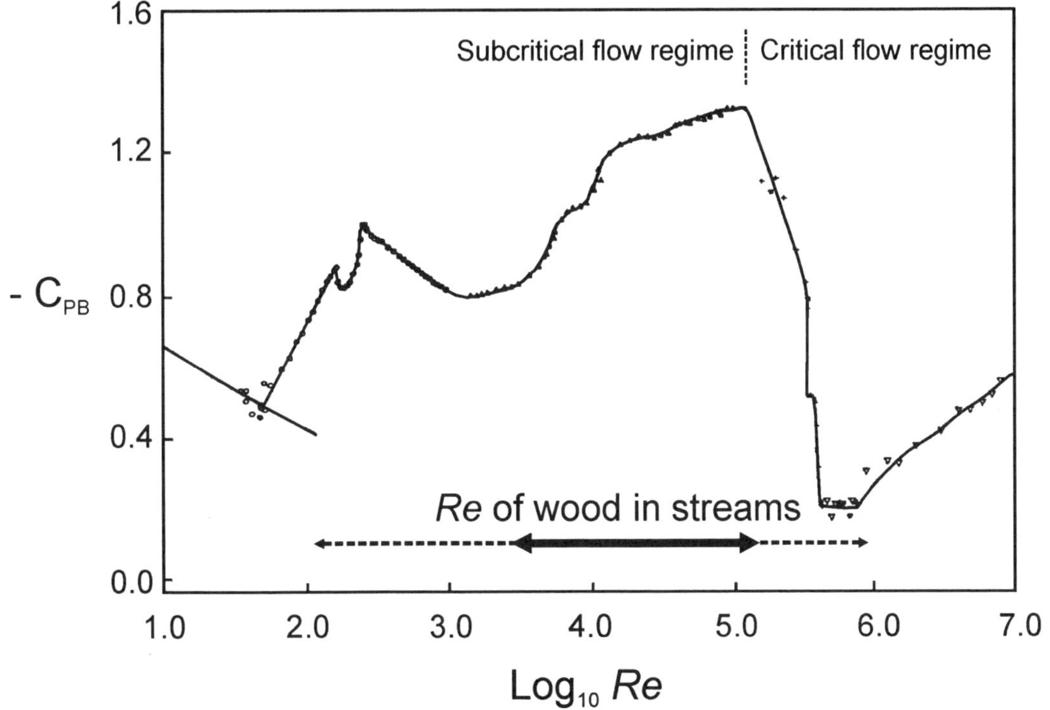

FIGURE 2. Variation of the base suction coefficients ($-C_{PB}$) with Reynolds number. The base suction coefficient is closely related to the process of vortex formation and flow pattern in the near wake. Division between subcritical and critical flow regime is indicated. The arrows show the range of potential Reynolds numbers of wood in streams and rivers: solid arrow marks the range of most probable Reynolds numbers. Adapted from Williamson (1997).

out its circumference until the separation, which is called the subcritical flow regime.

If the boundary layer at the cylinder surface becomes turbulent at Reynolds numbers above 10^5, the distance from stagnation point to separation becomes longer, and the wake behind the cylinder becomes narrower, creating what is called the critical flow regime. If Reynolds numbers are increased further, the flow regime shifts to supercritical, and the wake widens again. Only very large logs in fast-flowing rivers cause a supercritical flow regime (Figure 2).

The transition of flow with changing Reynolds numbers can be seen from the variation of the base suction, the negative static pressure acting at the rear surface of the cylinder (Figure 2). This force is closely coupled to the flow characteristics of the wake. Its variability with varying Reynolds numbers shows a complex sequence and overlay of different types of vortices. In addition to the vortex sequence, there are two- and three-dimensional instabilities of the boundary layer, the separating shear layer, and the wake, which occur at certain Reynolds numbers (Lin et al. 1995; Williamson 1997). The flow field changes drastically with minor variation of the Reynolds number, particularly in the critical flow regime.

The relation between Reynolds numbers and the cross-flow regime sequence, as shown in Figure 2, is valid for smooth cylinders. Rough surfaces, such as on a log with bark, are known to promote turbulent boundary-layer flow. Thus, rough cylinders show a shift of the flow-regime sequence towards lower Reynolds numbers (Achenbach 1971; Ribeiro 1991). This hydraulic effect of roughness depends largely on the shape and spacing of the roughness elements (Ribeiro 1991), with spacing having greater influence than roughness height (Leung et al. 1992). All types of roughness tested so far were regular, however, and no information is available on the hydraulic effect of irregular roughness, such as that caused by different types of rough bark or a decaying log.

Now, I concentrate on the subcritical flow regime, for which most information is available. I assume that, despite the effect of roughness, most wood in streams and rivers with Reynolds numbers in the range between 10^2 and 10^5 produces a subcritical cross-flow regime.

The distribution of the mean local forces at the cylinder surface relates to the boundary layer. Pressure is highest at the stagnation point, reduces as the boundary layer develops, and is lowest at flow separation. At the base of the cylinder, where the wake develops, static pressure is always negative; thus, this pressure is also known as base suction. An insect larva crawling around a log exposed perpendicularly to flow would be affected by considerable change in the static pressure and skin friction along its way. The pattern of both forces varies with the Reynolds number (Figure 3).

As logs are not perfectly cylindrical and log surfaces are irregular, flow patterns are generalized approximations of the natural environment. Shedding of vortices at edges or surface irregularities, and instabilities of the cross-flow field caused by the unsteady and highly turbulent approaching flow must be common.

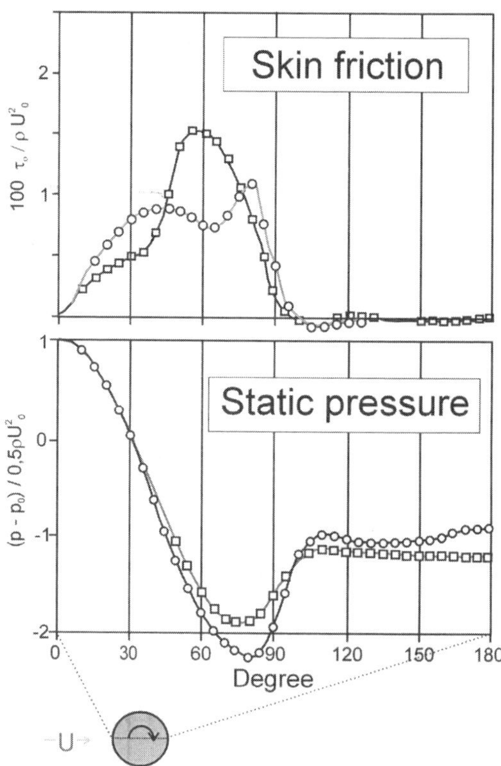

FIGURE 3. Variation of skin friction and local pressure along the surface of a rough circular cylinder exposed perpendicular to flow: *x*-axis is distance from stagnation point given in degree. Circles indicate data for Reynolds number of $1.7 \cdot 10^5$, squares indicate data for Reynolds number of $1 \cdot 10^5$. Relative roughness at the surface of the cylinder (k_s/D, k_s = height of roughness projection along the surface of the cylinder) was $4.5 \cdot 10^{-6}$. Adapted from Achenbach (1971).

Logs in vertically confined flow

The water body of streams and rivers is not of infinite vertical extent; it has distinct boundaries with the atmosphere and the streambed. Both boundaries are flexible or deformable and interact with the cross flow of logs. In such an environment, the relative size of a log compared to the water depth and the position of a log relative to the boundaries are important to the flow pattern produced.

The relative size of an obstruction to the streams cross-sectional area—the blockage ratio—determines the intensity of the deflection and of the acceleration of the flow and the general resistance to it. All three effects of wood are well reported (Keller and Swanson 1979; Lisle 1986; Smith et al. 1992; Abbe and Montgomery 1996; Manga and Kirchner 2000; Hygelund and Manga 2002). Manga and Kirchner (2000) found that wood blocking 56% of the cross-sectional area provided roughly half of the total flow resistance in a gravel-bed stream even though the wood only covered 2% of the streambed area. With a high blockage ratio, a single piece of wood can affect the thread of the flow for a long way downstream. For a large lowland river, Gippel (1995) reported flow deflection at bank-full discharge clearly visible more than 100 m downstream of a debris dam blocking 50% of the 35-m-wide river. A small blockage seems to have surprisingly little effect on the total conveyance of channels. In a flume study and field measurements, wood obstructing less than 10% of the cross-sectional area did not cause any detectable upstream afflux (Gippel 1995; Gippel et al. 1996).

Even small obstructions can cause local blockage, local flow deflection, and local flow acceleration because of the decrease in cross-sectional area. It is well known that removing wood reduces the heterogeneity of the current and significantly reduces low velocity habitats (Smith et al. 1992). Abbe and Montgomery (1996) measured the flow around a debris jam in a large river. The jam was induced by a large log parallel to the flow, with its rootwad facing upstream. Smaller pieces of wood had been deposited in front of the rootwad. The structure obstructed the total water depth for a small portion of the river width, and the flow pattern was the classical horseshoe vortex. The flow was directed down towards the streambed at the blunt surface of the rootwad, and a flow reversal developed along the streambed in front of the obstruction (Figure 4). The circulating flow surrounded

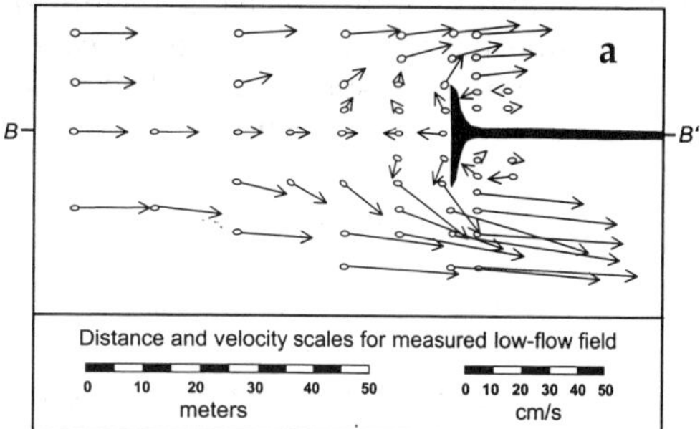

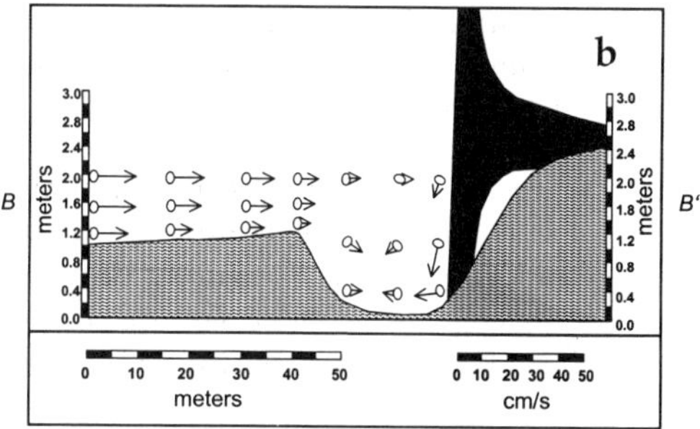

FIGURE 4. Flow pattern in the proximity of a small wood jam racked up against the rootwad of a large tree. Length of arrows indicates velocity; orientation of arrows indicates the flow direction: (a) plan view of near-bed flow in 0.2 flow depth; (b) profile of streambed and flow pattern through section indicated in plan view (*B* to*B'*). From Abbe and Montgomery (1996).

the object at both sides and separated with high velocity from the edges of the obstruction. Such horseshoe vortex systems can cause intensive local scour (Raudkivi 1990) and lead to excavations of the streambed around the obstacles (Cherry and Beschta 1989). Behind blunt wooden obstacles is an area of rapid flow deceleration. In this area of deceleration, sediments can be deposited and bars form (Keller and Swanson 1979; Lisle 1986). Upstream of the wood, the current velocity can be locally reduced and sediments deposited. As shear and sediment mobility correlate positively with current velocity and water depth, wood induces excavation and deposition of sediments mostly during high flow. Such created pools and bars are significant features of a natural streambed related to the hydraulics of wood. They influence the flow pattern and significantly contribute to the diversity of the current in particular at low discharge. In the example given by Abbe and Montgomery (1996), the excavation in front of the blunt surface as well as the deposition of sediments upstream of the excavation and behind the obstacle clearly affected the flow field at low discharge (Figure 4). For more information on the relation between wood, bed scour, and resulting streambed structures, see Montgomery et al. 2003.

Logs near the water surface

Sheridan et al. (1997) conducted a series of flume studies in a subcritical flow regime with circular cylinders close to the water surface. I describe selected results of this study because they demonstrate the complexity of cross-flow fields and because I assume that the effect of the water surface in the field is comparable to laboratory findings.

A cylinder, perpendicular to the flow and close to the water surface, generates a third vortex layer, in addition to the two symmetrical vortices shedding from the cylinder. This third layer separates from the water surface just behind the cylinder. It determines the nature of the cross-flow field and the nature of the water surface above and behind the cylinder. The flow pattern varies with the relative submergence (h^*) of the cylinder

$$h^* = h/D,$$

and the Froude number

$$\mathrm{Fr} = U/(g \cdot D)^{1/2},$$

where h is the distance from the top of the cylinder to the water surface. Small changes in relative submergence induce radical alterations of the flow pattern (Figure 5). At $h^* = 0$–0.16, a low-velocity flow discharged into a region of calm recirculating flow behind the cylinder. An increase of the relative submergence to 0.24 produced an abrupt onset of a fast, jet-like flow framed by high vorticity originating from the water surface and the top of the cylinder. This jet was directed downward along the cylinder base. At further increases in the cylinder submergence ($h^* > 0.31$), the jet detached from the cylinder and the angle of its downward deflection decreased until the jet was attached to the water surface at $h^* = 0.75$. At the water surface, the effect of this attached jet depended on the Froude number. For constant submergence of $h^* = 0.4$, the jet velocity was below critical and no wave motion appeared below Fr = 0.22. With rising Froude numbers, a sequence of surface waves of different amplitude, wavelength, and regularity developed downstream of the cylinder (Figure 6). At Fr = 0.72, the jet detached from the water surface and deflected downward and

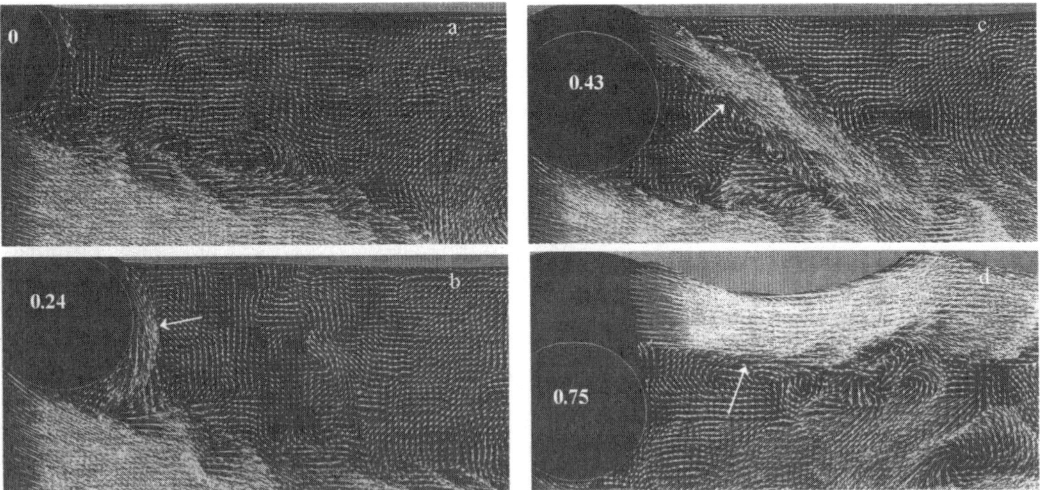

FIGURE 5. Instantaneous velocity fields downstream of a circular cylinder showing the effect of the relative submergence h^*. Thin arrows are velocity vectors; thick arrows point to the jet-like flow framed by the two vortex streets shedding from the water surface and from the cylinder. Numbers are relative submergence of cylinder. Froude numbers based on submergence are a = ∞, b = 1.22, c = 0.91, d = 0.69. Adapted from Sheridan et al. (1997).

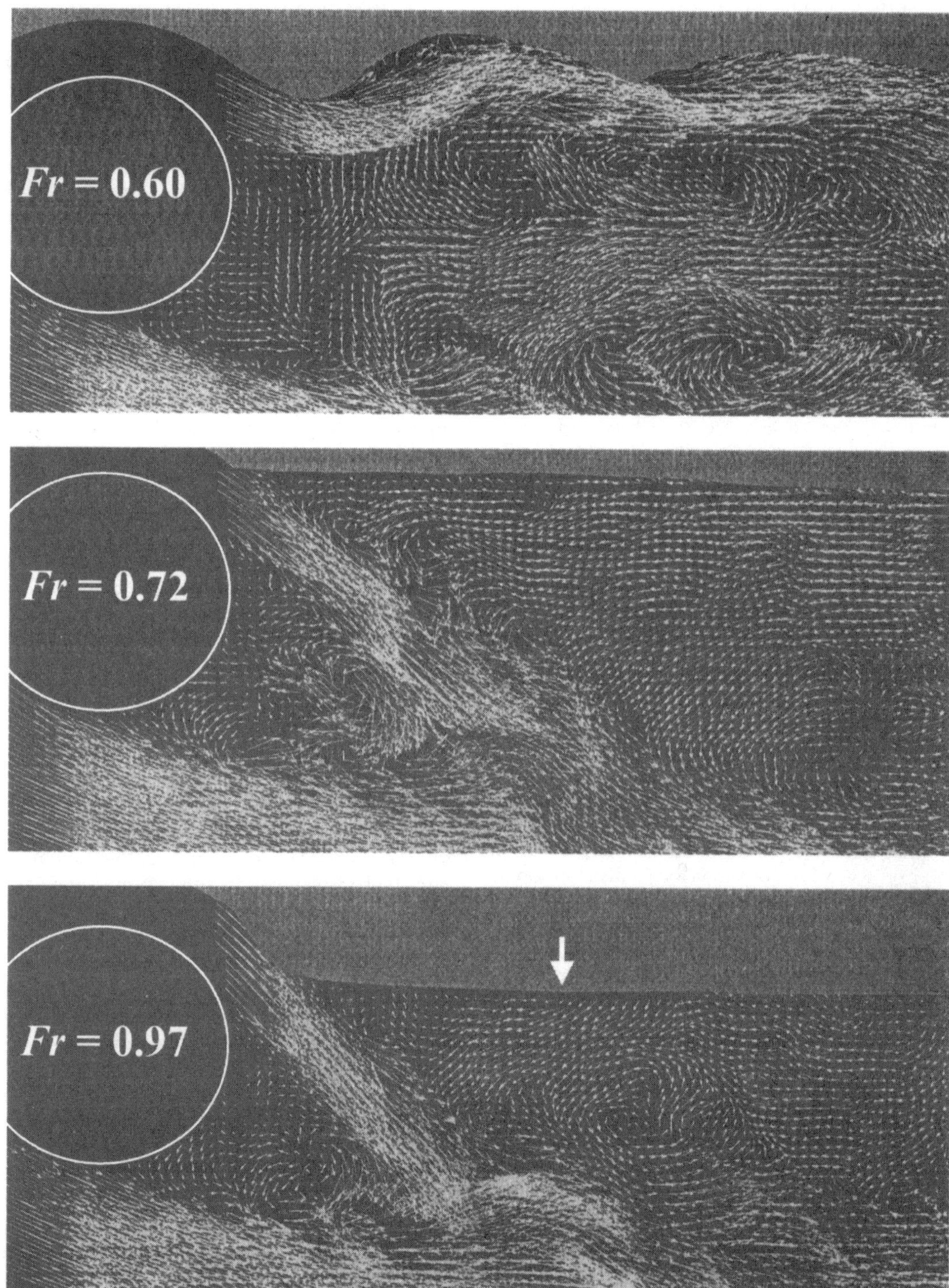

FIGURE 6. Variation of instantaneous velocity fields and water surface shapes downstream of a circular cylinder at constant submergence ($h^* = 0.4$) with varying Froude number (Fr). Thin arrows are velocity vectors; thick arrow points to the lowered water surface. Adapted from Sheridan et al. (1997).

the surface became undistorted. As Froude numbers increased further (Fr = 0.97), a hydraulic drop developed, and the elevation of the water surface was substantially lower behind the cylinder than upstream. Waves broke at the water surface even at large submergence of $h^* = 2$ and Froude number 0.72.

The sequence of different wake states showed instabilities because the flow field flipped irregularly between different states for certain ranges of the relative submergence and the Froude number. The detailed relation between the two codependent factors on the cross flow for a wide range of Froude numbers and relative submergence still has to be resolved. I assume, however, that the effect of the free water surface in the field must be comparable to the laboratory findings of Sheridan et al. (1997), even though, with the unsteady and highly turbulent flow in the field situation, the irregularity must be larger. No detailed field data on the near cross-flow field of logs close to the water surface are currently available.

Little field data exist on the downstream extent of the effect of wooden obstructions close to the water surface in sand-bed streams. The enlarged turbulence downstream of logs close to the water surface was reduced to undisturbed turbulence within a distance of about 8 times the log diameter at $Re \approx 3 \cdot 10^5$ (Beebe 1997) or less than 10 times the log diameter at $Re \approx 4.5 \cdot 10^4$ (Mutz 2000). These measurements were carried out during low to mean flow. The data given are based on the means of the log diameters, as the logs narrowed, and on the means of the velocity distribution at stream cross-sections. The findings fit well with the extent of the pools scoured downstream of cylindrical obstructions near the water surface in a flume study (Beschta 1983). Pool length was smaller than eight times the diameters of the obstructing cylinders. The study by Beschta was conducted at discharges with the potential to transport the sediments in the flume and at Reynolds numbers ranging from $3 \cdot 10^4$ to $9 \cdot 10^4$; hence, the pool length marked the zone of enlarged shear and turbulence.

In the field situations assessed, as well as in Beschta's flume experiments, both the water surface and the sediment surface certainly affected the flow pattern. Based on the results of Sheridan et al. (1997) and considering the vertical velocity distribution with higher speed near the free surface, I assume, however, that the effect of the free water surface, mostly ignored in the literature, might often override the effect of the streambed on the cross flow of logs in a position near both boundaries.

Logs near the streambed

The relation between the hydraulic roughness caused by the bed sediments and the size of the wood certainly has a large influence on the cross-flow field of logs at or near the streambed. Hydraulic roughness creates a vertical velocity gradient of the flow and, depending on the distance between individual roughness elements, different types of near-bed flow regimes (Morris 1955; Davis and Barmuta 1989; Young 1992). If a log is within the roughness projection of the bed sediments, such as large stones or boulders, it is just an additional roughness element with minor significance to the general flow pattern determined by the nature of the overall roughness. If the diameter of wood is much larger than the sediment grain roughness of the bed, it acts as individual flow obstruction with large hydraulic significance. Even small wood can thus be effective in gravel or sand-bed streams.

I will now concentrate on the effect of wood much larger than the grain roughness of the bed sediments. From flume studies, for a cylinder close to a plane bed, the cross-flow field and the forces acting on the cylinder in the subcritical flow regime are known to depend on the gap ratio (GR)

$$GR = G/D,$$

where G is the gap between the bottom of the cylinder and the bed (Lei et al. 1999). At a gap ratio larger than 2, the influence of the bed is negligible. As the cylinder moves farther toward a plane bed, the stagnation point shifts downward, and the pressure distribution at the cylinder surface becomes asymmetric (Bearman and Zdravkovich 1978). The lower separation point moves downstream along the cylinder surface, the upper separation point shifts upstream, and the size of the wake increases. At decreasing gap ratios below 2, the strength of the vortex shedding becomes weaker until, at $GR \approx 0.3$–0.2, the vortex shedding at the bottom of the cylinder is suppressed.

To what extent these findings from plane beds apply to the natural environment is unknown because the streambed is not level and can be hydraulically rough. Often, the streambed reacts to the flow field created by wood with a change of bed form. The jet-like flow underneath logs produces strong scour (Cherry and Beschta 1989), and pools are frequently found underneath wood with small gap ratios in streams with fine and coarse sediments (Beschta and Platts 1986; Mutz 2000).

In a lowland sand-bed stream, 48% of the total length of the large wood below the water surface had no direct contact to the streambed and gap ratios below two (Mutz and Daniel 1999). Although this type of wood position is common in streams and rivers, no detailed data on the near cross-flow field around logs with small gaps to natural bed forms are available at present.

Logs at the streambed

If a log is in close contact with the streambed or partly embedded in the sediments, recirculating vortices develop in front of and behind the log. In settings with level and smooth beds, the form, spatial extent, and number of vortex tubes produced and their variation with Reynolds number can be predicted (Savory and Toy 1986; Tamai et al. 1987). In the natural environment, the streambed near embedded logs is never level and seldom smooth, so the transferability of this knowledge is uncertain. As a rule of thumb, the diameter of the recirculating vortices in front and behind a log is about half its diameter. Sometimes this estimate can be verified in sand-bed streams by observing a line of fine particulate organic matter, which settles at the streambed parallel to the front of partly embedded logs at about that distance. This line indicates the saddle point where the near-bed flow is close to zero because the approaching flow meets the local flow reversal, and organic material can settle.

Although logs positioned on the streambed produce less scour than partially elevated logs do (Cherry and Beschta 1989), the vortex tubes with high turbulence and the locally upward directed flow can still cause enough scour to excavate under the wood (Beschta 1983; Beebe 1997). Moreover, this flow pattern applies significant pressure variations at the streambed, which can induce pumping of stream water through the bed sediments. Such pumping is documented for sand-bed environments in flumes (Huettel et al. 1996; Hutchinson and Webster 1998), and the significance of this process in a natural environment is supported by the strong vertical water exchange rates assessed in a sand-bed stream rich in wood (Mutz and Rohde 2003).

Using an acoustic Doppler velocimeter (ADV, NORTEC), I assessed the flow in the far wake of logs in sand-bed streams. The logs had direct contact with or were embedded in the sediments. For details of the method, see Mutz (2000, 2001). As expected, zones of reduced mean velocity were found near the bed in front and behind the logs (see Figure 7). The turbulence pattern showed clearly the enlarged turbulence intensity of the vortex street shedding from the logs. The fact that the zones of reduced near-bed velocity in front and behind the logs, as well as the zones of enlarged turbulence, extended high above the streambed and a long way up and downstream from the logs was unforeseen. The enlarged turbulence intensity in the upstream near-bed distance of the logs was caused by the shear between the cushion of reduced near-bed velocity and faster free-flowing water. The unexpected large upstream extent of this reduced near-bed flow might have been caused by slight elevations of the streambed in this area (Figure 7). Behind the logs, the velocity reduction extended farther downstream than predicted from a laboratory-based model (Okamoto 1981; Figure 8). The turbulence increase in the wake was stronger in the field than in Okamoto's model, but it diminished comparable to the model in a distance of about 10 times the log diameter for all Reynolds numbers assessed. Although these field data are still sparse, the flow in the natural environment seems to have comparable basic patterns to the laboratory findings with level beds, but the spatial extent differs significantly. I assume that it was the flow-related bed form that caused the differences. In all the field studies, the shear stress acting at the streambed was above critical for the transport of the fine sediments. Therefore, the large spatial extent of the hydrodynamic effects of wood found in sand-bed streams might have resulted from a self-amplifying interaction between the flow and the streambed-form.

Importance of angular orientation

The patterns described so far are cross-flow fields of logs more or less perpendicular to the flow. With increasing rotation away from perpendicular, the flow deflection along the longitudinal axis of the log superimposes and displaces the cross-flow patterns. Besides the diversion of parts of the flow, the shedding at the edges, created by the log's ends, and the development of the boundary layer in the longitudinal direction along the surface gain increasing importance. No detailed flow data from such angular orientations are available so far, but two studies recorded the drag coefficient of cylinders while rotating them from perpendicular to parallel (Gippel et al. 1996; Hygelund and Manga 2002). As drag is closely related to the flow field

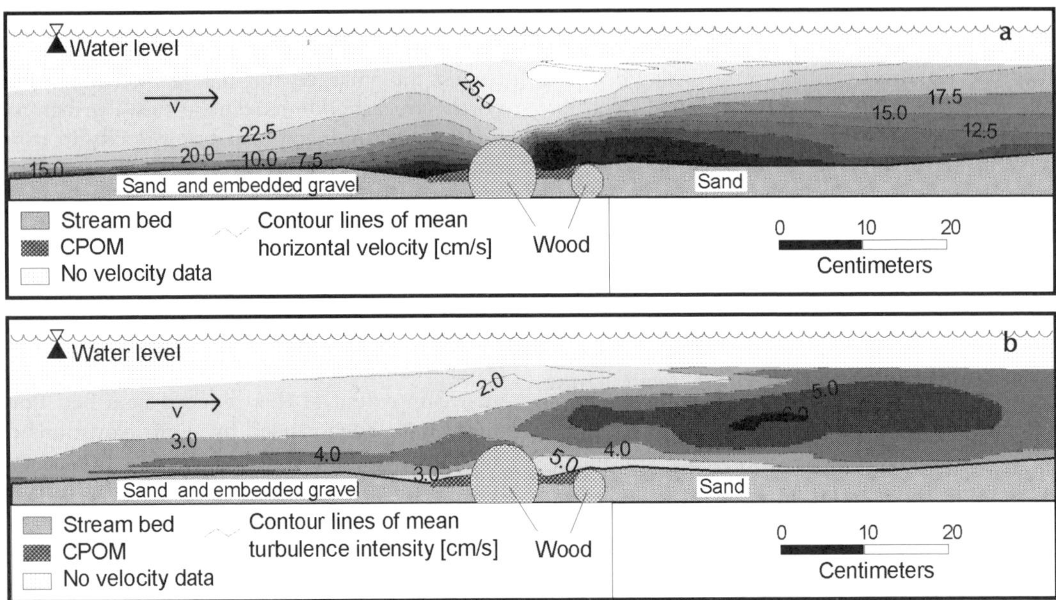

FIGURE 7. Flow pattern on a longitudinal section cutting wood embedded in a natural sand-bed stream: (a) isolevels of mean streamwise velocity; (b) isolevels of turbulence intensity in streamwise direction. Flow was recorded by small-scaled point measurements by use of an acoustic Doppler velocimeter.

around the cylinders, I present briefly the findings of these papers.

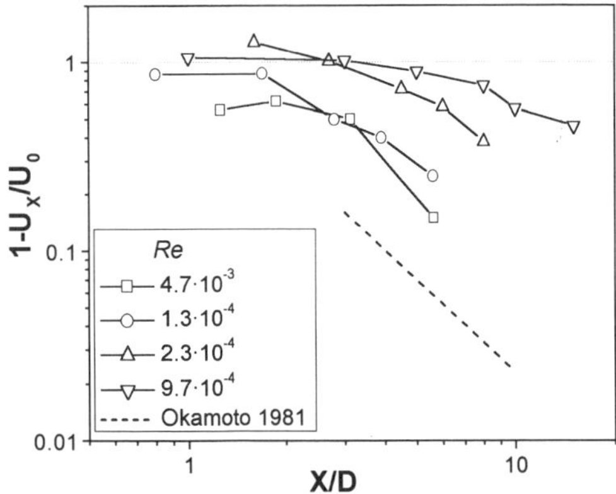

FIGURE 8. Comparison of the spatial extent of the velocity reduction downstream of logs partly embedded in a natural sand-bed stream with the laboratory-based model: Solid lines show field data at different Reynolds numbers (Re). Dotted line shows Okamoto's model (1981). ($1-U_x/U_0$) = relative reduction of mean flow at 0.5 times the height of the obstruction above the streambed. U_x = mean velocity at X, U_0 = mean velocity approaching the log, X = distance from center of log, D = log diameter.

In a flume with a smooth bed, Gippel et al. (1996) showed a significant decrease in the drag coefficient at an orientation of less than about 65 degrees (90 degrees was perpendicular to flow; Figure 9). The change in the drag coefficient in the data from Gippel et al. (1996) indicates the superimposition of the perpendicular cross-flow field with secondary components of the flow directed along the cylinder's longitudinal axis. Hence, according to these data, flow deflection along the longitudinal axis of a log seems to be of increasing significance at orientations below 65 degrees. Contrary to Gippel et al. (1996), Hygelund and Manga (2002) found no significant change in the drag coefficient of cylinders with angular orientation. Their experiments were conducted in a gravel-bed stream, and they explained that the difference to the findings of Gippel et al. (1996) was due to the effect of the rough bed, which produced a far-reaching vertical velocity gradient. In such a velocity gradient, the fluid acceleration of the cross flow under the log must be strong, and Hygelund and Manga hypothesized that because of this acceleration angled logs resemble bluff bodies. Further data from logs in natural streambeds or logs in flumes with vertical velocity gradient are needed to clarify the effect of angular orientation.

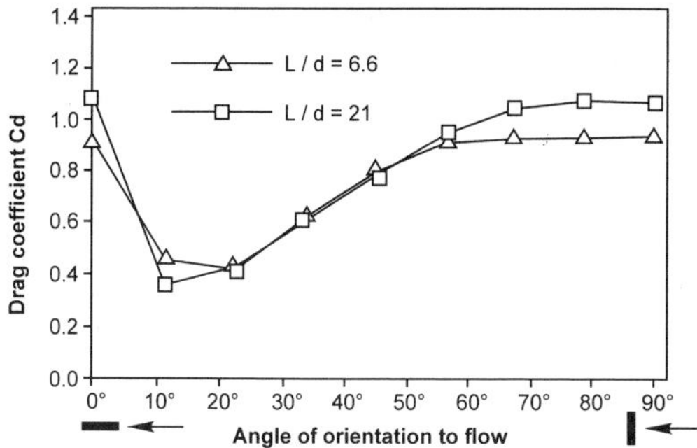

FIGURE 9. Variation in drag coefficient of circular cylinders with angle of orientation to flow. Data for two cylinders with various lengths (*L*) and diameter (*d*). From Gippel et al. (1996).

Complex Wooden Objects and Wood Accumulations

Many wooden elements in streams and rivers have more complex shapes than individual logs do. They can be bent or forked branches, trunks with rootwads and branches, or wood loosely accumulated in various positions, and dense wood jams.

Detailed data on the flow pattern caused by wooden elements with complex shape are few. From theoretical considerations, the flow field of the object is largely determined by its bluff surface and the free ends with an orientation obtuse to flow because the flow separation here is most probable. In their field study, Hygelund and Manga (2002) registered no change of drag force after adding branches to a cylinder. As the blunt surface area of the cylinder with branches was larger, the drag coefficient had decreased. Hygelund and Manga explained the decrease of the drag coefficient by flow separation and generation of turbulence at the branches, which changed the flow characteristics impinging on the cylinder.

Gippel et al. (1996) reported that, for a tree-shaped model, the orientation of the object to the flow had less effect on the drag coefficient, and hence on the flow pattern, compared with single cylinders. For such loose objects, there seems to be a replacement of particularly flow relevant structures with rotation. Abbe and Montgomery (1996) assessed the velocity pattern around a dense complex wood structure (Figure 4). It confirms that such structure can be modeled as one simplified solid obstruction with good results, at least as long as the focus is on the morphological effect of wood. If such wood is frequent, it contributes considerably to the form roughness of the channel.

The flow through wood jams that protrude high above the streambed is complex because of interference between the various elements. From the engineering literature, knowledge about the flow pattern caused by parallel cylinders in staggered arrangement is available (Gu and Sun 1999), but it can hardly be expanded to wood jams because wood is irregular in size, shape, and orientation.

Near-bed flow regimes

Wood is frequently spread out over the streambed without protruding high above the bed. It then blocks only a small portion of the cross-sectional area (Mutz 2000). Such wood contributes considerably to the skin roughness of the streambed. Distinct near-bed flow regimes are produced if the wooden elements protrude less than 30% of the water depth. The flow regimes can be categorized by the separation distance (S) of the wooden elements in the flow direction and the flow depth above them (y) (Davis and Barmuta 1989; Young 1992). The separation distance is defined as $S = j/k$, where j is the groove width in the flow direction and k is the height of the roughness projection.

At a low separation distance, skimming flow is produced. In this near-bed flow regime, the flow skims over the crests of the logs and stable vortices with reduced velocity developed in the grooves. With isolated roughness caused by very large separation distances, the vortices dissipate before the flow reaches the next log. The transition between these two flow regimes is wake-interference flow. In this regime, the trail of the vortex from the first log hits the next log with high turbulence, causing unstable and irregular flow patterns.

Based on flume studies with uniform roughness, threshold criteria for the spacing of roughness elements can be given to distinguish between these near-bed flow regimes. Skimming flow develops at separation distances below 1. The transition between wake interference and isolated roughness is a function of the roughness height and the depth of flow above the roughness elements (Young 1992; Figure 10).

Another approach to differentiate between the near-bed flow regimes is based on the roughness density (Novell and Church 1979). The roughness density is defined as

$$N \cdot A_e / A_t \, ,$$

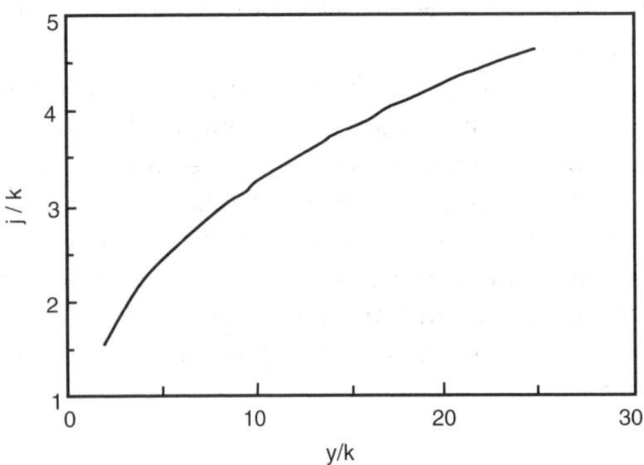

FIGURE 10. Critical spacing for the transition between wake interference flow (below curve) and isolated roughness flow given in nondimensional groove width (j/k) as a function of nondimensional depth (y/k). The function is valid for roughness element drag coefficient of cobbles. From Young (1992).

where N is the number of roughness elements, A_e is the total streambed area, and A_t is the area of roughness elements. Roughness densities below one-twelfth produce skimming flow, between one-fifteenth and one-twentieth wake interference flow, and above one-fiftieth isolated roughness flow (Davis and Barmuta 1989).

These thresholds are based on the drag coefficients and spatial dimensions of spheres or stones as roughness elements. Wood has a cylindrical shape, and the accumulations are usually created by a variety of wooden elements. They probably do not produce uniform roughness. The velocity distribution registered at a cross-section of a sand-bed stream is shown in Figure 11. The section cut a channel bend with wood distributed loosely and in all angular orientations at the outside of the bend. Where the wood was present, the vertical velocity distribution clearly showed wake-interference flow. The wood accumulation had a variety of groove widths and roughness heights. The mean separation distance in the accumulation was two, and the nondimensional depth (y/k) was four. According to the relation of the separation distance to the nondimensional depth (Figure 10), wake interference flow should result; hence, the observation supports the model given by Young (1992). These data are from a single field study, however, and because the flow pattern registered at individual logs partially embedded in the streambed showed a larger spatial extent of the wake (Figure 9), these thresholds should be used with great care.

The roughness created by wood accumulations and the resulting near-bed flow regime can also affect the pattern of the free-flowing water because they determine the overall frictional resistance. The friction is highest at the transition between wake interference and isolated roughness flow (Young 1992). In the cross-section shown in Figure 11, the velocity was reduced until high above the crest of the wood elements; hence, the wood obviously caused very strong frictional resistance. It even forced the thread of flow towards the inside bank of the bend, which was free of wood. As the stream had the competence for sediment transport, the flow pattern caused by the wood also explains the unexpected cross-sectional morphology. The outside bank, where at least parts of the sediments were protected by wood, had a shallow slope, whereas a deep pool was scoured close to the inside bank.

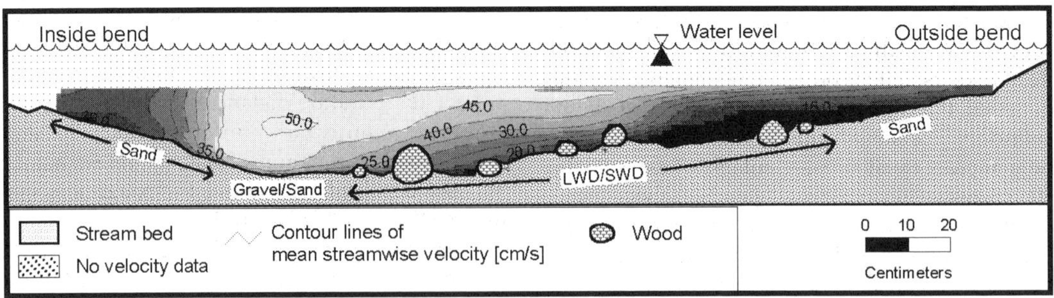

FIGURE 11. Velocity distribution on a cross-section cutting an accumulation of wood loosely distributed at the streambed. The section cuts a bend of the sand-bed stream. Wooden elements had various angular orientations to streamwise. LWD/SWD = large and small wood.

Conclusions

Information on hydraulic effects of wood in streams and rivers is sparse. Apart from observations on the general diversion of flow, so far, few data are available describing flow fields near wood.

Knowledge about hydraulic effects of circular cylinders (Williamson 1997; Munson et al. 1998) can be expanded, with caution, to the hydraulics of wood in streams and rivers. The Reynolds number (Re) related to the diameter of the flow-obstructing object determines the principal cross-flow fields. The findings from the technical environment also suggest that the cross-flow patterns produced by logs in streams and rivers depend on the logs' distance to the water surface (Sheridan et al. 1997). If the water surface is within two diameters, the combination of the logs' relative submergence (h^*) and the Froude number (Fr), related to the depth of submergence, determine largely the highly variable nature of the cross-flow field. Field measurements indicate that the hydraulic effects downstream of a log oriented near the water surface disappear in less then 10 times the log diameter (Beebe 1997; Mutz 2000).

For logs near the streambed, the gap ratio to the bed (GR) affects the hydraulics at the log (Bearman and Zdravkovich 1978). In this orientation, the flow fields of the natural environment must drastically differ from a model based on laboratory findings. The model is valid for level and smooth boundaries, but roughness and a complex three-dimensional shape are typical for natural beds. In the natural environment, submerged jet-like flows and local scour are typical (Beschta and Platts 1986; Cherry and Beschta 1989). Knowledge about detailed flow patterns around wood with a gap to a natural-shaped streambed is missing so far.

Few field data are available for logs with tight contact to sand beds or logs partly embedded in sand beds. They show cross-flow patterns comparable to the findings from flume studies (Savory and Toy 1986; Tamai et al. 1987). The spatial extent of the hydraulic effect seems to be larger in the natural environment than in the laboratory. I assume the uneven streambed to be the reason for this difference.

Depending on the angular orientation and the blockage ratio, the flow deflection along the longitudinal axis superimposes the cross-flow pattern. From the change of the drag coefficient with rotation to flow, which was recorded in a flume with a smooth bed (Gippel et al. 1996), it seems that flow deflection along cylinders is relevant at orientations of less than 65 degrees to flow parallel. However, such a change of drag was not observed for cylinders in the vertical velocity of a natural gravel-bed stream (Hygelund and Manga 2002).

Dense and complex wooden objects and dense wood accumulations can be modeled as solid structures (Abbe and Montgomery 1996). The flow near such objects is determined by the size of bluff surfaces and the shedding from edges or free ends with orientations obtuse to the flow. In dense wood accumulations, which extend into the current, the flow pattern can hardly be modeled; the composition and arrangement of the wooden elements are too variable.

Wood spread out over the streambed acts as streambed roughness. Models based on technical roughness can approximate the near-bed flow regimes (Davis and Barmuta 1989; Young 1992). The few field data available so far support these models.

The simplified laboratory conditions and the natural environment, however, differ in many of the factors that determine the cross flow of obstructions. Therefore, uncertainty remains as to what extent the findings from technical settings can be applied to wood in streams and rivers.

To reduce this uncertainty and to check the validity of existing hydraulic models for wood in streams and rivers, flow data from the field and data from flume experiments that simulate the natural complexity are needed.

Despite vagueness about the validity of the current models, the main factors that determine the hydraulic effects of wood are—to a large extent—evident; they are the Reynolds number, Froude number, blockage ratio, relative submergence, gap ratio to the streambed, angular orientation to the flow, and, for accumulations spread out at the streambed, the roughness parameters (such as separation distance and nondimensional depth). These factors can be easily assessed in the field and give at least a rough impression of the flow pattern produced by the wood in streams and rivers. They allow the evaluation and comparison of the hydraulic effects of wood in different locations and at varying velocities and discharges. I suggest that these factors should be assessed in future studies about wood in streams and rivers. These would not only help to evaluate the hydraulic significance of wood, but also contribute to better insight into the ecological and morphological processes related to the hydraulic effects of wood in streams and rivers.

Acknowledgments

I acknowledge the helpful comments of two anonymous reviewers on an earlier draft of the manuscript. I am grateful to Martha H. Brookes, Corvallis, for linguistic help with the manuscript and technical edit.

References

Abbe, T. B., and D. R. Montgomery. 1996. Large woody debris jams, channel hydraulics and habitat formation in large rivers. Regulated Rivers Research and Management 12:201–221.

Achenbach, E. 1971. Influence of surface roughness on cross flow around a circular cylinder. Journal of Fluid Mechanics 46:321–335.

Allan, D. J. 1995. Stream ecology: structure and functioning of running waters, 1st edition. Chapman and Hall, London.

Ambühl, H. 1959. Die Bedeutung der Strömung als ökologischer Faktor. Schweizerische Zeitschrift für Hydrologie 21(2):133–264.

Bearman, P. W., and M. M. Zdravkovich. 1978. Flow around a circular cylinder near a plane boundary. Journal of Fluid Mechanics 89:33–47.

Beebe, J. T. 1997. Fluid patterns, sediment pathways and woody obstructions in the Pine River, Angus, Ontario. Doctoral dissertation. Wilfrid Laurier University, Waterloo, ON.

Beschta, R. L. 1983. The effect of large organic debris upon channel morphology: a flume study. Pages 8.63–8.78 in D. B. Simons, editor. Symposium on erosion and sedimentation. Simons, Li, and Associates, Fort Collins, Colorado.

Beschta, R. L., and W. S. Platts. 1986. Morphological features of small streams: significance and function. Water Resources Bulletin 22:369–379.

Cherry, J., and R. L. Beschta. 1989. Coarse woody debris and channel morphology: a flume study. Water Resources Bulletin 25:1031–1036.

Davis, J. A., and L. A. Barmuta. 1989. An ecologically useful classification of mean and near-bed flows in stream and rivers. Freshwater Biology 21:271–282.

Gippel, C. J. 1995. Environmental hydraulics of large woody debris in stream and rivers. Journal of Environmental Engineering 128:388–395.

Gippel, C. J., I. C. O´Neill, B. L. Finlayson, and I. Schnatz. 1996. Hydraulic guidelines for the re-introduction and management of large woody debris in lowland rivers. Regulated Rivers Research and Management 12:223–236.

Gu, Z., and T. Sun. 1999. On the interference between circular cylinders in staggered arrangement at high subcritical Reynolds numbers. Journal of Wind Engineering and Industrial Aerodynamics 80:287–309.

Harmon, M. E., J. F. Franklin, F. J. Swanson, P. Sollins, S. V. Gregory, J. D. Lattin, N. H. Anderson, S. P. Cline, N. G. Aumen, J. R. Sedell, K. Cromack Jr., K. W. Cummins, and G. W. Lienkaemper. 1986. Ecology of coarse woody debris in temperate ecosystems. Advances in Ecological Research 15:133–302.

Huettel, M., W. Ziebis, and S. Forster. 1996. Flow-induced uptake of particulate matter in permeable sediments. Limnology and Oceanography 41:309–322.

Hutchinson, P. A., and I. T. Webster. 1998. Solute uptake in aquatic sediments due to current-obstacle interactions. Journal of Environmental Engineering 124:419–426.

Hygelund, B., and M. Manga. 2002. Field measurements of drag coefficients for model large woody debris. Geomorphology 1297:1–11.

Keller, E. A., and F. J. Swanson. 1979. Effects of large organic material on channel form and fluvial processes. Earth Surface Processes 4:361–380.

Lei, C., K. Cheng, and K. Kavanagh. 1999. Re-examination of the effect of a plane boundary on force and vortex shedding of circular cylinder. Journal of Wind Engineering and Industrial Aerodynamics 80:263–286.

Leung, Y. C., N. W. M. Ko, and K. M. Tang. 1992. Flow past circular cylinder with different surface configurations. Journal of Fluids Engineering - Transactions of the ASME 114:170–177.

Lin, J. -C., J. Towfighi, and D. Rockwell. 1995. Instantaneous structure of the near wake of a circular cylinder: on the effect of Reynolds number. Journal of Fluids and Structures 9:409–418.

Lisle, T. E. 1986. Effects of woody debris and anadromous salmonid habitat, Prince of Wales Island, south Alaska. North American Journal of Fisheries Management 6:538–550.

Lisle, T. E. 1995. Effects of woody debris and its removal on a channel affected by the 1980 eruption of Mount St. Helens, Washington. Water Resources Research 31:1797–1808.

Manga, M., and J. W. Kirchner. 2000. Stress partitioning in streams by large woody debris. Water Resources Research 36:2373–2379.

Montgomery, D. R., B. D. Collins, J. M. Buffington, and T. B. Abbe. 2003. Geomorphic effects of wood in rivers. Pages 21–47 *in* S. V. Gregory, K. L. Boyer, and A. M. Gurnell, editors. The ecology and management of wood in world rivers. American Fisheries Society, Symposium 37, Bethesda, Maryland.

Morris, H. M. 1955. Flow in rough conduits. Transactions of the American Society of Civil Engineers 120:373–398.

Munson, B. R., D. F. Young, and T. H. Okiishi. 1998. Fundamentals of fluid mechanics, 3rd edition. Wiley, Chichester, New York, Brisbane, Toronto, Singapore.

Mutz, M. 2000. Influences of wood on flow patterns

and channel morphology in a low energy sand-bed stream reach. International Review of Hydrobiology 85:107–121.

Mutz, M. 2001. Near-bed flow of a quasi-natural lowland stream. Verhandlungen der Internationalen Vereinigung der Limnologie 27:2444–2447.

Mutz, M., and S. Daniel. 1999. Totholz in einem naturnahen tieflandbach - art, menge und räumliche verteilung. Deutsche Gesellschaft für Limnologie Tagungsberichte 1999, 235–240. Tutzing, Eigenverlag der DGL, Tutzing 2000. (Woody debris in a natural lowland stream – type amount and spatial distribution. German Association of Limnology, Proceedings 1999: 235–240. Tutzing 2000.)

Mutz, M., and A. Rohde. In press. Processes of surface-subsurface water exchange in a low energy sand-bed stream. International Review of Hydrobiology.

Novell, A. R. M., and M. Church. 1979. Turbulent flow in a depth-limited boundary layer. Journal of Geophysical Research 84:4816–4824.

Okamoto, S. 1981. Turbulent shear flow behind hemisphere-cylinder placed on ground plane. Pages 171–185 *in* Proceedings of the Third International Symposium on Turbulent Shear Flows. Springer-Verlag, Berlin, Germany.

Raudkivi, A. J. 1990. Loose boundary hydraulics. Pergamon Oxford, UK.

Ribeiro, J. L. D. 1991. Effects of surface-roughness on the 2-dimensional flow past circular cylinders - 1. mean forces and pressures. Journal of Wind Engineering and Industrial Aerodynamics 37:299–309.

Savory, E., and N. Toy. 1986. Hemispheres and hemisphere-cylinders in turbulent boundary layers. Journal of Wind Engineering and Industrial Aerodynamics 23:344–364.

Sheridan, J., J. -C. Lin, and D. Rockwell. 1997. Flow past a cylinder close to a free surface. Journal of Fluid Mechanics 330:1–30.

Smith, R. H., F. D. Shields, E. A. Dardeau, T. E. Schaefer, and A. C. Gibson. 1992. Incremental effects of large woody debris removal on physical aquatic habitat. Technical Report EL-92–35, Pages 1–67. 5285 Port Royal Road, Springfield, Virginia, National Technical Information Service.

Statzner, B., J. A. Gore, and V. H. Resh. 1988. Hydraulic stream ecology: observed patterns and potential applications. Journal of the North American Benthological Society 7:307–360.

Tamai, N., T. Aseada, and N. Tanaka. 1987. Vortex structures around a hemispheric hump. Boundary-Layer Meterology 39:301–314.

Williamson, C. H. K. 1997. Advances in our understanding of vortex dynamics in bluff body wakes. Journal of Wind Engineering and Industrial Aerodynamics 69- 71:3–32.

Young, W. C. 1992. Clarification of criteria used to identify near-bed flow regimes. Freshwater Biology 28:383–391.

American Fisheries Society Symposium 37:109–133, 2003

Dynamics of Wood in Large Rivers

HERVE PIÉGAY

CNRS - UMR 5600 "Environnement, Ville, Société", 18 rue chevreul, 69362 Lyon cedex 07, France

Abstract.—A large river, in relation to wood dynamics, has a width several times greater than the height of the trees in its riparian area. Large rivers undergo particular physical and biological processes related to wood that vary according to their condition (pristine or managed) and their locations in the landscape (upland area or downstream, tropical or temperate climates).

In this chapter, several topics are developed to illustrate how large wood plays a significant role in the functioning of large rivers: production and transfer of wood, interactions between wood and channel geometry (accumulation sites, trapping efficiency, effects of wood on channel forms), and interactions between wood and human activities and management (effects of wood on fish resources, damages associated with the wood, restoration and maintenance strategies).

Bank erosion, wind, beavers, floodplain forest clearing by overbank flow, meteorological events (snow, wind), and humans (dump areas, residuals of timber harvest) are the main vectors of wood introduction to large rivers. The amount of wood stored in the channel is usually low compared to what is introduced annually. Wood is mainly located at the edge of the floodplain (islands, concave banks, side channels), but differences exist in accumulation sites, according to the channel pattern and geometry. The effects of wood on island formation as well as its effects on meander cut-off have been observed but are variable from one river to another according to the size of the wood and the character of the floodplain.

Fish abundance and diversity are influenced by wood structures, even in rivers with low amounts of wood. Wood obstructions and their associated geomorphic facies provide rearing sites and habitats for aquatic invertebrates. Because the amount of wood in large rivers is usually low, it rarely increases flooding risks locally. Wood affects navigation where unobstructed channels are favored; wood transport has then a cost because it must be removed. When the wood in rivers is enhancing fish resources, there is a need to define a balance between the ecological benefit of the wood and the need for removing it for safety purposes. In the 1980s, research was focused on habitat questions at a local spatial scale, linking hydraulic and geomorphic conditions, and characters of fish species. A research priority is now necessary to understand wood mobility and its downstream impacts.

Introduction

Pioneering research on wood in rivers was conducted as far back as the early 1970s (Heede 1972; Swanson et al. 1976), but the importance of wood as a structural element in river networks has received wider attention in the scientific community only over the past two decades (Vannote et al. 1980; Gregory et al. 1985; Harmon et al. 1986; Maser et al. 1988; Gurnell et al. 1995). Most research on wood has focused on montane streams in northern Pacific regions of North America where old-growth forests (dominated by Douglas-fir and western redcedar) produce large pieces of wood (more than 60 m long and 1.5 m in diameter) that decay slowly and move downstream over decades to centuries. Because threshold conditions above which the wood is transported are infrequently passed in such an environment, by the sheer volume, great sizes, and stability of the wood pieces, it is not surprising that wood affects, in many ways, the flow of water and sediment, channel geometry, and associated living communities.

Large wood has been studied throughout the world, but this research has focused chiefly on river systems that are strongly influenced by human activities and produce smaller-sized wood.

These studies have mostly been pursued in the temperate zone (England, Spain, Germany, France, Japan, Australia, and eastern parts of North America). Headwater streams with limited management (Gregory et al. 1985; Inoue et al. 1997; Elosegi et al. 1999) and large rivers have been less investigated because managers and scientists (1) assume that wood does not critically affect the functioning of such systems, (2) can design studies more easily for small streams, or (3) assume that social uses of large rivers are so important economically and culturally that natural functions of wood are irrelevant in these large systems. For example, Keller and Swanson (1979) found that the amount of wood in a sixth-order stream (McKenzie River) was negligible compared to the amount observed in first- to third-order streams of Oregon. Over the past two decades, however, studies in Australia, the eastern U.S. coastal plains, France, and Italy have demonstrated that the dynamics and management of large wood along large rivers are potentially key topics of research (Wallace and Benke 1984; Gippel et al. 1996; Downs and Simon 2001; Gurnell et al. 2002). The volume of large wood can be substantial, sometimes comparable to amounts observed in headwater streams.

This chapter assesses and provides examples of both the physical and biological dynamics of wood in large rivers. Major gaps in knowledge and unanswered questions are highlighted. Areas of research that will improve management decisions about river systems are identified. Five topics will be discussed: (1) geographical and historical diversity of wood in large rivers, (2) production and transfer of wood, (3) morphological and biological effects of wood once in the rivers, (4) management options to be investigated, and (5) problems to be solved.

Geographical and Historical Diversity of Large Rivers

Definition of a large river

The literature contains few studies of large wood in large rivers; large rivers, in the traditional sense, are generally considered to have a basin area of several hundred thousand square kilometers and a mean annual discharge of several hundred cubic meters per second. Only one such river is known to have been studied, and this is a historical account of wood in the Red River in Louisiana (Triska 1984). Consequently, river size here will be considered in a relative sense. A "large river" will be considered to have a channel width (w) several times greater than the maximum length (l) of pieces of wood that are actively transported downstream. Several authors have noted that size ratio, S_r (l/w), is a key variable potentially dictating the amount of time wood remains in a given reach (Swanson et al. 1984; Heede 1985). Along a stream in southeast Alaska, these same authors observed that wood pieces with lengths less than 0.5 times the bankfull channel width tended to float downstream during storms. Where the size ratio is greater than one, wood is fairly stable and affects local channel geometry, flow velocity, and patterns of sedimentation. Where size ratios are less than one and the wood is more mobile, physical effects are transmitted downstream.

Most of this complexity arises because wood affects reaches far from its sources, and it is transitory, often moving and changing form with every flood (Piégay et al. 1998, 1999). In large rivers, floods are critical to wood transport regardless of channel size. Large wood accumulations often have moderate effects on flood conveyance in the channel, notably in lowland rivers (Gippel et al. 1994). In headwater streams of old-growth forests, S_r ratios are expected to range from 15 to 30, but they range from 2 to 0.05 in large rivers (Table 1). In most large rivers, a basin of 1,000–5,000 km^2 is drained and encompasses contrasting landforms, ranging from mountainous reaches to lowland plains, with geomorphic patterns varying from braided, wandering channels to meandering, sinuous, single-thread channels.

Summarizing the typical features of wood in large rivers is difficult because rivers are so diverse and influenced differently in each case by human activities and the characteristics of the river basin, such as forest cover, channel depth, and climate. Distinct physical and biological processes are associated with large wood in rivers. These processes vary, however, based on how much a river has been managed, whether it is in upland or lowland areas, and its climatic regime (tropical or temperate).

Large wood in pristine temperate rivers

Historically, pristine temperate large rivers were not so different from unmanaged streams observed today in the Pacific Northwest. As de-

TABLE 1. General characteristics of studied large rivers.

	Channel slope m/km	Channel width (m)	Basin area (km^2)	Dominant riparian species	Size ratio S_r	Reference
Lookout Creek, Oregon	30	24	61	Old-growth forest (*Pseudotsuga menziesii*)	2.5	Keller and Swanson 1979
Queets River, Washington	1–20	30–83	1,164	Old-growth forest (*Pseudotsuga menziesii, Tsuga heterophylla*)	2.3–1.2	Abbe and Montgomery 1996
McKenzie River, Oregon	6	40	1,024	Old-growth forest (*Pseudotsuga menziesii*)	1.5	Keller and Swanson 1979
Willamette River, Oregon	0.9	108	29,138	*Populus trichocarpa, Fraxinus oregana, Pseudotsuga menziesii*	0.20	Sedell and Frogatt 1984
Ain River, France	1.3	100–150	3,640	*Populus nigra, Fraxinus excelsior*	0.1–0.15	Piégay and Marston 1998
Drôme River, France	3–8	150–200	1,620	*Populus nigra, Fraxinus excelsior*	0.07–0.10	Piégay et al. 1999
Tagliamento River, Italy	1–3	300–800	2,500	*Alnus incana, Populus nigra*	0.01–0.07	Gurnell et al. 2000a
South Fork Obion River, Tennessee	Lowland river	18–23	927	*Acer negundo, Betula nigra*	*1.7–2*	Shields and Smith 1992
Mallar Creek, North Carolina	1	12		Hardwood species	1–1.25	Keller and Swanson 1979
Ogeechee River, Georgia	0.2	33	7,000	*Taxodium distichum, Quercus nigra, Liquidambar* spp.	1–1.1 1–1.1	Wallace and Benke 1984
Thomson River, Australia	0.028	48	3,540	*Eucalyptus camaldulensis*	0.8–1	Gippel et al. 1996
Red River, Louisiana	Lowland river	215–365	236,000	*Populus* spp., *Quercus* spp.	0.05–0.1	Triska 1984

scribed by Triska (1984) on the Red River in Louisiana, and also by Sedell and Froggatt (1984) on the Willamette River in Oregon, large rivers of North America were full of wood before European settlement, and large channels were blocked in places by stable wood jams. Channel roughness was very high as a consequence of wood accumulations, which also changed river morphology, increased connectivity between the floodplain and the river channel, and influenced channel geometry. The upper Willamette corridor, in its pristine state, was a complex network of multiple active channels, oxbow lakes, sloughs, and backwaters, where wood created and maintained complex habitats in the outside bends of the river (Gregory et al. 2002). Sedell and Luchessa (1982) compiled a nonexhaustive list of rivers in Oregon and Washington that contained substantial wood jams blocking channels over distances of 100–1,500 m before river managers began removing the wood in the second half of the 19th century. The "great raft" on the Red River was a complex series of wood jams caused by congested transport of wood; it affected a 390–480-km length of river for 300–400 years (Triska 1984; Harvey et al. 1988). This example demonstrates how the channel of a large river can be completely blocked, even if the length of wood pieces is much less than the channel width; the overwhelming number of wood fragments forms extensive accumulations of wood.

Historical examples illustrate how channel geometry in the temperate zone is significantly influenced by human activities and how the characteristics of wood transport may be more closely associated with human actions than with channel size. Large temperate rivers today typically drain large valleys that are managed by people, especially for agricultural production and residential use and urbanization—all land uses that have reduced wood inputs and transport. In this context, volumes of wood in human-dominated rivers are often too small to have the same effects on either hydrologic or geomorphic variables as in streams or pristine large rivers.

Gradients in the lowlands and uplands and in the temperate and tropical regions

Tropical and lowland rivers, the least studied rivers in the world, may have unique features associated with the wood they contain. Lowland rivers usually have limited stream power and, thus, are less efficient in wood production, principally because of lower rates of bank erosion. Nevertheless, though the capacity of the river to generate wood is low, the amount of wood stored in the channel can be relatively high compared with streams (Wallace and Benke 1984; Gippel et al. 1994). A combination of a deep channel, low stream velocity, small sediment grain sizes, and local inputs of wood, contributes to producing wood with unique features. Along lowland rivers, fragments of wood are less physically worked by the river, and frequently, significant sections of the original tree architecture remain intact.

In tropical regions, we may expect that greater humidity and higher temperatures accelerate the decomposition of wood deposited along bank edges. As a result, wood should have a reduced capacity to structure its environment. In much cooler climates where decomposition is slow, logs buried some 5,000–7,000 years ago in floodplain sediments can eventually be reintroduced into the river (Australia, Tasmania). Carbon dating on fragments of wood from the Hoh River, in western Washington State, USA, determined some to be about 2,000 years old (Bilby and Bisson 1997). Not all wood in rivers behaves in the same way, however. Different tree species clearly have different wood densities, influencing both the buoyancy of the wood and its relative resistance to decomposition (Harmon et al. 1986; Gurnell 2003, this volume).

Increases of in-channel wood quantities and transfer due to human influences

Historical studies have demonstrated that human activities tend to decrease the amount of wood in large rivers primarily because of clearing for improved navigation and flood risk management. Although this trend is observed worldwide, some regions have also witnessed an increase in wood along large rivers. Most European rivers today store and generate more wood than they did over the first half of the 20th century because European riparian forests have expanded since the Second World War. Dendrochronological analysis of riparian trees in some French rivers has shown that most of the trees became established after 1945, after changes in agricultural practices (Liébault and Piégay 2001). Before 1945, only pioneer stages of riparian forest and isolated trees grew in the riparian zones, largely because these areas were used extensively for grazing and collecting wood for domestic fires or to make baskets. With the modernization of agriculture, these areas were abandoned by farmers and gradually became forested again, increasing wood in rivers. Wood observed in French rivers today is therefore a reflection of widespread afforestation of the landscape in response to human abandonment of both riparian zones and mountain hillslopes (Figure 1).

In the Mississippi delta, flow regulation and, straightening of channels have contributed to channel degradation and widening. The Schumm conceptual model of river degradation (Schumm et al. 1984; Simon and Hupp 1987) describes how bank destabilization increases with channel degradation, such that an increase in bank height above a critical threshold causes mass-wasting of riparian zones and an associated loading of large wood from riverbanks (Downs and Simon 2001). This problem is becoming so important that research has recently been initiated to estimate, at the basin scale, the amount of wood that could be introduced into the river before causing a flooding increase (Downs and Simon 2001). The degree to which bridges are affected by wood transport and the ways this risk can be managed are questions being explored in several studies (Diehl 1997; Wallerstein 1999).

A wide range of physical processes in large rivers is associated with large wood. They vary according to the characteristics of the riparian vegetation bordering the river, such as (1) size, species, and the condition of the riparian woodland (itself related to other influences, such as incidence of fire or human actions on the channel reach); (2) channel characteristics and form (such as channel slope, stream power, or a straight, stable channel versus a less predictable freely meandering channel); and (3) human pressure on the river system and its floodplain (such as modification of riverside vegetation, river diversions, and destabilization of the river channel by gravel extraction).

The amounts of wood stored in the channel differ greatly among rivers. Compared with

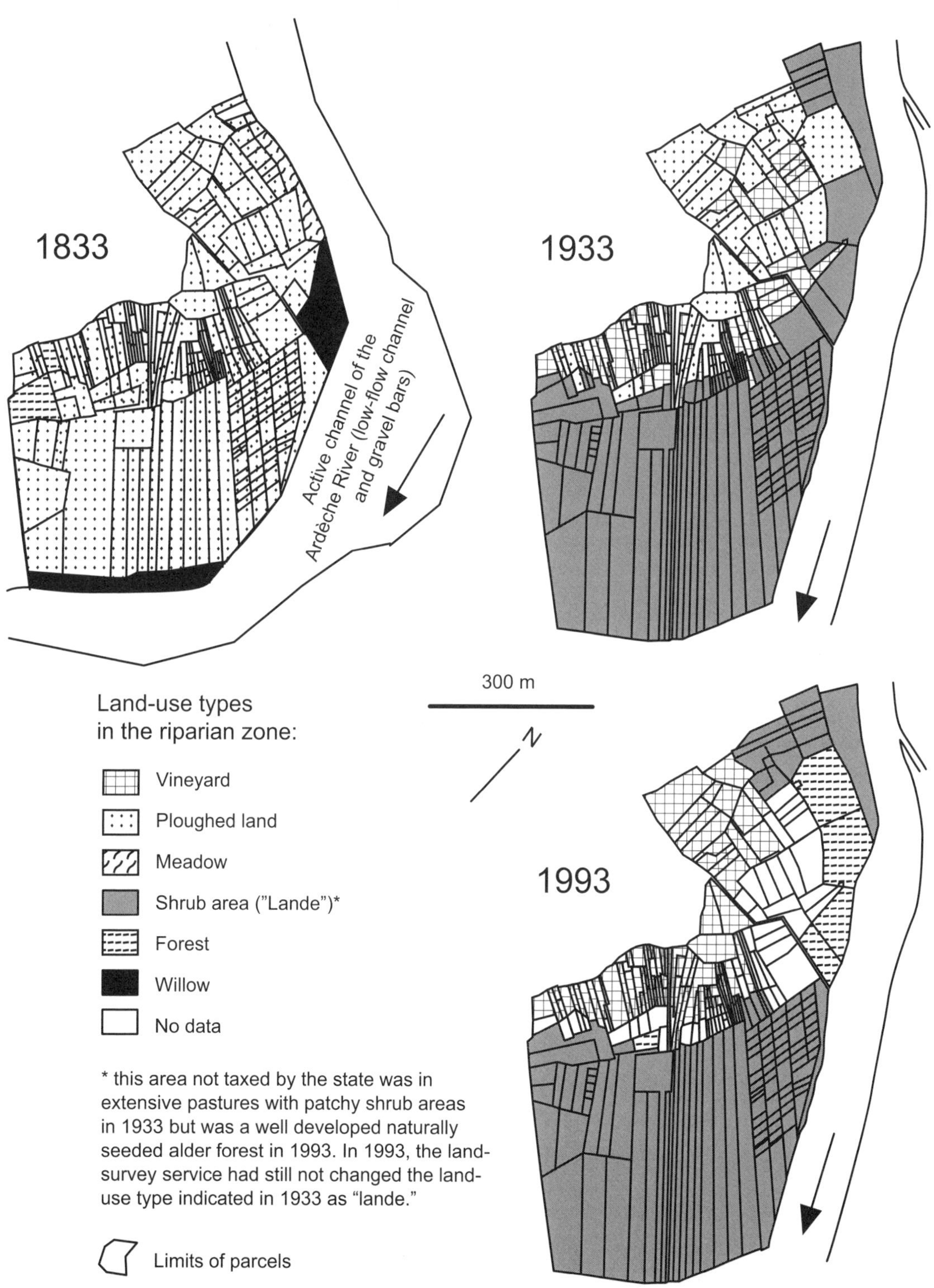

FIGURE 1. Riparian afforestation along the Ardèche River between 1833 and 1993.

unmanaged upland streams (for example, the Redwood Creek basin, California, where wood storage can range from 150 to 2,500 t/ha; Keller and Macdonald 1995), some large river reaches, notably those in lowland areas, can support substantial volumes of wood: 260–540 t/ha in the South Fork Obion River (Shields and Smith 1992), 90–110 t/ha on the Ogeechee (Wallace and Benke 1984) and Thompson rivers (Gippel et al. 1996). Research on piedmont rivers indicates that total volumes of wood stored in river channels are usually low: 6 t/ha in the McKenzie River, Oregon; 8–32 t/ha in the active channel of the Drôme, France; and 1–6 t/ha along the Tagliamento River, Italy. Nevertheless, much higher amounts of wood can accumulate in local portions of floodplain margins where vegetation may trap it. For example, Gurnell et al. (2000a) found an average of 80 t/ha of wood trapped on established islands along the Tagliamento River, and Piégay and Marston (1998) documented an average of 56 t/ha of wood in forest units along the concave bank of Mollon, along the Ain River, France. These amounts of wood are commonly observed in large temperate rivers if managers do not disrupt critical natural processes, such as succession to mature riparian forest, delivery of trees into the river through bank erosion, and wood storage in the channel. People influence these riverine processes differently in upland versus lowland areas, however. Conclusions about relative amounts of wood in upland river reaches (which are more often relatively unmanaged) versus lowland river reaches may be weak or erroneous.

Wood Production and Transfer

When wood dynamics are studied in large rivers, it is important to first understand patterns of "wood flux" and the origin of the wood. Where is the wood coming from? What amounts of wood are being introduced? What is the variability of inputs at different time scales?

Wood sources and conditions

According to Keller and Swanson (1979), several different vectors transfer wood into the river: ice flows, mass wasting, debris from torrential flows and avalanches, wind-throw, bank erosion, flooding (which may clear a floodplain of dead wood produced through natural tree mortality or live trees unable to remain standing during intense flows), meteorological events such as snow or wind, human actions such as dumping wood from cuttings, dam-building by beaver, and floating from upstream reaches. The first three vectors are mainly in headwater streams, but the rest can also be found downstream in wider reaches.

Wood sources also can vary from upstream to downstream reaches in large rivers, and it is important to recognize relative contributions of wood from uplands, floodplain avulsion, and tree fall along river margins. A critical question is, "What is the reach input rate from wind-throw, bank erosion, and flooding compared with the upstream input rate?" Data collected on upland systems, such as the Asusa River in the Kamikochi Valley of the Japanese Alps (1,500 m in elevation; 175 km^2) (data collected in 1999 with H. Shimazu, T. Oguchi), and the Rhône River downstream of Geneva (a basin area of 10,910 km^2 across which only 2400 km^2 is actually producing wood) (Moulin and Piégay 2003), have demonstrated that significant amounts of wood can originate in the riparian corridor. The analysis of wood fragments deposited in a braided channel of the Asusa River showed that more than 50% of the pieces were introduced from the adjacent riparian forest (*Salix* spp., *Fraxinus mandshurica, Toisusu urbanian*), which is less than 1 km from steep torrential tributaries draining the valley sides.

Along the Rhône River, some distance downstream from the mountain summits, wood fragments (>7.5 cm in diameter and >20 cm long) trapped in the Genissiat Reservoir over 10 months (February to November 1998) were comprised of 44% softwood species (*Alnus glutinosa, Salix* spp., *Populus nigra*), 18% hardwood species (*Fraxinus excelsior, Juglans regia*) and 25% conifers (*Abies, Picea*, and *Pinus* spp.). Most of the large wood contributions comprised of hardwood species were probably from the local riparian area, but coniferous wood fragments traveled farther, originating in the hillslopes lining the river basin (Moulin 1999).

Observations on piedmont rivers such as the Ain, the Loire, the Drôme and the Eygues in France, but also the Willamette upstream from Corvallis, Oregon, indicate that most of the large pieces of wood deposited in the active channel originate from the riparian corridor. Similarly, findings documented along a lowland river, the Thompson River, Australia (Gippel et al. 1994) and along several U.S. coastal plain rivers, such as the Roanoke and the Lumber rivers in North Carolina, have linked large wood found in the river channels with the surrounding riparian tree populations (Figure 2). Wood inputs vary longitudinally from upstream to downstream reaches, reflecting changes in abrasion, biological decomposition, and total volumes of wood introduced along the length of the river. As expected for sediment during its transport, a gradual disintegration of wood, or "fining" of wood occurs as it makes its way downstream. Change in organic matter from upstream to downstream reaches is one of the key components of the River Continuum Concept (Vannote et al. 1980). It is based on the assumption that invertebrates in upstream reaches play a principal role in breaking down organic material into finer pieces. Physical processes also contribute to the gradual disintegration of wood during transport, such as during movement across boulders and cobbles in shallow, fast-flowing water.

In montane river channels, physical breakdown is substantial because of high stream velocities, and abrasion against boulders and shallow channels which fragments wood. High incidence of branchless trunks with discontinuous bark is therefore expected. Thus, these systems supply mostly small pieces of wood to downstream reaches (Figure 3). Analysis of the condition of the wood confirms this. Wood fragments in upstream reaches are more often missing roots, branches, and bark than wood fragments in downstream reaches. Such clear contrasts have been observed between an intra-mountain river (Ubaye River, France) and two piedmont rivers (Ain and Drôme rivers, France) (Piégay and Gurnell 1997), the former exporting larger volumes of finer pieces of wood. In the Genissiat Reservoir along the Rhône River, wood fragments greater than 20 cm in diameter were evaluated, and 86% had no roots, 82% had no branches, and 77% had no bark (Moulin 1999). A survey completed on wood deposited in the channel of the Asusa River in 1999 recorded 86% of all trunks without branches, although—in contrast to ob-

FIGURE 2. Most of the wood fragments stored in the channel of the Lumber River, a blackwater river in North Carolina, originate from the riparian zone; many of the fragments retain their branches and bark.

FIGURE 3. Abrasive effects of shallow streambeds and high velocity flows on wood in upland rivers (Waipo River, New Zealand).

servations along the Rhône—45% of wood fragments still had roots and 55% were found with their bark intact. These differences can be explained by differences in the sampling strategies of the two studies. The sampling site of the Rhône River was 20–40 km downstream of where the wood was introduced into the river, but the study site along the Asusa River was in the active channel, very close to the sources of the wood.

Differences in wood production among large rivers

Large river basins differ greatly in floodplain features, human population densities, and wood delivery (Figure 4). As a result, size, species, and volumes of wood produced differ greatly among rivers. Nevertheless, if unmanaged wooded corridors are compared, they can be ranked according to their capacity to produce wood. One of the major processes is bank erosion, which controls the number and size of wood and wood fragments introduced to the channel and, consequently, the total amount of wood. Freely meandering channels produce larger pieces of wood than braided channels, which is explained by bank erosion in the concave banks of meander bends with mature forest (Citterio 1996). Along braided channels, banks erode in a band parallel to the main channel, in zones frequently rejuvenated,

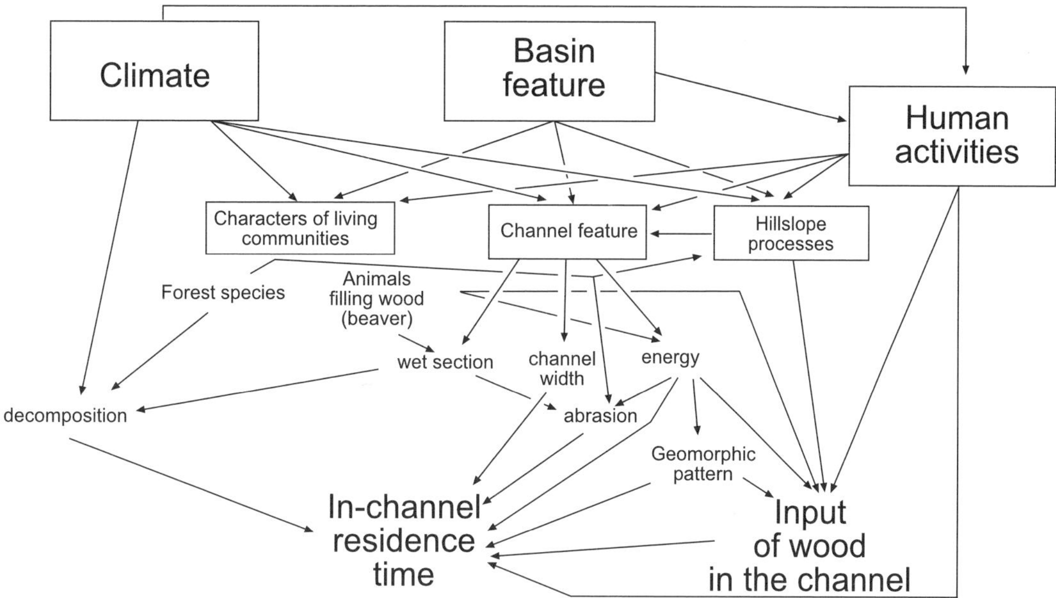

FIGURE 4. Conceptual model of the factors controlling wood input and residence time in large rivers.

which results in higher turnover rates of woody vegetation. Standing riparian tree biomass was evaluated relative to their successional stages and erosion rate along two contrasting river floodplains, the Ain, a meandering river, and the Drôme, a braided river. The study supports the difference in wood input according to river type (Citterio 1996). The meandering river averaged a bank erosion rate of 0.24 ha/year per kilometer of river length (1983–1991), with a correlative wood input rate of 39 t/year/km, and the braided channel averaged a bank erosion rate of 0.15 ha/year per kilometer of river length (1983–1991), with a correlative wood input rate of 12 t/year/km. The slightly higher bank erosion rate on the Ain River does not in itself explain differences in wood input. More significant is that 50% of the eroded floodplain along the Ain River was characterized by mature forest units compared with only 10% along the Drôme River.

A conceptual model summarizing variations in wood size, residence time, and volume of input along the river continuum relative to channel size and geomorphic pattern has been established based on the research just described (Figure 5). In mountain streams, where the size ratio S_r is higher than 1, wood tends to be stable in the channel, and residence times are therefore relatively long. Wood size and input are also high in these reaches, and related to the characteristics of the trees lining mountain streams. As the channel becomes wider downstream, the wood gradually disintegrates and is trapped along its course. The trend of increasing fragmentation and abrasion of wood as it moves downstream is counteracted by large amount of wood inputs at junctions with smaller tributaries. In mountain valleys and piedmont areas, wood is added through bank erosion and mostly exported downstream ($S_r > 1$). This local contribution of wood is significant and increases where channel pattern evolves from a braided to a meandering system. Along lowland rivers, bank erosion and flow velocity are very slow, explaining the strong decrease in the volume of wood input and its residence times, both contributing to high amounts of wood storage in the channel. Wood fragments here are also fairly large because they originate principally from mature trees established along the edge of the riverbanks.

Events that contribute wood

With the exception of wind, most of the factors that add wood to rivers are a function of duration and magnitude of floods (bank erosion, clearing of the floodplain, upstream input rates). Extreme weather, such as a hurricane, is important in lowland rivers of the eastern United States, where the residence time of wood is long, annual inputs of wood are low, and flow velocity during floods is not always sufficient to erode banks and intro-

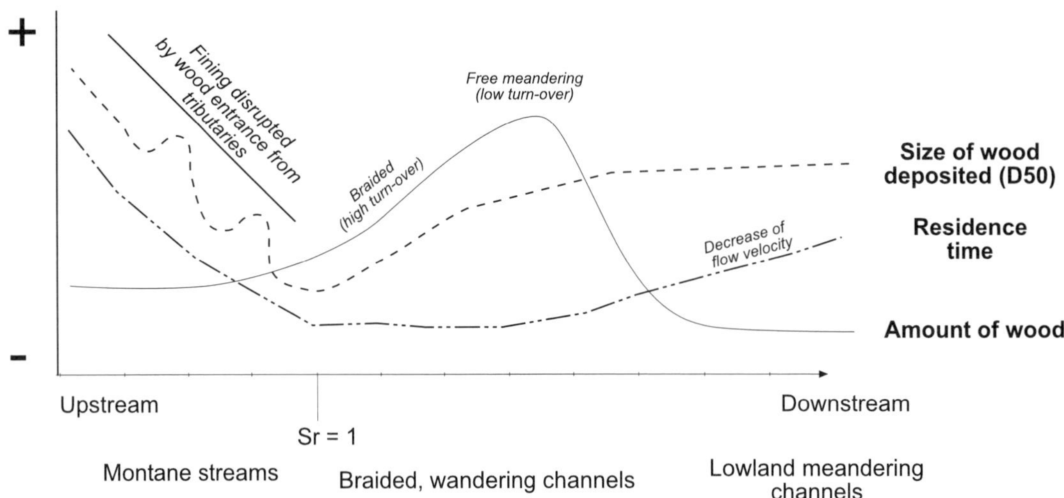

FIGURE 5. Theoretical changes in volumes of wood input, residence times, and median size of wood fragments from upland to lowland reaches.

duce wood into the channel. For example, I observed a large amount of wood in spring 2000 along the Roanoke River, North Carolina, 10 km upstream from the ocean. The source of these accumulations was associated with the toppling of trees on terrace slopes during a hurricane that occurred just a few months earlier.

Sedell et al. (1988) hypothesized that wood input and output are irregular over space and time. At the Genissiat Reservoir, the volume (V) of wood inputs (m^3) over a given period is quantitatively related to peak flows over that period (m^3/s) (Moulin and Piégay, 2003):

$$V = 7.41\ Q - 3459,$$
$$\text{where } r^2 = 0.71,\ n = 18, \text{ and } p < 0.0001$$

Daily rates of discharge and monthly volumes of wood extracted have also been compared at the Beauvoir Dam along the Isère River, France (unpublished data provided by Electricité de France). This example demonstrates how the timing of a given flood event plays an important role in explaining the volume of wood extracted (Figure 6). Two lesser winter peak flows (730 and 770 m^3/s) imported high volumes of large wood (5,000 and 8,500 m^3) along the Isère River. However, a peak of 1000 m^3/s, only 3 months later introduced less than 5,000 m^3 of large wood. The earlier events enlarged the channel and displaced riparian vegetation, leaving relatively little material to be displaced during the second flood event. The seasonality of flooding events is also important. During the growing season, floods can wash away floodplain undergrowth, which may explain why small accumulations of large wood are so abundant after summer floods compared with those in winter.

Most wood is delivered and transported during floods, but periods of low flow may also contribute some wood (for example, see May or August in Figure 6). Other factors contribute to wood input in addition to floods. Monthly volumes of wood extracted at the Saint-Egrève Dam, along the Isère River (France), between 1994 and 1998, reveal inter-annual and seasonal variations in wood output (Figure 7). Over the study period, wood volumes removed from the reservoir ranged from 896 m^3 in 1997 to 6,144 m^3 in 1994 and from 561 m^3 per month during periods of high flow (May to July) to 70 m^3 per month in March–April and 99 m^3 in August–September, when discharge was low. Wood transport can occur at high rates over a long period. For example, in 1994–1995, at least 100 m^3 were extracted each month from September 1994 to July 1995 (unpublished data provided by Electricité de France).

Because critical transport and delivery events, such as floods, are not distributed evenly in space and time, we can expect discontinuous rates of wood being generated in downstream reaches of the main river channel, but also variable in-channel volumes of wood through time. Wood delivery rates along riverbanks are difficult to predict. River basins act as complex reservoirs of wood with various roughness characters and transfer

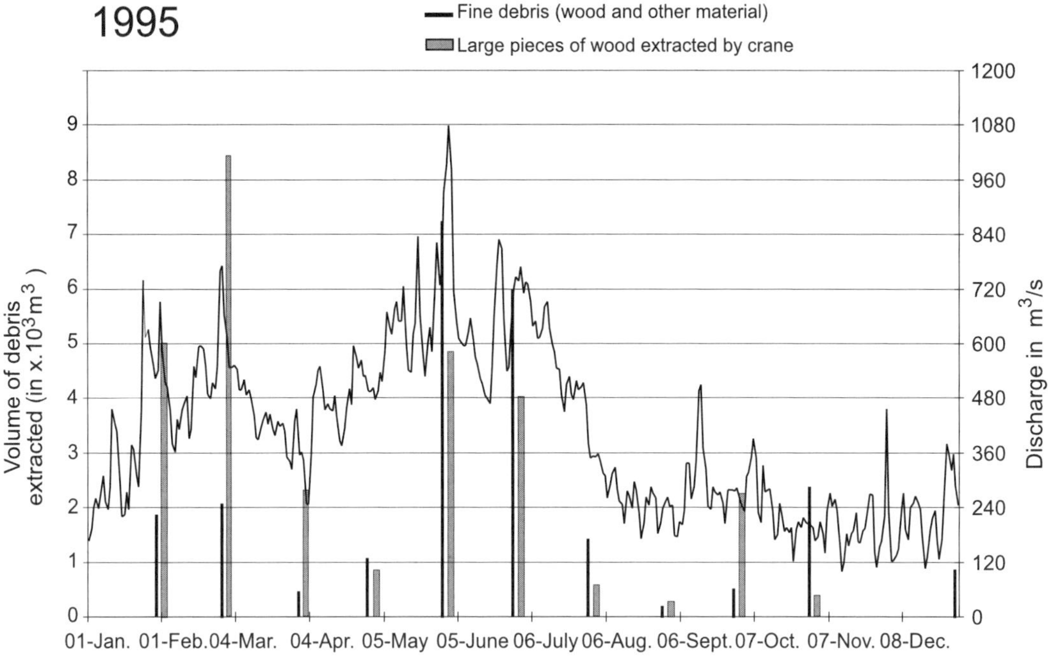

FIGURE 6. Rate of discharge and volume of wood extracted monthly in 1995 at the Beauvoir Reservoir along the Isère River near Grenoble, France (source: Electricité de France).

distances between sites of wood input and the study sites where wood "discharge" is recorded. Patterns of river discharge can be represented by complex flood hydrographs adjusted for basin area, retention capacity of the river basin, and the duration and intensity of flooding events. Similarly, researchers can develop graphs of wood discharge through time to reflect flood patterns. For example, a single peak in wood output is likely to be linked to a single flood event, and two peaks in wood output might be linked to storms in different sections of the river basin, one being closer than the other to the site where wood discharge is recorded. On the other hand, a long smooth peak in wood output can typically be linked to flooding after a long and widespread rainfall event occurring throughout the river basin. Further research is needed to reveal discharge thresholds above which wood transport increases abruptly and the relationship between the wood discharge curve and the water discharge curve. We should expect that the wood transport peak occurs later than, or simultaneously with, the flood peak (in contrast to the peaks of solute concentration or suspended sediments, which are sooner) because the wood input in large rivers is mainly derived from bank erosion. Wood transport then decreases slowly because of multiple roughness factors all along the channel. Moreover, the timing and interval length between critical events merits further study because the amounts of wood generated are not only associated with the magnitude of a flood event, but also with the timing of a flood in relation to other critical events such as season (for example, growing season versus dormant period).

Volumes of wood deposited in the active channel are spatially variable. Quantities of wood generally decline with increasing distance from their sites of origin. In upstream braided reaches of the Drôme River, more wood is trapped in the channel than in downstream reaches because of the disintegration of wood as it travels downstream (Piégay et al. 1999). This process helps to explain why isolated wood fragments in downstream reaches are smoother and more abraded (highly fractured root systems, fewer branches, smaller average sizes). The decline in wood accumulations originating from upstream reaches in channels is also associated with fewer mobile wood fragments downstream compared to upstream. Wood deposited in the active channel also decreases after a critical event because gradual decomposition and fragmentation of individual pieces reduces resistance to flow. Wood additionally will become remobilized by lower magnitude flooding events

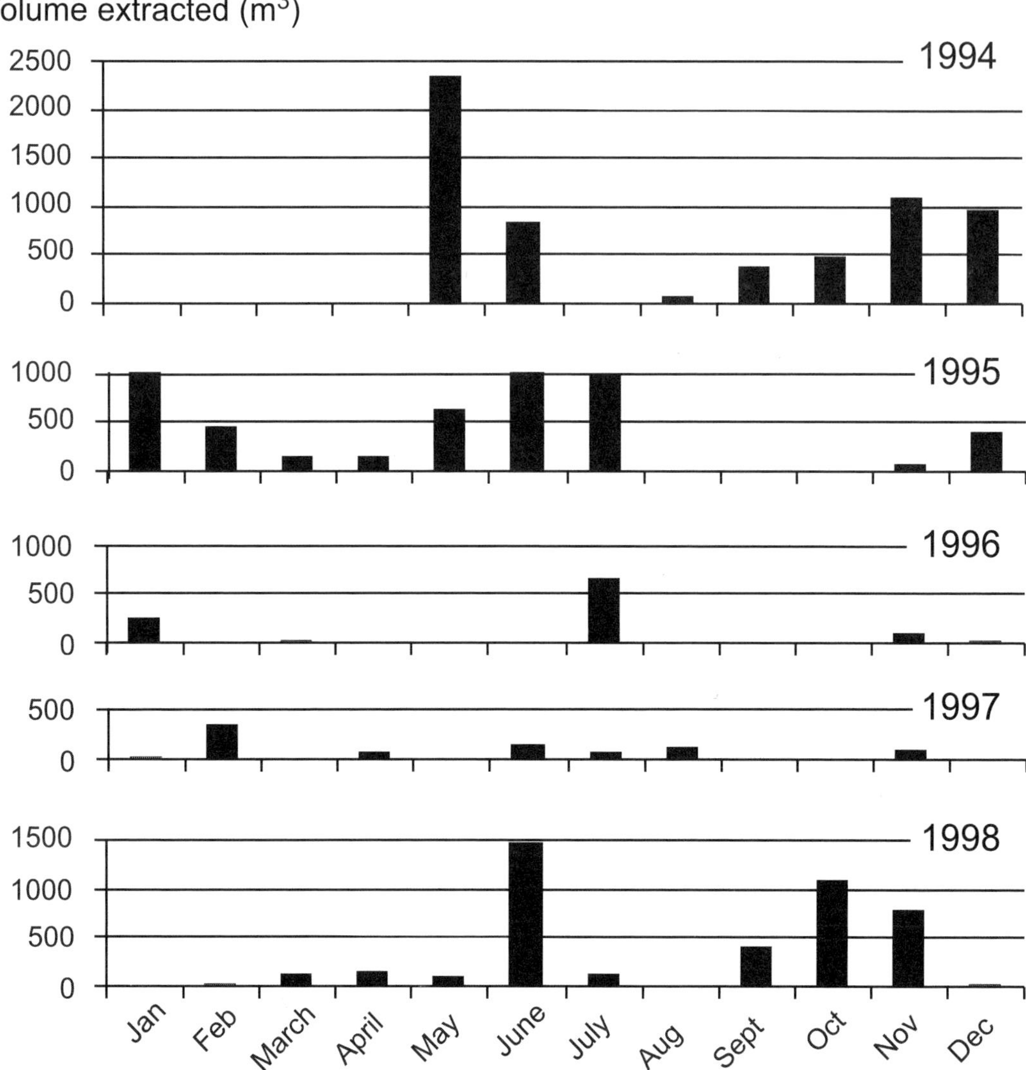

FIGURE 7. Annual and seasonal variation in quantities of large wood extracted at the Saint Egrève Dam along the Isère River near Grenoble, France from 1994 to 1998 (source: Electricité de France).

after a critical event because any wood deposited on bar surfaces during the falling limb of a flood hydrograph can be moved in these lesser flows. Delivery and transport can also leave vulnerable accumulations, easily moved at lower discharge events.

Interactions between Large Wood and Channel Form

Influences of large wood on channel form in large rivers are complex. The effects can be observed at the site scale over a few square meters or at the reach scale over several kilometers. Moreover, the geomorphic effects of wood depend not only on the characteristics of the wood, but also on the local geomorphologic structure of the channel, which also dictates patterns of wood deposition.

Sites of deposition

Although large wood is distributed stochastically in upper sections of large rivers, where wood can move easily, the distribution of wood is strongly correlated with certain key roughness structures (Piégay 1993; Piégay et al. 1999). As a result, wood deposits can be found at predictable locations, such as bar surfaces where they define the strand line (Figure 8), points in the river where flow is diverted, concave banks, entrances to overbank flow channels, upstream tips of vegetated islands, or against bridges. On bar formations, most accumulations are formed by entrapment in shrubs or trees that are inundated during floods. Some stable pieces of large wood trap other wood and create wood accumulations perpendicular to the channel. Wood blocked against structures tends to form large accumulations instead of isolated fragments (Figure 9).

Distribution patterns of wood differ by channel type. Wood in braided channels tends to be deposited on bars. Along meandering and anastomosing channels, wood is frequently deposited on concave banks. Wood is located primarily at the channel margins or along the edges of the floodplain on islands, along concave banks, or in side-channels. On the Tagliamento River, Italy, limited quantities of large wood are deposited on unvegetated gravel surfaces (1–21 t/ha, compared with tips of pioneer islands, which trap 293–1,664 t/ha, (Gurnell et al. 2000b). As the number of vegetated islands increases going downstream, the volume of large wood and encounter rates of wood accumulations increase. Along lowland rivers, flow velocity is less efficient at dislodging and transporting wood; wood therefore is more randomly distributed along riverbanks in these rivers and often is still anchored to the bank, confirming the local origin of many wood fragments (Gippel et al. 1994). Where wood occurs in these rivers, isolated fragments are more frequent than wood accumulations. On the Thompson River in Australia, 83% of the wood fragments found were oriented in the direction of river flow with the rootwad upstream, suggesting that they originated through transportation during a flood.

FIGURE 8. Large wood deposits along the Loire River, France. Pieces are isolated from one another but define the strand line of flood flows in the river landscape.

FIGURE 9. Large wood deposited along a concave bank of the Ain River, France; wood fragments accumulate in discontinuous lines along the bank.

Many authors consider that concave banks are the most common sites for wood deposition (Keller and Swanson 1979; Hickin 1984). However, each meander bend differs in terms of its bank height, location and number of overbank flow channels, and degree of sinuosity (Piégay and Marston 1998), all of which affect rates of wood deposition.

Trapping efficiency

Some proportion of the wood in a reach is always mobile, thus the amount of wood trapped in the channel is not an accurate indication of the total amounts of wood introduced into the river. In large rivers, we might expect that most of the wood is transported and only a small proportion is stored in the channel.

In montane and peri-montane environments, the amount of wood stored in the channel is usually low compared to the amounts introduced annually. In braided channels of the Drôme River, France, we quantified the amounts of wood stored in the channel and generated from the surrounding floodplain to evaluate the trapping efficiency of the system (Piégay et al. 1999). The amount of wood found in the channel corresponded to 1.3–3.1 times the amount of wood produced annually, most of which was trapped on bars. Only 0.12–0.32 times the amount of wood produced annually (0.7–1.9 t/km) was stored in the low flow channels and was potentially accessible to aquatic communities at low flow. A 4-year study (1992–1995) in a 4,000-m^2 floodplain plot in a concave bank of the Ain River showed that wood deposits moved each year, indicating that the storage capacity of the site is low. The amount of wood produced by bank erosion was also greater than the amount of wood stored from one year to the next. This finding suggests that wood deposition on the concave bank is not a cumulative process and that most of the wood deposits are derived instead from annual inputs (Figure 10A). Moreover, the line of wood migrated as the sinuosity rate was modified with a change in the orientation of high velocity flows. From 1992 to 1995, as the concave bank shifted by 10–20 m per year, the orientation of wood accumulations in the sampling plot changed direction, moving E–W to SW–NE, a response to changes in the orientation of overbank flows (Figure 10B).

Mobile wood fragments vary in distances they travel downstream and transfer rates. Some wood fragments are not exported out of the reach from

A.

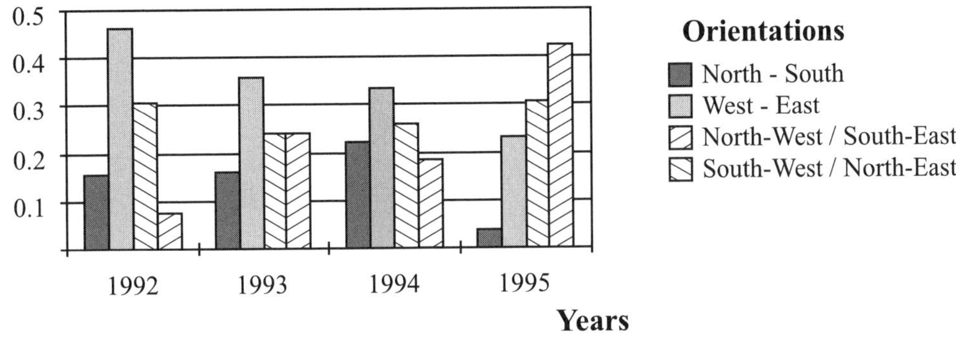

Years

B.

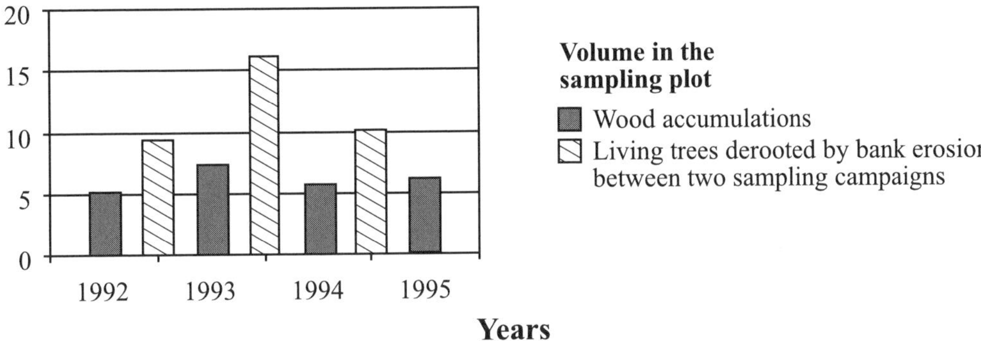

FIGURE 10. Evolution of (B) volumes of annual wood accumulations and wood mass of standing trees that fell down, and (A) wood accumulations according to their orientation, between 1992 and 1995 in a 4,000 m^2 sampling plot in the concave bank of Mollon (Ain River, France) (Piégay et al. 1998).

which they originate, but others are never trapped. Some authors described possible re-exposure of ancient wood buried in floodplain sediments hundreds or thousands of years ago (Nanson et al. 1995; Bilby and Bisson 1997; Hyatt and Naiman 2001). Moreover, most of the research, notably in lowland rivers, has focused on wood fragments that are visible in the channel during low flow. Little is known about the characteristics and geomorphic role of large wood below the water surface. In North America, during the active period of harvesting and log floating in the 19th century, many logs sank (called *"sinkers"*) initially to the bottom of the channel soon after entry into the river, where they may now affect transport along the riverbed, geomorphic features, and the sensitivity of the riverbed to degradation (C. R. Hupp, USGS, personal communication). Although wood accumulations change in time and space and some wood always moves downstream during each flood, different generations or cohorts of wood (sets of pieces introduced at the same date during a given flood event) are nevertheless always stored in river channels. On bar formations along the Asusa River, two generations of wood pieces dominated the river landscape. One of them was in an advanced stage of disintegration (decomposed wood, branchless trees without roots), and the other was evidently more recent (trees with branches, leaves). Both generations of wood had similar species composition, suggesting that two distinct events contributed to the deposition of wood in the study reach, one from only one year before and the other some time earlier.

Along the Loire River, the homogenous and advanced stages of decomposition among wood fragments imply that they originated simultaneously during a single flooding event 6–10 years earlier (Figure 8). The temporal variability in the wood might be expected to reflect changes in morphological and biological features over time (Figure 4). More attention should be focused on temporal variations in large wood input rates, both as a research area and as a context for specific research projects. A better understanding of the complex changes in wood over time in large rivers will contribute to our knowledge of the dynamics of aquatic ecosystems.

Effects of wood on floodplain landforms

Several authors (Piégay 1993; Abbe and Montgomery 1996) have described local effects of large wood accumulations on fluvial landforms, both in the channel and on the floodplain. Modification of flow velocity creates associated wood-formed pools and affects the distribution of sediment in the channel and on the floodplain. If the scale of geomorphic features (a few square meters) has been well studied, more controversial questions about interactions between large wood and river geomorphology are posed at the corridor scale. For example, does large wood contribute to floodplain development and the encroachment of vegetation onto the floodplain? Does large wood slow meander migration rates?

Effects of wood on island formation differ from one river to the next and vary according to the size of the large wood. Wood accumulations are roughness elements in the riverine landscape that promote forest establishment and development of floodplains and islands (Swanson and Lienkaemper 1982; Fetherston et al. 1995). In montane gravel-bed rivers, large wood creates sites for vegetation colonization and initiates developing forested islands (Swanson et al. 1998). Large wood also promotes formation of mid-channel bars and a corresponding increase in channel width. The morphological effects of these bars can be significant, especially where they are large compared to the channel width and have long residence times. In the Pacific Northwest, plant propagules colonize the active channel and floodplain in spring and early summer as discharge gradually subsides. Some wood fragments act as nurse logs during this period by providing woody species such as conifer seedlings with ideal microsites for regeneration (Harmon et al. 1986). Other species can also sprout from broken stems, such as cottonwood and willow. On the Tagliamento River, Gurnell et al. (2000b) noted high proportions of living wood caught in wood accumulations, mostly in downstream reaches, where conditions favor the survival and regrowth of live woody propagules. These reaches have relatively homogenous topography, and finer sediment grain sizes improve the water retention capacity of deposits and therefore encourage adventitious rooting.

Vegetation establishment associated with large wood is thought to be relatively infrequent on French piedmont rivers and apparently also on other rivers, such as the Willamette River, Oregon and the Asusa River, Japan. On the Drôme River, vegetated islands are not always characterized by wood accumulations at the upstream tips, and many wood accumulations show little evidence of vegetation establishment below them. Moreover, though we saw several live uprooted trees lying on bars and sprouting fresh shoots, they mostly failed to survive into a second growing season (Figure 11). On the Drôme, wood accumulations are important micro-sites for regenerating *Populus nigra* from seed at higher, more protected sections of the floodplain, thus producing higher survival rates (Barsoum 2001). However, the large wood is so unstable that it does not allow a long-term environment for establishing vegetation and forming islands.

The influence of large wood on tree establishment and floodplain development is far from clear. We can hypothesize that stand establishment depends on wood size and the timing of critical flood events related to the growing season. Survivorship of uprooted plants is lower than that of seedlings, and rates of regrowth are poor, which may explain why wood accumulations in the Drôme River do not lead to forming vegetated islands in the active channel but are instead trapped by previously established shrubs.

Several authors have posed the following question: "Does wood favor meander cut-offs, or reduce their frequency?" Keller and Swanson (1979) observed that wood can block the main channel, diverting river flow in the inner section of a meander bend, and favoring chute cut-off. This reference is often cited in the literature (Gurnell et al. 1995; Montgomery and Piégay 2003) as an example of large river processes, but it happens only where wood fragments are

FIGURE 11. Isolated cottonwood tree deposited on a bar of the Willamette River, Oregon, USA. New shoots have emerged from the top of the trunk.

shorter than channel width or where the volume of wood is sufficient to affect hydraulic conditions in an entire reach. Hickin (1984) described the function of a line of wood along a concave meander bank as a barrier preventing flooding onto the floodplain. Contrasting results were documented along the Ain River, where the line of wood in a concave bank was discontinuous, concentrating overbank flows along a particular axis where floodplain soil was eroded and overbank flow channels were found. In this context, wood does not accelerate meander cut-offs. Because of a change in channel sinuosity, positions of previous overbank flow channels change before floodplain soil erosion is sufficiently advanced to result in channel chute cut-off. Moreover, roots of standing trees in the floodplain armor the surfaces and slow soil erosion. Meander cut-offs in forested river corridors evolve over a longer time than in open corridors because of the greater floodplain soil protection afforded by lines of large wood on inner banks and by live tree roots. Frequency of channel cut-off in the lower Ain River changed in response to land-use changes on the floodplain. Before 1950, when the floodplain was grazed by cattle, chute cut-offs were found at sinuosity rates of approximately 2.5–3. The present sinuosity rate of 4.5 has still not provoked chute cut-off in the Ain corridor, which is now forested and has well-developed wood accumulations along concave banks (Figure 12).

Ecology, River Management, and Wood

Fish and wood

From mountains to coastal plains, large wood plays important ecological roles, even where the river system has been modified (Zalewski et al. 2003, this volume). Bisson et al. (1987) demonstrated that channel widening and bar formation associated with wood obstructions allow development of the short, braided reaches and secondary channels that are important spawning grounds for salmon and trout in the rivers of the Pacific Northwest. In the sand-bed coastal plain rivers of the southeastern United States, wood also provides important habitats for invertebrates and, thus,

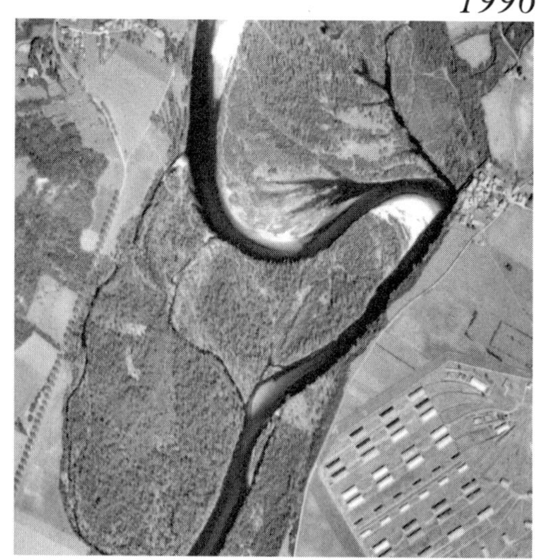

FIGURE 12. Two examples of meander features on the Ain River: the Planet meander, cut-off in 1954 through a grazed floodplain (sinuosity rate: 2.8) and the Mollon meander, still not cut off in 1996 because of afforestation and the high resistance of floodplain soils afforded by a dense network of roots (sinuosity rate: 4.5)(source: Institut Géographique National, France).

indirectly provides certain fish species with a source of food (Wallace and Benke 1984). On the Satilla River, Benke et al. (1985) noted that, in spite of wide habitat availability (snags representing only 4% of the total available area compared with 80% of the area comprising sandy habitat), snags contained 60% of total invertebrate biomass per meter along the river.

In large river systems with limited volumes of wood, fish abundance and diversity are also influenced by the structure of wood accumulations. Research on large French rivers has shown that fish abundance and diversity depend on accumulations of large wood as witnessed along the Drôme River (a braided river system) but also along the Loire and Rhône rivers, which have deeper channels and shallower gradients than the Drôme (Thévenet 1998). The reaches studied along these rivers were mostly occupied by cyprinids: 80% of fish captured are *Leuciscus cephalus* European chub, gudgeon *Gobio gobio,* and Eurasian minnow *Phoxinus phoxinus.* The Drôme River is characterized by heterogeneous lotic habitats with a rheophilous fish assemblage, but the Loire comprises a homogeneous network of lentic habitats in a degraded channel, influenced by marl outcrops (Thévenet 1998).

Sites with large wood have a more diverse fish community, with 1.3–2 times more species than sites without wood. Areas with wood have also been found to have 1.6–7 times more fish, both in the fall and in winter (Thévenet 1998). These results are confirmed by similar findings from the Mississippi where 2–50 times more fish were observed in habitats with large wood compared to areas without large wood (Lehtinen et al. 1997). Moreover, abundances of most of the common species (*Leucistus cephalus, Phoxinus phoxinus,* or spirlin *Alburnoides bipunctatus*) were greater with larger wood accumulations, with the exception of barbel *Barbus barbus, Gobio gobio,* bitterling *Rhodeus sericeus*, and stone loach *Barbatula barbatula* (Figure 13). Although large wood cover may benefit the species cited above, the small species like *Gobio gobio, Rhodeus sericeus,* or *Barbatula barbatula* become vulnerable to several predators such as trout and chub. As a result, their numbers are not directly influenced by wood cover. *Barbus barbus* is slightly different, however. This species only uses wood when flow velocity is high, and its abundance therefore is not influenced by the amount of large wood. Fish-tracking experiments on adult chub in the upper Rhône (Allouche et al. 1999) determined that the presence of protective structures along banks was a key factor in the fish distributions. In the summer, fish spent most of their time near large wood or boulders provided by rip-rap bank protections. In winter, chub assembled exclusively around clusters of boulders. Because fish reduce their activities and movement during winter, fish favor habitats closer to feeding areas (for example, areas of high flow velocity), which is more often true near boulders than near wood. The position of large wood in a channel reach potentially explains fish distribution apart from simply size and structure of the large roughness features.

Risks associated with wood

The volumes of wood in managed rivers are usually sufficiently small compared to the channel cross-section areas, such that the wood generally does not increase the risk of flooding. In France, river managers increasingly allow wood accumulations to remain in large rivers as a measure for mitigating flood risks, particularly when no infrastructures sensitive to wood build-ups and associated flooding are present. Maintenance efforts are concentrated instead on smaller streams, where wood accumulations have become more frequent since the 1950s because of widespread afforestation.

Several activities on large rivers may be disrupted by wood movement, such as commercial navigation (the Fraser River, British Columbia) and leisure activities (such as canoeing along the Snake River in Grand Teton National Park, USA). Wood clearing costs must also be considered in fluvial systems: the main channel of a river receives wood from the rest of the river network, and this material invariably must be cleared near reservoirs, dams, bridges, and weirs. Agencies in charge of maintaining roads and bridges are obliged to spend funds to keep rivers clear, at least in the vicinity of sensitive infrastructure. Hydroelectric companies, such as Electricité de France, also clear the river channel periodically to protect their dams. Wood is usually mixed with garbage and different measures are needed to treat this material. At the Génissiat Dam on the Rhône River, annual expenditure for extracting and treating trapped material between 1989 and 1998 was about $65,000. In this context, several questions can be raised about managing wood for environmental purposes: must all of the material be extracted from reservoirs or only the garbage? Can wood be allowed to move downstream, and

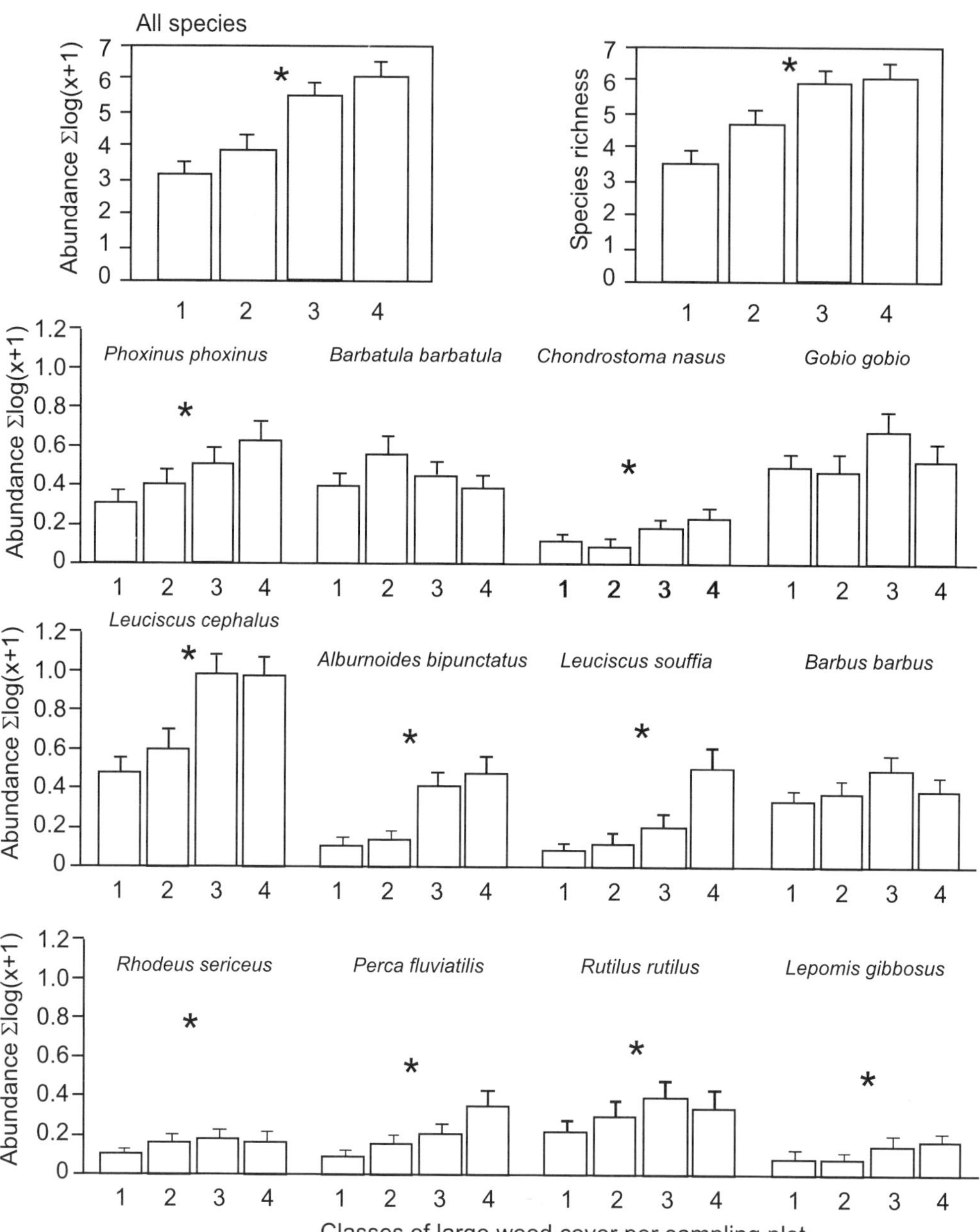

Large wood cover (in % of sampling plot area) :
1 : 0 ; 2 :]0-5] ; 3 :]5-15] ; 4 : >15
Mean +/- 1 standard error (n = 251)
Classes of equal population
* : significant effect of large wood cover on fish abundance or richness (Kruskal-Wallis test, p<0.01)

FIGURE 13. Fish species richness and abundance relative to degrees of wood cover in the fishing plots (Thévenet 1998).

would such a decision increase the risk of downstream flooding?

Uses of wood for improved fish habitat

As described by Sedell and Luchessa (1982) and Triska (1984), wood is removed from large rivers to improve conditions for navigation and to improve drainage of the adjacent floodplain to develop agriculture. Because wood provides habitat for fish, however, they would benefit if at least some wood was allowed to accumulate in large rivers, especially where wood is not abundant. Approaches for finding a balance between environmental needs, river stakeholder needs, and risk management are needed.

A first attempt at sustainable large wood management has been proposed for the Drôme River (Piégay and Landon 1997), and a wood management plan has been implemented. Along embanked reaches where numerous sensitive structures exist (such as bridges with narrow arches), wood accumulations are removed from the channel annually. In other reaches less sensitive to flooding and infrastructure damage, either all of the wood jams are retained or only those that create fish habitat, and again, depending on the sensitivity of the reach to wood movement (Figure 14).

Conclusions

Wood in large rivers is a critical research area of ecological, geomorphic, and social importance. More research is needed for a better understanding of the influences of large wood on geomorphic processes at different temporal and spatial scales, as well as its role in both aquatic and floodplain ecosystems. Although much attention has been focused on upland temperate rivers over the past two decades, more research is needed on lowland rivers, notably on wood input processes and residence times. Equally needed are studies of wood functions in biological and physical processes of large tropical rivers, a topic little is known about.

Integration of wood dynamics into river management plans remains a challenge. Applied research will be necessary to improve evaluation of river networks and the human infrastructures (bridges, weirs) vulnerable to damage by mobile wood (Diehl 1997; Wallerstein 1999). Further research into how infrastructure is damaged and possible solutions to reduce the likelihood of damage will also be useful. Engineers could provide solutions in the form of structures designed to trap wood during floods upstream of urban reaches or for designing infrastructures less affected by wood accumulations. A few of these experiments have been initiated in Germany, Switzerland, Japan, Canada, and the United States. Effectiveness of trash racks, as described by Diehl (1997), and iron or concrete pins inserted in the channel or the floodplain floor perpendicular to the direction of flow need to be tested, and more research must be conducted to improve their design, identify the best locations to trap wood, and mitigate effects to the infrastructure and river users (Hartlieb and Bezzola 2000; Bezzola 2001).

Most research in the 1980s was focused on habitat-related questions at the local scale, linking hydraulic conditions with aspects of the trophic chain, fish diversity, and geomorphology. Now, a substantial investment in research is needed to better understand wood mobility in the river network and possible damage it may induce downstream. Critical periods, critical discharges, and the duration of the events during which wood is transported must be identified by using "observation windows," such as reservoirs which store all of the wood generated, even during floods. Such sources of information can tell us about amounts of wood moving into a river but also the relationships between wood and the frequency, geographic origin, and magnitude of flood events.

From an environmental perspective, it is important to resolve how and where wood-formed habitats might be maintained. Because of the ephemeral nature of wood in some reaches, the best option is to sustain the processes that transport and store wood throughout the river network (including bank erosion), allowing each flood event to destroy some wood accumulations but at the same time create others. This process will create some conflicts among users (such as increased risk of erosion and flooding in urban reaches for example). In the same context, applied research is also needed to evaluate interactions between fish and wood and to better understand how the most effective wood accumulations might be maintained while minimizing the aforementioned risks to managed lands. For example, wood accumulations close to feeding habitats (sections of the channel with high flow velocities) may provide a more beneficial habitat than wood farther from a food supply. In rivers where channels are

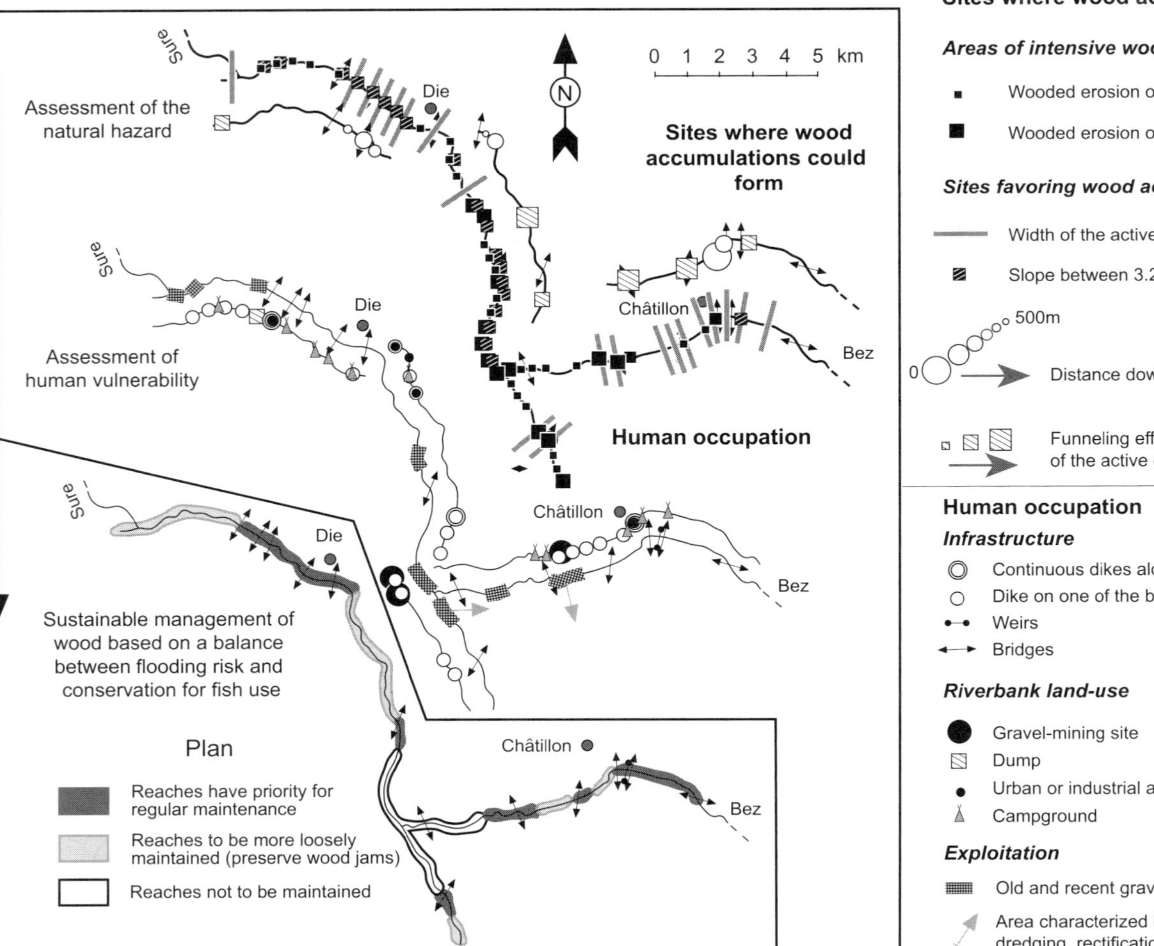

FIGURE 14. Finding a balance between conserving in-channel wood and flood risk management: example of mapping along the Drôme River (France) to define an intervention strategy for managing wood accumulations (Piégay and Landon 1997).

relatively stable and where wood poses local risks or has potential effects downstream, wood transport is unlikely to be acceptable. Thus, research is needed to propose restoration measures (such as permanent structures attractive to fish both in terms of their architecture and location).

Acknowledgments

I gratefully acknowledge A. Citterio, B. Moulin, and A. Thévenet, who have contributed to the large wood research in the Rhône River basin over the past 10 years. My sincere thanks also go to T. Oguchi, H. Shimazu, and C. Hupp, who broadened my experience on the relation between wood and river dynamics both in upland and lowland rivers, and S. Allouche for useful discussions about fish behavior in wood accumulations. C. Bracq from Electricité de France provided the Isère River data. Most of the research on French rivers was supported by the Région Rhône-Alpes (XI° Contrat de Plan Etat – Région). I also acknowledge N. Barsoum for revising the English and improving the text as well as S. V. Gregory, K. Boyer, M. H. Brookes, and anonymous reviewers for their advisable comments.

References

Abbe, T. B., and D. R. Montgomery. 1996. Large woody debris jams, channel hydraulics and habitat formation in large rivers. Regulated Rivers: Research and Management 12:201–221.

Allouche, S., A. Thévenet, and P. Gaudin. 1999. Habitat use by chub (*Leuciscus cephalus* L. 1766) in a large river, the French upper Rhône, as determined by radiotelemetry. Archives für Hydrobiologie 145(2):219–236.

Barsoum, N. 2001. Relative contributions of sexual and asexual regeneration strategies in *Populus nigra* and *Salix alba* during the first years of establishment on a braided gravel bed river. Evolutionary Ecology 15:4–6.

Benke, A. C., R. L. Henry, D. M. Gillespie, and R. J. Hunter. 1985. Importance of snag habitat for animal production in southeastern streams. Fisheries 10(5):8–13.

Bezzola, G. R. 2001. Schwemmholz-Rückhalt oder Weiterleitung? Wasser, Energie, Luft 9–10:247–252.

Bilby, R. E., and P. A. Bisson. 1997. Function and distribution of large woody debris in Pacific coastal streams and rivers. Pages 324–346 *in* R. J. Naiman and R. E. Bilby, editors. River ecology and management lessons from the Pacific coastal ecoregion. Springer, New York.

Bisson, P. A., R. E. Bilby, M. D. Bryant, C. A. Dolloff, G. B. Grette, R. A. House, M. L. Murphy, K. V. Kosky, and J. R. Sedell. 1987. Large woody debris in forested streams in the Pacific Northwest: past, present and future. Pages 143–190 *in* E. O. Salo and T. W. Cundy. Streamside management: forestry and fishery interactions. College of Forest Resources, University of Washington, Seattle.

Citterio, A. 1996. Dynamique de dépôts et de prise en charge du bois mort sur deux hydrosystèmes. Mémoire de maîtrise, University Lyon 3, Lyon, France.

Diehl, T. H. 1997. Potential drift accumulation at bridges. U.S. Department of Transportation, Federal Highway Administration Research and Development, Turner-Fairbank Highway Research Center, Publication FHWA-RD-97–028, McLean, Virginia. http://tn.water.usgs.gov/pubs/FHWA-RD-97–028/drfront1.htm

Downs, P. W., and A. Simon. 2001. Fluvial geomorphological analysis of the recruitment of large woody debris in the Yalobusha River network, Central Mississippi, USA. Geomorphology 37(1–2):65–91.

Elosegi, A., J. R., Diez, and J. Pozo. 1999. Abundance, characteristics, and movement of woody debris in four Basque streams. Archives für Hydrobiologie 144(4):455–471.

Fetherston, K. L., R. J. Naiman, and R. E. Bilby. 1995. Large woody debris, physical process, and riparian forest development in mountain river networks of the Pacific Northwest. Geomorphology 13:133–144.

Gippel, C. J., I. C. O'Neill, B. L. Finlayson, and I. Schnatz. 1994. Hydraulic guidelines for the reintroduction and management of large woody debris in degraded lowland rivers. 1st International Symposium on Habitat Hydraulics, Trondheim, Norway.

Gippel, C. J., B. L. Finlayson, and I. O'Neill. 1996. Distribution and hydraulic significance of large woody debris in a lowland Australian river. Hydrobiologia 318:179–194.

Gregory, K. J., A. M. Gurnell, and C. T. Hill. 1985. The permanence of debris dams related to river channel processes. Hydrological Sciences Journal 30(3):371–381.

Gregory, S. V., H. Li, and J. Li. 2002. The conceptual basis for ecological responses to dam removal. Bioscience 52(8):713–723.

Gurnell, A. M., G. E. Petts, and C. T. Hill. 1995. The role of coarse woody debris in forest aquatic habitats: implications for management. Journal of Aquatic Conservation: Marine and Freshwater Ecosystems 5:1–24.

Gurnell, A. M., G. E. Petts, N. Harris, J. V. Ward, K. Tockner, P. J. Edwards, and J. Kollmann. 2000a. Large wood retention in river channels: the case of the Fiume Tagliamento, Italy. Earth Surface Processes and Landforms 25:255–275.

Gurnell, A. M., G. E. Petts, D. M. Hannah, B. P. G. Smith, P. J. Edwards, J. Kollmann, J. V. Ward, and K. Tockner. 2000b. Wood storage within the active zone of a large European gravel-bed river. Geomorphology 34:55–72.

Gurnell, A. M., H. Piégay, F. J. Swanson, and S. V. Gregory. 2002. Large wood and fluvial processes. Freshwater Biology 47:601–619.

Gurnell, A. M. 2003. Wood storage and mobility. Pages 75–91 *in* S. V. Gregory, K. L. Boyer, and A. M. Gurnell, editors. The ecology and management of wood in world rivers. American Fisheries Society, Symposium 37, Bethesda, Maryland.

Harmon, M. E., J. F. Franklin, F. J. Swanson, P. Sollins, S. V. Gregory, J. D. Lattin, N. H. Anderson, S. P. Cline, N. G. Aumen, J. R. Sedell, G. W. Lienkaemper, K. Cromack, and K. W. Cummins. 1986. Ecology of coarse woody debris in temperate ecosystems. Pages 133–302 *in* A. MacFadayen and E. D. Ford, editors. Advances in ecological research 15. Academic Press, London.

Hartlieb, A., and G. R. Bezzola. 2000. Ein Uberblick zur Schwemmholzproblematik. Wasser, Energie, Luft 1/2:1–5.

Harvey, M. D., D. S. Bidenharn, and P. Combs. 1988. Adjustments of Red River following removal of the Great Raft in 1873. EOS Transactions in Geophysical Union 69(18):567.

Heede, B. H. 1972. Influence of a forest on the hydraulic geometry of two mountains streams. Water Resources Bulletin 8:523–530.

Heede, B. H. 1985. Interactions between streamside vegetation and stream dynamics. U.S. Forest Service, General Technical Report RM-120:54–58, Fort Collins, Colorado.

Hickin, E. J. 1984. Vegetation and river channel dynamics. Canadian Geographer 28(2):110–126.

Hyatt, T. L., and R. J. Naiman. 2001. The residence time of large woody debris in the Queets River, Washington, USA. Ecological Applications 11:191–202.

Inoue, M., S. Nakano, and F. Nakamura. 1997. Juvenile masu salmon abundance and stream habitat relationships in northern Japan. Canadian Journal of Fisheries and Aquatic Sciences 54(6):1331–1341.

Keller, E. A., and F. J. Swanson. 1979. Effects of large organic material on channel form and fluvial processes. Earth Surface Processes and Landforms 4:361–380.

Keller, E. A., and A. Macdonald. 1995. River channel change: the role of large woody debris. Pages 217–235 *in* A. M. Gurnell and G. E. Petts, editors. Changing river channels. Wiley, Chichester, UK.

Lehtinen, R. M., N. D. Mundhal, and M. Madelczyck. 1997. Autumn use of woody snags by fishes in backwater and channel border habitats of a large river. Environmental Biology of Fishes 49:7–19.

Liébault, F., and H. Piégay. 2001. Assessment of channel changes due to long-term bedload supply decrease, Roubion River, France. Geomorphology 36(3–4):167–186.

Maser, C., R. F. Tarrant, J. M. Trappe, and J. F. Franklin. 1988. From the forest to the sea: a story of fallen trees. U.S. Forest Service, General Technical Report PNW-GTR 229, Portland, Oregon.

Montgomery, D. R., and H. Piégay. 2003. Wood in rivers: interactions with channel morphology and processes. Geomorphology 51:1–5.

Moulin, B. 1999. Evaluation du transit des débris ligneux à partir de certains barrages EDF. Mémoire de DEA Interface Nature - Société, UMR 5600 CNRS and University of Saint-Etienne, Saint-Etienne, France.

Moulin, B., and H. Piégay. 2003. Characteristics and temporal variability of LWD stored in the reservoir of Genissiat (Rhone): elements for river basin management. River Research and Application 19:1–19.

Nanson, G. C., M. Barbetti, and G. Taylor. 1995. River stabilization due to changing climate and vegetation during the late Quaternary in western Tasmania, Australia. Geomorphology 13:145–158.

Piégay, H. 1993. Nature, mass and preferential sites of coarse woody debris deposits in the lower Ain Valley (Mollon Reach), France. Regulated Rivers Research and Management 8:359–372.

Piégay, H., and A. M. Gurnell. 1997. Coarse woody debris and river geomorphological style: examples from European rivers. Geomorphology 19:99–116.

Piégay, H., and N. Landon. 1997. Promoting an ecological management of riparian forests on the Drôme River, France. Aquatic Conservation: Marine and Freshwater Ecosystems 7:287–304.

Piégay, H., A. Citterio, and L. Astrade. 1998. Ligne de débris ligneux et recoupement de méandres, exemple du site de Mollon sur l'Ain (France). Zeitzchrift für Geomorphologie 42(2):187–208.

Piégay, H., and R. A. Marston. 1998. Distribution of coarse woody debris along the concave bank of a meandering river (the Ain River, France). Physical Geography 19(4):318–340.

Piégay, H., A. Thévenet, and A. Citterio. 1999. Input, storage and distribution of large woody debris along a montain river continuum, the Drôme River, France. Catena 35:19–39.

Schumm, S. A., M. D. Harvey, and C. C. Watson. 1984. Incised channels: initiation, evolution, dynamics, and control. Water Resources Publication, Littleton, Colorado.

Sedell, J. R., and K. J. Luchessa. 1982. Using the historical record as an aid to salmonid habitat enhancement. Pages 210–223 *in* N. B. Armantrout, editor. Aquisition and utilization of aquatic habitat inventory information. Western Division, American Fisheries Society, Portland, Oregon.

Sedell, J. R., P. A. Bisson, F. J. Swanson, and S. V.

Gregory. 1988. What we know about large trees that fall into streams and rivers. Pages 47–122 *in* C. Maser, R. F. Tarrant, J. M. Trappe, and J. F. Franklin, editors. From the forest to the sea: a story of fallen trees. U.S. Forest Service, General Technical Report PNW-GTR-229, Portland, Oregon.

Sedell, J. R., and J. L. Froggatt. 1984. Importance of streamside forests to large rivers: the isolation of the Willamette River, Oregon, USA, from its floodplain by snagging and streamside forest removal. Verhandlungen International Vereinigung Limnologie 22:1828–1834.

Shields, F. D., and R. H. Smith. 1992. Effects of large woody debris removal on physical characteristics of a sand-bed river. Aquatic Conservation: Marine and Freshwater Ecosystems 2:145–163.

Simon, A., and C. R. Hupp. 1987. Geomorphic and vegetative recovery processes along modified Tennessee streams: an interdisciplinary approach to disturbed fluvial systems. International Association of Hydrological Sciences 167:251–262.

Swanson, F. J., G. W. Lienkaemper, and J. R. Sedell. 1976. History, physical effects and management implications of large organic debris in western Oregon streams. U.S. Forest Service, General Technical Report PNW-56, Portland, Oregon.

Swanson, F. J., and G. W. Lienkaemper. 1982. Interactions among fluvial processes, forest vegetation and aquatic ecosystems, South Fork Hoh River, Olympic National Park. Ecological Research in National Parks of the Pacific Northwest. Oregon State University, Corvallis.

Swanson, F. J., G. W. Lienkaemper, and M. D. Bryant. 1984. Organic debris in small streams, Prince of Wales Island, southeast Alaska. U.S. Forest Service, Pacific Northwest Forest and Range Experimental Station, General Technical Report PNW-166, Portland, Oregon.

Swanson, F. J., S. L. Johnson, S. V. Gregory, and S. A. Acker. 1998. Flood disturbance in a forested mountain landscape: interactions of land use and floods. BioScience 48:681–689.

Thévenet, A. 1998. Intérêt des débris ligneux grossiers pour les poissons dans les grands cours d'eau: pour une prise en compte de la dimension écologique des débris ligneux grossiers dans la gestion des cours d'eau. Doctoral dissertation, University of Lyon I, Lyon, France.

Triska, F. J. 1984. Role of wood debris in modifying channel geomorphology and riparian areas of a large lowland river under pristine conditions: a historical case study. Verhandlungen International Vereinigung Limnologie 22:1876–1892.

Vannote, R. L., G. W. Minshall, K. W. Cummins, J. R. Sedell, and C. E. Cushing. 1980. The river continuum concept. Canadian Journal of Fisheries and Aquatic Sciences 37:130–137.

Wallace, J. B., and A. C. Benke. 1984. Quantification of wood habitat in subtropical coastal plain streams. Canadian Journal of Fisheries and Aquatic Sciences 41:1643–1652.

Wallerstein, N. 1999. Impact of large woody debris on fluvial processes and channel geomorphology in unstable sand-bed rivers. Unpublished Doctoral dissertation., University of Nottingham, Nottingham, UK.

Zalewski, M., M. Lapinska, and P. B. Bayley. 2003. Fish relationships with wood in large rivers. Pages 195–211 *in* S. V. Gregory, K. L. Boyer, and A. M. Gurnell, editors. The ecology and management of wood in world rivers. American Fisheries Society, Symposium 37, Bethesda, Maryland.

American Fisheries Society Symposium 37:135–147, 2003

Decomposition and Nutrient Dynamics of Wood in Streams and Rivers

Robert E. Bilby

Weyerhaeuser Co., WTC 1A5, P.O. Box 9777, Federal Way, Washington 98063–9777, USA

Abstract.—Wood degradation in freshwater aquatic ecosystems differs from degradation in the terrestrial environment. Oxygen concentration in the interior of wet pieces of wood is not high enough to support fungi, the primary decomposing agents of wood in terrestrial environments. The primary decomposing agents of wood in streams and rivers are bacteria and actinomycetes limited to a thin layer on the surface of the piece. As a result, microbial decomposition of wood proceeds much more slowly in water, and changes in wood properties occur only on the surface. Fragmentation is an important degradation process in streams. Flowing water or abrasion from sediment removes wood softened by microbial action, and several types of invertebrates ingest or construct burrows in the softened, outer layer. Removing the outer layer exposes new surfaces to bacterial colonization. Pieces of wood more than 1,000 years old have been identified in rivers by using dendrochronological and radioisotope techniques, although most wood degrades within 100 years after it enters a stream. Nutrient content of wood is low relative to other types of organic matter. Wood abundance in streams is often an order of magnitude greater than that of other types of organic matter, however. Thus, wood does contain a significant proportion of the nutrients associated with organic matter. Wood also affects the nutrient dynamics of streams by controlling the rate of transport of materials downstream. Wood primarily affects nutrients being transported in a particulate form. Removing wood from a 200-m section of a headwater stream in New Hampshire resulted in a sevenfold increase in the transport of sediment and particulate organic matter the next year. The increase was the combined result of mobilizing stored particulate material and increasing the efficiency with which material in the channel was routed downstream. The effect of wood on material transport influences nutrient availability in many Pacific Northwest streams by retaining salmon carcasses, which contribute significant amounts of N and P to the systems where these fish spawn and die.

Introduction

The dynamics of wood decay in streams differs from the better understood processes of wood decomposition in terrestrial environments. Flowing water accelerates disappearance of wood through mechanical abrasion, but inundation of wood also slows biological decay by reducing oxygen in the wood. Because the rate at which wood is removed directly changes the abundance of wood in channels (Welty et al. 2002), decomposition can have a major influence on wood abundance and function in stream systems.

Wood also affects nutrient dynamics in streams. Although the concentration of most biologically significant elements in wood is low, wood may still contain a significant proportion of the standing stock of certain nutrients because of the large amount of wood in some stream channels. Wood decay gradually releases the nutrients stored in wood to the stream. Wood also influences nutrient dynamics in streams by forming sites of particulate matter deposition. Microbial processing of these stored materials releases nutrients to the system. In the absence of wood, much of this particulate material and its associated nutrients may be flushed from the stream. In this chapter, I review the current state of our knowledge regarding wood decay in streams and rivers and the role wood plays as a source of nutrients and a control of nutrient movement in lotic ecosystems.

Wood Decomposition

In this discussion, wood decomposition will be considered to include all the physical and biological processes that can contribute to the degradation of a piece of wood in a stream channel. The basic processes include leaching, microbial degradation, and both biological and physical fragmentation. The relative importance of these processes varies with characteristics of the wood and the site. Wood degradation in streams differs from degradation in terrestrial environments in a variety of ways. Chief among them is the much greater influence of physical fragmentation in streams and the composition of the microbial community responsible for biological decay.

Processes of wood degradation

Biological factors influencing wood degradation in streams include microbial decomposition and fragmentation by invertebrates. Fungi are the primary microbial agents of wood decomposition in terrestrial environments (Savory 1954; Rayner and Todd 1979). In aquatic environments, however, oxygen concentrations are very low in wood except at the outer surface of the piece. Under these conditions, fungi play a lesser role in wood decay because insufficient oxygen is available to support the penetration of fungal hyphae into the piece. Microbial decomposition in water-saturated wood is caused primarily by single-celled bacteria and actinomycetes, which break down the structural components of wood much slower than many other decay-promoting organisms (Crawford and Sutherland 1979; Aumen et al. 1983). Saturated wood decomposes in a thin layer on the surface of the piece.

Several taxa of macroinvertebrates contribute to the degradation of wood in streams, including some beetles, stoneflies, caddiflies, mayflies, and true flies (Anderson et al. 1978; Benke and Wallace 2003, this volume). Some of these animals ingest wood, and others burrow into the wood for shelter. The macroinvertebrates work in concert with the microbial flora in wood decomposition. The invertebrates typically ingest or burrow in wood tissue previously colonized and softened by microbes (Harmon et al. 1986). The microbial flora increases the nutritional value of wood, and wood softened by decay is easier to excavate for burrows. As the invertebrates remove wood, new surfaces are exposed to oxygen, allowing microbial colonization. Thus, the rate of wood decay is accelerated by the complimentary functions of microbes and invertebrates.

Physical fragmentation is an important process in wood degradation in streams (Harmon et al. 1986). Physical fragmentation can be caused directly by the force of the flowing water or secondarily as sediment carried by the water abrades the wood (Aumen 1985). Physical fragmentation is also augmented by microbial activity on the wood surface. Once the wood has been softened by microbial decomposition, it is much more likely to be fragmented by flowing water or transported sediment. Fragmentation of softened wood from the surface of a piece exposes fresh wood surface for microbial colonization, generating a feedback loop as described for macroinvertebrate fragmentation.

Leaching of soluble materials from wood plays only a very small role in wood decomposition. A small amount of dissolved organic matter is rapidly released by freshly fallen wood, especially if the wood is from a live tree (Buchanan et al. 1976). Leaching loss can increase as microbes transform the insoluble materials forming wood into more easily dissolved compounds (Harmon et al. 1986). This soluble material forms a very small fraction of the total wood mass, however, so this process plays only a minor role in the degradation of wood in streams.

Numerous studies have documented the changes in wood properties with time in terrestrial environments (Harmon et al. 1986), but relatively few studies have documented these changes for wood in an aquatic system. Evaluation of the changes in properties of wood submerged in a stream for five years was evaluated in western Washington (Bilby et al. 1999). This study examined the changes in 20-cm diameter logs from five tree species: Douglas-fir *Pseudotsuga menzesii*, western redcedar *Thuja plicata*, western hemlock *Tsuga heterophylla*, bigleaf maple *Acer macrophyllum*, and red alder *Alnus rubra*. Logs were cut green and submerged in a small stream draining the foothills of the Cascade Mountains. Samples were collected annually. Changes through time in the structural integrity of the wood were evaluated by measuring breaking strength and flex, using standard tests for properties of lumber. Densities of the wood at the piece surface and within the interior of the piece were also measured. In addition, microbial activity on the surface of the pieces was evaluated by measuring rate of oxygen consumption.

All the changes in wood characteristics noted over the 5-year period of exposure were at the surface of the piece. No changes were seen in either strength or flexibility of the wood. Density of the interior wood did not change, but all five species exhibited decreased density in wood within 12 mm of the log surface (Figure 1). Differences were found among species in the loss of wood over the period of exposure (Table 1). Diameter of the bigleaf maple pieces decreased the most; they lost more than 20 mm in diameter. The other four species exhibited similar decreases in diameter, all less than 15 mm.

This slow rate of decomposition of wood in water also has been noted in an experiment in the southeastern United States (Webster et al. 1999). Logs from 20 cm to 32 cm in diameter from a live, yellow popular tree *Liriodendra tulipifera* were placed in a stream at the Coweeta Hydrological Laboratory in July 1988. Samples from each log were collected in 1989 and 1996 to determine changes in wood density. After 4 or 5 years, bark losses on the upper surfaces of the logs were noted, and some evidence of decay was seen on the ends of the logs. However, no evidence of decay was noted more than 5 mm from the surface, and no changes occurred in wood density in the interior of the logs.

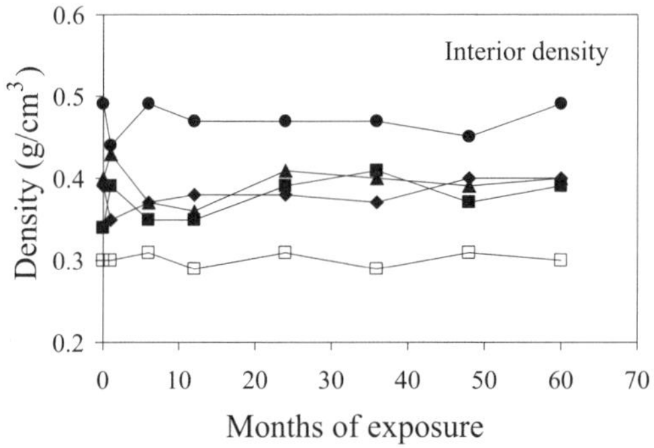

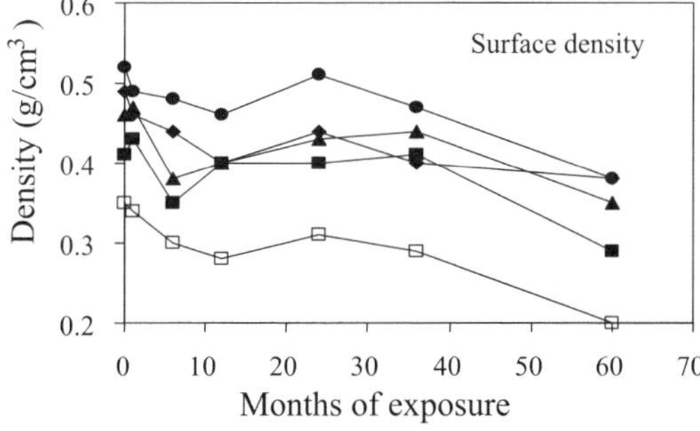

FIGURE 1. Changes in the density of wood from five species of trees common in the Pacific Northwest during 5 years of submersion in a small stream. Panel A represents the average change in density for wood more than 12 mm from the wood surface (n = 5 for each species on each sample date). Panel B represents average density changes for wood within 12 mm of the piece surface (n = 5 for each species on each sample date). Symbols represent the species: ■- western hemlock, ▲- Douglas fir, ●- bigleaf maple, ◆- red alder, □- western redcedar. (Bilby et al. 1999).

Factors influencing degradation rates

Two factors dictate the rate of wood decomposition: the characteristics of the wood and the characteristics of the wood's environment. Piece size is one attribute of wood that influences decay rate. This relationship is caused, at least in part, by the reduced ratio of surface area to volume (Abbott and Crossley 1982). The effect may be especially pronounced in freshwater environments where microbial processing is restricted primarily to the wood surface. The relative proportion of different wood tissues also changes with increased piece size. Smaller pieces of wood typically have a relatively high proportion of their mass as bark and sapwood with little heartwood. Heartwood comprises a much larger proportion of larger pieces of wood, and heartwood decomposes slower than the sapwood and bark (Harmon et al. 1986).

Heartwood contains compounds that retard decay (Scheffer and Cowling 1966). The nature and concentration of these decay-retarding compounds varies among species. For example, western redcedar contains a particular class of compounds called thujaplicins, which confer a high degree of decay resistance to the heartwood of this species (Scheffer and Cowling 1966). Heartwood of species containing lower concentrations or less effective decay-resisting compounds decompose more rapidly. Sapwood and bark do not contain these compounds and decay at much higher rates. The susceptibility of sapwood to decay also varies among species, based on the concentrations of starches, sugars, and proteins. Higher concentrations of these materials promote more rapid decomposition (Harmon et al. 1986).

The highly labile materials found in bark are used rapidly after a piece of wood enters a stream. Therefore, microbial decomposition rates decline through time as these more desirable materials are depleted. The decline can be rapid, as indicated by measuring of the rate of microbial action

TABLE 1. Decrease in diameter and decay rate constants (*k*) for wood from five species of trees submerged in water for 5 years (Bilby et al.1999).

Species	Diameter loss after 5 years[a](mm)	Decay rate constant (*k*)[b]
Douglas-fir	13.0 ± 2.7	0.026
Western redcedar	13.2 ± 4.2	0.026
Western hemlock	10.6 ± 4.9	0.031
Red alder	14.3 ± 4.0	0.033
Bigleaf maple	21.8 ± 6.3	0.038

[a] Diameter loss values are given ± one standard error.
[b] The derivation of the decay rate constant is explained in the text.
Note: Because these data are for small logs and cover a relatively brief period of exposure, values are likely higher than if wood degradation were followed longer or for larger pieces of wood.

on wood from five species of trees common in the Pacific Northwest over a 5-year period (Bilby et al. 1999; Figure 2). Microbial activity, as indexed by rate of oxygen consumption, declined for all species over time, but the initial rate and the extent of the decline differed among species. Rates of oxygen consumption were 5–10-fold greater shortly after introduction into the stream than after 5 years of exposure. This elevated oxygen consumption persisted for about a year, at which time most of the bark had been removed. Rates of consumption declined thereafter, with the conifer species exhibiting the most rapid declines. After 5 years, the two hardwood species consumed about twice as much oxygen as the conifer species, indicating a more rapid rate of decomposition.

Decay rate is also influenced by the availability of nutrients in the wood. Although all wood tends to contain low concentrations of biologically significant nutrients like nitrogen and phosphorus, species and tissue types vary considerably in these concentrations (Whittaker et al. 1979). Species and tissues with higher concentrations of those elements required by the wood-decaying microbes decompose more rapidly than species and wood tissues with lower concentrations. Angiosperms typically have higher concentrations of most nutrients than do gymnosperms (Harmon et al. 1986). Bark has higher concentrations of most nutrients than sapwood or heartwood. Sapwood generally contains more phosphorus, magnesium, and iron than heartwood, but heartwood has more manganese (Woodwell et al. 1975). The difference in decay rate between heartwood and sapwood is caused primarily by the presence of decay-retarding compounds in the heartwood, not by differences in nutrient content (Scheffer and Cowling 1966).

Wood is composed primarily of cellulose, hemicellulose, and lignin. The molecular properties of these materials influence the rate at which they decay. Cellulose, the chemically simplest of the components, decomposes most rapidly, and the complex lignin compounds are the slowest (Swift 1977; Mellilo et al. 1983). Thus, a higher proportion of lignin in wood is related to slower rates of decomposition. Gymnosperms generally have more lignin (26–30%) than angiosperms (16–32%) (Cote 1977), contributing to their slower rate of decay.

The environment of a piece of wood also influences its rate of decomposition. In turbulent streams where oxygen is always near saturation, the oxygen is sufficient to support microbial action at the wood surface. In slow-flowing streams with high organic loads, however, oxygen levels can be low enough to retard microbial action and greatly reduce decomposition rates (Zeikus 1980). The effect of oxygen on wood decomposition rates is emphasized by the extremely great age achieved by pieces buried in streambanks or beds, where oxygen is often very low. Wood excavated from floodplains of Australian rivers has been dated at more than 5,000 years (Nanson et al. 1995), and wood unearthed from the floodplain of a river in Washington was 1,400 years old (Hyatt and Naiman 2001).

Temperature also plays an important role in wood decomposition. Many wood-decaying microbes have a thermal optimum of 25–30°C (Kaarik 1974). At lower temperatures, metabolic rates decrease two–threefold with every 10°C decrease in temperature. Temperatures also can influence the rate of biological fragmentation. The larvae of the wood-ingesting beetle *Lara avara* doubled the production of fecal matter with an increase in water temperature from 5° to 15°C (Steedman 1983).

Nutrients available from external sources can augment those available from the wood and accelerate microbial activity and decay (Cowling

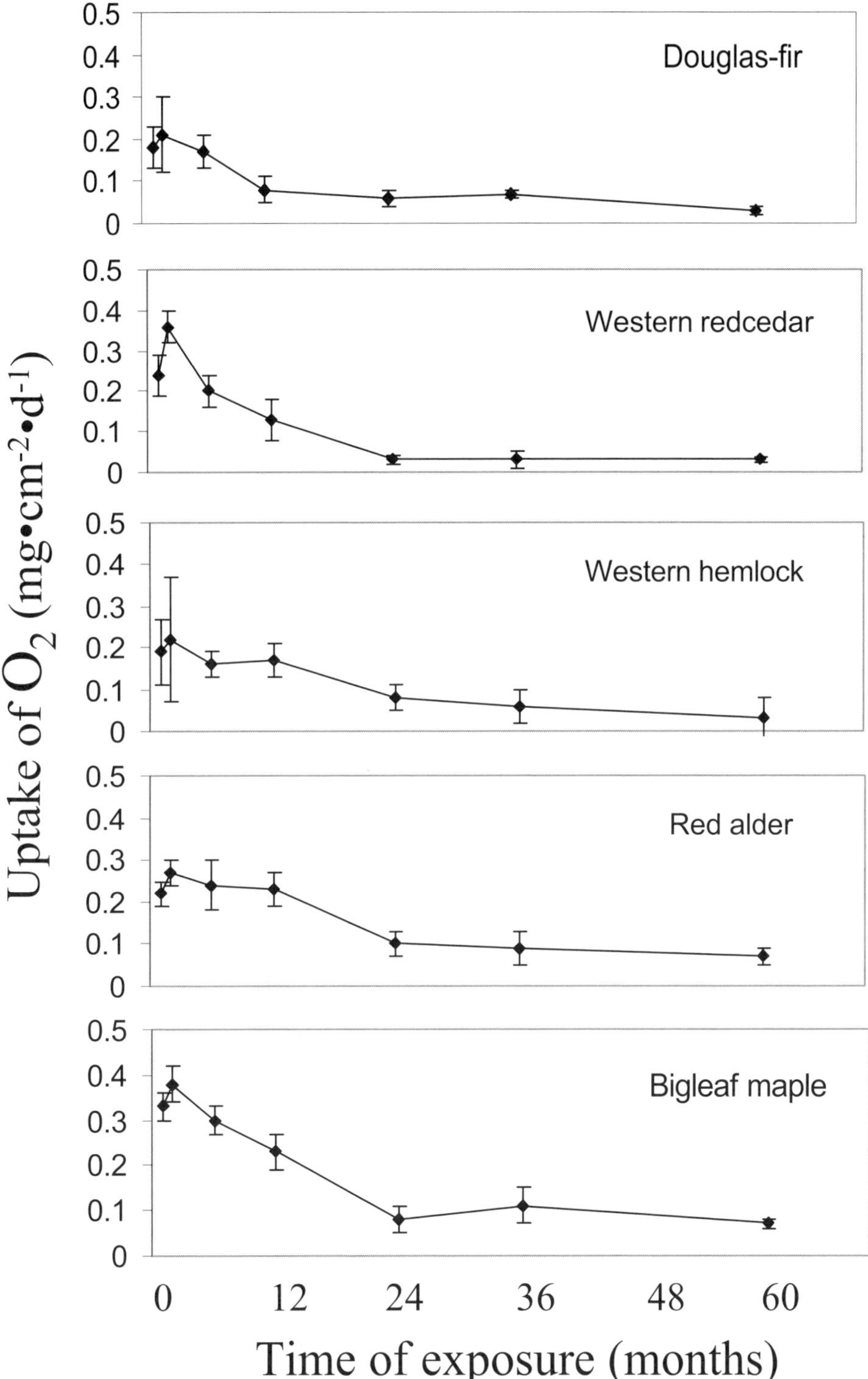

FIGURE 2. Changes in oxygen uptake by the microbial community on wood from five species of trees common in the Pacific Northwest during 5 years of submersion in a small stream. Oxygen use rates represent uptake per cm^2 of wood surface area. Values are shown ± one standard error (Bilby et al. 1999).

and Merrill 1966). Experimental addition of nitrogen to stream water increased the breakdown of both cellulose and lignin (Aumen 1985). In these experiments, the nitrogen rapidly transformed from an inorganic form in solution to an organic form on the wood surface. Phosphorus availability is also positively related to wood decay rates in streams (Mellilo et al. 1983). Experimental enrichment of stream water with phosphorus suggests that the effect of this nutrient on wood decay rate is most pronounced for species that have low lignin content (Mellilo et al. 1985).

Wood decay rates

The most common models used to express the loss of mass from a piece of wood through time are exponential models that take the form:

$$V_t = V_0 e^{-kt},$$

where V_t is the dimension of a piece of wood at time t, V_0 is the initial dimension of the piece, and k represents the decay-rate constant. This model provides a more realistic representation of loss in wood mass over time than do other models that have been used to describe this process (Wieder and Lang 1982). Because k indicates the rate at which a piece of wood is losing mass, it is a useful parameter for comparing rates of wood decomposition among different wood tissues or different species. These values are usually obtained in experiments where changes in the dimensions of pieces of wood are monitored through time (Olson 1963). Comparison of k values for wood of the same species in terrestrial and aquatic environments confirms the points made above about the slower rate of decomposition in water. Values for k calculated from data on 5 years of exposure in a Pacific Northwest stream for five species of wood ranged from about 0.026 to 0.038 with the angiosperm species exhibiting the higher values (Table 1; Bilby et al. 1999). Far more estimates of k have been generated for wood in terrestrial environments (Harmon et al. 1986), and the terrestrial values are generally considerably higher than for the same species in water.

If a wood decomposition experiment continues over sufficient time to determine the k value for the various components of the wood, including the heartwood, an estimate of piece longevity can be made by calculating the inverse of k. Rarely does a study continue long enough to accurately determine piece longevity using this approach. Therefore, various alternative methods have been devised to estimate wood longevity in aquatic systems. Both dendrochronological methods and radiocarbon dating have proved useful for aging wood. Wood longevity has also been estimated by measuring the rate of change in wood abundance in a stream over time.

Various dendrochronological methods have been used to age wood in streams. Aging the young trees growing on a log in a channel provides a minimum estimate of piece age (Triska and Cromack 1980), but variability in the length of time required for a piece of wood to provide a suitable germination site for a tree seed can introduce considerable error. Murphy and Koski (1989) used this method on streams in southeastern Alaska to age wood pieces and used average age by piece-size-class to determine a rate of wood depletion from these channels (Figure 3). Depletion rates were estimated as the inverse of the average age of wood in four diameter classes. The results were reassuring in that they showed an increase in piece longevity (decrease in depletion rate) with increasing piece diameter. Dating of scars on standing trees near a piece of wood in the channel has also been used to estimate piece age (Swanson and Lienkaemper 1978) by counting the annual rings produced since the tree was scarred. This method assumes the fallen log was

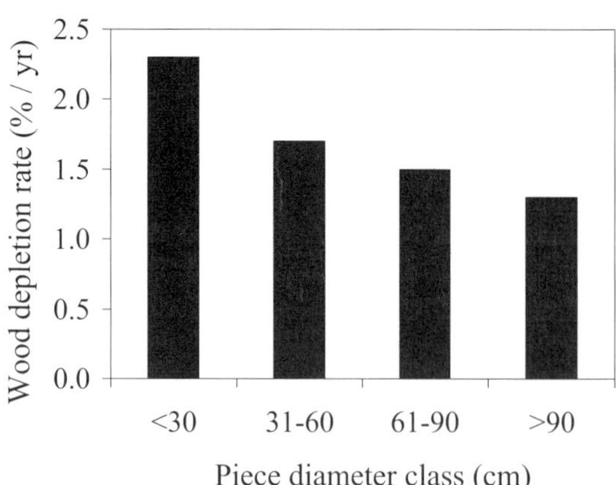

FIGURE 3. Wood-depletion rates for streams in southeast Alaska. These values were estimated by determining a minimum age for 252 pieces wood in the channel by dating the young trees growing on the wood. Depletion rate was then estimated by calculating the inverse of average piece age for each size-class (Murphy and Koski 1989).

responsible for the scar on the standing tree. Verification of this assumption is often difficult, and many pieces of wood fall without scarring nearby trees. Thus, this technique cannot be used for all pieces.

Tree-ring chronologies have been used more recently to date wood in channels. The year of the death of a tree producing a piece of wood is estimated by determining the year when the outer ring was produced. There may be some lag between tree death and entry into the channel; thus, this method may overestimate the time a piece of wood has been in the channel. Hyatt and Naiman (2001) used this technique to date wood in the channel and on the floodplain of the Queets River, a large, relatively undisturbed river on Washington's Olympic Peninsula. More than half the pieces of wood were less than 20 years old, and nearly 90% were less than 50 years old (Figure 4). Several pieces were found that could not be dated with the ring chronology developed for the site. Radiocarbon dating of the wood on the outer surface of these pieces indicated that they were very old and may have been delivered to the system as long as 1,400 years ago. Hyatt and Naiman (2001) attributed the existence of these very old pieces to burial in the bed or bank, which retards decay and prevents flushing from the system, followed by excavation of the piece by scour sometime in the recent past. Generally, using multiple dating techniques provides the most complete picture of wood age.

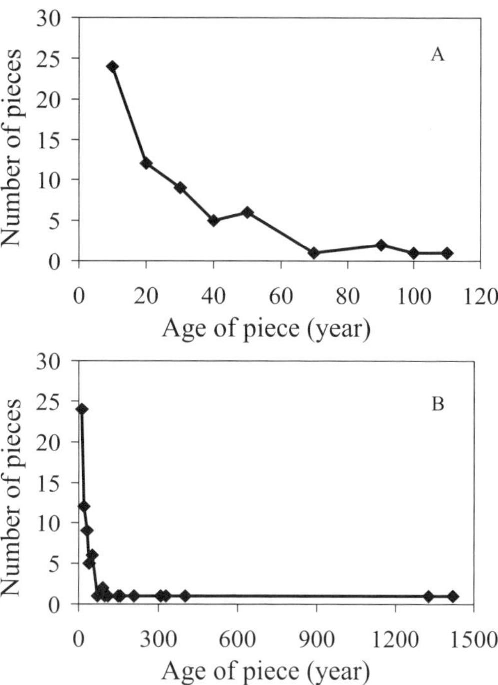

FIGURE 4. Age of wood in the channel and floodplain of the Queets River, Washington. These values were determined by using two techniques: tree-ring dating and radiocarbon dating. Panel A illustrates the frequency distribution of ages for those pieces less than 100 years old. Panel B includes all the wood dated in the study and illustrates the extreme age of a few of the pieces. The tree-ring chronology dated most of the pieces, but the very old pieces required carbon dating. Despite the few, remarkably old pieces discovered, most of the wood in this system had been delivered within the past 50 years (Hyatt and Naiman 2001).

Wood longevity in streams has also been estimated by measuring changes in the standing stock of wood in a channel through time. Because these techniques are based on rates of change in wood input, abundance in the channel, or both, they provide an estimate of the average rate at which both degradation and fluvial transport are removing wood from the sites being examined. Two basic approaches to this technique have been used: monitoring changes in wood abundance after trees are removed from the streamside area (McHenry et al. 1998) and measuring both input rate and wood abundance in the channel and estimating residence time as abundance/input rate (Lienkaemper and Swanson 1987). The estimates from the second measure require a relatively long record because wood input and fluvial export tend to be episodic. Because most records of wood in streams are relatively short (<20 years), estimates of wood longevity made with this technique have been variable. Wood residence time estimated with this technique for stream reaches at the H. J. Andrews Experimental Forest in Oregon ranged from 12 to 83 years (Lienkaemper and Swanson 1987). Nonetheless, the most accurate estimates of wood degradation rate will ultimately be provided by such long-term studies of wood behavior in streams.

Measurement of wood abundance in channels from which the riparian canopy was removed at various times in the past can be used to estimate of the average rate of wood depletion. This technique can only be used if the new wood added to the channel since the riparian trees were removed can be distinguished from the old wood. Generally wood provided by regrowth of the riparian forest can be distinguished from that added to the stream from the preceding stand by piece size or species. Grette (1985) used this technique at a series of sites on Washington's Olympic Peninsula. Trees bordering these stream reaches had been logged from 10 to more than 60 years before the study. The original forest at these sites had consisted of large conifer trees, but the replacement forest was primarily red alder so that wood added since logging could be easily identified. Rate of wood depletion from the sites was estimated from the change in abundance of conifer wood in the channel with time since logging (Figure 5). This relationship varied considerably among sites, reflecting differences in the dynamics of wood delivery and depletion. The analysis suggested that the depletion rate of conifer wood pieces at these sites was about 1%/year, indicating a residence time of 100 years.

Wood and Nutrient Dynamics in Streams

Wood influences the nutrient dynamics of aquatic systems in two ways: the wood contains various nutrient elements released during decomposition, and wood influences the rate and timing of material transport in the system. In small stream channels, both of these processes play a key role in determining nutrient cycling rates and flux.

Nutrients contained in wood

Wood is composed primarily of carbon and contains very low concentrations of most other elements (Harmon et al. 1986), but there are differences in the elemental content of various wood tissues (Figure 6). Bark contains higher quantities of virtually all biologically significant elements than wood. Elemental content also differs somewhat between sapwood and heartwood (Woodwell et al. 1975). In addition, nutrient content of wood and bark differs among tree species. Generally, angiosperms contain higher concentrations of most elements than gymnosperms do.

Despite the low concentration, wood often contains a significant proportion of the nutrient capital in a stream because of its abundance. Other types of organic matter in streams—leaves, needles, twigs and fruit—are nutrient rich relative to large wood. For example, concentration of nitrogen in fine organic matter (<1 mm) stored in the channel of a small stream in New Hampshire was 10 times the concentration in wood in that system (Bilby 1979). The abundance of wood in

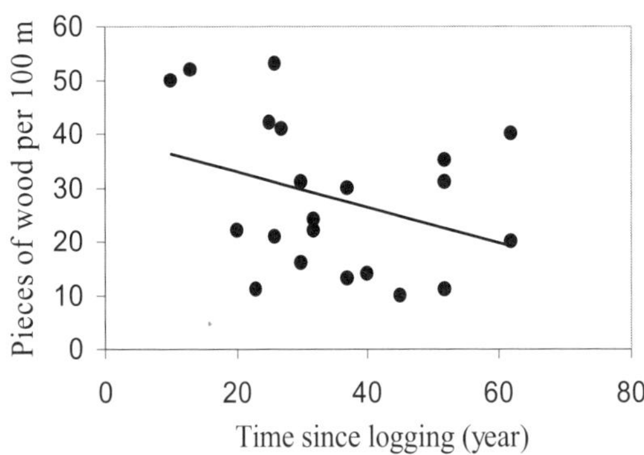

FIGURE 5. The relationship between the abundance of wood from conifer trees and time since logging. Wood abundance values represent the number of individual pieces of wood greater than 10 cm in diameter per 100 m of channel length. These data are for streams on the Olympic Peninsula of Washington. Because no new conifer wood was supplied to these channels after logging, the change in the amount of wood with time since logging provides an indication of the rate of wood depletion (Grette 1985). Wood abundance = –0.32 time since logging + 38.35; r^2 = 0.12; $p > 0.05$.

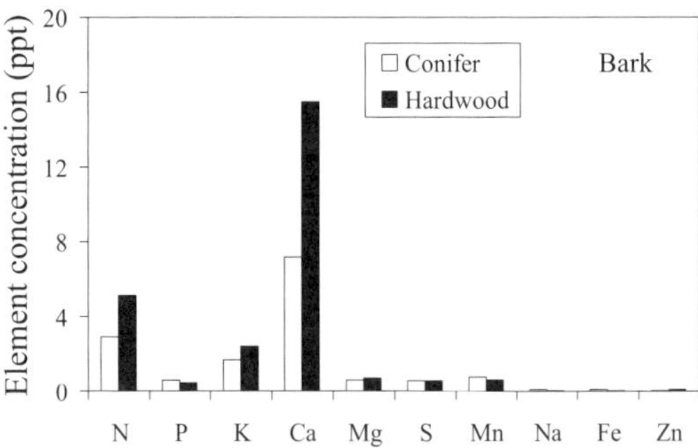

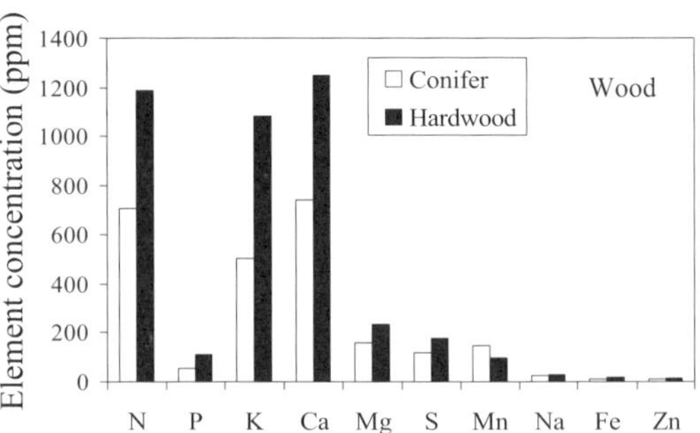

FIGURE 6. The nutrient content in the bark and wood of conifer and hardwood trees. These data are from multiple sources and compiled in Harmon et al. (1986).

the channel was far greater than that of finer organic matter, however, and the proportion of nutrients stored in wood and in finer organic matter was about equal for all elements except calcium (Figure 7). Because calcium is relatively abundant in wood, it was responsible for storing about five times as much of this element than that stored in the finer organic matter. A similar nutrient distribution was found in a first-order channel in the Cascade Range of Oregon where large wood contained 32% of the stored nitrogen in the system and the fine wood 18%; the remaining 50% was associated with finer organic matter (Triska et al. 1984).

Availability of the nutrients held by wood is generally low. These nutrients are released only as the wood decomposes. Because this process is slow in aquatic systems, the nutrients are released very slowly (Grier 1978). In contrast, the nutrients held in finer organic matter often decomposes relatively rapidly, making these nutrients more readily available to the stream ecosystem (Triska et al. 1975). The processes of fragmentation, leaching, and invertebrate processing of wood in streams, however, can accelerate the release of nutrients from wood (Harmon et al. 1986).

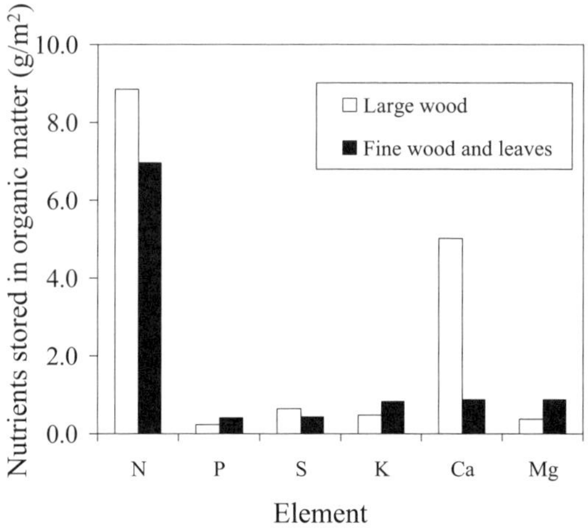

FIGURE 7. The nutrients contained in wood and in fine organic matter in streams in the White Mountains of New Hampshire. Despite the very low concentrations of nutrients in wood relative to fine organic matter, both constituents account for comparable proportions of the nutrients because of the much greater quantity of wood in these channels. Wood abundance was 1.65 kg/m^2, and fine organic debris abundance was 0.14 kg/m^2 (Bilby 1981).

Influence of wood on nutrient transport

Wood also influences nutrient availability and cycling in streams by providing sites for storing of fine organic matter and sediment in the channel. As a result, nutrients associated with these small particles can be used on site rather than transported from the system in a particulate form. The influence of wood on material transport has been evaluated by experimental removal of wood from small stream channels (Beschta 1979; Bilby 1981). Wood removal influences transport in two ways. In the short term, wood removal mobilizes sediment and organic matter stored by the wood, resulting in a rapid increase in transport during the next several periods of elevated discharge. The removal of wood also increases the efficiency of transport of materials delivered to the channel in the future by reducing the frequency of both depositional sites and steps (Bilby 1981). Channels with many steps are inefficient at transporting material (Heede 1972); thus, in the absence of wood, particulate material is rapidly moved through the system with little opportunity for biological processing.

Examples of increased particulate matter transport after wood removal are numerous. Wood removal from a 250-m stream reach in the Coast Range mountains of Oregon resulted in the release of 5,250 m^3 of particulate material during the following winter (Beschta 1979), and a similar treatment in a northern California stream reduced sediment storage by 60% (MacDonald and Keller 1983). Removing wood from a 175-m long stream reach in the White Mountains of New Hampshire doubled the particulate matter export rate (Bilby 1979). Conversely, adding wood to stream channels has been shown to dramatically increase the storage of particulate matter. Addition of wood to a small stream in the southern Appalachian Mountains, North Carolina produced nearly an 18-fold increase in the amount of particulate organic matter in the immediate vicinity of the added piece (Wallace et al. 1995).

The materials flushed from streams after wood removal contain significant amounts of nutrients. Nonwood organic matter tends to be nutrient rich when compared with wood, and a significant proportion of the total nutrient standing stock in a stream may be associated with this material. Export of particulate organic matter after wood removal from a White Mountain stream increased 446% during the following year (Bilby and Likens 1980). The export of dissolved organic matter changed little. Inorganic material export also increased as a result of wood removal, and total export (organic matter + sediment) increased by 531%. Change in nutrient export after wood removal varied by element depending on the relative proportion of that element in dissolved and particulate form (Figure 8). Therefore, elements existing primarily in a particulate form, like phosphorus, aluminum, iron, and manganese, exhibited large increases in export. In contrast, elements primarily in a dissolved form showed only small increases in export. These elements included sulfur, calcium, sodium, and nitrogen.

The influence of wood on nutrient dynamics in streams is greater in smaller channels where wood is more abundant and the influence of wood on material routing is more pronounced. Wood abundance decreases with stream size (Bilby and Ward 1989); this decline corresponds to a decrease in the nutrients held by wood (Table 2). These data suggest an 80% decline in nutrients contained in wood in Pacific Northwest coastal streams as channel size increases from 5 to 15 m. Fine organic matter held by wood also declines with stream size in response to the decline in depositional sites formed by wood (Naiman and Sedell 1979; Bilby and Ward 1989). In a series of streams in the White Mountains of New Hampshire, the standing stock of nutrients associated with fine organic matter retained by wood decreased more than 10-fold as stream size increased from first to third order (Figure 9; Bilby 1979). The reduction in wood abundance with increasing channel size was the primary cause of this decline because the proportion of this fine organic matter retained by wood in these systems decreased with increasing stream order: 74.5% in first-order, 58.4% in second-order, and 20.0% in third-order channels. Thus, the influence of wood on nutrient retention is far greater in small channels.

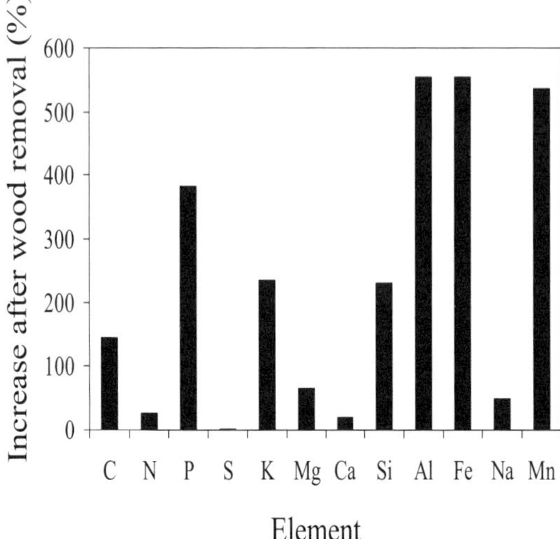

FIGURE 8. Increase in the transport of nutrients by a stream after wood was removed from the channel. Variation in the extent of increase among the nutrients is related to the degree to which that element is transported in a dissolved or particulate form. Nutrients transported in a particulate form increased most dramatically; those showing only small increases were primarily dissolved (Bilby 1981).

A special case of nutrient retention by wood occurs in streams where semelparous (fish that die after spawning), anadromous fishes spawn. These fishes have been shown to make significant contributions of nutrients to the freshwater systems where they spawn and die (Gresh et al. 2000). Tagged salmon carcasses were monitored in a series of streams on Washington's Olympic Peninsula to evaluate their distribution (Cederholm et al. 1989). About half of the carcasses were removed from the stream by scavengers, and about 60% of those that remained were held by large wood. In the absence of wood, a high proportion of these

TABLE 2. The amount of wood and the amount of nitrogen, phosphorus and potassium contained in it in Pacific Northwest streams with channel widths of 5 m, 10 m, and 15 m.

Channel width (m)	Wood volume[a] (m^3/m^2)	Wood mass[b] (kg/m^2)	N[c] (g/m^2)	P[c] (g/m^2)	K[c] (g/m^2)
5	.081	32.4	22.8	1.7	16.3
10	.031	12.4	8.7	0.7	6.2
15	.016	6.4	4.5	0.3	3.2

[a] Values for wood volumes are from Bilby and Ward (1989).
[b] To convert volume to wood mass, a wood density of 0.4 g/cm^3 was assumed (Figure 1).
[c] The nutrient values are for conifer wood; they are taken from Harmon et al. (1986): N = 704 ppm, P = 54 ppm, K = 505 ppm.

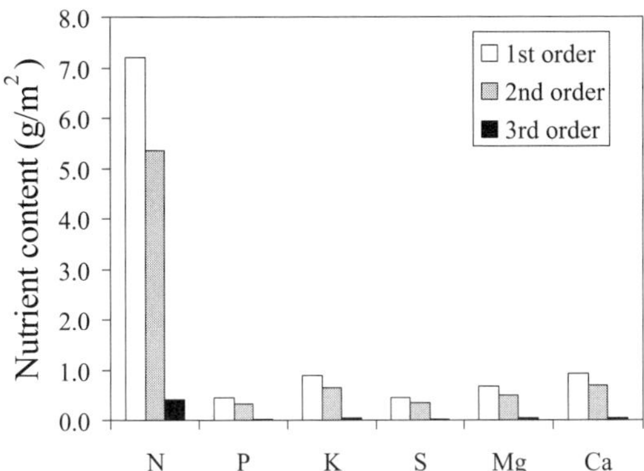

FIGURE 9. Change in the amount of nutrients in fine organic matter stored by wood with changing stream size. Decrease in wood-stored nutrients with increasing stream size is caused by a corresponding decrease in the abundance of wood (Bilby 1979).

carcasses likely would be flushed downstream, potentially affecting the biological productivity of the system (Bilby et al. 1998).

Wood plays an important role in the nutrient dynamics of stream ecosystems. By slowing water flow, enhancing particulate matter storage, and reducing transport efficiency, downstream movement of nutrients is reduced. Reducing the rate of transport increases nutrient availability as stored particulate matter accumulates. The affect of this function of wood on productivity of stream systems has not been well quantified. Nonetheless, the quantity of nutrients accumulating in channels with abundant wood clearly indicates that this role of large wood may be very significant in supporting system production.

References

Abbott, D. T., and D. A. Crossley. 1982. Woody litter decomposition following clearcutting. Ecology 63:35–42.

Anderson, N. H., J. R. Sedell, L. M. Roberts, and F. J. Triska. 1978. The role of aquatic invertebrates in processing wood debris in coniferous forest streams. American Midland Naturalist 100:64–82.

Aumen, N. G. 1985. Characterization of lignocellulose decomposition in stream wood samples using ^{14}C and ^{15}N techniques. Doctoral dissertation. Oregon State University, Corvallis, Oregon.

Aumen, N. G., P. J. Bottomley, G. M. Ward, and S. V. Gregory. 1983. Microbial decomposition of wood in streams: distribution of microfauna and factors affecting ^{14}C lignocellulose mineralization. Applied Environmental Microbiology 46:1409–1416.

Benke, A. C., and J. B. Wallace. 2003. Influence of wood on invertebrate communities in streams and rivers. Pages 149–177 *in* S. V. Gregory, K. L. Boyer, and A. M. Gurnell, editors. The ecology and management of wood in world rivers. American Fisheries Society, Symposium 37, Bethesda, Maryland.

Beschta, R. L. 1979. Debris removal and its effect on sedimentation in an Oregon coast range stream. Northwest Science 53:71–77.

Bilby, R. E. 1979. The function and distribution of organic debris dams in forest stream ecosystems. Doctoral dissertation. Cornell University, Ithaca, New York.

Bilby, R. E. 1981. Role of organic debris dams in regulating the export of dissolved and particulate matter from a forested watershed. Ecology 62:1234–1243.

Bilby, R. E., and G. E. Likens. 1980. Importance of organic debris dams in the structure and function of stream ecosystems. Ecology 61:1107–1113.

Bilby, R. E., and J. W. Ward. 1989. Changes in characteristics and function of woody debris with increasing size of streams in western Washington. Transactions of the American Fisheries Society 118:368–378.

Bilby, R. E., B. R. Fransen, P. A. Bisson, and J. K. Walter. 1998. Response of juvenile coho salmon and steelhead to the addition of salmon carcasses to two streams in southwest Washington, USA. Canadian Journal of Fisheries and Aquatic Sciences 55:1909–1918.

Bilby, R. E., J. T. Heffner, B. R. Fransen, J. W. Ward, and P. A. Bisson. 1999. Effect of immersion in stream water on the deterioration of wood from 5 tree species commonly used for habitat enhancement projects. North American Journal of Fisheries Management 19:687–695.

Buchanan, D. V., P. S. Tate, and J. R. Moring. 1976. Acute toxicities of spruce and hemlock bark extracts to some estuarine organisms in southeastern Alaska. Journal of the Fisheries Research Board Canada 33:1188–1192.

Cederholm, C. J., D. B. Houston, D. L. Cole, and W. J. Scarlett. 1989. Fate of coho salmon (*Oncorhynchus kisutch*) carcasses in spawning streams. Canadian Journal of Fisheries and Aquatic Sciences 46:1347–1355.

Cote, W. A. 1977. Wood ultrastructure in relation to chemical composition. Recent Advances in Phytochemistry 11:1–14.

Cowling, E. B., and W. Merrill. 1966. Nitrogen in wood and its role in wood deterioration. Canadian Journal of Botany 44:1539–1554.

Crawford, D. L., and J. B. Sutherland. 1979. The role of actinomycetes in the decomposition of lignocellulose. Developments in Industrial Microbiology 20:143–151.

Gresh, T., J. Lichatowich, and P. Schoonmaker. 2000. An estimation of historic and current levels of salmon production in the northeast Pacific ecosystem: evidence of a nutrient deficit in the freshwater systems of the Pacific Northwest. Fisheries 25(1):15–21.

Grette, G. B. 1985. The abundance and role of large organic debris in juvenile salmonid habitat in streams in second growth and unlogged forests. Master's dissertation. University of Washington, Seattle.

Grier, C. C. 1978. A *Tsuga heterophylla-Picea sitchensis* ecosystem of coastal Oregon: decomposition and nutrient balances of fallen logs. Canadian Journal of Forest Research 8:198–206.

Harmon, M. E., J. F. Franklin, F. J. Swanson, P. Sollins, S. V. Gregory, J. D. Lattin, N. H. Anderson, S. P. Cline, N. G. Aumen, J. R. Sedell, G. W. Lienkaemper, K. Cromack, Jr., and K. W. Cummins. 1986. Ecology of coarse woody debris in temperate ecosystems. Advances in Ecological Research 15:133–302.

Heede, B. H. 1972. Influences of a forest on the hydraulic geometry of two mountain streams. Water Resources Bulletin 8:523–530.

Hyatt, T. L., and R. J. Naiman. 2001. The residence time of large woody debris in the Queets River, Washington, USA. Ecological Applications 11:191–202.

Kaarik, A. A. 1974. Decomposition of wood. Pages 129–174 *in* C. H. Dickson and G. E. Pugh, editors. Biology of plant litter decomposition, volume 1. Academic Press, London.

Lienkaemper, G. W., and F. J. Swanson. 1987. Dynamics of large woody debris in streams in old-growth Douglas-fir forests. Canadian Journal of Forest Research 17:150–156.

MacDonald, A., and E. A. Keller. 1983. Large organic debris and anadromous fish habitat in the coastal redwood environment: the hydrologic system. Water Resources Center, Technical Completion Report OWRT Project B-213-CAL, University of California, Davis.

McHenry, M. L., E. Schott, R. H. Conrad, and G. B. Grette. 1998. Changes in the quantity and characteristics of large woody debris in streams of the Olympic Peninsula, Washington, U.S.A. (1982–1993). Canadian Journal of Fisheries and Aquatic Sciences 55:1395–1407.

Mellilo, J. M., R. J. Naiman, J. D. Aber, and K. N. Eshleman. 1983. The influence of substrate quality and stream size on wood decomposition dynamics. Oecologia 38:281–285.

Mellilo, J. M., R. J. Naiman, J. D. Aber, and A. E. Linkins. 1985. Factors controlling mass loss and nitrogen dynamics of plant litter decaying in northern streams. Bulletin of Marine Science 35:341–356.

Murphy, M. L., and K. V. Koski. 1989. Input and depletion of woody debris in Alaska streams and implications for streamside management. North American Journal of Fisheries Management 9:427–436.

Naiman, R. J., and J. R. Sedell. 1979. Benthic organic matter as a function of stream order in Oregon. Archiv fur Hydrobiologie 87:404–422.

Nanson, G. C., M. Barbetti, and G. Taylor. 1995. River stabilization due to changing climate and vegetation during the late Quaternary in western Tasmania, Australia. Geomorphology 13:145–158.

Olson, J. S. 1963. Energy storage and the balance of producers and decomposers in ecological systems. Ecology 44:322–331.

Rayner, A. D. M., and N. K. Todd. 1979. Population and community structure and dynamics of fungi in decaying wood. Advances in Botanical Research 7:333–420.

Savory, J. G. 1954. Breakdown of timber by ascomycetes and fungi imperfecti. Annals of Applied Biology 41:336–347.

Scheffer, T. C., and E. B. Cowling. 1966. Natural resistance of wood to microbial deterioration. Annual Review of Phytopathology 4:147–170.

Steedman, R. J. 1983. Life history and feeding role of the xylophagous aquatic beetle *Lara avara* LeConte (Dryopoidea: Elmidae). Master's thesis. Oregon State University, Corvallis.

Swanson, F. J., and G. W. Lienkaemper. 1978. Physical consequences of large organic debris in Pacific Northwest streams. USDA Forest Service, Pacific Northwest Forest and Range Experiment Station, General Technical Report PNW-69, Portland, Oregon.

Swift, M. J. 1977. The ecology of wood decomposition. Science Progress 64:175–199.

Triska, F. J., J. R. Sedell, and B. M. Buckley. 1975. The processing of conifer and hardwood leaves in two coniferous forest streams. II. Biogeochemical and nutrient changes. Verhandlungen Internationale Vereinigung fur Theoretische und Angewandte Limnologie 19:1628–1639.

Triska, F. J., and K. Cromack. 1980. The role of wood debris in forests and streams. Pages 171–190 *in* R. H. Waring, editor. Forests: fresh perspectives from ecosystem analysis. Proceedings of the 4th Biological Colloquium. Oregon State University, Corvallis.

Triska, F. J., J. R. Sedell, K. Cromack, S. V. Gregory, and F. M. McCorison. 1984. Nitrogen budget for a small coniferous forest stream. Ecological Monographs 54:119–140.

Wallace, J. B., J. R. Webster, and J. L. Meyer. 1995. Influence of log additions on physical and biotic characteristics of a mountain stream. Canadian Journal of Fisheries and Aquatic Sciences 52:2120–2137.

Webster, J. R., E. F. Benfield, T. P. Ehrman, M. A. Schaeffer, J. L. Tank, J. J. Hutchens, and D. J.

D'Angelo. 1999. What happens to allochthonous material that falls onto streams? A synthesis of new and published information from Coweeta. Freshwater Biology 41:687–705.

Welty, J. J., T. Beechie, K. Sullivan, D. M. Hyink, R. E. Bilby, C. Andrus, and G. Pess. 2002. Riparian Aquatic Interaction Simulator (RAIS): a model of riparian forest dynamics for the generation of large woody debris and shade. Forest Ecology and Management 162:299–318.

Wieder, R. K., and G. E. Lang. 1982. A critique of the analytical methods used in examining decomposition data. Ecology 63:1636–1642.

Whittaker, R. H., G. E. Likens, F. H. Bormann, J. S. Eaton, and T. G. Siccama. 1979. The Hubbard Brook ecosystem study: forest nutrient cycling and element behavior. Ecology 60:203–220.

Woodwell, G. M., R. H. Whittaker, and R. A. Houghton. 1975. Nutrient concentration in plants in the Brookhaven oak-pine forest. Ecology 56:318–332.

Zeikus, J. G. 1980. Fate of lignin and related aromatic substances in anaerobic environments. Pages 101–109 *in* T. K. Kirk, T. Higuchi, and H. M. Chang, editors. Lignin biodegradation, microbiology, chemistry and potential applications, volume 1. CRC Press, Boca Raton, Florida.

American Fisheries Society Symposium 37:149–177, 2003

Influence of Wood on Invertebrate Communities in Streams and Rivers

ARTHUR C. BENKE

Aquatic Biology Program, Box 870206, Department of Biological Sciences University of Alabama, Tuscaloosa, Alabama 35487-0206, USA

J. BRUCE WALLACE

Department of Entomology and Institute of Ecology University of Georgia, Athens, Georgia 30602, USA

Abstract.—Wood plays a major role in creating multiple invertebrate habitats in small streams and large rivers. In small streams, wood debris dams are instrumental in creating a step and pool profile of habitats, enhancing habitat heterogeneity, retaining organic matter, and changing current velocity. Beavers can convert sections of free-flowing streams into ponds and wetlands by killing trees and building dams. In low-gradient rivers, undercut trees that fall into the main channel (snags) are often the only stable habitat for invertebrates and provide a refuge and food resource for fishes as well. Invertebrates may use or require wood as food, but many species simply occupy wood as habitat. Although some species adapt to the woody environment by gouging and tunneling into the wood surface, others obtain their food from allochthonous and autochthonous resources that accumulate on the wood surfaces or are directly filtered from the water column. In streams of all sizes, accumulations of wood are often the hot spots of invertebrate diversity, and snags in Coastal Plain rivers of the southeastern United States support invertebrate production that is among the highest in lotic systems. The distribution of biomass and production among functional groups on wood varies greatly depending on the type of system. Loose streambed wood is colonized especially by shredders (gougers), and stable snags in larger streams and rivers are dominated by filterers and gatherers. High diversity of snag predators can result in complex food-web pathways, and snag taxa can be the major components of invertebrate drift in low-gradient rivers. Snag sampling is becoming a standard part of bioassessment, particularly in low-gradient systems, because snags are recognized as a major site of invertebrate diversity and production. Re-introduction of wood in streams and rivers is becoming an important aspect of restoration and management strategies around the world, as attempts are made to increase biodiversity and refuges both for fishes and their invertebrate prey.

Introduction

> It is interesting to contemplate an entangled bank ... with various insects flitting about, and worms crawling through damp earth ... the production of higher animals, directly follows. [Darwin 1859.]

Charles Darwin chose an entangled bank to illustrate the complexity of life in the final paragraph of his landmark book. What he may not have realized was that this entangled habitat probably continued beneath the surface of the stream or pond in the form of tree roots and dead wood. Associated with this submerged entanglement would have been an entirely different community of aquatic insects and other invertebrates. And similarly, the production of higher life forms (crayfishes, fishes, and so on) would directly follow in the aquatic as well as the terrestrial realm. The value of submerged wood as an important habitat for aquatic invertebrates, particularly in streams and rivers, has been known for at least 50 years (see early references in Hynes 1970), but it has received increased attention in the past 25 years. Although recognition has been slow, that wood can influence invertebrate communities in many ways is now apparent: creating habitats such as pools, providing substrate on which invertebrates can live, and serving as food. Consequently, wood and its associated invertebrates can be important in the functioning of aquatic systems.

The general study of invertebrates on wood in streams and rivers has coincided with an abundance of related work on the importance of wood to stream structure and function and as fish refuge and food source (for example, Harmon et al. 1986; Dolloff and Warren 2003; Wondzell and Bisson 2003; both this volume).

Much of the work on wood-dwelling invertebrates has been concentrated at specific sites in the United States: Midwestern U.S. rivers (Berner 1951; Morris et al. 1968; Modde and Schmulbach 1973; Nilsen and Larimore 1973; Nord and Schmulbach 1973), Oregon's Coast and Cascade range streams (Anderson et al. 1978, 1984; Dudley and Anderson 1982), southeastern U.S. Coastal Plain rivers (Benke et al. 1979, 1984, 1985, 2001; Cudney and Wallace 1980; Wallace et al. 1987; Benke and Wallace 1997), southeastern U.S. Coastal Plain streams (Smock et al. 1985, 1989, 1992; Thorp et al. 1985), southeastern U.S. Appalachian Mountain streams (Golladay and Webster 1988; Wallace et al. 1995, 1996, 1999), and south-central U.S. streams (Golladay and Hax 1995; Hax and Golladay 1998; Phillips 1997a, 1997b). Several relatively recent studies have been conducted in streams and rivers in Australia (O'Connor 1991, 1992; Humphries et al. 1996, 1998; McKie and Cranston 1998; Sheldon and Walker 1998; Growns et al. 1999), New Zealand (Tank and Winterbourn 1995, 1996; Quinn et al. 1997; Collier et al. 1998; Collier and Halliday 2000), and central Europe (Weigelhofer and Waringer 1999; Hoffmann and Hering 2000; Warmke and Hering 2000).

Some of the earliest hints of the importance of wood-dwelling invertebrates were related to their being a source of fish food (Berner 1951; Morris et al. 1968). Subsequent studies, however, have focused on a wide variety of questions: the role of wood-eating (xylophagous) taxa in processing wood, using wood as invertebrate habitat and food by nonxylophages, wood food webs, invertebrate production on wood, the role of debris dams in creating other habitats and affecting resource availability, effects of flow disturbance on wood invertebrates, and using wood-dwelling species in bioassessment.

Previous reviews of invertebrates on wood have focused on identifying species associated with wood (Dudley and Anderson 1982), how invertebrates use wood for habitat and food (Hoffmann and Hering 2000), and how wood contributes to invertebrate biodiversity in the southeastern United States (Wallace et al. 1996). We attempt to review and synthesize a wide variety of approaches that suggest wood is important for invertebrates in habitats ranging from small streams to large rivers. Unfortunately, wood has been removed from lotic systems of all sizes, often being associated with channelization and navigation in medium-to-large rivers. Consequently, much of the invertebrate diversity and production supported by woody habitats have disappeared, and their ecological functions can be studied and appreciated only in relatively undisturbed systems.

Wood-Created Habitat

The physical presence of wood creates invertebrate habitats in any freshwater ecosystem because it provides a solid substrate that decays slowly when submerged, and large wood lasts from decades to centuries (Harmon et al. 1986; Wallace et al. 1996, 2001a; Hyatt and Naiman 2001; Bilby 2003, this volume). Wood habitat often is found along the shoreline of natural lakes, where it blows in from surrounding forest, or in newly created reservoirs, where entire trees may be submerged (McLachlan 1970; Boon 1984; Bowen et al. 1998). More commonly, however, wood is a major natural feature of streams and rivers, where it accumulates from mortality of riparian trees or branches (Anderson and Sedell 1979; Triska and Cromack 1980; Wallace and Benke 1984). The major types of wood-created habitats in lotic systems are loose stream wood, represented by individual branches that lay on the stream bottom; wood dams, formed when trees fall across small-to-medium streams or when smaller branches fall and accumulate at specific stream sites; beaver ponds and wetlands, formed behind dams built by beavers (*Castor canadensis* in North America) on small-to-medium streams; and snags (fallen trees and branches), accumulating and often anchored along the banks of medium-to-large rivers (Figure 1).

Wood accumulations

Wood accumulations dramatically alter a stream's longitudinal profile by disrupting the flow of wa-

FIGURE 1. Wood-created habitats in lotic systems: (A) wood dam created by a log in a small stream in the southern Appalachian Mountains (note the plunge pool to the left and sediment accumulation to the right); (B) beaver dam built on a small stream in the southeastern U.S. Coastal Plain (note dam to the left consisting of logs and mud, and wetland with stumps of killed trees to the right); (C) snags along bank of a medium-sized river in the southeastern U.S. Coastal Plain at a very low river stage (note, all wood in the photograph will be submerged and colonized by aquatic invertebrates when the river rises.

ter and influencing an array of habitat features that affect the invertebrate assemblages and their role in the structure and function of stream communities (Swanson and Lienkaemper 1978; Harmon et al. 1986; Huryn and Wallace 1987; Trotter 1990; Gregory 1992; Wallace et al. 1995, 1999). In small, high-gradient streams, such obstructions create a stair-step profile that dramatically alters the streams' physical nature (Bilby and Likens 1980). Upstream of the dam, streams are deeper, current velocities lower, retention of particulate inorganic and organic matter higher, and substrate particles smaller. This depositional zone creates an invertebrate habitat different from that provided by erosional zones (such as riffles), with higher current velocities, larger substrate, and so on. A turbulent plunge pool may develop downstream of the dam, followed by riffle areas. All of these factors influence the composition of invertebrate assemblages, such as the distribution among functional feeding groups, and the role that they play in trophic and nutrient dynamics of stream ecosystems (Bilby 1981; Molles 1982; Newbold et al. 1982; Melillo et al. 1983; Smock et al. 1989).

Wood accumulations also are important features in low-gradient streams. Although the stair-step profile is far less pronounced, these obstructions still create depositional areas upstream that retain organic matter and represent a more pool-like environment (Smock et al. 1989). Alternatively, wood accumulations can develop along the margins of larger low-gradient streams and increase the diversity of benthic habitat patches (Palmer et al. 1996). The wood-formed dams of low-gradient streams, however, appear to be more of a hot spot for invertebrate diversity, abundance, biomass, and secondary production than those of high-gradient systems, given the unstable nature of sediments both upstream and downstream of the dam (Roeding and Smock 1989; Smock et al. 1989, 1992; Weigelhofer and Waringer 1999).

Beaver dams

In addition to debris dams formed by tree and limb fall and physical processes, dam-building by beavers can be a major force in creating ponds and wetlands, typically on first- to fourth-order streams (McDowell and Naiman 1986; Naiman et al. 1988; Clifford et al. 1993; Gurnell 1997; Pollock et al. 2003, this volume). These ecosystem engineers (Jones et al. 1994) topple trees, cut wood, and build and repair their dams. In doing so, they create pond and wetland habitats, many of which last for decades. Rather than simply creating a pool and riffle sequence like those with most dams formed by fluvial accumulations of wood, multiple beaver dams along a stream can create a major shift in the balance between lotic and lentic habitats for invertebrates (Naiman et al. 1988). Such lentic environments are colonized by an entirely different invertebrate assemblage than found in lotic habitats (McDowell and Naiman 1986; Benke et al. 1999; Wissinger and Gallagher 1999; Rolauffs et al. 2001). Beaver dams create pond-like environments in regions of relatively high gradient or shallow wetland environments in areas of low gradient. The beaver dam itself represents a habitat that can be colonized by invertebrates (Clifford et al. 1993; Rolauffs et al. 2001). Like most dams, beaver dams tend to modulate discharge regimes and alter stream chemistry downstream, which strongly affect invertebrate diversity and abundance (Smith et al. 1991).

River snags

In addition to the profound effect of wood in forming impoundments on first- to fourth-order streams, stabilized wood can itself be a major habitat, especially in medium-to-large rivers and particularly in Coastal Plain regions where the gradient is low (Wallace and Benke 1984; Maser and Sedell 1994). Riparian trees are undercut and fall along the bank of the channel, creating snag habitats that can remain from decades to centuries because the rate of decomposition of inundated wood is so slow. Historical records show that wood was abundant in many rivers of the United States (Sedell and Luchessa 1982; Sedell et al. 1982, 1990; Sedell and Froggatt 1984; Wallace and Benke 1984). During much of the 19th century and early 20th century, however, extensive snag removal eliminated most wood from all large rivers in the United States, including the Mississippi, and we can only speculate about the importance of this wood to these historical aquatic habitats and to the invertebrate species that must have colonized them. Today, wood is still seen in considerable abundance in some of the medium-sized low-gradient rivers of the U.S. Coastal Plain (Wallace and Benke 1984; Benke and Wallace 1990), and its important role for invertebrate communities has become apparent (Benke et al. 1979, 1984; Cudney and Wallace 1980).

In some large rivers, the accumulation of wood in the channel rivaled the influence of wood dams and beaver dams in altering the character of small streams. A well-documented example was the formation of the "Great Raft" on the Red River in northwestern Louisiana, USA (Triska 1984; McCall 1988). Over a period from the late 1400s to the mid-1800s, wood from riparian trees formed enormous debris dams that blocked the channel for more than 400 km and formed a series of large lakes along the Red River (Triska 1984). About 70 years (1833–1904) of wood dam and snag removal, levee projects, and dredging ultimately resulted in a cleared, wide meandering channel (Triska 1984). We do not know how many lowland rivers had wood similar to the Great Raft (Triska 1984). The Great Raft and studies of wood in smaller rivers only provide a clue to what ecological conditions and invertebrate assemblages of larger rivers must have been. Even in desert rivers, where snags are uncommon, driftwood can provide an important substrate that expands the diversity of niches available to invertebrates, as has been shown by recent work in the Colorado River (Haden et al. 1999).

In summary, wood plays a major role in creating habitats in streams and rivers of many sizes and in many geographic areas, and this role is often in forming dams that create lentic depositional environments along the river's course. In spite of this very important habitat-creating role of wood, much of the remainder of our review will primarily address the invertebrates that actually colonize wood surfaces and use it for habitat and food.

Wood as Habitat and a Potential Food Source

That invertebrates colonize any solid substrate in flowing water is widely known (Berner 1951; Fremling 1960). For example, artificial substrate samplers made of mineral (such as a rock basket) or organic (such as a Hester-Dendy multiplate sampler) materials have been used for many years to assess water quality. Similarly, that invertebrates readily colonize the branches of inundated trees in both lotic (Berner 1951; Nilsen and Larimore 1973; Benke et al. 1979, 1984; Cudney and Wallace 1980) and lentic systems (McLachlan, 1970; Boon 1984) has been known for decades. Invertebrates are also known to colonize leaves and small branches deposited on stream bottoms (Anderson et al. 1978; Anderson and Sedell 1979). Although mineral substrates have received much more attention, wood sometimes can provide a superior invertebrate habitat. Like mineral substrates, wood can be relatively stable if anchored in the stream, can provide good sites for invertebrate attachment, and can allow colonization of biofilms and plants that provide additional habitat and food. Unlike mineral substrates, however, wood can be eaten by some species, is soft enough to be gouged to create refuges, and is particularly effective in trapping other organic matter with its branching structure. Thus, wood can provide access to a wider variety of food than is available on mineral substrates. This same complexity also creates habitats and provides refuges for motile animals that do not necessarily colonize the wood itself, a well known phenomenon for fishes (Fausch and Northcote 1992; Lehtinen et al. 1997; Monzyk et al. 1997; Quist and Guy 2001; Dolloff and Warren 2003; Zalewski et al. 2003, this volume) but a less well-established one for large invertebrates such as shrimp (Everett and Ruiz 1993; Pyron et al. 1999).

Invertebrates use wood in many ways during all stages of their life cycle, regardless of their food source (Table 1). Adults use wood for resting and reproductive activities, and they deposit eggs on wood both above and below the water line. Many invertebrates, particularly insect larvae, use wood as a refuge by hiding under loose bark, by tunneling, and by occupying crevices or creating them by gouging (such as the filtering caddisfly *Macrostemum*). Some insect larvae use small pieces of bark and twigs to build mobile cases, such as the caddisflies *Heteroplectron* and *Pycnopsyche*. Perhaps the major reason for invertebrates being found on submerged wood is that they can exploit numerous food resources by using a wide variety of feeding strategies. Finally, many insects enter the pupal stage firmly attached to the wood surface or in the wood or use partially submerged wood to climb out of the water for emergence.

Dudley and Anderson (1982) were among the first to draw attention to the potential for wood as an invertebrate food source by surveying various wood-associated invertebrates in North America. More recently, Hoffmann and Hering (2000) classified the fauna associated with wood into three categories based on the relative degree of wood-feeding (xylophagy) exhibited by various species of macroinvertebrates. Some species are found *only* on wood and restricted to feeding on it (obligate xylophages). Others feed on wood to some degree (facultative xylophages) but also eat leaf litter. And some species have an affinity for wood substrates, but little of their diet is contributed by wood itself (nonxylophage). A more traditional classification would place xylophages into the shredder functional feeding group (Cummins and Merritt 1996; Wallace and Webster 1996). Nonxylophagous species would fall into any of the other functional groups (scrapers, gatherers, filterers, and predators).

Obligate xylophagous macroinvertebrates are dominated by miners, gougers, and tunnelers (Hoffmann and Hering 2000). These authors recognized some caddisflies (Trichoptera), beetles (Coleoptera), and true flies (Diptera) as obligate xylophages in central Europe (their Table 3), but many examples come from other parts of the world. Examples common to Europe and North America include caddisflies such as several *Heteroplectron* spp., elmid beetles, chironomids such as *Stenochironomus* spp., and tipulids such as *Lipsothrix* spp. Many species of chironomids can be especially common as wood miners and tunnelers (Cranston and Oliver 1988; Borkent 1984; Kaufman and King 1987; Anderson 1989; Cranston and Hardwick 1996). The chironomid *Xylopus par* tunneled into soft (decayed) wood to depths of 2 cm, and attained densities of $7,600/m^2$ on natural wood and more than $20,000/m^2$ on wood baits (Kaufman and King 1987). *Lipsothrix* spp. are associated with wood in advanced stages of decomposition, but the elmid *Lara avara* and the caddisfly *H. californicum* are gougers of firm waterlogged wood (Anderson et al. 1978; Anderson 1989). Presumably, the wood gouged and consumed by *L. avara* is very low in food quality because the beetle's larvae have the slowest growth rates and the longest life cycles (5–6 years) published for a stream insect (Steedman and Anderson 1985). On the other hand, many species of chironomids, such as *X. par*, can complete development in a year or less (Kaufman and King 1987). Collier and Halliday (2000) used gut and stable isotope analyses to demonstrate that 66% of the body carbon of *Pycnocentria funera* was derived from wood, with lesser amounts coming from wood for other co-existing species, in several New Zealand streams.

Facultative xylophagous macroinvertebrates are common in small streams where wood forms dams or lies on the streambed, rather than as a fixed element in the current. Dudley and Anderson (1982) identified several invertebrates that were associated with wood in North America, but the degree of xylophagy was not specified. In a companion paper, however, Pereira et al. (1982) examined gut contents of 108 taxa of lotic insects and found that 45 of them contained a large amount of wood; some of the insects probably required wood as food and others did not. Other early studies suggested that xylophagy was relatively common in streams in New Zealand (Anderson 1982) and Australia (Chessman 1986). Hoffmann and Hering (2000) recognized several species of facultative xylophages in Europe (their Table 2), but many additional taxa were suspected

Table 1. Potential uses of wood by stream invertebrates (modified from Wallace et al. 1996).[a]

Invertebrate use of wood	Comments	References
Adult activities	Resting sites; copulation; emergence	Dudley and Anderson 1982 Steedman and Anderson 1985 Dudley and Anderson 1987
Oviposition site	Eggs deposited above or below waterline	Dudley and Anderson 1982 Hoffman 2000
Direct food source		
Xylophagy (obligate or facultative)	Feeding directly on wood and epixylic biofilm coating it	Pereira et al. 1982 Anderson et al. 1984 Anderson et al. 1978 Hoffmann 2000 Steedman and Anderson 1985 Chergui and Pattee 1991 Roeding and Smock 1989 Collier and Halliday 2000
Predators of primary consumers	Invertebrate predators feeding on other snag invertebrates	Wallace et al. 1987 Benke et al. 2001 Smith and Smock 1992
Habitat and indirect food source		
Retention of coarse particulate organic matter as food	Litter trapped by wood and consumed by shredders	Wallace et al. 1987 Bilby and Likens 1980 Wallace et al. 1995 Smock et al. 1989 Roeding and Smock 1989 Díez et al. 2000
Retention of fine particulate organic matter as food	Seston (microbially enriched) trapped by wood and consumed by gatherers	Wallace et al. 1987 Edwards and Meyer 1990 Benke et al. 1992 Bilby and Likens 1980 Wallace et al. 1995 Díez et al. 2000
Biofilm on wood as food	Biofilm growth on snags consumed by grazers or gatherers	Wallace et al. 1987 Couch and Meyer 1992 Wallace et al. 1996 Tank and Winterbourn 1995, 1996 Golladay and Sinsabaugh 1991 Tank and Webster 1998 Tank et al. 1998
Filter-feeding of seston	Insect attachment to snags (such as black flies) or net building (hydropsychid caddisflies) for filtering	Wallace et al. 1987 Benke and Wallace 1997 Edwards and Meyer 1987 Edwards 1987 Wallace et al. 1977
Refuge from predators		
Woody material for cases	Caddisfly cases	Anderson et al. 1978 Wiggins 1998
Retreat construction	Gouging to build retreats (hydropsychid caddisflies)	Wallace and Sherberger 1974 Wallace et al. 1977
Hiding places	Crevices, loose bark	Anderson et al. 1978
Pupation site	Stable substratum or refuge under loose bark	Dudley and Anderson 1982 Hoffman 2000

[a] See also Hoffman and Hering (2000).

of requiring wood (their Table 4). One interesting example of a facultative xylophagous insect is the lepidostomatid caddisfly *Lasiocephala basalis* (Kol.), described from a third-order German stream by Hoffmann (2000). This "shredding" caddisfly's gut contents varied according to the substrate with which larvae were associated, consisting largely of wood fragments when the insect was found on wood or alder roots, leaf material and amorphous detritus when it was found in leaf litter, and amorphous detritus when it was collected from stone substrates. Similarly, Hall et al. (2000) showed that when litter was experimentally excluded from a southern Appalachian stream, *Tipula* spp. shifted from a diet of predominantly leaf litter to one of mostly wood.

Among the nonxylophagous invertebrates are scrapers and gatherers that feed on epixylic biofilms that often coat submersed wood. These biofilms consist of algae, fungi, bacteria, protists, amorphous detritus, and extracellular polysaccharides, which can develop from in situ production or from the deposition or flocculation of seston from the water column (Couch and Meyer 1992; Tank and Webster 1998). Virtually all components of epixylic biofilms are consumed by scrapers and gatherers; when gut contents are examined, what is found is distinguished only as amorphous detritus (Wallace et al. 1987). In some systems, fungi may develop more extensively on woody substrates than on leaf detritus (Golladay and Sinsabaugh 1991), and in New Zealand streams, ^{14}C-glucose incorporated into fungi and bacteria was at least partially assimilated by grazing amphipods (Tank and Winterbourn 1995). In U.S. Coastal Plain rivers, radio-labeled bacteria, deposited on snags from the seston, were shown to be readily assimilated by gathering mayflies (Edwards and Meyer 1990). Epixylic biofilms in Coastal Plain rivers may be enhanced by flocculated dissolved organic carbon and fine seston particles from the water column that provides the bulk of food for highly productive gathering invertebrates (Wallace et al. 1987; Benke et al. 1992; Couch and Meyer 1992; Benke and Jacobi 1994). Biofilms are known to develop within two weeks after exposure of wood to river water (Couch and Meyer 1992), which corresponds well with rapid invertebrate colonization on introduced woody substrates (Van Arsdall 1977; Thorp et al. 1985). Macroinvertebrate colonization and species richness has been positively correlated with development of biofilm on organic substrates (Hax and Golladay 1993).

Filterers also can be an important component of nonxylophagous invertebrates on wood, particularly on snags in Coastal Plain rivers where the benthic habitats are dominated by shifting sand and stable substrate is limited (Cudney and Wallace 1980; Benke et al. 1984; Smock et al. 1985; Benke and Wallace 1997). Filterers do not necessarily depend on biofilms because their food (drifting organic particles and drifting animals) is produced elsewhere in the system and delivered by the current. Gut analyses of snag-dwelling filterers from the Ogeechee River showed that they consumed large quantities of amorphous detritus from the seston (Wallace et al. 1987). Analyses of seston and feeding studies using labeled foods supplied to filtering black flies suggest that bacteria, bacterial extracellular polysaccharide, and protists may all be important components in the diet of filterers in the Ogeechee River (Edwards 1987; Edwards and Meyer 1987; Carlough and Meyer 1989, 1991; Carlough 1994; Couch et al. 1996). Other dominant filtering groups, such as chironomid midges and caddisflies (Wallace et al. 1987), are also likely to depend heavily on these microbial components.

Predators are a nonxylophagous group inhibiting wood substrates that have received relatively little attention. Most studies of wood-dwelling predators have been conducted in the southeastern United States; they suggest that predaceous species (dragonflies, damselflies, perlid stoneflies, megalopterans, ceratopogonids, and tanypodine chironomids) could be found on any solid substrate (Benke et al. 1984, 2001; Smock et al. 1985; Smith and Smock 1992). Nonetheless, the flattened bodies of several perlid stoneflies and the hellgrammite *Corydalus cornutus* are well adapted to hiding under loose bark of well-conditioned wood. Not surprisingly, these wood-dwelling predators consume coexisting primary consumers, but detailed analyses suggest that predators of differing sizes and species have specific preferences for prey (Smith and Smock 1992; Benke et al. 2001). In addition to insects generally considered to be predators, hydropsychid caddisflies from Ogeechee River snags are omnivores that consume substantial numbers of chironomids and mayflies (Benke and Wallace 1997; Benke et al. 2001).

Invertebrate Assemblages on Wood

Quantitative studies are often necessary for understanding the structural and functional characteristics of invertebrate assemblages found on wood. Such studies require both sampling invertebrates and quantifying the wood itself. Invertebrate assemblages have been assessed in various ways: by determining invertebrate life history, density, biomass, population dynamics, functional groups, secondary production, and food webs; and by contributions to drift, fish prey, and stream biodiversity.

Quantifying invertebrate assemblages on wood

Sampling wood-dwelling invertebrates in lotic systems can be difficult, particularly in large rivers. Access to wood in large rivers is virtually impossible from depths greater than 1 m, especially with swift currents and high turbidity. Scuba diving or snorkeling in swift currents near submerged trees can be dangerous. Therefore, unless waters are shallow (<1 m), wood samples have usually been collected from a boat where portions of the wood are just below the surface or partially sticking out of the water. Large wood is especially difficult to sample because it cannot be lifted from the water by hand. Sampling wood in small streams can be somewhat easier, but it is fraught with problems—such as whether the wood is normally submerged or inaccessible to aquatic fauna because it is buried in sediments or out of the water, or obtaining a "representative" sample from a wood-formed dam.

To assess the importance of invertebrates on wood, not only must wood samples be retrieved, but their contributions must be compared to those found in traditional habitats on the streambed (that is, sand, mud, gravel, cobble, and bedrock). Such a comparison may include at least two major steps: developing a quantitative sampling approach in which population variables (density, biomass, production) can be calculated per unit area of habitat surface and quantifying the wood habitat itself to convert habitat-specific values to streambed values.

The general sampling approach is to remove the wood from the stream in a way that minimizes loss of invertebrates. Once the wood is removed from the stream, invertebrates are typically brushed, picked, and washed from the wood surface. Then, the wood surface area is calculated so that density, biomass, or production can be presented per square meter of wood surface. In the case of wood-mining chironomids, however, a combination of multiple washings and emergence over several months has been necessary to estimate density (Anderson 1989).

Removing wood from streams and rivers can be as simple as carefully lifting the wood from the water and cutting it into transportable pieces (Nilsen and Larimore 1973; Cudney and Wallace 1980; Golladay and Hax 1995; Phillips 1997b; Hax and Golladay 1998; Spänhoff et al. 2000). Alternatively, some investigators have surrounded the snag with a collecting device so that organisms are not lost during retrieval. Smock et al. (1985) sampled snags in small streams by placing a bucket around the wood before cutting it. Others have collected snags by placing a mesh net around the snag before cutting off a section with a saw (O'Connor 1992; Phillips and Kilambi 1994a, 1994b).

Some investigators have built devices specifically for snag sampling. Benke et al. (1984) used a 45-cm longitudinal sieve with handles and a notch at each end (Figure 2). It was placed under an intact snag (submerged tree branch) and then lifted out of the water so that the wood at each end could be cut off with clippers or a saw. Delong et al. (1993) designed a more elaborate sampler consisting of plastic tubing and Nitex mesh that could be closed around the snag before it is cut. Unlike previously described samplers, the one designed by Growns et al. (1999; snag bag) could be used for large wood (>10-cm diameter) impossible to retrieve from lowland Australian rivers. Their mesh bag was wrapped around the snag and held in place with Velcro strips as invertebrates were brushed from the snag into the net.

Once invertebrates are removed from individual snags, the surface area sampled must be

FIGURE 2. Longitudinal sieve sampler (45 cm long) for retrieving snags up to 8 cm in diameter. The sampler is placed under snag before it is pulled out of the water. Excess wood is clipped or sawed off each end, and the remaining piece is placed in a plastic bag, along with any invertebrates that dropped into the sieve.

measured if numbers are to be converted to square meter of snag surface. Most investigators measure the length and diameter of wood in the laboratory after the invertebrates have been removed (Cudney and Wallace 1980; Benke et al. 1984), but Growns et al. (1999) measured wood in place after they brushed invertebrates into their snag bag. Spänhoff et al. (2000) estimated wood surface area in the laboratory by covering the wood with aluminum foil after invertebrates were removed, then weighing it and determining an area-to-weight relation.

Sampling wood from accumulations can be more difficult than sampling individual branches because wood pieces are jammed together tightly and attempts to dislodge single pieces would likely lose many of the invertebrate colonists. Smock and others (Roeding and Smock 1989; Smock et al. 1989, 1992) sampled entire wood dams by placing a 5-mm seine downstream of each dam to collect large invertebrates swept away as the dam was disassembled. Numbers were converted to square meter of streambed, based on the surface areas covered by the dam. Weigelhofer and Waringer (1999) sampled the vertical distribution in invertebrates in wood dams by driving a 10-cm-diameter plastic tube (with a 100-mm mesh net at the bottom end) horizontally into the dam in an upstream direction. Protruding twigs were cut from the anterior opening with clippers as the tube was withdrawn from the dam. Their volumetric estimates of wood could also be converted to numbers per square meter of streambed. Quantitative sampling of invertebrates from beaver dams has proved extremely difficult because they are typically too large to be sampled as a unit as Smock et al. (1989) did with small wood accumulations. Clifford et al. (1993) collected qualitative samples by disturbing and disassembling small sections of dam so that invertebrates were swept into a net. Rolauffs et al. (2001) obtained indirect estimates of invertebrate abundance and production in stream, beaver dam, and beaver pond through the use of emergence traps.

Because natural snags are often difficult to retrieve from rivers, several investigators have used artificial snags in attempts to characterize and quantify the wood assemblages (Nilsen and Larimore 1973; Thorp et al. 1985; Anderson 1989; O'Connor 1991; Hax and Golladay 1993; Tank and Winterbourn 1995, 1996; Magoulick 1998; McKie and Cranston 1998; Spänhoff et al. 2000). Humphries et al. (1998) compared various methods for sampling invertebrates in large rivers and found that artificial snags were superior to onion bag baskets, airlift sampling of sediments, and sweep-net sampling of edges for both diversity and quantitative analyses. For some xylophagous taxa, however, it may be necessary to condition wood in the stream for several years before it becomes a suitable habitat (Anderson 1989). Although artificial snags can be useful for various purposes (such as estimating colonization fluxes), assessing how representative they might be is impossible without quantitative sampling of natural snags.

If numbers of invertebrates per snag surface area are to be converted to a common unit for an entire lotic system, the amount of wood in the stream or river must be quantified. Wallace and Benke (1984) modified a line-intersect technique used by foresters so that surface area of wood could be calculated per square meter of riverbed as a function of river height in the Ogeechee River, a Coastal Plain river in the southeastern United States. Then, the number of invertebrates per square meter of snag surface could be converted to number per square meter of riverbed (Benke and Parsons 1990). O'Connor (1992) and Humphries et al. (1996) used the same method to estimate invertebrate densities on wood in lowland Australian streams and rivers. In contrast, Smock et al. (1985) estimated the surface area of all inundated pieces of wood (a census) along a 50-m reach of a second-order Coastal Plain stream in South Carolina. Gippel et al. (1996) compared the line-intersect method with census estimates and aerial photography in a lowland Australian river and concluded that the line-intersect method overestimated wood abundance, and offered suggestions about how overestimation could be reduced. Similarly, when Wallace et al. (2001a) compared the line-intersect method with direct wood removal in a mountain stream, they also found the method produced overestimates for large wood. In contrast, small wood was slightly underestimated by the line-intersect method compared to actual wood removal (Wallace et al. 2001b). Because the line-intersect method is the only alternative means of quantifying wood, aside from a complete wood census or aerial photography, investigators should take steps to assess its accuracy in their specific situation.

Abundance, biomass, production, and biodiversity

Invertebrates on wood have been quantified in studies from around the world, but the variation

TABLE 2. Quantitative studies of invertebrates on wood in streams and rivers; studies are grouped by location (state or country).[a]

Stream and rivers (no. streams or sites)	Location (state or country)	Stream order	Discharge (m^3/s)	Taxa included	Type of analysis	Reference
Coast Range (3)	Oregon	1, 3, 6	0.3, 0.8, 38	All	B	Anderson et al. 1978 Anderson et al. 1984
Cascade Range (4)	Oregon	1, 3, 5, 7	0.15, 2.2, 20, 200	All	B	Anderson et al. 1978
Cascade and Coast ranges (13)	Oregon			Tipulidae	N	Dudley and Anderson 1987
Berry Creek	Oregon	2	<0.015	All	N	Anderson 1989
San Joaquin River (25)	California			All	N, R	Brown and May 2000
Satilla River (2)	Georgia	6	28-62	All	N, B, P, R	Benke et al. 1979, 1984
Ogeechee River	Georgia	6	67	Simuliidae	N, B, P, R	Benke and Parsons 1990
Ogeechee River	Georgia	6	67	Ephemeroptera	N, B, P, R	Jacobi and Benke 1991 Benke and Jacobi 1994
Ogeechee River	Georgia	6	67	Trichoptera	N, B, P, R	Benke and Wallace 1997
Ogeechee River	Georgia	6	67	Chironomidae	N, B, P, R	Benke 1998
Ogeechee River	Georgia	6	67	Coleoptera	N, B, P, R	Benke 2002
Ogeechee River	Georgia	6	67	Predators	N, B, P, R	Benke et al. 2001
Savannah River	Georgia			Trichoptera	N, B, P, R	Cudney and Wallace 1980
Cedar Creek	South Carolina	2	1.2	All	N, B, P, R	Smock et al. 1985
Steel Creek	South Carolina		1.3	All	N, R	Thorp et al. 1985
Little Tennessee River	North Carolina	6		All	B	Wallace et al. 1996
Buzzard's Branch and Colliers Creek	Virginia	1	<0.1	All	N, B, P, R	Smock et al. 1989, 1992

TABLE 2. Continued.

Stream and rivers (no. streams or sites)	Location (state or country)	Stream order	Discharge (m^3/s)	Taxa included	Type of analysis	Reference
Buzzards' Branch	Virginia	1	<0.1	predators	N, B, P, R	Smith and Smock 1992
Fleming Creek	Michigan	3		All	N, R	Magoulick 1998
Chippewa River	Michigan	4		Chironomidae	N, B	Kaufman and King 1987
St. Regis River	New York	4		All	N, R	Hax and Golladay 1993
3 streams	Arkansas	1, 2, 3, 4		Trichoptera	N, R	Phillips 1994
3 streams	Arkansas	1, 2, 3, 4		Coleoptera	N, R	Phillips 1995
3 streams	Arkansas	1, 2, 3, 4		Plecoptera	N, R	Phillips and Kilambi 1994a
3 streams	Arkansas	1, 2, 3, 4		Ephemeroptera	N, R	Phillips and Kilambi 1994b
3 streams	Arkansas	1, 2, 3, 4		Diptera	N, R	Phillips and Kilambi 1994c
3 streams	Arkansas and Texas	2, 3, 4		Coleoptera	N, B, P	Phillips 1997a
3 streams	Arkansas and Texas	2, 3, 4		Coleoptera	N, B, P	Phillips 1997b
Sister Grove Creek	Texas	Intermittent		All (macro)	N, R	Hax and Golladay 1998
Sister Grove Creek	Texas	Intermittent		Meiofauna	N	Golladay and Hax 1995
Elm Fork	Texas			Trichoptera	N, B, P	Johnson et al. 1998
Marmaton River	Missouri	7	8.8	All	N, B	Rabeni and Hoel 2000
Marais des Cygnes River	Missouri	7	61	All	N, B	Rabeni and Hoel 2000
Pranjip-Creightons Creek	Australia		2	15 most common	N, B	O'Connor 1992
Ladberger Mühlenback and Ibbenbürener Aa	Germany	2, 3		All	N, R	Spänhoff et al. 2000
Weidlingbach (Danube)	Austria	3	0.045	All	N	Weigelhofer and Waringer 1999

[a]N = density, B = biomass, P = production, R = taxonomic richness.

in approaches and lack of consistent units of measure make comparisons and generalities difficult (Table 2). Sampling has sometimes been done with natural snags and wood accumulations, and at other times, colonization of introduced woody substrates has been measured. Invertebrates on wood have been quantified by using density, biomass, or production, sometimes apportioning one of more of these measures into functional feeding groups (Cummins and Merritt 1996; Wallace and Webster 1996). Sometimes, entire invertebrate assemblages have been analyzed, but other studies have focused on single species, small species groups, or orders (Table 2).

Density (number per square meter of snag surface area) for entire invertebrate assemblages has been estimated in several studies and varies greatly (Table 3). Invertebrate density estimates on snags from Germany, Australia, Texas, and Missouri were relatively low (<5,000/m^2), but most other estimates approached or exceeded 10,000/m^2. In an early study on the small (about 1 m^3/s) Kaskaskia River, Illinois, densities averaged more than 22,000/m^2 on three natural logs but reached 96,000/m^2 on introduced wood substrates heavily colonized by oligochaetes (Nilsen and Larimore 1973). Average density on snags collected monthly from two sites on the Satilla River, a medium-sized Coastal Plain river in Georgia, exceeded 26,000/m^2 (Benke et al. 1984). Average density from the Ogeechee River, also in the Georgia Coastal Plain, was greater than 97,000/m^2, about 70% of which were chironomids (see Benke 1998 and other references for Ogeechee River in Table 2). Average density from Cedar Creek, a second-order Coastal Plain stream in South Carolina was somewhat lower (14,900/m^2) than that found in larger rivers (Smock et al. 1985). Mean density for wood accumulations in Buzzard's Branch, a small Coastal Plain stream in Virginia, was greater than 22,000/m^2, but this estimate was per square meter of streambed directly under the dam, not surface area of wood (Smock et al. 1989). High density also has been found in studies that addressed only a single group (for example, Cudney and Wallace (1980) found mean

TABLE 3. Density, biomass, and production for entire invertebrate assemblages on wood surfaces from streams and rivers throughout the world from sampling natural snags, wood dams, or from colonization studies; values from colonization studies are shown in parentheses. In some cases, mean values were calculated where there was more than one site, stream year, or sampling approach. For river size and literature sources, see Table 2.

Stream and location	Density (No./m^2)	Biomass (g dry mass/m^2)	Production (g dry mass/m^2/year)
Upper Satilla River, Georgia	33,315	3.4	72.2
Lower Satilla River, Georgia	26,207	5.8	57.4
Ogeechee River, Georgia	97,704	6.4	147.6
Collier Creek, Virginia	8,915	0.4	2.8
Buzzards Branch, Virginia (DD)[a]	22,302	5.2	27.4
Cedar Creek, South Carolina	14,900	1.1	7.4
Steel Creek, South Carolina	(6,777)		
Little Tennessee River, North Carolina		2.7	
Marmaton River, Missouri	977	0.5	
Marais des Cygnes River, Missouri	471	0.3	
Coastal Mountains streams, Oregon		1.0	
Cascade Mountains streams, Oregon		0.7	
Berry Creek, Oregon	2,475		
San Joaquin River, California	4,700		
Kaskaskia River, Illinois	22,110	0.7	
	(96,060)	(0.3)	
Sister Grove Creek, Texas	2,780		
Fleming Creek, Michigan	(14,000)		
Pranjip-Creightons Creek, Australia	3,086	1.6	
Ladberger Mühlenback and Ibbenbürener Aa, Germany	1,464		
	(3,382)		

[a] Wood dam

density for net-spinning caddisflies varied from almost 6,000/m^2 to more than 22,000/m^2 in the Savannah River, Georgia, depending on current velocity). Low density can sometimes be explained, particularly by sampling procedures. For example, the low density of macroinvertebrates (mean <2,800/m^2) sampled in a Texas stream surely resulted from using a coarse mesh (500-µm) sieve (Hax and Golladay 1998). The same authors estimated meiofauna density of 22,000–224,000/m^2 using a fine mesh (45-µm) sieve (Golladay and Hax 1995). In German streams (Spänhoff et al. 2000), low invertebrate density were found on wood pieces lifted from the stream bottom where they were sometimes partially buried in sand. They probably did not experience the same type of environment as in those studies with extremely high density (such as in the Ogeechee and Satilla rivers). In summary, invertebrate density was usually very high on wood substrates, particularly those that were anchored and exposed to the current, suggesting that wood is a highly attractive habitat site for invertebrates in many lotic systems. If wood simply lies on the streambed where it may become partially buried, it will be attractive to shredders, but it is unlikely to be as attractive to filterers and gatherers that might otherwise reach high density.

Biomass (grams of dry mass per square meter of snag surface) for entire invertebrate assemblages also has been estimated in several studies but not necessarily for the same streams as for density (Table 3). Biomass estimates also varied greatly, possibly for some of the same reasons as suggested for density. In general, invertebrate biomass was relatively high, usually approaching or exceeding 1 g/m^2. The highest values were found on snags in the two Coastal Plain rivers in Georgia (Ogeechee and Satilla rivers) and for the wood accumulations in Buzzards Branch, Virginia. Invertebrate biomass on wood lying on the bottom of Oregon streams was generally lower than found on stabilized snags of the southeastern U.S. Coastal Plain.

Annual production (grams dry mass per square meter each year) studies for entire invertebrate assemblages have been done only in southeastern U.S. Coastal Plain streams and rivers (Table 3). Studies on the Satilla River were the first to establish that production on snags is often higher than in benthic habitats (Benke et al. 1979, 1984). Cudney and Wallace (1980) showed that production of net-spinning caddisflies on snag was similarly high in the Savannah River. The highest invertebrate production on snags was found in the Ogeechee River, a large fraction of which was from chironomids (Benke 1998; see other sources in Table 2), which function as the base of a complex invertebrate food web. Production on snags in the much smaller Cedar Creek (Smock et al. 1985) was substantially lower than on snags in the Satilla and Ogeechee rivers. Subsequent studies by Smock et al. (1989, 1992) showed that debris dams were a site of relatively high production in some (Buzzards Branch), but not all (Colliers Creek), Coastal Plain streams.

High invertebrate production on snags in southeastern Coastal Plain rivers results from both high biomass and high biomass turnover rate. In the Ogeechee River, growth rates were particularly high for chironomids (up to 90% per day; Stites and Benke 1989; Hauer and Benke 1991), black flies (up to 43% per day; Hauer and Benke 1987), and mayflies (up to 35% per day; Benke and Jacobi 1986; Benke et al. 1992), in part because of relatively high temperatures (>20°C) most of the year. High biomass and high growth rates are possible for invertebrates that colonize stable snags because they are able to exploit materials produced in other habitats and carried to them by the current (Cudney and Wallace 1980). Furthermore, the amorphous detritus carried to these insects by the current is microbially enriched and thus high in food value (Edwards 1987; Edwards and Meyer 1987, 1990; Carlough and Meyer 1989; Benke et al. 1992; Carlough 1994; Couch et al. 1996).

Several studies have shown that when invertebrate production is considered by specific habitat, snags can be a production "hot spot" compared to other habitats in some streams (Wallace et al. 1996). This phenomenon was particularly true for rivers in the southeastern U.S. Coastal Plain (Satilla and Ogeechee), where production on snags was much higher than was found in the shifting sand of the riverbed (Figure 3). It was also true of Cedar Creek, although total channel production was lower than in the larger rivers. Similarly, production in the wood dams of Buzzards Branch was higher than in sediments, but the reverse was true of Colliers Creek because of the relatively stable organic matter in the sediments. Some additional exceptions to the pattern in Figure 3 are notable. In the Ogeechee River, production was much higher on snags than in sediments, but biomass in sediments was almost as high on snags because of the nonnative mollusk *Corbicula fluminea* in sediments. In the Little Tennessee River, biomass was lower on snags than on macrophyte-covered cobbles that provide a stable three-dimensional habitat in the streambed.

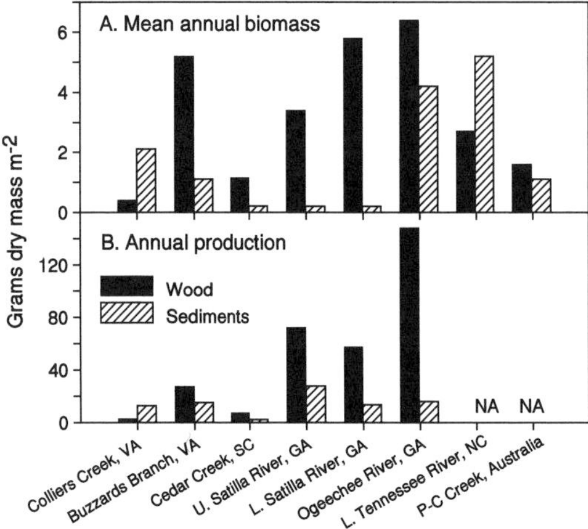

FIGURE 3. Mean annual biomass and production of invertebrates on wood surfaces compared to sediments for several streams. All units are per square meter of habitat surface area. Most wood surfaces are of snags, but Colliers Creek and Buzzards Branch are per square meter of streambed under wood dams. Sources: Colliers Creek and Buzzards Branch, Virginia (Smock et al. 1992); Cedar Creek, South Carolina (Smock et al. 1985); upper and lower Satilla River, Georgia (Benke et al. 1984); Ogeechee River, Georgia (from several papers in Table 2); Little Tennessee River, North Carolina (Wallace et al. 1996); and Pranjip-Creightons Creek, Australia (O'Connor 1992).

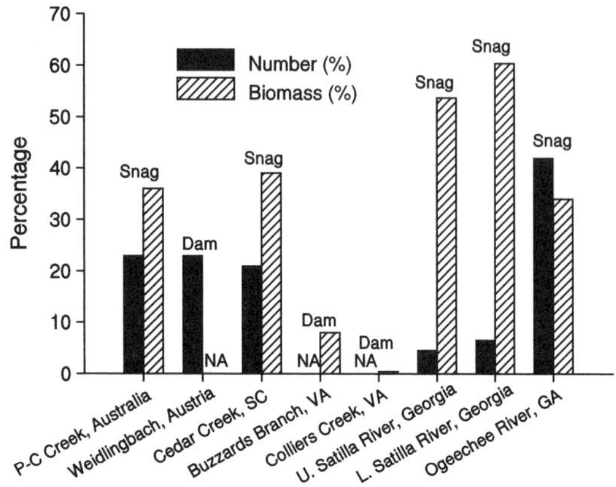

FIGURE 4. Percentage of snag invertebrates that contribute to total numbers and biomass in several streams. Some contributions are from invertebrates found in wood dams (Austria, Virginia) rather than snags. Sources: Pranjip-Creightons Creek, Australia (O'Connor 1992); Weidlingbach, Austria (Wiegelhofer and Waringer 1999); Cedar Creek, South Carolina (Smock et al. 1985); Buzzards Branch and Colliers Creek, Virginia (Smock et al. 1989, 1992); upper and lower Satilla River, Georgia (Benke et al. 1984); Ogeechee River, Georgia (Benke 2001). NA means values were not available.

Clearly, when wood is the only stable substrate in lotic systems, it will be a center of invertebrate biomass and production. When other stable substrates are abundant, however, wood may only be a hot spot for xylophagous species or shredders commonly associated with detritus (also see Humphries et al. 1996).

Although wood can be a site of intense invertebrate activity (Table 3; Figure 3), wood does not necessarily contribute significant invertebrate numbers, biomass, or production for the stream as whole. In studies that included attempts to quantify wood habitat, however, the numbers per habitat surface area can be converted to numbers per unit area of streambed and thus make direct comparisons among habitats possible (Wallace and Benke 1984; Smock et al. 1992; Benke 2001). Percentages of abundance and biomass on snag habitats compared to other habitats are shown for selected streams in Figure 4. Clearly, snags often contain more than 20% of total invertebrate numbers and more than 30% of invertebrate biomass when values are converted to area of streambed. The most extreme case was for the Satilla River, where biomass of snag-dwelling invertebrates was more than half of the total invertebrate biomass (Benke et al. 1984). Thus, snag invertebrates can comprise not only a habitat-specific concentration of invertebrates in streams and rivers (Figure 3), but also can make important contributions to the entire riverine ecosystem (Figure 4). On the other hand, if snags are only a tiny fraction of benthic habitats, perhaps because of wood removal at some time in the past, the contribution of invertebrates from snags may be minor (Rabeni and Hoel 2000).

In addition to being a site of high invertebrate biomass and production, wood can also be a "hot spot" of biodiversity. Wallace et al. (1996) compared invertebrate richness among habitats from six streams in the southeastern United States and found that half of the streams had more species associated with wood than with streambed habitats. Invertebrate richness on wood is particularly pronounced in Coastal Plain systems where few other stable habitats exist. For example, Benke et al. (1984) found 63 species inhabiting snags, but only 31 in sandy habitats and 41 in muddy backwaters in the Satilla River. A more intensive study on the Ogeechee River has found more than 108 invertebrate species on snags (Benke, author's unpublished data) and 70 in the sandy sediments (Stites 1986). The snags included 52 species of Ephemeroptera, Plecoptera, and Trichoptera, however, and the sediments had none from these orders. Collier and Halliday (2000) reported that species richness was greater on wood than on inorganic substrates in 11 of 12 pumice-bed streams in New Zealand, with a total of 81 species on wood and 64 on mineral substrates for all streams. Similarly, twice as many taxa were collected from wood (23) as from shifting sand and pumice (11) during the recovery of Clearwater Creek from the eruption of Mt. St. Helens (Dudley and Anderson 1982; Anderson 1992). Somewhat similar results have been found in other streams (Smock et al. 1985, 1992; O'Connor 1991; Hax and Golladay 1998). Wallace et al. (1996) noted that in lotic systems where snags are not prominent and extensive mineral substrates exist (such as cobble or cobble covered with macrophytes), biodiversity on snags may be less than on mineral substrates.

Wood colonization and drift

The availability of wood habitat to aquatic invertebrates is generally more variable than mineral substrates located on the riverbed. Unlike mineral substrates, wood is continuously supplied and lost to streams and rivers, even though it may be stabilized for decades. The greatest instability associated with wood is when it has a vertical orientation and is subjected to variable degrees of inundation (Figure 1C). Depending on where wood is positioned in the stream, it may be inundated for only a single day or for an entire year. Variable degrees of wetting and drying suggest the importance of two interrelated phenomena for invertebrates: colonization and drift. Colonization determines how quickly snags become populated as water rises to inundate previously dry habitat. Drift is probably a major source of this colonization (in addition to reproduction) because snag invertebrates cannot crawl easily across unstable habitats. Furthermore, if invertebrates found on snags are distinctly different from those in other habitats, quantification of drift can provide insight to both colonization and the origin of the drift assemblage.

Several studies have shown that invertebrate colonization of wood begins quickly, as does biofilm development, but it can take anywhere from one to several weeks to achieve maximum biomass and diversity. Nilsen and Larimore (1973) found that at least four weeks were required before maximum density (of mostly oligochaetes and chironomids) was reached on small logs in a slow-flowing section of the small Kaskaskia River, Illinois. In fast-flowing riffles, the assemblage shifted to caddisflies and black flies, although densities were not so high. Nord and Schmulbach (1973) used multiplate (masonite) samplers as surrogates for snags in an unchannelized section of the Missouri River; they found a diverse assemblage dominated by filter-feeding caddisflies in fast-flowing water after 32 d. Van Arsdall (1977), focusing only on caddisflies in the Satilla River, found substantial colonization in less than 10 d, with similar density near the water surface and at 2-m depth. Fremling (1960) had noticed earlier that hydropsychid caddisflies colonizing buoy chains (a snag surrogate) in the upper Mississippi River had similar abundance at the water surface and at 8 m deep. O'Connor (1991) found that 2 weeks was sufficient for natural colonization density in an Australian lowland stream, and Drury and Kelso (2000) found that abundance peaked on tree branches in Coastal Plain streams in Louisiana after two to three weeks. Thorp et al. (1985) examined the patterns of colonization over eight weeks at three sites in a low-gradient stream-swamp system in South Carolina. They found highest density with the fastest current; snags were well-colonized within a week but shifted from a predominance of filterers to gatherers in that period. Similarly, Hax and Golladay (1993) found more rapid colonization in higher velocity. Although colonization studies can be extremely useful in understanding population dynamics and succession on snags, caution should be exercised in interpreting patterns over various periods. The invertebrate assemblage found on an artificial snag at any given time is not only due to its length of submergence, but also to the stream's biological and environmental history (for example, species-specific life history patterns can cause a shift in assemblage structure from one week to the next, regardless of the colonization period).

Wood conditioning, surface texture, and gouging by invertebrates has been suggested to influence colonization. For example, O'Connor (1991) found that grooved woody substrates were colonized by more invertebrate species than was smooth wood. McKie and Cranston (1998) have even gone so far as to suggest that certain elmid beetles (particularly *Notriolus* spp.) function as "keystone" species because their gouging of *Eucalyptus* spp. wood facilitates colonization by other invertebrates and microbes. In a Michigan stream, however, Magoulick (1998) found no difference in colonization of smooth and rough wood, although species richness was greater on conditioned (previously submerged) wood than on unconditioned wood. Furthermore, invertebrate colonization was greater on smooth wood than on smooth Plexiglas, indicating that natural organic substrates are more suitable than artificial substrates (Magoulick 1998). Spänhoff et al. (2000) found much higher invertebrate density on conditioned than on fresh wood in German lowland streams, but not on different wood species over 8-weeks exposure.

Stream drift is a well-known phenomenon that has been better studied in small streams than large rivers (Waters 1972; Benke et al. 1986). If wood-dwelling invertebrates in lotic systems are sufficiently distinct from those in other habitats, such as sandy or muddy riverbeds, then examining drift can be used to measure snag importance. Early studies in the Missouri River suggested that most of the drifting insects originated from brush piles (snags) rather than benthic habitats (Morris et al. 1968; Modde and Schmulbach 1973). Similarly, Kovalak (1978) found more drift leaving than entering a pool area of a small stream in Michigan and suggested that the pool was a drift source because of the logs and twigs along the pool margin. Benke et al. (1986) compared drift density with habitat density in the Satilla River, Georgia, and showed that snags contributed 72–81% of the numbers and 78–82% of the biomass to drift. A subsequent study in the Ogeechee River demonstrated similar contributions from snags (Benke et al. 1991), but snag abundance—and thus drift density—were substantially higher (>20 individuals/m^3 and >2.4 mg dry mass/m^3) than in the Satilla. Although invertebrates can drift from different habitats at different rates, these findings provide an independent demonstration of the importance of snag habitats in low-gradient rivers and a potentially important source of food for filtering invertebrates and drift-feeding fishes.

Although snags can be the major source of drifting invertebrates in natural streams, they also represent a stable refuge, particularly in low-gradient streams during periods of high discharge. For example, Borchardt (1993) showed that accumulated wood substantially reduced drift losses under experimentally induced hydraulic stress. Palmer et al. (1996) also conducted experiments suggesting that wood dams conferred increased resistance of chironomids and copepods to floods in certain habitats. Invertebrates are known to drift from snags to escape desiccation as water levels fall, and drift is important in the recolonization of newly inundated snags, but the understanding of the relation between drift density, hydrologic regime, and colonization of snags remains weak.

Functional feeding groups on wood

The number of wood-eating species appears to decrease from western to eastern North America (Dudley and Anderson 1982), suggesting that degree of xylophagy may differ among geographic regions. Determining the importance of xylophagy (or any other feeding mode) for entire invertebrate assemblages from different geographic regions, however, requires not only determination of invertebrates' trophic status (Dudley and Anderson 1982; Hoffmann and Hering 2000), but also a quantitative approach. Although quantitative data are limited, we have used published analyses from streams in Oregon, Virginia, South Carolina, and Georgia to summarize the distribution of either biomass or production among functional feeding groups (Figures 5 and 6). Although wood gougers are considered part of the shredder feeding group (Cummins and Merritt 1996; Wallace and Webster 1996), we have separated gougers from shredders (when possible) for our analysis of wood-associated invertebrates.

The distribution of biomass among functional groups clearly distinguishes two patterns (Figure 5). Those studies that identified wood as "snags" (Figure 5C through 5D) contained invertebrates that primarily fell into gatherer, filterer, and predator functional groups, with few contributions from gougers or shredders. All of these studies were from eastern U.S. Coastal Plain streams or rivers. In contrast, the Oregon streams were heavily weighted toward the gougers and shredders, consistent with the higher diversity of obligate xylophages, with no filterers or predators. Scrapers did not comprise a significant proportion of invertebrate biomass on wood from eastern streams, but they were important in some Oregon streams (Anderson et al. 1978, 1984).

The distribution of production among functional groups was limited to only the eastern U.S. Coastal Plain streams and rivers and included the two Virginia streams in which wood dams were sampled instead of snags (Figure 6). Production on snags from the South Carolina stream and the Georgia rivers were similar to their biomass distributions (Figure 5), except that the relative proportions shifted in the gatherer, filterer, and predator groups as a function of their turnover rates. The wood dams from Virginia streams, however, contained invertebrates that were primarily shredders and predators with low fractions of gatherers and filterers, suggesting a significantly different microhabitat than for snags.

Based on the distributions of biomass and production among functional feeding groups in this limited number of streams (Figures 5 and 6), we suggest that these differences are caused largely by the habitat in which the wood was sampled and the manner in which the wood was collected. Snags from Coastal Plain systems in South Carolina and Georgia typically were branches of fallen trees well anchored in the current above the sediments. In contrast, wood accumulations from the Oregon streams were typically loose or tethered sticks lying on the stream bottom. Wood accumulations in Virginia streams, on the other hand, were packed tightly, trapped other detrital material, and thus attracted a high fraction of shredders. Other studies using abundance of functional groups are reasonably consistent with the biomass and production data. Colonization of fixed snags in another South Carolina Coastal Plain stream again demonstrated high gatherer and filterer density, but also included scrapers (Thorp et al. 1985). Snags from an Australian stream consisted primarily of gatherers (O'Connor 1992). On the other hand, stick packs placed on the bottom of small Appalachian streams in North Carolina attracted only 11% shredders; most of the rest were collectors and scrapers (Golladay and Webster 1988). Similarly, stick packs placed on the bottom of streams in New York were composed of only 6% shredders and piercers with the majority gatherers and filterers (Hax and Golladay 1993). In contrast to the wood dam results from Virginia (Smock et al.

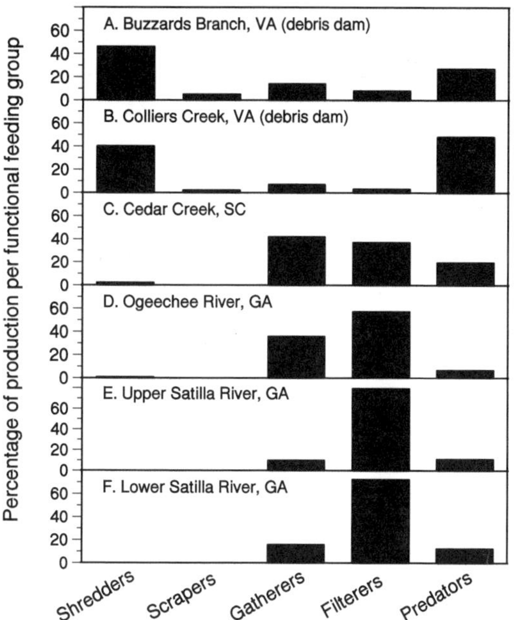

FIGURE 6. Distribution of invertebrate production percentages in functional feeding groups for streams and rivers. Sources: Virginia streams (Smock et al. 1989, 1992); Cedar Creek, South Carolina (Smock et al. 1985); Ogeechee River, Georgia (Benke et al. 2001, see other references in Table 2); Satilla River sites, Georgia (Benke et al. 1984).

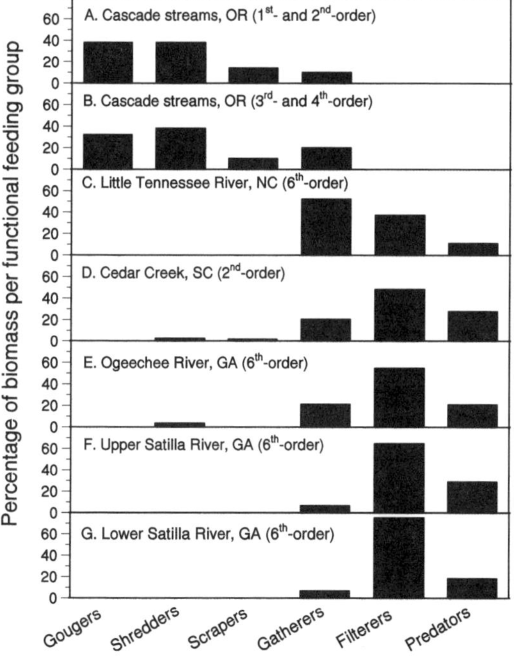

FIGURE 5. Distribution of invertebrate biomass percentages in functional feeding groups for streams and rivers. Sources: Cascade streams (first- and second-order and third- and fourth-order from Anderson and Sedell 1979); Little Tennessee River, North Carolina (Wallace et al. 1996); Cedar Creek, South Carolina (Smock et al. 1985); Ogeechee River, Georgia (see sources in Table 2), Satilla River, Georgia (upper and lower sites, Benke et al. 1984). Note: for the Cascade streams, we have arbitrarily placed those organisms identified as "other" into the gatherer functional group.

1989), invertebrates from a beaver dam in Alberta, Canada consisted primarily of filtering black flies and more closely resembled the fauna of a lake outlet (Clifford et al. 1993). In summary, quantitative studies suggest that nonxylophagous taxa (particularly filterers and gatherers) can dominate wood habitat in a wide variety of conditions.

Although the geographical distribution of obligate xylophages across North America may play a role in affecting differences in functional feeding groups that colonize snags (Dudley and Anderson 1982), we believe that the limited quantitative data (relative density, biomass, and production) presented here support the view that habitat and sampling approaches are of overriding importance in determining assemblage composition. Xylophages (gougers and other shredders) and gatherers appear to be important on wood that has settled in depositional habitats, but filterers and gatherers dominate on snags anchored in erosive environments. Filterers and gatherers have their highest production as stream size and water velocity increases, and more enriched fine particulate organic matter is transported.

Food webs on wood: beyond functional groups

Identification of invertebrate food habits and classification into functional feeding groups can provide considerable insight into the role of wood-dwelling invertebrates in streams. Such analysis, however, is only a crude characterization of wood food webs, particularly beyond the primary consumers. Detailed analyses for the Ogeechee River snag assemblage (Wallace et al. 1987; Benke and Wallace 1997; Benke et al. 2001) demonstrated trophic pathways between individual groups and species that have been realized in few other stream habitats (Smock and Roeding 1986; Smith and Smock 1992). Most of the primary consumers on Ogeechee snags are filterers and gatherers (chironomids and mayflies) that depend on microbially enriched amorphous detritus collected from the seston (Figure 7). In spite of a complex feeding hierarchy, most of the chironomids and mayflies are consumed by omnivorous net-spinning caddisflies, possibly after they enter the current, and are later recaptured from the drift by caddisflies on another snag. The predatory portion of the food chain also includes larger, strictly predaceous insects such as perlid stoneflies, dragonflies, and the hellgrammite *Corydalus cornutus* as the top invertebrate predator (Benke et al. 2001). Production of these larger predators totaled 7.1 g dry mass/m^2 (wood surface)/year, which required a high consumption of about 52% of total prey production. Smith and Smock (1992) also estimated production of the predator component on the wood dam of Buzzards Branch (8.4 g dry mass · m^{-2} · year) and found that predators consumed about 94% of invertebrate prey production for the stream as a whole. Food webs on wood from other lotic systems (such as high-gradient streams) have yet to be described, but they would probably be more difficult to separate from coexisting food webs on mineral habitats.

In addition to the complexity of feeding levels in the invertebrate assemblage on the snag habitat in Coastal Plain rivers, invertebrates provide the foundation for higher trophic levels. In the Satilla River, snag invertebrates are a major source of food for insectivorous fishes; up to 60% of the diet of several fish species originated from snags (Benke et al. 1979, 1985). The insectivorous fishes are, in turn, consumed by fish-eating species (Benke et al. 1985). In spite of many studies demonstrating the importance of wood in creating habitat and as a refuge for fishes (Dolloff and Warren 2003; Zalewski et al. 2003), relatively little other research documents the importance of wood-dwelling invertebrates as a food source to fishes (Angermeier and Karr 1984; Lehtinen et al. 1997; Crook and Robertson 1999). Although direct evidence is sparse, Crook and Robertson (1999) provide an excellent review on the potential use of snags as foraging sites by fishes; they conclude that, at least in low-gradient rivers throughout the world, fish do indeed rely on snag-dwelling invertebrates, whether they forage directly or whether they obtain their prey from the drift. Quist and Guy (2001) have found that growth rates of some prairie stream fishes are related to the amount of instream wood, presumably because it is the source of their invertebrate prey.

Influence of invertebrates on wood: structure and process

Given that many xylophagous invertebrates are found on stream wood, whether they play a major role in wood decomposition is of paramount importance. In terrestrial environments, wood decomposition is relatively rapid, enhanced by invertebrates that create deep galleries into the interior of the wood, promoting fragmentation

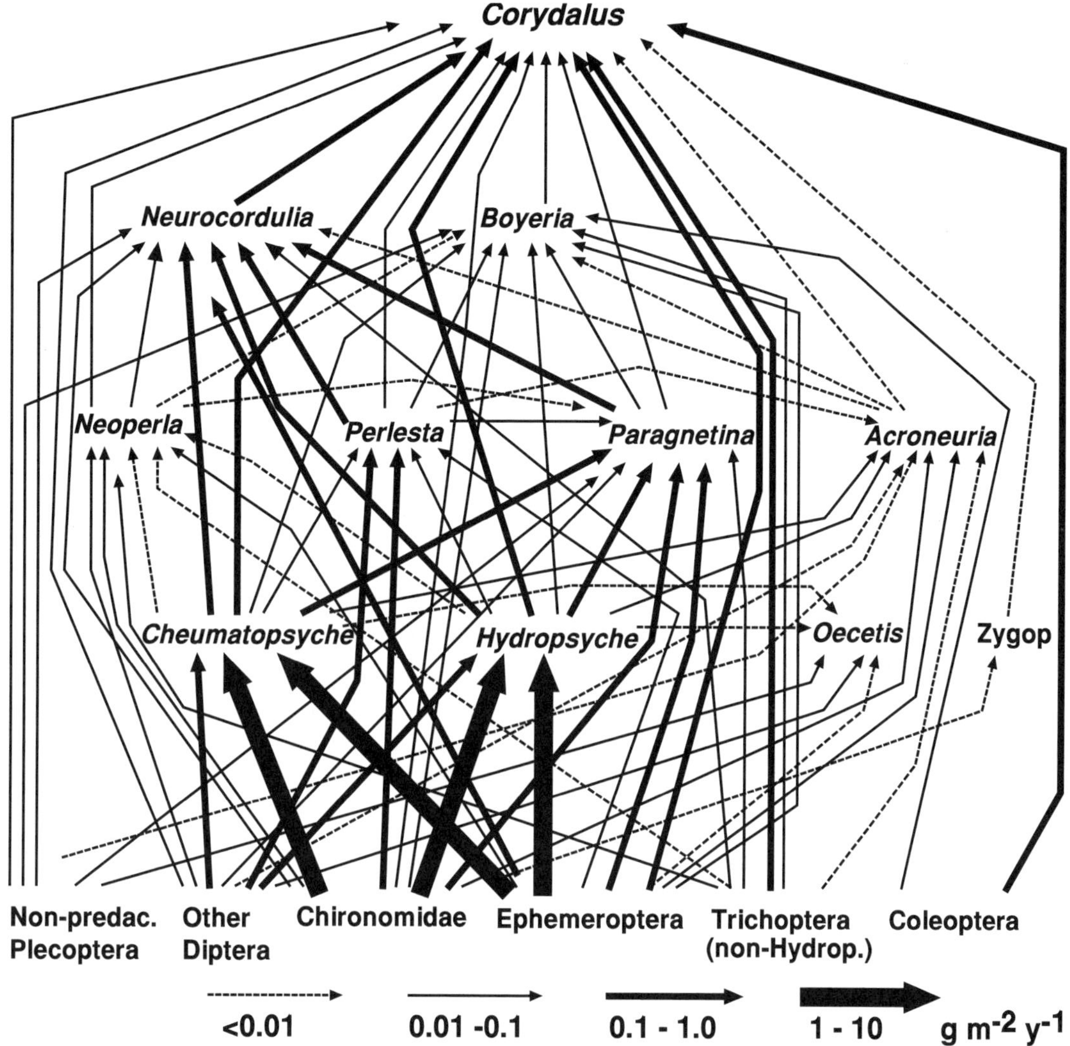

FIGURE 7. Quantitative food web on the snag habitat in the Ogeechee River, emphasizing the predator component (from Benke et al. 2001). Line thickness indicates the ingestion flux. See original paper for actual values. Note that *Cheumatopsyche* spp. and *Hydropsyche* spp. are omnivores, and their ingestion of basal food resources is not shown (see Figure 3 in Benke and Wallace 1997). Abbreviations: Zygop = Zygoptera, Non-predac. = non-predaceous, non-Hydrop. = non-Hydropsychidae. (reproduced with permission from *Freshwater Biology*).

and increased surface area for microbial decomposition (Harmon et al. 1986). In contrast, wood decomposition is much slower in freshwater environments, probably because oxygen concentrations in the interior of wetted wood are very low, serving as a barrier to bacteria, fungi, and invertebrates. Hence, wood decomposition in lotic systems is primarily a surface phenomenon, but the possibility still exists that invertebrates play an important role in this slow decomposition.

Although most invertebrates are limited to the wood surface, some evidence shows that they promote decomposition by scraping, gouging, and tunneling wood colonized by bacteria and fungi: these activities expose additional wood to further microbial decomposition (Anderson et al. 1984; Dudley and Anderson 1982). Unfortunately, relatively few quantitative studies exist, making comparative analysis difficult. Some species are incredibly effective in consuming wood; for example, the tipulid *Lipsothrix* spp. can produce feces from wood at a rate of 88% to 223% of its body weight per day in Oregon streams (Anderson et al. 1984), and the caddisfly *Pycrocentria funerea* can produce feces 22% to 168% of its body weight per day in New Zealand streams (Collier and Halliday 2000). Because wood has a low assimilation efficiency, these fecal production rates should be comparable to (but slightly less than) consumption rates. Determination of the invertebrate effects on wood requires not only consumption rates per individual, but also quantitative estimates of consumer biomass or production. Fecal production fluxes by three major xylophagous species in Oregon streams have been estimated to be 17.5 g of feces per kilogram of wood per year or more than 20 g/m^2 of streambed per year. (Anderson et al. 1978, 1984; Steedman and Anderson 1985). These authors estimate that such fecal production rates would process as much as 2% of existing wood per year, which would be significant considering the otherwise slow rate of decomposition. We are unaware of other estimates of wood disappearance rates caused by invertebrates. Nonetheless, Collier and Halliday (2000) estimated that *Pycnocentria*, the dominant xylophagous species in New Zealand streams, consumed about 29 mg $\cdot$ $m^2 \cdot d^{-1}$ (11 g $\cdot$ $m^{-2} \cdot year^{-1}$), comparable to that found in Oregon streams. Hoffmann (2000) estimated annual fecal production for the caddisfly *Lasiocephala basalis* of 100 g dry mass/m^2 of wood surface, but this cannot be directly compared to the other estimates. Roeding and Smock (1989) estimated wood consumption, rather than fecal production rates, by using secondary production analysis. Their consumption values of 12.6 g $\cdot$ $m^{-2} \cdot year^{-1}$ (or 34.5 mg $\cdot$ $m^{-2} \cdot d^{-1}$) for the dominant shredders in Virginia Coastal Plain streams were surprisingly similar to fecal production fluxes in Oregon and New Zealand. Using a similar approach, Hall et al. (2000) estimated wood consumption for shredders in a southern Appalachian stream of 6 mg to 23 mg ash free dry mass $\cdot$ $m^{-2} \cdot d^{-1}$ in a reference stream and a higher wood consumption of 34 mg to 48 mg ash free dry mass $\cdot$ $m^{-2} \cdot d^{-1}$ in a litter-excluded stream, again reasonably similar to previous estimates.

Nonxylophagous species also have the potential to gouge and stimulate wood decomposition, even though they do not necessarily eat the wood. Several species of net-spinning caddisflies (Hydropsychidae) are known to gouge oval depressions in snags, particularly in Coastal Plain rivers of the southeastern United States (Figure 8), but we are unaware of any attempts to measure this process. The significance of such gouging has been observed from underwater damage to wooden structures, however. In 1988, a section of bridge spanning the Pocomoke River in Maryland collapsed (NTSB 1989). Subsequent investigation showed that the untreated timber pilings had a 53% to 58% reduction in cross-sectional area, resulting in the collapse. The National Transportation Safety Board attributed the reduction to the combined effects of microbial decomposition and many decades of gouging by larvae of *Hydropsyche incommoda*, which had colonized the pilings

FIGURE 8. Submerged log from the Sipsey River, Alabama. The deep gouges were apparently made by hydropsychid caddisflies, particularly *Macrostemum carolina*, which is abundant on snags in this river. Diameter of log is about 10 cm.

(NTSB 1989; Flint 1996). In conclusion, both xylophagous invertebrates and nonxylophagous gougers probably play a major role in wood decomposition in lotic ecosystems, although further research in this area is clearly needed.

Wood Invertebrates in Bioassessment

The importance of wood as invertebrate habitat in streams and rivers has not escaped the attention of investigators developing protocols for bioassessment. Lenat (1988) described wood sampling as one of the standardized qualitative collection techniques used for water quality assessment in North Carolina. Cuffney et al. (1993) summarized methods for collecting macroinvertebrates in the U.S. Geological Survey's National Water-Quality Assessment Program and described woody snags as an important habitat to be sampled. Similarly, the rapid bioassessment protocol developed for the U.S. Environmental Protection Agency for macroinvertebrates has recently recognized snags as one of the five basic habitat types that should be sampled in streams of any type (Barbour et al. 1999). The USEPA (1997) and Maxted et al. (2000) developed field and laboratory methods specifically for macroinvertebrate and habitat assessment in low gradient, nontidal streams in the eastern U.S. Coastal Plain. Because the mineral substrate of such streams is often sand or silt, which typically have few species that are pollution-sensitive, they recognized that snag sampling should be essential to their protocol.

Various investigators have now conducted field studies in which they have sampled snags as part of water quality studies. Rabeni (2000) sampled six common habitat types, including snags, from 45 streams in three ecoregions in Missouri and calculated a variety of metrics for assessing water quality. He found distinctive assemblages associated with each habitat type and concluded that bioassessments need to use habitat-specific sampling, including snags, if they are a common feature of streams. Brown and May (2000) effectively assessed the relation between wood-dwelling macroinvertebrates (by family) and environmental variables in the lower Sacramento and San Joaquin River basins, California, as part of the National Water Quality Assessment Program. In New Zealand, Collier et al. (1998) sampled mineral substrates, macrophytes, and wood in 18 lowland streams, and they applied various biotic indices to each habitat. They found that wood was relatively common in most streams and an important habitat to sample in bioassessments. Quinn et al. (1997) measured wood accumulation in 11 hill-country streams in New Zealand and collected invertebrates with a Surber sampler. They found that potential differences in invertebrate communities based on differences in pollution were ameliorated by the presence of wood habitat in all the streams. In short, now that wood substrates are well recognized as invertebrate habitat through basic research, stream biologists have incorporated this important finding into bioassessment protocols.

Managing and Restoring Streams

Given that wood plays an important role in the ecological characteristics of streams and rivers and given that invertebrate colonists are an essential functional component, such considerations should obviously be an integral part of managing streams and planning to restore them. We have known for many years that protecting riparian zones is essential for natural functioning of lotic systems, particularly as a source of allochthonous inputs for aquatic consumers and in maintaining healthy river-floodplain interactions (Gregory 1992; Sweeney 1993; Benke 2001). Increased knowledge of the importance of wood to streams and river channels reinforces this understanding, particularly from the viewpoint of wood as essential habitat for both invertebrate and vertebrate animals. Not only is protecting riparian zones important, but also, the destructive practices of channelization and "snagging" should be eliminated or reduced in many systems, now that their harm to natural wood habitats is appreciated. Nonetheless, many streams and rivers have already been seriously damaged, and attempts to restore natural channels and floodplains should consider reintroducing wood to channels. While research in this area is still in its infancy, management and restoration strategies need to balance boating safety and other perceived liability issues with the importance of natural riverine habitats.

Although applied aspects are dealt with elsewhere in this volume, we will briefly mention a few restoration attempts that deal with invertebrates. In an experiment related to the concept of restoration, Wallace et al. (1995) added logs to a small mountain stream in North Carolina and demonstrated that such manipulations can result in significant functional shifts in the abundance, biomass, and production of the invertebrate assemblages in newly created depositional habitats. The addition of wood to a low-gradient third-order stream in Virginia caused an increase in pool habitat and a subsequent shift in functional group biomass (Hilderbrand et al. 1997; Lemly and Hilderbrand 2000). None of the Virginia streams in these studies were degraded, however, but the researchers were not intending to assess the success of restoration. On the other hand, Gerhard and Reich (2000) studied changes in microhabitats and invertebrates in a regulated and straightened third-order stream in Germany to which wood had been added as an accidental effect of forest management. Although they sampled on only a single date, they found an increase in microhabitats, an increase in invertebrate species, and an almost fivefold increase in invertebrate density in "restored" sections. The paucity of invertebrate assessments in attempts to re-introduce wood for restoration illustrates the clear need for work on this topic. The possibility for the greatest benefits of wood introduction will probably be found in systems where streambed substrate is unsuitable for many invertebrate species (shifting sand, pumice) and where wood has historically provided the only major stable substrate.

Summary

Research on wood-dwelling invertebrates can be divided into three basic approaches. The first deals with wood as a substrate from which invertebrates can obtain food resources and play a role in decomposition. Such studies were in smaller streams in which wood is part of the detrital pool. The second approach treats wood as a major habitat or attachment site for a wide variety of functional feeding groups (rather than just a detrital food source), and such habitats are referred to as snags. Although snags (as habitat) can be important in smaller streams, their relative importance seems to increase as stream size increases, as gradient decreases, and as alternative hard substrates disappear. The third considers the important role of wood in creating lentic habitats for invertebrates along the course of streams, either by enhancing pool and riffle sequences with wood accumulations or by forming major ponds and wetlands from beaver dams. Thus, wood has a profound effect on the diversity, abundance, production, and function of invertebrate assemblages in lotic systems of all sizes. The habitats and food sources created by wood and their associated invertebrates, in turn, form the foundation for higher trophic levels, particularly fishes, but including all vertebrate predators. Although much has been accomplished in understanding the role of invertebrates on wood in streams and rivers, much remains to be learned in quantifying that role and in using this information in bioassessment and restoration protocols for streams and rivers of all sizes.

Acknowledgments

We thank Stan Gregory and Kathryn Boyer for organizing the International Conference on Wood in World Rivers from which this chapter was developed. We also are grateful to Norm Anderson and an anonymous reviewer for many helpful comments and insights on the manuscript. We appreciate the outstanding editing job provided by Martha H. Brookes.

References

Anderson, N. H. 1982. A survey of aquatic insects associated with wood debris in New Zealand streams. Mauri Ora 10:21–33.

Anderson, N. H. 1989. Xylophagous Chironomidae from Oregon streams. Aquatic Insects 11:33–45.

Anderson, N. H. 1992. Influence of disturbance on insect communities in Pacific Northwest streams. Hydrobiologia 248:79–92.

Anderson, N. H., and J. R. Sedell. 1979. Detritus processing by macroinvertebrates in stream ecosystems. Annual Review of Entomology 24:351–377.

Anderson, N. H., J. R. Sedell, L. M. Roberts, and F. J. Triska. 1978. The role of aquatic invertebrates in processing of wood debris in coniferous forest streams. American Midland Naturalist 100:64–82.

Anderson, N. H., R. J. Steedman, and T. Dudley. 1984. Patterns of exploitation by stream invertebrates of wood debris (xylophagy). Internationale Vereinigung für Theoretische und Angewandte Limnologie 22:1847–1852.

Angermeier, P. L., and J. R. Karr. 1984. Relationships between woody debris and fish habitat in a small warmwater stream. Transactions of the American Fisheries Society 113:716–726.

Barbour, M. T., J. Gerritsen, B. D. Snyder, and J. B. Stribling. 1999. Rapid bioassessment protocols for use in streams and wadeable rivers: periphyton, benthic macroinvertebrates and fish. 2nd edition. U.S. Environmental Protection Agency, Office of Water, EPA 841-B-99–002, Washington, D.C.

Benke, A. C. 1998. Production dynamics of riverine chironomids (Diptera): extremely high biomass turnover rates of primary consumers. Ecology 79:899–910.

Benke, A. C. 2001. Importance of flood regime to invertebrate habitat in an unregulated river-floodplain ecosystem. Journal of the North American Benthological Society 20:225–240.

Benke, A. C. 2002. Secondary production of riffle beetles on the snag habitat of a Coastal Plain river in the southeastern USA. Internationale Vereinigung für Theoretische und Angewandte Limnologie 28:958–961.

Benke, A. C., D. M. Gillespie, F. K. Parrish, T. C. Van Arsdall, Jr., R. J. Hunter, and R. L. Henry III. 1979. Biological basis for assessing impacts of channel modifications: invertebrate production, drift, and fish feeding in a southeastern blackwater river. Georgia Institute of Technology, Environmental Resources Center Publication ERC 06–79, Atlanta.

Benke, A. C., F. R. Hauer, D. L. Stites, J. L. Meyer, and R. T. Edwards. 1992. Growth of snag-dwelling mayflies in a blackwater river: the influence of temperature and food. Archiv für Hydrobiologie 125:63–81.

Benke, A. C., R. L. Henry III, D. M. Gillespie, and R. J. Hunter. 1985. Importance of snag habitat for animal production in southeastern streams. Fisheries 10:8–13.

Benke, A. C., R. J. Hunter, and F. K. Parrish. 1986. Invertebrate drift dynamics in a subtropical blackwater river. Journal of the North American Benthological Society 5:173–190.

Benke, A. C., and D. I. Jacobi. 1986. Growth rates of mayflies in a subtropical river and their implications for secondary production. Journal of the North American Benthological Society 5:107–114.

Benke, A. C., and D. I. Jacobi. 1994. Production dynamics and resource utilization of snag-dwelling mayflies in a blackwater river. Ecology 75:1219–1232.

Benke, A. C., and K. A. Parsons. 1990. Modelling black fly production dynamics in blackwater streams. Freshwater Biology 24:167–180.

Benke, A. C., K. A. Parsons, and S. M. Dhar. 1991. Population and community patterns of invertebrate drift in an unregulated Coastal Plain river. Canadian Journal of Fisheries and Aquatic Sciences 48:811–823.

Benke, A. C., T. C. Van Arsdall, Jr., D. M. Gillespie, and F. K. Parrish. 1984. Invertebrate productivity in a subtropical blackwater river: the importance of habitat and life history. Ecological Monographs 54:25–63.

Benke, A. C., and J. B. Wallace. 1990. Wood dynamics in Coastal Plain blackwater streams. Canadian Journal of Fisheries and Aquatic Sciences 47:92–99.

Benke, A. C., and J. B. Wallace. 1997. Trophic basis of production among riverine caddisflies: implications for food web analysis. Ecology 78:1132–1145.

Benke, A. C., J. B. Wallace, J. W. Harrison, and J. W. Koebel. 2001. Food web quantification using secondary production analysis: predaceous invertebrates of the snag habitat in a subtropical river. Freshwater Biology 46:329–346.

Benke, A. C., G. M. Ward, and T. D. Richardson. 1999. Beaver-impounded wetlands of the southeastern coastal plain: habitat-specific composition and dynamics of invertebrates. Pages 217–245 *in* D. P. Batzer, R. B. Rader, and S. A. Wissinger, editors. Invertebrates in freshwater wetlands of North America: ecology and management. Wiley, New York.

Berner, L. M. 1951. Limnology of the lower Missouri River. Ecology 32:1–12.

Bilby, R. E. 1981. Role of organic debris dams in regulating export of dissolved and particulate matter from a forested watershed. Ecology 62:1234–1243.

Bilby, R. E. 2003. Decomposition and nutrient dynamics of wood in streams and rivers. Pages 135–147 *in* S. V. Gregory, K. L. Boyer, and A. M. Gurnell, editors. The ecology and management of wood in world rivers. American Fisheries Society, Symposium 37, Bethesda, Maryland.

Bilby, R. E., and G. E. Likens. 1980. Importance of organic debris dams in the structure and function of stream ecosystems. Ecology 61:1107–1113.

Boon, P. J. 1984. A re-examination of the submerged tree fauna in Lake Kariba (South Central Africa), 24 years after inundation. Archiv für Hydrobiologie 101:569–576.

Borchardt, D. 1993. Effects of flow and refugia on drift loss of benthic macroinvertebrates: implications for habitat restoration in lowland streams. Freshwater Biology 29:221–227.

Borkent, A. 1984. The systematics and phylogeny of the *Stenochironomus* complex (*Xestochironomus*, *Harrisius*, and *Stenochironomus* (Diptera: Chironomidae). Memoirs of the Entomological Society of Canada 128:1–269.

Bowen, K. L., N. K. Kaushik, and A. M. Gordon. 1998. Macroinvertebrate communities and biofilm chlorophyll on woody debris in two Canadian oligotrophic lakes. Archiv für Hydrobiologie 141:257–281.

Brown, L. R., and J. T. May. 2000. Macroinvertebrate assemblages on woody debris and their relations with environmental variables in the lower Sacramento and San Joaquin river drainages, California. Environmental Monitoring and Assessment 64:311–329.

Carlough, L. A. 1994. Origins, structure, and trophic significance of amorphous seston in a blackwater river. Freshwater Biology 31:227–237.

Carlough, L. A., and J. L. Meyer. 1989. Protozoans in two southeastern blackwater rivers and their importance to trophic transfer. Limnology and Oceanography 34:163–177.

Carlough, L. A., and J. L. Meyer. 1991. Bacterivory by sestonic protists in a southeastern blackwater river. Limnology and Oceanography 36:873–883.

Chergui, H., and E. Pattee. 1991. The breakdown of wood in the sidearm of a large river: preliminary investigations. Internationale Vereinigung für Theoretische und Angewandte Limnologie 24:1785–1788.

Chessman, B. C. 1986. Dietary studies of aquatic insects from two Victorian rivers. Australian Journal of Marine and Freshwater Research 37:129–146.

Clifford, H. F., G. M. Wiley, and R. J. Casey. 1993. Macroinvertebrates of a beaver-altered boreal stream of Alberta, Canada, with special reference to the fauna on the dams. Canadian Journal of Zoology 71:1439–1447.

Collier, K. J., and J. N. Halliday. 2000. Macroinvertebrate-wood associations during decay of plantation pine in New Zealand pumice-bed streams: stable habitat or trophic subsidy? Journal of the North American Benthological Society 19:94–111.

Collier, K. J., R. J. Wilcock, and A. S. Meredith. 1998. Influence of substrate type and physico-chemical conditions on macroinvertebrate faunas and biotic indices of some lowland Waikato, New Zealand, streams. New Zealand Journal of Marine and Freshwater Research 32:1–19.

Couch, C. A., and J. L. Meyer. 1992. Development and composition of the epixylic biofilm in a blackwater river. Freshwater Biology 27:43–51.

Couch, C. A., J. L. Meyer, and R. O. Hall, Jr. 1996. Incorporation of bacterial extracellular polysaccharide by black fly larvae (Simuliidae). Journal of the North American Benthological Society 15:289–299.

Cranston, P. S., and R. A. Hardwick. 1996. The immature stages and phylogeny of *Imparipecten* freeman, an Australian endemic genus of wood-mining chironomid (Diptera). Aquatic Insects 18:193–207.

Cranston, P. S., and D. R. Oliver. 1988. Aquatic xylophagous Orthocladiinae – systematics and ecology (Diptera: Chironomidae). Spixiana(Supplement 14):143–151.

Crook, D. A., and A. I. Robertson. 1999. Relationships between riverine fish and wood debris: implications for lowland rivers. Marine and Freshwater Research 50:941–953.

Cudney, M. D., and J. B. Wallace. 1980. Life cycle, microdistribution, and production dynamics of six species of net-spinning caddisflies in a large southeastern (U.S.A.) river. Holarctic Ecology 3:169–182.

Cuffney, T. F., M. E. Gurtz, and M. R. Meador. 1993. Methods for collecting benthic invertebrate samples as part of the National Water-Quality Assessment Program. U.S. Geological Survey, Open-File Report 93–406, Raleigh, North Carolina.

Cummins, K. W., and R. W. Merritt. 1996. Ecology and distribution of aquatic insects. Pages 74–86 *in* R. W. Merritt and K. W. Cummins, editors. An introduction to the aquatic insects of North America. 3rd edition. Kendall/Hunt Publishing Company, Dubuque, Iowa.

Darwin, C. 1859. On the origin of species by means of natural selection. J. Murray, London.

Delong, M. D., J. H. Thorp, and K. H. Haag. 1993. A new device for sampling macroinvertebrates from woody debris (snags) in nearshore areas of aquatic systems. American Midland Naturalist 130:413–417.

Díez, J. R., S. Larrañaga, A. Elosegi, and J. Pozo. 2000. Effect of removal of wood on streambed stability and retention of organic matter. Journal of the North American Benthological Society 19:621–632.

Dolloff, C. A. and M. L. Warren, Jr. 2003. Fish relationships with large wood in small streams. Pages 179–193 *in* S. V. Gregory, K. L. Boyer, and A. M. Gurnell, editors. The ecology and management of wood in world rivers. American Fisheries Society, Symposium 37, Bethesda, Maryland.

Drury, D. M., and W. E. Kelso. 2000. Invertebrate colonization of wood debris in coastal plain streams. Hydrobiologia 434:63–72.

Dudley, T., and N. H. Anderson. 1982. A survey of invertebrates associated with wood debris in aquatic habitats. Melanderia 39:1–21.

Dudley, T. L., and N. H. Anderson. 1987. The biology and life cycles of *Lipsothrix* spp. (Diptera: Tipulidae) inhabiting wood in western Oregon streams. Freshwater Biology 17:427–451.

Edwards, R. T. 1987. Sestonic bacteria as a food source for filtering invertebrates in two southeastern blackwater rivers. Limnology and Oceanography 32:221–234.

Edwards, R. T., and J. L. Meyer. 1987. Bacteria as a food source for black fly larvae in a blackwater river. Journal of the North American Benthological Society 6:241–250.

Edwards, R. T., and J. L. Meyer. 1990. Bacterivory by deposit-feeding mayfly larvae (*Stenonema* spp.). Freshwater Biology 24:453–462.

Everett, R. A., and G. M. Ruiz. 1993. Coarse woody debris as a refuge from predation in aquatic communities: an experimental test. Oecologia 93:475–486.

Fausch, K. D., and T. G. Northcote. 1992. Large woody debris and salmonid habitat in a small coastal British Columbia stream. Canadian Journal of Fisheries and Aquatic Sciences 49:682–693.

Flint, O. S., Jr. 1996. Caddisflies do count: collapse of S. R. 675 bridge over the Pocomoke River, Pocomoke City, Maryland. Bulletin of the North American Benthological Society 13:376–383.

Fremling, C. R. 1960. Biology and possible control of nuisance caddisflies on the upper Mississippi

River. Iowa Agricultural Experiment Station, Research Bulletin 483, Ames, Iowa.

Gerhard, M., and M. Reich. 2000. Restoration of streams with large wood: effects of accumulated and built-in wood on channel morphology, habitat diversity and aquatic fauna. International Review of Hydrobiology 85:123–137.

Gippel, C. J., B. L. Finlayson, and I. C. O'Neill. 1996. Distribution and hydraulic significance of large woody debris in a lowland Australian river. Hydrobiologia 318:179–196.

Golladay, S. W., and C. L. Hax. 1995. Effects of an engineered flow disturbance on meiofauna in a north Texas prairie stream. Journal of the North American Benthological Society 14:404–423.

Golladay, S. W., and R. L. Sinsabaugh. 1991. Biofilm development on leaf and wood surfaces in a boreal river. Freshwater Biology 25:437–450.

Golladay, S. W., and J. R. Webster. 1988. Effects of clear-cut logging on wood breakdown in Appalachian mountain streams. American Midland Naturalist 119:143–155.

Gregory, K. J. 1992. Vegetation and river channel process interactions. Pages 255–269 *in* P. J. Boon, P. Calow, and G. E. Petts, editors. River conservation and management. Wiley, Chichester, UK.

Growns, J. E., A. J. King, and F. M. Betts. 1999. The snag bag: a new method for sampling macroinvertebrate communities on large woody debris. Hydrobiologia 405:67–77.

Gurnell, A. M. 1997. Analysis of the effects of beaver dam building activities on local hydrology. Scottish Natural Heritage Review Series 85.

Haden, G. A., D. W. Blinn, J. P. Shannon, and K. P. Wilson. 1999. Driftwood: an alternative habitat for macroinvertebrates in a large desert river. Hydrobiologia 397:179–186.

Hall, R. O., J. B. Wallace, and S. L. Eggert. 2000. Organic matter flow in stream food webs with reduced detrital resource base. Ecology 81:3445–3463.

Harmon, M. E., J. F. Franklin, F. J. Swanson, P. Sollins, S. V. Gregory, J. D. Lattin, N. H. Anderson, S. P. Cline, N. G. Aumen, J. R. Sedell, G. W. Lienkaemper, K. Cromack, Jr., and K. W. Cummins. 1986. Ecology of coarse woody debris in temperate ecosystems. Advances in Ecological Research 15:133–302.

Hauer, F. R., and A. C. Benke. 1987. The influence of temperature and river hydrograph on the growth rates of black fly larvae in a subtropical blackwater river. Journal of the North American Benthological Society 6:251–261.

Hauer, F. R., and A. C. Benke. 1991. Rapid growth of snag-dwelling chironomids in a blackwater river: the influence of temperature and discharge. Journal of the North American Benthological Society 10:154–164.

Hax, C. L., and S. W. Golladay. 1993. Macroinvertebrate colonization and biofilm development on leaves and wood in a boreal river. Freshwater Biology 29:79–87.

Hax, C. L., and S. W. Golladay. 1998. Flow disturbance of macroinvertebrates inhabiting sediments and woody debris in a prairie stream. American Midland Naturalist 139:210–223.

Hilderbrand, R. H., A. D. Lemly, C. A. Dolloff, and K. L. Harpster. 1997. Effects of large woody debris placement on streams channels and benthic macroinvertebrates. Canadian Journal of Fisheries and Aquatic Sciences 54:931–939.

Hoffmann, A. 2000. The association of the stream caddisfly *Lasiocephala basilis* (Kol.) (Trichoptera: Lepidostomatidae) with wood. International Review of Hydrobiology 85:79–93.

Hoffman, A., and D. Hering. 2000. Wood-associated macroinvertebrate fauna in central European streams. International Review of Hydrobiology 85:25–48.

Humphries, P., P. E. Davies, and M. E. Mulcahy. 1996. Macroinvertebrate assemblages of littoral habitats in the Macquarie and Mersey rivers, Tasmania: implications for the management of regulated rivers. Regulated Rivers Research & Management 12:99–122.

Humphries, P., J. E. Growns, L. G. Serafini, J. H. Hawkins, A. J. Chick, and P. S. Lake. 1998. Macroinvertebrate sampling methods for lowland Australian river. Hydrobiologia 364:209–218.

Huryn, A. D., and J. B. Wallace. 1987. Local geomorphology as a determinant of macrofaunal production in a mountain stream. Ecology 68:1932–1942.

Hyatt, T. L., and R. J. Naiman. 2001. The residence time of large woody debris in the Queets River, Washington, USA. Ecological Applications 11:191–202.

Hynes, H. B. N. 1970. The ecology of running waters. University of Toronto Press, Toronto, Canada.

Jacobi, D. I, and A. C. Benke. 1991. Life histories and abundance patterns of snag-dwelling mayflies in a blackwater Coastal Plain river. Journal of the North American Benthological Society 10:372–387.

Johnson, Z. B., A. K. Riggs, and J. H. Kennedy. 1998. Microdistribution and secondary production of *Cyrnellus fraternus* (Trichoptera: Polycentropodidae) from snag habitats in the Elm Fork of the Trinity River, Texas. Annals of the Entomological Society of America 91:641–646.

Jones, C. G., J. H. Lawton, and M. Shachak. 1994. Organisms as ecosystem engineers. Oikos 69:373–386.

Kaufman, M. G., and R. H. King. 1987. Colonization of wood substrates by the aquatic xylophage *Xylotopus par* (Diptera: Chironomidae) and a description of its life history. Canadian Journal of Zoology 65:2280–2286.

Kovalak, W. P. 1978. Effects of a pool on stream invertebrate drift. American Midland Naturalist 99:119–127.

Lehtinen, R. M., N. D. Mundahl, and J. C. Madejczyk. 1997. Autumn use of woody snags by fishes in backwater and channel border habitats of a large river. Environmental Biology of Fishes 49:7–19.

Lemly, A. D., and R. H. Hilderbrand. 2000. Influence of large woody debris on stream insect communities and benthic detritus. Hydrobiologia 421:179–185.

Lenat, D. R. 1988. Water quality assessment of streams using a qualitative collection method for benthic macroinvertebrates. Journal of the North American Benthological Society 7:222–233.

Magoulick, D. H. 1998. Effect of wood hardness, condition, texture and substrate type on community structure of stream invertebrates. American Midland Naturalist 139:187–200.

Maser, C., and J. R. Sedell. 1994. From the forest to the sea: the ecology of wood in streams, rivers, estuaries and oceans. St. Lucie Press, Delray Beach, Florida.

Maxted, J. R., M. T. Barbour, J. Gerritsen, V. Poretti, N. Primrose, A. Silvia, D. Penrose, and R. Renfrow. 2000. Assessment framework for mid-Atlantic coastal plain streams using benthic macroinvertebrates. Journal of the North American Benthological Society 19:128–144.

McCall, E. 1988. The attack on the great raft. Invention & Technology 3(3):10–16.

McDowell, D. M., and R. J. Naiman. 1986. Structure and function of a benthic invertebrate stream community as influenced by beaver (*Castor canadensis*). Oecologia 68:481–489.

McKie, B. G. L., and P. S. Cranston. 1998. Keystone coleopterans? Colonization by wood-feeding elmids of experimentally immersed woods in southeastern Australia. Australian Journal of Marine and Freshwater Research 49:79–88.

McLachlan, A. J. 1970. Submerged trees as a substrate for benthic fauna in the recently created Lake Kariba (central Africa). Journal of Applied Ecology 7:253–266.

Melillo, J. M., R. J. Naiman, J. D. Aber, and K. N. Eshleman. 1983. The influence of substrate quality and stream size on wood decomposition dynamics. Oecologia (Berlin) 58:281–285.

Modde, T. C., and J. C. Schmulbach. 1973. Seasonal changes in the drift and benthic macroinvertebrates in the unchannelized Missouri River in South Dakota. Proceedings of the South Dakota Academy of Science 52:118–126.

Molles, M. C., Jr. 1982. The Trichoptera of streams associated with aspen and spruce-fir forests: long-term community change. Ecology 63:1–6.

Monzyk, F. R., W. E. Kelso, and D. A. Rutherford. 1997. Characteristics of woody cover used by brown madtoms and pirate perch in Coastal Plain streams. Transactions of the American Fisheries Society 126:665–675.

Morris, L. A., R. N. Langemeier, T. R. Russell, and A. Witt, Jr. 1968. Effects of main stem impoundments and channelization upon the limnology of the Missouri River, Nebraska. Transactions of the American Fisheries Society 97:380–388.

Naiman, R. J., C. A. Johnston, and J. C. Kelley. 1988. Alteration of North American streams by beaver. BioScience 38:753–762.

Newbold, J. D., R. V. O'Neill, J. W. Elwood, and W. VanWinkle. 1982. Nutrient spiralling in streams: implications for nutrient limitation and invertebrate activity. American Naturalist 120:628–652.

Nilsen, H. C., and R. W. Larimore. 1973. Establishment of invertebrate communities on log substrates in the Kaskaskia River, Illinois. Ecology 54:366–374.

Nord, A. E., and J. C. Schmulbach. 1973. A comparison of the macroinvertebrate aufwuchs in the unstabilized and stabilized Missouri River. Proceedings of the South Dakota Academy of Science 52:127–139.

NTSB (National Transportation Safety Board). 1989. Collapse of the S. R. 675 bridge spans over the Pocomoke River near Pocomoke City, Maryland. PB89–916205. National Transportation Safety Board/HAR-89/04, Washington, D.C.

O'Connor, N. A. 1991. The effects of habitat complexity on the macroinvertebrates colonising wood substrates in a lowland stream. Oecologia 85:504–512.

O'Connor, N. A. 1992. Quantification of submerged wood in a lowland Australian stream system. Freshwater Biology 27:387–395.

Palmer, M. A., P. Arensburger, A. P. Martin, and D. W. Denman. 1996. Disturbance and patch-specific responses: the interactive effects of woody debris and floods on lotic invertebrates. Oecologia 105:247–257.

Pereira, C. R. D., N. H. Anderson, and T. Dudley. 1982. Gut content analysis of aquatic insects from wood substrates. Melanderia 39:23–33.

Phillips, E. C. 1994. Habitat preference and seasonal abundance of Trichoptera larvae in Ozark streams, Arkansas. Journal of Freshwater Ecology 9:91–95.

Phillips, E. C. 1995. Associations of aquatic Coleoptera with coarse woody debris in Ozark streams, Arkansas. The Coleopterists Bulletin 49:119–126.

Phillips, E. C. 1997a. Life cycle, growth, and production of *Macronychus glabratus* (Coleoptera: Elmidae) in northwest Arkansas and southeast Texas streams. Invertebrate Biology 116:134–141.

Phillips, E. C. 1997b. Life history and energetics of *Ancyronyx variegata* (Coleoptera: Elmidae) in northwest Arkansas and southeast Texas. Annals of the Entomological Society of America 90:54–61.

Phillips, E. C., and R. V. Kilambi. 1994a. Habitat type

and seasonal effects of the distribution and density of Plecoptera in Ozark streams, Arkansas. Annals of the Entomological Society of America 87:321–326.

Phillips, E. C., and R. V. Kilambi. 1994b. Utilization of coarse woody debris by Ephemeroptera in three Ozark streams of Arkansas. The Southwestern Naturalist 39:58–62.

Phillips, E. C., and R. V. Kilambi. 1994c. Use of coarse woody debris by Diptera in Ozark streams, Arkansas. Journal of the North American Benthological Society 13:151–159.

Pollock, M., M. Heim, and D. Werner. 2003. Hydrologic and geomorphic effects of beaver dams and their influence on fishes. Pages 213–233 *in* S. V. Gregory, K. L. Boyer, and A. M. Gurnell, editors. The ecology and management of wood in world rivers. American Fisheries Society, Symposium 37, Bethesda, Maryland.

Pyron, M., A. P. Covich, and R. W. Black. 1999. On the relative importance of pool morphology and woody debris to distributions in a Puerto Rican headwater stream. Hydrobiologia 405:207–215.

Quinn, J. M., A. B. Cooper, R. J. Davies-Colley, J. C. Rutherford, and R. B. Williamson. 1997. Land use effects on habitat, water quality, periphyton, and benthic invertebrates in Waikato, New Zealand, hill-country streams. New Zealand Journal of Marine and Freshwater Research 31:579–597.

Quist, M. C., and C. S. Guy. 2001. Growth and mortality of prairie stream fishes: relations with fish community and instream habitat characteristics. Ecology of Freshwater Fish 10:88–96.

Rabeni, C. F. 2000. Evaluating physical habitat integrity in relation to the biological potential of streams. Hydrobiologia 422/423:245–256.

Rabeni, C. F., and S. M. Hoel. 2000. The importance of woody debris to benthic invertebrates in two Missouri prairie streams. Internationale Vereinigung für Theoretische und Angewandte Limnologie 27:1499–1502.

Roeding, C. E., and L. A. Smock. 1989. Ecology of macroinvertebrate shredders in a low-gradient sandy-bottomed stream. Journal of the North American Benthological Society 8:149–161.

Rolauffs, P., D. Hering, and S. Lohse. 2001. Composition, invertebrate community and productivity of a beaver dam in comparison to other stream habitat types. Hydrobiologia 459:201–212.

Sedell, J. R., F. H. Everest, and F. J. Swanson. 1982. Fish habitat and streamside management: past and present. Pages 244–255 *in* Proceedings of the Society of American Foresters Annual Meeting (September 27–30, 1981). Society of American Foresters, Bethesda, Maryland.

Sedell, J. R., and J. L. Froggatt. 1984. Importance of streamside forests to large rivers: the isolation of the Willamette River, Oregon, U.S.A., from its floodplain by snagging and streamside forest removal. Internationale Vereinigung für Theoretische und Angewandte Limnologie 22:1828–1834.

Sedell, J. R., and K. J. Luchessa. 1982. Using the historical records as an aid to salmonid habitat enhancement. Pages 210–223 *in* N. B. Armantrout, editor. Acquisition and utilization of aquatic habitat inventory information. American Fisheries Society, Western Oregon Division, Bethesda, Maryland.

Sedell, J. R., G. H. Reeves, F. R. Hauer, J. A. Stanford, and C. P. Hawkins. 1990. Role of refugia in recovery from disturbances: modern fragmented and disconnected river systems. Environmental Management 14:711–724.

Sheldon, F., and K. F. Walker. 1998. Spatial distribution of littoral invertebrates in the lower Murray-Darling River system, Australia. Marine and Freshwater Research 49:171–182.

Smith, L. C., and L. A. Smock. 1992. Ecology of invertebrate predators in a Coastal Plain stream. Freshwater Biology 28:319–329.

Smith, M. E., C. T. Driscoll, B. J. Wyskowski, C. M. Brooks, and C. C. Cosentini. 1991. Modification of stream ecosystem structure and function by beaver (*Castor canadensis*) in the Adirondack Mountains, New York. Canadian Journal of Zoology 69:55–61.

Smock, L. A., E. Gilinsky, and D. L. Stoneburner. 1985. Macroinvertebrate production in a southeastern United States blackwater stream. Ecology 66:1491–1503.

Smock, L. A., J. E. Gladden, J. L. Riekenberg, L. C. Smith, and C. R. Black. 1992. Lotic macroinvertebrate production in three dimensions: channel surface, hyporheic, and floodplain environments. Ecology 73:876–888.

Smock, L. A., G. M. Metzler, and J. E. Gladden. 1989. Role of debris dams in the structure and functioning of low-gradient headwater streams. Ecology 70:764–775.

Smock, L. A., and C. E. Roeding. 1986. The trophic basis of production of the macroinvertebrate community of a southeastern U.S.A. blackwater stream. Holarctic Ecology 9:165–174.

Spänhoff, B., C. Alecke, and E. I. Meyer. 2000. Colonization of submerged twigs and branches of different wood genera by aquatic macroinvertebrates. International Review of Hydrobiology 85:49–66.

Steedman, R. J., and N. H. Anderson. 1985. Life history and ecological role of the xylophagous aquatic beetle, *Lara avara* LeConte (Dryopoidea: Elmidae). Freshwater Biology 15:535–546.

Stites, D. L. 1986. Secondary production and productivity in the sediments of blackwater rivers. Doctoral dissertation. Emory University, Atlanta.

Stites, D. L., and A. C. Benke. 1989. Rapid growth rates of chironomids in three habitats of a subtropical blackwater river and their implications for P:B ratios. Limnology and Oceanography 34:1278–1289.

Swanson, F. J., and G. W. Lienkaemper. 1978. Physical consequences of large organic debris in Pacific Northwest streams. USDA Forest Service Technical Report PNW-69, Pacific Northwest Research Station, Portland, Oregon.

Sweeney, B. W. 1993. Effects of streamside vegetation on macroinvertebrate communities of White Clay Creek in Eastern North America. Proceedings of the Academy of Natural Sciences of Philadelphia 144:291–340.

Tank, J. L., and J. R. Webster. 1998. Interaction of substrate availability and nutrient distribution on wood biofilm development in streams. Ecology 79:2168–2179.

Tank, J. L., and M. J. Winterbourn. 1995. Biofilm development and invertebrate colonization of wood in four New Zealand streams of contrasting pH. Freshwater Biology 34:303–315.

Tank, J. L., and M. J. Winterbourn. 1996. Microbial activity and invertebrate colonization of wood in a New Zealand forest stream. New Zealand Journal of Marine and Freshwater Research 30:271–280.

Thorp, J. H., E. M. McEwan, M. F. Flynn, and F. R. Hauer. 1985. Invertebrate colonization of submerged wood in a cypress-tupelo swamp and blackwater stream. American Midland Naturalist 113:56–68.

Triska, F. J. 1984. Role of wood debris in modifying channel geomorphology and riparian areas of a large lowland river under pristine conditions: an historical case study. Internationale Vereinigung für Theoretische und Angewandte Limnologie 22:1876–1892.

Triska, F. J., and K. Cromack, Jr. 1980. The role of wood debris in forests and streams. Pages 171–190 *in* R. H. Waring, editor. Forests: fresh perspectives from ecosystem analysis. Proceedings of the 40th Annual Biology Colloquium. Oregon State University Press, Corvallis.

Trotter, E. H. 1990. Woody debris, forest-stream succession, and catchment geomorphology. Journal of the North American Benthological Society 9:141–156.

USEPA (U. S. Environmental Protection Agency). 1997. Field and laboratory methods for macroinvertebrate and habitat assessment of low gradient nontidal streams. Mid-Atlantic Coastal Streams Workshop, Environmental Services Division, Region 3. Wheeling, West Virginia.

Van Arsdall, T. C., Jr. 1977. Production and colonization of the snag habitat in a southeastern blackwater river. Master's thesis. Georgia Institute of Technology, Atlanta.

Wallace, J. B., and A. C. Benke. 1984. Quantification of wood habitat in subtropical Coastal Plain streams. Canadian Journal of Fisheries and Aquatic Sciences 41:1643–1652.

Wallace, J. B., A. C. Benke, A. H. Lingle, and K. Parsons. 1987. Trophic pathways of macroinvertebrate primary consumers in subtropical blackwater streams. Archiv für Hydrobiologia(Supplement 74):423–451.

Wallace, J. B., S. L. Eggert, J. L. Meyer, and J. R. Webster. 1999. Effects of resource limitation on a detrital-based ecosystem. Ecological Monographs 69:409–442.

Wallace, J. B., J. W. Grubaugh, and M. R. Whiles. 1996. Influence of coarse woody debris on stream habitats and invertebrate biodiversity. Pages 119–129 *in* J. W. McMinn and D. A. Crossley, editors. Biodiversity and coarse woody debris in southern forests (Proceedings of the workshop on coarse woody debris in southern forests: effects on biodiversity). USDA Forest Service, Southern Research Station, Asheville, North Carolina.

Wallace, J. B., and F. F. Sherberger. 1974. The larval retreat and feeding net of *Macronema carolina* Banks (Trichoptera: Hydropsychidae). Hydrobiologia 45:177–188.

Wallace, J. B., and J. R. Webster. 1996. The role of macroinvertebrates in stream ecosystem function. Annual Review of Entomology 41:115–139.

Wallace, J. B., J. R. Webster, S. L. Eggert, J. L. Meyer, and E. R. Siler. 2001a. Large woody debris in a headwater stream: long-term legacies of forest disturbance. International Review of Hydrobiology 86:501–513.

Wallace, J. B., J. R. Webster, S. L. Eggert, and J. L. Meyer. 2001b. Small wood dynamics in a headwater stream. Internationale Vereinigung für Theoretische und Angewandte Limnologie 27:1361–1365.

Wallace, J. B., J. R. Webster, and J. L. Meyer. 1995. Influence of log additions on physical and biotic characteristics of a mountain stream. Canadian Journal of Fisheries and Aquatic Sciences 52:2120–2137.

Wallace, J. B., J. R. Webster, and W. R. Woodall. 1977. The role of filter-feeders in flowing waters. Archiv für Hydrobiologie 79:506–532.

Warmke, S., and D. Hering. 2000. Composition, microdistribution and food of the macroinvertebrate fauna inhabiting wood in low-order mountain streams in Central Europe. International Review of Hydrobiology 85:67–78.

Waters, T. F. 1972. The drift of stream insects. Annual Review of Entomology 17:253–272.

Weigelhofer, G., and J. A. Waringer. 1999. Woody debris accumulations – important ecological components in a low order forest stream (Weidlingbach, lower Austria). International Review of Hydrobiology 84:427–437.

Wiggins, G. B. 1998. Larvae of the North American caddisfly genera (Trichoptera), 2nd edition. University of Toronto Press, Toronto.

Wissinger, S. A., and L. J. Gallagher. 1999. Beaver pond wetlands in northwestern Pennsylvania: modes of colonization and succession after drought. Pages 333–262 *in* D. P. Batzer, R. B. Rader, and S. A. Wissinger, editors. Invertebrates in freshwater wetlands of North America: ecology and management. Wiley, New York.

Wondzell, S. M. and P. A. Bisson. 2003. Influence of wood on aquatic biodiversity. Pages 249–263 *in* S. V. Gregory, K. L. Boyer, and A. M. Gurnell, editors. The ecology and management of wood in world rivers. American Fisheries Society, Symposium 37, Bethesda, Maryland.

Zalewski, M., M. Lapinska and P. B. Bayley. 2003. Fish relationships with wood in large rivers. Pages 195–211 *in* S. V. Gregory, K. L. Boyer, and A. M. Gurnell, editors. The ecology and management of wood in world rivers. American Fisheries Society, Symposium 37, Bethesda, Maryland.

American Fisheries Society Symposium 37:179–193, 2003

Fish Relationships with Large Wood in Small Streams

C. Andrew Dolloff

USDA Forest Service, Southern Research Station, Department of Fisheries and Wildlife Virginia Tech, Blacksburg, Virginia 24060, USA

Melvin L. Warren, Jr.

USDA Forest Service, Southern Research Station 1000 Front Street, Oxford, Massachusetts 38655, USA

Abstract.—Many ecological processes are associated with large wood in streams, such as forming habitat critical for fish and a host of other organisms. Wood loading in streams varies with age and species of riparian vegetation, stream size, time since last disturbance, and history of land use. Changes in the landscape resulting from homesteading, agriculture, and logging have altered forest environments, which, in turn, changed the physical and biological characteristics of many streams worldwide. Wood is also important in creating refugia for fish and other aquatic species. Removing wood from streams typically results in loss of pool habitat and overall complexity as well as fewer and smaller individuals of both coldwater and warmwater fish species. The life histories of more than 85 species of fish have some association with large wood for cover, spawning (egg attachment, nest materials), and feeding. Many other aquatic organisms, such as crayfish, certain species of freshwater mussels, and turtles, also depend on large wood during at least part of their life cycles.

Introduction

Large wood can profoundly influence the structure and function of aquatic habitats from headwaters to estuaries. In streams and rivers flowing through forests worldwide, large wood slows, stores, and redirects surface water and sediments and provides cover, substrate, and food used by fish and other vertebrates, aquatic invertebrates, and microbes (Table 1). We here review the role of large wood in small streams and highlight specific examples of how wood influences the distribution, abundance, and life history of various organisms. Our focus in this chapter will be on small streams, defined here as fourth-order or lower and having a bank-full width of 10 m or less.

Physical Effects in Stream Channels

Woody material of all sizes—from tiny fragments to intact trees—plays a role in stream systems. Because decay rate and probability of displacement are a function of size, large pieces have a greater influence on habitat and physical processes than small pieces. In general, rootwads, branches, snags, detached boles, and other pieces 1.0–1.5 m long and wider than 10 cm in diameter are defined as large wood. Although somewhat arbitrary, this size approximates the wood that is entrained rather than simply passed through channels of small perennial streams. Smaller pieces and fragments, even those too small to count, also play a role in forming habitat (for example, as part of the matrix of material that forms debris dams; Bilby and Likens 1980; Dolloff and Webster 2000). Depending on the characteristics of the stream channel (for example, the presence of hydraulic controls, channel constrictions, channel roughness) and relative size and quality (for example, surface area and resistance to decay) of the wood, wood accumulations or dams of both large and small wood create habitat diversity and may persist for a few months to hundreds of years (Bilby and Likens 1980).

TABLE 1. The physical and biological roles of large wood in small streams.

Physical		Biological	
Water	Substrate	Food	Cover
storage	storage	substrate	protect from predators
quality	routing	abundance	isolate competitors
		availability	protect from displacement
Channel shape		delivery	
width, depth		substrate	
habitat type		type	

Nearly all wood that impinges on stream channels has the capacity to influence habitat and aquatic communities. Large wood oriented perpendicular to the thalweg is often associated directly with pool formation (Cherry and Beschta 1989; Richmond and Fausch 1995; Hauer et al. 1999). Even wood outside or above the wetted portion of the channel influences pool formation by directing patterns of scour at bank-full flows. Wood bridges eventually weaken with decay, break into two or more pieces, and exert entirely different influences on channel shape and habitat formation.

The amount of large wood in streams (wood loading) varies with age of the riparian forest, species of riparian vegetation, size of stream, time since disturbance, and land-use history. All of these factors interact with the frequency and magnitude of natural disturbances as well as human perturbations. Of these human alterations, history of land use, particularly urbanization, agriculture, and logging, has changed not only the physical and biological characteristics of streams but also our perception of what is natural. Natural disturbances like fire, wind, and floods do not typically affect all parts of a landscape equally. A flood-producing storm in one watershed may not similarly affect an adjacent watershed. Because of the dynamic spatial effects of natural disturbance regimes, large wood loading and hence stream habitat features across natural landscapes vary greatly. At any one time, some stream corridors may have large amounts of large wood configured into highly complex habitats, but others—perhaps even in the same watershed—may lack wood and have greatly simplified habitats (Sedell et al. 1990).

Today, the lack of large wood in streams near urban centers or in those draining agricultural regions is readily apparent, but even streams flowing through forested regions often lack significant amounts of wood (Maser and Sedell 1994). Nearly all of the riparian forests across the eastern United States have been harvested at least once, if not multiple times, in the last 150 years. Most have not had time to produce the large trees that are not only the source of instream wood, but are also critical in forming complex, long-lasting habitat configurations (Dolloff and Webster 2000). The loss of old or late-successional riparian forests in the last century constitutes a new disturbance regime. The drastically reduced recruitment of large wood coupled with natural processes of decay and downstream transport have resulted in unnaturally low accumulations of large wood in streams across entire landscapes. Recovery of stream ecosystems through formation of wood dams may not occur for many decades (Golladay et al. 1989; Webster et al. 1990). The absence of large wood in riparian forests over much of North America has greatly reduced habitat complexity in streams and also reduced variability among streams draining a given ecoregion or geoclimactic setting. Without adequate large wood, these systems become more homogeneous with reduced habitat complexity, species diversity, and productivity.

Habitats in small streams are particularly sensitive to changes in large wood loading, and the full implications of reduced amounts of wood over large areas are unknown but potentially far-reaching for certain groups and life stages of organisms. Risk to the persistence, distribution, and abundance of pool-dependent species may be compounded after a natural disturbance, particularly if the abundance of suitable alternative habitats is already low or fragmented because of a previous disturbance.

Directly removing wood typically results in loss of pools and habitat complexity; in sand-bed streams, it may trigger or exacerbate destructive channel incision (Shields and Smith 1992). Removing wood from the channel or floodplain reduces numbers, average size, and biomass for both warmwater (Angermeier and Karr 1984; Shields et al. 1994) and coldwater (Elliott 1986; Dolloff 1986) fish species. Adding wood typically increases total pool area and depth, but simply adding wood may not produce the same result in all streams. The quality of habitat produced by large wood is a function of geomorphic characteristics, such as channel size, channel form, gradient, and substrate size. For example, pool habitat more than doubled within 1 year of adding large wood to a small, low gradient (1.0% slope) Virginia stream (Hilderbrand et al. 1997). In contrast, pool area remained essentially unchanged after similar amounts of large wood were added to a moderate gradient (3–6%) stream where boulders were the dominant pool-forming element.

Wood is deposited and accumulates in small streams by a combination of chronic and episodic processes (Table 2). Inputs of the greatest amounts of large wood usually can be traced to specific, weather-related events; piles of logs deposited by floodwaters along stream sides are obvious indicators of the importance of episodic events. But for perspective, the annual input of organic material to small upland streams from seasonal litter fall (leaves, twigs, small branches) typically exceeds the input of large wood (Webster et al. 1990). In general, large wood input rates vary with tree species and age, slope and shape of surrounding land forms, health and history of surrounding forest (insects, disease, land use), and size of the watershed. Recruitment rates are generally lower in second-growth forests; in eastern North America, input rates tend to be very low because many riparian forests have been completely removed or have been harvested repeatedly (three or more timber rotations).

Hydraulic processes control the particular arrangement of large wood after it enters a stream channel, but the mechanism of entry can also be influential. Stream size, which incorporates differences in flow magnitudes, has a major influence on the fate and distribution of large wood (Bisson et al. 1987). In small streams (first- and second-order), the dominant spatial pattern appears random: dams may form at predictable intervals from leaves and litter; however, large pieces tend to stay where they fall because transport power is low. Accumulations of wood in intermediate-size streams (second to fourth-order) tend to be larger, more variable in size, and clumped at or near bends or changes in channel gradient. The mechanism of entry also influences the arrangement of large wood accumulations. In contrast to the relatively loose architecture of wood jams that form incrementally from the transport and entrainment of individual pieces in a stream, large wood pieces from landslides, debris flows, and snow avalanches tend to deposit in interlocking wood accumulation's (see Benda et al. 2003, this volume). Although formed by different processes, both types of accumulations make important contributions to habitat.

Some large wood accumulations in small streams last for hundreds of years. However, the total amount of large wood is dynamic, changing, sometimes dramatically, over the course of a year. As an illustration of this variability, the half-life of individual pieces of large wood ranges from 2.3 to 230 years (Harmon et al. 1986). The half-life of wood dams in very small streams tends to be less, about 6–24 months, depending on size and composition of the material in the dam and characteristics of flow. Whether limbs or whole trees, decay begins with death. The longevity of intact pieces of large wood is thus influenced by decay rates (resistance to abrasion, breakage, and rot) and transport processes. Softwoods generally decay faster than hardwoods, but certain softwood species, like hemlock, that contain high levels of decay-resistant tannins and other substances, last longer than many "soft" hardwoods, such as tulip poplar and aspen. Temperature of both the air and water, exposure to sunlight and humidity, and frequency and duration of flow all affect the rate of wood decay and its longevity at a particular site. Wood buried in the streambed or continually submerged resists decay better than wood that is only partly or periodically submerged, alternating between being wet and dry.

The loading of large wood in streams flowing through forests varies greatly, from 2.5 to 4,500 m^3/ha (Harmon et al. 1986). In general, streams flowing through forests undisturbed by human

TABLE 2. Mechanisms of large wood input to small streams.

Chronic processes	Episodic processes
Litterfall, self-pruning from riparian areas	Tree fall from stream bank failures
Mortality of trees from insects and disease	Tree fall from windthrow
Tree fall from streambank undercutting	Snow and ice accumulation and melting
	Avalanches, debris flows

activities have the greatest volumes or density of large wood. High loads are well documented in the Pacific Northwest, but small streams flowing through forests in eastern North America also have substantial loads (Hedman et al. 1996). By inference, historical amounts of large wood in many small eastern streams probably were high. In the southern Appalachians, streams draining old-growth forests contained amounts of large wood comparable to those in northwestern old-growth forests and up to four times more large wood than in eastern wilderness streams draining forests cleared early in the century (Flebbe and Dolloff 1995). Likewise, small streams in unlogged watersheds of the Great Smoky Mountains National Park contained 4 times the volume of wood and 10 times the material in wood dams than streams logged early in the 20th century (Silsbee and Larson 1983; Table 6 in Harmon et al. 1986). In small streams of southern New Zealand, researchers determined that the amount of large wood in older forest streams was greater than in young forests, with overall amounts much smaller relative to those found in North American forested streams (Evans et al. 1993).

Habitat Relations

For many aquatic organisms, particularly fishes, large wood is most important in creating and maintaining deepwater or pool habitat. Pools and other habitats associated with large wood are attractive to fish for a variety of reasons. Conspicuous among them are the lower water velocities and greater depths associated with pools during low-flow periods. Salmonids and drift-feeding minnows inhabit areas with low velocity water and make forays into fast water to capture food items in the drift (Matthews 1998). Trout and minnows of eastern England were also found to be strongly associated with large wood (Punchard et al. 2000). Several American species also seek the lower velocities behind logs to avoid swift, cold winter flows. Refuges from high velocities, such as those found in the lee of logs, may significantly affect overwintering survival of the federally listed bayou darter *Etheostoma rubrum* (Ross et al. 1992) and many other small benthic fishes, particularly in sand-bottomed coastal plain streams. Still other fishes use the cover of logs in pools for nest sites and spawning (for example, centrar-chids) or attach their eggs to logs (for example, shiners *Cyprinella* spp. and relict darter *Etheostoma chienense*). Pools typically harbor more and larger fish than shallower areas because of the greater space (volume) of habitat available, particularly during times of exceptionally low flow, such as in late summer or winter. Streams with poorly developed, shallow pools have a low residual pool volume (Lisle and Hilton 1992) and consequently tend to support fewer species, have simple trophic structure, and show higher variability in fish abundance (Schlosser 1987).

Pools can form around any obstruction that creates friction and resists displacement by flowing water, and large wood is often the dominant pool-forming element. In small streams in forested areas, large wood can be responsible for 50–90% of all pools (Andrus et al. 1988; Hedman et al. 1996). The proportion of pools formed by large wood tends to be much lower (<10%) in streams flowing through second-growth and younger forests.

Large wood is an important habitat that is included in the definitions of many fluvial habitat types (Bisson et al. 1982; Hawkins et al. 1993; Armantrout 1998). Large wood can contribute to pool formation in a variety of ways depending on factors such as the size of the piece, the mechanism of entry into a stream channel, channel size and gradient, and streamflow. The most commonly envisioned orientation is channel spanning and perpendicular to the channel, which often results in forming plunge pools in the downstream and dam pools in the upstream direction. Perpendicularly oriented large wood that juts into but does not span a channel may create eddies and backwater pools. The deepest pools form behind pieces that span the entire width of the channel and are oriented perpendicular to flow. Large wood helps maintain the diversity of physical habitat by delimiting pool position and increasing depth variability.

Although the role of large wood in deep pool formation is important, its contribution as an element of channel roughness in shallow stream reaches should not be overlooked. In many coastal plain streams of the eastern United States, riffles and cobble substrate are absent and gravel is rare (Meffe and Sheldon 1988). Here, large wood is often the only element contributing to channel roughness and, hence, to the forming of complex, flowing habitats (Smock and Gilinsky 1992). The presence of wood accumulations, even relatively small-diameter pieces, in shallow, sandy, flowing areas creates a heterogeneous zone of variable velocities and depths. These wood-formed "riffles and runs" in turn support a significant proportion of the stream fish diversity in coastal plain streams and likely are critical to the persistence of many darters (*Percina* spp. and *Etheostoma* spp.) and madtom catfishes *Noturus* spp. (Chan and Parsons 2000; Warren et al. 2002).

Large wood creates sediment storage sites (floodplains and terraces), the wetted portions of which contribute to producing food (algae and macroinvertebrate) and spawning habitat. Forming new sediment terraces is particularly important in drainages where natural floodplain deposits and terraces have been lost, such as in many small watersheds in the Cascade Range of Oregon and across much of eastern North America. Conversely, large wood also encourages scour and promotes transportation of fine sediments, thereby exposing gravel where fish spawn and benthic macroinvertebrates thrive. In general, streamflow determines the influence of large wood on routing and rates of sediment scour and fill. At low flow, pools formed by wood tend to fill and riffles tend to scour, but when flow is high, pools scour and riffles are depositional.

Although many researchers have focused on how large wood affects physical processes in streams, large wood also has a major impact on many, if not all, biological processes, such as substrate type, water depth, and velocity. These physical processes are influenced by large wood. Large wood also facilitates primary production by providing attachment sites for microbes and algae, sources of nutrients, and storage zones for organic matter. Biofilms, composed of microorganisms and the mucoidal matrix that surrounds them, develop readily on wood, a substrate that also provides a portion of the nutrients required to maintain them (Aumen et al. 1990; Golladay and Sinsabaugh 1991; Tank and Winterbourn 1995; Benke and Wallace 2003, this volume). Wood also creates conditions favorable to the entrainment and processing of terrestrially derived fine particulate organic material (FPOM) from riparian vegetation. Microbes and macroinvertebrates process this material more readily than wood. Large wood also traps and retains carcasses of spent fish in streams that are home to runs of anadromous fish (Cederholm and Peterson 1985). Nutrients, particularly nitrogen, from decomposing carcasses boost primary production by enhancing both algal and biofilm development (Richey et al. 1975; Wold and Hershey 1999).

Large wood enhances secondary production both directly, such as by increasing the surface area available to macroinvertebrate grazers and scrapers (Benke et al. 1985), and indirectly, by physical habitat partitioning. Angermeier and Karr (1984) experimentally removed large wood from one side of a reach of a small Illinois stream and found that, in addition to more and larger fish, the side of the stream with wood had more litter and benthos. In coastal plain headwater streams of Virginia, invertebrate densities were 10 times and biomass 5 times higher in wood dams than on sandy substrates (Smock et al. 1989). In pumice-bed streams of New Zealand, summer densities of macroinvertebrates were higher in streams where wood was present, providing either increased habitat or greater food resources or a combination of both (Collier and Halliday 2000). Wood snags in southeastern Australian lowland streams serve as important habitat for macroinvertebrates, influencing not only the diversity of species, but also their abundance (O'Connor 1991). Invertebrates associated with wood substrates form essential elements of the diet of many fishes in these streams (Felley and Felley 1987).

Aside from its role in creating pools and velocity breaks, large wood can be an important source of cover and habitat complexity for aquatic species. Wood cover can be simply an overhanging log or intricute accumulation of wood and other material, often referred to a wood jam. In northern Japanese streams, masu salmon *Oncorhynchus masou* density was correlated with wood cover in both pool and nonpool habitats (Inoue and Nakano 1998). In general, streams with complex habitats tend to have more fish and fish species than streams lacking complexity. Complexity provides a measure of security from predators (Harvey and Stewart 1991), isolation from competitors, and points of refuge from severe environmental conditions, such as flood or drought. Wood provides shadow, which not only makes fish harder for predators to see, but also aids fish in seeing approaching predators (Helfman 1981). Large wood protects fishes from aquatic predators (Everett and Ruiz 1993), and complexity thwarts diving and wading predators (Power 1987; Harvey and Stewart 1991. Complexity is critical for many fishes, particularly aggressive species like salmonids, which do not tolerate conspecifics in close proximity. Submerged branches and other wood partition habitat and visually isolate individual fish, allowing more fish per unit of available space (Dolloff 1986). Field observation of this "condominium effect" provides part of the rationale for enhancing habitat by using discarded Christmas trees, brush bundles, and tops from softwood trees (Boussu 1954; Reeves et al. 1991).

Complex cover and habitat, like that created by large wood, is especially important during times of stress, such as when streamflows are exceptionally low or high (Swales et al. 1986; Pearsons et al. 1992). In small, drought-prone streams, large wood and other channel obstructions may provide the only refuge during low flows. White pines planted near Benson's Run, Virginia toppled into the stream by the laterally migrating channel caused local scour and increased numbers and distribution of pools that are used as low-flow refugia by a host of fish species, including brook trout *Salvelinus fontinalis* (Figure 1). Species other than fish also frequent these pools; predators such as raccoons, herons, water snakes, and anglers are known to exploit the dense congregations of fish whose only escape lies in the tangled roots of the toppled trees. Adult cutthroat trout *O. clarki* in northern California small streams occupied habitat influenced by large wood during winter floods and showed a strong fidelity to wood-formed pools during normal fall and winter flows (Harvey et al. 1999).

In individual habitat units, fish tend to segregate by species and size. Particularly among fish with aggressive natures, physical habitat partitioning is an important element of coexistence. Larger or more aggressive individuals tend to occupy the most efficient positions in small streams (Fausch and White 1981), which often are characterized by structure at the heads of pools, near but out of high-velocity currents. Juvenile salmonids of different species and sizes are known to segregate in small streams where large wood is the dominant pool-forming element. For example, Dolloff and Reeves (1990) observed that coho salmon *O. kisutch* and Dolly Varden *S. malma* occupied distinct regions in small (2–10-m^2) pools in Alaska streams. One-year-old and older coho salmon typically occupied positions at the heads of pools near the water surface, but similar-sized Dolly Varden were associated more closely with the stream bottom and dense (frequently wood) cover. Young-of-the-year coho salmon and Dolly Varden occupied positions similar to the larger fish but farther down stream in the pools and associated less closely with cover.

Multivariate studies of warmwater, stream fish communities often demonstrate habitat partitioning based on factors involving amount of

FIGURE 1. Refuge habitat created by the rootwads of toppled streamside trees in Benson's Run, Virginia.

wood and cover after depth and velocity are taken into account (Baker and Ross 1981; Felley and Hill 1983; Felley and Felley 1987; Taylor et al. 1993; Warren et al. 2002). The actual association of species-rich, warmwater fish assemblages with wood and cover is often a complex mix of velocity preferences, behavioral responses (such as predator avoidance), or feeding responses, which can only be partially resolved in fish assemblage–habitat studies (Meffe and Sheldon 1988). In small northern Mississippi streams, many of which are affected by stream incision, the relative abundance of sunfishes was positively associated with wood in relatively deep habitats, a response attenuated by stream incision (Warren et al. 2002). In flowing habitats, large wood was associated positively with the relative abundance of darters (genera *Etheostoma* and *Percina*) and catfishes (mostly madtoms *Noturus* spp. and bullheads *Ameiurus* spp; Warren et al. 2002). In Ozark streams, stream-dwelling rock bass and smallmouth bass showed high association with wood, but habitat use showed distinct differences in segregation by species, size, and relation to current velocities (Probst et al. 1984). Rock bass *Ambloplites rupestris* were more often associated with rootwads, regardless of size, but larger smallmouth bass *Micropterus dolomieu* (>350 mm) used log complexes in higher velocities than velocities associated with rock bass cover (Probst et al. 1984).

As fish mature, their use of large wood in streams can shift in relation to their size and food preferences. For example, in a prairie stream in Kansas, growth increments of some stream fishes were associated strongly with amounts of large wood (Quist and Guy 2001). Creek chubs *Semotilus atromaculatus* were strongly associated with amounts of large wood for only the first year of growth, but all growth stages of green sunfish *Lepomis cyanellus* were associated with wood (Quist and Guy 2001). The authors attributed the response of creek chubs to shifts from eating invertebrates while small to eating fish when they were larger. Green sunfish likely exploited invertebrates on wood at small sizes but, at large sizes, continued to use large wood as cover (Quist and Guy 2001).

Many other fish species are also associated with accumulations of large wood. In Kentucky's Beaver Creek, for example, divers found the highest concentrations of the federally threatened blackside dace *Phoxinus cumberlandensis* in pools with a complex mixture of large and small wood. Two nocturnally active fishes, the brown madtom *Noturus phaeus* and the pirate perch *Aphredoderus sayanus*, are strongly associated with complex wood habitats in small coastal plain streams (Monzyk et al. 1997; Chan and Parsons 2000) of the southeastern United States. Large wood is a requisite feature of the diurnal cover of these species, but specific characteristics (for example, cavity space, entrained leaves, adjacent flow) also influence fish use (Monzyk et al. 1997). Brush bundles, leaf packs, and faux rootwads experimentally placed in northern Mississippi streams were used by 32 species of small-stream fishes. Strong position effects on fish numbers and identity reflected depth and velocity at individual samplers. Brush bundles and leafpacks held equal numbers of fish; smaller root samplers held fewer fish, but darters used them regularly. In January (water 2–5°C), up to 70 lethargic cyprinids occupied a single bundle, suggesting that such refuges are critical winter habitat (A. J. Sheldon, University of Montana, M. L. Warren, Jr. and W. R. Haag, USDA Forest Service, unpublished data). Even in the arid Southwest, large wood forms important habitat for native fishes. Presence of pools with root masses of standing or uprooted trees was the best predictor of the occurrence and abundance of the federally threatened Gila chub *Gila intermedia* (Probst and Stefferud 1994).

A cursory compilation of southeastern U.S. fishes that have some association with large wood is instructive (Table 3.). For many of these species, presence of large wood is facultative, particularly where rocky substrates or other elements can substitute for cover. For other species, large wood is the primary element of habitat complexity, especially in streams of predominantly fine substrates. Although not exhaustive, our compilation of fish associated with large wood habitat includes 12 families and more than 86 species of warmwater fishes. For these species, association with large wood includes diurnal cover, cover for adult or juvenile stages, food sources (such as algae or macroinvertebrates on wood), and reproductive uses (such as spawning and nesting cover, egg attachment). These species, and likely others that live in small streams, find protection from predators, as well as resting and foraging areas and spawning sites among the branches and pieces that comprise the matrix of wood jams.

Much information has accumulated showing the relation of large wood to fish populations, particularly of salmonids in western North America. Clear demonstration of similar relations in other regions and for other species, however, has often been problematic. For example, though Flebbe and

TABLE 3. Warmwater fish species associated with large wood in the southeastern United States

Species	Common name	Association with large wood	Reference
Ichthyomyzon castaneus	Chestnut lamprey	Spawning cover	Becker 1983
Lampetra aepyptera	Least brook lamprey	Cover for transforming larvae	Walsh and Burr 1981
Amia calva	Bowfin	Cover, nesting materials, nesting cover	Scott and Crossman 1973
Esox niger	Chain pickerel	Cover	Scott and Crossman 1973
Cyprinella analostana	Satinfin shiner	Egg attachment	Gale and Buynak 1978
Campostoma spp.	Stonerollers	Feeding on algae	Matthews 1998
Cyprinella callitaenia	Bluestripe shiner	Probable egg attachment	Wallace and Ramsey 1981
Cyprinella galactura	Whitetail shiner	Egg attachment	Pflieger 1997
Cyprinella lutrensis	Red shiner	Feeding, egg attachment	Quist and Guy 2001; Pflieger 1997
Cyprinella spiloptera	Spotfin shiner	Egg attachment	Pflieger 1965
Cyprinella venusta	Blacktail shiner	Egg attachment	Pflieger 1997
Cyprinella whipplei	Steelcolor shiner	Egg attachment	Pflieger 1965
Exoglossum laurae	Tonguetied minnow	Nesting cover	Trautman 1981; Jenkins and Burkhead 1994
Exoglossum maxillingua	Cutlips minnow	Nesting cover	Jenkins and Burkhead 1994
Opsopoeodus emiliae	Pugnose minnow	Probable egg attachment	Johnston and Page 1990
Nocomis leptocephalus	Bluehead chub	Nesting cover	Jenkins and Burkhead 1994
Phoxinus cumberlandensis	Blackside dace	Cover	Starnes and Starnes 1978
Phoxinus tennesseensis	Tennessee dace	Cover	Starnes and Jenkins 1988
Pimephales notatus	Bluntnose minnow	Nesting cover, egg attachment	Hubbs and Cooper 1936; Scott and Crossman 1973
Pimephales promelas	Fathead minnnow	Nesting cover, egg attachment	Wynne-Edwards 1932; Scott and Crossman 1973
Pimephales vigilax	Bullhead minnow	Nesting cover, egg attachment	Parker 1964
Pteronotropis euryzonus	Broadstripe shiner	Cover	Mettee et al. 1996
Semotilus atromaculatus	Creek chub	Feeding (juvenile), diurnal cover	Pflieger 1997; Quist and Guy 2001
Erimyzon oblongus	Creek chubsucker	Cover	Mettee et al. 1996
Erimyzon tenuis	Sharpfin chubsucker	Cover	Mettee et al. 1996
Ameiurus melas	Black bullhead	Nesting cover	Breder and Rosen 1966
Ameiurus natalis	Yellow bullhead	Nesting cover	Carlander 1969
Ameiurus nebulosus	Brown bullhead	Nesting cover	Blumer 1985
Ameiurus serracanthus	Spotted bullhead	Cover	Mettee et al. 1996
Noturus flavater	Checkered madtom	Cover	Pflieger 1997
Noturus flavipinnis	Yellowfin madtom	Diurnal cover	Dinkins and Shute 1996
Noturus funebris	Black madtom	Cover	Mettee et al. 1996
Noturus gyrinus	Tadpole madtom	Cover, nesting cover	Burr and Warren 1986; Burr and Stoeckel 1999
Noturus gilberti	Orangefin madtom	Nesting cover	Burr and Stoeckel 1999
Noturus hildebrandi	Least madtom	Cover, nesting cover	Mayden and Walsh 1984; Etnier and Starnes 1993
Noturus leptacanthus	Speckled madtom	Cover, probably nesting cover	Mettee et al. 1996
Noturus miurus	Brindled madtom	Cover, nesting cover	Burr and Warren 1986; Mettee et al. 1996; Burr and Stoeckel 1999
Noturus munitus	Frecklebelly madtom	Cover	Etnier and Starnes 1993
Noturus nocturnus	Freckled madtom	Cover	Burr and Warren 1986; Mettee et al. 1996

TABLE 3. Continued.

Species	Common name	Association with large wood	Reference
Noturus phaeus	Brown madtom	Diurnal cover, probably nesting cover	Monzyk et al. 1997
Noturus stigmosus	Northern madtom	Cover	Etnier and Starnes 1993
Pylodictis olivaris	Flathead catfish	Cover, nesting cover	Jackson 1999
Aphredoderus sayanus	Pirate perch	Cover, feeding	Benke et al. 1985; Monzyk et al. 1997
Fundulus olivaceus	Blackspotted topminnow	Cover	Mettee et al. 1996
Lucania goodei	Bluefin killifish	Cover	Mettee et al. 1996
Cottus bairdi	Mottled sculpin	Nesting cover, egg attachment	Rohde and Arndt 1982
Ambloplites ariommus	Shadow bass	Cover	Mettee et al. 1996
Ambloplites rupestris	Rock bass	Cover	Probst et al. 1984; Matthews 1998
Centrarchus macropterus	Flier	Cover	Mettee et al. 1996
Enneacanthus gloriosus	Bluespotted sunfish	Cover	Mettee et al. 1996
Lepomis spp.		Cover (various life stages), feeding	Benke et al. 1985; Matthews 1998
Micropterus dolomieu	Smallmouth bass	Cover	Probst et al. 1984
Etheostoma asprigene	Mud darter	Cover, probable egg attachment	Page et al. 1982; Burr and Warren 1986; Pflieger 1997
Etheostoma artesiae	Redfin darter	Cover	Mettee et al. 1996
Etheostoma boschungi	Slackwater darter	Cover	Etnier and Starnes 1993
Etheostoma chlorosomum	Bluntnose darter	Cover, egg attachment	Page et al. 1982; Etnier and Starnes 1993
Etheostoma chienense	Relict darter	Cover, nesting cover, egg attachment	Piller and Burr 1999
Etheostoma collis	Carolina darter	Cover	Jenkins and Burkhead 1994
Etheostoma colorosum	Coastal darter	Cover	Mettee et al. 1996
Etheostoma coosae	Coosa darter	Egg attachment	Mettee et al. 1996
Etheostoma corona	Crown darter	Nesting cover, egg attachment	Mettee et al. 1996
Etheostoma cragini	Arkansas darter	Cover	Pflieger 1997
Etheostoma davisoni	Choctawhatchee darter	Cover	Mettee et al. 1996
Etheostoma gracile	Slough darter	Cover, egg attachment	Braasch and Smith 1967
Etheostoma histrio	Harlequin darter	Cover, juvenile cover	Warren 1982; Pflieger 1997
Etheostoma lachneri	Tombigbee darter	Cover	Mettee et al. 1996
Etheostoma neopterum	Lollipop darter	Cover, nesting cover, probable egg attachment	Etnier and Starnes 1993
Etheostoma nigrum	Johnny darter	Nesting cover, egg attachment	Jenkins and Burhead 1994
Etheostoma olmstedi	Tessellated darter	Cover, nesting cover, probable egg attachment	Jenkins and Burhead 1994
Etheostoma oophylax	Guardian darter	Cover, nesting cover, probable egg attachment	Etnier and Starnes 1993
Etheostoma parvipinne	Goldstripe darter	Cover, egg attachment	Robison 1977; Johnston 1994
Etheostoma proeliare	Cypress darter	Cover, egg attachment	Burr and Page 1978; Pflieger 1997
Etheostoma punctulatum	Stippled darter	Cover	Pflieger 1997
Etheostoma pyrrhogaster	Firebelly darter	Probable egg attachment	Carney and Burr 1989; Etnier and Starnes 1993

TABLE 3. Continued.

Species	Common name	Association with large wood	Reference
Etheostoma ramseyi	Alabama darter	Cover	Mettee et al. 1996
Etheostoma swaini	Gulf darter	Cover	Etnier and Starnes 1993
Etheostoma tallapoosae	Tallapoosa darter	Spawning cover, probable egg attachment	Mettee et al. 1996
Etheostoma trisella	Trispot darter	Cover	Etnier and Starnes 1993
Etheostoma vitreum	Glassy darter	Egg attachment	Winn and Picciolo 1960
Etheostoma zonale	Banded darter	Juvenile cover	Pflieger 1997
Percina cymatotaenia	Bluestripe darter	Cover	Pflieger 1997
Percina macrocephala	Longhead darter	Cover	Etnier and Starnes 1993
Percina maculata	Blackside darter	Cover, particularly for juveniles	Etnier and Starnes 1993; Pflieger 1997
Percina nigrofasciata	Blackbanded darter	Cover	Etnier and Starnes 1993
Percina sciera	Dusky darter	Cover	Page and Smith 1970; Etnier and Starnes 1993; Pflieger 1997
Percina stictogaster	Frecklebelly darter	Cover	Burr and Page 1993; Etnier and Starnes 1993

Dolloff (1995) demonstrated high use by trout of habitat created by large wood, they were unable to document a direct relation of trout density to the amount of wood in three streams draining wilderness areas in the eastern United States.

Unpublished information regarding the role of large wood as habitat for fishes of the Amazon Basin provides preliminary insights about the role of wood in tropical streams (P. Petry, Fish Division, Field Museum of Natural History, Chicago, personal communication). Many species of fishes are believed to use large wood as their habitat in the Amazon River and its tributaries. This includes several groups of catfishes (families Loricariidae, Pseudopimelodidae, Auchenipteridae, and Ageneosidae). Species of the family Auchenipteridae live among dead trees and branches; they are called driftwood catfishes because they are always found inside logs. Several species of electric knifefishes (Order Gymnotiformes) are also frequently found inside logs. In addition to those fish that actually live inside the logs, many species are associated with the log jams, including many cichlids and characins.

Large wood can be a vital resource for completing life history phases for a host of fish species. Almost half the species associations with large wood are related to reproduction or juvenile rearing (Table 3). Fishes living in streams of the coastal plain, where streambed materials tend to be fine-grained and highly mobile (Felley 1992), often use large wood for attaching their eggs, spawning cover, or nest cover. For example, logs with intact, deeply ridged bark provide suitable spawning habitat for minnows of the genus *Cyprinella* (Pflieger 1997), which deposit their eggs in crevices. The large range of the blacktail shiner *Cyprinella venusta* across southeastern U.S. coastal plain streams is thought to be at least partially attributable to its ability to use large wood for egg attachment. Madtom catfishes *Noturus* spp. provide extensive care to nests, eggs, and young and several species nest on the undersides of large wood (Burr and Stoeckel 1999). The federally endangered relict darter attaches its eggs to the underside of sticks or logs, and the male remains to guard the nest site; paucity of woody spawning substrates in its sandy coastal plain habitat is a primary factor limiting recruitment of the species (Piller and Burr 1999). A similar case can be made for other egg attaching darters inhabiting coastal plain and piedmont streams (Table 3).

Beyond Fish

Adding wood to Puerto Rican headwater streams for the purpose of increasing cover for freshwater shrimp did not appear to affect total numbers of shrimp per pool area (Pyron et al. 1999). However, large wood in streams around the world has been found to influence a broad array of aquatic and riparian dependent organisms in addition to fishes and insects. One of the most spectacular examples of dependence on large wood is the giant freshwater crayfish of Tasmania. These animals, known to live for 30 years and reach weights of 4 kg, depend on wood both for food (they peel and eat layers of rotting eucalyptus logs) and for habitat (Hamr 1996). Instream habitat for the Tasmanian giant crayfish includes submerged logs, root masses, and wood accumulations, all of which are generally available and abundant in streams flowing through intact riparian areas. Another equally unique example is the interaction of large wood and the "fishing mussels" (*Lampsilis australis*, *L. perovalis*, and *L. subangulata*) of the southeastern United States (Haag et al. 1995). Females in this group release their larvae, which are obligate parasites on fishes, in two discrete masses that resemble a pair of small fishes in shape and color. After release, the lures are tethered to the female by a long, transparent mucous strand. As it undulates in the current, the tethered lure mimics movement of small fishes (Haag et al. 1995). After some time, the strand detaches from the female, drifts downstream, catches on wood, and continues "fishing" for the host fish (*Micropterus* spp.) (Haag et al. 1995; Haag and Warren 1999). Although the importance of this "snag and fish" strategy to recruitment has not been documented empirically, the strategy obviously extends temporal and spatial exposure of the lure (and, hence, the larvae) to the target host-fish species.

Perhaps less spectacular, but nevertheless important, is the association of aquatic turtles and wood. A frequently cited cause of decline of turtles in streams and rivers is the loss or deliberate removal of large wood from the channel (Felley 1992; Buhlmann and Gibbons 1997). Although few studies have focused on the relationship of turtles to large wood, snags and logs provide them with basking sites (Conant and Collins 1998), particularly the map and false map turtles (genus *Graptemys*). Other turtle species seem to prefer habitats where wood is present. In Whiskey Chitto Creek, a tributary of the Calcasieu River, Louisiana, Shively and Jackson (1985) found that the Sabine map turtle was limited to stream reaches with high numbers of snags. The snags provided both basking areas and a substrate for algae used as food by the turtle (Shively and Jackson 1985; Felley 1992).

In the last 20 years, the list of species associated with, if not dependent on, large wood has grown dramatically. New relationships between wood in the water and many taxa of animals emerge regularly, and there clearly is much to be learned (see also Steel et al. 2003, this volume). Additional research is needed to provide managers with a more complete understanding of the role of wood in streams including variables such as tree species, amounts, distribution, relations to biota (not only fishes but other vertebrates and invertebrates), and influence on habitat features in small streams. Knowledge gaps notwithstanding, our current knowledge provides a rigorous basis for resource managers to protect, maintain, and begin the complex task of restoring the functional character and features of small streams in forested regions.

References

Angermeier, P. L., and J. R. Karr. 1984. Relationships between woody debris and fish habitat in a small warmwater stream. Transactions of the American Fisheries Society 113:716–726.

Andrus, C. W., B. A. Long, and H. A. Froelich. 1988. Woody debris and its contribution to pool formation in a coastal stream 50 years after logging. Canadian Journal of Fisheries and Aquatic Sciences 45:2080–2086.

Aumen, N. G., C. P. Hawkins, and S. V. Gregory. 1990. Influence of woody debris on nutrient retention in catastrophically disturbed streams. Hydrobiologia 190:183–192.

Armantrout, N. B., compiler. 1998. Glossary of aquatic habitat inventory terminology. American Fisheries Society, Bethesda, Maryland.

Baker, J. A., and S. T. Ross. 1981. Spatial and temporal resource partitioning by southeastern cyprinids. Copeia 1981:178–184.

Becker, G. C. 1983. Fishes of Wisconsin. The University of Wisconsin Press, Madison.

Benda, L., D. Miller, J. Sias, D. Martin, R. Bilby, C. Veldhuisen, and T. Dunne 2003. Wood recruitment processes and wood budgeting. Pages 49–73 *in* S. V. Gregory, K. L. Boyer, and A. M. Gurnell, editors. The ecology and management of wood in world rivers. American Fisheries Society, Symposium 37, Bethesda, Maryland.

Benke, A. C., R. L. Henry, III, D. M. Gillespie, and R. J. Hunter. 1985. Importance of snag habitat for animal production in southeastern streams. Fisheries 10:8–13.

Benke, A. C., and J. B. Wallace. 2003. Influence of wood on invertebrate communities in streams and rivers. Pages 149–177 *in* S. V. Gregory, K. L. Boyer, and A. M. Gurnell, editors. The ecology and management of large wood in world rivers. American Fisheries Society, Symposium 37, Bethesda, Maryland.

Bilby, R. E., and G. E. Likens. 1980. Importance of organic debris dams in the structure and function of stream ecosystems. Ecology 61:1107–1113.

Bisson, P. A., J. L. Nielson, R. A. Palmason, and L. E. Grove. 1982. A system of naming habitat types in small streams, with examples of habitat utilizations by salmonids during low streamflow. Pages 62–73 *in* N. B. Armantrout, editor. Acquisition and utilization of aquatic habitat inventory information. American Fisheries Society, Western Division, Bethesda, Maryland.

Bisson, P. A., R. E. Bilby, M. D. Bryant, C. A. Dolloff, G. B. Grette, R. A. House, M. L. Murphy, K. V. Koski, and J. R. Sedell. 1987. Large woody debris in forested streams in the Pacific Northwest: past, present, and future. Pages 143–190 *in* E. O. Salo and T. W. Cundy, editors. Streamside management: forestry and fishery interactions. University of Washington, Seattle.

Blumer, L. S. 1985. Reproductive natural history of the brown bullhead *Ictalurus nebulosus* in Michigan. American Midland Naturalist 114:318–330.

Boussu, M. F. 1954. Relationship between trout production and cover on a small stream. Journal of Wildlife Management 18:229–239.

Braasch, M. E., and P. W. Smith. 1967. The life history of the slough darter, *Etheostoma gracile* (Pisces, Percidae). Illinois Natural History Survey Biological Notes 58:1–12.

Breder, C. M., Jr., and D. E. Rosen. 1966. Modes of reproduction in fishes. American Museum of Natural History, The Natural History Press, Garden City, New York.

Buhlmann, K. A., and J. W. Gibbons. 1997. Imperiled aquatic reptiles of the southeastern United States: historical review and current conservation status. Pages 201–231 *in* G. W. Benz and D. E. Collins, editors. Aquatic fauna in peril: the southeastern perspective. Southeast Aquatic Research Institute, Special Publication 1, Lenz Design & Communications, Decatur, Georgia.

Burr, B. M., and L. M. Page. 1993. A new species of *Percina* (*Odontopholis*) from Kentucky and Tennessee with comparisons to *Percina cymatotaenia* (Teleostei: Percidae). Bulletin of the Alabama Museum of Natural History 16:15–28.

Burr, B. M., and L. M. Page. 1978. The life history of the cypress darter, *Etheostoma proeliare*, in Max Creek, Illinois. Illinois Natural History Survey Biological Notes 106:1–15.

Burr, B. M., and J. N. Stoeckel. 1999. The natural history of madtoms (genus *Noturus*) North America's diminutive catfishes. Pages 51–101 *in* E. R. Irwin, W. A. Hubert, C. F. Rabeni, H. L. Schramm, Jr., and T. Coon, editors. Catfish 2000: proceedings of the International Ictalurid Symposium. American Fisheries Society, Symposium 24, Bethesda, Maryland.

Burr, B. M., and M. L. Warren, Jr. 1986. A distributional atlas of Kentucky fishes. Kentucky Nature Preserves Commission Scientific and Technical Series 4.

Carlander, K. D. 1969. Handbook of freshwater fishery biology. Volume 1. The Iowa State University Press, Ames.

Carney, D. A., and B. M. Burr. 1989. Life histories of the bandfin darter, *Etheostoma zonistium*, and the firebelly darter, *Etheostoma pyrrhogaster*, in western Kentucky Illinois. Illinois Natural History Survey Biological Notes 134:1–16.

Cederholm, C. J., and N. P. Peterson. 1985. The retention of coho salmon (*Oncorhynchus kisutch*) carcasses by organic debris in small streams. Canadian Journal of Fisheries and Aquatic Sciences 42:1222–1225.

Chan, M. D., and G. R. Parsons. 2000. Aspects of brown madtom, *Noturus phaeus*, life history in northern Mississippi. Copeia 2000:757–762.

Cherry, J., and R. L. Beschta. 1989. Coarse woody debris and channel morphology: a flume study. Water Resources Bulletin 25:1031–1036.

Collier, K. J., and J. N. Halliday. 2000. Macroinvertebrate-wood associations during decay of plantation pine in New Zealand pumice-bed streams: stable habitat or trophic subsidy? Journal of the North American Benthological Society 19:94–111.

Conant, R., and J. T. Collins. 1998. A field guide to reptiles and amphibians, eastern and central North America (3rd edition, expanded). Houghton Mifflin Company, New York.

Dinkins, G. R., and P. W. Shute. 1996. Life histories of *Noturus baileyi* and *N. flavipinnis* (Pisces: Ictaluridae), two rare madtom catfishes in Citico Creek, Monroe County, Tennessee. Bulletin of the Alabama Museum of Natural History 18:43–69.

Dolloff, C. A. 1986. Effects of stream cleaning on juvenile coho salmon and Dolly Varden in southeast Alaska. Transactions of the American Fisheries Society 115:743–755.

Dolloff, C. A., and G. H. Reeves. 1990. Microhabitat partitioning among stream-dwelling juvenile coho salmon, *Oncorhynchus kisutch*, and Dolly Varden, *Salvelinus malma*. Canadian Journal of Fisheries and Aquatic Sciences 47:2297–2306.

Dolloff, C. A., and J. R. Webster. 2000. Particulate organic contributions from forests to streams: debris isn't so bad. Chapter 6 *in* E. S. Verry, J. Hornbeck, and C. A. Dolloff, editors. Riparian management in forests of the eastern continental United States. CRC-Lewis Publishers, Boca Raton, Florida.

Elliott, S. T. 1986. Reduction of a Dolly Varden population and macrobenthos after removal of logging debris. Transactions of the American Fisheries Society 15:392–400.

Evans, B. F., C. R. Townsend, and T. A. Crowl. 1993. Distribution and abundance of coarse woody debris in some southern New Zealand streams from contrasting forest catchments. New Zealand Journal of Marine and Freshwater Research 27:227–239.

Etnier, D. A., and W. C. Starnes. 1993. The fishes of Tennessee. The University of Tennessee Press, Knoxville.

Everett, R. A., and G. M. Ruiz. 1993. Coarse woody debris as a refuge from predation in aquatic communities. Oeocologia 93:475–486.

Fausch, K. D., and R. J. White. 1981. Competition between brook trout (*Salvelinus fontinalis*) and brown trout (*Salmo trutta*) for positions in a Michigan stream. Canadian Journal of Fisheries and Aquatic Sciences 386:1220–1227.

Felley, J. D. 1992. Medium-low-gradient streams of the Gulf coastal plain. Pages 233–269 *in* C. T. Hackney, S. M. Adams, and W. H. Martin, editors. Biodiversity of the southeastern United States, aquatic communities. Wiley, New York.

Felley, J. D., and S. M. Felley. 1987. Relationships between habitat selection by individuals of a species and patterns of habitat segregation among fishes: fishes of the Calcasieu drainage. Pages 61–68 *in* W. J. Matthews and D. C. Heins, editors. Community and evolutionary ecology of North American stream fishes. University of Oklahoma Press, Norman.

Felley, J. D., and L. G. Hill. 1983. Multivariate assessment of environmental preferences of cyprinid fishes of the Illinois River, Oklahoma. American Midland Naturalist 109:209–221.

Flebbe, P. A., and C. A. Dolloff. 1995. Trout use of woody debris and habitat in Appalachian wilderness streams of North Carolina. North American Journal of Fisheries Management 15:579–590.

Gale, W. F., and G. L. Buynak. 1978. Spawning frequency and fecundity of satinfin shiner (*Notropis analostanus*)—a fractional, crevice spawner. Transactions of the American Fisheries Society 107:460–463.

Golladay, S. W., and R. L. Sinsabaugh. 1991. Biofilm development on leaf and wood surfaces in a Boreal river. Freshwater Biology 25:437–450.

Golladay, S. W., J. R. Webster, and E. F. Benfield. 1989. Changes in stream benthic organic matter following watershed disturbance. Holarctic Ecology 12:96–105.

Haag, W. R., and M. L. Warren, Jr. 1999. Mantle displays of freshwater mussels elicit attacks from fish. Freshwater Biology 42:35–40.

Haag, W. R., R. S. Butler, and P. D. Hartfield. 1995. An extraordinary reproductive strategy in freshwater bivalves: prey mimicry to facilitate larval dispersal. Freshwater Biology 34:471–476.

Hamr, P. 1996. A giant's tale: the life history of *Astacopsis gouldi* (Decapoda: Parastacidae) a freshwater crayfish from Tasmania. Freshwater Crayfish XI:13–34.

Harmon, M. E., J. E. Franklin, F. J. Swanson, P. Sollins, S. V. Gregory, J. D. Lattin, N. H. Anderson, S. P. Cline, N. G. Aumen, J. R. Sedell, G. W. Lienkaemper, K. Cromack, Jr., and K. W. Cummins. 1986. Ecology of coarse woody debris in temperate ecosystems. Advances in Ecological Research 15:133–302.

Harvey, B. C., R. J. Nakamoto, and J. L. White. 1999. Influence of large woody debris and a bankfull flood on movement of adult resident coastal cutthroat trout (*Oncorhynchus clarki*) during fall and winter. Canadian Journal of Fisheries and Aquatic Sciences 56:2161–2166.

Harvey, B. C., and A. J. Stewart. 1991. Fish size and habitat depth relationships in headwater streams. Oecologia 87:336–342.

Hauer, F. R., G. C. Poole, J. T. Gangemi, C. V. Baxter. 1999. Large woody debris in bull trout (*Salvelinus confluentus*) spawning streams of logged and wilderness watersheds in northwest Montana. Canadian Journal of Fisheries and Aquatic Sciences 56:915–924.

Hawkins, C. P., J. L. Kershner, P. A. Bisson, M. D. Bryant, L. M. Decker, S. V. Gregory, D. A. McCullough, C. K. Overton, G. H. Reeves, R. J. Steedman, and M. K. Young. 1993. A hierarchical approach to classifying stream habitat features. Fisheries 18(6):3–12.

Hedman, C. W., D. H. Van Lear, and W. T. Swank. 1996. In-stream large woody debris loading and riparian forest seral stage associations in the southern Appalachian mountains. Canadian Journal of Forest Research 26:1218–1227.

Helfman, G. S. 1981. The advantage to fishes of hovering in shade. Copeia 1981:392–400.

Hilderbrand, R. H., A. D. Lemly, C. A. Dolloff, and K. L. Harpster. 1997. Effects of large woody debris placement on stream channels and benthic macroinvertebrates. Canadian Journal of Fisheries and Aquatic Sciences 54:931–939.

Hubbs, C. L., and G. P. Cooper. 1936. Minnows of Michigan. Cranbook Institute of Science Bulletin 8.

Inoue, M., and S. Nakano. 1998. Effects of woody debris on the habitat of juvenile masu salmon (*Oncorhynchus masou*) in northern Japanese streams. Freshwater Biology 40:1–16.

Jackson, D. L. 1999. Flathead catfish: biology, fisheries, and management. Pages 23–25 *in* E. R. Irwin, W. A. Hubert, C. F. Rabeni, H. L. Schramm, Jr., and T. Coon, editor. Catfish 2000: proceedings of the International Ictalurid Symposium. American Fisheries, Symposium 24, Bethesda, Maryland.

Jenkins, R. E., and N. M. Burkhead. 1994. Freshwater fishes of Virginia. American Fisheries Society, Bethesda, Maryland.

Johnston, C. E. 1994. Spawning behavior of the goldstripe darter (*Etheostoma parvipiine* Gilbert and Swain) (Percidae). Copeia 1994:823–825.

Johnston, C. E., and L. M. Page. 1990. The breeding behavior of *Opsopoedus emiliae* (Cyprinidae) and its phylogenetic implications. Copeia 1990:1176–1180.

Lisle, T. E., and S. Hilton. 1992. The volume of fine sediment in pools: an index of sediment supply in gravel-bed streams. Water Resources Bulletin 28(2):371–383.

Maser, C., and J. R. Sedell. 1994. From the forest to the sea: the ecology of wood in streams, rivers, estuaries, and oceans. St. Lucie Press, Delray Beach, Florida.

Matthews, W. J. 1998. Patterns in freshwater fish ecology. Chapman and Hall, New York.

Mayden, R. L., and S. J. Walsh. 1984. Life history of the least madtom, *Noturus hildebrandi* (Siluriformes: Ictaluridae) with comparisons to related species. American Midland Naturalist 112:349–368.

Meffe, G. K., and A. L. Sheldon. 1988. The influence of habitat structure on fish assemblage composition in southeastern blackwater streams. American Midland Naturalist 120:225–240.

Mettee, M. F., P. E. O'Neil, and J. M. Pierson. 1996. Fishes of Alabama and the Mobile Basin. Oxmoor House, Inc., Birmingham, Alabama.

Monzyk, F. R., W. E. Kelso, and D. A. Rutherford. 1997. Characteristics of woody cover used by brown madtoms and pirate perch in Coastal Plain streams. Transactions of the American Fisheries Society 126:665–675.

O'Connor, N. A. 1991. The effects of habitat complexity on the macroinvertebrates colonizing wood substrates in a lowland stream. Oecologia 85:504–512.

Page, L. M., and P. W. Smith. 1970. The life history of the dusky darter, *Percina sciera*, in the Embarras River, Illinois. Illinois Natural History Survey Biological Notes 69:1–15.

Page, L. M., M. A. Retzer, and R. A. Stiles. 1982. Spawning behavior in seven species of darters. Brimelyana 8:135–143.

Parker, H. L. 1964. Natural history of *Pimephales vigilax* (Cyprinidae). Southwestern Naturalist 8:226–235.

Pearsons, T. N., H. W. Li, and G. A. Lamberti. 1992. Influences of habitat complexity on resistance to flooding and resilience of stream fish assemblages. Transactions of the American Fisheries Society 121:427–436.

Pflieger, W. L. 1997. The fishes of Missouri (revised edition). Missouri Department of Conservation, Jefferson City.

Pflieger, W. L. 1965. Reproductive behavior of the minnows, Notropis spilopterus and *Notropis whipplii*. Copeia 1965:1–8.

Piller, K. R., and B. M. Burr. 1999. Reproductive biology and spawning habitat supplementation of the relict darter, *Etheostoma chienense*, a federally endangered species. Environmental Biology of Fishes 55:145–155.

Power, M. E. 1987. Predator avoidance by grazing fishes in temperate and tropical streams: importance of depth and prey size. Pages 333–351 *in* W. C. Kerfoot and A. Sih, editors. Predation. University Press of New England, Hanover, New Hampshire.

Probst, W. E., C. F. Rabeni, W. G. Covington, and R. E. Martinez. 1984. Resource use by stream-dwelling rock bass and smallmouth bass. Transactions of the American Fisheries Society 113:283–294.

Probst, D. L., and J. A. Stefferud. 1994. Distribution and status of the Chihuahua chub (Teleostei: Cyprinidae: *Gila nigrescens*), with notes on its ecology and associated species. Southwestern Naturalist 39:224–234.

Punchard, N. T., M. R. Perrow, and A. J. D. Jowitt. 2000. Fish habitat associations, community structure, density and biomass in natural and channelised lowland streams in the catchment of the River Wensum, UK. Pages 143–157 *in* I. G. Cowx, editor. Management and ecology of river fisheries. Fishing News Books, Blackwell Scientific Publications Ltd., Oxford, UK.

Pyron, M., A. P. Covich, and R. W. Black. 1999. On the relative importance of pool morphology and woody debris to distributions of shrimp in a Puerto Rican headwater stream. Hydrobiologia 405:207–215.

Quist, M. C., and C. S. Guy. 2001. Growth and mortality of prairie stream fishes: relations with fish community and instream habitat characteristics. Ecology of Freshwater Fish 10:88–96.

Reeves, G. H., J. D. Hall, T. D. Roelofs, T. L. Hickman, and C. O. Baker. 1991. Rehabilitating and modifying stream habitats. Pages 519–557 *in* W. R. Meehan, editor. Influences of forest and rangeland management on salmonid fishes and their habitats. American Fisheries Society, Special Publication 19, Bethesda, Maryland.

Richey, J. E., M. A. Perkins, and C. R. Goldman. 1975. Effects of kokanee salmon (*Oncorhynchus nerka*) decomposition on the ecology of a subalpine stream. Journal of the Fisheries Research Board of Canada 32:817–820.

Richmond, A. D., and K. D. Fausch. 1995. Characteristics and function of large woody debris in subalpine Rocky Mountain streams in northern Colorado. Canadian Journal of Fisheries and Aquatic Sciences 52:1789–1802.

Robison, H. R. 1977. Distribution, habitat, variation, and status of the goldstripe darter, *Etheostoma parvipinne* Gilbert and Swain, in Arkansas. Southwestern Naturalist 22:435–442.

Rohde, F. C., and R. G. Arndt. 1982. Life history of a coastal plain population of the mottled sculpin, *Cottus bairdi* (Osteichthyes: Cottidae), in Delaware. Brimleyana 7:69–94.

Ross, S. T., J. G. Knight, and S. D. Wilkins. 1992. Distribution and microhabitat dynamics of the threatened bayou darter, *Etheostoma rubrum*. Copeia 1992(3):658–671.

Schlosser, I. J. 1987. A conceptual framework for fish communities in small warmwater streams. Pages 17–24 *in* W. J. Matthews and D. C. Heins, editors. Community and evolutionary ecology of North American stream fishes. University of Oklahoma Press, Norman.

Scott, W. B., and E. J. Crossman. 1973. Freshwater fishes of Canada. Fisheries Research Board of Canada Bulletin 184.

Sedell, J. R., G. H. Reeves, F. R. Hauer, J. A. Stanford, and C. P. Hawkins. 1990. Role of refugia in recovery from disturbances: modern fragmented and disconnected river systems. Environmental Management 14:711–724.

Shields, F. D., Jr., and R. H. Smith. 1992. Effects of large woody debris removal on physical characteristics of a sand-bed river. Aquatic Conservation: Marine and Freshwater Ecosystems 2:145–163.

Shields, F. D., Jr., S. S. Knight, and C. M. Cooper. 1994. Effects of channel incision on base flow stream habitats and fishes. Environmental Management 18:43–57.

Shively, S. H., and J. F. Jackson. 1985. Factors limiting the upstream distribution of the Sabine map turtle. American Midland Naturalist 114:292–303.

Silsbee, D. G., and G. L. Larson. 1983. A comparison of streams in logged and unlogged areas of Great Smoky Mountains National Park. Hydrobiologia 102:99–111.

Smock, L. A., and E. Gilinsky. 1992. Coastal plain blackwater streams. Pages 271–313 *in* C. T. Hackney, S. M. Adams, and W. H. Martin editors. Biodiversity of the southeastern United States, aquatic communities. Wiley, New York.

Smock, L. A., G. M. Metzler, and J. E. Gladden. 1989. Role of debris dams in the structure and functioning of low-gradient headwater streams. Ecology 70:764–775.

Starnes, W. C., and L. B. Starnes. 1978. A new cyprinid fish of the genus *Phoxinus* endemic to the upper Cumberland River drainage. Copeia 1978:508–516.

Starnes, W. C., and R. E. Jenkins. 1988. A new cyprinid fish of the genus *Phoxinus* (Pisces: Cypriniformes) from the Tennessee River drainage with comments on relationships and biogeography. Proceedings of the Biological Society of Washington 101:517–529.

Steel, E. A., W. H. Richards, and K. A. Kelsey. 2003. Wood and wildlife: benefits of river wood to terrestrial and aquatic vertebrates. Pages 235–247 *in* S. V. Gregory, K. L. Boyer, and A. M. Gurnell, editors. The ecology and management of wood in world rivers. American Fisheries Society, Symposium 37, Bethesda, Maryland.

Swales, S., R. B. Lauzier, and C. D. Levings. 1986. Winter habitat preferences of juvenile salmonids in two interior rivers in British Columbia. Canadian Journal of Zoology 64:1506–1514.

Tank, J. L., and M. J. Winterbourn. 1995. Biofilm development and invertebrate colonization of wood in four New Zealand streams of contrasting pH. Freshwater Biology 34:303–315.

Taylor, C. M., M. R. Winston, and W. J. Matthews. 1993. Fish species-environment and abundance relationships in a Great Plains river system. Ecography 16:16–23.

Trautman, M. B. 1981. The fishes of Ohio (revised edition). Ohio State University Press, Columbus.

Wallace, R. K., and J. S. Ramsey. 1981. Reproductive behavior and biology of the bluestripe shiner (*Notropis callitaenia*) in Uchee Creek, Alabama. American Midland Naturalist 106:197–200.

Walsh, S. J., and B. M. Burr. 1981. Distribution, morphology and life history of the least brook lamprey, *Lampetra aepyptera* (Pisces: Petromyzontidae), in Kentucky. Brimleyana 6:83–100.

Warren, M. L., Jr., 1982. Rediscovery of *Etheostoma histrio* and *Percina ouachitae* in Green River, Kentucky, with distribution and habitat notes. Transactions of the Kentucky Academy of Science 43:21–26.

Warren, M. L., Jr., W. R. Haag, and S. B. Adams. 2002. Forest linkages to diversity and abundance in lowland stream fish communities. Pages 168–182 *in* M. M. Holland, M. L. Warren, Jr., and J. A. Stanturf, editors. Proceedings of a Conference on Sustainability of Wetlands and Water Resources: How Well Can Riverine Wetlands Continue to Support Society into the 21st Century? U.S. Department of Agriculture, Forest Service, Southern Research Station, General Technical Report SRS-50, Asheville, North Carolina.

Webster, J. R., S. W. Golladay, E. F. Benfield, D. J. D'Angelo, and G. T. Peters. 1990. Effects of forest disturbance on particulate organic matter budgets of small streams. Journal of the North American Benthological Society 9:120–140.

Winn, H. E., and A. R. Picciolo. 1960. Communal spawning of the glassy darter *Etheostoma vitreum* (Cope). Copeia 1960:186–192.

Wold, A. K. F., and A. E. Hershey. 1999. Effects of salmon carcass decomposition on biofilm growth and wood decomposition. Canadian Journal of Fisheries and Aquatic Sciences 56:767–773.

Wynne-Edwards, V. 1932. The breeding habits of the black-headed minnow (*Pimephales promelas* Raf.). Transactions of the American Fisheries Society 62:382–383.

American Fisheries Society Symposium 37:195–211, 2003

Fish Relationships with Wood in Large Rivers

Maciej Zalewski[1,2] and Malgorzata Lapinska[1]

[1]International Centre for Ecology, Polish Academy of Sciences, 1 Marii Konopnickiej Str. 05–092 Lomianki, Warsaw, Poland

[2]Department of Applied Ecology, University of Lodz, 12/16 Banacha Str., 90–237 Lodz, Poland

Peter B. Bayley

Department of Fisheries and Wildlife, Oregon State University, Corvallis, Oregon 97330, USA

Abstract.—The importance of wood for fish as individuals, populations, and communities and its role in the formation of habitats in large rivers is reviewed. Small-scale effects of wood on fish dominate the literature, but effects on populations and communities at the basin scale are not additive because increasing development of wood-influenced habitats modifies the transport of water, nutrients, and sediments. This, in turn, creates and maintains habitat resulting in increased spatial and temporal diversity and system resilience, which subsequently affect different life cycles of riverine fishes. Such effects depend on location; the relative influence of instream wood on fish and physical processes tends to decrease downstream in large basins, but the converse may be true in floodplains as their influence increases downstream. Because climatic factors control hydrology and temperature throughout the basin, and fish responses to management depend on large-scale, longer term processes as well as smaller ones, the research results reviewed here are assessed in the context of a basin-scale framework, the Ecohydrology Concept (EHC). A restoration strategy based on a basin-scale framework fosters a natural disturbance regime, increases the ability of the system to absorb impacts, and utilizes natural processes in the development of management tools. Reintroduction of wood in large rivers is an essential strategic component but will not, on its own, achieve significant benefits for fish and other biota without at least partial restoration of hydrological cycles, riparian vegetation, selected floodplains, and the associated natural disturbance regime.

Introduction

Increasing human impacts on most of the world's major river systems has been dominated by protection against flooding, enhancement of navigation, development of agriculture, and urbanization. These processes have lead to fragmentation of the riverscape (sensu Fausch et al. 2002) and alterations of hydrological regimes (Poff et al. 1997) and thus to degradation of ecological integrity (Karr et al. 1986) of river-floodplain ecosystems (for example, Petts et al. 1989a, 1989b; Schmutz et al. 2000; Schiemer et al. 2001). Restoration attempts to mitigate negative human impacts and improve fish habitat have been conducted for several decades, but research that evaluates restoration projects has concentrated on small streams (Newbury and Gaboury 1993; White 1996). Since the mid-1980s, rehabilitation, conservation, and more sustainable management of large rivers has increased (for example, Dodge 1989; Sedell et al. 1989; Boon et al. 1992; Armantrout 1995; Arthington and Welcomme 1995; Gore and Shields 1995; Buijse and Vriese 1996; Dokulil 1996). Although holistic approaches are occasionally considered, they are rarely applied, and, with rare exceptions (Cowx and Welcomme 1998), practical guidance on large river restoration is wanting, and approaches for managing wood in large rivers are nonexistent.

Restoration of most physically degraded riverine fish habitats needs to include the addition of, or capacity for the production of, large wood. Although re-introduction of wood is widely recognized as a management tool to improve fish habitat quality, most work has been done in North American small streams and rivers (Harmon et al. 1986; Murphy and Koski 1989; Armantrout 1991; Crispin et al. 1993; Bryant and Sedell 1995; Slaney and Martin 1997; Abt et al. 1998; Hilderbrand et al. 1998; Koning et al. 1998; Bisson et al. 2003, this volume). In the past 10 years, research on the importance of instream structure, including wood, has increased in Europe (Gregory and Davis 1992; Piégay 1993; Gurnell et al. 1995, 2000, 2001; Mann 1996; Piégay and Marston 1998; Hering et al. 2000; Mutz 2000), Japan (Inoue and Nakano 1998; Nagasaka and Nakamura 1999), South America (Jepsen et al. 1997), Australia (Gippel et al. 1996a, 1996b; Koehn 1996; Koehn and Nicol 1997), and Africa (Jacobson et al. 1999, 2000), but most studies still assess small-scale effects.

The role of wood in controlling processes in rivers is only beginning to be understood, (Piégay 1993; Crook and Robertson 1999; Gurnell et al. 2000; Gurnell 2003, this volume). A significant gap lies in understanding its role in large, lowland rivers where minimal quantities result from cumulative human alterations basinwide and locally from de-snagging operations against flood and navigation hazards (Triska 1984; Smith et al. 1992; Gippel et al. 1996a, 1996b). Consequently, natural processes of wood accumulation are largely defunct in most temperate systems (Sedell and Froggatt 1984; Maser and Sedell 1994; Gurnell 1997; Piégay and Gurnell 1997). What little we know about the role of wood has been based mainly on perennial, temperate river channels. Conversely, information about wood and its function in perennial arid-zone rivers (Black and Crowl 1994), ephemeral rivers (Jacobson et al. 1999, 2000) and rivers of the humid tropics are scarce (Bryant and Sedell 1995). Most river management generally focuses on river channels, while the importance of and need for wood in floodplain habitats is frequently overlooked (Bayley et al. 2000).

The role of wood in affecting the development of adaptations in riverine fish species, potential coevolution shaping their communities, and the subsequent survival and production of populations is barely understood. Of necessity, studies have to be conducted at a scale small enough for quantitative sampling, and inferences about patterns and processes at larger scales may be distorted by nonlinear effects (Allen and Starr 1982). While this is a problem in relatively well-studied small temperate river basins, it is an enormous challenge in large river systems due to their scale, lack of replicate and natural systems, difficulties of sampling in deeper habitats, and increased influence of floodplain habitats. Management actions at smaller scales are needed to begin the restoration process before we can obtain a more complete understanding of the scale-dependent mechanisms involving wood and fish. Therefore, a holistic perspective is essential for achieving sustainable changes in the context of competing river and land uses (Vannote et al. 1980; Dodge 1989; Petts et al. 1989b; Ward 1989; National Biological Service 1995; Frissell et al. 1986; Décamps 1996; Dokulil 1996; Zalewski et al. 1997; Harper et al. 1999; Zalewski 2000, 2002a, 2002b).

The need for management solutions to sustain or restore temperate ecosystems has already led to research on links between river channel dynamics and riparian vegetation (Chauvet and Decamps 1989; Naiman et al. 1989; Naiman and Decamps 1990; Piégay and Maridet 1994; Persat et al. 1995; Tabacchi et al. 1998; Zalewski and Frankiewicz 1998; Zalewski et al. 1998, 2001; Ward et al. 1999). There is an associated interest in large wood as a management tool (Gurnell et al. 1995, 1996a, 1996b; Ptolemy 1997; Gerhard and Reich 2000) and in its potential supply to river channels (Van Sickle and Gregory 1990).

Here, we attempt to assess current knowledge of the influence of large wood on fish in large river systems. Processes governing wood production depend on physical conditions from basin to microhabitat scale. The subsequent distribution and retention of wood is determined by physical conditions associated with position on the stream continuum and associated floodplains (see Gurnell 2003; Piégay 2003, this volume). Responses of aquatic biota, particularly local populations of fishes, are initially influenced by physical and biological processes mainly at the habitat or microhabitat scale. Relationships between wood and fish should play a key role in long-term management strategies to restore system function, which should involve the restoration of all the foregoing processes up to the scale of the river basin.

Influence of Wood and Forests on Biological Processes and Fish Communities

Climatic and forest impacts on hydrological processes in rivers should be considered as large scale-drivers controlling habitats and fish populations (Tonn 1990; Zalewski 1995). However, fish and habitat studies normally can only be undertaken at microhabitat and habitat scales. Each of the following five subsections considers effects of wood and forest on fish and fish habitats in order of decreasing spatial scale.

1. Effects of forests on fish

While this paper emphasizes the effects of wood on fish, sources of wood from adjacent riparian forests, floodplain forests, and upstream sources need to be considered. There are several direct and indirect effects of the living component of riparian and floodplain forests. A direct link between floodplain trees and fish is exemplified in neotropical rivers where several fish species are adapted to feed on fruits, seeds, and/or leaves (Goulding 1980). These activities represent consumption of autochthonous production because the inundated trees and the fish share the same habitat (Junk et al. 1989). Floodplain forests produce large amounts of organic matter (Brinson 1990) and may support fish through the detrital cycle. In the central Amazon floodplain, when it was already 50% deforested, estimated production of forest litter (excluding wood production) from the dominant secondary forest comprised about 24% of primary production from all sources (Bayley 1989).

The canopy of riparian forest may have an indirect effect on riverine fish as it determines solar energy access to the river channel. Zalewski et al. (1994) predicted an increase in production of algae and higher trophic level organisms when vegetative canopy was reduced by 40–60%. A test of this prediction in medium-sized rivers (Lapinska 1996; Lapinska et al. 2002) indicated that biomass and diversity of fish was maximized when the channel received 300–700 $mE/cm^2/s$ in upland and lowland reaches (Figure 1).

The foregoing relationship may not be applicable to large river reaches where the ratio of bank length to water surface area is low. However, fish tend to be more common in inshore habitats with riparian vegetation than those without vegetation. In the large, Warta River in Poland, Penczak (1995) observed declines in fish species caught from 17 to 11 and in standing crop from 31.9 kg/ha to 7.9 kg/ha when bankside vegetation was removed.

Riparian and floodplain forests dissipate kinetic energy of floods partly because of their high hydraulic roughness (Bren 1993; Piegay 1997; Tabacchi et al. 2000). This effect, along with that of distributing overbank discharges into floodplains, increases the smoothness and duration of the flood pulse. This produces favorable conditions for fish habitats and growth in floodplains (Welcomme 1985; Junk et al. 1989) and moderates bank erosion downstream. Extensive reduction of riparian forest, wood removal, and channelization may cause an increase in flood peaks by 1.5–2.5 times, reduce peak duration by 7 h (Nagasaka and Nakamura 1999), and raise water temperature above tolerance levels of fishes. Flash floods can be harmful for fish instream when they can cause downstream displacements that are not part of their life cycle (Harvey 1987). However, instream wood decreases the severity of floods and fish more rapidly recolonize habitats maintained by wood (Pearsons et al. 1992).

2. Indirect effects of wood on the fish energy budget

Large wood influences flow, which, in turn, influences the food supply of fish and the amount of energy they expend when swimming. Large wood dissipates energy of fast flows and increases habitat diversity by producing a variety of flows (hydraulic complexity) in addition to the direct effect of the wood itself (Harmon et al. 1986; Gurnell et al. 1995; Tabacchi et al. 1998, 2000). Instream wood can reduce shear stress fourfold (Shields and Smith 1992).

Wood in larger rivers has less hydraulic significance than in streams because flows are usually reduced and logs are not large enough to bridge the channel. Although these effects reduce energy dissipation compared to discharge, important effects occur at the microhabitat scale. Habitat/microhabitat preferences of riverine fish are to a great extent determined by bioenergetics (Benke et al. 1988). The energy-expenditure/energy-gain conceptual model proposed for drift-feeding fish (Bachman 1982; Figure 2) illustrates an important microhabitat effect of wood that could occur across several stream orders. Wood provides a stable substrate for invertebrate communities (Phillips and Kilambi 1994; Phillips 1995; Bis and Higler 2001; Benke and Wallace 2003, this volume), influences the supply of drifting macroinvertebrates (Borchardt 1993), and provides velocity refuges for invertebrates during floods (Palmer et al. 1996). The quality of woody material entering the channel—which depends on riparian forest species composition—influences biofilm development and the abundance of invertebrates (Anderson 1984). Up to 50% of the bio-

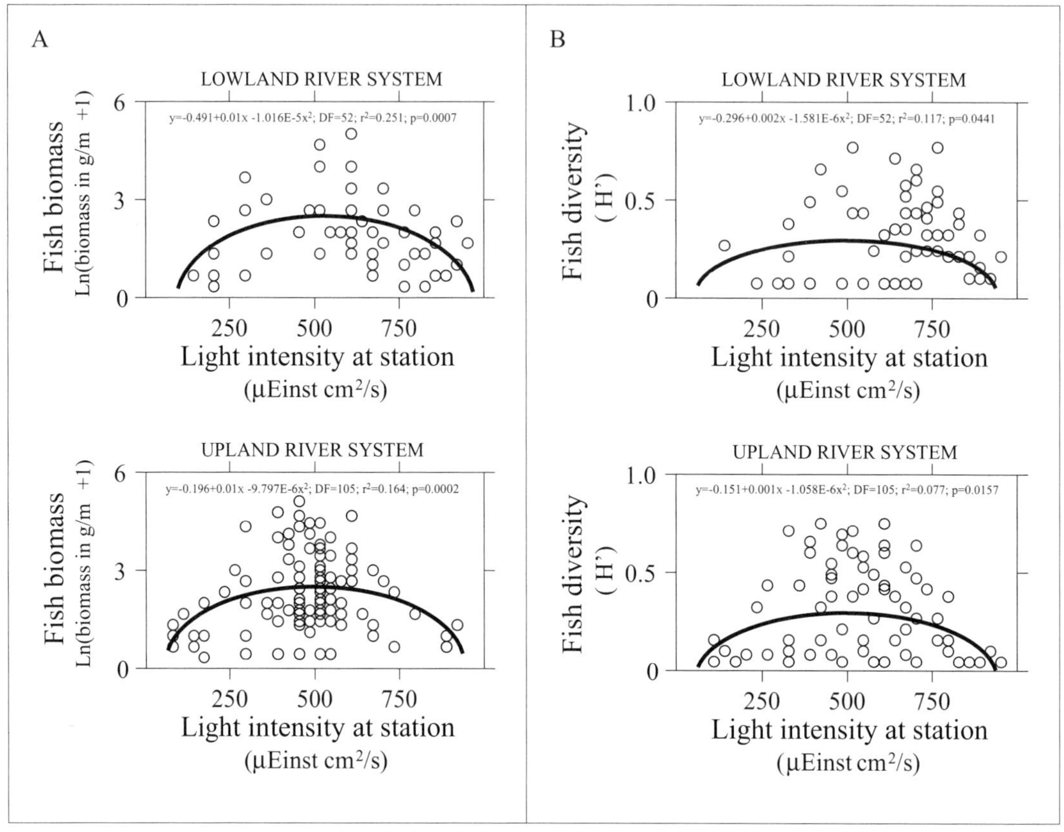

FIGURE 1. Effect of riparian canopy on fish biomass (A) and fish species diversity (B) in different river systems (based on Lapinska 1996; Lapinska et al. 2002). Diversity was calculated using the Shannon (H′) Index (Odum 1980): $H' = -Sp_i \ln p_i$, where p_i is the number of individuals of species *i* divided by number of individuals of all species in the sample, with the products summed over all species (transformed to range 0–1). Fish biomass (transformed as $\ln(x + 1)$ where x = biomass density in g/m²), fish diversity, and light intensity at the station are the average annual values estimated for three habitat types combined (pool, run, riffle). Number of samples were 53 for lowland and 106 for upland river.

mass of riverine invertebrates can occur on snags and about 15% on logs (Benke et al. 1984; Benke and Wallace 2003).

Fish respond on a diel basis to hydraulic complexity associated with wood. Winter use of complex refuges by cutthroat trout may be 2.5 times higher at night than during the day (Simondet 1997). Summer use of submerged logs and rootwads by smallmouth bass *Micropterus dolomieu* is greater during the day (Todd and Rabeni 1989). Many species use refuges during the day and actively forage in open water during the night when invertebrate drift is most abundant (Kwak et al. 1992; Baxter and McPhail 1997).

Although the main channels of large rivers typically lack structure, an abundance of lower velocity habitats occur laterally, including shoreline wood and overhanging banks associated with riparian vegetation, anabranches, backwaters, and seasonal floodplain habitats (Mann and Penczak 1984; Stalnaker et al. 1989; Schiemer and Zalewski 1992; Agostinho and Zalewski 1995; Penczak 1995, 2001; Agostinho et al. 2001; Schiemer et al. 2001). Koehn (1996) observed Murray cod moving closer to Ovens and Murray river shorelines during floods where they occupied microhabitats with wood and submerged vegetation. The number of juvenile fish species caught was strongly correlated with the diversity of shoreline and floodplain microhabitats in the Danube River (Schiemer et al. 1991, 2001). Riparian and floodplain habitats, which usually contain wood, often contain high densities of macroinvertebrates capable of supporting fish

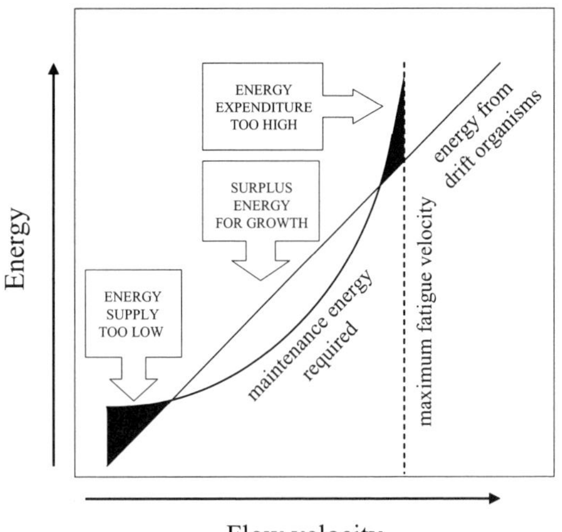

FIGURE 2. The energy-expenditure/energy-gain relationship for drift-feeding fish as a function of water velocity (based on Bachman 1982).

(Grzybkowska et al. 1990; Boulton and Lloyd 1991; Smock et al. 1992; Penczak 2001).

3. Effect of wood microhabitats as cover (refuge, shelter) for fish

The function of wood as microhabitat for salmonids (McMahon and Hartman 1989; Bjorn and Reiser 1991; Simondet 1997; Inoue and Nakano 1998; Sundbaum and Naeslund 1998) and non-salmonids (Angermeier and Karr 1984; Todd and Rabeni 1989; Scott and Angermeier 1998) has been well established in small and medium-size rivers. Lapinska (1996) compared fish biomass (Figure 3A) and diversity (Figure 3B) estimates before and after removal of wood in an experimental stream channel. Both fish community measures were approximately halved following wood removal. Community composition also changed. The larger, open-water roach *Rutilus rutilus* was dominant with wood (Figure 4A) while the smaller, bottom-dwelling stone loach *Nemacheilus barbatulus* was dominant after wood removal (Figure 4B). When wood was restored, the former community structure, in terms of dominance and estimated densities, was restored after 3 weeks (Figure 4C).

Experimental placement of large wood in a large river (Rhode River, Chesapeake Bay) resulted in significant increases in common epibenthic species compared to a control site (Everett and Ruiz 1993). A summary of other studies (Table 1) indicates that wood as fish microhabitat in large rivers generally is important, but fish responses can be negative (Hesse 1994) as well as positive (Koehn 1996; Jepsen et al. 1997; Koehn and Nicol 1997; Lehtinen et al. 1997). While these studies imply some functional relationship due to wood in larger rivers as well as streams, several mechanisms may be responsible for these associations. There is no evidence that such habitat/microhabitat scale mechanisms depend on the

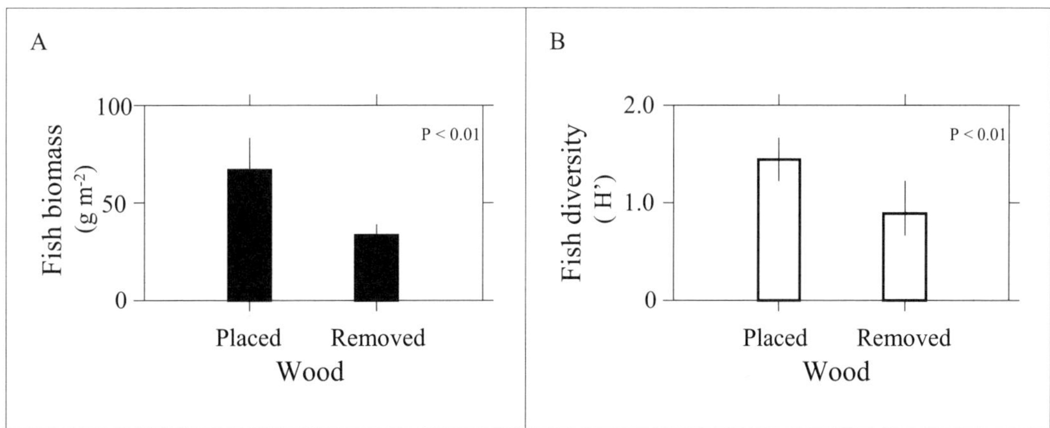

FIGURE 3. The effect of wood experimental placement and removal from river channel on fish community biomass (A) and fish species diversity (B) (untransformed Shannon Index; based on Lapinska 1996; Lapinska et al. 2002). Error bar is standard deviation. *P*-values based on Tukey tests from analysis of variances.

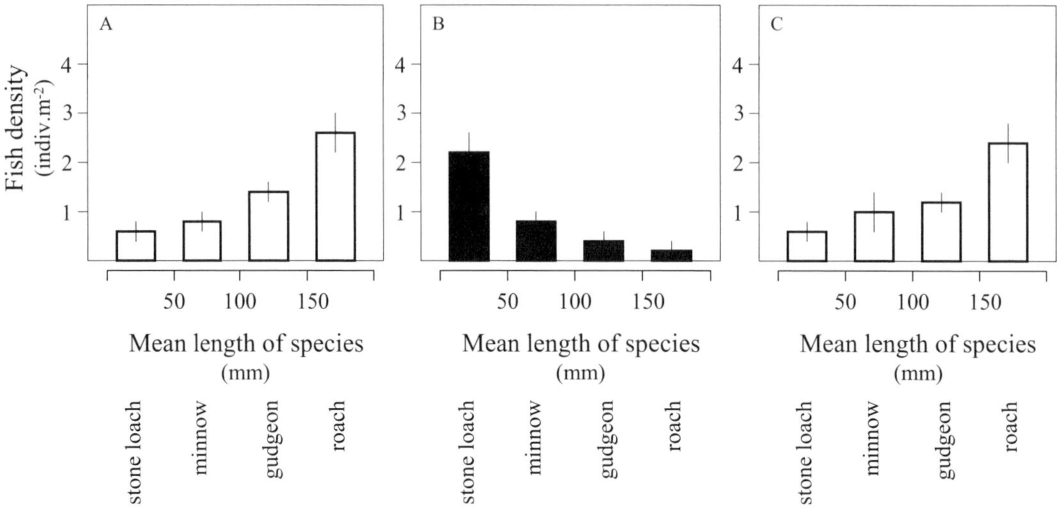

FIGURE 4. The effect of experimental wood placement (A), removal (B), and restoration (C) on density of four co-existing fish species in a river channel (based on Lapinska 1996; Lapinska et al. 2002). Error bar is standard deviation. Tukey tests were performed on ANOVAs. Type I probabilities for: stone loach A versus B $p = 0.001$, B versus C $p = 0.002$; minnow A versus C $p = 0.029$; gudgeon *Gobio gobio* A versus B $p = 0.009$, B versus C $p = 0.016$; roach A versus B $p = 0.002$, B versus C $p = 0.003$.

TABLE. 1. Reports on influence of large wood microhabitat on fish in large rivers.

River	Species	Response comments	Citation
Mississippi River (North America)	Smallmouth bass Bluegill *Lepomis macrochirus* Rock bass *Ambloplites rupestris* Spotted sucker *Minytrema melanops* Bigmouth buffalo *Ictiobus cyprinellus* Largemouth bass *Micropterus salmoides* Black crappie *Pomoxis nigromaculatus* Emerald shiner *Notropis atherinoides* Common carp *Cyprinus carpio* Walleye *Stizostedion vitreum* Sauger *S. vitreum*	• fish biomass and abundance were higher (2 to 50 times) in habitat with wood • fish biomass was higher in more complex accumulations of wood than in less complex acculations	Madejczyk et al. 1998
Missouri River (North America)	Blue catfish *Ictalurus furcatus*	• population density was reduced by the presence of wood in habitat	Hesse 1994
Cinaruco River (South America)	Speckled pavon *Cichla temensis* *Cichla orinocensis*	• fish species distribution was strongly associated with wood microhabitats	Jepsen et al. 1997
Ovens River and Murray Darling River (Australia)	Murray cod *Maccullochella peeli peeli* Trout-cod *Maccullochella macquariensis* Golden perch *Macquaria ambigua* Common carp	• fish species distribution was strongly associated with wood	Koehn 1996; Koen and Nicol 1997

scale of the river channel, but that does not imply that larger scale responses of populations and community structure would be similar.

4. Role of wood microhabitats in intra- and interspecific fish interactions

The influence of microhabitats created or influenced by wood on intra- and interspecific relationships among fish appears to be critical at small scales, while such relationships may be indirectly influenced by trophic and physical processes governed by large-scale effects of climate and hydraulic variability (Schlosser and Ebel 1989; Schlosser et al. 2000). The importance of instream and overhead cover obstructing the visual detection of fish by fish has been already verified but mostly in microhabitat experiments using artificial structures (Valdimarsson and Metcalfe 1998). Studies testing natural wood structures are rare (for example, McMahon and Hartman 1989; Lapinska 1996, 2001; Sundbaum and Naeslund 1998). Wood may reduce inter- and intraspecific interference competition among fish and consequently energy expenditure to the benefit of enhanced growth (McMahon and Hartman 1989; Sundbaum and Naeslund 1998). This may be true for fish in lower trophic levels, but negative relationships between habitat complexity associated with wood and foraging success of predatory fish species have been demonstrated (Everett and Rhuiz 1993; Lapinska 2001). Lapinska (2001) compared the influence of four littoral zone types, including one with wood, on growth rates of pikeperch *Stizostedion lucioperca* and northern pike *Esox lucius* juveniles together (total length of both species less than 130 mm) and pikeperch juveniles alone (total length less than 60 mm). Mean growth rates of both predators appeared to be lower in woody habitats (Figure 5), with the possible exception of pike, compared with gravel habitat (Figure 5B).

Other negative associations between structural complexity in littoral zones and predator efficiency have been reported in simulated and natural environments (Eklov and Diehl 1994; Persson and Eklov 1995; Chick and McIvor 1997). The lower mean growth rate of pikeperch in woody habitat (Figure 5) is consistent with predator efficiency being impaired by decreased visual acuity and/or increased ability of prey to hide (Savino and Stein 1982, 1989a, 1989b; Anderson 1984; Diehl 1988; Eklov and Diehl 1994; Eklov 1997) or by limited maneuverability of predators in complex structures (Vince et al. 1976). Conversely, the efficiency of an ambush predator, such as pike, tends to increase as structural complexity increases (Eklov 1997). Also, pike prefer vegetation and may have adaptations that result in the observed higher growth rates in that habitat (Eklov and Diehl 1994; Figure 5B).

Recent studies suggest that fish may respond to different levels of habitat complexity that is influenced by wood (Black and Crowl 1994; Shields et al. 1994; Monzyk et al. 1997). Complexity in wood accumulations may be expressed as the joint effect of cavity space, stem diameter, suspended and benthic leaves, depth, inside and outside flow, undercut bank, and lateral position (Monzyk et al. 1997). Species and life cycle stage of the fish involved are important in influencing habitat choice (Table 2).

Summary and Conclusions

Wood and carrying capacity of large river systems for fish

Because of its deliberate removal, artificial impoundment, or interference with its source, large wood has been conspicuous by its absence in most temperate and some tropical systems. The successional stage of riparian forests determines the quality and quantity of wood supply to the river, through processes that may span several centuries and extend over whole basins. The climatic conditions favoring forest types and other primary production sources also function at large scales. However, hydrological conditions change over several shorter time periods but are still effective at large spatial scales. Temporal habitat diversity is an important expression of the resilience and resistance of the riverscape to more extreme climatic events. Natural hydrological disturbances in the aquatic component of the water mesocycle, that have been excessively reduced by most current management practices, are essential in providing the diversity of habitats at several temporal scales, as well as maintaining spatial diversity across stream orders and floodplains.

Such spatial and temporal heterogeneity in habitats impacts individual fish, life history stages, interactions among species, and subsequently community structure (Sections 3, 4, Table 2). Wood supply operates at smaller scales to form and maintain pool mesohabitats in lower order streams, stabilizes riverbanks and pro-

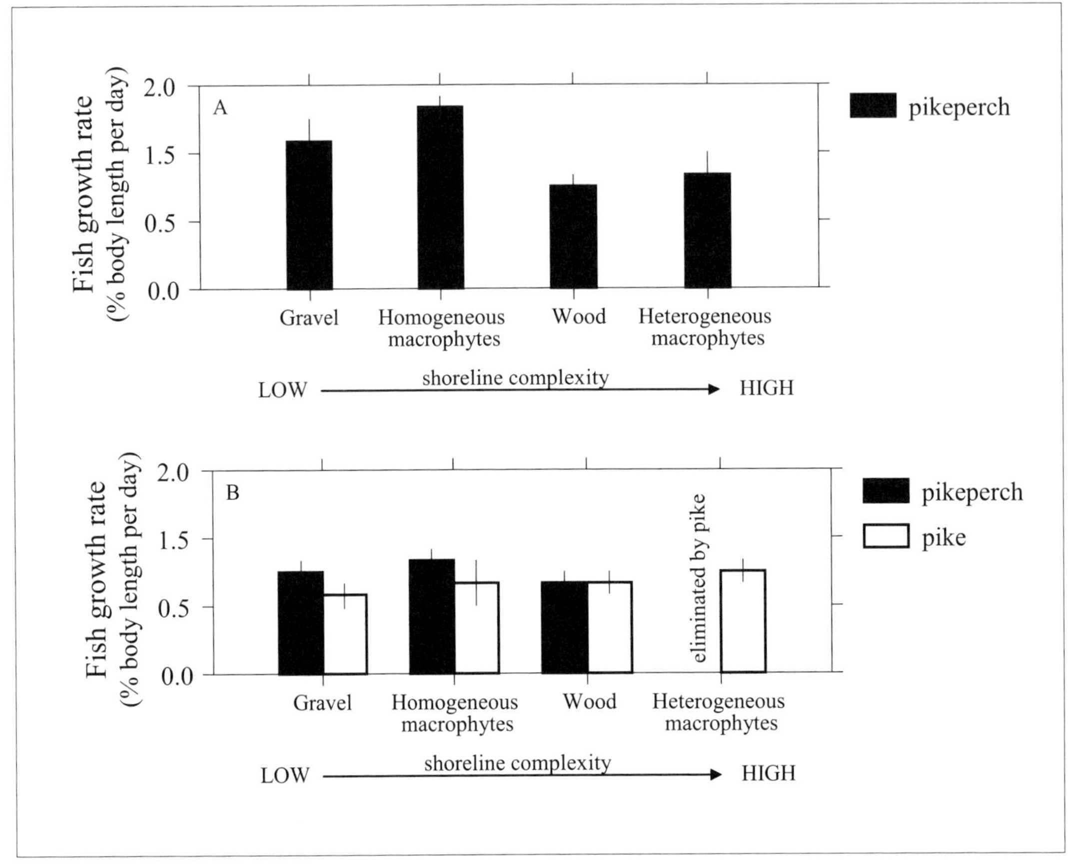

FIGURE 5. The role of wood as a river shoreline structure in decreasing predator fish efficiency (A) and stabilizing interspecific interactions of predators (B) (based on Lapinska 2001). Homogeneous macrophytes (only *Juncus* sp.). Heterogeneous macrophytes (*Acorus calamus, Potamogeton* sp., *Elodea canadensis, Glyceria* sp., *Nuphar luteum, Lemna* sp.). Error bar is standard deviation. Tukey tests were performed on ANOVAs. Type I probabilities for A for pikeperch (n = 24): gravel versus homogeneous macrophytes p = 0.01; gravel versus wood p = 0.049, homogeneous macrophytes versus wood p = 0.000; homogeneous versus heterogeneous macrophytes p = 0.000. Type I probabilities for B for pikeperch (n = 8): gravel versus wood p = 0.033, homogeneous macrophytes versus wood p = 0.017. Type I probabilities for B for pike (n = 12): gravel versus heterogeneous macrophytes p = 0.041.

duces microhabitats for fish in small and large streams, and helps the diversion of river flow and sediment to form side channels and maintain seasonal wetland habitats for aquatic and terrestrial biota in floodplains, particularly those formed by larger rivers. If these natural, small-scale mechanisms are allowed to function throughout the drainage basin, larger scale effects that control the transport of water, nutrients, and inorganic and organic sediment will in turn magnify these processes of habitat development and diversity.

Life cycle stages of riverine fish populations require habitats specific to different seasons and river discharges and, for migratory species, different locations in the drainage basin. Therefore any consideration of the population of a species migrating longitudinally or laterally relative to the river channel necessitates a consideration of seasonal connectivity among habitats and the creation and maintenance of such habitats by climatic forcing factors such as precipitation and temperature. Even if a riverine fish species that completed its whole life cycle in one habitat were discovered, that habitat would still be subject to changes in hydrology and temperature and delivery of wood, sediment, nutrients, and food that are dependent on processes that operate at least on the scale of the basin upstream.

TABLE. 2. Reports on influence of wood complexity on fish.

Characteristic investigated	Goal of experiment	Fish responses to complexity	Citation
Habitat choice	To examine the debris characteristics at sites used as diurnal cover by adult brown madtoms *Noturus phaeus* and pirate perch *Aphredoderus sayanus*	• wood used by brown madtoms has greater cavity space, structural complexity, and more suspended leaves than that used by pirate perch	Monzyk et al. 1997
Predation and habitat choice	To examine selection of interstices of different sizes by bluegills in artificial ponds containing polyethylene pipes and wooden rods, in the presence and absence of largemouth bass predation	• the presence of predation caused increased use of smaller interstices	Johnson et al. 1988
Size-structure	To examine differences in use of structurally complex debris by different sizes of pirate perch	• larger individuals occupied more structurally complex debris than smaller ones	Monzyk et al. 1997
	To examine effects of habitat structure types on fish assemblage structure	larger fish were present at locations with structure (wing dikes, woody snags) than at sites without (bare shore)	Madejczyk et a al. 1998
Life stages	To examine differences in use of structurally complex debris by different life stages of largemouth bass	• juveniles preferred complex patches with combination of wood, leaf packs, and aquatic vegetation • non-nesting adults were found regularly in areas lacking structure	Annet et al 1996
Life cycle	To examine differences in use of structurally complex debris by different sexes and life stages of largemouth bass	• non-nesting males did not use patches with wood • nesting mates used patches associated with the large, but simple, woody structure	Annet et al. 1996
Seaward migration	To examine behavior, habitat use, and movements of coho salmon smolts during seaward migration	• smolts formed large aggregations in pools with large wood during seaward migration	McMahon and Holtby 1992
Homing abilities	To examine migration of radiotagged Murray cod relative to wood	• cod migrated up to 100 km upstream during the spawning season and returned to the same woody complex that they had inhabited before spawning	Koehn 1996

Fish as populations and communities are maintained over generations—as opposed to individuals within a day—that respond to a mosaic of seasonal habitats that changes spatially but maintains its integrity so that adaptive characteristics, including those resulting from co-evolution, can develop among the biota. An important large-scale spatial and temporal result of these micro- and mesohabitat effects—provided that they are maintained by a natural disturbance regime of seasonal flows—is to control carrying capacity, life history and species diversity, and other group properties of fish species in the whole river ecosystem (Schiemer and Zalewski 1992). We propose that the magnitudes of such basinwide properties would be enhanced if the processes mediated by wood in naturally functioning floodplains were restored.

Limiting factors and modeling concepts for understanding large river systems

Currently, there are limited, ecologically oriented concepts that address the functioning of river channels (Vannote et al. 1980) and river-floodplains (Junk et al. 1989). Neither addresses the whole basin from a physical and biological viewpoint and, in particular, does not adequately account for the role of wood or the trade-off between abiotic and biotic controlling factors on fish. Zalewski and Naiman (1985) and Bayley and Li (1992: Figure 12.2b) independently proposed that abiotic processes dominate (particularly hydrology) except when they are stable and/or predictable, in which case, biotic controls on fish populations start to manifest themselves. Zalewski and Naiman (1985) proposed the Concept of Abiotic-Biotic Regulatory Continuum in which this trade-off varies along the river continuum and according to different geographical regions. The influence of abiotic factors on fish populations gradually decreases with increases in spatial heterogeneity and stability of habitats, which typically occurs with increasing stream order, and/or with a parallel increase in water temperature in temperate rivers. Conversely, in arid-zone rivers, abiotic controls of even higher temperatures (Zalewski and Naiman 1985) and low hydrological predictability (Bayley and Li 1992) can reassert themselves as regulators of fish and their community structure. Conversely, seasonal and interannual variation in ice formation and flows may dominate the hierarchy of abiotic controls in arctic rivers.

Whether wood is a potential biotic or abiotic control depends on the context. Wood that functions as microhabitat cover, or as a component of pool habitat or bank protection, acts as a potential abiotic control. The potential influence of wood in river networks tends to decrease downstream in large basins and is generally less influential in affecting physical processes and producing habitat in arctic and arid zone rivers. This hypothesized decrease of the role of wood along the length of streams and rivers contrasts with its role in floodplains, which tends to increase with stream order. Spatial stability of wood in lotic and floodplain habitats over time has not been addressed, but in floodplains with predictable flood pulses, it is expected to be relatively stable and conform to the concept of increased biotic controls, at least within hydrological seasons (Bayley and Li 1992). However, other abiotic effects may be less predictable in floodplains, such as temperature and dissolved oxygen, which can limit fish populations. Converse effects of these abiotic candidates may be true in the adjacent river channel, and controls on the populations of the many fish species that occupy channels and floodplains during their life cycles are not self evident. However, the degree of annual flooding within and among systems exerts a major control on fishery yields (Welcomme 1985; Bayley 1995).

Management options and public perceptions in large river systems

One management option is to restore the natural process of wood supply, which depends on riparian and, where appropriate, floodplain forests in which natural succession is allowed (Boyer et al. 2003, this volume). Dynamic interactions between wood, hydrology, and sediments may also help restore some of the natural disturbance regime, providing intermediate disturbance levels and habitat heterogeneity that promote increased species diversity (Connell 1978). As hypothesized in the Ecohydrology Concept (EHC), such a dynamic riverscape should also increase system resilience to some man-made disturbances. These processes depend also on some degree of restoration towards a natural hydrograph, given that flows in most large river systems have been made more homogeneous, with higher minimum levels, controlled moderate floods, but more catastrophic floods.

Restoration of wood supply in lotic habitats by the reestablishment of riparian forest restoration and in selected floodplains by land-use and forestation is the preferred management solution. This process becomes more cost-effective on a large scale but requires a consistent policy and investment over several decades. The alternative engineering approach of placing and periodically replacing wood may be useful for small-scale demonstration purposes but is not a viable long-term, large-scale solution and would not provide other ecological benefits obtainable from restored forests.

Public perceptions of landscape aesthetics should also be considered. Gregory and Davis (1993) indicated that in-channel large wood negatively affects the perception of riverscape aesthetics. Management strategies that make long-term sense in terms of natural restoration and low-cost maintenance may require demonstration projects to build a sense of familiarity in a changed riverscape and an understanding of its functions that improve tangible benefits, such as angling.

In conclusion, reintroduction of wood will not, on its own, achieve benefits without at least partial restoration of hydrological cycles, riparian vegetation, selected floodplains, and the associated natural disturbance regime. An evolving basinwide concept, such as the EHC, is necessary to guide these processes so that larger-scale, less-tangible benefits such as increased resilience and biological production can be achieved.

Acknowledgments

This work was supported under Grant No. DHR-5600-G-00-1045-00: "Restoration Ecology of Two Pericid Fishes," Program in Science and Technology Co-operation, Office of Science Advisor, U.S. Agency for International Development, Washington, USA, Ohio State University, and in the framework of a Research Project Contract No. EVK1 - CT-2001-00094: "Development, Evaluation and Implementation of a Standardised Fishbased Assessment Method for the Ecological Status of European Rivers. A Contribution to the Water Framework Directive (acronym: FAME)," supported by the European Commission under the Fifth Framework Programme and contributing to the implementation of the Key Action - Sustainable Management and Quality of Water within the Energy, Environment and Sustainable Development Programme.

References

Abt, S., S. J. Dudley, and J. C Fischenich. 1998. Woody debris influence on flow resistance. Engineering approaches to ecosystem restoration. Pages 42–47 *in* Proceedings of the 1998 Wetlands Engineering and River Restoration Conference, Denver, Colorado. American Society of Civil Engineers, Reston, Virginia.

Agostinho, A. A., L. C. Gomes, and M. Zalewski. 2001. The importance of floodplains for the dynamics of fish communities of the upper river Parana. Ecohydrology & Hydrobiology 1:209–217.

Agostinho, A. A., M. Zalewski. 1995. The dependence of fish community structure and dynamics on floodplain and riparian ecotone zone in Parana River, Brazil. Pages 141–148 *in* F. Schiemer, M. Zalewski, and J. Thorpe, editors. The importance of aquatic-terrestrial ecotones for freshwater fish. Developments in Hydrobiology 105. Kluver Academic Publishers, Dordrecht, Boston, London.

Allen, T. F. H., and T. B. Starr. 1982. Hierarchy: perspectives for ecological complexity. University of Chicago Press, Chicago.

Anderson, O. 1984. Optimal foraging by largemouth bass in structured environments. Ecology 65:851–861.

Angermeier, P. L., and J. R. Karr. 1984. Relationships between woody debris and fish habitat in a small warmwater stream. Transactions of the American Fisheries Society 113:716–726.

Annet, C., J. Hunt, and E. D. Dibble. 1996. The compleat bass: habitat use patterns of all stages of the life cycle of largemouth bass. Pages 306–314 *in* L. E. Miranda, and D. R. DeVries, editors. Multidimensional approaches to reservoir fisheries management. American Fisheries Society, Symposium 16, Bethesda, Maryland.

Armantrout, N. B. 1991. Restructuring streams for anadromous salmonids. Pages 136–149 *in* J. Colt, and R. J. White, editors. Fisheries Bioengineering Symposium. American Fisheries Society, Symposium 10, Bethesda, Maryland.

Armantrout, N. B., editor. 1995. Condition of the world's aquatic habitats. Proceedings of the World Fisheries Congress, Theme 1. Oxford & IBH Publishing Co. Pvt. Ltd., New Delhi.

Arthington, A., and R. L. Welcomme. 1995. The conditions of large river systems of the world. Pages 16–43 *in* N. B. Armantrout (1991).

Bachman, R. A. 1982. A growth model for drift feeding salmonids: a selective pressure for migration. Pages 128–134 *in* E. L. Brannon and E. O. Salo, editors. Salmon and Trout Migratory Behaviour Symposium. University of Washington, Seattle.

Baxter, J. S., and J. D. McPhail. 1997. Diel microhabitat preferences of juvenile bull trout in an artificial stream channel. North American Journal of Fisheries Management 17:975–980.

Bayley, P. B. 1989. Aquatic environments in the Amazon Basin, with an analysis of carbon sources, fish production, and yield. Special Publication of the Canadian Journal of Fisheries and Aquatic Sciences 106:399–408.

Bayley, P. B. 1995. Understanding large river-floodplain ecosystems. Bioscience 45:153–158.

Bayley, P. B., and H. W. Li. 1992. Riverine fishes. Pages 252–281 *in* P. Calow and G. E. Petts, editors. The rivers handbook: hydrological and ecological principles. Volume 1. Blackwell Scientific Publications, Oxford, UK.

Bayley, P. B., K. O'Hara, and R. Steel. 2000. Defining and achieving fish habitat rehabilitation in large, low-gradient rivers. Pages 279–289 *in* I. G. Cowx, editor. Management and ecology of river fisheries. Fishing News Books, Blackwell Scientific Publications, Oxford, UK.

Benke, A. C., C. A. S. Hall, C. P. Hawkins, R. H. Lowe-McConnell, J. A. Stanford, K. Suberkropp, and J. A. Ward. 1988. Bioenergetic considerations in the analysis of stream ecosystem. Journal of North American Benthological Society 7:480–502.

Benke, A. C., T. C. van Arsdall, Jr., D. M. Gillespie, and F. K Parrish. 1984. Invertebrate productivity in a subtropical blackwater river. Ecological Monographs 54:25–63.

Benke, A. C., and J. B. Wallace. 2003. Influence of wood on invertebrate communities in streams and rivers. Pages 149–177 *in* S. V. Gregory, K. L. Boyer, and A. M. Gurnell, editors. The ecology and management of wood in world rivers. American Fisheries Society, Symposium 37, Bethesda, Maryland.

Bis, B., and L. W. G. Higler. 2001. Riparian vegetation of streams and the macroinvertebrate community structure. Ecohydrology & Hydrobiology 1:253–260.

Bisson, P. A., S. M. Wondzell, G. H. Reeves, and S. V. Gregory. 2003. Trends in using wood to restore aquatic habitats and fish communities in western North American rivers. Pages 391–406 *in* S. V. Gregory, K. L. Boyer, and A. M. Gurnell, editors. The ecology and management of wood in world rivers. American Fisheries Society, Symposium 37, Bethesda, Maryland.

Bjorn, T. C., and D. W. Reiser. 1991. Habitat requirements of salmonids in streams. Pages 83–138 *in* W. R. Meehan, editor. Influences of forest and rangeland management on salmonid fishes and their habitats. American Fisheries Society, Special Publication 19, Bethesda, Maryland.

Black, R. W., and T. A. Crowl. 1994. Effects of instream woody debris and complexity of the aquatic community in a high mountain, desert stream community. In D. A. Hendrickson, editor. Annual Symposium of the Desert Fishes Council, Furnace Creek, California. Proceedings of the Desert Fishes Council. ISSN 1068-0381.

Boon, P. J., P. Calow, and G. E. Petts, editors. 1992. River conservation and management. Wiley, Chichester, UK.

Borchardt, D. 1993. Effects of flow and refugia on drift loss of benthic macroinvertebrates: implications for habitat restoration in lowland streams. Freshwater Biology 29(3):221–227.

Boulton, A. J., and L. N. Lloyd. 1991. Macroinvertebrate assemblages in floodplain habitats of the lower river Murray, South Australia. Regulated Rivers: Research and Management 6(3):183–201.

Boyer, K. L., D. R. Berg, and S. V. Gregory. 2003. Riparian management for wood in rivers. Pages 407–420 *in* S. V. Gregory, K. L. Boyer, and A. M. Gurnell, editors. The ecology and management of wood in world rivers. American Fisheries Society, Symposium 37, Bethesda, Maryland.

Bren, L. J. 1993. Riparian zone, stream, and floodplain issues: a review. Journal of Hydrology 150:227–299.

Brinson, M. M. 1990. Riverine forests. Pages 87–141 *in* A. E. Lugo, M. M. Brinson and S. Brown, editors. Forested wetlands. Ecosystems of the world, 15. Elsevier Publishing, New York.

Bryant, M. D., and J. R. Sedell. 1995. Riparian forests, wood in the water, and fish habitat complexity. Pages 202–224 *in* N. B. Armantrout (1991).

Buijse, A. D., and F. T. Vriese. 1996. Assessing potential fish stocks in new nature developments in floodplains of large rivers in the Netherlands. Large Rivers 10, Archiv fur Hydrobiologie, Supplement 113:339–343.

Chauvet, E., and H. Decamps. 1989. Lateral interactions in a fluvial landscape: the river Garonne, France. Journal of the North American Benthological Society 8(1):9–17.

Chick, J. H., and C. C. McIvor. 1997. Habitat selection by three littoral zone fishes: effects of predation pressure, plant density and macrophyte type. Ecology of Freshwater Fish 6:27–35.

Connell, J. H. 1978. Diversity in tropical rainforest and coral reefs. Science 199:1302–1310.

Cowx, I. G., and R. L. Welcomme editors. 1998. Rehabilitation of rivers for fish. Food and Agriculture Organization of the United Nations (FAO) and Fishing News Books, Oxford, UK.

Crispin, V., R. House, and D. Roberts. 1993. Changes in instream habitat, large woody debris, and salmon habitat after the reconstructing of a coastal Oregon stream. North American Journal of Fishery Management 13(1):96–102.

Crook, D. A., and A. I. Robertson. 1999. Relationships between riverine fish and woody debris: implications for lowland rivers. Marine and Freshwater Research 50:941–953.

Decamps, H. 1996. The ecology of large rivers: a symposium in perspective (Kerms, Austria, 1995). Archiv fur Hydrobiologie Supplement 113(1–4), Large Rivers 10(1–4):593–598.

Diehl, S. 1988. Foraging efficiency of three freshwater fish: effects of structural complexity and light. Oikos 53:207–214.

Dodge, P. D., editors. 1989. Proceedings of the International Large River Symposium (LARS). Canadian Special Publication of Fisheries and Aquatic Sciences 106:1–629.

Dokulil, M., editor. 1996. Archiv fur Hydrobiologie Supplement 113(1–4), Large Rivers 10(1–4).

Eklov, P. 1997. Effects of habitat complexity and prey abundance on the spatial and temporal distributions of perch (*Perca fluviatilis*) and pike (*Esox lucius*). Canadian Journal of Fisheries and Aquatic Sciences 54(7):1520–1531.

Eklov, P., and S. Diehl. 1994. Piscivore efficiency and refuging prey: the importance of predator search mode. Oecologia 98:344–353.

Everett, R. A., and G. M. Ruiz. 1993. Coarse woody debris as refuge from predation in aquatic communities. An experimental test. Oecologia 93(4):475–486.

Fausch, K. D., C. E. Torgeson, C. V. Baxter, and H. W. Li. 2002. Landscapes to riverscapes: bridging the gap between research and conservation of stream fishes. BioScience 52:483–498.

Frissell, C. A., W. J. Liss, C. E. Warren, and M. D. Hurley. 1986. A hierarchical framework for stream classification: viewing streams in a watershed context. Environmental Management 10:199–214.

Gerhard, M., and M. Reich. 2000. Restoration of streams with large wood: effects of accumulated and built-in wood on channel morphology, habitat diversity and aquatic fauna. Internationale Revue der gesamten Hydrobiologie 85(1):123–137.

Gippel, C. J., B. L. Finlayson, and I. C. O'Neill. 1996b. Distribution and hydraulic significance of large woody debris in a lowland Australian river. Hydrobiologia 318:179–194.

Gippel, C. J., I. C. O'Neill, B. L. Finlayson, I. Schnatz, and S. J. Saltveit, editors. 1996a. Hydraulic guidelines for the re-introduction and management of large woody debris in lowland rivers. Regulated Rivers: Research and Management 12(2–3):223–236.

Gore, J. A., and F. D. Shields, Jr. 1995. Can large rivers be restored? BioScience 45:142–152.

Goulding, M. 1980. The fishes and the forest: explorations in Amazonian natural history. University of California Press, Berkeley.

Gregory, K. J., and R. J. Davis. 1992. Coarse woody debris in stream channels in relation to river channel management in woodland areas. Regulated Rivers: Research and Management 7(2):117–136.

Gregory, K. J., and R. J. Davis. 1993. The perception of riverscape aesthetics: an example from two Hampshire rivers. Journal of Environmental Management 39(3):171–185.

Grzybkowska, M., J. Hejduk, and P. Zielinski. 1990. Seasonal dynamics and production of *Chironomidae* in a large lowland river upstream and downstream from a new reservoir in Central Poland. Archiv fur Hydrobiologie 119:439–455.

Gurnell, A. 1997. The hydrological and geomorphological significance of forested floodplains. Global Ecology and Biogeography Letters 6(3–4):219–229.

Gurnell, A. M. 2003. Wood storage and mobility. Pages 75–91 *in* S. V. Gregory, K. L. Boyer, and A. M. Gurnell, editors. The ecology and management of wood in world rivers. American Fisheries Society, Symposium 37, Bethesda, Maryland.

Gurnell, A. M., K. J. Gregory, and G. E. Petts. 1995. The role of coarse woody debris in forest aquatic habitats: implications for management. Aquatic Conservation: Marine and Freshwater Ecosystems 5(2):143–166.

Gurnell, A. M., G. E. Petts., D. M. Hannah, B. P. G. Smith, P. J., Edwards, J. Kollmann, J. V. Ward, and K. Tockner. 2001. Riparian vegetation and island formation along the gravel-bed River Tagliamento, Italy. Earth Surface Processes and Landforms 26:31–62.

Gurnell, A. M., G. E. Petts, N. Harris, J. V. Ward, K. Tockner, P. J. Edwards, and J. Kollmann. 2000. Large wood retention in river channels: the case of the River Tagliamento, Italy. Earth Surface Processes and Landforms 25:255–275.

Harmon, M. E., N. H. Anderson, J. F. Franklin, S. P. Cline, F. J. Swanson, N. G. Aumen, P. Sollins, J. R. Sedell, S. V. Gregory, G. W. Lienkaemper, J. D. Lattion, K. Cromack, Jr., and K. W. Cummins. 1986. Ecology of coarse woody debris in temperate ecosystems. Advances in Ecological Research 15:133–302.

Harvey, B. C. 1987. Susceptibility of young-of-the-year fish to downstream displacement by flooding. Transactions of the American Fisheries Society 116:851–855.

Harper, D. M., M. Ebrahimnezhad, E. Taylor, S. Dickinson, O. Decamp, G. Verniers, and T. Balbi. 1999. A catchment-scale approach to the physical restoration of lowland UK rivers. Aquatic Conservation: Marine and Freshwater Ecosystems 9:141–157.

Hering, D., J. Kail, S. Eckert, M. Gerhard, E. I. Meyer, M. Mutz, M. Reich, and I. Weiss. 2000. Coarse woody debris quantity and distribution in Central European streams Internationale Revue der gesamten Hydrobiologie 85:5–23.

Hesse, L. W. 1994. The status of Nebraska fishes in the Missouri River, 4. Flathead catfish *Pylodictis olivaris*, and blue catfish, *Ictalurus furcatus* (Ictaluridae). Transactions of Nebraska Academic Sciences 21:89–98.

Hilderbrand, R. H., A. D. Lemly, C. A Dolloff, and K. L. Harpster. 1998. Design considerations for large woody debris placement in stream enhancement projects. North American Journal of Fisheries Management 18(1):161–167.

Inoue, M., and S. Nakano. 1998. Effects of woody debris on the habitat of juvenile masu salmon (*Oncorhynchus masou*) in northern Japanese streams. Freshwater Biology 40(1):1–16.

Jacobson, P. J., K. M. Jacobson, P. L. Angermeier, and D. S. Cherry. 1999. Transport, retention, and ecological significance of woody debris within a large ephemeral river. Journal of the North American Benthological Society 18:429–444.

Jacobson, P. J., K. M. Jacobson, P. L. Angermeier, and D. S. Cherry. 2000. Variation in material transport and water chemistry along a large ephemeral river in the Namib Desert. Freshwater Biology 44:481–491.

Jepsen, D. B., K. O. Winemiller, and D. C. Taphorn. 1997. Temporal patterns of resource partitioning among *Cichla* species in a Venezuelan blackwater river. Journal of Fish Biology 51:1085–1108.

Johnson, D. L., R. A Beaumier, and W. E. Lynch Jr. 1988. Selection of habitat structure interstice size by bluegills and largemouth bass in ponds. Transactions of the American Fisheries Society 117:171–179.

Junk, W. J., P. B. Bayley, and R. E. Sparks. 1989. The flood pulse concept in river-floodplain systems. Canadian Special Publication of Fisheries and Aquatic Sciences 106:110–127.

Karr, J., K. D. Fausch, P. L. Angermeier, P. R. Yant, and I. J. Schlosser. 1986. Assessing biological integrity in running waters: a method and its rationale. Illinois Natural History Survey, Special Publication 5, Champaign, Illinois.

Koehn, J. D. 1996. Habitats and movements of freshwater fish in the Murray-Darling Basin. Pages 27–32 *in* R. J. Banens and R. Lehane, editors. Proceedings of the 1995 Riverine Research Forum. October 1995, Attwood, Australia. Murray-Darling Basin Commission, Canberra, Australia.

Koehn, J. D., and S. Nicol. 1997. Habitat and movement requirements of fish. Pages 1–6 *in* R. J. Banens and R. Lehane. Proceedings of the Inaugural Riverine Environment Research Forum of MDBC Natural Resource Management Strategy Funded Projects. October 1996, Brisbane, Australia. Murray-Darling Basin Commission, Canberra, Australia.

Koning, C. W., M. N. Gaboury, M. D. Feduk, and P. A. Slaney. 1998. Techniques to evaluate the effectiveness of fish habitat restoration works in streams impacted by logging activities. Canadian Water Resources Journal 23(2):191–207.

Kwak, T. J., M. J. Wiley, L. L. Osborne, and R. W. Larimore. 1992. Application of diel feeding chronology to habitat sustainability analysis of warmwater stream fishes. Canadian Journal of Fisheries and Aquatic Sciences 49:1417–1430.

Lapinska, M. 1996. Space as a limiting factor for fish communities in lowland and upland river systems. Doctoral thesis, University of Lódz, Poland (in Polish).

Lapinska, M. 2001. The influence of littoral zone type and presence of YOY pike (*Esox lucius* L.) on growth and behaviour of YOY pikeperch, *Stizostedion lucioperca* (L.) – consequences for water quality in lowland reservoirs. Ecohydrology & Hydrobiology 1(3):355–372.

Lapinska, M., Z. Kaczkowski, and M. Zalewski. 2002. Restoration of streams for water quality improvement and fishery enhancement. Pages 113–125 *in* M. Zalewski, editor. Guidelines for the integrated management of the watershed – phytotechnology and ecohydrology. United Nations Environment Programme, Division of Technology, Industry and Economics, Freshwater Management Series No. 5.

Lehtinen, R. M., N. D. Mundahl, and J. C. Madejczyk. 1997. Autumn use of woody snags by fishes in backwater and channel border habitats of a large river. Environmental Biology of Fishes 49:7–19.

Madejczyk, J. C., N. D Mundahl, and R. M. Lehtinen. 1998. Fish assemblages of natural and artificial habitats within the channel border of the upper Mississippi River. American Midland Naturalist 139:296–310.

Mann, R. H. K. 1996. Environmental requirements of European non-salmonid fish in rivers. Hydrobiologia 323:223–235.

Mann, R. H. K., and T. Penczak. 1984. The efficiency of a new electrofishing technique in determining fish numbers in a large river in Central Poland. Journal of Fish Biology 24:173–185.

Maser, C., and J. R. Sedell. 1994. From the forest to the sea: the ecology of wood in streams, rivers, estuaries, and oceans. St. Lucie Press, Delray Beach, Florida.

McMahon, T. E., and G. F. Hartman. 1989. Influence of cover complexity and current velocity on winter habitat use by juvenile coho salmon (*Oncorhyncus kisutch*). Canadian Journal of Fisheries and Aquatic Sciences 46:1551–1557.

McMahon, T. E., and L. B. Holtby. 1992. Behaviour, habitat use, and movements of coho salmon (*Oncorhynchus kisutch*) smolts during seaward migration. Canadian Journal of Fisheries and Aquatic Sciences 49(7):1478–1485.

Monzyk, F. R., W. E. Kelso, and D. A. Rutherford. 1997. Characteristics of woody cover used by madtoms and pirate perch in coastal plain streams. Transactions of American Fisheries Society 126(4):665–675.

Murphy, M. L., and K. V. Koski. 1989. Input and depletion of woody debris in Alaska streams and implications for streamside management. North American Journal of Fisheries Management 9(4):427–436.

Mutz, M. 2000. Influences of woody debris on flow patterns and channel morphology in a low energy, sand-bed stream reach. Internationale Revue der gesamten Hydrobiologie 85(1):107–121.

Nagasaka, A., and F. Nakamura. 1999. The influences

of land-use changes on hydrology and riparian environment in a northern Japanese landscape. Landscape Ecology 14:543–556.

Naiman, R. J., and H. Décamps, editors. 1990. The ecology and management of aquatic-terrestrial ecotones. UNESCO, Paris, and Parthenon Publishing Group, Carnforth, UK.

Naiman, R. J., H. Decamps, and F. Fournier, editors. 1989. Role of land/inland water ecotones in landscape management and restoration, proposals for collaborative research. UNESCO, Vendome, France.

National Biological Service 1995. Sustaining the ecological integrity of large floodplain rivers: application of ecological knowledge to river management. Conference and workshop summary. Environmental Management Technical Center, Onalaska, Wisconsin.

Newbury, R. W. and M. N. Gaboury. 1993. Stream analysis and fish habitat design. Newbury Hydraulics Ltd., Gibsons, BC.

Odum, E. P. 1980. Ecology. Holt-Saunders, London.

Palmer, M. A., P. Arensburger, A. P. Martin, and D. W. Denman. 1996. Disturbance and patch-specific responses: the interactive effects of woody debris and floods on lotic invertebrates. Oecologia 105:247–257.

Pearsons, T. N., H. W. Li, and G. A. Lamberti. 1992. Influence of habitat complexity on resistance to flooding and resilience of stream fish assemblages. Transactions of the American Fisheries Society 121:427–436.

Penczak, T. 1995. Effects of removal and regeneration of bankside vegetation on fish population dynamics in the Warta River, Poland. Hydrobiologia 303:207–210.

Penczak, T. 2001. Populations of fish in relation to riparian ecotone development in the Narew river catchment. Ecohydrology & Hydrobiology 1:163–176.

Persat, H., J. M. Olivier, and J. P. Bravard. 1995. Stream and riparian management of large braided Mid-European Rivers, and consequences for fish. Pages 139–169 *in* N. B. Armantrout (1991).

Persson, L., and P. Eklov. 1995. Prey refuges affecting interactions between piscivorous perch and juvenile perch and roach. Ecology 76:70–81.

Petts, G. E., P. Armitage, and A. Gustard, editors. 1989a. Fourth International Symposium on Regulated Streams. Regulated Rivers 3:1–394.

Petts, G. E., H. Moller, and A. L. Roux, editors. 1989b. Historical change of large alluvial rivers: Western Europe. Wiley, Chichester, UK.

Phillips, E. C. 1995. Associations of aquatic *Coleoptera* with coarse woody debris in Ozark streams, Arkansas. Coleoptera Bulletin 49(2):119–126.

Phillips, E. C., and R. V. Kilambi. 1994. Habitat type and seasonal effects on the distribution and density of *Plecoptera* in Ozark streams, Arkansas. Annals of the Entomological Society of America 87(3):321–326.

Piégay, H. 1993. Nature, mass and preferential sites of coarse woody debris deposits in the lower Ain Valley (Mollon Reach), France. Regulated Rivers: Research and Management 8:359–372.

Piégay, H. 1997. Interactions between floodplain forests and overbank flows: data from three piedmont rivers of southeastern France. Global Ecology and Biogeography Letters 6(3–4):187–196.

Piégay, H. 2003. Dynamics of wood in large rivers. Pages 109–133 *in* S. V. Gregory, K. L. Boyer, and A. M. Gurnell, editors. The ecology and management of wood in world rivers. American Fisheries Society, Symposium 37, Bethesda, Maryland.

Piégay, H., and A. M. Gurnell. 1997. Large woody debris and river geomorphological pattern: examples from S. E. France and S. England. Geomorphology 19:99–116.

Piégay, H., and L. Maridet. 1994. Bibliographical review. Riparian forest of stream and fish management. Bulletin Francais de la Peche et de la Pisciculture 333:125–147.

Piégay, H., and R. A. Marston. 1998. Distribution of large woody debris along the outer bend of meanders in the Ain River, France. Physical Geography 19:318–340.

Poff, L. N., J. D. Allan, M. B. Bain, J. R. Karr, K. L. Prestegaard, B. D. Richter, R. E. Sparks, and J. C. Stromberg. 1997. The natural flow regime. BioScience 47:769–784.

Ptolemy, R. A. 1997. A retrospective review of fish habitat improvement projects in British Columbia: do we know enough to do the right thing? In J. D. Hall, P. A. Bisson, and R. E. Gresswell, editors. Sea run cutthroat trout: biology, management, and future conservation. American Fisheries Society, Oregon Chapter, Corvallis, Oregon.

Savino, J. F., and R. A. Stein. 1982. Predator-prey interactions between largemouth bass and bluegills as influenced by simulated, submersed vegetation. Transactions of the American Fisheries Society 11:255–266.

Savino, J. F., and R. A. Stein. 1989a. Behavioural interactions between predators and their prey: effects of plant density. Animal Behaviour 37:11–321.

Savino, J. F., and R. A. Stein. 1989b. Behaviour of fish predators and their prey: habitat choice between open water and dense vegetation. Environmental Biology of Fishes 24:287–293.

Schiemer, F., H. Keckeis, and L. Flore. 2001. Ecotones and hydrology: key conditions for fish in large rivers. Ecohydrology & Hydrobiology 1(1–2):49–55.

Schiemer, F., T. Spindler, H. Wintersberger, A. Schneider, and A. Chovanec. 1991. Fish fry associations: important indicators for the ecological status of large rivers. Verhandlungen der

Internationalen Vereinigung fur theoretische und angewandte Limnologie 24:2497–2500.

Schiemer, F., and M. Zalewski. 1992. The importance of riparian ecotones for diversity and productivity of riverine fish communities. Netherlanden Journal of Zoology 42:323–335.

Schlosser, I. J., and K. K. Ebel. 1989. Effects of flow regime and cyprinid predation on a headwater stream. Ecological Monographs 59:41–57.

Schlosser, I. J., J. D. Johnson, W. L. Knotek, and M. Lapinska. 2000. Climate variability and size-structured interactions among juvenile fish along a lake-stream gradient. Ecology 81:1046–1057.

Schmutz, S., M. Kaufmann, B. Vogel, M. Jungwirth, and S. Muhar. 2000. A multi-level concept for fish-based, river-type-specific assessment of ecological integrity. Hydrobiologia 422/423:279–289.

Scott, M. C., and P. L. Angermeier. 1998. Resource use by two sympatric black basses in impounded and riverine sections of the New River, Virginia. North American Journal of Fisheries Management 18(2):221–235.

Sedell, J. R., and J. L. Froggatt. 1984. Importance of streamside forest to large rivers: the isolation of the Willamette River, Oregon, USA, from its floodplain by snagging and streamside forest removal. Verhandlungen der Internationalen Vereinigung fur Theoretische und Angewandte Limnologie 22:1828–1834.

Sedell, J. R., J. E. Richey, and F. J. Swanson. 1989. The river continuum concept: a basis for the expected ecosystem behavior of very large rivers? Canadian Special Publication of Fisheries and Aquatic Sciences 106:49–55.

Shields, F. D., Jr., S. S. Knight, and C. M. Cooper. 1994. Effects of channel incision on base flow stream habitats and fishes. Environmental Management 18(1):43–57.

Shields, F. D., Jr., and R. H. Smith. 1992. Effects of large woody debris removal on physical characteristics of a sand-bed river. Aquatic Conservation: Marine and Freshwater Ecosystems 2(2):145–163.

Simondet, J. A. 1997. Seasonal and diel response to habitat complexity by an allopatric population of coastal cutthroat trout. In J. D. Hall, P. A. Bisson, and R. E. Gresswell, editors. Sea run cutthroat trout: biology, management, and future conservation.. American Fisheries Society, Oregon Chapter, Corvallis, Oregon.

Slaney, P. A., and A. D. Martin. 1997. The watershed restoration program of British Columbia: accelerating natural recovery processes. Aquatic Ecosystem Restoration 32:325–346.

Smith, R. H., F. D. Shields, Jr., A. C. Gibson, T. E. Schaefer, and E. A. Dardeau. 1992. Incremental effects of large woody debris removal on physical aquatic habitat. Waterways Experiment Station, Corps of Engineers, Technical Report E1-92-35 (EIRP), Vicksburg, Massachusetts.

Smock, L. A., J. E. Gladden, J. L. Riekenberg, L. C. Smith, and C. R. Black. 1992. Lotic macroinvertebrate production in three dimensions: channel surface, hyporheic, and floodplain environments. Ecology 73:876–886.

Stalnaker, C. B., R. T. Milhous, and K. D. Bovee. 1989. Hydrology and hydraulics applied to fishery management in large rivers. Canadian Special Publication of Fisheries and Aquatic Sciences 106:13–30.

Sundbaum, K., and I. Naeslund. 1998. Effects of woody debris on the growth and behaviour of brown trout in experimental stream channels. Canadian Journal of Zoology 76(1):56–61.

Tabacchi, E., D. L. Correll, R. Hauer, G. Pinay, A. M. Planty-Tabacchi, and R. C. Wissmar. 1998. Development, maintenance and role of riparian vegetation in the river landscape. Freshwater Biology 40:497–516.

Tabacchi, E., L. Lambs, H. Guilloy, A. M. Planty-Tabacchi, E. Muller, and H. Décamps. 2000. Impacts of riparian vegetation on hydrological processes. Hydrological Processes 14:2959–2976.

Todd, B. L., and C. F. Rabeni. 1989. Movement and habitat use by stream-dwelling smallmouth bass. Transactions of the American Fisheries Society 118:229–242.

Tonn, W. 1990. Climate change and fish communities: A conceptual framework. Transactions of the American Fisheries Society 119:337–352.

Triska, F. J. 1984. Role of woody debris in modifying channel geomorphology and riparian areas of a large lowland river under pristine conditions: a historical case study. Verhandlungen der Inter-nationalen Vereinigung fur theo-retische und angewandte Limnologie 22:1876–1892.

Valdimarsson, S. K., and N. B. Metcalfe. 1998. Shelter selection in juvenile Atlantic salmon, or why do salmon seek shelter in winter? Journal of Fish Biology 52:42–49.

Van Sickle, J., and S. V. Gregory. 1990. Modeling inputs of large woody debris to streams from falling trees. Canadian Journal of Forest Research 20:1593–1601.

Vannote, R. L., G. W. Minshall, K. W. Cummins, J. R. Sedell, and C. E. Cushing. 1980. The river continuum concept. Canadian Journal of Fisheries and Aquatic Sciences 37:130–137.

Vince, S., I. Valiela, N. Backus, and J. M. Teal. 1976. Predation by the saltmarsh killifish *Fundulus heterocilitus* in relation to prey size and habitat structure: consequences for prey distribution and abundance. Journal of Experimental Marine Biology and Ecology 23:225–266.

Ward, J. V. 1989. The four-dimensional nature of lotic ecosystems. Journal of North American Benthological Society 8:2–8.

Ward, J. V., K. Tockner, F. Schiemer, and J. B. Layzer, editors. 1999. Biodiversity of floodplain river ecosystems: ecotones and connectivity. Regulated Rivers: Research and Management 15:125–139.

Welcomme, R. L. 1985. River fisheries. FAO Fisheries Technical Paper 262, Rome.

White, R. J. 1996. Growth and development of North American stream habitat management for fish. Canadian Journal of Fisheries and Aquatic Sciences 53 (Supplement 1):342–363.

Zalewski, M. 1995. Freshwater habitat management and restoration in face of the global changes. Pages 170–194 *in* N. B. Armantrout (1991).

Zalewski, M. 2000. Ecohydrology – the scientific background to use ecosystem properties as management tools toward sustainability of water resources. Ecological Engineering 16:1–8.

Zalewski, M. 2002a. Ecohydrology - the use of ecological and hydrological processes for sustainable management of water resources. Hydrological sciences. Journal des Sciences Hydrologiques 47(5):825–834.

Zalewski, M. editor. 2002b. Guidelines for the integrated management of the watershed – phytotechnology and ecohydrology. United Nations Environment Programme, Division of Technology, Industry and Economics, Freshwater Management Series No. 5.

Zalewski, M., B. Bis, M. Lapinska, P. Frankiewicz, and W. Puchalski. 1998. The importance of the riparian ecotone and river hydraulics for sustainable basin-scale restoration scenarios. Aquatic Conservation: Marine and Freshwater Ecosystems 8:287–307.

Zalewski, M., and P. Frankiewicz. 1998. The influence of riparian ecotones on the dynamics of riverine fish communities. Pages 87–96 *in* L. C. de Waal, A. R. G. Large, and P. M. Wade, editors. Rehabilitation of rivers, principles and implementation. Wiley, Chichester, UK.

Zalewski, M., G. A. Janauer, and G. Jolankaj. 1997. Ecohydrology: a new paradigm for the sustainable use of aquatic resources. In Conceptual background, working hypothesis, rationale and scientific guidelines for the implementation of the IHP-V projects 2.3/2.4. UNESCO, Paris Technical Documents in Hydrology No. 7.

Zalewski, M., and R. J. Naiman. 1985. The regulation of riverine fish communities by a continuum of abiotic-biotic factors. Pages 3–9 *in* J. S. Alabaster, editor. Habitat modification and freshwater fisheries. Pages 3–9 in J. S. Alabaster, editor. Habitat modification and freshwater fisheries. FAO UN, Butterworths, London.

Zalewski, M., W. Puchalski, P. Frankiewicz, and B. Bis. 1994. Riparian ecotones and fish communities in rivers – intermediate complexity hypothesis. Pages 152–160 *in* I. G. Cowx, editor. Rehabilitation of freshwater fisheries. Fishing News Book, Blackwell, Oxford, UK.

Zalewski, M., F. Schiemer, and J. Thorpe, editors 2001. International Journal of Ecohydrology & Hydrobiology. Special issue on: Catchment processes, land-water ecotones and fish communities 1(1–2).

American Fisheries Society Symposium 37:213–233, 2003

Hydrologic and Geomorphic Effects of Beaver Dams and Their Influence on Fishes

Michael M. Pollock, Morgan Heim, and Danielle Werner

National Oceanic and Atmospheric Administration, Northwest Fisheries Science Center 2725 Montlake Boulevard E., Seattle, Washington 98112, USA

Abstract.—Beaver dams alter the hydrology and geomorphology of stream systems and affect habitat for fishes. Beaver dams measurably affect the rates of groundwater recharge and stream discharge, retain enough sediment to cause measurable changes in valley floor morphology, and generally enhance stream habitat quality for many fishes. Historically, beaver dams were numerous in small streams throughout most of the Northern Hemisphere. The cumulative loss of millions of beaver dams has dramatically affected the hydrology and sediment dynamics of stream systems. Assessing the cumulative hydrologic and geomorphic effects of depleting these millions of wood structures from small and medium-sized streams is urgently needed. This is particularly important in semiarid climates, where the widespread removal of beaver dams may have exacerbated effects of other land use changes, such as livestock grazing, to accelerate incision and the subsequent lowering of groundwater levels and drying of streams.

Introduction

In most of the temperate Northern Hemisphere, beaver historically altered low-gradient, small-stream ecosystems by constructing millions of dams made primarily of wood. Almost every northern temperate ecosystem that had trees or shrubs growing along streams also once had beaver dams. In Eurasia, evidence of beaver has been found in streams as far south as Iraq and Turkey, in the Arctic, and stretching from Scotland in the west to Kamchatka in the east (Figure 1; Halley and Rosell 2002). In North America, beaver were once found far south into the arid environments of Arizona and northern Mexico along rivers such as the San Pedro, Colorado, and the Rio Grande (Pattie 1833; Leopold 1972; Fredlake 1997) and occupied all biomes north of the border from coast to coast, except for the Arctic, peninsular Florida, and the dry Great Basin and desert country of Nevada and southern California (Figure 1; Jenkins 1979).

Historically, beaver dams created streams systems with slow, deep water and floodplain wetlands dominated by emergent vegetation and shrubs. Geomorphology and plant communities of small low-gradient streams were much changed throughout much of the Northern Hemisphere after reduction of beaver populations (Rea 1983; Naiman et al. 1988). In both Eurasia and North America, beaver populations have generally declined as human populations have increased. In both continents, only small populations survived by the end of the 19th century (Naiman et al. 1988; MacDonald et al. 1995; Nolet and Rosell 1998; Halley and Rosell 2002). The primary reasons for the declines were that people trapped beavers either because they were resources for fur or oil or competitors for productive valley bottom lands (MacDonald et al. 1995; Mackie 1997; Halley and Rosell 2002). More recently, however, there has been widespread recognition that beaver dams play a vital role in maintaining and diversifying stream and riparian habitat (Naiman et al. 1988; Pollock et al. 1994; Gurnell 1998; Collen and Gibson 2001). In the past century, land managers throughout the Northern Hemisphere have attempted to reintroduce beaver in areas where they have been extirpated. Today, beaver populations are rebounding throughout North America, with the population estimated to be about 10 million and reoccupying most of its former range (Naiman et al. 1988). Throughout Eurasia, recovery has been slower, with the Eurasian beaver population be-

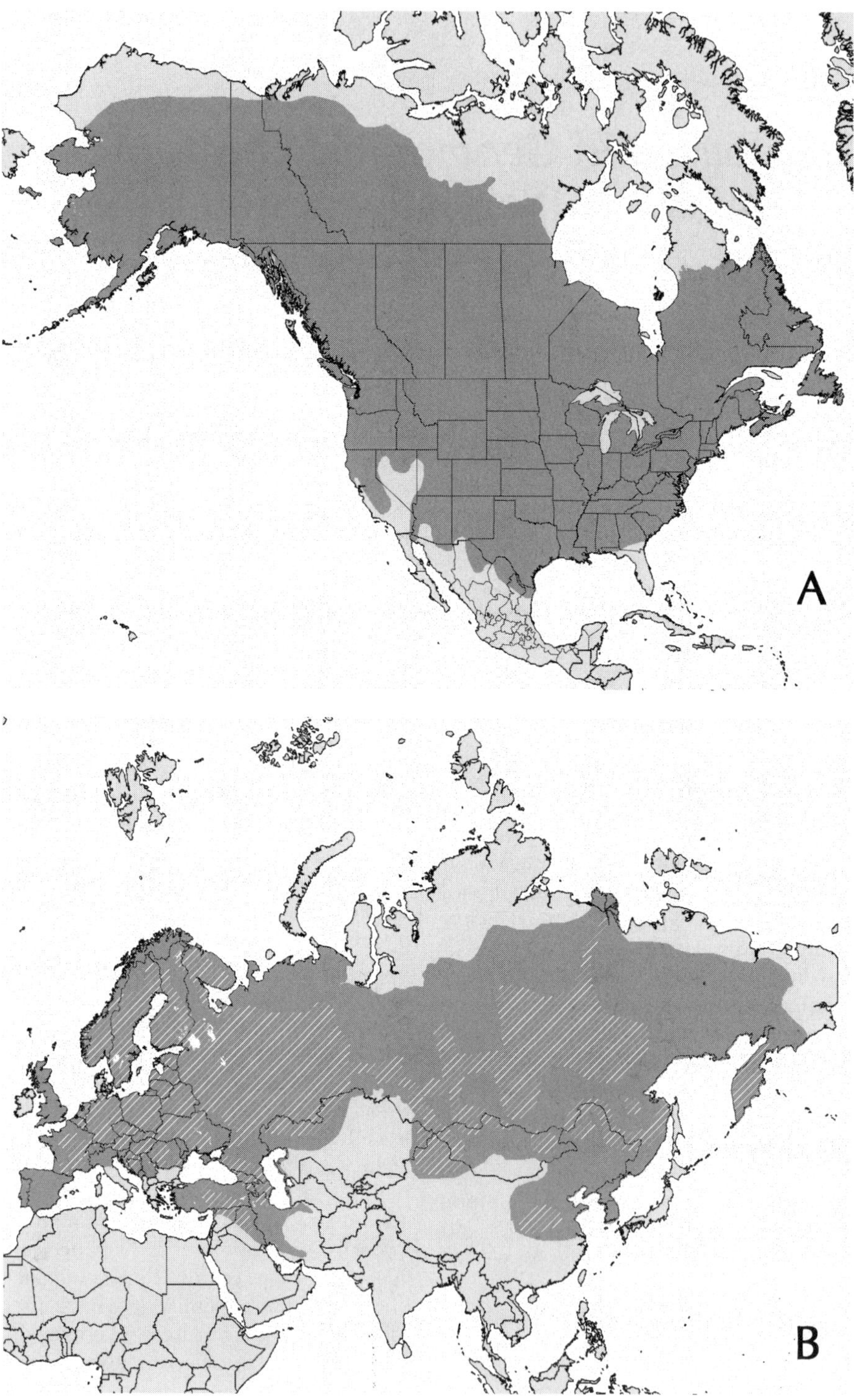

FIGURE 1. Estimated current and historic distribution of beaver in North America (A) and Eurasia (B). Isolated populations in peninsular Florida and Southern California are not shown. In Eurasia, cross hatching delineates current distribution. In North America, the current and historic distributions are approximately coincidental. (Based on Jenkins 1979; Halley and Rosell 2002; MacDonald et al. 1995).

tween a half to one million, with large geographic areas where no beaver are present or where the current status is unknown (Halley and Rosell 2002).

North American beaver have also become established in regions outside their geographic range, such as the Tierra del Fuego region of Argentina and Chile where they are spreading rapidly (Jaksic et al. 2002). The current beaver population in Tierra del Fuego is estimated to be around 70,000 (Jaksic et al. 2002). Several North American beaver populations have also been established in Europe prior to the recognition of the Eurasian and North American beavers as separate species (Figure 1; Nolet and Rosell 1998; Halley and Rosell 2002). In such areas, beaver are exotic species and may have undesirable impacts on native species and ecosystem processes.

The primary instream habitat value of beaver dams is that they impound water to form large pools and ponds. These impoundments trap sediment, help to create productive and diverse wetland environments on adjacent floodplains, improve water quality, and facilitate groundwater recharge. All these functions are ultimately the result of the dams reducing stream velocities and spreading water over a large surface area.

Here, we review how beaver dams have altered stream hydrology and morphology, both in the past when dams were much more abundant and under current conditions. We estimate the historic abundance and location of beaver dams in watersheds throughout the Northern Hemisphere, assess the cumulative hydrologic effects of multiple beaver dams, evaluate the geomorphic consequences of those effects, and examine what influence these physical effects have on fishes. Most of the scientific literature on the ecological effects of beaver dams come from studies of the North American beaver while literature on the ecological effects of the Eurasian beaver dams are rarer, if for no other reason than that the Eurasian beaver itself is still quite rare throughout much of its former range. However, available studies indicate that the two species are quite similar in most respects, though some evidence suggests the Eurasian beaver builds fewer dams and lodges and has a lower reproductive rate (Danilov and Kan'shiev 1983; Collen and Gibson 2001; Halley and Rosell 2002). Studies of the hydrologic and geomorphic effects of North American beaver dams should be applicable to Eurasian beaver, but studies of the effects on fishes may not be directly applicable, depending on what fish species were studied. Reviews of the many additional ecosystem impacts of beaver that are not directly related to their dams, such as their effects on riparian communities, can be found elsewhere (Naiman et al. 1988, 1994; Pollock et al. 1994; Gurnell 1998; Collen and Gibson 2001).

The Historical Abundance of Beaver Dams

Contemporary studies of protected or remote beaver populations support the contention that, historically, dams were very common in most small, low-gradient streams, throughout North America and Eurasia, but that the frequency of these dams varied considerably. Where beaver populations are undisturbed, localized dam frequencies range from 7.5 per km to as high as 74 per km, with frequencies of around 10 dams per km being more typical in low-gradient streams (Table 1; Warren 1926; Scheffer 1938). Frequencies may decrease when larger areas are considered. Two studies examining dam occurrence across entire, multiple watersheds found frequencies of 2.5 per km for the 750 km^2 Kabetogama Peninsula in Minnesota and 9.6 dams per km for an 85 km^2 area encompassing two watersheds in Wyoming (Skinner et al. 1984; Johnston and Naiman 1990b).

Estimates of historic dam densities over large geographical areas can be calculated if colony densities are known and the average dams built per colony can be estimated. Reported colony densities of remote or protected populations show a trend of lower densities in subarctic regions and higher densities in more temperate regions, with an overall average of a little less than 0.5 colonies per km^2 (Table 2). Johnston and Naiman (1990b) used remote sensing on the Kabetogama Peninsula and estimated a colony density of 0.92 km^2 and a beaver pond density of 2.96 km^2, which gives an average of 3.2 dams per colony. Recognizing both the high spatial variation and uncertainty of estimates of both colony density and dams per colony, these studies suggest that, before the arrival of Europeans, at least 25 million beaver dams spanned small to medium-sized streams throughout the 15.5 × 10^6 km^2 of the North American continent where beaver once existed (Figure 1; Jenkins 1979). Historic dam densities were probably similar throughout Eurasia, though there is some evidence to suggest that the remnant Eurasian beaver build fewer dams than their North American counterparts (Danilov and Kan'shiev 1983).

TABLE 1. Beaver dam density reported in the literature for pristine, remote, or protected areas.

Source	Dams /km	Range dams/km	Surveyed length/km	Location	Gradient	Comments
Naiman et al. 1986	10.6	8.6–16.0	4.3	Quebec	low	Pristine
Smith 1950	12.0	n/a	4.4	Colorado	1–3%	Remote
Smith 1950	19.1	n/a	4.4	Colorado	1–3%	Includes inactive dams
Warren 1926	73.7	n/a	0.3	Colorado	12.5%	Pristine
Warren 1926	41.6	n/a	1.2	Colorado	3.5%	Pristine
Scheffer 1938	7.5	n/a	8.0	Washington	low	Transplanted population
Scheffer 1938	35.4	n/a	0.6	Washington	low	Transplanted population
Skinner et al. 1984	9.6	2.9–41.1	43.3	Wyoming	85 km^2 area	Remote
Woo and Waddington 1990	14.3	5-19	n/a	Ontario	low	Remote, pristine
Naiman et al. 1988	2.5	2.0–2.9	n/a	Minnesota	750 km^2 area	Protected population
Hering et al. 2001	27.8	n/a	0.9	W. Germany	n/a	Recovering population
Hering et al. 2001	12.3	n/a	1.3	S. Germany	n/a	Recovering population
Medwecka-Kornas and Hawro 1993	20.0	n/a	1.0	Poland	n/a	Recovering population

Using Collen and Gibson's (2001) estimated overall mean of 5.2 individuals per colony of North American beaver, and using a continental mean of 0.5 colonies per km^2 (based on Table 2), provides an historic population estimate of North American lodge-building beaver of around 40 million individuals. Only about 75% of beaver live in lodges, and the rest burrow into banks (Danilov and Kan'shiev 1983); we estimate a total historic population of around 55 million. This estimate is on the low end of, but not inconsistent with, Seton's (1929) often cited population estimate of 60–400 million, which was derived primarily from the qualitative data. However, we will never be certain whether current estimates of individuals per colony or colony densities are applicable to historic populations. Because colony density is affected by habitat quality, it is not unreasonable to assume that, historically, overall colony densities were much higher, and thus populations much greater, when the entire vast, productive lowland river bottom habitat throughout North America had not yet been altered by humans and was available for beavers

TABLE 2. Colony density estimates for North American and Eurasian beaver for protected, unprotected, and recovering populations.

Source	Location	Density km^2	Status
Howard and Larson 1985	Massachusetts	0.92	Protected
Johnston and Naiman 1990a	Minnesota	0.92	Protected
McCall et al. 1996	Maine	0.15–0.32	Managed
Voigt et al. 1976	Ontario	0.38–0.76 (3.0)	Remote
Bergerud and Miller 1977	Newfoundland	0.2–0.3 (4.2)	Remote
Aleksiuk 1970	Northwest Territories	0.38	Remote
MacDonald et al. 1995	Latvia	0.37	Recovering
Hartman 1994	Sweden	0.25	Recovering
Zurowski and Kasperczyk 1986	Poland	0.15	Recovering
Jaksic et al. 2002	Argentina	1.9	Introduced

to use. The Eurasian beaver has a slower reproductive rate and a lower average colony size (3.8); so, historic estimates of populations and densities cannot be directly extrapolated from studies of the North American beaver (Danilov and Kan'shiev 1983). However, recovering populations of European beaver suggest that high densities may be achievable in the long term if management authorities desire such a goal (Halley and Rosell 2002).

Location of Dams in Watersheds

Beaver prefer to dam small, low-gradient streams with unconfined valleys, but they can also dam both large and high-gradient streams. Retzer et al. (1956) studied 365 reaches in 61 streams throughout Colorado to determine the physical factors determining beaver pond location. Beaver built dams on 82% of all the low-gradient (1–3%) streams surveyed, 73% of reaches with 4–6% gradients, and 61% of reaches with 7–9% gradients (Figure 2). Use of streams with a slope greater than 9% dropped off dramatically. On streams with gradients greater than 15%, just one active dam was found on a 16% stream slope and one abandoned dam was found on a 21% stream slope. Beaver overwhelmingly preferred to dam streams in valleys wider than 46 m (150 ft); they used 85% of such streams, but only 35% of streams with narrower valleys, regardless of their slope. These results are confounded because wide valleys tend to have low-gradient streams.

Consistent with these results, Pollock and Pess (1998) studied 341 beaver ponds in the 1,741 km^2 Stillaguamish watershed, Washington and found that 91% were in low-gradient (≤4%) streams with unconfined (>4 channel widths) valleys, and almost all of them were in watersheds less than 15 km^2. In this study, 16% was the steepest gradient where a beaver pond was found. Similarly, Suzuki and McComb (1998) studied 170 beaver dams in the Drift Creek basin, Oregon and consistent with these results found that more than 90% were on stream gradients of less than 6%. Beier and Barrett (1987) also found that geomorphic and hydrologic conditions were the best predictors of dam-site suitability, with gradient, stream depth, and stream width the most important factors. They found that biological factors, such as food availability, did not provide additional explanatory power as to the location of dam sites.

The maximum size of streams that beaver can dam is not well documented, and it will certainly vary from region to region, depending on hydrologic conditions. In Washington, historical

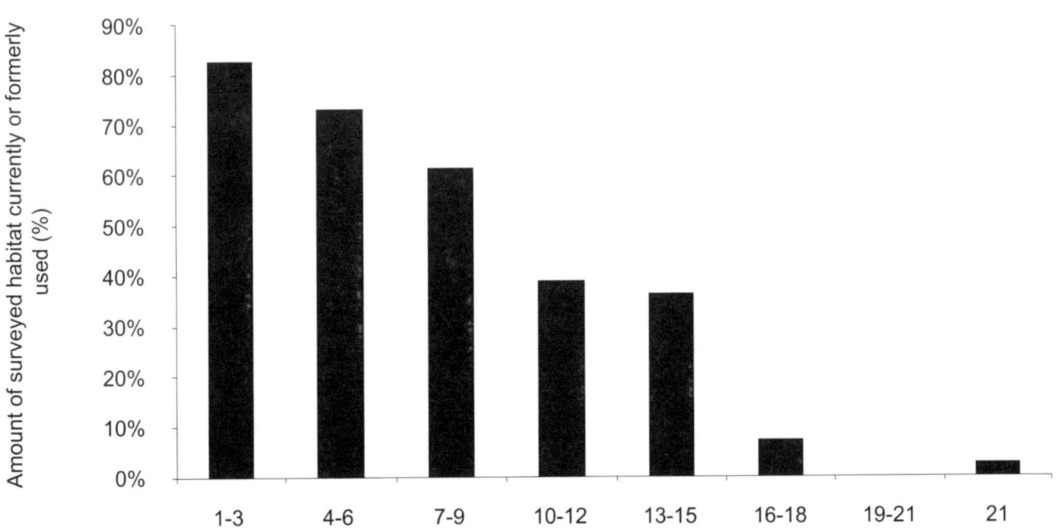

FIGURE 2. Beaver–dam frequency based on stream gradient from a survey of 356 stream reaches in the mountainous regions of Colorado, showing that beaver strongly prefer to dam low-gradient streams (adapted from Retzer et al. 1956).

records note a beaver dam on a stream with a drainage area of 107 km^2. Naiman et al. (1984) report seasonal dams being built on the lower reaches of the 207 km^2 Muskrat River watershed, Quebec. In 1998, one of us (Pollock) noted a stable beaver dam crossing the entire South Fork of the Little Colorado River with an upstream drainage area of 280 km^2. We suspect that unreported dams exist on larger and more powerful streams, but stream size or drainage area are generally not reported in the literature. The size of streams and rivers that could be dammed by the bear-sized Pleistocene beaver is a matter of interesting speculation (Mai 1978). We are not aware of any published literature that has identified prehistoric beaver dams.

Effects of Beaver Dams on Stream Hydrology

Even the casual observer notices that beaver dams retain water and thus slow its downstream movement. What is surprising is how few studies have attempted to quantify what happens to the water retained by beaver dams and what the hydrologic effects are. The limited available evidence suggests that key hydrologic functions of beaver dams are to dissipate stream energy, attenuate peak flows, and increase groundwater recharge and retention, which increases summer low flows and elevates groundwater levels in stream valleys, thus expanding the extent of riparian vegetation (Stabler 1985; Lowry 1993).

Beaver dams slow the velocity of water, thereby reducing energy and increasing retention time. During floods, energy is dissipated as the water works its way through a tortuous path of small branches on the downstream side of the dam. Finally, energy is further dissipated by floodplain vegetation below the dam that must be overcome as the water works its way back to the stream channel (Li and Shen 1973; Woo and Waddington 1990; Dunaway et al. 1994).

Because beaver dams slow stream velocity, they should also attenuate flood peaks. Research on the effects of wood dams in small (third-order) streams suggests that they can retain water at least 50% longer than streams where such dams are absent (Ehrman and Lamberti 1992). Given the much lower permeability of beaver dams compared to large wood jams, it is reasonable to expect them to retain water for much longer periods of time (Hering et al. 2001).

Beedle (1991) used the dimensional characteristics of beaver dams in the glacially carved valleys on a southeastern Alaska island to evaluate the hydrologic effects of these structures on peak flows. Using simulated peak-flow routing, Beedle estimated that a single full beaver pond reduced peak flows by more than 5%, but that a series of five large ponds in series could reduce peak flows of a 2-year event by 14% and peak flows of a 50-year event by 4%. The simulation also suggested that beaver dams did not greatly alter the shape of outflow hydrographs, resulting in a 10–15-min delay in the peak and a slight increase in flood duration. Beedle's simulation assumed that all ponds were at full storage capacity before the storm and that water pouring over dams instantly returned to the channel. Although these assumptions were necessary for the simulation, under natural conditions, where dams are not filled to capacity and flood water pouring over dams is spread out across the floodplain, peak flows could likely be further reduced. Because many streams may have dozens of dams, expecting stronger cumulative effects is reasonable (Scheffer 1938; Smith 1950; Naiman et al. 1986).

Effects of Beaver Dams on Aquifer Recharge and Low Flow Discharges

Some researchers have suggested that beaver dams and other instream obstructions that retain water contribute to groundwater recharge and thus help to increase summer low flows. Anecdotal reports in the literature support this contention, though few quantitative data exist. In Massachusetts, Wilen et al. (1975) compared changes in summer low flows caused by beaver dams when beaver colonized one of their study sites. In a paired stream study, beaver dams and large wood were removed from one stream, and the other stream was retained as a control. During the first year of the experiment, both streams remained perennial, but the undisturbed stream had more flow. In spring of the next year, beaver recolonized the experimental reach, building a series of dams. That summer, the experimental stream remained perennial below the newly built dams, but the "control" stream with no beaver dams went dry.

Tappe (1942) noted summer flows in several streams in northern California increased after beaver colonized upstream reaches. In Missouri,

Dalke (1947, cited in Stabler 1985) reported that the return of beaver to small streams restored perennially flowing water that could support fish. Likewise, Finley (1937) reported that stream flow in Silver Creek, in the Ochoco National Forest of eastern Oregon, decreased substantially after the loss of beaver dams. Trappers removed about 600 beaver from the headwaters of Silver Creek, and several creeks dried up in following years as water levels dropped. According to Finley, groundwater levels dropped sufficiently that pastures near streams could no longer support grass, resulting in an annual loss of approximately 15,000 tons of pasturage. Further downstream, water was no longer available to irrigate farmland, resulting in additional economic losses.

In eastern Oregon, groundwater levels in the floodplain near a beaver dam substantially increased as compared to a reach lacking any dams, suggesting that dams do help increase water availability to riparian areas at least 50 m from the pond (Lowry 1993). Additionally, Lowry estimated that about 90 m^3 of groundwater could be obtained if the beaver dam was breached.

Human-built structures have yielded analogous results. Brown (1963), studying a conservation program in Flat Top Ranch, Texas noted that small lakes and ponds created in sandy soils allowed for rapid infiltration of stormwater and helped to convert all major drainages from intermittent to perennial streams. In coastal British Columbia, a small check dam no larger than a beaver dam was used to retain stormwater in a 27-ha pond and ensure that the stream flowed throughout the summer (Wood 1997).

In Alkali Creek, Colorado, Heede and DeBano (1984) observed that perennial flow developed in downstream reaches 7 years after installation of 132 small check dams in ephemeral, gullied streams. Old beaver dams were observed in the gully walls while the check dams were being built, and 14 years after the project began, beaver have begun recolonizing some of the reaches with perennial flow (Stabler 1985). Stabler also noted the work of Jester and McKirdy (1966) who installed check dams in Taos Creek, New Mexico. Within 10 years, they obtained perennial flow that could sustain trout populations. Similarly, Ponce and Lindquist (1990) noted that perennial flow was an unexpected result of a small (5-m) dam built to retain sediment on Sheep Creek, Utah.

Lowry's (1993) study suggests that the water storage value of beaver dams does not lie solely in the water stored behind the dam because 90 m^3 is only enough water to keep even a small stream flowing for a very short period. If, as the previous examples suggest, beaver dams change stream hydrology so streams that formerly went dry in the summer now flow year round, their ability to recharge groundwater must be substantial. Simple calculations indicate that recharge from the dams of even a few beaver colonies can contribute substantially to summer low flows. As an example, a small stream flowing at a modest 0.1 m^3/s over a period of 100 d (the approximate length of summer in much of the western United States) will need about 8.64×10^5 m^3 of water above and beyond what is lost to evapotranspiration. Assuming that a series of beaver dams provides 1×10^5 m^2 of pond surface area (20 dams, assuming that a typical pond has 5×10^4 m^2 of surface area), aquifer hydraulic conductivities of about 4×10^{-7} m/s are needed to accept this much water during the other 265 d of the year. Because 4×10^{-7} m/s is well within the range of hydraulic conductivities typical of the silts and sands that would be found at the bottom of beaver ponds and fine-grained alluvial deposits (Dunne and Leopold 1978; Lowry 1993), even a small number of beaver impoundments should be able to augment low flow discharges during long periods without rain, provided an aquifer is present to accept the recharge.

A stream may remain perennial if the amount of valley alluvium is sufficient to store waters infiltrating from beaver ponds. Assuming a specific yield of 20%, typical of fine-grained alluvial deposits (Dunne and Leopold 1978; Lowry 1993), an aquifer volume of 4.32×10^6 m^3 is needed to store the 8.64×10^5 m^3 of water needed to maintain flow at 0.01 m^3/s for 100 d. As an example, sustaining this flow would require valley alluvium 150 m wide by 2.4 km long by 12 m deep. By comparison, to retain this much water behind typical beaver dams (0.5 m high on a 1% stream gradient and a 50-m-wide dam, and assuming they are not storing any sediment) would require the impoundment of 35 km of low-gradient streams. Together, these numbers suggest that if beaver impoundments augment summer low flows in small streams, they do so primarily by recharging aquifers, and the direct storage of water behind dams probably plays a minor role in low-flow augmentation.

Whether a series of beaver dams is actually able to recharge an alluvial aquifer, and whether that water is ultimately available to augment streamflows in the summer, depends upon the geometry and hydraulic properties of the aquifer. Aquifer geometry determines the size of the aquifer. Hydraulic properties, such as hydraulic conductivity and the coefficient of storage, control the rate of recharge and discharge, which determines whether an aquifer will drain throughout the summer. For example, a highly porous aquifer may quickly recharge but also quickly discharge, thereby precluding the slow release of water throughout the summer.

Effects of Beaver Dams on Valley Floor Morphology

The slow velocity of water behind beaver dams creates extensive depositional areas for sediment and organic material transported from upstream reaches. The shallow waters behind the dams allow highly productive emergent vegetation to grow, allowing in situ development of organic material, much of which is ultimately deposited on the (submerged) valley floor. Together, the onsite creation of organic material and the deposition of sediment and organic material from upstream raise the surface of the stream bed and inundate the adjacent valley floor. Sediment storage behind beaver dams can be substantial.

In a boreal forest ecosystem, Naiman et al. (1986) found that sediment storage behind beaver dams ranged from 35 to 6,500 m^3, yet they found no relation between dam size and sediment retained. Butler and Malanson (1995) studied sediment deposition behind five young (<6 years) and three "old" (>10 years) beaver dams in Glacier National Park, Montana. They found more sediment stored behind the older ponds. Sediment volumes behind the more recent dams ranged from 9 to 165 m^3 (average = 56 m^3); while behind the older dams, sediment storage ranged from 77 to 267 m^3 (average = 203 m^3). In parts of the arid western United States, beaver dams have been used to trap sediment in eroding streams with good results (Scheffer 1938; Apple 1985). In Mission Creek, Washington, beaver were successfully "employed" to control sediment losses resulting from poor land-use practices. In just 2 years, 4,863 m^3 of sediment were trapped behind 22 dams along a 622-m reach for an average of 7.8 m^3 of sediment stored per linear meter of stream (Scheffer 1938). The average sediment retention behind beaver dams was 221 m^3 and ranged from 34 m^3 to 586 m^3. A reanalysis of Scheffer's data shows a significant relationship between the surface area of the dam face and the amount of sediment stored ($r^2 = 0.30$, $p < 0.001$, $n = 22$). Removing the three statistical outliers greatly increases the strength of the correlation ($r^2 = 0.82$, $p < 0.001$, $n = 19$; Figure 3). These data suggest a general trend of increasing sediment storage with increasing dam size, with some notable exceptions. These exceptions could be explained by the age of the dams or the recent failure of a dam upstream.

These studies indicate that sedimentation rates behind beaver dams vary widely. Factors influencing sedimentation rates include growth rates of the emergent vegetation (which itself is determined by species composition, climate, and site productivity), upstream sediment loads (which is determined by geology, watershed land-use history, and disturbance history), the number of beaver dams upstream, and the frequency of dam failure for both the dam in question and any upstream dams. Given the high spatial and temporal variation for all of these factors, general predictions of sedimentation rates will be inaccurate and site-specific measurements will be required.

Sediment behind beaver dams is often a combination of fluvially transported material and organic material produced by the emergent vegetation growing on the pond edge. As ponds fill in with sediment, emergent vegetation is able to grow towards the pond center, helping to trap more sediment and producing more organic material, thereby accelerating the rate of filling. Before being completely filled, most ponds are abandoned, and dams often breach (Pastor et al. 1993), thus ending the accumulation of sediment. Although inorganic sediment accumulation may cease, however, wetland graminoids (sedges, rushes, grasses) colonize the exposed sediments on the pond edges, typically turning the former pond into a highly productive wet meadow and creating and depositing yet more organic matter on the former pond floor.

The observation that beaver dams trap sediment have led some scientists to conclude that the accumulation of such sediment over long periods can cause permanent changes in valley floor morphology. Rudemann and Schoonmaker (1938), Ives (1942), and Rutten (1967) argued that the numerous small, broad, relatively level surfaces at the bottom of recently deglaciated, U-shaped valleys were the result of sediment accumulation behind the numerous beaver dams built and maintained since the last glacial recession.

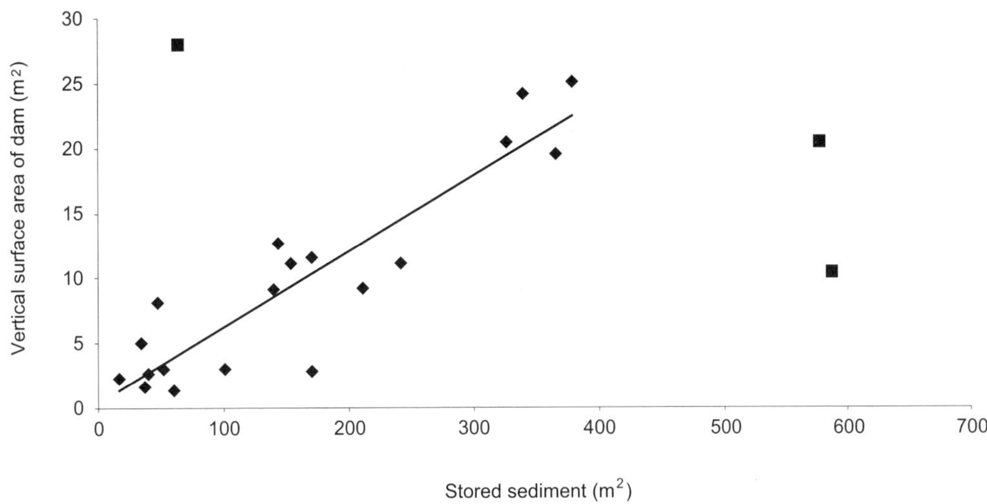

FIGURE 3. The relation between sediment stored and vertical dam face area is weak but significant ($r^2 = 0.30$, $p < 0.01$, $n = 22$). With three statistical outliers removed (large square symbols), the relation becomes much more apparent ($r^2 = 0.83$, $p < 0.01$, $n = 19$) (adapted from Scheffer 1938).

Ives observed that a competing hypothesis, that postglacial lake deposits formed the bottoms, was unlikely because the valleys were, in fact, gently sloping over long distances. Ives also dismissed the hypothesis that the flatness of the valleys was the result of the meandering of postglacial braided rivers on the grounds that such action would have left a more uneven topography because braided reaches differentially incise and because the valley fill was fine-textured, suggesting a depositional environment rather than the coarse-textured sediment that would be typical of a braided river transporting glacial outwash. Finally, Ives observed slight steps in some valley floors coincidental with the presence of abundant buried wood similar in size and shape to the wood of beaver dams.

Others, such as Rutherford (1964), have argued that although beaver dams have contributed to the accumulation of fine sediments on valley floors, the presence of underlying geologic features, such as glacial moraine or dykes, are the underlying control that allows flat valleys to form. We believe that both hypotheses have merit and are not mutually exclusive. However, the extent to which beaver dams control valley floor morphology (for example, slope) over long periods is still largely unknown.

Over short periods in small reaches, beaver dams create a stair-step valley and stream profile consisting of flat areas with abrupt gradient changes at dam sites. We suggest that, over long periods, beaver dams can change the longitudinal profile of streams such that they effectively change valley slope over long distances. This change could be accomplished if beaver dams at the mouth of a valley were consistently built higher or were more closely spaced to act as a series of step dams. Although the total gradient would stay the same, most of the valley would be gently sloping, with a steeply graded but stepped gradient near the mouth. Simple calculations suggest that, over long periods, significant morphological changes are possible.

Geometric relations can be used to estimate maximum potential sediment storage behind beaver dams:

$$V_m = 0.5H^2W/S ,$$

where V_m = maximum sediment storage volume, H = dam height, W = dam width, and S = stream slope. Thus, as an example, a dam rising 1.0 m above a 50-m-wide valley floor with a slope of 0.01 could store a maximum of 2,500 m^3 of sediment, or about 25 m^3/m of stream length, which on average would elevate the valley floor 0.5 m. Assuming that on average beaver build 1.0-m-high dams and that they are half filled every 50 years, a valley floor could be raised at a rate of 0.5 cm/year, or accumulate about 12 m (vertical) of sediment and organic material, over a period of 10,000 years. Measured sediment accumulation rates behind dams range from 4 to 30 cm/year over short time frames and from 0.25 to 6.5 cm/

year over decadal time scales (Scheffer 1938; Devito and Dillon 1993; Butler and Malanson 1995), suggesting that our calculations are conservative. On the other hand, sediment will compact under the weight of additional sediment, and a certain amount of sediment behind abandoned dams is exported if the stream incises into the deposits behind the former dam. Also, the cycle of beaver habitat colonization and abandonment can be at shorter intervals than 50 years (Warren 1932; Neff 1959). As a rough approximation, however, these calculations illustrate the theoretical capacity of beaver dams to substantially raise valley floor elevations over long periods. Studies of the cumulative geomorphic effects of beaver dams (and their widespread removal) would be worthwhile. Observations of meters-thick accumulations of compressed organic material exposed on the valley walls of incised streams, where beaver dams are known to have existed until recently, provide additional physical evidence to support the idea that a series of beaver dams built over long periods can raise valley floors (Pattie 1833, Fredlake 1997; Pollock, author's personal observation).

Other researchers have observed streams that historically contained beaver and well-vegetated floodplains and are now incised in many places (Winegar 1977; Parker et al. 1985; Elmore and Beschta 1987). The rapid incision that has occurred throughout much of the western United States and other parts of the world in recent centuries suggests that base elevations were not controlled by hard geological features, such as bedrock, but by "soft" grade-control structures like beaver dams, large wood, or the dense roots of streamside vegetation (for example, Zierholz et al. 2001). In the western United States, the coincidence of widespread incision with the arrival of the first white settlers suggests that one of their early land-use activities may have been the causal agent. Removal of beaver by trappers was one of the first major land-use changes to occur in the United States following European settlement (Mackie 1997). Rapidly removing millions of beaver resulted in the loss of millions of functioning dams throughout the West. For small streams, this loss must have resulted in a series of catastrophic floods, as dams ruptured during high flows and destroyed downstream dams that were acting as "roughness elements" (sensu Lisle 1982) slowing stream velocities and dissipating energy.

The effects of catastrophic beaver dam failures are documented (Butler 1989; Stock and Schlosser 1991). Often channel incision occurs just downstream of the dam and fine sediment is deposited farther downstream, along with uprooted plant material and organic matter. Large wood from streams has also resulted in the loss of sediment and the rapid downcutting of channel beds (Smith et al. 1993). Recent restoration efforts using large wood have elevated the base of streambeds (Slaney and Zaldokas 1997).

These studies taken as a whole suggest that beaver dams can substantially elevate and maintain wide, low-gradient valley floors over long periods and that these geomorphic features should persist as long as healthy populations of beaver are available to maintain the dams. The available evidence also suggests that the repeated building of beaver dams, the subsequent backfilling with sediment and organic matter, and the building of new dams on top of old fill can create valley and stream-slope elevational profiles that differ from the underlying slope.

Whether the widespread elimination of beaver dams is the ultimate cause of the recent incision that has occurred in many streams in the western United States has not been determined. But this hypothesis is consistent with available information: indirect support for it comes from the work of land managers who are using beaver dams—or small artificial dams—as a tool to aggrade incised streams.

In Wyoming, Apple et al. (1983) provided cut cottonwood branches to recently relocated beaver who subsequently built dams that aggraded incised channels. The initial dams quickly backfilled with sediment, and the beaver continued to build additional structures upstream. In Mission Creek, Washington, relocated beaver built dams and trapped thousands of cubic meters of sediment in just a few years (Scheffer 1938). Similar human-built structures have also yielded good results. In Sheep Creek, Utah, a 5-m-high retaining dam with a storage volume of 1.09×10^5 m^3 of sediment was built in 1961 and filled up in just 1 year. The dam continues to cause aggradation upstream of the storage area and, by 1984, had retained a total volume of 4.39×10^5 m^3 of sediment (Haveren et al. 1987). A formerly incised xeric valley is now supporting abundant riparian vegetation on a gently sloping valley floor.

On a smaller scale, in Alkali Creek, Colorado, Heede (1978) built small concrete "check" dams on small ephemeral streams containing no riparian vegetation. The barriers quickly filled with sediment and now support substantial amounts of riparian vegetation because the dams create localized areas of groundwater recharge. The vegetation itself helps to trap additional sediment, thus creating a positive feedback of valley aggradation. Together, these studies suggest that, where small dams exist, whether built by beavers or people, they can cause localized aggradation of valley floors and infilling of incised valleys, provided a sufficient supply of sediment is upstream. Additional studies to determine the feasibility of using beaver dams to initiate aggradation of incised valleys would be useful.

Beaver Dams and Fish

Effect of dams on communities and populations

When beaver impound streams by building dams, they substantially alter stream hydraulics that may benefit many fish species (Murphy et al. 1989; Snodgrass and Meffe 1998). More than 80 North American fishes have been documented in beaver ponds, with 48 species commonly using them (Table 3). Fish use of beaver ponds in Eurasia is not as well documented. In Sweden, Hagglund and Sjoberg (1999) observed that minnow *Phoxinus phoxinus* and brown trout *Salmo trutta* used beaver ponds, while burbot *Lota lota* and northern pike *Esox lucius* were found in the ponds occasionally. In general however, studies of the effects of the Eurasian beaver on fish populations are not common, and data are often extrapolated from studies of the North American beaver (for example, see Collen and Gibson 2001). Because beaver ponds usually have slow current velocities and large edge-to-surface-area ratios, they provide extensive cover to fish and a productive environment for both vegetation and aquatic invertebrates, affording fish foraging opportunities not found in unimpounded stream habitat (Hanson and Campbell 1963; Keast and Fox 1990). Fish expend less energy when foraging in slow water. Thus sections of streams impounded by beaver dams are often more productive than unim-pounded reaches in both the number and size of fish (for example, Gard 1961b; Hanson and Campbell 1963; Murphy et al. 1989; Leidholt Bruner et al. 1992; Schlosser 1995). Fish are not the only beneficiaries of beaver dams. Relative to unimpounded reaches, increases in either biomass or diversity have also been observed for a wide range of taxa, including birds, mammals, amphibians, plants, and insects in areas affected by beaver (see reviews in Naiman et al. 1988; Pollock et al. 1994).

The physical components of beaver dams affect a variety of fish species. Much of the research on beaver–fish interactions has been concerned with how salmonids are affected by beaver ponds, primarily brook trout *Salvelinus fontinalis* and juvenile coho salmon *Oncorhynchus kisutch* (however, see Hanson and Campbell 1963; Keast and Fox 1990; Schlosser 1995; Snodgrass and Meffe 1999). Some early observations of the increased siltation in beaver ponds, along with increased temperatures caused by the loss of shade-producing vegetation, led some to conclude that beaver ponds could be detrimental to salmonid populations (Salyer 1935; Reid 1952; Knudsen 1962). These opinions were based on subjective observations and assumptions that fine sediment deposited behind beaver dams smothered redds and decreased benthic invertebrate production, that higher summer temperatures found in ponds were detrimental, and that the flooding of vegetation led to water quality problems such as low dissolved oxygen levels.

Although these researchers expressed concern that beaver dams are harmful to trout, no study has ever demonstrated a detrimental population-level effect of dams on salmonids, nor has a study shown that beaver dams are more than a seasonal barrier to fish movement. On the contrary, most studies support the contention that the pond habitat formed by beaver dams is highly beneficial to many fishes and that species regularly cross dams in both upstream and downstream directions (Rupp 1954; Huey and Wolfrum 1956; Gard 1961b; Hanson and Campbell 1963; Call 1970; Bustard and Narver 1975; Swales et al. 1988; Murphy et al. 1989; Keast and Fox 1990; Leidholt Bruner et al. 1992; Schlosser 1995; Snodgrass and Meffe 1998, 1999).

Comparing salmonid productivity in stream reaches above beaver dams with reaches where beaver dams are absent generally demonstrate that the reaches above beaver dams produce either more fish or larger fish or both. In a detailed study of the effects of beaver dams on trout in Sagehen Creek in the Sierra Nevada of California, Gard (1961b) found that the number of trout (brook, rainbow *O. mykiss*, and brown) were about the same over a 3-year period for a dam-affected reach and a "control" reach of a similar length (67 m), but that the size of the trout in the reach above the beaver dam were much larger.

TABLE 3. Common and scientific names of fish species known to use beaver ponds. The relevant scientific papers are also noted.

Species	Common name	Abundance[a]	References[b]
Aphredoderus sayanus	Pirate perch	C	1
Campostomus anomalum	Central stoneroller	C	2
Catostomus commersoni	White sucker	C	2,7,9
Centrarchus macropterus	Flier	C	1
Phoxinus eos	Northern redbelly dace	C	3,4,7,15
Culaea inconstans	Brook stickleback	C	3,4
Enneacanthus chaetodon	Blackbanded sunfish	C	1
Enneacanthus gloriosus	Bluespotted sunfish	C	1
Erimyzon oblongus	Creek chubsucker	C	1
Esox americanus americanus	Redfin pickerel	C	1
Esox niger	Chain pickerel	C	1
Etheostoma serrifer	Sawcheek darter	C	1
Fundulus diaphanus	Banded killifish	C	3
Fundulus lineolatus	Lined topminnow	C	1
Gambusia holbrooki	Eastern mosquitofish	C	1
Hybognathus hankinsoni	Brassy minnow	C	4
Ameiurus melas	Black bullhead	C	2,4,5
Lepomis auritus	Redbreast sunfish	C	1
Lepomis cyanellus	Green sunfish	C	2
Lepomis gibbosus	Pumpkinseed sunfish	C	3,4
Lepomis gulosus	Warmouth	C	1
Lepomis marginatus	Dollar sunfish	C	1
Lepomis punctatus	Spotted sunfish	C	1
Notropis cummingsae	Dusky shiner	C	1
Notropis heterodon	Blackchin shiner	C	3,4
Notropis lutipinnis	Yellowfin shiner	C	1
Luxilus umbratilis	Redfin shiner	C	2
Oncorhynchus clarkii	Cutthroat trout	C	6, 9
Oncorhynchus kisutch	Coho salmon	C	6, 10,12,13[c]
Oncorhynchus nerka	Sockeye salmon	C	6
Phoxinus neogaeus	Finescale dace	C	4
Phoxinus phoxinus	Minnow	C	16
Pimephales promelas	Fathead minnow	C	2,3,4,16
Pungitius pungitius	Ninespine stickleback	C	7,15
Salmo trutta	Brown trout	C	8,17
Salvelinus fontinalis	Brook trout	C	8,9,12,16[c]
Salvelinus malma	Dolly Varden char	C	6,10,12
Semotilus atromaculatus	Creek chub	C	1,2,4,5,7,16
Semotilus coporalis	Fallfish	C	7
Cottus spp.	Sculpins	LC	6
Esox lucius	Northern pike	LC	4,14,17
Gasterosteus aculeatus	Threespine stickleback	LC	6, 12
Micropterus salmoides	Largemouth bass	LC	1,2,4,5
Notemigonus crysoleucas	Golden shiner	LC	1,2,3,16
Luxilus cornutus	Common shiner	LC	2,4
Notropis heterolepis	Blacknose shiner	LC	4
Notropis topeka	Topeka shiner	LC	2
Perca flavescens	Yellow perch	LC	3,4,16
Prosopium spp.	Whitefish	LC	6
Ameiurus natalis	Yellow bullhead	U	1,2

TABLE 3. Continued.

Species	Common name	Abundance[a]	References[b]
Ameiurus platycephalus	Flat bullhead	U	1
Acantharchus pomotis	Mud sunfish	U	1
Ameiurus nebulosus	Brown bullhead	U	3
Lepomis humilis	Orangespotted sunfish	U	2
Lota lota	Burbot	U	16
Cyprinella lutrensis	Red shiner	U	2
Noturus leptacanthus	Speckled madtom	U	1
Notropis petersoni	Coastal shiner	U	1
Oncorhynchus mykiss	Steelhead	U	6, 10
Oncorhynchus mykiss	Rainbow trout	U	8, 9
Oncorhynchus tshawytscha	Chinook salmon	U	6, 10
Pimephales notatus	Bluntnose minnow	U	2,3
Prosopium williamsoni	Mountain whitefish	U	10
Margariscus margarita	Pearl dace	O	16
Amia calva	Bowfin	O	1
Anguilla rostrata	American eel	O	1
Elassoma zonatum	Banded pygmy sunfish	O	1
Etheostoma exile	Iowa darter	O	3
Etheostoma fricksium	Savannah darter	O	1
Etheostoma fusiforme	Swamp darter	O	1
Etheostoma nigrum	Johnny darter	O	2
Etheostoma olmstedi	Tessellated darter	O	1
Etheostoma spectabile	Orangethroat darter	O	2
Lepomis macrochirus	Bluegill	O	2
Minytrema melanops	Spotted sucker	O	1
Nocomis leptocephalus	Bluehead chub	O	1
Notropis dorsalis	Bigmouth shiner	O	2
Noturus gyrinus	Tadpole madtom	O	1
Oncorhynchus gorbuscha	Chum salmon	O	6
Percina caprodes	Logperch	O	2
Umbra limi	Central mudminnow	O	3
Umbra pygmaea	Eastern mudminnow	O	1

[a] Species sorted by frequency of use: C = common, LC = locally common, U = uncommon, O = occasional.
[b] Reference key: 1 = Snodgrasss and Meffee 1998; 2 = Hanson and Campbell 1963; 3 = Keast and Fox; 1990, 4 = Schlosser 1995; 5 = Stock & Shlosser 1991; 6 = Murphy et al. 1989; 7 = Rupp 1954; 8 = Gard 1961a, 1961b; 9 = Huey and Wolfrum 1956; 10 = Swales et al. 1988; 11 = Bryant 1983; 12 = Call 1970; 13 = Bustard and Narver 1975; 14 = Knudsen 1962; 15 = Rupp 1954; 16 = Balon and Chadwick 1979; Hagglund and Sjoberg 1999.
[c] For brook trout and coho salmon use, there is an abundance of additional literature, referenced in the text.

The average length in the dam-affected reach was 147 mm, and the average weight was 117 g; trout lengths in the control reach averaged 110 mm, and their average weight was 22 g. The brook trout and brown trout generally benefited more from the beaver ponds because they could feed on the pond's bottom fauna. After floods destroyed the dams, the brown trout population crashed and rainbow trout became the dominant species (Gard and Seegrist 1972).

Similarly, Huey and Wolfrum (1956) compared summer trout populations in three stream reaches upstream of beaver dams with three control reaches in the upper Red River, New Mexico and found that the dam reaches contained many more trout (mostly cutthroat *O. clarki* and brook trout, but some rainbow). Average trout populations in the three control reaches ranged from 12 to 80 (mean = 40) and, in the dam reaches, ranged from 131 to 218 (mean = 167). All the reaches were about 46 m long. Other studies have noted the abundance of trout upstream of beaver dams but did not directly compare them with unimpounded reaches. Call (1970), studying the effects of beaver on headwater streams in Wyoming, noted that beaver created trout habitat, where previously none existed, by damming very small streams and seeps, substantially increasing available brook trout habitat and allowing for the development of a productive fishery. Similarly, Gard (1961a) built small artificial dams to mimic beaver dams in the headwaters of Sagehen Creek, California and created a productive brook trout fishery where none had previously existed. Some of the increased salmonid productivity may be a result of the increased numbers of forage fish in beaver ponds. Rupp (1954) observed in a beaver pond in Maine that increased numbers of small fish, primarily three-spine stickleback *Gasterosteus aculeatus* and northern redbelly dace *Phoxinus eos*, were used as forage by brook trout.

Beaver ponds may be particularly important habitat for overwintering resident char and trout. Both Chisholm et al. (1987) and Cunjak (1996) observed that brook trout had a strong tendency to move into the slow water habitat of beaver ponds to overwinter. Likewise, Jakober (1995) observed that, in Montana streams, bull trout *Salvelinus confluentus* and cutthroat trout aggregated in large numbers to overwinter in beaver ponds. As temperatures drop in the autumn, other fishes appear to aggregate and seek out slow water habitat in which to overwinter, but it is not clear whether or not they utilize or prefer beaver ponds (Craig 1978; Paragamian 1981).

In anadromous-fish streams along the west coast of North America, comparisons of the growth and survival of juvenile coho salmon and other salmonids between reaches upstream of beaver dams and nonimpounded reaches have yielded similar results. Swales et al. (1988) compared populations of juvenile coho, chinook *O. tshawytscha*, and steelhead *O. mykiss* in side-channels impounded by beaver with the mainstem of the Coldwater River in British Columbia and found that the side channels upstream of beaver dams were heavily used by overwintering coho, but less so by chinook and steelhead. Over the course of 3 years, these researchers captured and measured coho in both habitats monthly and found that the coho upstream of the beaver dam were consistently larger, more abundant, and grew faster than those downtream. Juvenile coho were not only using the ponds for overwintering habitat, but as important refuge and rearing areas throughout the year. Murphy et al. (1989) studied summer use of mainstem and off-channel habitat in the valley floor of the Taku River, Alaska and found the highest densities of juvenile coho in reaches upstream of beaver dams (0.59 per m^2) and virtually all the larger coho were in beaver ponds (Figure 4). These reaches accounted for just 0.7% of the total available habitat; yet, 34% of all the juvenile coho were found there. Consistent with other studies, the coho in these reaches were longer than coho found in other habitat types. Murphy et al. also found that juvenile sockeye used reaches upstream of beaver dams, averaging 0.48 per m^2, and that these fish too were larger and grew faster than fish using other instream habitats. Similarly, Leidholt-Bruner et al. (1992) found that summertime densities of juvenile coho in "pools" upstream of beaver dams (0.34 per m^2) were higher than in pools formed by other obstructions (0.26 per m^2). Studying three small coastal streams in the islands of southeast Alaska, Bryant (1983) found that populations of coho juveniles in summer were significantly greater in impounded reaches compared with reaches just upstream and downstream, but that the densities in the impounded reach were lower because the beaver dams had greatly expanded the surface area of the stream. The highest population of Dolly Varden *Salvelinus malma* char was found in one of the impounded reaches, but no consistent pattern among all streams was found. In Carnation Creek, British Columbia, Bustard and Narver (1975) found that survival rate of overwintering juvenile coho in old beaver ponds was about twice as high as the 35% estimated for the entire stream system.

Other studies have found that many fish species, besides salmonids, are abundant in beaver ponds. Hanson and Campbell (1963), in a study of headwater streams in north Missouri, determined that the presence of beaver dams greatly increased the productivity of fishes compared to unimpounded reaches. They used rotenone to sample three beaver ponds and one "natural" pool to determine biomass and species composition for each of the habitats. The three beaver ponds had an average biomass of 2.8 g/m^2, and the natural pool had a biomass of 1.1 g/m^2. Hanson and Campbell also found 21 species in

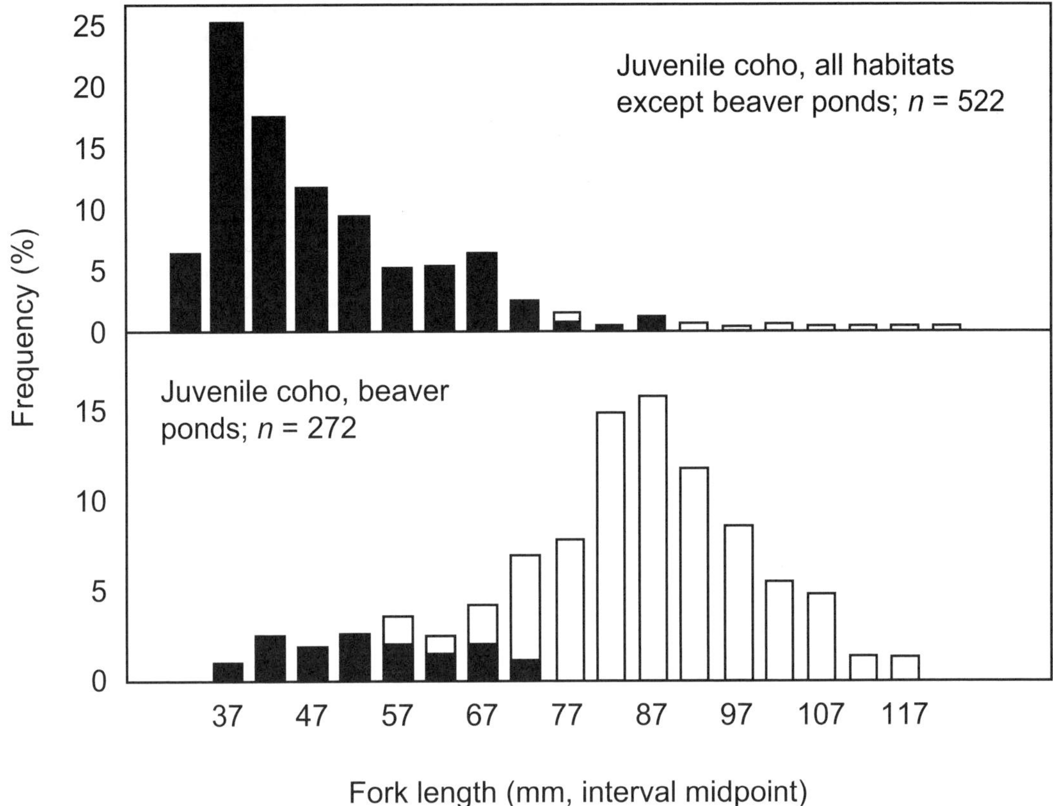

FIGURE 4. Size-frequency distribution of juvenile coho in main channel and off-channel habitat in the Taku River, southeast Alaska, showing that larger coho (age-1 light columns, age-0 dark columns) overwhelmingly prefer beaver ponds over any other habitat. Beaver ponds account for just 0.7% of the total instream habitat area in the Taku River floodplain (adapted from Murphy et al. 1989).

the beaver ponds, a 50% increase in number over the 14 species they found in an unimpounded reach nearby.

Snodgrass and Meffe (1998), in a study of fish community use of beaver ponds in the southeastern United States, found that beaver dams in first- and second-order streams had higher fish diversity (32 and 38 species, respectively), than in third-order streams (26 species). They concluded that, by creating pool habitat in small headwater streams where it is generally absent, beaver dams increase fish diversity. They also concluded that the generally positive relation observed between species richness and drainage area is a recent phenomenon resulting from the extirpation of beavers from their historical range. Finally, they found that headwater stream reaches upstream of beaver dams tended to be dominated by a few predators, suggesting that the deeper, more open water created more predation opportunities.

Schlosser (1995) concluded that beaver ponds in a headwater stream in northern Minnesota provided a source for fish populations, while adjacent stream environments acted as "sinks." Keast and Fox (1990) found a high amount of habitat and dietary specialization among the common fish species observed in an Ontario beaver pond. For example, Iowa darters were found only over bare (unsilted) substrate; one of the preferred habitats of the blackchin shiner *Notropis heterodon* was the open, silted area near the pond inlet; and fathead minnows *Pimephales promelas* were generally found in the old creek channel. They also determined that fish species richness and body size were smaller in a beaver pond relative to more permanent lentic water bodies in the area.

Impact of beaver dams on spawning

A number of observers have speculated that beaver dams can affect fish populations by covering spawning sites with silt or deep, slow water or by blocking access to upstream spawning grounds (Knudsen 1962; Call 1970; Swanston 1991; Cunjak and Therrien 1998). While beaver dams have led to the siltation of spawning habitat and probably restrict access to spawning grounds for some species, there is little evidence of negative population-level effects. Because beaver ponds trap sediments and dampen floods, siltation and scouring of spawning gravels further downstream may be reduced, making determination of an overall negative population effect problematic (Scheffer 1938; Beedle 1991; Dunaway et al. 1994; Butler and Malanson 1995; Hering et al. 2001). Further, some species that require clean spawning gravels, such as brook trout, cutthroat trout, and coho salmon, extensively use beaver ponds as rearing habitat (Gard 1961b; Call 1970; Murphy et al. 1989). However, where spawning habitat is limiting, beaver ponds may affect fish populations. As an example, Rabe (1970) found that beaver ponds with little accessible spawning sites nearby contained fewer, larger brook trout than those ponds where nearby spawning habitat was abundant. Similarly, Balon and Chadwick (1979) observed that three lithophilic fish species, two *Semotilus* species, and brook trout, creek chub *Semotilus atromaculatus*, and *Semotilus margarita*, needing gravel or rock substrate for their early development (for example, spawning and egg incubation), did not reproduce in an Ontario lake after construction of a beaver dam that raised the lake level. Three other species, yellow perch *Perca flavescens*, fathead minnow, and golden shiner *Notemigonus crysoleucas*, were not lithophilic and were able to successfully reproduce.

Beaver dams as a barrier to fish movement

Studies examining the permeability of beaver dams to fish passage suggested that many species are able to gain passage, but they are seasonally restricted by low flows. Gard (1961b) found that all three species that they studied (brown, brook, and rainbow trout) were able to move both upstream and downstream across a series of 14 dams. Brown trout crossed dams more frequently than the others; rainbow trout showed greater ability to cross a series of dams, with some crossing an entire series of 14 dams. Movement upstream and downstream was equal for all species. In a less detailed study, Rupp (1954) monitored the movement of brook trout in a series of five beaver dams in the Sunkhaze Stream, Maine. Rupp set up two monitoring stations, one above all the dams and one below, and found substantial movement across the entire dam series. In contrast to Gard, Rupp found more downstream movement than upstream movement, and he could not correlate movement with streamflow. In Gould Creek, Minnesota, Schlosser (1995) studied the movement of a 12-species fish assemblage in beaver ponds and observed that downstream movement of fish was weakly correlated with elevated stream flows that coincided with the approximate stage at which water began moving over and around a dam. Upstream movement, however, was not correlated with streamflow, and fish moved up over dams across a broad range of flow conditions.

On the west coast of North America, the question of whether beaver dams are a barrier to anadromous salmonid movement has long been debated, and even today, some natural resource agencies remove beaver dams to "enhance" salmonid habitat. Personal observations and published literature suggest that both adults and juveniles of coho, steelhead, sea-run cutthroat, Dolly Varden, and sockeye are able to cross beaver dams (Bryant 1983; Swales et al. 1988; Murphy et al. 1989). In a study of out-migration from a floodplain beaver pond on the Coldwater River, British Columbia, Swales et al. (1988) monitored 1,257 coho salmon smolts, 62 steelhead trout (mean length = 129 mm), and 3 Dolly Varden char moving downstream over a period of 2.5 months in the spring. Bryant (1983) found abundant populations of young-of-the-year and age-one coho juveniles in the slow water areas behind beaver-dammed streams in southeast Alaska. He concluded that the abundance of young of the year indicated that adult coho were able to cross the dams (including one dam 2.1 m high), on the presumption that such small fish (<52 mm) do not move upstream across beaver dams in great numbers. Little information has been published about whether the adults of chum salmon are able to cross beaver dams, but the general consensus among salmon fisheries managers is that beaver dams can be an obstacle to upstream chum salmon movement. Anecdotal evidence also suggests that beaver dams can be an obstacle to the upstream movement of adult Atlantic salmon *Salmo salar* (Cunjak and Therrien 1998).

Conclusions

The widespread removal of beaver dams across small and medium-sized streams throughout most of the Northern Hemisphere has caused substantial changes to stream hydrology and geomorphology, resulting in dramatic shifts in the composition of communities. Because beaver build dams on small streams, their presence creates slow water pool habitat in a portion of the network where such habitat is often uncommon or nonexistent. This allows fish depending on pool habitat to move farther up the system, sometimes to reaches previously inhospitable. The presence of beaver dams often, though not always, creates fish habitat with higher productivity or diversity. Certain species, such as brook trout and juvenile coho salmon, rely heavily on beaver ponds, and the loss of such habitat has undoubtedly affected populations.

Some studies suggest that beaver dams can restore perennial flow to intermittent streams, but the conditions under which such a transformation can be expected have not been well described. Our calculations suggest that the ability of beaver dams to increase aquifer recharge depends largely on the hydraulic and geometric characteristics of the aquifer. Manmade structures similar in size to beaver dams that have been shown to improve stream flow lend support to our contention that beaver dams can also increase stream flows. Based on our observations of stream systems in the western United States, it is likely that many intermittent streams could have perennial flow restored if beaver colonize them.

Observations of extensive sediment accumulation behind beaver dams lend support to the hypothesis that beaver, acting over long periods, can significantly change valley floor morphology. The potential for sediment accumulation behind beaver dams over the course of millennia is tremendous, especially considering that, until recently, few checks existed on their population outside of habitat or food limitations. There is some evidence to suggest that much of the recent stream incision throughout the western United States is at least partially a result of the widespread loss or removal of beaver dams. The historical records and the banks of incised streams provide evidence that beaver inhabited many of these streams at one time. The cause of this recent incision is still unclear, but the hypothesis that it might be the cumulative result of removing millions of beaver dams is at least consistent with available scientific information about the timing, location, and extent of incision. Studies of incision and the possible use of beaver in restoring incised streams in other arid parts of the beaver's former range, such as in southern Europe and parts of Asia, would be useful.

For all the scientific studies of beaver, very few have addressed the cumulative effects of the widespread dam-removal "experiment" conducted across North America over the past few centuries and across Eurasia over the past millennium. It is likely that the hydrologic, geomorphologic, and biological cumulative effects are, and continue to be, substantial. Watershed-scale experiments to assess the effects of restoring beaver dams to move historical numbers would greatly aid our understanding of how stream systems historically functioned and perhaps lead to innovative large-scale stream-restoration strategies.

References

Aleksiuk, M. 1970. The seasonal food regime of arctic beavers. Ecology 51:264–270.

Apple, L. L. 1985. Riparian habitat restoration and beavers. In Riparian ecosystems and their management: reconciling conflicting uses. USDA Forest Service, Rocky Mountain Forest & Range Experiment Station, General Technical Report RM-120, Fort Collins, Colorado.

Apple, L. L., B. H. Smith, J. D. Dunder, and B. W. Baker. 1983. The use of beavers for riparian/aquatic habitat restoration of cold desert, gully-cut stream systems in southwestern Wyoming. Pages 123–130 *in* Annual proceedings of the American Fisheries Society/Wildlife Society joint chapter meeting, February 8–10, Logan, Utah.

Balon, E. K., and E. M. P. Chadwick. 1979. Reclamation of a perch lake: a case study using density estimates and the guild concept. Archiv fuer Hydrobiologie 85.

Beedle, D. 1991. Physical dimensions and hydrologic effects of beaver ponds on Kuiu Island in southeast Alaska. Master's thesis. Oregon State University, Corvallis.

Beier, P., and R. H. Barrett. 1987. Beaver habitat use and impact in Truckee River basin. California Journal of Wildlife Management 51:794–799.

Bergerud, A. T., and D. R. Miller. 1977. Population dynamics of Newfoundland beaver. Canadian Journal of Zoology 55:1480–1492.

Brown, J. B. 1963. The role of geology in a unified conservation program, Flat Top Ranch, Bosque

County, Texas. Baylor Geological Studies Bulletin 5:1–29.

Bryant, M. D. 1983. The role of beaver dams as coho salmon habitat in southeast Alaska streams. Pages 183–192 *in* J. M. Walton and D. B. Houston, editors. Proceedings of the Olympic Wild Fish Conference. Olympic Wild Fish Conference, Port Angeles, Washington.

Bustard, D. R., and D. W. Narver. 1975. Aspects of the winter ecology of juvenile coho salmon (*Oncorhynchus kisutch*) and steelhead trout (*Salmo gairdneri*). Journal of the Fisheries Resource Board of Canada 32:667–680.

Butler, D. R. 1989. The failure of beaver dams and resulting outburst flooding: a geomorphic hazard of the southeastern Piedmont. Geographical Bulletin 31:29–38.

Butler, D. R., and G. P. Malanson. 1995. Sedimentation rates and patterns in beaver ponds in a mountain environment. Geomorphology 13(1–4):255–269.

Call, M. W. 1970. Beaver pond ecology and beaver-trout relationships in southeastern Wyoming. Doctoral dissertation. University of Wyoming, Laramie.

Chisholm, I. M., W. A. Hubert, and T. A. Wesche. 1987. Winter stream conditions and use of habitat by brook trout in high-elevation Wyoming streams. Transactions of the American Fisheries Society 116:176–184.

Collen, P., and R. J. Gibson. 2001. The general ecology of beavers (*Castor* spp.) as related to their influence on stream ecosystems and riparian habitats, and the subsequent effects on fish - a review. Reviews in Fish Biology and Fisheries 10:439–461.

Craig, P. C. 1978. Movements of stream resident and anadromous arctic char (*Salvelinus alpinus*) in a perennial spring of the Canning River, Alaska. Journal of the Fisheries Resource Board of Canada 35:48–52.

Cunjak, R. A. 1996. Winter habitat of selected stream fishes and potential impacts from land-use activity. Canadian Journal of Fisheries and Aquatic Sciences 53:267–282.

Cunjak, R. A., and J. Therrien. 1998. Inter-stage survival of wild juvenile Atlantic salmon, *Salmo salar* L. Fisheries Management and Ecology 5:209–223.

Dalke, P. D. 1947. The beaver in Missouri. Missouri Conservationist 8:1–3.

Danilov, P. I., and V. Y. Kan'shiev. 1983. The state of populations and ecological characteristics of European (*Castor fiber* L.) and Canadian (*Castor canadensis* Kuhl.) beavers in the northwestern USSR. Acta Zool Fenn. Helsinki: Finnish Zoological Publishing Board 174:95–97.

Devito, K. J., and P. J. Dillon. 1993. Importance of runoff and winter anoxia to the P and N dynamics of a beaver pond. Canadian Journal of Fisheries and Aquatic Sciences 50:2222–2234.

Dunaway, D., S. R. Swanson, J. Wendel, and W. Clary. 1994. The effect of herbaceous plant communities and soil textures on particle erosion of alluvial streambanks. Geomorphology 9:47–56.

Dunne, T., and L. B. Leopold. 1978. Water in environmental planning. Freeman, New York.

Ehrman, T. P., and G. A. Lamberti. 1992. Hydraulic and particulate matter retention in a 3rd-order Indiana stream. Journal of the North American Benthological Society 11:341–349.

Elmore, W., and R. L. Beschta. 1987. Riparian areas: perceptions in management. Rangelands 9:260–265.

Finley, W. L. 1937. Beaver - conserver of soil and water. Transactions of the American Wildlife Conference 2:295–297.

Fredlake, M. 1997. Re-establishment of North American Beaver (*Castor canadensis*) into the San Pedro Riparian National Conservation Area. U.S. Bureau of Land Management, Environmental Assessment EA no. AZ-060–97-004, Tucson, Arizona.

Gard, R. 1961a. Creation of trout habitat by constructing small dams. Journal of Wildlife Management 52:384–390.

Gard, R. 1961b. Effects of beaver on trout in Sagehen Creek, California. Journal of Wildlife Management 25:221–242.

Gard, R., and D. W. Seegrist. 1972. Abundance and harvest of trout in Sagehen Creek, California. Transactions of the American Fisheries Society 3:463–477.

Gurnell, A. M. 1998. The hydrogeomorphological effects of beaver dam-building activity. Progress in Physical Geography 22:167–189.

Hagglund, A., and G. Sjoberg. 1999. Effects of beaver dams on the fish fauna of forest streams. Forest Ecology and Management 115:259–266.

Halley, D. J., and F. Rosell. 2002. The beaver's reconquest of Eurasia: status, population development and management of a conservation success. Mammalian Reviews 32:153–178.

Hanson, W. D., and R. S. Campbell. 1963. The effects of pool size and beaver activity on distribution and abundance of warm-water fishes in a north Missouri stream. American Midland Naturalist 69:137–149.

Hartman, G. 1994. Long-term population development of a reintroduced beaver (*Castor fiber*) population in Sweden. Conservation Biology 8:713–717.

Haveren, B. P., W. L. Jackson, and G. C. Lusby. 1987. Sediment deposition behind Sheep Creek Barrier Dam, southern Utah. Journal of Hydrology (NZ) 26:185–196.

Heede, B. H. 1978. Designing gully control systems for eroding watersheds. Environmental Management 2:509–522.

Heede, B. H., and L. F. DeBano. 1984. Gully rehabilitation–a three-stage process in a sodic soil. Journal of the Soil Society of America 48:1416–1422.

Hering, D., M. Gerhard, E. Kiel, T. Ehlert, and T. Pottgiesser. 2001. Review: study on near-natural conditions of central European mountain streams, with particular reference to debris and beaver dams: results of the REG meeting 2000. Limmologica 31:81–92.

Howard, R. J., and J. S. Larson. 1985. A stream habitat classification system for beaver. Journal of Wildlife Management 49:19–25.

Huey, W. S., and W. H. Wolfrum. 1956. Beaver-trout relations in New Mexico. Progressive Fish-Culturist 18:70–74.

Ives, R. L. 1942. The beaver-meadow complex. Journal of Geomorphology 5:191–203.

Jakober, M. J. 1995. Autumn and winter movement and habitat use of resident bull trout and westslope cutthroat trout in Montana. Master's thesis. Montana State University, Bozeman.

Jaksic, F. M., J. A. Iriarte, J. E. Jimenez, and D. R. Martinez. 2002. Invaders without frontiers: cross-border invasions of exotic animals. Biological Invasions 4:157–173.

Jenkins, S. H. 1979. Castor canadensis. Mammalian Species 120:1–8.

Jester, D. B., and J. H. McKirdy. 1966. Evaluation of trout stream improvement in New Mexico. Annual Conference of the Western Association of State Game and Fish Commissioners 26:79–81.

Johnston, C. A., and R. J. Naiman. 1990a. Aquatic patch creation in relation to beaver population trends. Ecology 71:1617–1621.

Johnston, C. A., and R. J. Naiman. 1990b. The use of a geographic information system to analyze long-term landscape alteration by beaver. Landscape Ecology 4:5–19.

Keast, A., and M. G. Fox. 1990. Fish community structure, spatial distribution and feeding ecology in a beaver pond. Environmental Biology of Fishes 27:201–214.

Knudsen, G. J. 1962. Relationship of beaver to forests, trout and wildlife in Wisconsin. Wisconsin Department of Natural Resources Technical Bulletin 25:1–50.

Leidholt Bruner, K., D. E. Hibbs, and W. C. McComb. 1992. Beaver dam locations and their effects on distribution and abundance of coho salmon fry in two coastal Oregon streams. Northwest Science 66:218–223.

Leopold, A. S. 1972. Wildlife in Mexico. University of California Press, Berkeley.

Li, R. M., and H. W. Shen. 1973. Effects of tall vegetation on flow and sediment. Journal of the Hydraulics Division, Proceedings of the American Society of Civil Engineers 91:793–814.

Lisle, T. E. 1982. Roughness elements: a key resource to improve anadromous fish habitat. Pages 93–98 *in* T. J. Hassler, editor. Proceedings: Propagation, Enhancement, and Rehabilitation of Anadromous Salmonid Populations and Habitat in the Pacific Northwest Symposium. California Cooperative Fisheries Research Unit, Humboldt State University, Arcata, California.

Lowry, M. M. 1993. Groundwater elevations and temperature adjacent to a beaver pond in central Oregon. Master's thesis. Oregon State University, Corvallis.

MacDonald, D. W., F. H. Tattersall, E. D. Brown, and D. Balharry. 1995. Reintroducing the European beaver to Britain: nostalgic meddling or restoring biodiversity? Mammalian Reviews 25:161–200.

Mackie, R. S. 1997. Trading beyond the mountains. UBC Press, Vancouver.

Mai, H. 1978. Studies on the jaws of the Pleistocene beavers Trogontherium- and Castor- and their stratigraphic classification. Schriften des Naturwissenschaftlichen Vereins fuer Schleswig-Holstein 48:35–39.

McCall, T. C., P. Hodgman, D. R. Diefenbach, and R. B. Owen. 1996. Beaver populations and their relation to wetland habitat and breeding waterfowl in Maine. Wetlands 16:163–172.

Medwecka-Kornas, A., and R. Hawro. 1993. Vegetation on beaver dams in the Ojcow National Park (southern Poland). Phytocoenologia 23:611–618.

Murphy, M. L., J. Heifetz, J. F. Thedinga, S. W. Johnson, and K. V. Koski. 1989. Habitat utilization by juvenile Pacific salmon (*Oncorhynchus*) in the glacial Taku River, southeast Alaska. Canadian Journal of Fisheries and Aquatic Sciences 46:1677–1685.

Naiman, R. J., C. A. Johnston, and J. C. Kelley. 1988. Alteration of North American streams by beaver. Bioscience 38:753–761.

Naiman, R. J., D. M. McDowell, and B. S. Farr. 1984. The influence of beaver (*Castor canadensis*) on the production dynamics of aquatic insects. Proceedings of the Congress of the International Association of Limnology 22:1801–1810.

Naiman, R. J., J. M. Melillo, and J. E. Hobbie. 1986. Ecosystem alteration of boreal forest streams by beaver (*Castor canadensis*). Ecology 67:1254–1269.

Naiman, R. J., G. Pinay, C. A. Johnston, and J. Pastor. 1994. Beaver influences on the long-term biogeochemical characteristics of boreal forest drainage networks. Ecology 75:905–921.

Neff, D. J. 1959. A seventy-year history of a Colorado beaver colony. Journal of Mammalogy 40:381–387.

Nolet, B. A., and F. Rosell. 1998. Comeback of the beaver *Castor fiber*: an overview of old and new conservation problems. Biological Conservation 83:165–173.

Paragamian, V. L. 1981. Some habitat characteristics that affect abundance and winter survival of smallmouth bass in the Maquoketa River, Iowa. Pages 45–53 *in* L. A. Kruholz, editor. The Warmwater Streams Symposium: a national symposium on fisheries aspects of warmwater streams. American Fisheries Society, Southern Division, Bethesda, Maryland.

Parker, M., F. J. Wood, Jr., B. H. Smith, and R. G. Elder. 1985. Erosional downcutting in lower order riparian ecosystems: have historical changes been caused by removal of beaver? Pages 35–38 *in* R. R. Johnson, editor. Riparian ecosystems and their management: reconciling conflicting uses. USFS, RM GTR 120, Fort Tucson, Arizona.

Pastor, J., J. Bonde, C. Johnson, and R. J. Naiman. 1993. Markovian analysis of the spatially dependent dynamics of beaver ponds. Lectures on Mathematics in the Life Sciences 23:5–27.

Pattie, J. O. 1833. The personal narrative of James Ohio Pattie of Kentucky. Timothy Flint, editor. John H. Wood, Cincinnati.

Pollock, M. M., R. J. Naiman, H. E. Erickson, C. A. Johnston, J. Pastor, and G. Pinay. 1994. Beaver as engineers: influences on biotic and abiotic characteristics of drainage basins. Pages 117–126 *in* C. G. Jones and J. H. Lawton, editors. Linking species to ecosystems. Chapman and Hall, New York.

Pollock, M. M., and G. R. Pess. 1998. The current and historical influence of beaver (*Castor canadensis*) on coho (*Oncorhynchus kisutch*) smolt production in the Stillaguamish River Basin. 10, 000 Years Institute, Seattle.

Ponce, V. M., and D. S. Lindquist. 1990. Management of baseflow augmentation: a review. Water Resources Bulletin 26:259–268.

Rabe, F. W. 1970. Brook trout populations in Colorado beaver ponds. Hydrobiologia 35:431–448.

Rea, A. M. 1983. Once a river: bird life and habitat change on the Middle Gila. University of Arizona Press, Tucson.

Reid, K. A. 1952. The effect of beaver on trout waters. Maryland Conservationist 29:21–23.

Retzer, J. L., H. M. Swope, J. D. Remington, and W. H. Rutherford. 1956. Suitability of physical factors for beaver management in the Rocky Mountains of Colorado. State of Colorado, Department of Game and Fish Technical Bulletin No. 2, Denver.

Rudemann, R., and W. J. Schoonmaker. 1938. Beaver dams as geological agents. Science 88:523–524.

Rupp, R. S. 1954. Beaver-trout relationships in the headwaters of Sunkhaze Stream, Maine. Transactions of the American Fisheries Society 84:75–85.

Rutherford, W. H. 1964. The beaver in Colorado: its biology, ecology, management and economics. Colorado Game, Fish and Parks Department, Technical Publication 17, Denver.

Rutten, M. G. 1967. Flat-bottomed glacial valleys, braided rivers and the beaver. Geologie en Mijnbouw 46:356–360.

Salyer, J. C. 1935. Preliminary report on the beaver trout investigation. American Game 24:6–15.

Scheffer, P. M. 1938. The beaver as an upstream engineer. Soil conservation 3:178–181.

Schlosser, I. J. 1995. Dispersal, boundary processes, and trophic-level interactions in streams adjacent to beaver ponds. Ecology 76:908–925.

Seton, E. T. 1929. Lives of game animals. Doubleday, Doran and Co., Inc., Garden City, New York.

Skinner, Q. D., J. E. Speck, Jr., M. Smith, and J. C. Adams. 1984. Stream water quality as influenced by beavers within grazing systems in Wyoming. Journal of Range Management 37:142–146.

Slaney, P. A., and D. Zaldokas, editors. 1997. Fish habitat rehabilitation procedures. Ministry of the Environment, Lands and Parks, Vancouver.

Smith, A. E. 1950. Effects of water runoff and gradient on beaver in mountain streams. Master's thesis. University of Michigan, Ann Arbor.

Smith, R. D., R. C. Sidle, and P. E. Porter. 1993. Effects on bedload transport of experimental removal of woody debris from a forest gravel-bed stream. Earth Surface Processes and Landforms 18:455–468.

Snodgrass, J. W., and G. K. Meffe. 1998. Influence of beavers on stream fish assemblages: effects of pond age and watershed position. Ecology 79:928–942.

Snodgrass, J. W., and G. K. Meffe. 1999. Habitat use and temporal dynamics of blackwater stream fishes in and adjacent to beaver ponds. Copeia 3:628–639.

Stabler, D. F. 1985. Increasing summer flow in small streams through management of riparian areas and adjacent vegetation - a synthesis. USDA Forest Service General Technical Report RM-120:206–210, Fort Collins, Colorado.

Stock, J. D., and I. J. Schlosser. 1991. Short-term effects of a catastrophic beaver dam collapse on a stream fish community. Environmental Biology of Fishes 31:123–129.

Suzuki, N., and W. C. McComb. 1998. Habitat classification models for beaver (*Castor canadensis*) in the streams of the central Oregon Coast Range. Northwest Science 72:102–110.

Swales, S., F. Caron, J. R. Irvine, and C. D. Levings. 1988. Overwintering habitats of coho salmon (*Oncorhynchus kisutch*) and other juvenile salmonids in the Keogh River system, British Columbia. Canadian Journal of Zoology 66:254–261.

Swanston, D. N. 1991. Natural processes. Pages 139–179 *in* W. R. Meehan, editor. Influences of forest and rangeland management on salmonid fishes and their habitats. American Fisheries Society, Special Publication 19, Bethesda, Maryland.

Tappe, D. T. 1942. The status of beavers in California. State of California Department of Natural Resources Division of Fish and Game. Game Bulletin 3:1–60.

Voigt, D. R., G. B. Kolenosky, and D. H. Pimlott. 1976. Changes in summer food of wolves in central Ontario. Journal of Wildlife Management 40:663–668.

Warren, E. R. 1926. A study of the beaver in the Yancy Region of Yellowstone National Park. Roosevelt Wildlife Annals 1:5–191.

Warren, E. R. 1932. The abandonment and reoccupation of pond sites by beavers. Journal of Mammalogy 13:343–346.

Wilen, B. O., B. P. MacConnell, and D. L. Mader. 1975. The effects of beaver activity on water quality and water quantity. Proceedings of the Society of American Foresters:235–240.

Winegar, H. H. 1977. Camp Creek channel fencing–plant, wildlife, soil and water response. Rangeman's Journal 4:10–12.

Woo, M. K., and J. M. Waddington. 1990. Effects of beaver dams on subarctic wetland hydrology. Arctic 43:223–230.

Wood, J. A. 1997. Augmenting minimum streamflows in sediment deposition reaches. Chapter 14 *in* P. Slaney and D. Zaldokas, editors. Fish habitat rehabilitation procedures. Ministry of Environment, Lands and Parks, Watershed Restoration Program, Vancouver.

Zierholz, C., I. P. Prosser, P. J. Fogarty, and P. Rustomji. 2001. In-stream wetlands and their significance for channel filling and the catchment sediment budget, Jugiong Creek, New South Wales. Geomorphology 38(3–4):221–235.

Zurowski, W., and B. Kasperczyk. 1986. Characteristics of a European beaver population in the Suwalki Lakeland. Acta Theriologica 13:311–325.

American Fisheries Society Symposium 37:235–247, 2003

Wood and Wildlife: Benefits of River Wood to Terrestrial and Aquatic Vertebrates

E. Ashley Steel

NW Fisheries Science Center, 2725 Montlake Boulevard E, Seattle, Washington 98112, USA

William H. Richards[1]

The Pacific Forest Trust, 157 Yesler Way, Suite 419, Seattle, Washington 98104, USA

Kathryn A. Kelsey[2]

University of Washington, Seattle, Washington 98195, USA

Abstract.—Wood in rivers, or wood deposited from fluvial processes, provides unique habitat for terrestrial and aquatic wildlife species. Many wildlife species utilize riparian areas for some portion of their life history primarily due to the universal need for water, the presence of unique plant assemblages, and the diversity of microhabitats produced by the dynamics of river systems. Wood in rivers provides four primary functions for aquatic and terrestrial wildlife species: habitat structure, shelter, patchiness of habitat, and increased food resources. Abundance and diversity of wildlife species are enhanced by wood in rivers, and they, in turn, shape and maintain aquatic and riparian habitats. Though there is a clear link between wood in rivers and riparian wildlife communities, knowledge about their interactions and interdependence is sparse.

Introduction

Riparian areas are notedly rich in species diversity (Motroni 1980; O'Connell et al. 1993; Kelsey and West 1998; Kauffman et al. 2001), though in many ecosystems they represent a relatively small and fragile fraction of the landscape. Wood is one of the primary attributes of riparian habitats, enhancing wildlife abundance and diversity. Complexity, including the presence of wood and high frequency of natural disturbances in riparian areas, is, in part, responsible for the diversity of both habitat patches and wildlife species (Gregory et al. 1991). Many wildlife species are found exclusively or more commonly in riparian areas (Anthony et al. 1987, 1996; Gomez and Anthony 1998). For example, more than 60% of the bird species in the western United States nest primarily in riparian forest habitats (Szaro 1980). In the Blue Mountains of Oregon, 98% (278 of 285) of the terrestrial wildlife species are found either exclusively or more commonly in riparian areas (O'Connell et al. 1993). The importance of riparian habitats for multiple wildlife species has also been documented in the Midwestern United States (Stauffer and Best 1980), the Rocky Mountains (Knopf 1985), and in the desert Southwest (England et al. 1984). In the Pacific Northwest, most wildlife species may occur in both upland and riparian areas, but a few are found only in riparian areas (Anthony et al. 1987; Doyle 1990; McComb et al. 1993a, 1993b; O'Connell et al. 2000). These species are considered riparian obligates, which depend on riparian areas for all or part of their life histories (Gomez and Anthony 1998; Loegering and Anthony 1999). In western Washington and Oregon, where 87% of the terrestrial wildlife species are found either exclusively or more commonly in riparian areas (O'Connell et al. 1993), well-known riparian wildlife species include the tailed frog *Ascaphus truei*,

[1] Current Address: Seattle Public Utilities, Cedar River Watershed, 19901 Cedar Falls Road SE, North Bend, Washington 98045, USA

[2] Current Address: Seattle Public Schools, P.O. Box 34165, Seattle, Washington 98124, USA

Pacific giant salamander *Dicamptodon tenebrosus*, great blue heron *Ardea herodias*, American dipper *Cinclus mexicanus*, harlequin duck *Histrionicus histrionicus*, water shrew *Sorex palustris*, and river otter *Lutra canadensis*. Worldwide, riparian wildlife species represent an array of diversity, including turtles, monkeys, terrestrial and aquatic snakes, bears, hippopotami, alligators, and crocodiles.

The role of large wood in upland and riparian habitats has been extensively studied (Maser et al. 1979; Carey and Johnson 1995; Butts and McComb 2000; Lohr et al. 2002), as has the importance of wood to stream dynamics, fish, and aquatic invertebrate habitat (Harmon et al. 1986; Bisson et al. 1987; Maser et al. 1988; Bilby and Bisson 1998). In upland areas, many species of wildlife use dead wood, either standing or fallen (Johnson and O'Neil 2001). In the northeastern United States, 28 birds, 18 mammals, 23 reptiles and amphibians, and hundreds or thousands of insect species use wood (Hagan and Grove 1999). One recent study of southeastern Australian floodplains was able to associate variation in the density of wood with the abundance and diversity of native mammals but not of amphibians or reptiles (Mac Nally et al. 2001). Yet, the role of wood in rivers in supporting wildlife populations has seldom been specifically researched.

Instream wood serves many of the same purposes for wildlife as it does for fish: increasing pool density, dissipating stream energy, increasing the abundance and diversity of aquatic food resources, and providing cover from predation. Exposed river wood—piles and pieces of wood deposited along riverbanks, in the riparian forest, and on cobble islands during high water—is a resource that many properly functioning river systems provide adjacent terrestrial ecosystems. Some river-deposited wood remains dry for much of the year but is submerged during periods of high water. As the river migrates and meanders, much of the wood deposited by rivers becomes integrated into the riparian forest and floodplains and may remain dry for years or decades.

In this chapter, we synthesize research on the role of river wood in supporting wildlife communities. We draw from upland and instream findings to suggest hypotheses about river wood and wildlife that might be tested in the future. In the first of three sections, we summarize what is known about wood and wildlife, exclusive of fish and beaver, which are considered elsewhere in this book. Throughout the chapter, we use the phrase riparian wildlife to include all species that have been found at times in riparian areas: aquatic, obligate riparian dwellers, riparian associates, generalists, and primarily upland species. We consider four primary functions of instream and river-deposited wood for aquatic and terrestrial wildlife: habitat structure, shelter, patchiness of habitat, and food resources. Although the exact mechanisms by which river wood supports wildlife differ across species and ecoregions, we emphasize the many common relationships or patterns. The second section of the chapter describes feedback loops between wildlife and riparian ecosystems. We describe in detail one example to support the theory that healthy riparian wildlife communities are a key ingredient of processes that shape and sustain natural river and riparian systems. In the final section of the chapter, we summarize gaps in our current understanding of wood-wildlife to encourage future research on the role of wood in rivers for supporting diverse wildlife assemblages.

Four Functions of Wood in Supporting Riparian Wildlife

Habitat structure

In the most basic sense, wood in rivers provides habitat structure for riparian wildlife, including structure that supports perching and basking above the water and in the riparian corridor. Many bird species are known or have been observed to use wood for perching. Belted kingfishers *Ceryle alcyon*, a common riparian obligate in the Pacific Northwest, are most commonly seen perching on shoreline or suspended wood as they search for prey in the water (Ehrlich et al. 1988). In a study of the value of accumulations of wood in riparian areas, Mason and Koon (1985) observed violet-green swallows *Tachycineta thalassina*, Say's phoebe *Sayornis saya*, Townsend's solitaire *Myadestes townsendi*, and western flycatchers *Empidonax difficilis* resting on wood between aerial feedings. In estuarine areas, on mudflats, or along the coast, great blue herons and great egrets *Casmerodium albus* use emergent wood, floating logs, and log rafts for perching during high water. Only able to wade in water up to about 20 cm, these visual predators use wood to maximize their hunting time while conserving energy (Maser et al. 1988).

On exposed cobble bars, river-deposited piles of wood may provide a particularly important resource as the only available aboveground structure. During 160 h of bird observations along the North Fork Skykomish River in western Washington, 125 individual birds (16 species) were observed on wood accumulations (Table 1; Steel et al. 1999). More than half the bird activities (54%) around wood required only its physical structure for perching or hawking. Birds were also observed foraging or eating on the wood accumulations, using the internal spaces of the wood accumulations, and displaying territorial behaviors on the wood accumulations (Figure 1).

Reptiles, amphibians, and aquatic mammals benefit from the structure provided by wood as sites for basking and resting. Turtles and many species of lizards are often observed motionless in the sun on emergent wood. Partially submerged logs have been documented as basking sites for western pond turtles *Clemmys marmorata* and painted turtles *Chrysemys picta* (Nussbaum et al. 1983). Melquist and Hornocker (1983) found that river otters *Loutra canadensis* frequently use logjams as resting sites.

Shelter

River wood provides complex structure for nesting and denning while sheltering against predation and climatic extremes. Several species of birds and waterfowl have been observed to nest inside river wood accumulations. In the Skykomish River study, 15% of the bird activities included entering wood accumulations (Figure 1), and dark-eyed juncos *Junco hyemalis* used the wood as nesting sites (Steel 1993). Two nests of Canada geese *Branta canadensis* were documented on wood along the Stehekin River in eastern Washington, and winter wrens *Troglodytes troglodytes* were reported to nest inside wood accumulations (Mason and Koon 1985). On a salt plain in Oklahoma, Grover and Knopf (1982) found that 115 of 182 nests (63.2%) of American avocets *Recurvirostra americana*, snowy plovers *Charadrius alexandrinus*, and least terns *Sterna antillarum* were within 5 cm of river wood. And, in an observational study of associations between wood and terrestrial wildlife on southeastern Australian floodplains, avifauna abundance and diversity were greater in wood piles (Mac Nally et al. 2001). One postulated explanation was the importance of the shelter of wood accumulations in the selection of microhabitats.

Mammals use wood in rivers for shelter. Water shrews, for example, build nests of dry grass in burrows or under logs. The degree to which small mammals reside in accumulations of wood in streams is largely unstudied. Rector (1991) has suggested that small mammals captured on cobble bars reside in the forest and are captured during foraging expeditions. If this is true, the number of small mammals found on cobble bars would decrease with increasing distance from the forest edge. Steel et al. (1999) found that small-mammal abundance in wood on cobble bars did not de-

TABLE 1. Bird species seen on wood piles during 160 h of observation along the North Fork Skykomish River in western Washington, USA. Species list describes 125 individual birds. Modified from Steel et al. (1999).

Common name	Scientific name
Swainson's thrush	*Catharus guttatus*
Steller's jay	*Cyanocitta stelleri*
Dark-eyed junco	*Junco hyemalis*
White-crowned sparrow	*Zonotrichia leucophrys*
Rough-winged swallow	*Stelgidopteryx serripennis*
Epidonax flycatcher	*Empidonax* sp.
Western tanager	*Piranga ludoviciana*
Cedar waxwing	*Bombycilla cedrorum*
Winter wren	*Troglodytes troglodytes*
Varied thrush	*Ixoreus naevius*
Calliope hummingbird	*Stellula calliope*
Rufous hummingbird	*Selasphorus rufus*
American robin	*Turdus migratorius*
Blue-throated gray warbler	*Dendroica nigrescens*
Hairy woodpecker	*Picoides villosus*
Belted kingfisher	*Ceryle alcyon*

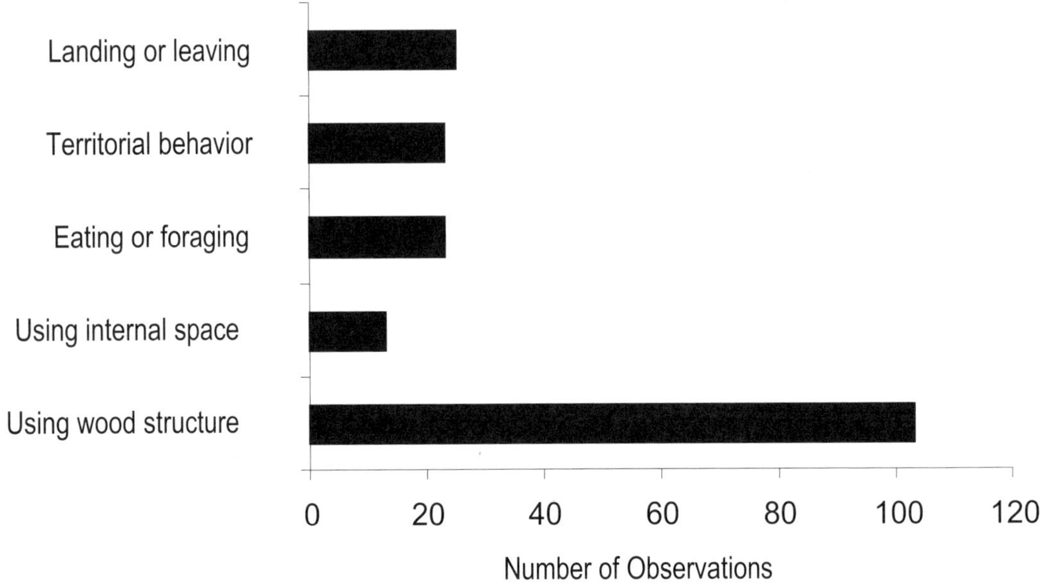

FIGURE 1. Bird activities. "Landing or leaving" includes birds that alighted momentarily or were seen only when flying away from the wood. "Territorial behavior" includes singing, calling, and chasing. "Using internal space" includes all birds hopping or flying inside the wood accumulation. "Using wood structure" includes all birds perching, hopping, preening, or hawking. Up to two activity codes were recorded for each bird sighting.

pend on the distance of the wood accumulations from the forest. Their findings indicate that either small mammals reside in wood on cobble bars or small mammals reside in the nearby forest, but river wood provides enough shelter for them to forage far from the forest edge (>107 m). In both the forested areas and on the cobble bars, total abundance and species richness of small mammals on the Skykomish River were greater on wood accumulations (nine species) than at nearby areas without wood (four species; Tables 2 and 3). In upland areas, there is much evidence that mammals use downed wood, indicating that river-deposited wood may be valuable habitat for these. For example, fishers *Martes pennanti* and other mustelids rest in dens made of river wood in riparian areas (Weir 1993). In upland areas, changes in behavior have been associated with the presence or absence of woody cover. White-footed mice *Peromyscus leucopus* in south-central Pennsylvania were more active in experimental plots with woody ground litter than in plots that did not contain it (Planz and Kirkland 1992).

Many species of amphibians are most commonly found under rocks or logs, though most studies of aquatic amphibians have not differentiated between the type of cover. Nussbaum et al. (1983) found that many amphibians are nocturnal and spend the day hiding under cover. In a large survey of amphibians in the Pacific Northwest, 65–93% (by species) of captures were found under rock or log cover (Bury et al. 1991). Structure provided by wood in rivers is important in sheltering the reproductive activities of amphibians. Nineteen of 24 (79%) of Washington's amphibian species use streams and ponds for egg deposition, often under, between, or in crevices of logs (Nussbaum et al. 1983). In upland communities, numerous studies have documented the relationship between wood and amphibian presence (Nussbaum et al. 1983). For example, ensatina *Ensatina eschscholtzi*, western redback salamander *Plethodon vehiculum*, and red-legged frog *Rana aurora* are associated with terrestrial wood (Aubry and Hall 1991).

Reptiles may also use river wood and other downed wood as protection from predation and climatic conditions. A study of wandering garter snakes *Thamnophis elegans vagrans* along the Rio de las Vacas in New Mexico found that snakes used wood as shelter from predation and when fleeing danger as well as for thermal shelter during times of inactivity (Szaro et al. 1985). No snakes were seen in the open during the hottest part of the day during warm months, and more than half of the overnight or inactive cover sites were accu-

TABLE 2. Mean catch (per unit effort) of three genera of small mammals for wood piles and reference sites on the cobble bar and in the forest. Modified from Steel et al. (1999).

	Cobble bar		Forest	
Genera	Wood piles	Reference sites	Wood piles	Reference sites
Peromyscus	8.0	3.9	4.1	9.5
Sorex	1.6	0.0	2.7	1.5
Microtus	2.7	0.0	0.0	0.0
Total	12.3	3.9	6.8	11.0

mulations of wood. Wood provides shelter from climatic variations on exposed cobble bars. Riparian lizards use the wood to avoid lethal midday temperatures in July and August (Sabo 2000). They can move back and forth between the sunny and shaded sides of exposed branches. Steel et al. (1999) found that the average maximum daily temperature was significantly cooler (3.2°C cooler) inside accumulations of wood on cobble bars than outside of them.

Creating patches of habitat

Many chapters in this book deal with the geomorphic and fluvial processes by which wood in rivers contributes to pool formation, sediment retention, cobble bar formation, and riparian forest succession (Montgomery et al. 2003; Mutz 2003; Piégay 2003; Swanson 2003; all this volume). In this chapter, we do not include additional information about the role of wood in these processes but focus on the importance of these wood-influenced habitat patches for aquatic and riparian wildlife. Many species of wildlife benefit from a mosaic of habitat patches, such as pools and riffles or cobble islands and emergent riparian forests, in part formed and maintained by wood in rivers.

Deposition and accumulation of wood are fundamental to river ecosystems, forming pools and shaping channel morphology (Bisson et al. 1987; Montgomery et al. 1995; Abbe and Montgomery 1996; Bilby and Bisson 1998). Sequences of pools and riffles contribute to habitat diversity and nutrient cycling. The importance of pool habitat and channel complexity is well known for fish (Bilby and Bisson 1998; Montgomery et al. 1999; Dolloff and Warren 2003; Zalewski et al. 2003; both this volume). Pools are also important for many species of amphibians (Bury et al. 1991; Wilkins and Peterson 2000).

Wood is a key ingredient in forming cobble bars and cobble islands (Bilby and Ward 1989; Buffington 1995; Abbe and Montgomery 1996) that benefit multiple riparian wildlife species. Wood thus indirectly affects the species that utilize these dynamic habitats. Pioneer islands that may initially appear lifeless often host large beetle and insect communities, providing an important food

TABLE 3. Small mammals captured at wood piles and reference sites by absolute number and catch per unit effort (calculated for all piles together, modified from Steel et al. 1999).

	Wood pile		Reference site	
Species	Number	CPUE	Number	CPUE
Deer mouse	10	2.2	5	2.2
Forest mouse	6	1.3	4	1.8
Peromyscus spp.	1	0.2		
Trowbridge shrew	6	0.9	1	0.3
Montane shrew	2	0.3	1	0.3
Water shrew	1	0.2		
Longtail vole	3	0.7		
Oregon vole	1	0.2		
Shrew mole	1	0.2		
Pacific mole	1	0.2		
Total	32	6.3	11	4.6

[a] Catch per unit effort was calculated for all piles together.

source for riparian bird and small mammal communities (Rector 1991). The rocky substrates found on cobble bars can also provide warm thermal retreats. In a study on the South Fork Eel River in Colorado, cobble bars provided thermal retreats that maintained optimal temperatures for Western fence lizard *Sceloporus occidentalis* digestion and may have been a factor in the female Western fence lizard's ability to produce more than one clutch of eggs per season (Sabo 2000).

Wood from rivers promotes stability of riparian habitats and enhances riparian forest growth (Boyer et al. 2003, this volume). Accumulations of wood can be maintained as part of the riparian ecosystem for long periods. In the Skykomish study (Steel et al. 1999), the largest wood accumulations were located on cobble bars and tended to be washed up against patches of riparian forest. Vegetation growing in and directly upstream of the wood was more than 30 years old in places, indicating that accumulations themselves had been on site for at least that long (Steel et al. 1999). Mason and Koon (1985) hypothesized that the wood on cobble areas leads to increased deposition of fine sediments and therefore complex and well-developed vascular plant communities that can be used by a wide variety of wildlife species.

Increased food resources

The presence of wood potentially increases the abundance of food resources for wildlife species. Instream wood, for example, increases the retention of leaf litter and other inputs and, thereby, increases their local or immediate availability as a food source (Bilby 1981, 2003, this volume). Leaf litter trapped in wood accumulations that are intermittently inundated breaks down quickly because of alternation between submerged and nonsubmerged states (Merritt et al. 1984; Mason and Koon 1985). Wood increases the availability of fungal spores as a food resource (Edmonds and Marra 1999). Higher humidity inside exposed wood accumulations on floodplains or riparian areas increases fungal growth (Harmon et al. 1986). Wood often contains abundant fungal spores when initially deposited (Shearer and Von Bodman 1983), and these fungal colonies may grow and flourish.

Insects are another important food resources that can be enhanced by the presence of river wood. (Benke and Wallace 2003, this volume). The relationship of wood in rivers to insect communities is treated thoroughly elsewhere in this book (Benke and Wallace 2003) and in a special issue of the *International Review of Hydrobiologia* (2000). Insects and other invertebrates support a food chain on which many species of riparian wildlife depend. Amphibian and reptile communities depend on aquatic insects as a primary food source. Sabo (2000) demonstrated that both growth rates and abundance of Western fence lizards are reduced when aquatic insect densities are artificially reduced. Mammals also rely on invertebrate food sources. For example, water shrews eat shrimp, water striders, and caddis fly larvae. River otters, hooded skunks *Mephitis macroura*, mink *Mustela vision*, and other mustelids rely to some degree on food resources potentially enhanced by wood in rivers. In a habitat restoration project of a spring-fed stream in Montreal, Canada, habitat modifications were associated with increased invertebrate food resources; structures of wood were associated with a more than twofold increase in crayfish biomass, the major component of mink diet in this area (Burgess 1980). Many species of bat also rely on emerging aquatic insects for food resources. For example, the Brazilian long-nosed bat *Rhynchonycteris naso* forages almost entirely over water and feeds exclusively on emergent insects, including chironomids, mosquitoes, and caddis flies (Plumpton and Jones 1992). Cobble bars, often formed downstream of river-deposited wood, can be abundant sources of food for birds and terrestrial animals. These areas support unusually high beetle and terrestrial invertebrate densities (Rector 1991).

River wood is clearly related to fish abundance and diversity (Bisson et al. 1987, 2003, this volume; Montgomery et al. 1999; Dolloff and Warren 2003; Zalewski et al. 2003). Although fish abundance and diversity are important for their own sake, the importance of fish as a staple for riparian wildlife species, such as kingfishers, bears, river otter, and mink, should not be overlooked.

Seeds are also a primary food resource for many small mammals and riparian birds. Submerged wood traps suspended particles such as seeds and sporocarps, and exposed accumulations of wood may create wind eddies that increase deposition of airborne seeds and sporocarps (Mason and Koon 1985). The wood serves as host to many riparian plant species, including lichens, mosses, ferns and fern allies, devil's club *Oplopanax horridus*, spruces, hemlocks, and cedars (Bragg and Kershner 1999). Increased diversity of plant communities supported by river wood eventually lead to localized increases in seed production.

Feedbacks between Riverine Processes and Riparian Wildlife

Riparian wildlife communities are not simply a result of riparian conditions but an integral component of the landscape. Many riparian species play valuable roles in shaping the riparian ecosystems on which they depend. Crocodilians, for example, maintain aquatic refugia during droughts, prey selectively on fish species, and increase nutrient recycling (Craighead, Sr. 1968; King 1988). Beavers *Castor canadensis* are a classic example of how animals can shape the habitats on which they depend by creating vast pond and marsh complexes (Pollock et al. 2003, this volume). Fish and other wildlife can move marine-derived nutrients into the streams. And, nesting birds bring phosphorus and ammonia upstream, stimulating phytoplankton production (Burton et al. 1979). Here, we detail one example of the feedback loop between stream habitat and wildlife populations.

The feedback loop between submerged wood, salmon carcass abundance, grizzly bear *Ursus arctos* activity, and riparian plant growth provides evidence that river wood, by increasing salmon production and therefore potential marine-derived nutrients, plays a role not only in stream function but also in riparian forest growth (Helfield and Naiman 2001). How does the marine-derived nitrogen get from the stream to the forest? Helfield (2001) examined concentrations of stable isotopes, ^{15}N, of marine-derived nitrogen in bear middens (areas with matted grass, salmon carcass remnants, and bear scat) in southeast Alaska and found high levels of marine-derived nitrogen in the vegetation surrounding these middens. Combined with results of recent research by Ben-David et al. (1998) and Hilderbrand et al. (1999), who also found spatial correlations between bear, river otter, mink, and marten activity and foliar ^{15}N concentrations, Helfield's findings provide evidence that wildlife bring marine-derived nutrients from the river or stream to the riparian forest. Increased riparian forest growth potentially benefits the stream ecosystem. If a 50-cm basal diameter is assumed as a minimum threshold size for wood to be retained in the stream during high water (Bilby and Ward 1989), the importance of wildlife to the stream environment can be calculated. Estimates of spruce growth rates within 25 m of salmon-bearing and nonsalmon-bearing streams indicate that trees enhanced by marine nitrogen (often through wildlife activity) may reach 50-cm basal diameter in less than 100 years; trees growing along streams without salmon may take more than 300 years to reach the same size (Helfield and Naiman 2001; Figure 2). Wildlife, such as bear, river, otter, mink, and marten are therefore beneficiaries of food resources that are enhanced by wood in rivers. And, by redistributing the nutrients from these food resources, increasing riparian tree growth, and increasing wood in rivers, these wildlife species may also play a part in engineering the amount of wood available to the river system.

Wood and Wildlife: Research Needs

Is wood in rivers or deposited by rivers essential for any wildlife species?

Research to date and the evidence presented in this chapter describing wildlife use of wood in rivers is primarily correlative. The value of wood to particular wildlife species cannot be assessed through such correlative studies alone. Whether a change in the amount of wood or type of wood in rivers would affect wildlife populations must be tested through controlled experiments or controlled comparisons between stream reaches with different types and amounts of wood. Studies could attempt to measure the importance of wood to particular wildlife species through changes in individual fitness, median population growth rate, or life stage specific survival parameters. Continued emphasis on animal use patterns and observed density may obscure important relationships between the quantity and distribution of river wood and wildlife use of that resource (van Horne 1983). Species that use wood on exposed cobble bars and in unvegetated areas would be of particular interest. Research to determine how well animals in these areas might survive in the absence of river wood will enhance our knowledge of both river ecosystems and wildlife populations and should lead to more effective management of these habitats for wildlife.

Which characteristics of wood in rivers are most important to wildlife populations?

All river wood may not be equally valuable to wildlife; wood of some species, sizes, shapes, de-

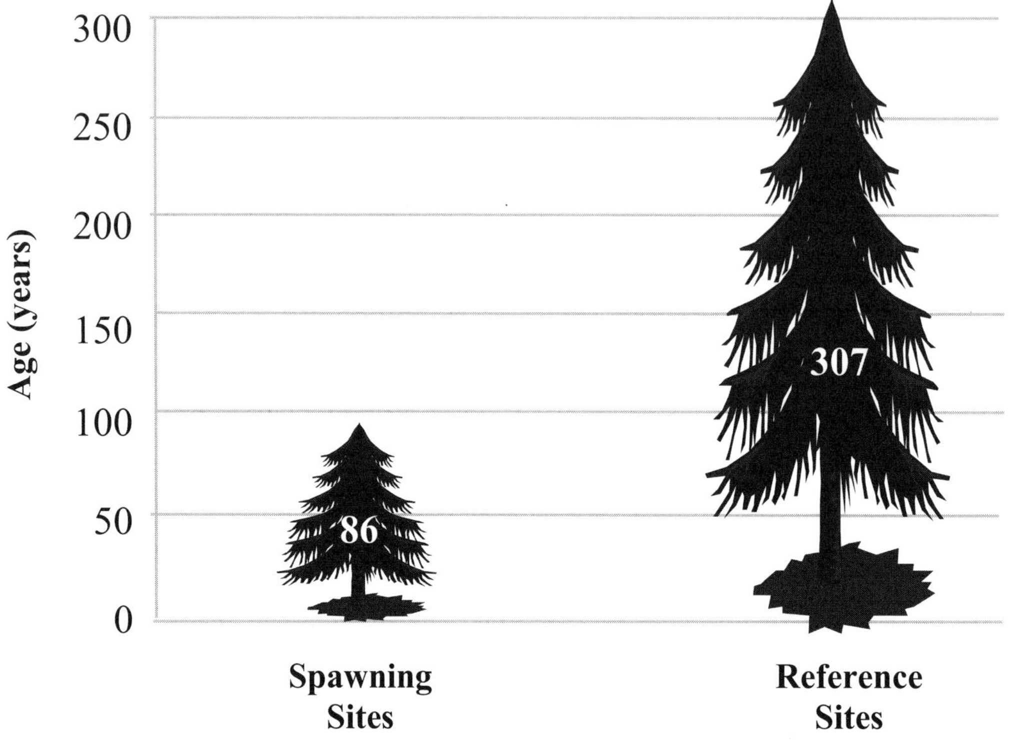

FIGURE 2. Expected age when riparian Sitka spruce trees, growing within 25 m of streams with and without salmon spawning, would reach 50 cm dbh, Tenakee Inlet, Alaska. Based on data from Helfield and Naiman 2001. Figure courtesy of James Helfield, University of Washington.

cay class, or even densities may be more valuable than others for particular wildlife species. Differences in small mammal abundances between types of wood accumulations were evident in the Skykomish River study (Steel 1993). Significantly more small mammals and higher species richness were found in association with newly deposited wood accumulations versus older wood accumulations. The findings suggest that green needles, sporocarps, or other fresh material may increase use of wood (Steel 1993). Tree species determines specific density and therefore whether or not the wood floats, how it is distributed, and patterns of retention (Abbe 2000; Webb and Erskine 2003). Tree species have inherently different bark characteristics, decay rates, and rootwads. Wood with attached rootwads, for example, is integral in forming logjams and promoting in-river stability of wood (Bisson et al. 1987). These rootwads may also entrap smaller pieces of wood, leaves, and twigs as food for wildlife or provide nesting sites for birds and small mammals.

In upslope communities, a great deal of research has demonstrated that characteristics of dead wood, such as mean log diameter, decay class, and log structure, can be important predictors of wildlife use (Hayes and Cross 1987; Bull et al. 1997). Upland salamanders appear to select wood according to decay class. Both ensatina and western red-backed salamander *Plethodon vehiculum* were captured more frequently in moderately decomposed wood than in either intact or fully decomposed material (Aubry et al. 1988). Similar patterns might be the case for wood deposited by rivers. Tree species are important for beavers in both building dams and selecting cache material. In Algonquin Provincial Park, Ontario, Canada, beavers preferentially used coniferous species and speckled alder in building dams (Doucet et al. 1994). Trembling aspen *Populus tremuloides*, beaked hazel *Corylus cornuta*, maple *Acer* sp., and serviceberry *Amelanchier canadensis* were used preferentially as cache material.

In what geographic regions or locations along the stream continuum are relationships between wildlife species and river wood likely to be strongest?

The importance of wood to wildlife will likely depend on the surrounding ecosystem, the isolation of the riparian corridor, and the location of the site within the river network. At the extremes, we would expect the relationship between wood and wildlife to differ between desert rivers with highly defined riparian corridors and sporadic movements of large accumulations of wood (Minckley and Rinne 1985) and tropical rivers with frequent, dramatic flooding and undefined riparian corridors (Dudgean 1992). Within an ecosystem, the importance of river wood to wildlife will be at least partially determined by location within the stream network both because of variation in wood accumulation and deposition patterns along the network and because of variation in wildlife abundance and diversity. The distribution of riparian-obligate species, species which might have the greatest dependence on wood in rivers, depends to a great extent on habitat characteristics associated with stream class (Kelsey and West 1998). For example, larger rivers with more distinct riparian forests may support more distinct riparian avifauna communities (Lock and Naiman 1998). In these areas, the presence of riverwood may be particularly critical. Identification of these broad relationships will aid in identifying areas where wood preservation or restoration might be most critical for wildlife.

While clear correlations between wood in rivers and wildlife diversity and abundance have been established, the lack of mechanistic or experimental studies limits our ability to identify definitive relationships. To provide information necessary for improving riverine or wildlife management, researchers must next identify the degree of species-specific dependence on wood in rivers for representative taxa or species of management importance and examine which types of wood and what locations provide the greatest resources to these wildlife species.

Acknowledgments

We thank James Helfield, Jonathon Sabo, and Robert Naiman for contributing content for this chapter. Two anonymous reviewers provided substantial contributions. We also thank Todd Bennett for his careful review of the manuscript and Karrie Hanson for assisting with manuscript preparation.

References

Abbe, T. B. 2000. Patterns, mechanics and geomorphic effects of wood debris accumulations in a forest river system. Doctoral dissertation. University of Washington, Seattle.

Abbe, T. B., and D. R. Montgomery. 1996. Large woody debris jams, channel hydraulics and habitat formation in large rivers. Regulated Rivers Research & Management 12:201–221.

Anthony, R. G., E. D. Forsman, G. A. Green, G. Witmer, and S. K. Nelson. 1987. Small mammal populations in riparian zones of different-aged coniferous forests. The Murrelet 68:94–102.

Anthony, R. G., G. A. Green, E. D. Forsman, and S. K. Nelson. 1996. Avian abundance in riparian zones of three forest types in the Cascade Mountains, Oregon. The Wilson Bulletin 108:280–291.

Aubry, K. B., and P. A. Hall. 1991. Terrestrial amphibian communities in the southern Washington cascades. Pages 327–338 *in* L. F. Ruggiero, K. B. Aubry, A. B. Carey, and M. H. Huff, editors. Wildlife and vegetation of unmanaged Douglas-fir forests. USDA, Forest Service, Pacific Northwest Research Station, General Technical Report PNW-GTR-285, Portland, Oregon.

Aubry, K. B., L. L. C. Jones, and P. A. Hall. 1988. Use of woody debris by plethodontid salamanders in Douglas-fir forests in Washington. Management of amphibians, reptiles, and small mammals in North America. USDA Forest Service, Rocky Mountain Forest and Range Experiment Station, General Technical Report GTR-RM-166, Fort Collins, Colorado.

Ben-David, M., T. A. Hanley, and D. M. Schell. 1998. Fertilization of terrestrial vegetation by spawning Pacific salmon: the role of flooding and predator activity. Oikos 83:47–55.

Benke, A. C., and J. B. Wallace. 2003. Influence of wood on invertebrate communities in streams and rivers. Pages 149–177 *in* S. V. Gregory, K. L. Boyer, and A. M. Gurnell, editors. The ecology and management of wood in world rivers. American Fisheries Society, Symposium 37, Bethesda, Maryland.

Bilby, R. E. 2003. Decomposition and nutrient dynamics of wood in streams and rivers. Pages 135–147 *in* S. V. Gregory, K. L. Boyer, and A. M. Gurnell, editors. The ecology and management of wood in world rivers. American Fisheries Society, Symposium 37, Bethesda, Maryland.

Bilby, R. E. 1981. Role of organic debris in regulating the export of dissolved and particulate matter from a forested watershed. Ecology 62:1234–1243.

Bilby, R. E., and P. A. Bisson. 1998. Function and distribution of large woody debris. Pages 324–346 *in* R. J. Naiman and R. E. Bilby, editors. River ecology and management: lessons from the Pacific Coastal Ecoregion. Springer-Verlag, New York.

Bilby, R. E., and J. W. Ward. 1989. Changes in characteristics and function of woody debris with increasing size of streams in western Washington. Transactions of the American Fisheries Society 118:368–378.

Bisson, P. A., S. M. Wondzell, G. H. Reeves, and S. V. Gregory. 2003. Trends in using wood to restore aquatic habitats and fish communities in western North American rivers. Pages 391–406 *in* S. V. Gregory, K. L. Boyer, and A. M. Gurnell, editors. The ecology and management of wood in world rivers. American Fisheries Society, Symposium 37, Bethesda, Maryland.

Bisson, P. A., R. E. Bilby, M. D. Bryant, C. A. Dolloff, G. B. Grette, R. A. House, M. L. Murphy, K. V. Koski, and J. R. Sedell. 1987. Large woody debris in forested streams in the Pacific Northwest: past, present, and future. Pages 143–190 *in* E. O. Salo and T. W. Cundy, editors. Streamside management: forestry and fishery interactions. College of Forest Resources, University of Washington, Seattle.

Boyer, K. L., D. R. Berg, and S. V. Gregory. 2003. Riparian management for wood in rivers. Pages 407–420 *in* S. V. Gregory, K. L. Boyer, and A. M. Gurnell, editors. The ecology and management of wood in world rivers. American Fisheries Society, Symposium 37, Bethesda, Maryland.

Bragg, D. C., and J. L. Kershner. 1999. Coarse woody debris in riparian zones: opportunity for interdisciplinary interaction. Journal of Forestry 97:30–35.

Buffington, J. M. 1995. Effect of hydraulic roughness and sediment supply on surface textures of gravel-bedded rivers. Master's thesis. University of Washington, Seattle.

Bull, E. L., C. G. Parks, and T. R. Torgersen. 1997. Trees and logs important to wildlife in the interior Columbia basin. USDA Forest Service, Pacific Northwest Research Station, General Technical Report PNW-GTR-391, Portland, Oregon.

Burgess, S. A. 1980. Effects of stream habitat improvements on invertebrates, trout populations, and mink activity. Journal of Wildlife Management 44:871–880.

Burton, T. M., M. T. Harriss, and D. Taylor. 1979. The influence of bird rookeries on nutrient cycling and organic matter production in the Shark River, Florida Everglades. Pages 73–79 *in* R. R. Johnson and J. F. McCormick, editors. Strategies for protection and management of the floodplain wetlands and other riparian ecosystems. USDA Forest Service, General Technical Report GTR-WO-12, Washington, D.C.

Bury, R. B., P. S. Corn, K. B. Aubry, F. F. Gilbert, and L. L. C. Jones. 1991. Aquatic amphibian communities in Oregon and Washington. Pages 353–362 *in* L. F. Ruggiero, K. B. Aubry, A. B. Carey, and M. H. Huff, editors. Wildlife and vegetation of unmanaged Douglas-fir forests. USDA Forest Service, Pacific Northwest Research Station, General Technical Report PNW-GTR-285, Portland, Oregon.

Butts, S. R., and W. C. McComb. 2000. Associations of forest-floor vertebrates with coarse woody debris in managed forests of western Oregon. Journal of Wildlife Management 64:95–104.

Carey, A. B., and M. L. Johnson. 1995. Small mammals in managed, naturally young and old-growth forests. Ecological Applications 5:336–352.

Craighead, Sr., F. C. 1968. The role of the alligator in shaping plant communities and maintaining wildlife in the southern Everglades. Florida Naturalist 41:2–94.

Dolloff, A., and M. L. Warren, Jr. 2003. Fish relationships with large wood in small streams. Pages 179–193 *in* S. V. Gregory, K. L. Boyer, and A. M. Gurnell, editors. The ecology and management of wood in world rivers. American Fisheries Society, Symposium 37, Bethesda, Maryland.

Doucet, C. M., I. T. Adams, and J. M. Fryxell. 1994. Beaver dam and cache composition: are woody species used differently? Ecoscience 1:268–270.

Doyle, A. T. 1990. Use of riparian and upland habitats by small mammals. Journal of Mammology 71:14–23.

Dudgean, D. 1992. Endangered ecosystems: a review of the conservation status of tropical Asian rivers. Hydrobiologia 248:167–191.

Edmonds, R. L. and J. L. Marra. 1999. Decomposition of woody material: nutrient dynamics, invertebrate/fungi relationships and management in Northwest forests. Pages 68–79 *in* R. T. Meurisse, W. G. Ypsilantis, and C. Seybold, editors. Proceedings: Pacific Northwest Forest and Rangeland Soil Organism Symposium 461. Corvallis, Oregon, March 17–19, 1998. USDA Forest Service, Pacific Northwest Research Station, General Technical Report PNW-GTR-461, Portland, Oregon.

Ehrlich, P. R., D. S. Dobkin, and D. Wheye. 1988. The birder's handbook: a field guide to the natural history of North American birds. Simon and Schuster, New York.

England, A. S., L. D. Foreman, and W, F. Laudenslayer Jr. 1984. Composition and abundance of bird populations in riparian systems of the California deserts. Pages 694–705 *in* R. E. Warner and K. M. Hendrix, editors. California riparian systems: ecology, conservation, and productive management. University of California Press, Berkeley.

Gomez, D. M., and R. G. Anthony. 1998. Small mammal abundance in riparian and upland areas of five seral stages in Oregon. Northwest Science 72:293–302.

Gregory, S. V., F. J. Swanson, W. A. McKee, and K. W. Cummins. 1991. An ecosystem perspective of riparian zones. BioScience 41:540–551.

Grover, P. B., and F. L. Knopf. 1982. Habitat requirements and breeding success of charadriiform birds nesting at Salt Plains National Wildlife Refuge, Oklahoma. Journal of Field Ornithology 53:139–148.

Hagan, J. M., and S. L. Grove. 1999. Coarse woody debris. Journal of Forestry 97:6–11.

Harmon, M. E., J. F. Franklin, F. J. Swanson, P. Sollins, S. V. Gregory, J. D. Lattin, N. H. Anderson, S. P. Cline, N. G. Aumen, J. R. Sedell, G. W. Lienkaemper, K. J. Cromack, and K. W. Cummins. 1986. Ecology of coarse woody debris in temperate ecosystems. Advances in Ecological Research 15:133–302.

Hayes, J. P., and S. P. Cross. 1987. Characteristics of logs used by western red-backed voles, *Clethrionomys californicus,* and deer mice, *Peromyscus maniculatus.* Canadian Field-Naturalist 101:543–546.

Helfield, J. M. 2001. Interactions of salmon, bear and riparian vegetation in Alaska. Doctoral dissertation. University of Washington, Seattle.

Helfield, J. M., and R. J. Naiman. 2001. Effects of salmon-derived nitrogen on riparian forest growth and implications for stream productivity. Ecology 82:2403–2409.

Hilderbrand, G. V., T. A. Hanley, C. T. Robbins, and C. C. Schwartz. 1999. Role of brown bears (*Ursus arctos*) in the flow of marine nitrogen into a terrestrial ecosystem. Oecologia 121:546–550.

Johnson, D. H., and T. A. O'Neil. 2001. Wildlife Habitat Relationships in Oregon and Washington. Oregon State University Press, Corvallis.

Kauffman, J. B., M. Mahrt, L. A. Mahrt, and W. D. Edge. 2001. Wildlife of riparian habitats. Pages 361–388 *in* D. H. Johnson and T. A. O'Neil, editors. Wildlife-habitat relationships in Oregon and Washington. Oregon State University Press, Corvallis.

Kelsey, K. A., and S. D. West. 1998. Riparian wildlife. Pages 235–252 *in* R. J. Naiman and R. E. Bilby, editors. River ecology and management: lessons from the Pacific Coastal Ecoregion. Springer-Verlag, New York.

King, F. W. 1988. Crocodiles: Keystone wetland species. Pages 18–19 in G. H. Dalyrymple, W. F. Loftus, and F. S. Bernardino, editors. Wildlife in the Everglades and Latin American wetlands. Abstracts of the proceedings of the First Everglades National Park Symposium, Miami 1985. Florida International University, the State University of Florida at Miami, and Everglades National Park Publishers.

Knopf, F. L. 1985. Significant riparian vegetation to breeding birds across an altitudinal cline. Pages 105–111 *in* R. R. Johnson, C. D. Ziebell, D. R. Patton, P. F. Folliot, and R. H. Hamre, editors. Proceedings of a Conference on Riparian Ecosystems and Their Management: Reconciling Conflicting Uses, April 16–18, Tucson, AZ. USDA Forest Service, Rocky Mountain Forest and Range Experiment Station, General Technical Report RM-GTR-120, Fort Collins, Colorado.

Lock, P. A., and R. J. Naiman. 1998. Effects of stream size on bird community structure in coastal temperate forests of the Pacific Northwest. Journal of Biogeography 25:773–782.

Loegering, J. P., and R. G. Anthony. 1999. Distribution, abundance, and habitat association of riparian-obligate and -associated birds in the Oregon Coast Range. Northwest Science 73:168–185.

Lohr, S. M., S. A. Gauthreaux, and J. C. Kilgo. 2002. Importance of coarse woody debris to avian communities in Loblolly pine forests. Conservation Biology 16:767–777.

Mac Nally, R., A. Parkinson, G. Horrocks, L. Conole, and C. Tzaros. 2001. Relationships between terrestrial vertebrate diversity, abundance and availability of coarse woody debris on south-eastern Australian floodplains. Biological Conservation 99:191–205.

Maser, C., R. G. Anderson, and K. J. Cromack. 1979. Dead and downed woody material. Pages 78–95 *in* J. W. Thomas, technical editor. Wildlife habitats in managed forests: the Blue Mountains of Oregon and Washington. U.S. Forest Service Agricultural Handbook No. 553. Wildlife Management Institute, Washington D.C. and Bureau of Land Management.

Maser, C., R. F. Tarrant, J. M. Trappe, and J. F. Franklin. 1988. From the forest to the sea: a story of fallen trees. USDA Forest Service, Pacific Northwest Research Station, General Technical Report PNW-GTR-229, Portland, Oregon.

Mason, D. T., and J. Koon. 1985. Habitat values of woody debris accumulations of the lower Stehekin River, with notes on disturbances of alluvial gravels. Final Report to the National Park Service, Contract CX-9000–3-EO66, Bellingham, Washington.

McComb, W. C., C. L. Chambers, and M. Newton. 1993a. Small mammal and amphibian communities and habitat associations in red alder stands, central Oregon coast range. Northwest Science 67:181–208.

McComb, W. C., K. McGarigal, and R. G. Anthony. 1993b. Small mammal and amphibian abundance in streamside and upslope habitats of mature Douglas-fir stands, western Oregon. Northwest Science 67:7–15.

Melquist, W. E., and M. G. Hornocker. 1983. Ecology of river otters in west central Idaho. Wildlife Monographs 83:1–60.

Merritt, R. W., W. Wuerthele, and D. L. Lawson. 1984. The effect of leaf conditioning on the timing of litter processing on a Michigan woodland floodplain. Canadian Journal of Zoology 62:179–182.

Minckley, W. L., and J. N. Rinne. 1985. Large woody debris in hot-desert streams: an historical review. Desert Plants 7:142–153.

Montgomery, D. R., B. D. Collins, J. M. Buffington, and T. B. Abbe. 2003. Geomorphic effects of wood in rivers. Pages 21–47 *in* S. V. Gregory, K. L. Boyer, and A. M. Gurnell, editors. The ecology and management of wood in world rivers. American Fisheries Society, Symposium 37, Bethesda, Maryland.

Montgomery, D. R., E. M. Beamer, G. R. Pess, and T. P. Quinn. 1999. Channel type and salmonid spawning distribution and abundance. Canadian Journal of Fisheries and Aquatic Sciences 56:377–387.

Montgomery, D. R., J. M. Buffington, R. D. Smith, K. M. Schmidt, and G. R. Pess. 1995. Pool spacing in forest channels. Water Resources Research 314:1097–1105.

Motroni, R. 1980. The importance of riparian zones to terrestrial wildlife: an annotated bibliography. Prepared for the Sacramento District's Sacramento River and Tributaries Investigation, U.S. Army Corps of Engineers, Sacramento, California.

Mutz, M. 2003. Hydraulic effects of wood in streams and rivers. Pages 93–107 *in* S. V. Gregory, K. L. Boyer, and A. M. Gurnell, editors. The ecology and management of wood in world rivers. American Fisheries Society, Symposium 37, Bethesda, Maryland.

Nussbaum, R. A., E. D. J. Brodie, and R. M. Storm. 1983. Amphibians and reptiles of the Pacific Northwest. University of Idaho Press, Moscow.

O'Connell, M. A., J. G. Hallett, and S. D. West. 1993. Wildlife use of riparian habitats: a literature review. Washington Department of Natural Resources, Timber, Fish, and Wildlife Program, TFW-WL1–93-001, Olympia.

O'Connell, M. A., J. G. Hallett, S. D. West, K. A. Kelsey, D. A. Manuwal, and S. F. Pearson. 2000. Effectiveness of riparian management zones in providing habitat for wildlife. Washington Department of Natural Resources, Final Report to Timber, Fish, and Wildlife Program, TFW-LWAG1–00-001, Olympia.

Piégay, H. 2003. Dynamics of wood in large rivers. Pages 109–133 *in* S. V. Gregory, K. L. Boyer, and A. M. Gurnell, editors. The ecology and management of wood in world rivers. American Fisheries Society, Symposium 37, Bethesda, Maryland.

Planz, J. V., and G. L. J. Kirkland. 1992. Use of woody ground litter as a substrate for travel by the white-footed mouse, *Peromyscus leucopus*. Canadian Field-Naturalist 106:118–121.

Plumpton, D. L., and J. K. J. Jones. 1992. *Rhynchonycteris naso*. Mammalian Species 413:1–5.

Pollock, M. M., M. Heim, and D. Werner. 2003. Hydrologic and geomorphic effects of beaver dams and their influence on fishes. Pages 213–233 *in* S. V. Gregory, K. L. Boyer, and A. M. Gurnell, editors. The ecology and management of wood in world rivers. American Fisheries Society, Symposium 37, Bethesda, Maryland.

Rector, M. 1991. Small mammal and carabid beetle communities on Mt. St. Helen's. Master's thesis. University of Washington, Seattle.

Sabo, J. L. 2000. River-watershed exchange: effects of rivers on the population and community dynamics of riparian lizards (*Sceloporus occidentalis*). Doctoral dissertation. University of California, Berkeley.

Shearer, C. A., and S. B. Von Bodman. 1983. Patterns of occurrence of ascomycetes associated with decomposing twigs in a midwestern stream. Mycologia 75:518–530.

Stauffer, D. F., and L. B. Best. 1980. Habitat selection by birds of riparian communities: evaluating effects of habitat alterations. Journal of Wildlife Management 44:1–15.

Steel, E. A. 1993. Woody debris piles: habitat for birds and small mammals in the riparian zone. Master's thesis. University of Washington, Seattle.

Steel, E. A., R. J. Naiman, and S. D. West. 1999. Use of woody debris piles by birds and small mammals in a riparian corridor. Northwest Science 73:19–26.

Swanson, F. J. 2003. Wood in rivers: a landscape perspective. Pages 299–313 *in* S. V. Gregory, K. L. Boyer, and A. M. Gurnell, editors. The ecology and management of wood in world rivers. American Fisheries Society, Symposium 37, Bethesda, Maryland.

Szaro, R. C. 1980. Factors influencing bird populations in southwestern riparian forests. Pages 403–418 *in* Proceedings of Conference on Management of Western Forests and Grasslands for Nongame Birds. USDA Forest Service, Intermountain Forest and Range Experiment Station, General Technical Report GTR-INT-86, Fort Collins, Colorado.

Szaro, R. C., S. C. Belfit, K. Aitkin, and J. N. Rinne. 1985. Impact of grazing on a riparian garter snake. Pages 359–363 *in* R. R. Johnson, C. D. Ziebell, D. R. Patton, P. F. Folliot, and R. H. Hamre, editors. Proceedings of a Conference on Riparian Ecosystems and Their Management: Reconciling Conflicting Uses, April 16–18, Tucson AZ. USDA Forest Service, Rocky Mountain Forest and Range Experiment Station, General Technical Report RM-GTR-120, Fort Collins, Colorado.

van Horne, B. 1983. Density as a misleading indicator of habitat quality. Journal of Wildlife Management 47:893–901.

Webb, A. A., and W. D. Erskine. 2003. Distribution, recruitment, and geomorphic significance of large woody debris in an alluvial forest stream: Tonghi

Creek, southeastern Australia. Geomorphology 51:109–126.

Weir, R. 1993. Use of riparian and riparian-associated habitats by Fishers. Pages 49–56 *in* K. H. Morgan1 and M. A. Lashmar, editors. Riparian habitat management and research. Proceedings of a workshop sponsored by Environment Canada and the British Columbia Forestry Continuing Studies Network held in Kamloops, B.C., 4–5 May, 1993. Fraser Action Plan Special Publication. Available at http://www.rem.sfu.ca/FRAP/rhmr.pdf.

Wilkins, R. N., and N. P. Peterson. 2000. Factors related to amphibian occurrence and abundance in headwater streams draining second-growth Douglas-fir forests in southwestern Washington. Forest Ecology and Management 139:79–91.

Zalewski, M., M. Lapinska, and P. B. Bayley. 2003. Fish relationships with wood in large rivers. Pages 195–211 *in* S. V. Gregory, K. L. Boyer, and A. M. Gurnell, editors. The ecology and management of wood in world rivers. American Fisheries Society, Symposium 37, Bethesda, Maryland.

American Fisheries Society Symposium 37:249–263, 2003

Influence of Wood on Aquatic Biodiversity

STEVEN M. WONDZELL AND PETER A. BISSON

USDA Forest Service, Pacific Northwest Research Station, Olympia Forestry Sciences Laboratory 3625 93rd Avenue SW, Olympia, Washington 98512-9193, USA

Abstract.—We review published literature examining the role of wood in mediating biodiversity in aquatic ecosystems, identifying the components of biodiversity, taxonomic groups, and scales that have been studied, and highlight gaps in existing knowledge. The components of biodiversity most frequently studied include species diversity (or richness) of macroinvertebrates and fishes, structural complexity within habitat units, and the diversity of habitats found in a stream reach. Many of these studies show that large wood increases biodiversity by providing stable, hard substrates for colonization by periphyton and macroinvertebrates; by increasing microhabitat complexity; and by shaping channel morphology by controlling patterns of erosion and deposition in stream reaches. The abundance of wood in channels, as well as its functional role, varies greatly in longitudinal, lateral, and vertical dimensions along the river corridor. The influence of wood on community structure and ecosystem processes also varies across these dimensions and from stream headwaters to river mouths and nearshore marine environments. Thus, wood can influence biodiversity at all of these scales. Numerous studies, however, have failed to show an effect of wood on biodiversity. These conflicting results illustrate that wood abundance, its functional role in streams, and its influence on biodiversity depend on a variety of factors, and it is the total effect of all these factors, not simply the presence of large wood, that determines patterns of biodiversity.

Introduction

The objective of this paper is to review published literature examining the role of wood in mediating biodiversity in aquatic ecosystems. We take a broad view of biodiversity, one that considers the multiple components of biodiversity that might be important in aquatic and riparian ecosystems, including genetic, age-class, and life history diversity of individual species; species richness or diversity; life-form or functional-group diversity; habitat diversity (or habitat complexity); ecosystem diversity; and landscape diversity.

A review of aquatic and riparian literature in a recently compiled bibliography of papers (http://riverwood.oregonstate.edu/html/intro.html v. 1) examining the role of large wood in the world's rivers turned up 54 papers that examined some component of biodiversity or species richness. These 54 papers comprise less than 5% of the publications cited. Further, none of these 54 studies focused solely on the relation between large wood (logs or snags) and species diversity or species richness in lakes, streams, or riparian zones. We realize that our literature search was limited or may have missed references. Even so, our examination identified several clear areas of past research emphasis and highlighted gaps in existing knowledge. First, the importance of wood as an element of physical habitat and its role in shaping channel morphology are the most widely studied aspects of the relation between biodiversity and large wood, accounting for the greater part of all the published literature (Table 1). Second, most studies examine only species diversity as it relates to the abundance of large wood or the role of wood in creating habitat or microhabitat diversity. Relatively few studies link the role of wood in creating habitat diversity to species diversity or species richness (Table 2). Third, we found no studies linking large wood to any of the other components of biodiversity (Table 2). Fourth, of the studies examining species diversity, the most commonly studied taxa are invertebrates, followed closely by fishes (Table 3). By comparison, studies of the role of large wood and biodiversity in riparian zones or of other types of organisms are quite rare. Finally, studies of large wood and biodiversity largely focus on individual channel-units (for example, pools or riffles) or short reaches (less than

TABLE 1. The 54 (of 1,192) literature citations examining the role of large wood in biodiversity of aquatic ecosystems, categorized by the topic of biodiversity studied; some citations fall into more than one category[a].

Topic	Number	Percentage
Source of food, energy, or nutrients	1	2
Wood as habitat		
Wood provides a substrate	20	37
Wood adds complexity to habitat unit	25	46
Added cover (predation)	4	7
Visual screening (competition)	1	2
Current velocity (refuge)	5	9
Disturbance (refuge)	2	4
Wood provides a unique habitat	8	15
Wood as a factor in habitat formation	21	39

[a] Literature source: International Wood in World Rivers, Wood Bibliography (http://riverwood.oregonstate.edu/html/intro.html v. 1).

500 m). Although large wood can influence migratory species in different ways and in different parts of the stream network depending on age-class and life form, we failed to find studies examining the influence of wood on biodiversity at the scale of an entire river network (Table 4). Clearly, the narrowly focused topic of large wood and biodiversity has received comparatively little attention from aquatic and riparian ecologists. Instead, most effort has been devoted to understanding the functional significance of large wood in aquatic ecosystems.

The Role of Large Wood Relative to Biodiversity

Large wood potentially influences biodiversity in aquatic ecosystems in many ways and at multiple spatial scales. Large wood is a source of organic matter and nutrients for aquatic insects (Benke and Wallace 2003, this volume). Thus, wood influences both the trophic structure and productivity of aquatic food webs and thereby influences both species composition and diversity. Wood also serves as habitat, for example, for wood-boring or mining invertebrates and for salamanders nesting in rotten logs. More typically, however, wood is viewed as an element of physical habitat, either providing a specific substrate, especially for biofilms and macroinvertebrates, or by adding complexity to existing habitats, such as pools. Also, large wood can control patterns of erosion and deposition of sediment, thus helping shape channel morphology and helping determine the types and relative abundance of physical habitats present in a stream reach. The role of wood in streams varies with location along the stream network. Wood recruitment to streams is often episodic, with

TABLE 2. The 54 (of 1,192) literature citations examining the role of large wood in biodiversity of aquatic ecosystems, categorized by the component of biodiversity examined[a].

Diversity component	Number	Percentage
Genetic	0	
Age-class	0	
Life-form	0	
Dietary	1	2
Species	29	54
Habitat	33	61
Ecosystem	0	
Landscape or network	0	

[a] Literature source: International Wood in World Rivers, Wood Bibliography (http://riverwood.oregonstate.edu/html/intro.html v. 1).

TABLE 3. The 54 (of 1,192) literature citations examining the role of large wood in biodiversity of aquatic ecosystems, categorized by the taxon of interest; some citations fall into more than one category[a].

Topic	Number	Percentage
Physical habitat studies (no other taxon)	9	17
Aquatic studies	44	81
Fungi	2	4
Macrophytes	1	2
Invertebrates	21	39
Crustaceans	1	2
Amphibians	2	4
Fish	17	31
Riparian studies	6	11
Moss	1	2
Plants	3	6
Other vertebrates	2	4

[a] Literature source: International Wood in World Rivers, Wood Bibliography (http://riverwood.oregonstate.edu/html/intro.html v. 1).

large pulses of wood entering streams during, or shortly after, major disturbances. Further, floods and debris flows can significantly restructure a stream's wood load. This high spatial and temporal variability in wood loading is likely to result in changes in biodiversity at multiple spatial and temporal scales.

Wood as a unique substrate or habitat

Several studies showed that wood provides a unique habitat to which some species are specially adapted or provides habitat critical for some aspect of a species' life history. Examples of wood specialists include the wood-boring isopods called gribbles *Limnoria* spp. and the wood-boring mollusks called shipworms (naval shipworm *Teredo navalis* and feathery shipworm *Bankia setacea*) that occupy bays and estuaries (Maser and Sedell 1994). Some freshwater macroinvertebrates are also obligate wood specialists, including the functional groups of miners and tunnelers, such as the larval cranefly *Lipsothrix* spp., the larval riffle beetle *Lara avara*, and species of caddisflies that gauge attachment sites on wood (Anderson et al. 1978; Wallace and Anderson 1996; Wallace et al. 1996; Benke and Wallace 2003). Although relatively few species require wood, many are facultative wood-using species. These include many aquatic macroinvertebrates that use wood as an attachment site when it is available (Anderson et al. 1978; Benke and Wallace 2003) and species like the Van Dyke's salamander *Plethodon vandykei*, a species that can nest in moderately decayed stumps and logs (Blessing et al. 1999). Wood also provides important basking habitat for several species of turtles in the southeastern United States, in environments where other basking habitats are not available (Lindeman 1999). Clearly, the

TABLE 4. The 54 (of 1,192) literature citations examining the role of large wood in biodiversity of aquatic ecosystems, categorized by the spatial scale of the study; some citations fall into more than one category[a].

Scale	Number	Percentage
Particle (or piece)	13	24
Subunit	25	46
Channel unit	21	39
Stream reach	31	57
Stream section	0	0
Stream network	0	0

[a] Literature source: International Wood in World Rivers, Wood Bibliography (http://riverwood.oregonstate.edu/html/intro.html v. 1).

presence of highly specialized, wood-obligate species depends on the presence of wood; similarly, the abundance of many facultative wood-using species is also dependent on wood, especially in environments that lack alternative hard substrates. Thus, wood contributes directly to the richness of species in these communities (Wallace et al. 1996; Benke and Wallace 2003).

Wood as a source of nutrients or energy

In general, wood does not provide a readily available form of energy for stream ecosystems. It is difficult to decompose, in part because lignin and cellulose are inherently resistant to decomposition, but also because lignin breakdown is dominantly an aerobic process (Suberkropp 1998; Bilby 2003, this volume). Water-saturated logs in streams and rivers become anaerobic, except near the wood surface where oxygen can diffuse into the log. Consequently, only the surface layers of the logs decompose readily (Harmon et al. 1986; Bilby et al. 1999). Because wood is more resistant to decomposition than leaves and other organic matter, it does not represent a significant energy source to most temperate stream food webs. Although many species of miners, gougers, scrapers, shredders, and collectors are abundant on wood, it is often the biofilm on the wood surface, not the wood substrate itself, that is the primary food for these macroinvertebrates (Bowen et al. 1998; Bisson and Bilby 1998). More recent studies do indicate, however, the use of wood as a primary food source for several aquatic insect taxa (Hoffman and Hering 2000). Although wood is a poor source of energy and nutrients, wood is often a key feature trapping other organic materials, forming what are commonly called "debris dams" (Angermeier and Karr 1984; Smock et al. 1989; Wallace et al. 1995; Murphy and Meehan 1991; Diez et al. 2000). Organic matter collected on or around large wood is an important substrate for heterotrophic organisms (Anderson and Sedell 1979; Triska et al. 1982; Gregory et al. 1987; Smock et al. 1989; Casas 1997). Thus, wood can strongly influence trophic structure, helping to determine the composition of macroinvertebrate communities and the relative abundance of functional-feeding groups in stream ecosystems (Anderson and Sedell 1979; Benke et al. 1985; Smock et al. 1989; Wallace et al. 1995; Wallace et al. 1996).

Wood as a substrate

One of the most studied aspects of wood and biodiversity is its influence as a substrate (Table 1). Studies have examined the importance of wood in providing a relatively stable, hard substrate in a variety of aquatic ecosystems, where it can support more species-rich communities of periphyton and benthic invertebrates (Benke et al. 1985). Research results on the role of large wood in biodiversity are often conflicting, however. The clearest evidence for increased biodiversity comes from studies of low-gradient streams, estuaries, and lakes with soft-textured bed sediments where wood is the only hard, stable substrate (Benke et al. 1985; Benke and Wallace 2003). If wood pieces are large or securely imbedded in the stream channel (including tree roots exposed in streambanks), they are less likely than the streambed sediment to be disturbed or moved during high flows (Wallace and Benke 1984; Benke and Wallace 2003). Similarly, because wood breaks down slowly, it lasts longer than leaves or other organic substrates (Wallace and Benke 1984; Wallace et al. 1996). Thus, in many low gradient sand or fine-sediment bedded streams, wood-dwelling macroinvertebrates comprise a substantial proportion of the total macroinvertebrate biomass and are an important food source for fish and other consumers (Angermeier and Karr 1984; Benke et al. 1985; Smock et al. 1989; Benke and Wallace 2003). Clearly, wood, when present, provides an important substrate for a variety of aquatic organisms, although its functional importance in stream ecosystems and its influence on biodiversity may depend on the availability of alternative, hard, stable substrates (also see Benke and Wallace 2003).

Wood surfaces also provide diverse microhabitats. Greater species richness of macroinvertebrates has been observed on rough-conditioned wood than on either unconditioned or smooth wood (Magoulick 1998). Wood properties change with exposure to water, physical abrasion, and decomposition; thus, surface complexity tends to increase with time. The complex microhabitats of decayed wood appear to contribute to increased biodiversity of macroinvertebrate communities (Hax and Golladay 1993). O'Connor (1991) found that rough-textured wood supported greater macroinvertebrate species richness, which was attributed to increased habitat complexity. Similarly, differences in bark texture among tree species can also contribute to microhabitat variation, apparently increasing microhabitat diversity and potentially leading to changes in species diversity of macroinvertebrates in aquatic ecosystems (Bowen et al. 1998). However, not all studies have shown relations between species and microhabitat diversity (France 1997; Magoulick 1998). In some cases, macroinvertebrates may simply be opportunistic colonizers of available hard substrates (France 1997). In other cases, analyses focused on species diversity—habitat complexity relations may miss more complex functional relations between macroinvertebrates and woody substrates (O'Connor 1991) or the effects of recent disturbances (Anderson 1992) that also affect community composition.

Wood creates complexity in habitat units

A second commonly studied aspect of wood and biodiversity is its influence in creating complexity in a habitat unit, such as a pool, riffle, or rapid (Table 1). A large body of research documents that wood-created habitat complexity serves important functions in many aquatic ecosystems. Wood can visually isolate individuals and thereby reduce competition (Dolloff and Reeves 1990; Crook and Robertson 1999; Dolloff and Warren, Jr. 2003, this volume); wood also provides a variety of refugia (Sedell et al. 1990; Crook and Robertson 1999), including hiding cover (Angermeier and Karr 1984; Everett and Ruiz 1993; Inoue and Nakano 1998). Log accumulations may also be good foraging sites for predatory macroinvertebrates and for fish feeding on macroinvertebrates (Benke et al. 1985; Wallace et al. 1996; Crook and Robertson 1999; Benke and Wallace 2003). Wood in pools also creates microhabitat patches (channel-subunit scale) with varying depths and flow velocities, which can contribute to microhabitat partitioning of co-occurring species (Reeves et al. 1997; Reeves et al. 1998; Thévenet and Statzner 1999) and can result in increased diversity of fish species and age-classes (Bisson et al. 1992). During floods, structural complexity created by wood provides refugia that can reduce disturbance to benthic macroinvertebrates (Borchardt 1993; Palmer et al. 1996) and contribute to the overwinter persistence of salmonids in a stream reach (McMahon and Hartman 1989; Quinn and Peterson 1996; Harvey et al. 1999). Wood is especially important in areas where alternative sources of structural complexity are not available, for example, in rivers and estuaries where streambed sediment is soft and rock, emergent vegetation, and reefs or reef-like structures are lacking (Everett and Ruiz 1993; Wallace et al. 1996; Benke and Wallace 2003) or in soft-bottomed lakes lacking macrophytes and other physical structures (France 1997).

We know of no examples in which wood decreases measures of diversity in habitat units. Linking the effects of increased habitat diversity to other components of biodiversity, however, is problematic. Some studies have shown that increased habitat diversity was correlated to increased diversity of fishes (Reeves et al. 1993; Thévenet and Statzner 1999). Other studies have failed to show this relationship. For example, the amount of wood in pools did not influence salmonid diversity in Oregon Coast Range streams (Chen 1999); coastal cutthroat trout *Oncorhynchus clarki clarki* density did not differ between simple pools and pools with wood-formed habitat diversity, which was attributed to a lack of interspecific competition (Simondet 1997). Also, adding wood to a tropical forest stream had little influence on populations of freshwater shrimp, probably because habitat structure was provided by cobbles and boulders (Pyron et al. 1999). Clearly, diversity at the microhabitat and habitat-unit scale can be functionally important, at least to some species and life forms, especially where wood-related habitat diversity contributes to functional processes in aquatic ecosystems that would otherwise be lacking (Dolloff and Warren 2003). However, separating the influence of diversity from other confounding effects or isolating the effect of diversity at the microhabitat scale from diversity at larger scales can be difficult.

Wood as a geomorphic element creating a diversity of habitats

A third commonly studied aspect of wood and biodiversity is its influence in creating a diversity of habitat units in a stream reach (Table 1). Wood-forced patterns of scour and deposition are often important in determining the type and number of channel units in alluvial reaches of streams. This relation between wood and channel morphology has been widely studied in streams of the northern Pacific coast of North America and in a wide variety of streams throughout the world (Heede 1972; Bilby 1981; Mosley 1981; Trotter 1990; Ebert et al. 1991; Shields and Smith 1992; Dose and Roper 1994; Assani and Petit 1995; Fetherston et al. 1995; Wallace et al. 1995; Abbe and Montgomery 1996; Dolloff 1996; Wood-Smith and Buffington 1996; Hilderbrand et al. 1997; Gurnell and Sweet 1998; Ward et al. 1999b; Gerhard and Reich 2000; Gurnell et al. 2002). The relation between wood and channel morphology is not present in all streams and rivers, however (Inoue and Nakano 1998).

Effects of large wood on channel morphology depend on the amount of wood, its size and orientation in the stream channel, and the size and type of stream (Keller and Swanson 1979; Swanson et al. 1982; Bilby and Bisson 1998; Montgomery and Buffington 1998; Beechie et al. 2000; Gurnell et al. 2002). These relations are well described for mountain stream networks. In small streams, single logs, as well as logjams, frequently form channel-spanning blockages and can alter channel morphology by obstructing the flow of water, forming pools, and storing sediment (Bilby 1981; Andrus et al. 1988; Bilby and Ward 1989; Nakamura and Swanson 1993; Richmond and Fausch 1995; Wallace et al. 1995; Gurnell et al. 2002). If logs are oriented parallel to the stream, or only partially block the stream, they tend to create scour pools.

In large streams and rivers, single logs are not large enough to control channel-forming processes. Instead, large logjams are required. These jams can block channels during floods, driving channel avulsions (Gottesfeld and Johnson-Gottesfeld 1990; Wondzell and Swanson 1999), helping form secondary channels and other off-channel habitat, which is important for many aquatic species (Sedell et al. 1984; Bryant et al. 1991; Crispin et al. 1993). Wood is also deposited on streambanks, gravel bars, and floodplains during floods, influencing channel morphology by stabilizing streambanks, creating scour pools, or helping form gravel bars (Malanson and Butler 1990; Fetherston et al. 1995; Richmond and Fausch 1995; Abbe and Montgomery 1996; Bilby and Bisson 1998; Montgomery and Buffington 1998; Edwards et al. 1999; Braudrick and Grant 2001; Gurnell et al. 2002). Large logjams can be long-lasting, protecting and stabilizing gravel bars in the active channel of braided rivers, initiating a successional sequence that will give rise to vegetated mid- channel islands (Fetherston et al. 1995; Abbe and Montgomery 1996; Edwards et al. 1999). All these processes help form complex, multiple-channel floodplains in large rivers, areas widely recognized for high biodiversity (Brown 1997; Ward et al. 1999a, 1999b). Because of the functional relations between wood and channel morphology, wood is widely recognized as contributing to habitat diversity in streams (Bisson et al. 1987).

Several studies have shown that the amounts of wood present in a stream reach are correlated to the number of pools and sometimes to either the area or volume of pool habitats in those streams (Murphy et al. 1986; House and Boehne 1987; Andrus et al. 1988; Carlson et al. 1990; Crispin et al. 1993; Richmond and Fausch 1995; Dolloff 1996; Gurnell and Sweet 1998). Management practices that reduce the amounts of wood in streams often lead to reductions in pool volume and area (Murphy et al. 1986; McIntosh et al. 1994) and reduce habitat diversity (Shields and Smith 1992). Changes in the relative abundance of different habitat units, especially pools and riffles, can have marked influence on the diversity of fish species and age-classes in streams. For example, comparing a wood-rich stream with a wood-poor stream in western Washington showed that the wood-poor stream was dominated by riffle habitats and the age-0 and age-1 steelhead trout *O. mykiss*, which prefer riffle environments. The wood-rich stream, however, was dominated by pools, and fish biomass was more evenly distributed among the three species present and among multiple age-classes of these species (Hicks et al. 1991). Studies of sand-bedded streams have shown similar results, with greater richness of fish communities in streams with more abundant wood and more pools (Ebert et al. 1991; Shields et al. 1994).

Several studies have compared composition and productivity of macroinvertebrate communities between forested and clear-cut reaches of streams (Murphy and Hall 1981; Anderson 1992) and among forest, plantation, and pasture reaches (Quinn et al. 1997), where the nonforest reaches have reduced amounts of instream large wood or wood is entirely lacking. Several of these studies found that the overall richness of macroinvertebrates differed little among reaches (Anderson 1992; Quinn et al. 1997) or was greater in nonforested reaches lacking large wood (Murphy and Hall 1981). Stream productivity, however, was limited by available light in the examples cited, so that forest removal increased primary production and resulted in large changes in macroinvertebrate community composition.

Many of the studies that have examined the role of large wood in streams have compared areas with different land uses. In the Pacific Northwest of the United States and Pacific coastal regions of Canada and Alaska, comparisons between clear-cut and unharvested areas dominate much of the published literature. Many of these studies focus on large wood, relating the amounts of large wood present in the stream to the attributes of interest, for example, numbers of pools, abundance of fish, or diversity of macroinvertebrate communities. However, wood abundance is only one of several stream attributes affected by land use patterns (Gregory et al. 1987; Hanchet 1990). For example, a study in New Zealand comparing native forest with pasture to examine effects of land use on native fish communities could not isolate the effect of wood abundance from the effects of shade, cover provided by instream vegetation, water temperature, and substrate coarseness (Hanchet 1990). Similarly, Quinn et al. (1997) concluded that several factors, including light, temperature, productivity, water quality, fine sediment, current velocity, and channel morphology—in addition to the presence and abundance of wood—may have contributed to differences in macroinvertebrate communities among native forest, plantations, and pasture. Collectively, the studies cited above show that the effect of wood on biodiversity depends strongly on the functional role of large wood in stream ecosystems. In some cases, wood strongly influences biodiversity; in other cases, wood appears to have little influence on biodiversity. In all cases, however, isolating the influence of wood from other confounding factors is difficult.

Organic-matter retention and sediment storage

Large wood effectively traps small pieces of wood and other organic materials, such as leaves, needles, and twigs (Angermeier and Karr 1984; Smock et al. 1989; Gregory et al. 1993; Wallace et al. 1995; Murphy and Meehan 1991; Diez et al. 2000) and even carcasses of dead fish such as salmon (Cederholm and Peterson 1985). Further, the abrasion or erosion of the surface of large wood can be a significant source of fine particulate organic matter (Ward and Aumen 1986). Wood thus influences the retention of organic matter in stream reaches (Bilby 1981; Harmon et al. 1986; Bisson and Bilby 1998). Because the organic matter collected on or around large wood is an important substrate for heterotrophic organisms and the primary consumers that feed on them, large wood can help determine stream trophic relations (Anderson and Sedell 1979; Triska et al. 1982; Gregory et al. 1987; Smock et al. 1989; Casas 1997), the diversity and composition of the macroinvertebrate communities, and the relative abundance of functional-feeding groups present in a stream reach (Anderson and Sedell 1979; Benke et al. 1985; Smock et al. 1989; Wallace et al. 1995; Wallace et al. 1996).

Large wood influences patterns of erosion, transport, and deposition of sediment at many different scales in stream networks (Palmer et al. 1996; Wallace et al. 1996; Buffington and Montgomery 1999; Lancaster et al. 2001). Wood can create low streamflow-velocity environments, characterized by deposition of fine-textured sediment that function as refugia for some macroinvertebrates during floods (Palmer et al. 1996). Also, patterns of sediment deposition create "textural patches" characterized by different grain-size distributions and can provide greater diversity of aquatic habitats than otherwise might be present (Gerhard and Reich 2000). For example, wood can create patches of sediment with grain-sizes suitable for fish that must spawn in substrates of a specific size (Buffington and Montgomery 1999). Large wood can also control sediment storage, especially in small mountain streams (Nakamura and Swanson 1993; Montgomery et al. 1996), where it forms pool-step sequences. The change in the longitudinal gradient of streams across pool-step sequences is the primary factor driving hyporheic exchange flow in small mountain streams (Harvey and Bencala 1993; Kasahara and Wondzell, in press), which creates unique physical, chemical, and hydrologic environments in streams and riparian zones that provide a diversity of habitats for many specially adapted macroinvertebrates (Stanford and Ward 1988). Similarly, large wood can control deposition of sediment in active stream channels, initiating formation of sand or gravel bars that can eventually become vegetated islands in a braided channel (Edwards et al. 1999; Ward et al. 1999b). Thus, wood controls sediment and organic matter storage in streams at scales ranging from the particle and microhabitat up to entire stream reaches, and it contributes to habitat diversity at these scales.

Wood and Biodiversity over Scales of Space and Time

Spatial scales

Spatial scale in streams has been defined geomorphologically, starting from the particle through the sub-unit, channel geomorphic unit, stream reach, river section, and up to the network scale (Frissell et al. 1986; Grant et al. 1990; Sedell et al. 1990; Gregory et al. 1991). We have examined the role of large wood in mediating biodiversity, organizing our discussion around the functional role of wood, and we have shown that the functional role of large wood and its effects on biodiversity have been studied at spatial scales ranging from the particle through the stream reach (Table 4). At larger spatial scales, however, the role of wood and its importance to maintaining biodiversity have received comparatively little attention, despite research showing that large wood influences habitat complexity from the headwater reaches of small streams through low-land sections of large rivers, in tidal reaches and estuaries, on ocean beaches, and even on the floor of the ocean (Maser and Sedell 1994). Further, the role of large wood, the way it is distributed in streams (Swanson et al. 1982; Gregory et al. 1993; Bilby and Bisson 1998; Martin and Benda 2001; Gurnell et al. 2002), and the way it influences streams and their biological communities changes with location in the stream network (Sedell et al. 1982; Bisson et al. 1987; Bilby and Ward 1989; Crook and Robertson 1999).

We draw from a study of fall chinook salmon *O. tshawytscha* in the Sixes River, Oregon (Reimers 1971) to illustrate the possible influence of large wood at the network scale. The life histories of anadromous salmonids include exposure to a sequence of habitats, from small tributary streams to the ocean and back again. Populations must evolve adaptations to the diversity of aquatic habitats available in the stream network, and often, this evolution leads to a strategy of spreading risk among different life history alternatives. Fall chinook salmon in the Sixes River, Oregon display five life history strategies (Table 5) defined by differences in rearing times in tributary streams, the mainstem river, and the estuary (Reimers 1971). The importance of wood in determining salmonid habitat quality in tributary and mainstem rivers has been widely studied (Bisson et al. 1987; McMahon and Hartman 1989; Bilby and Bisson 1998), and even in estuaries, most juvenile chinook were associated with wood (Reimers 1971). Presumably, the relative success of different life history strategies, each spending different lengths of time in tributary, mainstem, and estuary habitat, would be related, at least in part, to the abundance of large wood. Reimers (1971) examined this question in the Sixes River, finding that more than 90% of returning adults were of life history type 3. Reimers (1971) also hypothesized that improved tributary and mainstem habitat could lead to increased numbers of life history types 4 and 5. Despite the importance of large wood in streams and rivers of the north Pacific Coast of North America and the documented importance of wood to the habitat of salmon that inhabit those rivers, we know of no studies that have examined the role of large wood and habitat complexity in maintaining the relative abundance and long-term viability of alternative life histories of salmon at the scale of an entire stream network.

Schlosser and Angermeier (1995) pointed out that the lack of understanding of the links between the diversity of habitats available at a network scale and metapopulation dynamics are a critical gap in knowledge for fish conservation. An important topic for future conservation research will be to understand the importance of habitat diversity across scales in maintaining both genetic and life history diversity of individual species as well as species diversity. Further, there is a need to understand the role of biophysical processes in maintaining that habitat diversity and to understand how the spatial distribution and connectivity of habitat patches at river network, or larger scales, influences biodiversity.

Temporal scale

Wood abundance changes naturally over time at both site and watershed scales. Wood enters chan-

TABLE 5. Five different types of life-history strategies of fall chinook salmon in the Sixes River, Oregon for 1 year after emergence (em) of salmon fry from streambed gravels. Emergence occurs between March and May; peak emergence is in mid to late April, with some protracted emergence into June. Length of juvenile residence in tributary (T), mainstem river (R), and estuary habitat (E), and timing of out-migration to the ocean (O) are shown. Capital letters, above columns, denote months. From data in Reimers (1971).

Life history	M	A	M	J	J	A	S	O	N	D	J	F	M	A
Type 1	em	em	T	O										
Type 2	em	em	T	R	E	O								
Type 3	em	em	T	R	E	E	E	O						
Type 4	em	em	T	R	R	R	R	O						
Type 5	em	em	T	R	R	R	R	R	R	R	R	R	R	O

nels as the result of streambank erosion during major floods, from tree mortality caused by windstorms, fires, insects or disease, and from other disturbances, such as landslides (Harmon et al. 1986; Lienkaemper and Swanson 1987; Naiman et al. 1992; Dolloff et al. 1994; Reeves et al. 1995; Dolloff 1996; Wallace et al. 1996; Bragg 2000; Martin and Benda 2001; Benda et al. 2002, 2003, this volume; Gurnell et al. 2002; Gurnell 2003; Montgomery et al. 2003; Nakamura and Swanson 2003; all this volume). Once in the stream, mobilization of in-channel wood by fluvial transport during floods, or by debris flows, can cause significant restructuring of a stream's wood load. Reeves et al. (1995) hypothesized that centuries-long cycles of episodic inputs followed by long-term loss of wood and coarse sediment after wildfires drive long-term changes in both the composition and productivity of fish communities in Oregon Coast Range streams. Many of these changes are related to the abundance and physical characteristics of pools anchored by large wood; as wood disappears from stream channels—and in the absence of fresh inputs of wood and coarse sediment—the stream becomes dominated by shallow-water habitats that favor riffle-dwelling species (Hicks et al. 1991). Thus, patterns in wood-controlled biodiversity would be expected to change through time. Some recent studies have begun to examine long-term dynamics of large wood in streams and rivers at landscape scales (Lancaster et al. 2001; Martin and Benda 2001), but these dynamics have yet to be linked to long-term patterns of biodiversity.

Conclusions

Many studies have shown that wood in rivers and streams increases biodiversity, especially species diversity and richness of macroinvertebrates and fishes, structural complexity in habitat units, and the diversity of habitats in a stream reach. Many other studies, however, failed to show an effect of wood on biodiversity. These contrasting results highlight the importance of the functional role of wood in aquatic ecosystems. In some streams, wood provides unique functions: (1) providing the only hard and stable substrate available, (2) creating microhabitat complexity within channel units not provided by other structural elements such as large cobbles and boulders, and (3) controlling the morphology of large alluvial rivers or the number and type of channel units present in a smaller stream. In these cases, large wood often contributes significantly to biodiversity at multiple spatial scales. However, the relationships between large wood and biodiversity can be confounded by many factors (for example, number of fish species present, presence or absence of alternative hard substrates, availability of alternative cover, amounts of light and primary productivity). Further, both land use history and natural disturbances influence the amount of wood present in a stream both because natural disturbance is often related to patterns of wood delivery to streams and because the type of disturbance, time since disturbance, and the land use history are related to the successional age and species composition of the adjacent riparian forests. Further, the relative importance of all of these factors can vary with location within a stream network. Thus, the abundance of wood and its functional significance in a variety of biotic and abiotic processes depend on a variety of factors. It is the combined effect of all these factors, in addition to the presence of large wood, which determines patterns of biodiversity.

Acknowledgments

We thank two anonymous reviewers, Stan Gregory, and Deanna Stouder, for helpful comments. Support for this work was provided by the Interior Columbia Basin Ecosystem Management Project. Additional support was provided by the USDA Forest Service's Pacific Northwest Research Station and both the Aquatic and Land Interactions Program and the Managing Disturbance Regimes Program.

References

Abbe, T. B., and D. R. Montgomery. 1996. Large woody debris jams, channel hydraulics and habitat formation in large rivers. Regulated Rivers Research and Management 12:201–221.

Anderson, N. H. 1992. Influence of disturbance on insect communities in Pacific Northwest streams. Hydrobiologia 248:79–92.

Anderson, N. H., and J. R. Sedell. 1979. Detritus processing by macroinvertebrates in stream ecosystems. Annual Review of Entomology 24:351–377.

Anderson, N. H., J. R. Sedell, L. M. Roberts, and F. J. Triska. 1978. The role of aquatic invertebrates in processing wood debris from coniferous forest streams. American Midland Naturalist 100:64–82.

Andrus, C. W., B. A. Long, and H. A. Froehlich. 1988.

Woody debris and its contribution to pool formation in a coastal stream 50 years after logging. Canadian Journal of Fisheries and Aquatic Sciences 45:2080–2086.

Angermeier, P. L., and J. R. Karr. 1984. Relationships between woody debris and fish habitat in a small warmwater stream. Transactions of the American Fisheries Society 113:716–726.

Assani, A. A., and F. Petit. 1995. Log-jam effects on bed-load mobility from experiments conducted in a small gravel-bed forest ditch. Catena 25:117–126.

Beechie, T. J., G. Pess, P. Kennard, R. E. Bilby, and S. Bolton. 2000. Modeling recovery rates and pathways for woody debris recruitment in northwestern Washington streams. North American Journal of Fisheries Management 20:436–452.

Benda, L. E., P. Bigelow, and T. M. Worsley. 2002. Recruitment of wood to streams in old-growth and second-growth redwood forests, northern California, U.S.A. Canadian Journal of Forest Research 32:1460–1477.

Benda, L. E., D. Miller, J. Sias, D. Martin, R. Bilby, C. Veldhuisen, and T. Dunne. 2003. Wood inputs: quantitative theory, field practice, and modeling. Pages 49–73 *in* S. V. Gregory, K. L. Boyer, and A. M. Gurnell, editors. The ecology and management of wood in world rivers. American Fisheries Society, Symposium 37, Bethesda, Maryland.

Benke, A. C., R. L. Henry III, D. M. Gillespie, and R. J. Hunter. 1985. Importance of snag habitat for animal production in southeastern streams. Fisheries 10:8–13.

Benke, A. C., and J. B. Wallace. 2003. Influence of wood on invertebrate communities in streams and rivers. Pages 149–177 *in* S. V. Gregory, K. L. Boyer, and A. M. Gurnell, editors. The ecology and management of wood in world rivers. American Fisheries Society, Symposium 37, Bethesda, Maryland.

Bilby, R. E. 1981. Role of organic debris dams in regulating the export of dissolved and particulate matter from a forested watershed. Ecology 62:1234–1243.

Bilby, R. E. 2003. Decomposition and nutrient dynamics of wood in streams and rivers. Pages 135–147 *in* S. V. Gregory, K. L. Boyer, and A. M. Gurnell, editors. The ecology and management of wood in world rivers. American Fisheries Society, Symposium 37, Bethesda, Maryland.

Bilby, R. E., and P. A. Bisson. 1998. Function and distribution of large woody debris. Pages 324–346 *in* R. J. Naiman and R. E. Bilby, editors. River ecology and management. Springer-Verlag, New York.

Bilby, R. E., J. T. Heffner, B. R. Fransen, J. W. Ward, and P. A. Bisson. 1999. Effects of immersion in water on deterioration of wood from five species of trees used for habitat enhancement projects. North American Journal of Fisheries Management 19:687–695.

Bilby, R. E., and J. W. Ward. 1989. Changes in characteristics and function of woody debris with increasing size of streams in western Washington. Transactions of the American Fisheries Society 118:368–378.

Bisson, P. A., and R. E. Bilby. 1998. Organic matter and trophic dynamics. Pages 373–398 *in* R. J. Naiman and R. E. Bilby, editors. River ecology and management. Springer-Verlag, New York.

Bisson, P. A., R. E. Bilby, M. D. Bryant, C. A. Dolloff, G. B. Grette, R. A. House, M. L. Murphy, K. V. Koski, and J. R. Sedell. 1987. Large woody debris in forested streams in the Pacific Northwest: past, present, and future. Pages 143–190 *in* E. O. Salo and T. W. Cundy, editors. Streamside management: forestry and fishery interactions. Contribution 57, Institute of Forest Resources, University of Washington, Seattle.

Bisson, P. A., T. P. Quinn, G. H. Reeves, and S. V. Gregory. 1992. Best management practices, cumulative effects, and long-term trends in fish abundance in Pacific Northwest river systems. Pages 189–232 *in* R. J. Naiman, editor. Watershed management: balancing sustainability and environmental change. Springer-Verlag, New York.

Blessing, B. J., E. P. Phenix, L. L. C. Jones, and M. G. Raphael. 1999. Nests of Van Dyke's salamander (*Plethodon vandykei*) from the Olympic Peninsula, Washington Northwest Naturalist 80:77–81.

Borchardt, D. 1993. Effects of flow and refugia on drift loss of benthic macroinvertebrates: Implications for habitat restoration in lowland streams. Freshwater Biology 2:221–227.

Bowen, K. L., N. K. Kaushik, and A. M. Gordon. 1998. Macroinvertebrate communities and biofilm chlorophyll on woody debris in two Canadian oligotrophic lakes. Archiv Fur Hydrobiologie 141:257–281.

Bragg, D. C. 2000. Simulating catastrophic and individualistic large woody debris recruitment for a small riparian system. Ecology 81:1383–1394.

Braudrick, C. A., and G. E. Grant. 2001. Transport and deposition of large woody debris in streams: a flume experiment. Geomorphology 41:263–283.

Brown, A. G. 1997. Biogeomorphology and diversity in multiple-channel river systems. Global Ecology and Biogeography Letters 6:179–185.

Bryant, M. D., P. E. Porter, and S. J. Paustian. 1991. Evaluation of a stream channel-type system for southeast Alaska. United States Department of Agriculture, Forest Service, General Technical Report, Pacific Northwest Research Station GTR-PNW-267, Portland, Oregon.

Buffington, J. M., and D. R. Montgomery. 1999. Effects of hydraulic roughness on surface textures of gravel-bed rivers. Water Resources Research 35:3507–3521.

Carlson, J. Y., C. W. Andrus, and H. A. Froehlich. 1990. Woody debris, channel features, and macroinvertebrates of streams with logged and undisturbed riparian timber in Northeastern Oregon, U.S.A. Canadian Journal of Fisheries and Aquatic Sciences 47:1103–1111.

Casas, J. J. 1997. Invertebrate assemblages associated with plant debris in a backwater of a mountain stream: natural leaf packs vs. debris dam. Journal of Freshwater Ecology 12:39–49.

Cederholm, C. J., and N. P. Peterson. 1985. The retention of coho salmon (*Oncorhynchus kisutch*) carcasses by organic debris in small streams. Canadian Journal of Fisheries and Aquatic Sciences 42:1222–1225.

Chen, G. K. 1999. The relationship between stream habitat complexity and anadromous salmonid diversity and habitat selection, Doctoral dissertation. Oregon State University, Corvallis.

Crispin, V., R. House, and D. Roberts. 1993. Changes in instream habitat, large woody debris, and salmon habitat after the restructuring of a coastal Oregon stream. North American Journal of Fisheries Management 13:96–102.

Crook, D. A., and A. Robertson. 1999. Relationships between riverine fish and woody debris: implications for lowland rivers. Marine and Freshwater Research 50:941–953.

Diez, J. R., S. Larranaga, A. Elosegi, and J. Pozo. 2000. Effect of removal of wood on streambed stability and retention of organic matter. Journal of the North American Benthological Society 19:621–632.

Dolloff, C. A. 1996. Large woody debris, fish habitat, and historical land use. Pages 130–138 *in* J. W. McMinn and D. A. Crossley, Jr., editors. Biodiversity and coarse woody debris in southern forests. Proceedings of the workshop on coarse woody debris in southern forests: effects on biodiversity. USDA Forest Service, Southern Research Station, General Technical Report GTR-SE-094.

Dolloff, C. A., P. A. Flebbe, and M. D. Owen. 1994. Fish habitat and fish populations in a southern Appalachian watershed before and after Hurricane Hugo. Transactions of the American Fisheries Society 123:668–678.

Dolloff, C. A., and G. H. Reeves. 1990. Microhabitat partitioning among stream-dwelling juvenile Coho salmon, *Oncorhynchus kisutch*, and Dolly varden, *Salvelinus malma*. Canadian Journal of Fisheries and Aquatic Sciences 47:2297–2306.

Dolloff, C. A., and C. M. Warren, Jr. 2003. Fish relationships. Pages 179–193 *in* S. V. Gregory, K. L. Boyer, and A. M. Gurnell, editors. The ecology and management of wood in world rivers. American Fisheries Society, Symposium 37, Bethesda, Maryland.

Dose, J. J., and B. B. Roper. 1994. Long-term changes in low-flow channel widths within the South Umpqua Watershed, Oregon. Water Resources Bulletin 30:993–1000.

Ebert, D. J., T. A. Nelson, J. L. Kershner, J. L. Cooper, and R. H. Hamre. 1991. A soil-based assessment of stream fish habitats in coastal plain streams. Warmwater Fisheries Symposium 1. Pages 217–224 *in* United States Department of Agriculture, Forest Service, Rocky Mountain Forest and Range Experiment Station, General Technical Report, RM-GTR-207, Fort Collins, Colorado.

Edwards, P. J., J. Kollmann, A. M. Gurnell, G. E. Petts, K. Tockner, and J. V. Ward. 1999. A conceptual model of vegetation dynamics on gravel bars of a large alpine river. Wetlands Ecology and Management 7:141–153.

Everett, R. A., and G. M. Ruiz. 1993. Coarse woody debris as a refuge from predation in aquatic communities: an experimental test. Oecologia 93:475–486.

Fetherston, K. L., Naiman, R. J., and Bilby, R. E. 1995. Large woody debris, physical process, and riparian forest development in montane river networks of the Pacific Northwest. Geomorphology 13:133–144.

France, R. L. 1997. Macroinvertebrate colonization of woody debris in Canadian Shield lakes following riparian clearcutting. Conservation Biology 11:513–521.

Frissell, C. A., W. J. Liss, C. E. Warren, and M. D. Hurley. 1986. A hierarchical framework for stream classification: viewing streams in a watershed context. Environmental Management 10:199–214.

Gerhard, M., and M. Reich. 2000. Restoration of streams with large wood—effects of accumulated and built-in wood on channel morphology, habitat diversity and aquatic fauna. International Review of Hydrobiology 85:123–137.

Gottesfeld, A. S., and L. M. Johnson-Gottesfeld. 1990. Floodplain dynamics of a wandering river, dendrochronology of the Morice River, British Columbia, Canada. Geomorphology 3:159–179.

Grant, G. E., F. J. Swanson, and M. G. Wolman. 1990. Pattern and origin of stepped-bed morphology in high-gradient streams, western Cascades, Oregon. Geological Society of America Bulletin 102:340–352.

Gregory, K. J., R. J. Davis, and S. Tooth. 1993. Spatial distribution of coarse woody debris dams in the Lymington Basin, Hampshire, UK. Geomorphology 6:207–224.

Gregory, S. V., G. A. Lamberti, D. C. Erman, K. V. Koski, M. L. Murphy, and J. R. Sedell. 1987. Influence of forest practices on aquatic production. Pages 233–255 *in* E. O. Salo and T. W. Cundy, editors. Streamside management: forestry and fishery interactions. Contribution 57, Institute of Forest Resources, University of Washington, Seattle.

Gregory, S. V., F. J. Swanson, W. A. McKee, and K. W. Cummins. 1991. An ecosystem perspective of riparian zones. BioScience 41(8):540–550.

Gurnell, A. M. 2003. Wood storage and mobility. Pages 75–91 *in* S. V. Gregory, K. L. Boyer, and A. M. Gurnell, editors. The ecology and management of wood in world rivers. American Fisheries Society, Symposium 37, Bethesda, Maryland.

Gurnell, A. M., H. Piégay, F. J. Swanson, and S. V. Gregory. 2002. Large wood and fluvial processes. Freshwater Biology 47:601–619.

Gurnell, A. M., and R. Sweet. 1998. The distribution of large woody debris accumulations and pools in relation to woodland steam management in a small low-gradient stream. Earth Surface Processes and Landforms 23:1101–1121.

Hanchet, S. M. 1990. Effect of land use on the distribution and abundance of native fish in tributaries of the Waikato River in the Hakarimata Range, North Island, New Zealand. New Zealand Journal of Marine and Freshwater Research 24:159–171.

Harmon, M. E., J. F. Franklin, F. J. Swanson, P. Sollins, S. V. Gregory, J. D. Lattin, N. H. Anderson, S. P. Cline, N. G. Aumen, J. R. Sedell, G. W. Lienkaemper, K. Cromack, Jr., and K. W. Cummins. 1986. Ecology of coarse woody debris in temperate ecosystems. Advances in Ecological Research 15:133–302.

Harvey, B. C., R. J. Nakamoto, and J. L. White. 1999. Influence of large woody debris and a bankfull flood on movement of adult resident coastal cutthroat trout (*Oncorhynchus clarki clarki*) during fall and winter. Canadian Journal of Fisheries and Aquatic Sciences 56:2161–2166.

Harvey, J. W., and K. E. Bencala. 1993. The effect of streambed topography on surface-subsurface water exchange in mountain catchments. Water Resources Research 29:89–98.

Hax, C. L., and S. W. Golladay. 1993. Macroinvertebrate colonization and biofilm development on leaves and wood in a boreal river. Freshwater Biology 29:79–87.

Heede, B. H. 1972. Influences of a forest on the hydraulic geometry of two mountain streams. Water Resources Bulletin 8:523–530.

Hicks, B. J., J. D. Hall, P. A. Bisson, and J. R. Sedell. 1991. Responses of salmonids to habitat changes. Pages 483–518 in W. R. Meehan, editor. Influences of forest and rangeland management on salmonid fishes and their habitats. American Fisheries Society, Special Publication 19, Bethesda, Maryland.

Hilderbrand, R. H., A. D. Lemly, C. A. Dolloff, and K. L. Harpster. 1997. Effects of large woody debris placement on stream channels and benthic macroinvertebrates. Canadian Journal of Fisheries and Aquatic Sciences 54:931–939.

Hoffman, A., and D. Hering. 2000. Wood-associated macroinvertebrate fauna in central European streams. International Review of Hydrobiology 85:25–48.

House, R. A., and P. L. Boehne. 1987. The effect of stream cleaning on salmonid habitat and populations in a coastal Oregon drainage. Western Journal of Applied Forestry 2:84–87.

Inoue, M., and S. Nakano. 1998. Effects of woody debris on the habitat of juvenile masu salmon (*Oncorhynchus masou*) in northern Japanese streams. Freshwater Biology 40:1–16.

Kasahara, T., and S. M. Wondzell. In press. Geomorphic controls on hyporheic exchange flows in mountain streams. Water Resources Research.

Keller, E. D., and F. J. Swanson. 1979. Effects of large organic material on channel form and fluvial process. Earth Surface Processes and Landforms 4:361–380.

Lancaster, S. T., S. K. Hayes, and G. E. Grant. 2001. Modeling sediment and wood storage and dynamics in small mountainous watersheds. Pages 85–102 *in* J. M. Dorava, D. R. Montgomery, B. B. Palcsak, and F. A. Fitzpatrick, editors. Geomorphic processes and riverine habitat, water science and application, volume 4. American Geophysical Union, Washington D.C.

Lienkaemper, G. W., and F. J. Swanson. 1987. Dynamics of large woody debris in streams in old-growth Douglas-fir forests. Canadian Journal of Forest Research 17:150–156.

Lindeman, P. V. 1999. Surveys of basking map turtles *Graptemys* spp. in three river drainages and the importance of deadwood abundance. Biological Conservation 88:33–42.

Magoulick, D. D. 1998. Effect of wood hardness, condition, texture and substrate type on community structure of stream invertebrates. American Midland Naturalist 139:187–200.

Malanson, G. P., and D. R. Butler. 1990. Woody debris, sediment, and riparian vegetation of a subalpine river, Montana, U.S.A. Arctic and Alpine Research 22:183–194.

Martin, D. J., and L. E. Benda. 2001. Patterns of instream wood recruitment and transport at the watershed scale. Transactions of the American Fisheries Society 130:940–958.

Maser, C., and J. R. Sedell. 1994. From the forest to the sea: the ecology of wood in streams, rivers, estuaries and oceans. St. Lucie Press, Delray Beach, Florida.

McIntosh, B. A., J. R. Sedell, J. E. Smith, R. C. Wissmar, S. E. Clarke, G. H. Reeves, and L. A. Brown. 1994. Historical changes in fish habitat for select river basins of eastern Oregon and Washington. Northwest Science 68:36–53.

McMahon, T. E., and G. F. Hartman. 1989. Influence of cover complexity and current velocity on winter habitat use by juvenile coho salmon (*Oncorhynchus kisutch*). Canadian Journal of Fisheries and Aquatic Sciences 46:1551–1557.

Montgomery, D. R., T. B. Abbe, J. M. Buffington, N. P. Peterson, K. M. Schmidt, and J. D. Stock. 1996. Distribution of bedrock and alluvial channels in forested mountain drainage basins. Nature (London) 381:587–589.

Montgomery, D. R., and J. M. Buffington. 1998. Channel processes, classification, and response. Pages 13–42 *in* R. J. Naiman and R. E. Bilby, editors. River ecology and management. Springer-Verlag, New York.

Montgomery, D. R., B. D. Collins, J. M. Buffington, and T. B. Abbe. 2003. Geomorphic effects of wood in rivers. Pages 21–47 *in* S. V. Gregory, K. L. Boyer, and A. M. Gurnell, editors. The ecology and management of wood in world rivers. American Fisheries Society, Symposium 37, Bethesda, Maryland.

Mosley, M. P. 1981. The influence of organic debris on channel morphology and bedload transport in a New Zealand forest stream. Earth Surface Process and Landforms 6:571–579.

Murphy, M. L., and J. D. Hall. 1981. Varied effects of clear-cut logging on predators and their habitat in small streams of the Cascade Mountains, Oregon. Canadian Journal of Fisheries and Aquatic Sciences 38:137–145.

Murphy, M. L., J. Heifetz, S. W. Johnson, K. V. Koski, and J. F. Thedinga. 1986. Effects of clear-cut logging with and without buffer strips on juvenile salmonids in Alaskan streams. Canadian Journal of Fisheries and Aquatic Sciences 43:1521–1533.

Murphy, M. L., and W. R. Meehan. 1991. Stream ecosystems. Pages 17–46 *in* W. R. Meehan, editor. Influences of forest and rangeland management on salmonid fishes and their habitats. American Fisheries Society, Special Publication 19, Bethesda, Maryland.

Naiman, R. J., T. J. Beechie, L. E. Benda, D. R. Berg, P. A. Bisson, L. H. MacDonald, M. D. O'Connor, P. L. Olson, and E. A. Steel. 1992. Fundamental elements of ecologically healthy watersheds in the Pacific Northwest coastal ecoregion. Pages 127–188 *in* R. J. Naiman, editor. Watershed management: balancing sustainability and environmental change. Springer-Verlag, New York.

Nakamura, F., and F. J. Swanson. 1993. Effects of coarse woody debris on morphology and sediment storage of a mountain stream in western Oregon. Earth Surface Processes and Landforms 18:43–61.

Nakamura, F., and F. J. Swanson. 2003. Dynamics of wood in rivers in the context of ecological disturbance. Pages 279–297 *in* S. V. Gregory, K. L. Boyer, and A. M. Gurnell, editors. The ecology and management of wood in world rivers. American Fisheries Society, Symposium 37, Bethesda, Maryland.

O'Connor, N. A. 1991. The effects of habitat complexity on the macroinvertebrates colonising wood substrates in a lowland stream. Oecologia 85:504–512.

Palmer, M. A., P. Arensburger, A. P. Martin, and D. W. Denman. 1996. Disturbance and patch specific responses: the interactive effects of woody debris and floods on lotic invertebrates. Oecologia 105:247–257.

Pyron, M., A. P. Covich, and R. W. Black. 1999. On the relative importance of pool morphology and woody debris to distributions in a Puerto Rican headwater stream. Hydrobiologia 405:207–215.

Quinn, J. M., A. B. Cooper, R. J. Davies-Colley, J. C. Rutherford, and R. Williamson. 1997. Land use effects on habitat, water quality, periphyton, and benthic invertebrates in Waikato, New Zealand, hill-country streams New Zealand. Journal of Marine and Freshwater Research 31:579–597.

Quinn, T. P., and N. P. Peterson. 1996. The influence of habitat complexity and fish size on over-winter survival and growth of individually marked juvenile coho salmon (*Oncorhynchus kisutch*) in Big Beef Creek, Washington. Canadian Journal of Fisheries and Aquatic Sciences 53:1555–1564.

Reeves, G. H., L. E. Benda, K. M. Burnett, P. A. Bisson, and J. R. Sedell. 1995. A disturbance-based ecosystem approach to maintaining and restoring freshwater habitats of evolutionarily significant units of anadromous salmonids in the Pacific Northwest. Pages 334–349 *in* J. L. Nielsen, editor. Evolution and the aquatic ecosystem: defining unique units in population conservation. American Fisheries Society, Symposium 17, Bethesda, Maryland.

Reeves, G. H., P. A. Bisson, and J. M. Dambacher. 1998. Fish communities. Pages 200–234 *in* R. J. Naiman and R. E. Bilby, editors. River ecology and management. Springer-Verlag, New York.

Reeves, G. H., F. H. Everest, and J. R. Sedell. 1993. Diversity of juvenile anadromous salmonid assemblages in coastal Oregon basins with different levels of timber harvest. Transactions of the American Fisheries Society 122:309–317.

Reeves, G. H., J. D. Hall, and S. V. Gregory. 1997. The impact of land-management activities on coastal cutthroat trout and their freshwater habitats. Pages 138–144 *in* Sea-run cutthroat trout: biology, management and future conservation. Oregon Chapter, American Fisheries Society, Corvallis, Oregon.

Reimers, P. E. 1971. The length of residence of juvenile fall Chinook salmon in Sixes River, Oregon. Doctoral dissertation. Oregon State University, Corvallis.

Richmond, A. D., and K. D. Fausch. 1995. Characteristics and function of large woody debris in subalpine Rocky Mountain streams in northern Colorado. Canadian Journal of Fisheries and Aquatic Sciences 52:1789–1802.

Schlosser, I. J., and P. L. Angermeier. 1995. Spatial variation in demographic processes of lotic fishes: conceptual models, empirical evidence, and implications for conservation. Pages 392–401 *in* J. L. Nielsen, editor. Evolution and the aquatic ecosystem: defining unique units in population conservation. American Fisheries Society, Symposium 17, Bethesda, Maryland.

Sedell, J. R., P. A. Bisson, J. A. June, and R. W. Speaker. 1982. Ecology and habitat requirements of fish populations in south fork Hoh River, Olympic National Park. Pages 35–42 *in* E. E. Starkey, J. F. Franklin, J. W. Matthews, technical coordinators. Ecological research in national parks of the Pacific Northwest: proceedings, 2nd Conference on Scientific Research in the National Parks. Oregon State University Forest Research Laboratory, Corvallis.

Sedell, J. R., G. H. Reeves, F. R. Hauer, J. A. Stanford, and C. P. Hawkins. 1990. Role of refugia in recovery from disturbances: modern fragmented and disconnected river systems. Environmental Management 14:711–724.

Sedell, J. R., J. E. Yuska, and R. W. Speaker. 1984. Habitats and salmonids distribution in pristine, sediment-rich river valley systems: S. Fork Hoh and Queets River, Olympic National Park. Pages 33–46 *in* W. R. Meehan, T. R. Merrell, Jr., and T. A. Hanley, editors. Fish and wildlife relationships in old-growth forests—proceedings of a symposium. American Institute of Fishery Research Biologists, Asheville, North Carolina.

Shields, F. D. Jr., S. S. Knight, and C. M. Cooper. 1994. Effects of channel incision on base flow stream habitats and fishes. Environmental Management 18:43–57.

Shields, F. D. Jr., and R. H. Smith. 1992. Effects of large woody debris removal on physical characteristics of a sand-bed river. Aquatic Conservation: Marine and Freshwater Ecosystems 2:145–163.

Simondet, J. A. 1997. Seasonal and diel response to habitat complexity by an allopatric population of coastal cutthroat trout. Page 179 *in* J. D. Hall, P. A. Bisson, and R. E. Gresswell, editors. Sea-run cutthroat trout: biology, management and future conservation. American Fisheries Society, Oregon Chapter, Corvallis, Oregon.

Smock, L. A., G. M. Metzler, and J. E. Gladden. 1989. Role of debris dams in the structure and functioning of low-gradient headwater streams. Ecology 70:764–775.

Stanford, J. A., and J. V. Ward. 1988. The hyporheic habitat of river ecosystems. Nature (London) 335:64–66.

Suberkropp, K. F. 1998. Microorganisms and organic matter decomposition. Pages 120–143 *in* R. J. Naiman and R. E. Bilby, editors. River ecology and management. Springer-Verlag, New York.

Swanson, F. J., S. V. Gregory, J. R. Sedell, and A. G. Campbell. 1982. Land-water interactions: the riparian zone. Pages 267–291 *in* R. L. Edmonds, editor. Analysis of coniferous forest ecosystems in the western United States. Hutchinson Ross Publishing Company, US/IBP Synthesis Series 14, Stroudsburg, Pennsylvania.

Thévenet, A., and B. Statzner. 1999. Linking fluvial fish community to physical habitat in large woody debris: sampling effort, accuracy and precision. Archiv fur Hydrobiologie 145:57–77.

Triska, F. J., J. R. Sedell, and S. V. Gregory. 1982. Coniferous forest streams. Pages 292–332 *in* R. L. Edmonds, editors. Analysis of coniferous forest ecosystems in the western United States. Hutchinson Ross Publishing Company, US/IBP Synthesis Series 14, Stroudsburg, Pennsylvania.

Trotter, E. H. 1990. Woody debris, forest-stream succession, and catchment geomorphology. Journal of the North American Benthological Society 9:141–156.

Wallace, J. B., and N. H. Anderson. 1996. Habitat, life history, and behavioral adaptations of aquatic insects. Pages 41–73 *in* R. W. Merritt and K. W. Cummins, editors. An introduction to the aquatic insects of North America, 3rd edition. Kendall/Hunt, Dubuque, Iowa.

Wallace, J. B., and A. C. Benke. 1984. Quantification of wood habitat in subtropical coastal plain streams. Canadian Journal of Fisheries and Aquatic Sciences 41:1643–1652.

Wallace, J. B., J. W. Grubaugh, and M. R. Whiles. 1996. Influences of coarse woody debris on stream habitats and invertebrate biodiversity. Pages 119–129 *in* J. W. McMinn and D. A. Crossley, Jr., editors. Biodiversity and coarse woody debris in southern forests. Proceedings of the workshop on coarse woody debris in southern forests: effects on biodiversity. USDA Forest Service, Southern Research Station General Technical Report GTR-SE-094, Asheville, North Carolina.

Wallace, J. B., J. R. Webster, and J. L. Meyer. 1995. Influence of log additions on physical and biotic characteristics of a mountain stream. Canadian Journal of Fisheries and Aquatic Sciences 52:2120–2137.

Ward, G. M., and N. G. Aumen. 1986. Woody debris as a source of fine particulate organic matter in coniferous forest stream ecosystems. Canadian Journal of Fisheries and Aquatic Sciences 43:1635–1642.

Ward, J. V., K. Tockner, P. J. Edwards, J. Kollmann, G. Bretschko, A. M. Gurnell, G. E. Petts, B. Rossaro, and J. B. Layzer. 1999b. A reference river system for the Alps: the Fiume Tagliamento. Regulated Rivers Research and Management 15:63–75.

Ward, J. V., K. Tockner, and F. Schiemer. 1999a. Biodiversity of floodplain river ecosystems: ecotones and connectivity. Regulated Rivers Research and Management 15:125–139.

Wondzell, S. M., and F. J. Swanson. 1999. Floods, channel change, and the hyporheic zone. Water Resources Research 35:555–567.

Wood-Smith, R. D., and J. M. Buffington. 1996. Multivariate analysis of forest streams: implications for assessment of land use impacts on channel condition. Earth Surface Processes and Landforms 21:377–393.

American Fisheries Society Symposium 37:265–277, 2003

Dynamics and Ecological Functions of Wood in Estuarine and Coastal Marine Ecosystems

Charles A. Simenstad

Wetland Ecosystem Team, School of Aquatic and Fishery Sciences University of Washington, Seattle, Washington 98195, USA

Alicia Wick

Anchor Environmental, L. L. C., Seattle, Washington 98101, USA

Stan Van de Wetering

Confederated Tribes of the Siletz Indians, Siletz, Oregon 97380, USA

Daniel L. Bottom

NOAA-National Marine Fisheries Service Hatfield Marine Science Center, Newport, Oregon 97365, USA

Abstract.—The abundance, quality, and ecological fate of wood in estuarine and coastal ecosystems result from watershed and, ultimately, landscape-scale processes. Estuaries and coasts receive the wood exported from watersheds, in which fluvial import, retention, processing, and export mostly determine how much wood is available. The dynamics of wood in estuaries and coasts reflect somewhat different processes and rates than in rivers, however. Bidirectional and fluctuating fluvial-tidal flows, multiple sources, and retentive mechanisms generate complex patterns of wood movement through tidal ecosystems. Too often, wood is assumed to perform the same functions in estuaries as in rivers. With a few exceptions—such as wood acting as a disturbance in the rocky intertidal zone—these assumed similarities have not been demonstrated empirically. The dynamics and role of wood need to be examined across the continuum of watershed, estuarine, and marine ecosystems, with the coastal ocean serving as both sinks and sources of wood. Wood's function in juvenile salmonid habitat is a prime example of extrapolating the role of wood in river ecosystems, often without substantiation, to estuaries and oceans. Large wood is often designed into estuarine wetland restoration projects at significant cost without any real evidence of its importance to fish in tidal environments. The use of wood by juvenile salmonids may be scale-dependent, and it may vary with factors that affect availability of refugia and predation rates, such as tidal elevations, river flow, and diel changes in light intensity. Tides and wind can produce heterogeneous distributions of wood, which may help support the tidal and seasonal movements of juvenile salmon. Management and restoration plans should protect upstream wood sources and delivery processes to maintain and retain the supply of wood to estuaries and coasts. Research on large wood is still needed to evaluate its habitat functions in coastal and estuarine ecosystems to develop criteria for assessing and restoring habitat and to understand variation in the role of large wood across estuary-ocean landscapes.

Introduction

The ecological influence of wood does not cease once it is exported from rivers to the coastal margin. Estuaries and coasts are influenced by the dynamic passage of wood, even if these ecosystems are its ultimate repositories. In most estuaries and along coasts, wood is a dynamic source of organic matter, substrate, and disturbance. Although river processes—including fluvial import, retention, and fluvial-tidal export—control the delivery and movement of estuarine and coastal wood, other distinctly coastal processes play a major role in transporting, distributing, and functioning of large wood within these environments. Bidirectional and fluctuating fluvial-tidal flows, strong winds, waves, freshet flooding, and diverse geomorphic features of estuaries combine to generate complex patterns of wood movement, distribution, and residency. These interactions create a wide range of turnover rates for the wood imported to estuaries from watersheds, ranging from highly ephemeral material that pulses rapidly through the system with little impact to wood pieces that are deposited in the estuary and may remain for decades or centuries.

The functions ascribed to wood in estuaries and coasts are often assumed to parallel those described for rivers and streams. As illustrated by this volume's focus on rivers, the estuarine and coastal functions of wood have not been effectively documented. Too often, wood is merely assumed to perform similar functions in all the aquatic environments in which it is found. Here, we place watersheds and rivers in a broader coastal context by tracing the legacy of wood inherited by estuaries and coasts. We will not repeat the comprehensive syntheses of wood in estuaries and along the coast portrayed by Maser et al. (1988) and Maser and Sedell (1994). To some degree, the scientific knowledge about wood exported from watersheds and deposited in estuaries and along coasts has not changed appreciably from that described by Gonor et al. (1988). Our intent is to update these earlier accounts and highlight the distinct ecological functions of wood in coastal and estuarine environments. We devote particular attention to the role of estuarine wood as habitat for migrating juvenile salmon because of growing interest in placing wood in estuaries as a presumed method for restoring important habitat of at-risk salmon populations. During migration from freshwater to the ocean, some salmon species and life history types use estuaries extensively for feeding and growth (Groot and Margolis 1991; Simenstad et al. 1982), including chum salmon *Oncorhynchus keta* and ocean-type chinook salmon *O. tshawytscha*.

Origins and Dynamics

Wood originates and is modified, transported, and deposited across a continuum of watershed, estuary, and marine ecosystems (Figure 1). Most wood in estuaries probably originates from the watershed through the various processes described elsewhere in this volume (Boyer et al. 2003; Gurnell 2003; Nakamura and Swanson 2003; all this volume). Both the generation and input of wood are forced by storm and freshet events, but the characteristics and frequency of inputs have changed historically through industrial forestry practices, channelization, bank armoring, and other influences on riparian ecosystems, and hydrological changes that affect flooding frequencies and wood transport. In relatively unaltered watersheds, wood was likely exported into estuaries and to the ocean as a broad spectrum of species, sizes, frequencies, and intensities, as a result of chronic erosion of woody vegetation in adjacent riparian zones and as acute debris flows from larger segments of steep hillsides. In intensively modified watersheds, wood delivery has become more episodic, reducing the input of large pieces of stable wood, concentrating wood in large but infrequent accumulations, and removing sources of material for stream channels (Bisson et al. 1987; Hicks et al. 1991). These changes imply that wood import into estuaries may occur only during extreme floods, which deposit most wood at high elevations in or near estuaries. In severely channelized watersheds or where flows are regulated by dams, both input and flux of wood to estuaries are diminished significantly.

Estuarine wood can also originate from freshwater tidal floodplains, at the watershed-estuary ecotone, or be imported through tidal flow from lower segments of the estuarine tidal gradient or ocean. Flooded swamps may provide less material and be less vulnerable to either chronic or acute erosion than streamside riparian areas above tidewater because of the counteracting effects of the tides on fluvial currents. Moreover, because tidally influenced, forested wetlands are among the most intensively modified habitats in Pacific Northwest estuaries, much less woody material is now available from flooded swamps than during predevelopment periods (Thomas 1983). Nonetheless, bank erosion remains a common source of wood input at the upper reach of the tides.

The processes that distribute wood delivered to the estuary are complex because of interactions among tides, river flows, winds, and waves. Wood is continuously dispersed, deposited, and typically refloated and redispersed in estuaries

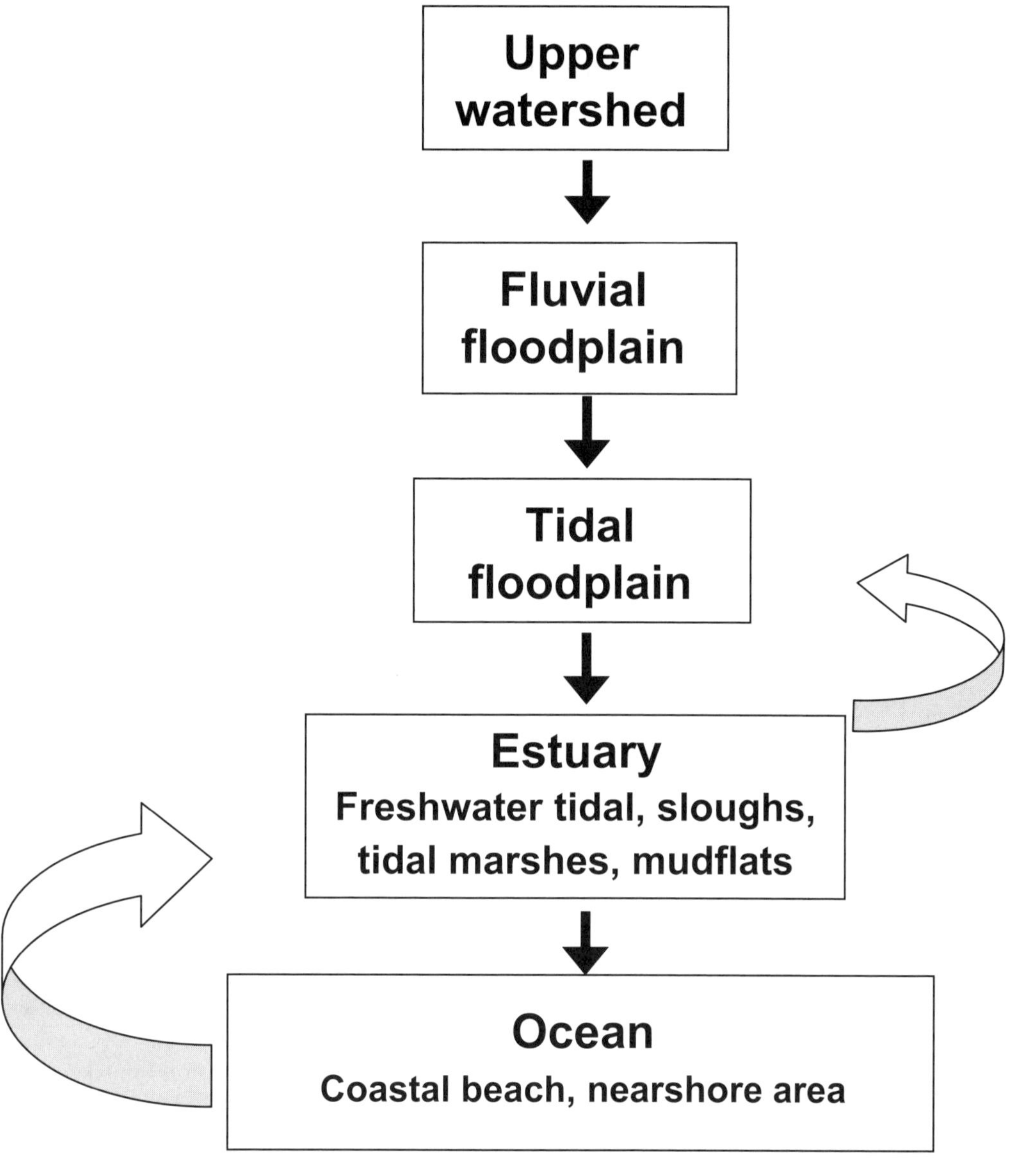

FIGURE 1. Cycling of wood among watershed, estuary, and ocean ecosystems.

by tidal action. These dynamics vary with the size and structure of wood retention features and the tidal range and strength of tidal currents associated with the stage of the tidal month. Thus, residence times for wood in estuaries and on ocean beaches are highly variable, especially where wood is actively harvested by people on beaches (Benner and Sedell 1987). Wood may become incorporated into sediment or periodically floated and relocated by tidal flushing and storms (Gonor et al. 1988).

The potential for wood to remain in the estuary also depends upon the availability of particular habitat classes and their relative complexity. Complex habitats like forested wetland sloughs, which may be choked with both live riparian and dead wood, are much more likely to trap new pieces of wood transported during flood tides than are gravel or sand bars. Low-energy, fine-sediment habitats like mudflats, however, can easily entrain wood. And complex pieces of wood are more likely to be trapped and retained in the estuary than are simple pieces that lack branches or rootwads.

Long-term persistence of wood is influenced by the species of decomposing colonizers, which vary with salinity and location. In more saline areas of the lower estuary, wood provides a substrate for wood-digesting fungi, cellulose-digesting bacteria (minor roles), and wood-boring gribbles and shipworms (major roles), which decompose wood to provide detrital energy (Maser and Sedell 1994; Jones et al. 1976). Upper-estuarine areas (such as freshwater and tide-influenced river), however, contain few degrader species, and biological decay may take many decades (Maser and Sedell 1994).

Ecological Functions

An array of functions is attributed to wood in estuaries and along other coastal regions (Gonor et al. 1988), including dissipation of wave energy, shoreline protection from erosion (Terich and Milne 1977; Stembridge 1979), and sediment trapping. Eilers (1975) found that piles of downed trees in a salt marsh of the Nehalem River (Oregon) trapped enough sediment to support vegetation. In addition, wood refloated by the tide can disturb marshes and increase their vegetative complexity (Eilers 1975).

With a few exceptions—including the role of wood as a disturbance factor in the rocky intertidal zone (Paine and Levin 1981)—the ecological functions of wood in estuarine and coastal ecosystems have not been evaluated (Table 1). Only a few studies have examined the association between wood and the plants and animals in estuaries. In their comprehensive review, Gonor et al. (1988) argued that wood could function as a major component of habitat complexity in estuaries, beaches, and the coastal ocean, where fine sediment is the dominant substrate. Everett and Ruiz (1993) considered wood an important refuge for fish and invertebrates in the Chesapeake Bay and hypothesized that this role was similar for other habitats where physical structure is limited. Large wood in the Nehalem River estuary creates shallow pool-type depressions that provide low-tide habitat for fish (Eilers 1975).

Restoration programs frequently promote the idea that additions of large wood to estuaries will benefit juvenile salmon, based on current understanding of wood as a critical salmon habitat component in streams and rivers. Yet, few studies have assessed the importance of estuarine wood to salmonids. The common assumption derived from stream research is that wood pieces in estuaries must be large to provide significant habitat or contribute to habitat-forming processes.

Most studies of the role of large wood as salmonid habitat have focused on coho salmon, one of the least estuarine-dependent salmonid species (Groot and Margolis 1991; but see Tschaplinski and Hartman 1983 and Miller and Simenstad 1997). During seaward migration of juvenile coho in the Carnation Creek (British Columbia) estuary, smolt abundance was correlated with large wood density (McMahon and Holtby 1992). In studies tracking juvenile coho in the Chehalis River, Washington, USA estuary, Moser et al. (1991) found that the fish congregated for extended periods in low-velocity areas created by log pilings. We are aware of no other studies documenting juvenile salmon associations with large wood in estuaries. Studies of the contributions of wood to invertebrate assemblages in Pacific Northwest estuaries are similarly rare.

The role of large wood in estuarine geomorphic processes supporting fish habitat is somewhat more developed. Gonor et al. (1988) note that large wood modifies local current and turbulence patterns and promotes sedimentation, accretion of bars and banks, and settling of organic matter. Terich and Milne (1977), Stembridge (1979), and Wolanski and Ridd (1986) also document the role of large wood in stabilizing shorelines, which, in turn, may allow riparian vegetation to become established in estuarine and ocean beach environments.

Based on these studies and our knowledge of juvenile salmon ecology in estuarine habitats with abundant large wood, such as the brackish sloughs in the Chehalis River estuary (Simenstad et al. 1992, 1993, 1997, 2000, 2001; Miller and Simenstad 1994a, 1994b, 1997), we propose that the role of wood in supporting juvenile salmon in estuaries depends on the retention, structure, size, and location of wood (Figure 2). Direct use of wood by juvenile salmon is associated with microhabitats where complex structure creates areas of reduced water velocity and turbulence, decreasing the metabolic costs for fish to maintain their positions and feed; complex cover affords refuge from predators. Large wood also indirectly supports juvenile salmon by trapping detrital organic matter and fine sediments, preferred by benthic or epibenthic prey organisms

TABLE 1. Inferred and documented functions of wood in estuarine and ocean ecosystems; * indicates that no sources were located.

Function	General reference	Primary study
Releases organic carbon	Maser and Sedell 1994	*
Harbors nitrogen-fixers	Harmon et al. 1986	*
Provides substrate for microalgae and macroinvertebrate colonization	Maser and Sedell 1994	Becker 1971 Gonor et al. 1988 Jones et al. 1976 Kodata 1958 Ray 1959 Turner 1971, 1977, 1981, 1984
Controls the movement of sediment and water	Maser and Sedell 1994	Gonor et al. 1988 (reference, not a study)
Controls the movement of organic matter	Harmon et al. 1986	*
Dissipates the energy of flow regimes	Maser and Sedell 1994	*
Provides habitat for fish, invertebrates, or both	Maser and Sedell 1994	Dudley and Anderson 1982 Eilers 1975 Levings et al. 1976 Levy et al. 1982
Provides cover habitat for fish, invertebrates, or both	Thom and Borde 1998	Daniel and Robertson 1990 Everett and Ruiz 1993 McMahon and Holtby 1992
Influences channel morphology	Hilderbrand et al. 1998	Gonor et al. 1988 (reference, not a study)
Creates hydraulic diversity that structures pools, riffles, and falls	Thom and Borde 1998	Gonor et al. 1988 (reference, not a study)
Creates hydraulic diversity that influences productivity	Harmon et al. 1986	*
Serves as an interface linking terrestrial and aquatic systems	Hilderbrand et al. 1998	Triska and Cromack 1980
Influences water column structure and complexity	Harmon et al. 1986	*
Serves as a source of disturbance that influences plant communities	Gonor et al. 1988	Eilers 1975
Provides wood directly consumed by invertebrates, fungi, and bacteria	Harmon et al. 1986	Becker 1971 Gonor et al. 1988 Jones et al. 1976 Kodata 1958 Ray 1959 Turner 1971, 1977, 1981, 1984

like amphipods (such as *Corophium* spp.) or mysids. Sediment trapping also promotes colonizing vegetation, such as the low brackish sedge *Carex lyngbyei*. In turn, this species forms habitat used by other preferred prey organisms, refuge from predators, and shade to channels that reduces water temperature. Large wood may be particularly important to fish as refugia during low tides, when shallow channels, high-elevation habitats, and vegetated banks are inaccessible.

Recent Investigations on the Role of Estuarine Wood for Juvenile Salmon

Siletz River (Oregon, USA) estuary

In a recent investigation, Van de Wetering (2001) explored the association between aggregations of

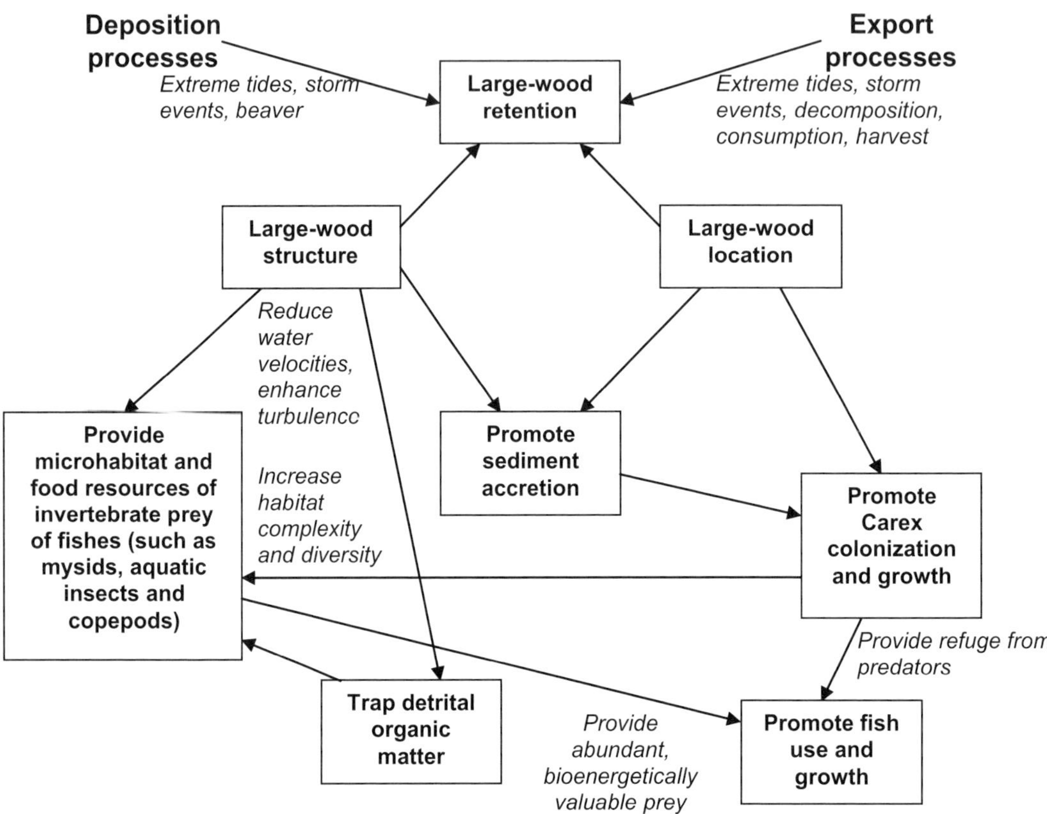

FIGURE 2. Conceptual model of the ecological functions of large wood in estuaries and other coastal ecosystems.

large wood and juvenile salmon migrating through and rearing in the Siletz River estuary. Using a mixture of videography and seining techniques during various stages of flood tides, he enumerated daylight fish densities among a series of estuarine habitats, including marsh, main-channel bank, eelgrass, open mud flat, and complex large wood. Wood habitats were located directly adjacent to main-channel bank and mudflat habitats. Results suggested that complex wood was used more frequently and by a greater density of salmonids than other habitats. Van de Wetering (2001) concluded that, during daylight flood tides, juvenile salmon preferred microhabitats with protective cover from predators, low water velocities, and optimal feeding lanes for energy-efficient capture of prey. Together with additional snorkeling observations, the camera observations suggest that large wood may provide an important habitat at low tide and during early flood tides, when water levels prevent fish from occupying complex intertidal marsh edges and tidal channels (Figure 3). To our knowledge, these results are the first to compare the occurrence and relative densities of salmon among estuarine habitats with different amounts of large wood.

Chehalis River (Washington, USA) estuary

A study recently completed in the Chehalis River estuary investigated fish, prey, and physical-habitat (sediments, large wood) associations in two (one created, one natural) brackish sloughs (Wick 2002). In each slough, fish abundance was related to measurements of wood for total fishes, salmonids only (chum, chinook), and other abundant fish species. Each slough was stratified into upper (head) and lower (mouth) areas. Once per month, at slack high tide, the head and mouth areas were partitioned with floating barrier nets into two sampling sections with different amounts of large wood (Figure 4; Table 2). Fish trapped after the tide receded

SUBTIDAL CHANNEL
INTERTIDAL MUDFLAT
MARSH AND TIDAL CHANNELS
Flood Tide/High Water
Expansion
Ebb Tide/Low Water
Contraction

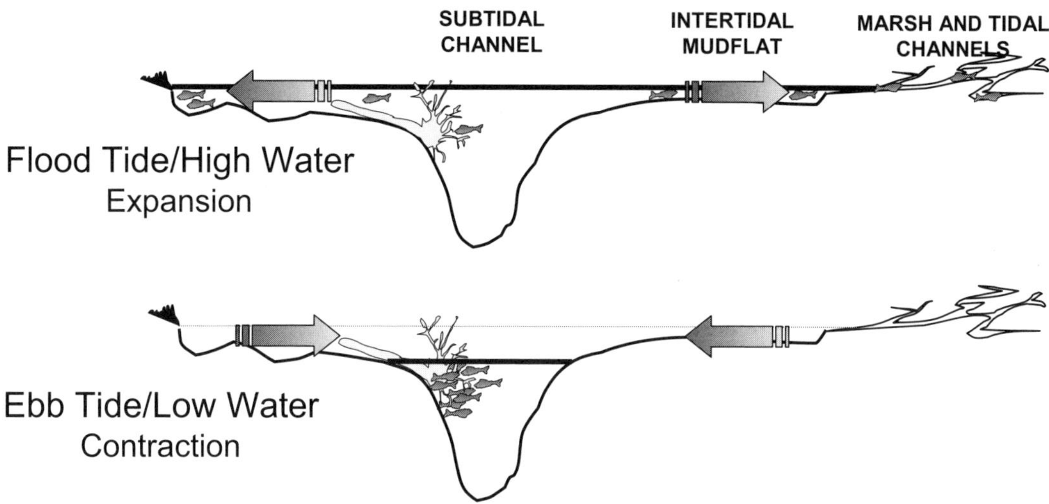

FIGURE 3. Landscape-scale interactions of juvenile salmon with estuarine wood.

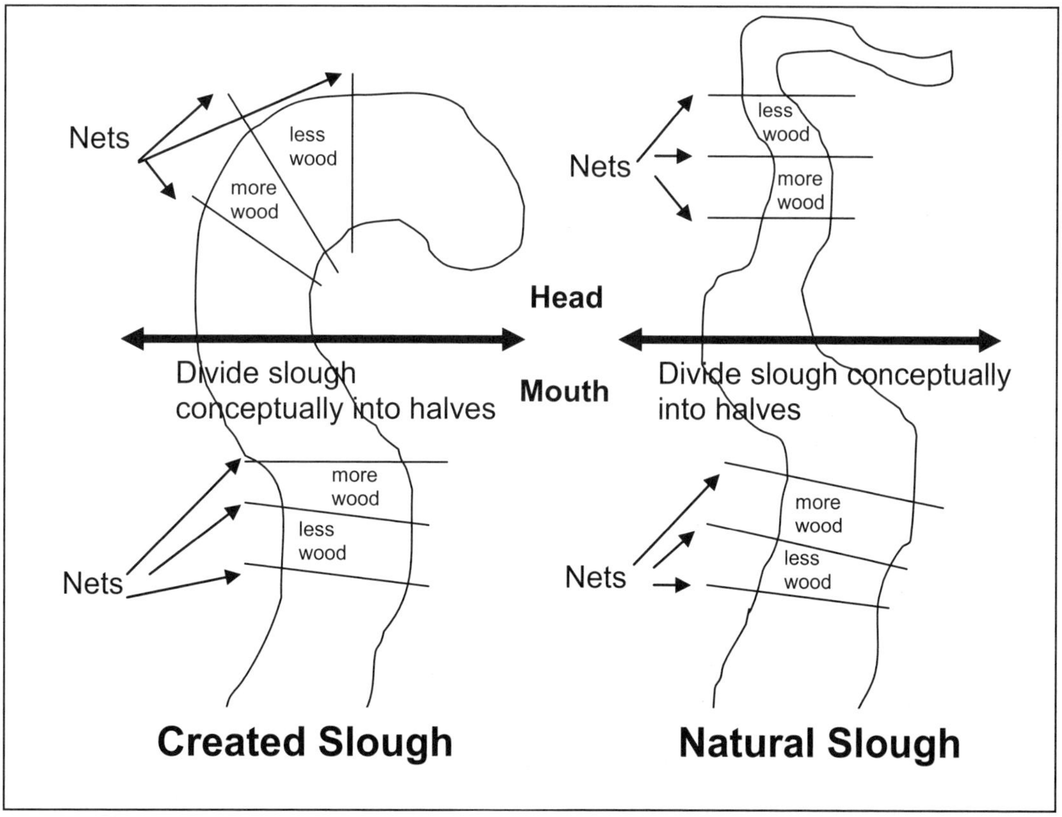

Nets
less wood
more wood
Nets
less wood
more wood
Head
Divide slough conceptually into halves
Mouth
Divide slough conceptually into halves
more wood
less wood
Nets
more wood
less wood
Nets
Created Slough
Natural Slough

FIGURE 4. Schmatic Illustration of the fish sampling regime in March–June 2001 in one created and one natural estuarine slough in the Chehalis River estuary, Washington.

TABLE 2. Wood surface areas (WSA m²), sampling section surface areas (FSSA m²), and wood surface areas as a percentage of sampling section surface areas (WSA % of SSSA m²) in created and natural sloughs; see Figure 4 for location of sampling sections.

Section	WSA (m²)	SSSA(m²)	WSA/SSSA (%)
Created slough			
1	6.44	484.50	1.33
2	13.31	464.30	2.87
3	80.27	1089.00	7.37
4	83.09	535.70	15.51
Natural slough			
1	5.49	837.70	0.66
2	9.03	1292.00	0.70
3	54.57	1250.30	4.36
4	139.41	831.40	16.77

in each sampling section were retrieved from the channel, identified, counted, and released. Wick (2002) found that fish densities in sampling sections were not correlated with wood surface area for any fish group, including salmonids.

Invertebrate fish prey and total invertebrate abundances were compared near (≤0.25 m) and far (>2 m) from large wood. Between these two categories of proximity to wood, no differences were found in benthic invertebrate densities, in invertebrate species composition, or in sediment deposition rates, grain size, or organic content. The composition of invertebrate assemblages on wood also was not significantly different from either benthic or epibenthic (invertebrates living on sediment or other surfaces) assemblages far from wood.

Wick (2002) also used GIS analysis to evaluate changes in large wood abundance, complexity, and movement based on three georeferenced aerial photographs of one natural and one created slough. Abundance and distribution of wood changed little during the 7-year time interval of the aerial photos. Wood type, slough, and channel bank location did not affect large-wood complexity, and the changes in the large wood did not differ between the natural and the created slough.

Implications of Recent Studies on Functions of Estuarine Wood for Juvenile Salmonids

Results of recent studies highlight the difficulties of interpreting the role of wood at the diverse scales over which it may function in estuarine and ocean environments, as illustrated by the investigations of juvenile salmon response to large wood. Studies in the Siletz River estuary find evidence, albeit preliminary, that juvenile salmon respond to large wood microhabitats at fine spatial scales and especially under early flood tides. These observations document fish reacting to different velocities and turbulence fields created by the wood on the scale of centimeters to meters. This finding implies, and Van de Wetering (2001) observations suggest, that the physical structure of large wood pieces or aggregates and their orientation to tidal currents may determine how much they contribute to bioenergetic efficiency of juvenile salmon positioned to use them and how they may concentrate prey. Thus, the primary function of large wood at this scale may be as a retention structure in tidal environments, interacting with currents to produce complex hydrological conditions (such as vortices) that concentrate prey resources and allow fish refuge from adjacent, relatively high-velocity waters without such structure.

Two other scales likely to be a factor in the role of estuarine large wood are the tidal elevation and physico-chemical gradients of the estuarine landscape. The location of wood at any position along the estuarine gradient may influence its function for fish migrating between low-tide refuges (such as main-stem and large distributary channels) and peripheral, high tide habitats (such as tidal channels in marshes and swamps). We hypothesize that subtidal wood positioned at the edges of estuarine distributary channels (Figure 5) may provide both low velocity or turbulence positions where fish can hide and feed efficiently during low tide and early flood tide periods. In this way, the complexity, cover, and velocity barrier afforded by large wood may offer important

FIGURE 5. Large wood joining a distributary channel and intertidal mudflat in the Siletz River estuary, Oregon. (S. Van de Wetering photograph)

connective structure for fish to readily access peripheral habitats preferentially occupied at high tide (Figure 3). Wood in the middle of low-gradient mudflats may not be as useful for positioning and feeding during high flow velocities but may provide refuge "islands" when the fish cross the flats on flooding tides. Large wood on marsh surfaces may not provide either function directly, though other indirect functions, such as maintaining diverse vegetation cannot be discounted.

Juvenile salmon may also respond to different spatial scales and locations of wood as they migrate through and rear in the estuary before entering the ocean. In the case of oligohaline-brackish sloughs of the Chehalis River estuary (Wick 2002), changes in current velocities or prey concentrations associated with large wood may be relatively inconsequential to salmon. In these wood-laden channels, the entire tidal slough may constitute a single, large wood complex to which juvenile salmon respond at scales of 10–100 km. Within these sloughs, finer-scale variations in velocity or prey concentration may be insignificant compared with those described in the main channel environment of the brackish Siletz River estuary (Van de Wetering 2001). These results suggest that wood may not function in a manner that is directly analogous to stream environments, where spatial scales are relatively confined by the linear structure of the stream-channel network.

The Chehalis and Siletz studies provide different views of fish use and complex wood habitats. What neither study fully addresses is how wood affects habitat complexity across estuaries. During the summer 2000, two of us (Van de Wetering and Bottom) documented marked differences in fish distribution during flood-tide underwater surveys at night in the Siletz River estuary. On one side of the main estuary channel, we observed higher juvenile salmon densities in shoreline mudflat habitats than in adjacent complex wood and open mudflat habitats (outer boundary = 300 m). On the other hand, we found no juvenile salmon in shoreline or open mudflat habitats on the opposite side of the main estuary channel where no complex wood habitats existed. This limited observation suggests wood aggregations may create a broad-scale habitat complex with other physical environments nearby.

From our Siletz observations above, it appears that juvenile salmon may select wood aggregations for a host of reasons that vary with tide and light conditions. For instance, low light levels may reduce the need for the complex cover of large wood and thereby permit tradeoffs between refuge and prey requirements. Thus, as light levels decreased, fish in the Siletz River estuary may have ventured from wood aggregations to take advantage of shoreline habitats where prey may have been more abundant. These results suggest possible salmon-habitat associations with estuarine wood over broad spatial and temporal scales not yet incorporated in scientific studies.

Conclusions and Recommendations

As should be evident from our synthesis of evidence for the role of wood at the watershed:ocean ecotone, there is little substantive documentation but considerable speculation on the role of large wood in estuaries. Most of this conjecture is based on extrapolating knowledge about the functional role of wood in rivers to assumptions about potential roles of wood in estuaries on other coastal ecosystems. Explicit research on the role of wood in estuaries and on ocean beaches is essential to separate the valid extrapolations from those that that cannot be validated. One of the most important current research needs is to evaluate the controls on wood dynamics within and through estuaries, particularly the hydrologic and geomorphic characteristics of estuaries that promote wood retention. Little is known about the influence of physical structure on the amount and configuration of imported and resident large wood. Specifically, understanding the role of wood across the broad landscape of this ecotone requires knowledge about the specific ecological and other roles of wood along the estuarine gradient and relative to tidal elevation.

We also need to know more about variation in the role of wood among estuaries with different histories of watershed development, physiographic setting, and structure (tidal prism, tidal slough and channel systems, and so on). For instance, most of the current large wood in the Siletz River estuary has apparently originated from bank fall either from upriver or locally, where people have created environments—like freshwater soils on estuarine dikes—which have allowed for conifer growth and later channel input. Historical observations, however, describe the Siletz as being "choked" with large wood during the 19th century (Maser and Sedell 1994). More wood could likely allow for more delta habitat, which, in turn, would allow better retention of all wood sizes as they pass through the estuary. In the Siletz, we infer that specific environmental conditions have to coincide at the time the wood is transported to and through the bay in order for estuarine wood to accumulate. The tide and river must form a unique eddy pushing the wood to shallower areas; a southwest wind can enhance this effect and also act as a final stabilizer or placer of retained wood by aligning the larger trees.

Another crucial component of the role of wood is its function as fish refuge or as habitat for fish predators. Fish have been associated in several studies with large wood in some estuarine areas. A leading question is whether the association of fish with wood habitats necessarily means that wood is high-quality habitat. If so, identifying the attributes of large wood important to fish is essential. And little is known about the mechanisms of fish use. For example, how closely and when do fish associate with wood? Camera observations indicate that wood provides both swimming and predator refugia areas in areas of increased velocity. The influence on fish feeding energetics is apparent from the Van de Wetering observations of fish lining up in eddy habitats behind wood, perhaps to maximize prey-capture rates. Such a mechanistic understanding of the observed patterns would significantly advance the appreciation of why and how wood may be important in estuarine food chains.

Despite the lack of past studies, sufficient evidence exists of the importance of estuarine wood and its historical prevalence in northwestern North America estuaries to recommend interim protection and prevent additional irreversible losses. In recent history, large-wood sources have shifted from natural to logging-related debris. Clearing the riparian corridor and isolating floodplains by diking has reduced source wood for watersheds and led to a general decline of wood in estuaries and coastal marine systems. Managers can help preserve sources by limiting the direct removal of local wood wherever possible and by preventing the clearing and harvesting of relocated, stranded wood from riparian, nearshore, estuarine and coastal areas. In restoration projects, planners and managers should think about promoting natural processes that will provide comparable amounts and a continuous supply of large wood like that in natural or reference systems. Conserving source wood and restoring natural hydrologic regimes and floodplain connectivity will help maintain the supply of wood to estuaries and coasts, both in amount and in rate of transport. Consideration should also be given to the characteristics and amounts of wood in reference sites or historical records, keeping in mind that reference amounts may not reflect actual past abundance.

Acknowledgments

This manuscript was prepared with support from the Washington Sea Grant Program, Grant R/ES-41, U.S. Army Corps-Seattle District, National Marine Fisheries Service-Northwest Fisheries Center, and the Siletz Tribe.

References

Becker, G. 1971. On the biology, physiology, and ecology of marine wood-boring crustaceans. Pages 303–326 *in* E. B. G. Jones and S. K. Eltringham, editors, Marine borers, fungi, and fouling organisms. Organization Economic Cooperative Development, Paris.

Benner, P. A. and J. R. Sedell. 1987. Chronic reduction of large woody debris on beaches at Oregon river mouths. Proceedings of the 8th Annual Meeting of the Society of Wetland Scientists, 1987 May 24–27, Seattle, Washington. Society of Wetland Scientists, Wilmington, North Carolina.

Bisson, P. A., R. E. Bilby, M. D. Bryant, C. A. Dolloff, G. B. Grette, R. A. House, M. L. Murphy, K. V. Koski, and J. R. Sedell. 1987. Large woody debris in forested streams in the Pacific Northwest: past, present, and future. Pages 143–190 *in* E. O. Salo and T. W. Cundy, editors. Streamside management: forestry and fishery interactions. University of Washington, Institute of Forest Resources Contribution 57, Seattle.

Boyer, K. L., D. R. Berg, and S. V. Gregory. 2003. Riparian management for wood in rivers. Pages 407–420 *in* S. V. Gregory, K. L. Boyer, and A. M. Gurnell, editors. The ecology and management of wood in world rivers. American Fisheries Society, Symposium 37, Bethesda, Maryland.

Daniel, P. A., and A. I. Robertson. 1990. Epibenthos of mangrove waterways and open embayments: community structure and the relationship between exported mangrove detritus and epifaunal standing stocks. Estuarine and Coastal Shelf Science 31:599–619.

Dudley, T., and N. H. Anderson. 1982. A survey of invertebrates associated with wood debris in aquatic habitats. Melanderia 39:1–21.

Eilers, H. P. 1975. Plants, plant communities, net production and tide levels: the ecological biogeography of the Nehalem salt marshes, Tillamook County, Oregon. Doctoral dissertation. Oregon State University, Corvallis.

Everett, R. A., and G. M. Ruiz. 1993. Coarse woody debris as a refuge from predation in aquatic communities: an experimental test. Oecologia 93:475–486.

Gonor, J. J., J. R. Sedell, and P. A. Benner. 1988. What we know about large trees in estuaries, in the sea, and on coastal beaches. Pages 83–112 *in* C. Maser, R. F. Tarrant, J. M. Trappe, and J. F. Franklin, technical editors. From the forest to the sea: a story of fallen trees. Pacific Northwest Research Station, USDA Forest Service General Technical Report GTR-PNW-229, Portland, Oregon.

Groot, C., and L. Margolis, editors. 1991. Pacific salmon life histories. UBC Press, Vancouver.

Gurnell, A. M. 2003. Wood storage and mobility. Pages 75–91 *in* S. V. Gregory, K. L. Boyer, and A. M. Gurnell, editors. The ecology and management of wood in world rivers. American Fisheries Society, Symposium 37, Bethesda, Maryland.

Harmon, M. E., J. F. Franklin, F. J. Swanson, P. Sollins, S. V. Gregory, J. D. Lattin, N. H. Anderson, S. P. Cline, N. G. Sumen, J. R. Sedell, G. W. Lienkaemper, K. Cromack, Jr., and K. W. Cummins. 1986. Ecology of coarse woody debris in temperate ecosystems. Advances in Ecological Research 15:133–302.

Hicks, B. J., J. D. Hall, P. A. Bisson, and J. R. Sedell. 1991. Responses of salmonids to habitat changes: influences of forest and rangeland management on salmonid fishes and their habitats. Pages 483–518 in W. R. Meehan, editor. Influences of forest and rangeland management on salmonid fishes and their habitats. American Fisheries Society, Special Publication 19, Bethesda, Maryland.

Hilderbrand, R. H., A. D. Lemly, C. A. Dolloff, and K. L. Harpster. 1998. Design considerations for large woody debris placement in stream enhancement projects. North American Journal of Fisheries Management 18:161–167.

Jones, E. B. G., R. Turner, S. E. Furtado, and H. Kuhne. 1976. Marine biodeteriogenic organisms. I. Lignicolous fungi and bacteria and wood boring mollusca and crustacea. International Biodeterioration Bulletin 12:120–134.

Kodata, H. 1958. Cellulose-decomposing bacteria in the sea. Pages 332–341 *in* D. L. Ray, editor. Marine wood boring and fouling organisms. University of Washington Press, Seattle.

Levings, C. D., N. G. McDaniel, and E. A. Black. 1976. Laboratory studies on the survival *of Anisogammarus confervicolus* and *Gammarus setosus* (Amphipoda, Gammaridea) in sea water and bleached kraft pulp mill effluent. Fisheries and Marine Service Canada, Technical Report 656:23.

Levy, D. A., T. G. Northcote, and R. M. Barr. 1982. Effects of estuarine log storage on juvenile salmon. University of British Columbia, Westwater Research Centre, Technical Report 26, Vancouver.

Maser, C., R. F. Tarrant, J. M. Trappe, and J. F. Franklin, technical editors. 1988. From the forest to the sea: a story of fallen trees. USDA Forest Service, Pacific Northwest Research Station, General Technical Report PNW-GTR-229, Portland, Oregon.

Maser, C., and J. R. Sedell. 1994. From the forest to the sea: the ecology of wood in streams, rivers, estuaries, and oceans. St. Lucie Press, Delray Beach, Florida.

McMahon, T. E., and L. B. Holtby. 1992. Behaviour, habitat use, and movements of coho salmon (*Oncorhynchus kisutch*) smolts during seaward migration. Canadian Journal of Fisheries and Aquatic Sciences 49:1478–1485.

Miller, J. A., and C. A. Simenstad. 1994a. Growth of juvenile coho salmon in natural and created estuarine habitats: a comparative study using otolith microstructure analysis. University of Washington, Fisheries Research Institute, FRI-UW-9405, Seattle.

Miller, J. A., and C. A. Simenstad. 1994b. Otolith microstructure preparation, analysis, and interpretation: a potential habitat assessment tool. University of Washington, Fisheries Research Institute, FRI-UW-9406, Seattle.

Miller, J. A., and C. A. Simenstad. 1997. A comparative assessment of a natural and created estuarine slough as rearing habitat for juvenile chinook and coho salmon. Estuaries 20:792–806.

Moser, M. L., A. F. Olson, and T. P. Quinn. 1991. Riverine and estuarine migratory behavior of coho salmon (*Oncorhynchus kisutch*) smolts. Canadian Journal of Fisheries and Aquatic Sciences 48:1670–1678.

Nakamura, F., and F. J. Swanson. 2003. Dynamics of wood in rivers in the context of ecological disturbance. Pages 279–297 *in* S. V. Gregory, K. L. Boyer, and A. M. Gurnell, editors. The ecology and management of wood in world rivers. American Fisheries Society, Symposium 37, Bethesda, Maryland.

Paine, R. T., and S. A. Levin. 1981. Intertidal landscapes: disturbance and the dynamics of pattern. Ecological Monographs 51:145–178.

Ray, D. L. 1959. Nutritional physiology of Limnoria. Pages 46–61 *in* D. L. Ray, editor. Marine wood boring and fouling organisms. University of Washington Press, Seattle.

Simenstad, C. A., J. R. Cordell, W. G. Hood, J. A. Miller, and R. M. Thom. 1992. Ecological status of a created estuarine slough in the Chehalis River estuary: report of monitoring in created and natural estuarine sloughs, January-December 1991. University of Washington, Fisheries Research Institute, FRI-UW-9206, Seattle.

Simenstad, C. A., J. R. Cordell, J. A. Miller, W. G. Hood, and R. M. Thom. 1993. Ecological status of a created estuarine slough in the Chehalis River estuary: assessment of created and natural estuarine sloughs, January-December 1992. University of Washington, Fisheries Research Institute, FRI-UW-9305, Seattle.

Simenstad, C. A., J. R. Cordell, W. G. Hood, B. E. Feist, and R. M. Thom. 1997. Ecological status of a created estuarine slough in the Chehalis River estuary: assessment of created and natural estuarine sloughs, January-December 1995. University of Washington, Fisheries Research Institute, FRI-UW-9621, Seattle.

Simenstad, C. A., K. L. Fresh, and E. O. Salo. 1982. The role of Puget Sound and Washington coastal estuaries in the life history of Pacific salmon: an unappreciated function. Pages 343–364 *in* V. S. Kennedy, editor. Estuarine comparisons. Academic Press, New York.

Simenstad, C. A., W. G. Hood, R. M. Thom, D. A. Levy, and D. L. Bottom. 2000. Landscape structure and scale constraints on restoring estuarine wetlands for Pacific Coast juvenile fishes. Pages 597–630 *in* M. P. Weinstein and D. A. Kreeger, editors. Concepts and controversies in tidal marsh ecology. Kluwer Academic Publishers, Dordrecht, Netherlands.

Simenstad, C. A., A. J. Wick, J. R. Cordell, R. M. Thom, and G. D. Williams. 2001. Decadal development of a created slough in the Chehalis River estuary: Year 2000 results. University of Washington, School of Aquatic and Fishery Sciences, SAFS-UW-0110, Seattle.

Stembridge, J. E., Jr. 1979. Beach protection properties of accumulated driftwood. Pages 1052–1068 *in* Proceedings of the Specialty Conference on Coastal Structures 1979. ASCE/Alexandria, Virginia, 14–16 March 1979. Coastal and Environmental Resources Institute, Salem, Oregon.

Terich, T., and S. Milne. 1977. The effects of wood debris and drift logs on estuarine beaches of northern Puget Sound. Department of Geography and Regional Planning, Western Washington University, Bellingham.

Thom, R. M., and A. B. Borde. 1998. Human intervention in Pacific Northwest coastal ecosystems. Pages 5–38 *in* G. R. McMurray and R. J. Bailey, editors. Change in Pacific Northwest coastal ecosystems. NOAA Coastal Ocean Program, Decision Analysis Series 11, NOAA, Washington, D.C.

Thomas, D. W. 1983. Changes in the Columbia River estuary habitat types over the past century. Columbia River Estuary Data Development Program, Astoria, Oregon.

Triska, F. J., and K. Cromack, Jr. 1980. The role of wood debris in forests and streams. Pages171–190 *in* R. H. Waring, editor. Forests: fresh perspectives from ecosystem analysis. Proceedings of the 40th Annual Biological Colloquium. Oregon State University Press, Corvallis.

Tschaplinski, P. J., and G. F. Hartman. 1983. Winter distribution of juvenile coho salmon (*Oncorhynchus kisutch*) before and after logging in Carnation Creek, British Columbia, and some implications for overwinter survival. Canadian Journal of Fisheries and Aquatic Sciences 40:452–461.

Turner, R. D. 1971. Australian shipworms. Australian Natural History 17:139–145.

Turner, R. D. 1977. Wood, mollusks, and deep-sea food chains. Bulletin of the American Malacological Union 1976:13–19.

Turner, R. D. 1981. "Wood islands" and "thermal vents" as center for diverse communities in the deep sea. Soviet Journal of Marine Biology 7:3–4. (1987 Translation of Biologiya Morya 7:3–10)

Turner, R. D. 1984. An overview of research on marine borers: past, progress, and future directions. Pages 3–16 *in* F. D. Costlow and R. C. Tipper, editors. Marine biodeterioration: an interdisciplinary study. Naval Institute Press, Annapolis, Maryland.

Van de Wetering, S. 2001. Juvenile salmonid densities across variable habitats within the Siletz River estuary. Confederated Tribes of the Siletz Indians, Technical Report 2001E3, Siletz, Oregon.

Wick, A. J. 2002. Ecological function and spatial dynamics of large woody debris in oligohaline-brackish estuarine sloughs for juvenile Pacific salmon. Master's thesis. University of Washington, Seattle.

Wolanski, E., and P. Ridd. 1986. Tidal mixing and trapping in mangrove swamps. Estuarine and Coastal Shelf Science 23:759–771.

American Fisheries Society Symposium 37:279–297, 2003

Dynamics of Wood in Rivers in the Context of Ecological Disturbance

FUTOSHI NAKAMURA

Department of Forest Science, Graduate School of Agriculture Hokkaido University, Sapporo 060-8589, Japan

FREDERICK J. SWANSON

Pacific Northwest Research Station, USDA Forest Service, Forestry Sciences Laboratory 3200 SW Jefferson Way, Corvallis, Oregon, 97331, USA

Abstract.—Disturbance relevant to dynamics of wood in rivers can take many forms. We consider effects of ecosystem disturbance related to wood in river systems in geographic settings, which include high-gradient, boulder-dominated streams, braided, gravel-bed streams, and low-gradient, sand-bed streams. Disturbances of forests affect delivery of wood to streams and rivers directly by causing wood input or wood removal and indirectly by limiting source material. Disturbance of the fluvial system, either the channel form or flow regime, alters the transport and standing crop of wood. Change in wood distribution by processes of deposition, transport, and removal can disturb riparian, benthic (streambed), hyporheic, and water-column habitat. Human actions, such as harvesting trees, building roads, and regulating water flow, can substantially alter the types, frequencies, spatial patterns, and severity of the natural disturbance regime. We summarize the current status of knowledge on these points and identify knowledge gaps in studies of wood in rivers within the context of ecological disturbance. Finally, we offer a framework for future work and management that integrates processes that shape the spatial and temporal dynamics of wood at a series of scales.

Introduction

Disturbance can be defined ecologically as "any relatively discrete event in time that disrupts ecosystem, community, or population structure and changes resources, substrate availability, or the physical environment" (Pickett and White 1985) and thus creates "space" for other organisms. Principal attributes of disturbances include timing, spatial aspects, intensity of the force that triggers change, and severity of response. This concept of disturbance is ecological and is not commonly used in geomorphology. Disturbance processes are so integral to the highly dynamic riverine and riparian ecosystems being considered here that absence of disturbance, such as elimination of peak flows by reservoir management, can itself be considered a disturbance (Resh et al. 1988).

Previous studies of wood in rivers have focused on the amounts and distribution of wood and its ecological and geomorphic effects. Few studies have considered wood explicitly in terms of disturbance processes and effects, though implications for interpreting disturbance appear in many published studies.

Disturbance relevant to dynamics of wood in rivers can take many forms. For example, the frequency and magnitude of flooding can be reduced by natural factors, such as drought, or management, such as dams and dikes, to regulate or divert flows. Human actions, such as cutting forest, building roads, and regulating water, can substantially alter the natural disturbance regime (that is, the types, frequencies, spatial patterns, and severity of disturbance). Human actions can increase or decrease the frequency and magnitude of natural processes, and they can introduce entirely new (exotic) factors.

Three effects of ecosystem disturbance related to wood in river systems can be significant. Disturbances of forests affect delivery of wood to streams and rivers directly or indirectly by limiting source material. Disturbance of the fluvial system, either the channel form or flow regime, alters the transport and standing crop of the wood. Change in wood distribution, including deposition, transport, and removal, can disturb riparian, benthic (streambed), hyporheic, and water-column habitat.

In this chapter, we first describe the geographic settings in which disturbance processes in river landscapes operate. We then summarize the current state of knowledge about wood-related processes, both as they are affected by disturbance (wood input) and as disturbance agents that affect the river landscape (wood transport and change in wood structure). For each of these wood-related processes, we discuss their spatial patterns and how they are influenced by management. We conclude by defining some major gaps in knowledge related to managing wood in rivers and present a framework for interpreting and applying results in different geographic and management situations.

Geographic Settings of Wood-Related Disturbance Processes

For many ecological phenomena, especially disturbance in stream and riparian networks, a geographic context is an important framework for analysis. Here, we consider relevant processes in terms of the size and type of the river landscape and the associated channel conditions. River landscapes are classified into three dominant morphologies. The relative importance of mechanisms that deliver wood to streams varies among these types: (1) high-gradient, boulder-dominated streams with very narrow valley floors, so hillslopes constrain the stream channel and hillslope processes directly influence channel morphology and wood dynamics; (2) braided, gravel-bed streams with a relatively steep channel bed and banks commonly dominated by a gravel-dominated stream, and also shows some development of multiple channels where wood dynamics are controlled mainly by fluvial processes; (3) low-gradient, sand-bed streams characteristic of low-gradient, meandering rivers (Figure 1). We refer to these three morphologies as mountain, boulder-bed streams; braided, gravel-bed streams; and meandering, sand-bed streams.

The concept of "disturbance cascades" is a complementary geographic perspective that considers the wood moving through stream networks by a sequence of transport processes (Nakamura et al. 2000). These sequences can pass from one of these stream types to another. A broader, landscape-ecology perspective also can be adopted to assessing disturbance effects on wood in rivers. For example, transport of wood in floods can contribute to forming patches of disturbed riparian vegetation, making a shifting patch-mosaic, landscape-dynamic model relevant (Bormann and Likens 1979). This landscape view of the river and riparian landscape is addressed in Swanson 2003, this volume.

Disturbance Effects on Wood Input to Rivers

Input processes include both geophysical and biotic processes, some highly influenced by vegetation structure and composition and others less so. Stage of vegetation succession can affect vulnerability of a stand to disturbance that leads to wood delivery to streams. Stage of succession of streamside forest also affects the size and spatial distributions and amount of wood available to be delivered to streams. Hedman et al. (1996), for example, showed that instream amounts of wood increased linearly with increasing forest age through a chronosequence of forests spanning a 165-year interval in a series of study sites in the southern Appalachian Mountains, USA. Similar observations have been made in conifer forests in the Pacific Northwest, USA (Harmon et al. 1986).

Mountain, boulder-bed stream type

Processes that deliver wood into the mountain, boulder-bed streams are dominated by disturbances to upland and riparian forests, such as windthrow, fire, and landslides. Lateral cutting by the stream may not significantly contribute wood to mountain streams because the channels are tightly constrained by the adjacent hillslopes.

Windthrow dominates the agents of delivery in many mountain streams. Lienkaemper and Swanson (1987), for example, observed that about 69% of the wood volume added to the stream was delivered by wind, possibly coupled with stem or root decay, which make trees more vulnerable to toppling. Instability imposed by the tilt of trees growing into the open canopy space over a chan-

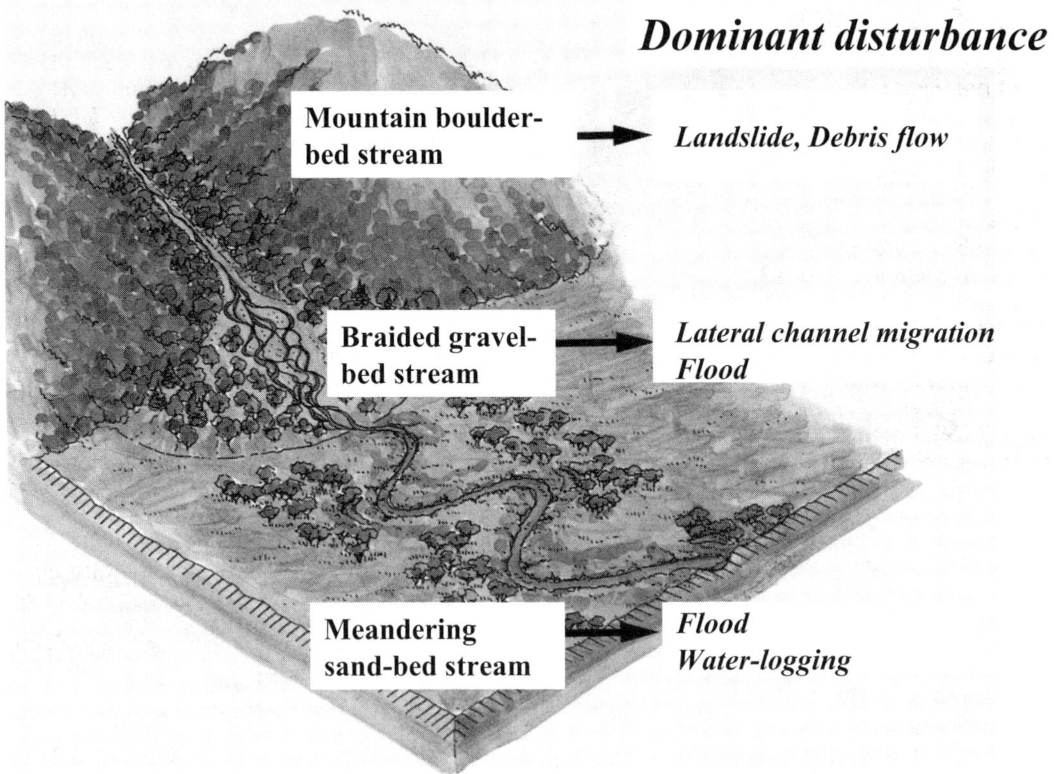

FIGURE 1. Classification of river landscape into the three dominant morphologies.

nel can increase the susceptibility of large trees to windthrow. Catastrophic windstorms, such as typhoons and hurricanes, are common in both temperate and tropical forests, and their effects are closely related to landform. Forest stands established on steep slopes facing the direction of strong winds and valley bottoms running parallel to the wind direction are particularly susceptible to windthrow (Tokyo Forestry Branch 1959). Strong typhoon winds in Japan create large patches of fallen trees (Figure 2) and abruptly supply large amounts of wood to streams in association with landslides (Miyabuchi et al. 1999). Stands planted in Japan after World War II now form densely stocked, 50-year-old stands on steep slopes; the trees are highly vulnerable to toppling into streams as large patches, perhaps in part by a domino effect under the influence of the high winds and intense rainfall of typhoons (Ishikawa 1989). Under these conditions, distinguishing influences of wind and landslides is difficult. In some stands, large trees are more susceptible to windthrow than small ones, creating individual small canopy gaps and occasional tree fall into streams.

Large wildfires, such as the Yellowstone fires of 1988, can burn forests over tens to thousands of square kilometers, affecting extensive areas of the mountain, boulder-bed and other stream types (Romme and Despain 1989). These fires substantially alter the amount and quality of the wood in streams over a long period. Minshall et al. (1989) evaluated the wood in stream channels before and after wildfire in a lodgepole pine forest in the Yellowstone area. Large amounts of wood accumulate in old-growth forest streams in periods leading up to fires. Because of its high moisture content, most of this material remained intact even after hot fires, and it was rapidly augmented by branches and trunks brought down by the fire (Figure 3). Thus, wood volume in the streams increases sharply immediately after fires (Spies and Franklin 1988). Although fallen, fire-killed snags continue to accumulate in streams for several decades after a fire; the new growth in the riparian forest contributes little large wood to the stream. Burned wood may readily break down into small pieces with greater stream instability, faster in the first few years after a fire than later. As forests

FIGURE 2. Catastrophic windthrow by the typhoon in September 1991, Ooita Prefecture, southern Japan (photo provided by Akira Shimizu).

mature and naturally thin, additional wood of increasing diameter and mass accumulates in streams. This process accelerates when old-growth forest conditions are attained. This process can take about 300 years in some of the forest types of the Yellowstone region (Romme and Despain 1989). Simulation modeling of forest stands replicates this temporal pattern of instream wood loading after disturbance (Bragg 2000; Benda et al. 2003, this volume).

Heavy rainfall and rapid snowmelt can initiate landslides, which deliver wood and sediment to mountain streams. The potential for landslides can be increased by windthrow and wildfire in steep landscapes. Typical landforms with landslides are bedrock hollows (zero-order basins; Tsukamoto 1973, Dietrich and Dunne 1978) and steep (>30 degrees), convergent slopes created by channel incision and then filled with colluvium after the last glacial period (Hatano 1974). Many of the wood pieces moved by landslides may be delivered directly to stream channels because there are few geomorphic surfaces, such as fans and terraces, that act to intercept wood delivery. The frequency of landslides at a particular site can range from a few decades (Shimokawa 1984; Yanai and Usui 1989) to 10,000 years or more (Reneau and Dietrich 1990), depending on the geological and geomorphological setting. Slope positions with very frequent landslides may supply smaller pieces of wood than slopes with rare landslides because of the limited time for accumulating live and dead wood for transport by the slides. Ishikawa (1989) studied three heavy rainfall disasters in Japan and concluded that most of the mobilized wood pieces were delivered by landslides adjacent to stream channels and also included old, remnant wood that was in the channel before the storm. Landslides seem to be an important process supplying wood to streams, especially in geologically unstable terrain, such as Japan and the Pacific Northwest, USA (Benda 1990; May 2001).

Volcanic eruptions and associated mass movements can dramatically increase wood volume in mountain streams in volcanic landscapes,

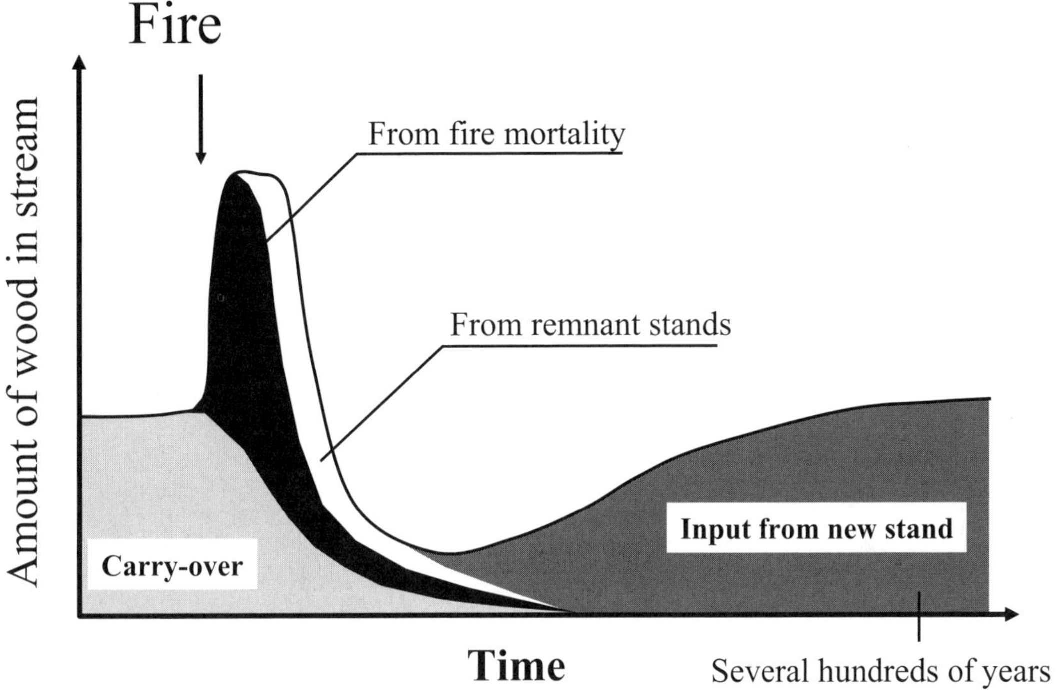

FIGURE 3. Postulated response of instream large wood in lodgepole pine forest after wildfire (modified from Spies and Franklin 1988 and Minshall et al. 1989).

such as in Japan and northwest USA. The timing of inputs of wood, however, can vary widely. The 1977 eruption of Usu Volcano—Hokkaido, Japan—ejected a large amount of ash and pumice, covering a forest landscape near the volcano with several meters of ash fall. Trees buried by deep ash fall were killed, but they were not instantly supplied to the deep gully systems created by rapid runoff after the eruption because most of the trees were standing dead. The wood pieces from broken limbs delivered right after the eruption were relatively short and thin and therefore contributed little to the large debris flows from the flanks of the volcano. In contrast, the 1980 eruption of Mount St. Helens included a much greater variety of volcanic processes that immediately delivered large volumes of wood to rivers draining the volcano and transported that material downstream in some of these rivers. A lateral blast leveled forest over 550 km^2, greatly increasing the wood in river networks (Figure 4). Massive volcanic mudflows raced down most major channels draining the mountain, entraining riparian forest and moving vast amounts of wood many tens of kilometers downstream. The magnitude of the Mount St. Helens 1980 eruption was far greater than recent activity at Mount Usu, and trees near Mount St. Helens were large conifers, up to 60 m (197 ft) in height and 2 m (6.6 ft) in diameter. Wood is still a very important component in streams and some lakes of Mount St. Helens 20 years after the eruption (Figure 4).

Braided, gravel-bed stream type

Forest disturbance processes that operate along mountain, boulder-bed stream types also affect wood in braided, gravel-bed streams, but the braided type additionally experiences disturbances resulting from lateral channel migration and the development and abandonment of secondary channels. Wood falls into streams where banks are eroded and live trees undermined, and wood enters primary channels where fallen trees in secondary channels and on floodplains are floated into the main channel. Multiple channels appear in this stream type at low water, and they may unite during flood flows.

An example of fluvial, wood-related disturbance in the braided, gravel-bed stream type has been described for the Saru River, Japan (Naka-

FIGURE 4. Wood pieces in an area where the 1980 eruption of Mount St. Helens blew down old-growth forest. The photograph was taken in the upper Green River 20 years after the eruption and after recent flooding caused some redistribution of wood in the channel.

mura and Kikuchi 1996). A major flood in 1992 inundated the entire valley floor with water, filling abandoned channels and knocking down the trees in its path. The degree of riparian forest disturbance by the 1992 flood varied according to the age of forest stand (Figure 5). Disturbance severity was described in three rankings: high severity, where trees and even soil were completely removed; moderate severity, where trees were toppled but not removed from the site; and low severity, where trees remained standing, but understory vegetation was removed. Younger riparian forests tended to be more extensively and severely disturbed than older forest stands. In general, mature forests are away from the active channel and on higher geomorphic surfaces, which limits the depth, velocity, and force of water on the trees. In contrast, younger forests are close to the stream channels and, therefore, disturbed frequently by flooding and impacts from floating logs. Thus, small wood pieces dominate braided, gravel-bed streams because the forest is frequently disturbed in the main zone of wood recruitment and the forest does not have time to grow trees of large size.

These observations in Japan parallel trends reported for braided rivers in southeast France by Piégay and Gurnell (1997) and in the Pacific Northwest of the United States (Acker et al. 2003). In alluvial channels in northwest Washington State, however, large "key member" logs (potentially >4 m in diameter, >70 m in length) initiate formation of stable bar-apex (head of gravel bar) and meander jams that alter local flow hydraulics, resulting in bar and pool formation. Individual jams are remarkably stable, providing habitat for trees to mature within a dynamic valley floor environment characterized by frequent channel migration and disturbances (Abbe and Montgomery 1996).

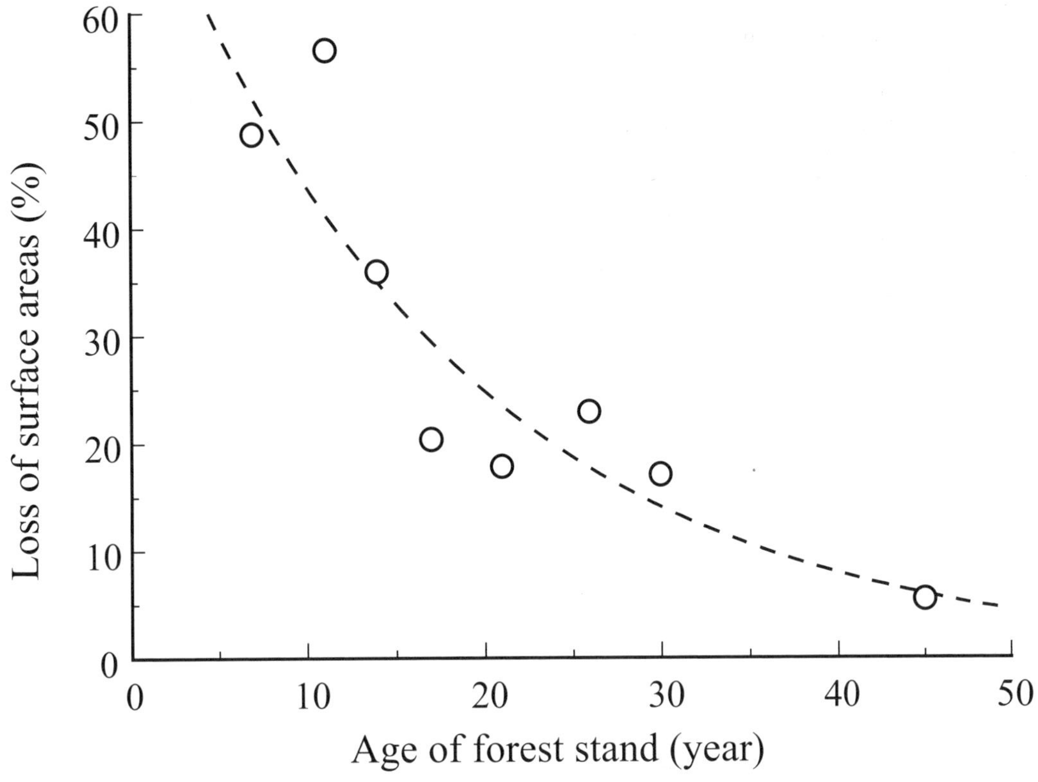

FIGURE 5. Areal percentage of eroded floodplain areas of the Saru River with respect to the stand age (from Nakamura and Kikuchi 1996).

Meandering, sand-bed stream type

River pattern dynamics along the meandering, sand-bed stream type strongly influence the spatial pattern and rate of wood input to channels. Point bars composed of fresh sediment deposits are formed one after another on the inside of meander bends, and progressively older plant communities are developed at increasing distance from the channel. As the entire meandering belt shifts downstream through time, however, older stands grow at the outside of bends where strong currents undercut the banks. This process produces large wood pieces that fall into the channel and may collect on downstream point bars. Thus, wood pieces are individually delivered from the outside banks of the channel, and fresh sand bars providing regeneration sites develop on the inside of meander bends.

Keller and Swanson (1979) suggested the importance of bank erosion in supplying wood to low-gradient, meandering streams. In a study of a meandering, sand-bed stream in northern Japan, Abe and Nakamura (1996) observed that about 78% of the wood volume was produced by bank erosion. Mortality and windthrow contributed only 3% of the total pieces for which supply processes could be identified in this study. Everitt (1968) documented wood entrainment and destruction of riparian forests in association with lateral migration of meandering, sand-bed channel in the Little Missouri River.

Management effects on wood inputs

Forestry activity has the potential to strongly influence wood in the mountain, boulder-bed stream type. Clearcutting, thinning, cleaning streams, and building roads, for example, can alter stand conditions that affect input of wood to streams. Removing large trees and using management schemes that operate on short rotation lengths can reduce the amount and size of wood in channels. If logging debris is left in channels, it may be more vulnerable to movement than natural wood, which includes large pieces, some with rootwads that can stabilize the entire wood supply of the channel. The amount of logging slash in a channel decreases over time as it is transported by debris flows or floods, however, and the wood source areas on adjacent hillslopes are limited to smaller pieces after intensive forestry (Murphy and Koski 1989). A rapid decline of the amount of wood in the first 7 years after harvesting has been reported (Bilby and Ward 1991). The selective removal of instream wood in association with timber harvesting and establishment of red alder trees in riparian zones shifts the size-class distribution of wood toward smaller pieces in low-order streams (Bilby and Ward 1991; Ralph et al. 1994).

Forest roads used to be built along the riparian areas in many countries in the 1950s and 1960s, resulting in removal of riparian trees and reduction of the source of wood input to streams. Even if the roads were removed, the influence of this activity could continue for a century until the re-established riparian trees reach maturity. More recently, forest roads are built away from the riparian zone and on the ridges, to minimize detrimental effects on streams.

Monocultural, even-aged plantations of commercial trees develop a uniform canopy layer. This simple structure may be vulnerable to windthrow, occasionally producing large and abrupt delivery of wood pieces to streams. Moreover, the edges of clearcut patches are particularly susceptible to windthrow where trees grown in fully stocked stands are exposed to strong winds moving through the new opening (Franklin and Forman 1987; Sinton et al. 2000).

In many areas, it is now common practice to leave buffer strips along streams in forest harvest sites to maintain stream and riparian habitat and processes, including wood delivery to streams. This practice has not been widely examined in terms of its long-term ability to meet these objectives, however. One concern is that narrow riparian buffer strips may be subject to substantial toppling of trees by wind and high levels of associated input of wood pieces because the buffer-strip edges are vulnerable to windthrow.

In some areas, such as Japan, regulations for management of both braided and meandering alluvial channel types require clearing of the riparian forests, and roads and agricultural lands are then developed to the edges of the river banks. The riparian forests, therefore, cannot regenerate on these sites, and grasses dominate the streambank vegetation. In Juyonsen Creek, for example, the length of streamside areas bordered by forested riparian zones decreased by half from 1947 (11.6 km [7.6 mi]) to 1989 (6.6 km [4.1 mi]). In the Toikanbetsu River basin, including the Juyonsen Creek tributary, the volume and number of wood pieces were different among stream reaches flowing through three forest-cover types—no forest cover, second-growth forest, and old-growth forest (Figure 6). The numbers of wood pieces in second-growth forest were significantly higher than those in reaches with no forest cover, and wood volume was significantly higher in old-growth forest (Nagasaka and Nakamura 1999). Field studies of the distribution of wood pieces and modeling studies have shown that the successional stage of riparian forests is an important factor limiting the amount of wood because these forests are the major source of wood for streams (Andrus et al. 1988; Murphy and Koski 1989; Bilby and Ward 1991).

After streamflow is regulated by dams, lateral channel change is suppressed and the frequency of modification of gravel bars and associated vegetation are limited. Effects of altered flow regimes on valley-floor inundation are exemplified by damming of the Satsunai River in Japan. Floods with a 1-year return period before the dam was built recurred only once every 5 years or so after the dam went into operation (Figure 7A). Because the maximum discharge below the Satsunai River dam is regulated at only 150 m^3/s (5,300 ft^3/s), flood disturbances as large as those every 2 years before the dam are not expected in the future. A large gravel bar 2.0 km (1.2 mi) downstream from the dam site was used to observe effects of a considerable drop of water level after the dam was built (Figure 7B). Before the dam, a flood once every 5 years would submerge the entire river bed, except for 8% of the higher elevations on the gravel bar. In contrast, the regulated flows resulted in only 55% of the river bed submerged by a flood occurring once in 20 years (Nakamura and Shin 2001). Thus, production of wood pieces by fluvial processes will be limited by the regulated streamflow regime.

Disturbances Caused by Transport of Wood

Once wood pieces are supplied to streams from hillside slopes and floodplains, they may be transported and redistributed by processes operating in the channel. The mode of transport,

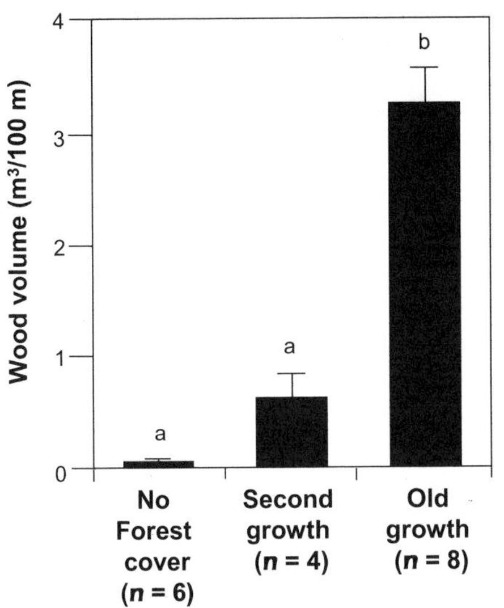

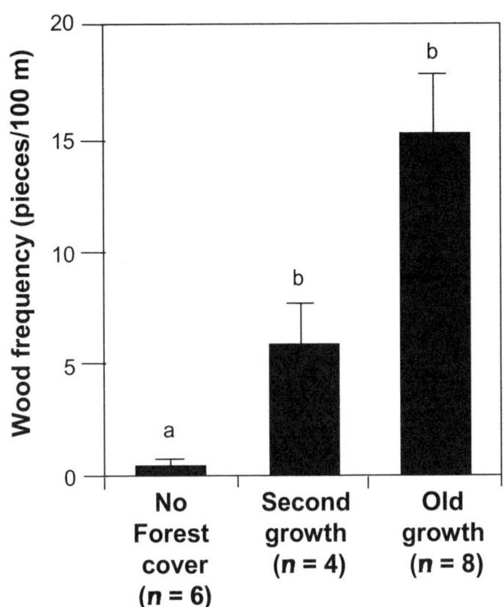

FIGURE 6. Comparison of wood volume and numbers of wood pieces among the three stand age-classes along tributaries in Toikanbetsu River basin, Japan. Values followed by the same letter are not significantly different using Mann-Whitney *U*-test ($p < 0.05$), conservatively adjusted with a Bonferroni procedure (from Nagasaka and Nakamura 1999).

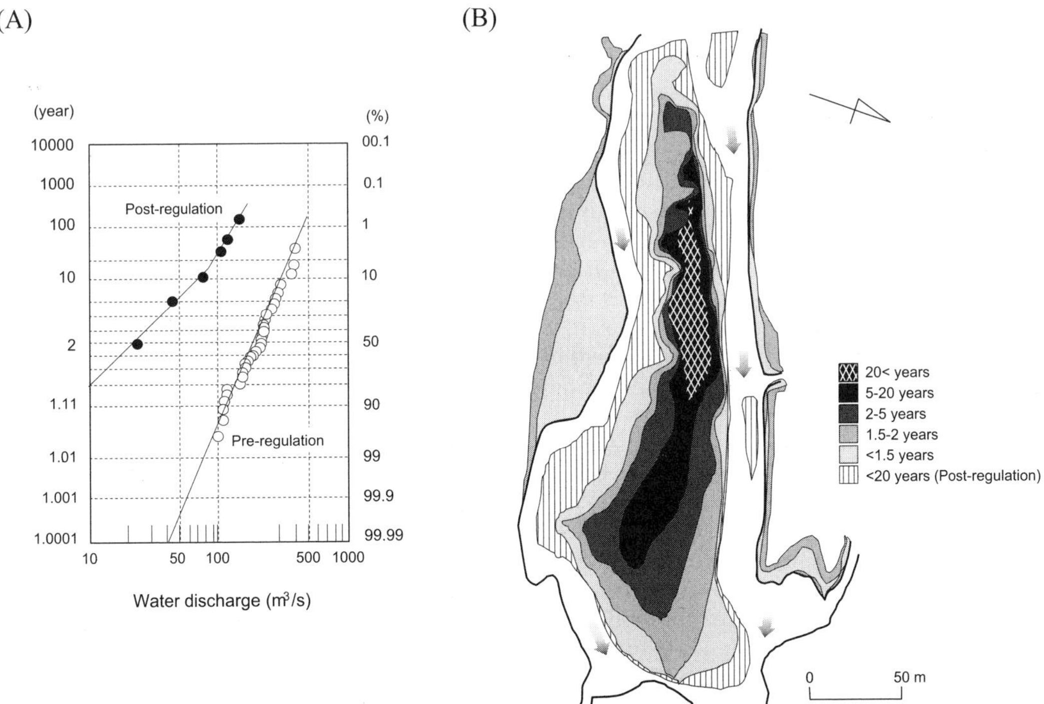

FIGURE 7. Return periods of peak discharge before and after Satsunai River dam (Japan) construction (A) and the resultant decrease in water level (B). The water level in flows with 20-year recurrence intervals was far lower than the water level during 1.5-year flows before flow regulation (Nakamura and Shin 2001).

including the abundance of wood and the type and amount of associated material in transit, can strongly influence the type and extent of disturbance it creates for in stream and riparian organisms and habitat structure (see Gurnell 2003, this volume).

Wood is transported by processes operating over a continuum of concentrations of wood and sediment in water. At least three classes of wood transport conditions are recognized: debris flows, floating wood moving in a congested manner in streamflow, and floating in an uncongested manner in streamflow (Braudrick et al. 1997). "Congested" transport refers to a high concentration of logs moving together as a single mass of pieces in contact with one another; "uncongested" transport refers to logs moving in sufficiently low densities that log–log contacts are infrequent. This spectrum of flow and transport conditions parallels to some extent the classification scheme of Scott (1988) for sediment transport, and we build on that framework by including wood.

Debris flows can transport large amounts of wood pieces as they flow down channels with abundant wood that can be entrained from the streambed and banks. This wood can move at the front of a debris flow as a pile of logs pushed forward by the force of the sediment-rich core of the debris flow. In steep mountain landscapes, debris flows pushing wood at their fronts can be 2 to more than 6 m (20 ft) thick and may therefore overflow the channel, exerting intense scour on streambed and banks. In steep channels with narrow valley floors, debris flows can maintain sufficient thickness to continue moving for many hundreds of meters into channels of lower gradient and wider valley floors. As debris flows spread laterally and become thin, frictional resistance to flow increases and standing trees interact with the moving boulders and wood to stop movement.

In larger channels, buoyant wood pieces can float during high flows in congested or uncongested modes of transport (Braudrick et al. 1997). Important mechanisms of disturbance by floating wood involve the impact force of a floated log against a standing tree and the "sail effect" of the transfer of force of flowing water on a horizontal, floated log lodged against a standing tree. In the latter case, the standing tree may gradually topple, making it possible for the floated log to slide up the tipping tree. This may cause the floated log to be lifted out of the water somewhat, thereby reducing the cross-sectional area affected by streamflow and reducing the force of streamflow against the toppling tree. This negative feedback mechanism may be responsible for formation of patches of tipped, still partially rooted saplings and small trees along channels where wood floatation events recently occurred (Johnson et al. 2000). Congested transport has the greater potential for disturbing riparian vegetation because groups of floating logs can have higher impact force on standing trees and may be less likely to pivot on impact and float away than is the case of a single piece of wood hitting a riparian tree. Floating wood is less likely to disturb benthic habitat than are debris flows because the floating wood is generally moving in the upper part of the water column and above the streambed.

The type of riparian vegetation disturbance depends on the vegetation itself and the amount and size distribution of transported wood. For example, young forests, such as 10–40-year-old alders, are flexible and have root systems that allow trees to be commonly found partially toppled but still rooted on gravel bars. This finding may be common where the channel has transported floated wood of sufficient size to knock the trees down, especially large, cylindrical pieces of conifer wood lacking massive, complex branch systems more typical of hardwoods. Large conifer trees in riparian zones may topple as a result of erosion of soil from their root systems. When they fall, their full length reaches the ground and may shatter on impact, but some pieces may not move because of their size relative to the amount of water available to float them. An important control on mobility of young, toppled hardwood trees, on the other hand, can be the stabilizing effect of partial rooting.

Johnson et al. (2000) investigated the role of floated wood on disturbance severity in riparian forests affected by a major flood in 1996 in the Cascade Range in Oregon, USA. Some riparian trees were removed by the flood, some were toppled, and others remained standing. The percentage of trees affected in these ways differed with transport processes, such as fluvial disturbance, uncongested transport, and congested transport. Stream reaches with uncongested wood transport had more toppled riparian trees than reaches experiencing high flow with very little wood transport and less than reaches with congested wood transport. Therefore, the availability of wood pieces to function as disturbance "tools" and the way they are transported influence the degree of disturbance in riparian forests.

Spatial perspectives of transport

Debris flows are a common wood transport and disturbance process in many steep, narrow channels of the mountain, boulder-bed stream type. On the other hand, floatation of wood pieces dominates transport in higher-order streams of the braided and meandering alluvial stream types. The factors controlling stream power (such as discharge, width, and gradient) and large obstructions or roughness elements (such as boulders, riparian trees, bars, and sinuosity) play major roles in redistributing wood pieces in streams (Nakamura and Swanson 1994; Piégay et al. 1999; Gurnell et al. 2000, 2002).

These different transport processes and their geomorphic settings can result in distinctive deposits of wood (Gurnell et al. 2002). Debris flows commonly produce tightly knit logjams at their snouts, which may accumulate at channel junctions (Benda and Cundy 1990), on roads, on valley floors, or in receiving channels. Floated wood transported in an uncongested mode commonly results in scattered deposits along the banks of rivers or in discrete trapping sites, such as at the heads of islands or mouths of secondary channels. Wood moved in a congested manner may form a series of log levees along the channel with pieces oriented parallel to it or pointing downstream and toward the channel (Johnson et al. 2000). Where streamside slides instantly deliver soil and many trees into a stream, they may dam the channel temporarily, followed by a dam-break flood (Coho 1993). This geomorphic process can transport logs in a congested manner. When the logs are deposited, they tend to be oriented perpendicular to the channel at the toe of the deposit and parallel to the channel at the lateral marginal levees. Floated wood pieces can be efficiently trapped on gravel bars in braided rivers, resulting in distinctive wood accumulations, bar forms, and patches of riparian vegetation (Nakamura and Swanson 1993; Abbe and Montgomery 1996). Wood may be preferentially deposited along floodplains adjacent to meandering streams (Piégay and Gurnell 1997; Gurnell et al. 2002).

Spatial and temporal interactions of transport processes

Sequences of geomorphic processes transport wood through stream networks. These sequences also function as a cascade of disturbances in mountain landscapes (Nakamura et al. 2000). Such disturbance cascades can be initiated by small, rapid landslides from hillslopes and channel head environments or by large, slow-moving earth-flows (Nakamura et al. 2000; Figure 8). Small landslides (1a) triggered on hillslopes may move into steep, headwater streams, move rapidly down those channels as debris flows (2), and ultimately deposit coarse sediment and logs in larger streams or on mainstem valley floors. Some debris flows may enter fourth- and fifth-order channels and be immediately entrained in part as rafts of wood floated by stream flow in the large channel. Alternatively, debris-flow deposits may form jams at the confluence with larger channels (A). These jams may break during floods, triggering a flood surge (3), which pushes a mass of floating logs in a congested mode of transport. Transport of wood accumulations from debris-flow runout or from jam break-up may end in distinct accumulations of wood or may simply dissipate in the downstream direction as the wood is left along the banks in a series of smaller accumulations, such as wood levees (Johnson et al. 2000).

Disturbance cascades can also begin where large, slow-moving earthflows gradually constrict channels (4a), increasing the potential for streambank erosion and streamside slides during high flows (5). These streamside slides can deliver sediment and trees, which form temporary dams (B) that can break up, triggering flood surges downstream for a kilometer or more (3). The associated logs may move in a congested manner, accentuating their ability to disturb riparian vegetation.

Transitions among these geomorphic processes are critical aspects of evaluating their potential to propagate through stream and riparian networks. The sequence of processes can be interrupted at any point along the flowpath. Landslides may come to rest where they reach a channel (1b) and form a debris jam (D) or, in areas of low gradient—such as slump benches or road surfaces—before reaching streams. Debris flows may stop at critical points where the channel slope is too low to sustain flow, channel direction changes abruptly by 70 degrees or more, valley floor width increases, or the debris flow encounters obstructions to flow, such as standing trees or road fills that can act as dams (C).

The extent of expression of the disturbance cascade can be assessed in terms of the probabil-

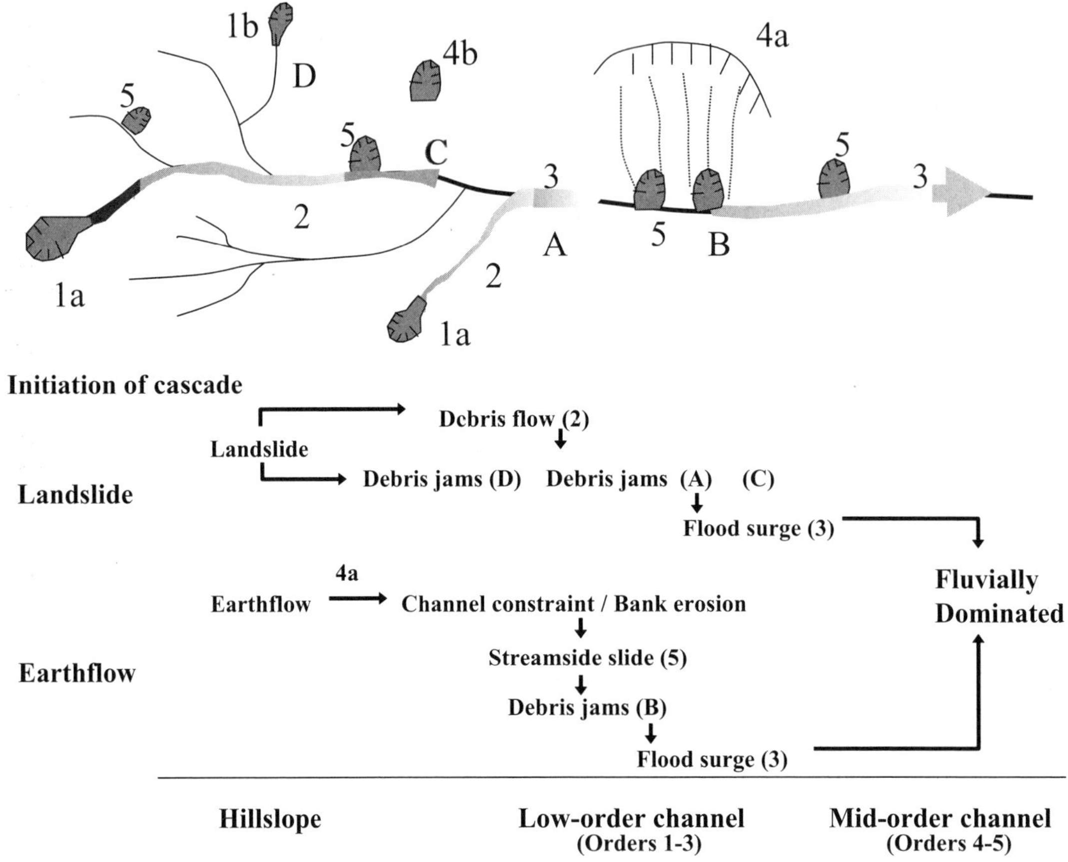

FIGURE 8. Representation of interaction among processes (numbered) operating in disturbance cascades. Letters denote depositional sites at the end of landslide and debris flow paths (from Nakamura et al. 2000).

ity of one process transforming to the next in the sequence, based on an observed history of events. To illustrate these relations, we examined linkages among the relevant processes in >50-year-recurrence-interval flood of February 1996 in the Oregon Cascade Range (Nakamura et al. 2000). The events were initiated mostly by shallow landslides unassociated with earthflows. About 30% of the landslides did not reach streams because they lodged on hillslopes or on roads. Nearly half of all the landslides transformed into debris flows down steep, first- through third-order channels. Thus, about 70% of the initial population of landslides and debris flows stopped at hillslope-channel intersections, in the channel at a downstream point, or on roads and alluvial fans before reaching large streams. Only 7 of 94 events inventoried were associated with earthflows. These observations suggest the relative importance of different flowpaths through the watershed in terms of wood transport by mass landslide-related processes and associated disturbance to stream and riparian habitat in this mountain stream network.

Management effects on wood transport

Management actions of various types—including forestry, erosion control, and recreation-related—can affect the rates, frequencies, and types of wood transport in watersheds and the size distribution of material moved. These factors affect disturbance regimes of stream and riparian systems. Forest clearcutting, for example, can increase the probability of landslides by a factor of two or more (Sidle et al. 1985), and the associated reduction of large trees in landslide deposits may increase the potential for these landslides to develop into debris flows. Forestry practices may also reduce the size of wood pieces in a stream, making the material more mobile during high flows.

Roads passing through headwater areas of mountain stream networks can be sources of landslides when roadfills collapse during heavy rainfall and snowmelt (Sidle et al. 1985, Wemple et al. 2001). Roads on valley floors of larger channels can serve as small dams that block downstream transport of wood in debris flows (Wemple et al. 2001). Roads may also increase peak flows in some situations by increasing the stream drainage density in a watershed (Wemple et al. 1996), thus potentially increasing the capacity of a stream to transport wood.

A comparative watershed study by Nakamura and Swanson (1993) showed that a third-order watershed with no landslides along its mainstem after clearcutting developed a wide valley floor with many wood pieces, including logging slash. A nearby watershed with many, large debris flows had relatively little instream wood remaining on the narrow, bedrock streambed.

Check dams for sediment control in Japan trap wood pieces on the flat areas of the sediment storage area behind the dam (Shimizu 1998). Slit dams designed to sieve large inorganic and organic materials from streamflow and debris flows capture wood pieces efficiently. Large reservoirs for power generation and flood control trap all wood pieces transported from the upstream drainage areas, resulting in substantial decrease in wood delivery to the downstream reaches.

Not only the wood supply from upstream, but also the entrainment of wood by lateral channel migration and high flow peaks can be greatly reduced by regulating water discharge. Straightening, revetment, and bed stabilization work in association with channelization projects can simplify channel margin features and bed roughness, increasing channel capacity for downstream wood transport, and leaving very few pieces after flooding in channelized reaches. The subsequent degradation of the riverbed by vertical erosion commonly forces a river into a single channel. Such hydrologic and geomorphic changes dramatically alter the frequency and magnitude of flood disturbances.

Disturbance Caused by Changes in Structure of Wood Accumulations

Because wood plays important roles in the habitat structure and ecological function of stream ecosystems, change in the arrangement of instream wood can amount to ecosystem disturbance. Effects of modifying wood have been examined in various ways, such as studies of processes and experimental manipulations of the wood, including both introducing and removing it.

Wood efficiently retains the particulate organic materials delivered from upstream and adjacent hillslopes, especially in low-order streams. This function is important in regulating energy flows in forested streams. In a study in Japanese low-gradient, small streams, hydrologic factors—such as velocity and depth—were shown not to regulate the amounts of leaf litter in streams; rather, wood volume best explains the standing stock of leaf litter in the three seasons studied (Kishi et al. 1999).

Wood pieces were experimentally removed from streams at Hubbard Brook Experimental Forest, New Hampshire, USA to investigate effects on particulate-matter storage and export and other factors (Bilby and Likens 1980). The results showed a striking increase in fine-particulate organic-matter export, especially when water discharge was high. Removing wood pieces not only decreased the ability of the channel to retain these materials, but also destroyed most of the pools in the stream, thereby removing depositional areas for particulate material. Wood also serves as a major mechanism for retention of salmon carcasses in the Pacific Northwest (Cederholm et al. 1989).

The abundance of dams not only regulates organic-matter storage, but also macroinvertebrate abundance. Macroinvertebrate density and biomass in wood dams of a headwater stream in Virginia were 5 to 10 times greater than on sandy sediments because of higher retention of particulate organic matter associated with wood dams (Smock et al. 1989). Thus, functional feeding groups also varied with wood dam abundance; for example, the density of leaf shredders was eight times higher in dams than on sediment.

Removing wood pieces initiates substantial changes in channel morphology and sediment transport (Beschta 1979, Abe and Nakamura 1999). Experimentally removing wood from a small, gravel-bed stream resulted in a fourfold increase in bedload transport at bank-full discharge (Smith et al. 1993).

Natural fluvial disturbance of wood can also greatly alter the characteristics and the extent of the hyporheic zone. Wondzell and Swanson (1999) observed changes in hyporheic zones before and after the major 1996 flood in the Oregon Cascade Range, USA (Figure 9). Before the flood, the active

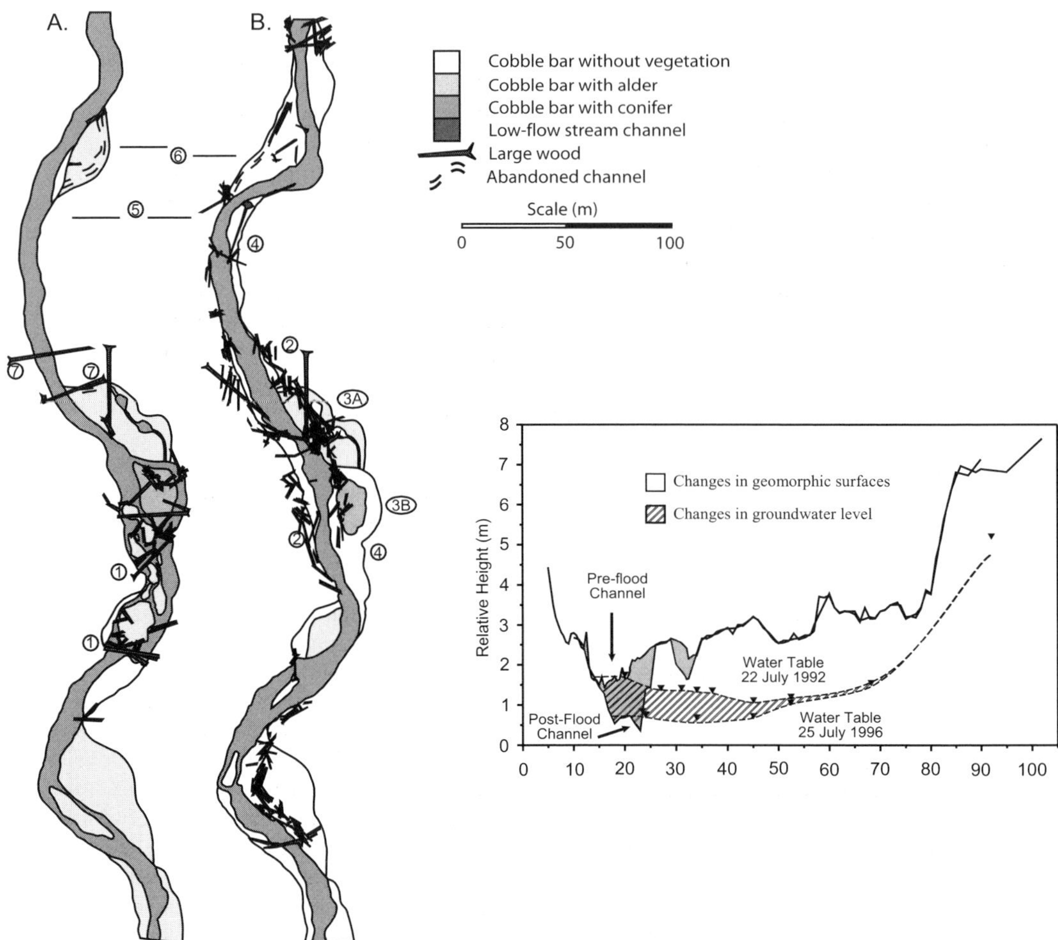

FIGURE 9. Lower McRae Creek showing the general planform of the active channel, locations of key large wood, secondary channels and gravel bars (A) before and (B) after the flood of February 1996. Cross-sectional profile indicates the drop in water table before and after the flood (from Wondzell and Swanson 1999).

channel was somewhat braided around small gravel bars and logjams. The flood of 1996 breached the logjams, and the channel cut down through the wedge of accumulated sediment. This downcutting altered subsurface flow paths and limited the extent of the hyporheic zone.

Wood pieces control instream habitat conditions for fish by creating pools and providing refuge during floods and cover in nonflood periods. Studies in the United States and Canada clearly indicate that wood provides salmonid habitat and influences fish abundance (Elliott 1986; Riley and Fausch 1995; Dolloff and Warren 2003, this volume). These results were experimentally confirmed by adding wood pieces or removing them. Observed increases in fish populations were attributed to an increase in pool volumes (Fausch and Northcote 1992). This function of wood pieces was not fully examined in Japanese streams where all riparian trees are small. Inoue and Nakano (1998) examined the relation between wood volume and juvenile salmon density in a small stream and concluded that wood affected fish more by providing cover than by creating pools. In the volcanic, pumice-bed streams of this study area, however, even small woody pieces could create pools and increase biomass of rainbow trout *Oncorhyunchus mykiss* (Urabe and Nakano 1998), probably because the low-density, pumice bed material could be easily scoured. Removal of wood by river regulation detrimentally affect fish populations and communities.

Spatial aspects of change in wood accumulations

Mountain boulder-bed channels, and especially bedrock channels, can be highly influenced by wood because these systems have high transport capacity, which wood can impede. Wood can trap sediment, creating benthic habitat, and retain floated organic litter in a system otherwise limited in substrates. Thus, removing wood from these systems is likely to strongly affect these processes and features. Disturbance of wood in this channel type is likely to have little effect on recruiting additional wood to the stream, except in the case of reduction of wood for inclusion in debris flows, which can affect their impact and extent. Modifying wood in this channel type may have limited effect on fish where the boulders are providing high channel complexity and fish populations are limited by high velocity and gradient and by limited habitat space.

Disturbance of wood structures in braided, gravel-bed streams is likely to affect both wood input and transport processes, which are sensitive to the presence of floated wood in these systems. Disturbance of wood structures in secondary channels can be particularly important in terms of loss of water-column habitat for fish. Secondary channels can trap small wood pieces that contribute to pool formation (Nakamura and Swanson 1994) and provide refugia during floods.

Disturbance of wood in the meandering, sand-bed channel type may have an impact on complexity of water-column habitat along the outside banks of bends in the channel where certain fish species may reside. In general, however, wood plays less significant roles in this channel type than in the other two, so disturbance of wood would be of less consequence. Experimentally removing the wood in the meandering, sand-bed channel decreased small pools associated with wood in one Japanese study but recreated large pools by lateral scouring, and total pool volume remained unchanged (Abe and Nakamura 1999).

Management effects on wood accumulations

Channelization projects common in developed countries can greatly reduce the ability of a channel to retain material transported by flowing water (such as organic matter, nutrients, and sediment) by removing wood pieces from stream channels. After channelization, further changes in channel structure are limited because of wood removal, straightening of the channel path, and revetment of stream banks. Thus, transport capacity in the channelized reaches, in general, is high, and wood pieces supplied from upstream are easily transported.

Check dams for sediment control create a large hyporheic zone behind (upstream of) the dams. Check dams, however, are commonly impermeable, and therefore, they can almost eliminate exchange flows because water cannot flow through the accumulated sediment behind dams. The residence time of groundwater in the dam-created hyporheic zone may be very long, so anaerobic conditions commonly occur in the stagnant groundwater. On the other hand, wedges of sediment accumulated behind permeable, step structures, such as logjams, permit complex subsurface water flow through the structure.

Knowledge Gaps and Framework for Future Work

Research on the influences of disturbance on patterns and processes of wood in rivers and stream is growing, but many knowledge gaps remain. Only little work on wood in rivers has taken a disturbance perspective, especially at the time scale of the disturbance regime and the spatial scale of whole watersheds. Few studies link wood production, storage, decay, and transport at the watershed scale (Snyder 2000; Benda et al. 2003). An important next step is to incorporate a disturbance perspective in the general view of the production and routing of wood pieces. Some components of the wood-routing system (Benda et al. 2003; Gurnell 2003) result in ecosystem disturbance and other components do not.

A further critical knowledge gap is the cumulative effects of management actions. Effects of management actions—intended and unintended—are poorly understood in terms of wood quantities, arrangements, movement, and effects. In general, management tends to reduce wood pieces in streams in the long term, though it may initially increase them. Many complexities of the temporal and spatial dynamics of wood in natural and managed systems remain to be unraveled, however. One valuable perspective would be an understanding of how the relative behavior of wood-related disturbance regimes compares with other disturbances in terms of types, degrees, and consequences. This information could benefit efforts to restore watersheds and biodiversity.

A disturbance-cascade perspective (Nakamura et al. 2000) may provide a good framework to advance understanding of the links between wood-transport processes and disturbance. We know little about effects of management activities on the frequency and intensity of disturbances and the continuity and connectivity of disturbances processes cascading through river networks. An assessment of management effects on wood-related disturbances would examine the roles of wood for each process and the transitions from one process to another.

Critical elements of a framework for future study and management of wood must include geographic, temporal, and process views at a series of scales. These elements are shown in a single figure encompassing the geographic sequence down the gravitational flow path from hillslopes to small streams to large rivers (Figure 10). Natural disturbances—such as fire, windthrow, and landslides—on hillslopes and in mountain creeks contribute wood pieces to streams, and fluvial processes like channel migration and flooding are responsible for producing wood in braided and meandering streams. Management activities, like timber harvesting and road building, influence both production and transport processes, and check dams and slit dams in mountain creeks interrupt transport of log pieces by debris flow. Embankment, revetment, and channelization constrain channel migration and flooding frequency. Time is another important factor affecting current conditions of wood in streams under a disturbance regime. From a watershed perspective, time can be expressed as a disturbance cascade providing a disturbance history in a stream network. Scaling reflects geographical differences, such as decay rate, tree size versus channel width, wood density, and bed material density, affect temporal changes in the standing stock of wood in the stream and residence time in a storage.

All of these factors eventually affect the stream and its riparian ecosystem, including habitat loss and creation, refugia and sources of propagules for recolonization, energy flows, and the dynamics of hyporheic zones.

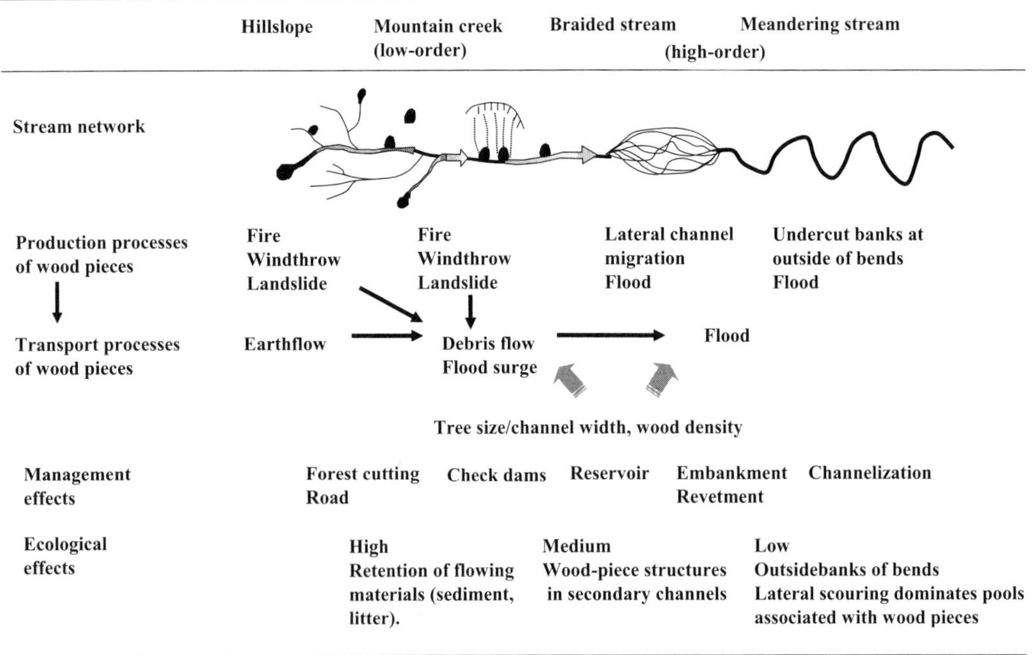

FIGURE 10. A framework for applying results in different geographic and management settings.

Acknowledgments

This work was supported in part by the USDA Forest Service and a National Science Foundation grant supporting the Long-Term Ecological Research program at the H. J. Andrews Experimental Forest (DEB-9632921), by the US-Japan Joint Program between National Science Foundation and the Japan Society for Promotion of Science, and by Grants in Aid for Scientific Research (Nos. 10460059, 13460061, 14506039,14380274) from the Ministry of Education, Japan.

References

Abbe, T. B., and D. R. Montgomery. 1996. Large woody debris jams, channel hydraulics and habitat formation in large rivers. Regulated Rivers: Research and Management 12:201–221.

Abe, T., and F. Nakamura. 1996. Pool and cover formation by coarse woody debris in a small low-gradient stream in northern Hokkaido. Journal Japanese Forestry Society 78:36–42 (in Japanese with English abstract).

Abe, T., and F. Nakamura. 1999. Effects of experimental removal of woody debris on channel morphology and fish habitats. Ecology and Civil Engineering 2:179–190 (in Japanese with English abstract).

Acker, S. A., S. V. Gregory, G. Lienkaemper, W. A. McKee, F. J. Swanson, and S. D. Miller. 2003. Composition, complexity, and tree mortality in riparian forests in the central western Cascades of Oregon. Forest Ecology and Management 173(2003):293–308.

Andrus, C. W., B. A. Long, and H. A. Froehlich. 1988. Woody debris and its contribution to pool formation in a coastal stream 50 years after logging. Canadian Journal of Fisheries and Aquatic Sciences 45:2080–2086.

Benda, L. 1990. The influence of debris flows on channels and valley floors in the Oregon Coast Range, U.S.A. Earth Surface Processes and Landforms 15:457–466.

Benda, L., and T. W. Cundy. 1990. Predicting deposition of debris flows in mountain channels. Canadian Geotechnical Journal 27:409–417.

Benda, L., D Miller, J. Sias, D. Martin, R. Bilby, C. Veldhuisen, and T. Dunne. 2003. Wood recruitment processes and wood budgeting. Pages 49–73 *in* S. V. Gregory, K. L. Boyer, and A. M. Gurnell, editors. The ecology and management of wood in world rivers. American Fisheries Society, Symposium 37, Bethesda, Maryland.

Beschta, R. L. 1979. Debris removal and its effects on sedimentation in an Oregon Coast Range stream. Northwest Science 53:71–77.

Bilby, R. E., and G. E. Likens. 1980. Importance of organic debris dams in the structure and function of stream ecosystems. Ecology 61:1107–1113.

Bilby, R. E., and J. W. Ward. 1991. Characteristics and function of large woody debris in streams draining old-growth, clear-cut, and second-growth forests in southwestern Washington. Canadian Journal of Fisheries and Aquatic Sciences 48:2499-2508.

Bormann, F. H., and G. E. Likens. 1979. Pattern and process in a forested ecosystem. Springer-Verlag, New York.

Bragg, D. C. 2000. Simulating catastrophic and individualistic large woody debris recruitment for a small riparian system. Ecology 81:1383–1394.

Braudrick, C. A., G. E. Grant, Y. Ishikawa, and H. Ikeda. 1997. Dynamics of woody transport in streams: a flume experiment. Earth Surface Processes and Landforms 22:669–683.

Cederholm, C. J., D. B. Houston, D. L. Cole, and W. J. Scarlett. 1989. Fate of coho salmon (*Oncorhynchus kisutch*) carcasses in spawning streams. Canadian Journal of Fisheries and Aquatic Sciences 46:1347–1355.

Coho, C. S. 1993. Dam-break floods in low order mountain channels of the Pacific Northwest. Master's thesis. University of Washington, Seattle.

Dietrich, W., and T. Dunne. 1978. Sediment budget for a small catchment in mountainous terrain. Zeitshrift fur Geomorphogie N. F. 29:191–206.

Dolloff, C. A., and M. L. Warren, Jr. 2003. Fish relationships with large wood in small streams. Pages 179–193 in S. V. Gregory, K. L. Boyer, and A. M. Gurnell, editors. The ecology and management of wood in world rivers. American Fisheries Society, Symposium 37, Bethesda, Maryland.

Elliott, S. T. 1986. Reduction of a Dolly Varden population and macrobenthos after removal of logging debris. Transactions of the American Fisheries Society 115:392–400.

Everitt, B. L. 1968. Use of the cottonwood in an investigation of recent history of a flood plain. American Journal of Science 266:417–439.

Fausch, K. D., and T. G. Northcote. 1992. Large woody debris and salmonid habitat in a small coastal British Columbia stream. Canadian Journal of Forest Research 49:682–693.

Franklin, J. F., and R. T. T. Forman. 1987. Creating landscape patterns by forest cutting: ecological consequences and principles. Landscape Ecology 1:5–18.

Gurnell, A. M., G. E. Petts, N. Harris, J. V. Ward, K. Tockner, P. J. Edwards, and J. Kollmann. 2000. Large wood retention in river channels: the case of the Fiume Tagliamento, Italy. Earth Surface Processes and Landforms 25:255–275.

Gurnell, A. M., H. Piégay, F. J. Swanson, and S. V. Gregory. 2002. Large wood and fluvial processes. Freshwater Biology 47:601–619.

Gurnell, A. M. 2003. Wood storage and mobility. Pages 75–91 *in* S. V. Gregory, K. L. Boyer, and A. M. Gurnell, editors. The ecology and management of wood in world rivers. American Fisheries Society, Symposium 37, Bethesda, Maryland.

Harmon, M. E., J. F. Franklin, F. J. Swanson, P. Sollins, S. V. Gregory, J. D. Lattin, N. H. Anderson, S. P. Cline, N. G. Aumen, J. R. Sedell, G. W. Lienkaemper, K. Cromack, Jr., and K. W. Cummins. 1986. Ecology of coarse woody debris in temperate ecosystems. Advances in Ecological Research 15:133–302.

Hatano, S. 1974. Landslide geomorphology (2). Tsuchi-To-Kiso 22(11):85–93 (in Japanese).

Hedman, C. W., D. H. Van Lear, and W. T. Swank. 1996. In-stream large woody debris loading and riparian forest seral stage associations in the southern Appalachian Mountains. Canadian Journal of Forest Research 26:1218–1227.

Inoue, M., and S. Nakano. 1998. Effects of woody debris on the habitat of juvenile masu salmon (*Oncorhynchus masou*) in northern Japanese streams. Freshwater Biology 40:1–16.

Ishikawa, Y. 1989 Studies on disasters caused by debris flows carrying floating logs down mountain stream. Ph.D. dissertation. Kyoto University, Kyoto, Japan.

Johnson, S. L., F. J. Swanson, G. E. Grant, and S. M. Wondzell. 2000. Riparian forest disturbances by a mountain flood—the influence of floated wood. Hydrological Processes 14:3031–3050.

Keller, E. A., and F. J. Swanson. 1979. Effects of large organic material on channel form and fluvial processes. Earth Surface Processes 4:361–380.

Kishi, C., F. Nakamura, and M. Inoue. 1999. Budgets and retention of leaf litter in Horonai Stream, southwestern Hokkaido, Japan. Japanese Journal of Ecology 49:11–20 (in Japanese with English abstract).

Lienkaemper, G. W., and F. J. Swanson. 1987. Dynamics of large woody debris in streams in old-growth Douglas-fir forests. Canadian Journal of Forest Research 17:150–156.

May, C. L. 2001. Spatial and temporal dynamics of sediment and wood in headwater streams of the Oregon Coast Range. Ph.D. Dissertation. Oregon State University, Corvallis.

Minshall, G. W., J. T. Brock, and J. D. Varley. 1989. Wildfires and Yellowstone's stream ecosystems—a temporal perspective shows that aquatic recovery parallels forest succession. BioScience 39:707–715.

Miyabuchi, Y., A. Shimizu, and Y. Ogawa. 1999. Storage of woody debris and sediment in a mountain stream, northern Kyusyu, southwestern Japan. Journal of the Japanese Society of Erosion Control Engineering 52(1):21–27 (in Japanese with English abstract).

Murphy, M. L., and K. V. Koski. 1989. Input and depletion of woody debris in Alaska streams and implications for streamside management. North American Journal of Fisheries Management 9:427–436.

Nagasaka, A., and F. Nakamura. 1999. The influences of land-use changes on hydrology and riparian environment in a northern Japanese landscape. Landscape Ecology 14:543–556.

Nakamura, F., and S. Kikuchi. 1996. Some methodological developments in the analysis of sediment transport processes using age distribution of floodplain deposits. Geomorphology 16:139–145.

Nakamura, F., and N. Shin. 2001. The downstream effects of dams on the regeneration of riparian tree species in northern Japan. Pages 173–181 *in* Geomorphic processes and riverine habitat. American Geophysical Union Monograph, Water Science and Application 4.

Nakamura, F., and F. J. Swanson. 1993. Effects of coarse woody debris on morphology and sediment storage of a mountain stream system in western Oregon. Earth Surface Processes and Landforms 18:43–61.

Nakamura, F., and Swanson, F. J. 1994. Distribution of coarse woody debris in a mountain stream, western Cascade Range, Oregon. Canadian Journal of Forest Research 24:2395–2403.

Nakamura, F., F. J. Swanson, and S. M. Wondzell. 2000. Disturbance regimes of stream and riparian systems—a disturbance-cascade perspective. Hydrological Processes 14:2849–2860.

Pickett, S. T. A., and P. S. White, editors. 1985. The ecology of natural disturbance and patch dynamics. Academic Press, San Diego, California.

Piégay, H., and A. M. Gurnell. 1997. Large woody debris and river geomorphological pattern: examples from S.E. France and S. England. Geomorphology 19:99–116.

Piégay, H., A. Thevenet, and A. Citterio. 1999. Input, storage and distribution of large woody debris along a mountain river continuum, the Drome River, France. Catena 35:19–39.

Ralph, S. C., G. C. Poole, L. L. Conquest, and R. J. Naiman. 1994. Stream channel morphology and woody debris in logged and unlogged basins of western Washington. Canadian Journal of Fisheries and Aquatic Sciences 51:37–51.

Reneau, S. L., and W. E. Dietrich. 1990. Depositional history of hollows on steep hillslopes, coastal Oregon and Washington. National Geographic Research 6:220–230.

Resh, V. H., A. V. Brown, A. P. Covich, M. E. Gurtz, H. W. Li, and G. W. Minshall. 1988. The role of disturbance in stream ecology. Journal of the North American Benthological Society 7:433–55.

Riley, S. C., and K. D. Fausch. 1995. Trout population response to habitat enhancement in six northern Colorado streams. Canadian Journal of Fisheries and Aquatic Sciences 52:34–53.

Romme, W. H., and D. G. Despain. 1989. Historical

perspective on the Yellowstone fires of 1988. BioScience 39(10):695–06.

Scott, K.M. 1988. Origins, behavior, and sedimentology of lahars and lahar-runout flows in the Toutle-Cowlitz River system. U.S. Geological Survey, Professional Paper 1447-A.

Shimizu, O. 1998. Sediment budgets to analyze sediment transport processes through drainage basins. Research Bulletin, Hokkaido University Forests 55(1):123–215 (in Japanese with English abstract).

Shimokawa, E. 1984. A natural recovery process of vegetation on landslide scars and landslide periodicity in forested drainage basins. Pages 99–107 *in* C. L. O'Loughlin and A. J. Pearce, editors. Proceeding of Symposium on Effects of Forest Land Use on Erosion and Slope Stability. East-West Center, Honolulu, Hawaii.

Sidle, R. C., A. J. Pearce, and C. L. O'Loughlin. 1985. Hillslope stability and land use. Water Resources Monograph 11. American Geophysical Union, Washington, D.C.

Sinton, D. S., J. A. Jones, J. L. Ohmann, and F. J. Swanson. 2000. Windthrow disturbance, forest composition and structure in the Bull Run Basin, western Oregon. Ecology 81(9):2539–2556.

Smith, R. D., R. C. Sidle, and P. E. Porter. 1993. Effects on bedload transport of experimental removal of woody debris from a forest gravel-bed stream. Earth Surface Processes and Landforms 18:455–468.

Smock, L. A., G. M. Metzler, and J. E. Gladden. 1989. Role of debris dams in the structure and functioning of low-gradient headwater stream. Ecology 70:764–775.

Snyder, K. U. 2000. Debris flows and flood disturbance in small, mountain watersheds. Master's thesis. Oregon State University, Corvallis.

Spies, T. A. and J. F. Franklin. 1988. Coarse woody debris in Douglas-fir forests of western Oregon and Washington. Ecology 69:1689–1702.

Swanson, F. J. 2003. Wood in rivers: a landscape perspective. Pages 299–313 *in* S. V. Gregory, K. L. Boyer, and A. M. Gurnell, editors. The ecology and management of wood in world rivers. American Fisheries Society, Symposium 37, Bethesda, Maryland.

Tokyo Forestry Branch. 1959. Survey of forest damage by the two typhoons in 1959. (in Japanese).

Tsukamoto, Y. 1973. Study on the growth of stream channel (I): relation between stream channel growth and landslides occurring during heavy storm. Journal of the Japanese Erosion Control Society (Shin-Sabo) 25(4):4–13 (in Japanese).

Urabe, H., and S. Nakano. 1998. Contribution of woody debris to trout habitat modification in small streams in secondary deciduous forest, northern Japan. Ecological Research 13:335–345.

Wemple, B. C, J. A. Jones, and G. E. Grant. 1996. Channel network extension by logging roads in two basins, western Cascades, Oregon. Water Resources Bulletin 32(6):1195–1207.

Wemple, B. C., F. J. Swanson, and J. A. Jones. 2001. Forest roads and geomorphic process interactions, Cascade Range, Oregon. Earth Surface Processes and Landforms 26:191–204.

Wondzell, S. M., and F. J. Swanson. 1999. Floods, channel change, and the hyporheic zone. Water Resources Research 35(2):555–567.

Yanai, S., and G. Usui. 1989. Measurement of the slope failures frequency on sediments with tephrochronological analysis. Journal of the Japanese Erosion Control Society (Shin-Sabo) 163:3–10 (in Japanese with English abstract).

American Fisheries Society Symposium 37:299–313, 2003

Wood in Rivers: A Landscape Perspective

FREDERICK J. SWANSON

USDA Forest Service, Pacific Northwest Research Station
3200 SW Jefferson Way, Corvallis, Oregon 97331, USA

Abstract.—A landscape perspective of wood in world rivers accounts for spatial and temporal patterns of sources of wood from streamside forests, processes of wood delivery to channels, transport of wood through river networks, and trapping sites of wood. Amounts of wood in a river system also depend on productivity of forests in source areas and decomposition rates. Collectively, these factors determine the amount and arrangement of individual pieces and accumulations of wood through a river network, which, in turn, affect ecological, geomorphic, social, and other features of rivers. Research to date deals with subsets of these components of wood in rivers, but there has been limited development of a general framework for wood in river networks. This chapter considers a framework for examining the arrangement of wood in river landscapes and how it may reflect the history of spatial patterns and timing of wood input and redistribution. Field studies provide examples of different spatial patterns and architectures of wood accumulations. Wood accumulations are shaped by input processes, trapping sites, and transport processes. Reaches in river networks may switch from wood patterns dominated by one set of controls to another because of gradual or abrupt input and redistribution. A framework for future studies and management includes interpretation of these different controls through time and over river networks.

Introduction

The physical dynamics and ecological significance of large pieces of wood (>1 m long and >0.1 m diameter) in river systems have strong landscape properties. "Landscapes" are land surface areas composed of units or patches of differing geophysical and biological properties (Risser et al. 1984; Forman and Godron 1986). Landscape studies commonly encompass an adequate sample of landscape units, examine spatial units within the study area explicitly, and consider time scales sufficiently long to reveal representative patterns in landscape conditions and change. Patterns and dynamics of wood in rivers include forest-stream interactions scaled up from local sites to full watersheds, effects of the network structure of streams and riparian zones, and interactions among input, transport, and accumulation of wood in different types of river systems. Furthermore, studies can focus on pattern-process relations of wood itself or on the ecological, environmental hazard, or other consequences of those patterns and processes.

Little work to date explicitly considers landscape aspects of wood in rivers, especially the arrangement of wood pieces and accumulations over widely differing stream sizes and types, although attention to these topics has grown in recent years. This paper addresses the current state of knowledge about landscape aspects of wood in rivers, identifies knowledge gaps, and offers a framework for further work to benefit science and management. The emergence of landscape perspectives in studies of wood in rivers is reflected in many chapters in this book.

State of Knowledge

Existing published and unpublished works touch only lightly on landscape perspectives of wood in rivers. Many chapters in this book provide a conceptual framework for analysis of sediment and wood delivery and routing through river networks. Most published work on standing crops of wood in streams has focused on relations to forest age and management history, and ecological effects (see reviews of Harmon et al. 1986; Bisson et al. 1987; Bilby and Bisson 1998). Existing work is largely at either a local scale, such as work in small streams, or adopts a generalized, broad perspective, such as compilations of samples of standing crop measurements over longitudinal profiles of rivers (such as Harmon et al. 1986; Bilby and Ward 1989). New studies, however, are taking a much more comprehensive view than existed a few years ago (Benda and Dunne 1997a, 1997b; Martin and Benda 2001).

In working at any particular scale, such as the landscape, drawing on information relevant to system behavior must be considered at the full range of scales. Integration of knowledge about wood in rivers from the local site to the full river network is essential (Frissell et al. 1986; Gregory et al. 1991) and in even broader geologic-tectonic (Montgomery 1999) and biogeographic (Harmon et al. 1986; Montgomery 1999) contexts. Temporal perspectives, including stochastic aspects of system behavior, are also integral to understanding system behavior (Reeves et al. 1995; Benda and Dunne 1997a, 1997b). Geologic context influences the amount and size distribution of sediment in rivers and the associated landforms, hence the geophysical dynamics of the system. Biotic context of the landscape determines forest composition, structure, productivity, and dynamics of a landscape. Together, the geologic and biotic components determine the amount, size distribution, and persistence of wood in river systems. In this discussion, wood dynamics in rivers are considered at the scales of constrained-unconstrained stream reaches (sensu Grant and Swanson 1995), longitudinal river profile (sensu Vannote et al. 1980), and the full river-riparian network in a forest landscape.

A general framework—wood in a landscape perspective

The basic elements of wood movement through landscapes are the river network, the surrounding forests, and the processes that link them (Figure 1). Forests are dynamic in response to disturbance and successional processes. Patterns of

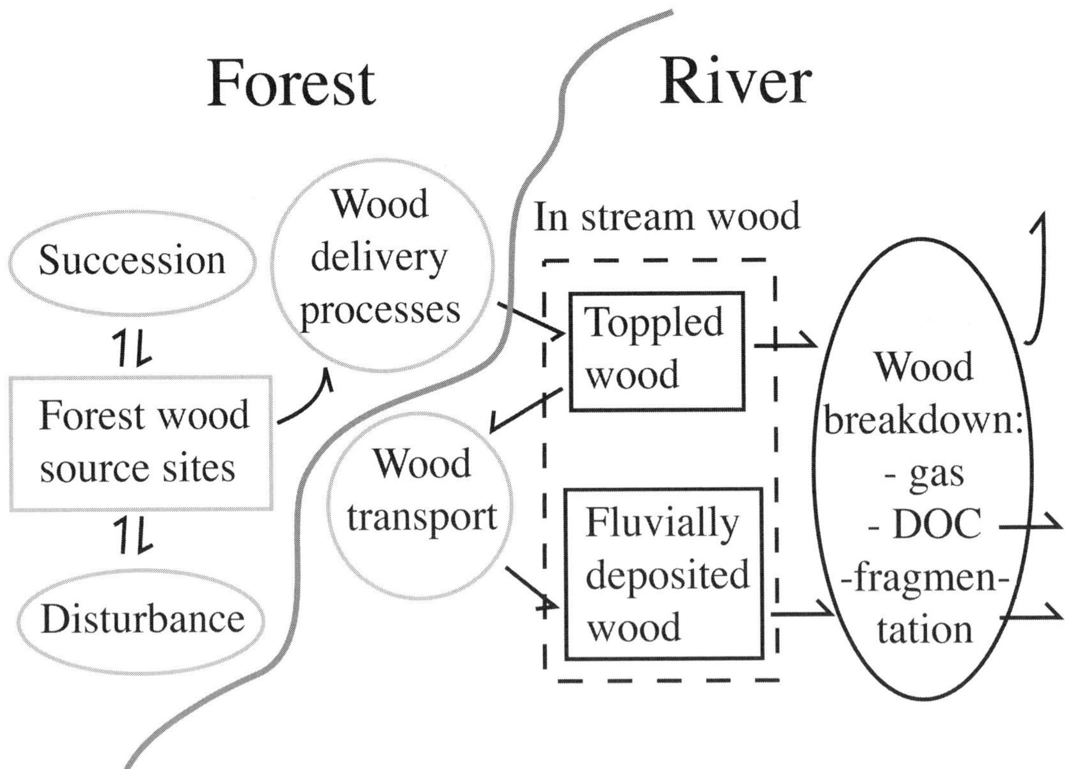

FIGURE 1. Stores (boxes) and transfer processes (circles) for routing of wood from forest sources to and through river systems.

wood movement follow gravitational flow paths from forest to stream by wood delivery processes and then potentially down rivers by fluvial wood transport processes. Wood is deposited at persistent or ephemeral accumulation sites along rivers, forming "standing stock." Wood in storage is subject to decomposition, resulting in release of gases to the atmosphere and dissolved organic carbon and other constituents that move downstream. Physical abrasion of wood produces fine fragments of organic matter that also move downstream, but are no longer in the large wood category.

A spatially explicit landscape approach considers landscape structure, including the types and arrangement of landscape units that affect wood supply (the forest patchwork) and transport and storage (the stream network); processes that vary in relation to position in the stream and riparian network; and concepts related to how these structures and processes change through time (Figure 2). The basis for these observations comes from conceptual, empirical, and modeling approaches (Keller and Swanson 1979; Gurnell et al. 2002; Abbe and Montgomery 2003; Benda et al. 2003, this volume).

The primary **landscape structure units** in this analysis are patches of vegetation that serve as sources of wood and the stream network itself where material is transported and deposited. Secondary features of potential significance include natural landforms, like alluvial fans, and engineered structures, such as roads, channelized stream reaches, and dams, that may alter the movement of wood through landscapes. Roads, for example, can be initiation sites of landslides that deliver wood to streams, and they may block passage of wood being transported by debris

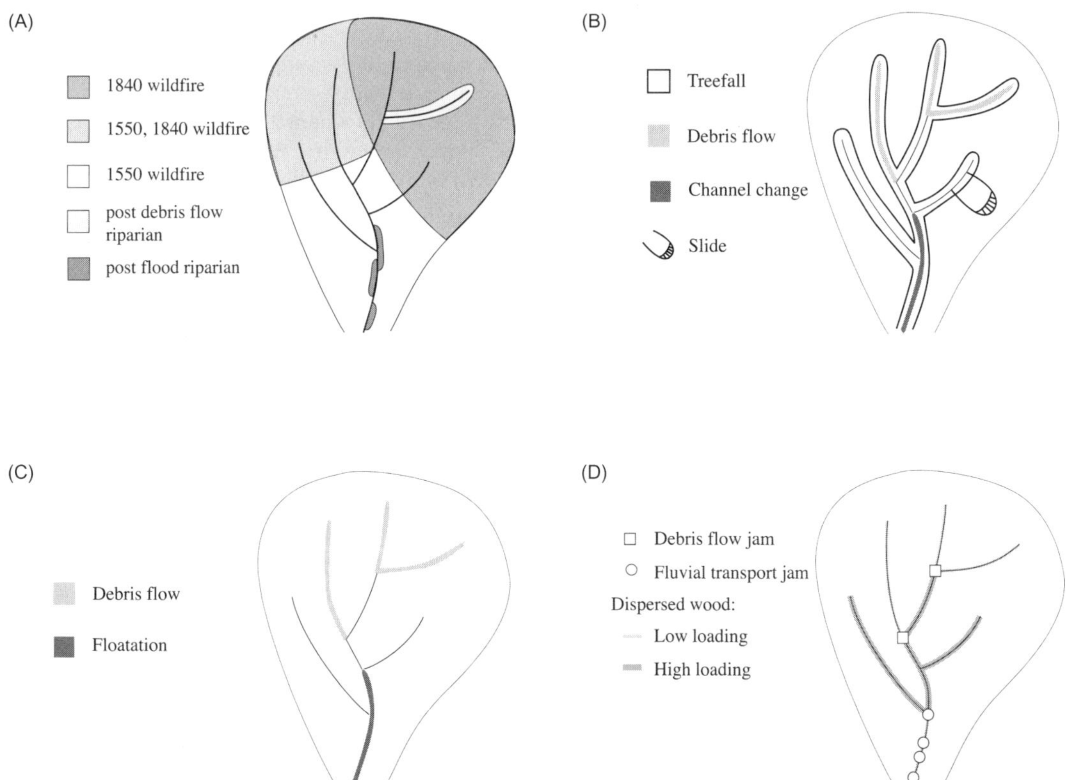

FIGURE 2. Schematic map of (A) forest disturbance history and resulting patchwork showing upland and riparian forest age-classes and stand-initiating processes; (B) zones influenced by different processes of wood delivery to channels; (C) zones of wood transport processes in streams; (D) patterns of wood accumulations in a stream network.

flows or flotation (Wemple et al. 2001). The **source area** of wood for rivers is the adjacent forest landscape whose structure reflects the history of terrestrial disturbance events, like fire and windstorm, and river-based events, such as debris flows, floods, and lateral channel change (Figure 2).

Processes that control the input and redistribution of wood vary with position in stream networks (Keller and Swanson 1979; Montgomery 1999; Martin and Benda 2001; Benda et al. 2003; Gurnell 2003, this volume). A landscape can be mapped as zones (or "process domains," in the language of Montgomery (1999)), in which different wood input and transport processes operate (Figure 2B and C). Some source areas and delivery processes disperse wood through time and space, such as chronic mortality in streamside forests in response to small-scale gap dynamics and competitive interactions among trees. Other processes deliver batches of wood from discrete source patches. Source patches and delivery processes may persist in locations or be transient, shifting locations through time. For example, wood delivery, both dispersed and in batches, comes from the streamside zone of a tree-height width where toppling of all or parts of live and dead trees is an important delivery mechanism (Figure 2B). Other processes, like landslides and snow avalanches, have the potential to deliver wood from areas farther upslope.

Transport processes move wood downstream through a river system, and different processes dominate in different parts of the network (Figure 3). Debris flows, for example, are relegated to small, steep channels. Flotation, a dominant transport process in large channels, can move wood in either a congested (batches of interacting wood pieces) or an uncongested (individual pieces) manner (Braudrick and Grant 2000). A potentially large, but unknown, fraction of wood in rivers leaves the system via decomposition as CO_2 or fine particulate matter, not as large wood pieces.

Patterns of wood sources and processes of input and transport also interact with channel and valley floor features to determine **wood accumulation sites**, which result from either lack of transport capacity or the presence of trapping structures (Gurnell et al. 2002; Abbe and Montgomery 2003). Trapping structures may be geomorphic, biotic, or hydraulic features. Wood can accumulate where a channel shallows or narrows, on obstructions in channels and along their margins, or stranded anywhere as flood flows drop.

The **standing stock**—the amount and arrangement—of wood in a river system is a reflection of this complex array of both dispersed and discrete input processes and events that redistribute wood (Gurnell et al. 2002; Abbe and Montgomery 2003). The standing stock of wood in a river reach is commonly composed of material that fell into place from the neighboring forest source area and other material that was transported into place by streamflow or other transport processes. This distinction can be useful in interpreting the history of processes affecting a site as well as the architecture of wood assemblages and their associated functions, such as pool formation, sediment trapping, and habitat for terrestrial and aquatic organisms. Given a certain amount of topographic and forest landscape control on source-sink relations, the pattern of standing crop of wood through a river network may be somewhat predictable (Figure 2D). An important temporal factor for interpreting standing crops of wood is the time since the last input and redistribution events. Immediately after a flood, for example, deposition of mobile wood by fluvial processes would be expected. In the same river reach after a wood movement event, some wood pieces will fall into the river and lie where they fell. This material may be more vulnerable to transport than wood pieces of the same size and shape that had been transported and deposited in more persistent depositional sites, such as lodged against massive boulders, bedrock outcrops, or stable trees.

The following illustrations of some of these points and conjectures come largely from the relatively well-studied rivers of the Pacific Northwest of the United States, but the phenomena are discussed in general terms. These examples reflect certain biases, such as large quantities of large pieces of wood and the transport processes characteristic of some montane environments. The approach of thinking in terms of patterns, processes, and controls is integral to landscape perspectives, however, and transferable to other geographic areas.

Landscape aspects of wood sources

The geography of forests as sources of wood to rivers is defined by the area potentially subject to wood input. Thus the map of wood delivery is the overlap of the forest patchwork map (Figure 2A) and the map of zones influenced by different modes of potential wood delivery (Figure 2B). Multiple processes, each with characteristic lat-

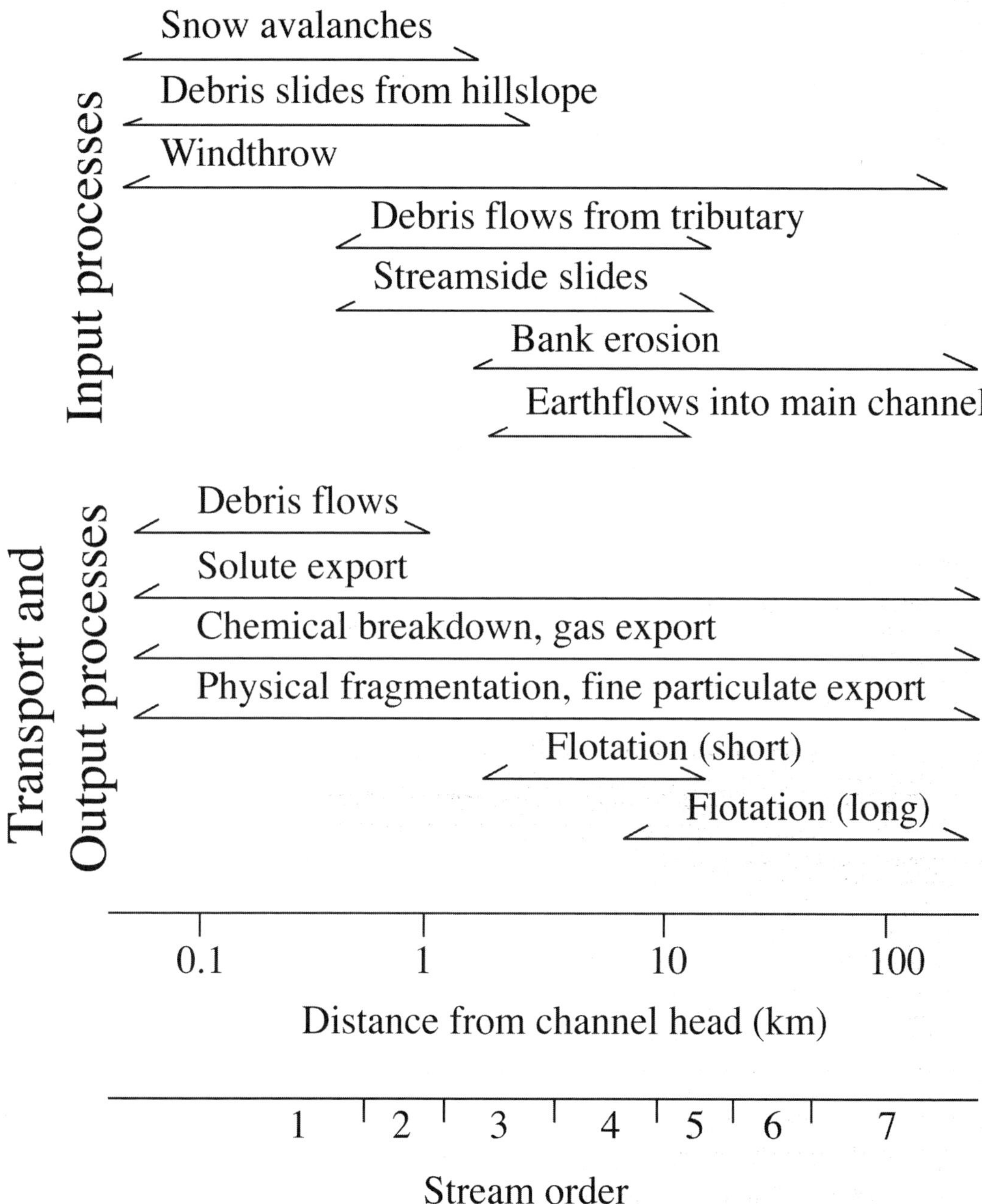

FIGURE 3. Schematic distribution of wood input, transport, and output processes in an Oregon Cascade Range river (adapted from Keller and Swanson 1979). Some processes, such as windthrow, are more significant in upstream areas, and other processes are more common in downstream areas. "Short" flotation transport refers to movement of pieces with lengths of 0.5–1.0 bank-full width that can commonly move distances of 10 s to a few hundred meters, which is the channel length that would include several wood accumulations. "Long" flotation transport refers to the potential for such wood pieces to move a kilometer or more as a single, floating piece.

eral extent away from the channel, can deliver wood to any particular stream reach (Figure 2C). The suite of processes that affect wood distribution in stream reaches varies down the stream system (Keller and Swanson 1979; Figure 3). Of course, bioregional and physiographic factors also affect the presence and absence and relative rates of input, such as snow avalanches and river ice movement operating in cold regions, and the undercutting of riverbanks, which is more common along large, low-gradient tropical rivers.

Commonly, riparian vegetation grows in narrow, linear patches along streams in response to the history of floods creating gravel bars on accreting portions of channel bends or sites of repeated disturbance and reestablishment of forests. Where the width of riparian vegetation patches is less than the height of adjacent forest, a form of lateral and vertical stacking of wood sources sometimes allows two or more types of riparian and upland forest to serve as wood sources for a given stream site (Swanson et al. 1990).

Several processes alter the rates of wood delivery from streamside forests (Naiman et al. 1993; Acker et al. 2003). Gradual processes of forest succession change the species composition, size distribution, and amount of wood delivered to streams. Abrupt disturbance of streamside forest may result in a pulse of wood input to streams, such as by wind-toppling of streamside forest, followed by reduced input while the stand is reestablishing. Clear-cutting of streamside forest removes wood sources without delivering a pulse of wood to the stream. Thus, negative feedback mechanisms can reduce input of wood to streams for decades or even centuries until the disturbed sites reestablish and produce large wood. In areas where disturbance is more frequent than the time required for trees to grow to the size-class of large wood, streamside vegetation may not produce large wood for long periods.

Landscape aspects of wood transport and accumulation sites

In many river systems, the significance of different processes of wood transport varies substantially through the river network as a result of geographic variation in size and amount of wood delivered to the channel and the capacity of the river system to transport wood. **Flotation**, for example, is constrained by the relations of channel width to wood-piece length and water depth to piece diameter—relatively small pieces are more mobile (Lienkaemper and Swanson 1987; Bilby and Ward 1989; Braudrick and Grant 2000). Therefore, based on these perspectives from ecosystems dominated by conifers, river systems with increasing flow in the downstream direction are expected to exhibit increased significance of flotation in lower river reaches. In deciduous forests, however, tree form is commonly broadly branching, so fallen deciduous trees with large limbs may be less susceptible to transport than cylindrical pieces with the same wood volume, particularly in low-velocity rivers (for example, Palik et al. 1998).

Debris flows, the rapid, downstream movement of 100 s to 1,000 s of m^3 of soil, sediment, and wood, provide a contrasting example of landscape influence on wood dynamics common in montane river systems. Debris flows flush sediment and large wood from small, steep, confined channels (Figure 3b in Swanson et al. 1998) and deposit it in lower gradient, less confined sites, such as alluvial fans, floodplains, and fourth-order and larger channels, or at channel junctions with abrupt change in flow direction (Benda and Cundy 1990). The resulting deposits can form large jams that greatly exceed channel dimensions or, in the case of delivery to large channels, the wood may be scattered in smaller accumulations down the main channel (Johnson et al. 2000).

A 50-year record of debris flows in the Blue River watershed of the Oregon Cascades provides examples of several aspects of debris flow influence on wood distribution across a stream network (Snyder 2000). The documented debris flows are restricted to first- through third-order channels and in low elevation parts of the landscape, which are characterized by steep slopes, slide-prone soils, and high water input to soils during rain-on-snow precipitation events (Swanson and Dyrness 1975; Swanson et al. 1998; Snyder 2000). The net effect is that a large percentage of the length of small streams in the debris flow-affected, lower-elevation half of this study area was flushed of wood accumulated over the previous 100 years. The upper-elevation half of the study area, however, has been much less affected by debris flows and wood transport.

Debris flows are episodic events in most landscapes. For example, most debris flows in the 50-year record for Blue River were in just two winters. About half of the inventoried events took place in the winter of 1965, and about 30% of the debris flows occurred during one flood in 1996. Some of the 1996 debris flows rescoured channels that had not yet completed reloading with wood after debris flows in the same channels in the winter of 1965. Presence of forest roads and clear-cut areas substantially increased the number of debris flows but did not alter the general pattern (Swanson and Dyrness 1975). Wood abundance in channels can take more than a century to recover, as indicated by dendrochronologically documented residence times of large wood in channels (Swanson et al. 1976; Keller and Tally 1979; Hyatt and Naiman 2001).

Some wood movement from hill slopes to small streams to large rivers can be viewed as a sequence of processes, that Nakamura et al. (2000) term a **disturbance cascade**. Wood transport begins as debris slides on steep hill slopes, transforming into debris flows down steep, headwater channels, and potentially traveling to larger channels where the wood moves by flotation. Events in the 1996 flood affecting the Blue River watershed provide examples of this sequence (Nakamura et al. 2000; Nakamura and Swanson 2003, this volume). In the 1996 flood in Blue River, only 22 of 39 debris flows delivered wood to fourth- and fifth-order channels, and the remainder stopped in smaller channels (Nakamura et al. 2000). A variety of geomorphic and engineered features and large, standing trees along the flow path can interrupt delivery of wood and sediment to large channels (Wemple et al. 2001). Roads, for example, were involved in stopping the movement of 28% of the debris flows in the 50-year record in Blue River (Snyder 2000).

Wood accumulation sites may be distinctive to the transport process that produced them and the location in the stream network (Gurnell et al. 2002; Abbe and Montgomery 2003). Development of effective trapping sites varies among river systems. For example, simple, straight channel form may facilitate efficient movement of wood downstream, but complex, braided river channels exhibit low wood transport efficiency. Some types of trapping sites persist for many millennia and others may be ephemeral, like single large trees. Wood transported by debris flows tends to accumulate in sites of decreased channel gradient and increased width—often associated with obstructions, such as large standing trees or road fills. In channels wide enough to transport most wood pieces present, a variety of channel margin and secondary channel features retain transported wood (Nakamura and Swanson 1994; Gurnell et al. 2002; Abbe and Montgomery 2003; Gurnell 2003). Along large rivers wood accumulates persistently on the heads of large bars or islands, in the mouths of secondary channels, along the riparian forest fringe where flow enters the floodplain, at the confluence of major rivers where the flood crest of one river creates a backwater effect in the mouth of the other, and in other sites of diminished transport capacity (Gurnell et al. 2000; Piégay and Gurnell 1997; Abbe and Montgomery 2003). River meandering and channel avulsion play critical roles in these processes of wood input and redistribution in low-gradient rivers (Piégay et al. 1999; Piégay 2003, this volume).

Properties of stream networks may influence the effects of certain processes on wood distribution. For example, the role of debris flows in delivering wood to fourth- and fifth-order channels and valley floors may be accentuated in networks that are elongate and have numerous debris flow-prone tributaries. In such basins, a higher proportion of channel junctions may have the potential to deliver wood by debris flows than more dendritic channel networks. The potential for wood transport by debris flows through a network depends in part on network properties, such as junction angles and channel gradient (Benda and Cundy 1990), and also on valley-floor landforms. Debris flow deposition of wood jams at the confluence of second- and fifth-order channels, for example, may result from landforms, such as alluvial fans, rather than the junction angle readily interpreted from topographic maps (Snyder 2000).

Landscape aspects of standing stock of wood in rivers

Patterns of the amount and locations of wood in rivers reflect interactions among input, redistribution, and loss. Spatial patterns of wood in river systems can be addressed at a series of nested scales—variation down the longitudinal profile of a river, variation with respect to position in the river network, distribution of accumulations in a river reach, and location of pieces within individual accumulations. At the finer scales, wood characteristics include the degree of aggregation, the architecture of accumulations, and degree of burial in sediment.

The River Continuum Concept (Vannote et al. 1980) provides a useful conceptual framework for understanding riverine ecosystem structure and function along a longitudinal gradient from headwaters to river mouth. Both reduced inputs and increased transport capacity may be responsible for observed downstream decreases in standing crop of wood (Naiman and Sedell 1979; Harmon et al. 1986; Bilby and Ward 1989). Few workers, however, have studied both small streams and large rivers in a single basin with compatible sampling methods, although Martin and Benda (2001) offer an interesting step forward on this issue. The simple, longitudinal depiction of wood along the river continuum reveals a relatively high standing stock per unit area in headwater streams in response to high input rates from the surrounding forest and the limited transport capacity of the stream.

Arrangement of wood in river networks has been hypothesized to exhibit increased aggregation in the downstream direction, based on assumptions about the controls of source area, transport capacity, and distribution of trapping sites (Swanson et al. 1982; Benda et al. 2003). In headwater streams, wood tends to lie where it fell from the adjacent forest because of the limited transport capacity of small streams, except when and where debris flows can mobilize large volumes of wood. A clumped pattern is produced by localized sources in sites of limited transport capacity, such as outside of meander bends undercutting forested riverbanks, resulting in wood accumulations on the heads of point bars. In large channels, transport by flotation can accumulate wood at trapping sites, such as the prows of islands and mouths of secondary channels (Gurnell et al. 2002; Abbe and Montgomery 2003).

Various features of wood pieces and their arrangement can be used to distinguish pieces that have fallen into place from those transported into place by stream flow (for example, Swanson et al. 1984). The distinction between pieces emplaced by fluvial transport and those located where they fell can be useful in interpreting recent processes and long-term dynamics of wood in stream channels.

The architecture of wood accumulations may be characterized by the degree of contact among pieces and their relative orientation (Abbe and Montgomery 1996, 2003). Trees that have toppled into a channel typically form open structures with abundant void space that can serve as habitat for terrestrial and aquatic species. Wood accumulations formed by fluvial deposition and debris flows are commonly assemblages of parallel, tightly packed wood pieces. In some situations, scour around wood pieces precludes burial in sediment, but at other times, such as with debris-flow deposits, wood structures facilitate their own burial. All these aspects of accumulation form affect function.

A typology of controls on wood amount and arrangement

Basic questions in the landscape analysis of wood in rivers are: What is the amount and arrangement of wood in a river system? What controls these features? How do they vary in time and space? A general framework, or typology, for examination of controls on arrangement of wood in rivers can be used in field and modeling studies and for planning stream habitat restoration projects. Such a typology of wood conditions can be based on distinguishing the relative influence of critical factors affecting the arrangement of wood in river reaches. Dominant critical factors are (1) input may be dispersed or patchy in time, space, or both; (2) discrete trapping sites may be strongly or weakly expressed; (3) locations of source and accumulation sites may be persistent or transient; and (4) transport distances may be long or short relative to the spacing of source or accumulation sites. Examination of forest and river systems with different wood dynamics provides examples of cases with different relative strengths of source areas, transport processes, and accumulation sites in determining the arrangement of wood in a particular river. Once we have defined a hypothetical typology of wood dynamics in rivers, it can be tested through field and modeling studies.

Four major types of controls on wood arrangement are identified:

Discrete-source-area control of pattern.—Arrangement of discrete source areas along a river dominates patterns of wood in the river where transport distances are much shorter than the spacing of source areas (Figure 4a).

Trapping-site control of pattern.—In systems with effective trapping sites, their arrangement dominates wood accumulation patterns where transport distances are long relative to spacing of source areas (Figure 4b, in the case where deposition sites are determined by presence of trapping sites).

Transport control of pattern.—In river reaches lacking discrete wood-trapping sites and where transport distances are long relative to source-area spacing, wood is randomly distributed, regardless of the pattern of

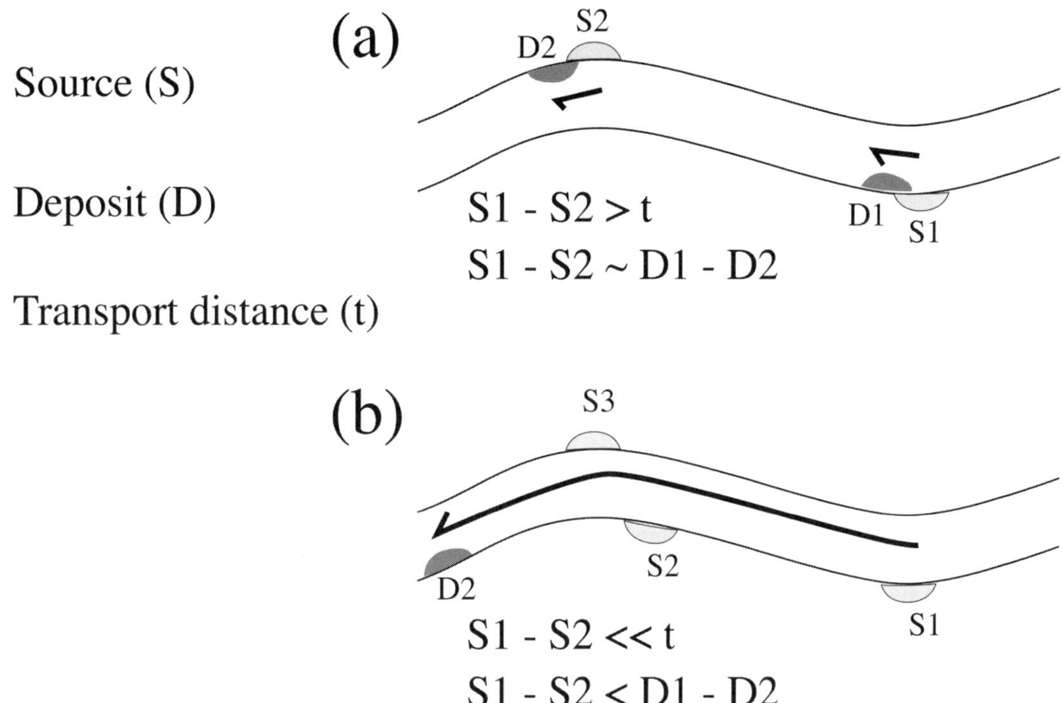

FIGURE 4. Hypothetical patterns of standing crop of wood in a river network when trapping sites are discrete: (a) transport distance is short relative to spacing of source areas, and (b) transport distance and spacing of deposition sites are long relative to spacing of source areas.

wood source areas and input processes (Figure 4b, in the case where deposition sites are determined by limits on transport unrelated to presence of a trapping site).

Dispersed-source control of pattern.—In areas of dispersed input and very limited transport capacity, wood is randomly distributed and amounts reflect forest stand and decomposition histories.

The history of wood input and redistribution is an important, additional dimension of controls on wood amounts and arrangements. Wood conditions in a site strongly reflect the relative and absolute timing of wood input, redistribution, and loss. The historical dimension of the arrangement of wood implies that individual sites may display different types of controls on wood conditions at different times. For example, a protracted period with dispersed wood input and no redistribution events may exhibit dispersed-input control. Immediately after a major flood, however, the site may appear to be an example of either transport or trapping-site control, depending on the strength of trapping-site influence. After an additional period of dispersed wood input, the site may exhibit a mixture of patterns produced by both input and transport. This history of wood dynamics may be reflected not only in the amount and arrangement of wood, but also in the distribution of wood-decay classes (Harmon et al. 1986) with the older, transported pieces in more advanced stages of decay than many of the recently input pieces. Clearly, placing the status of wood in a study reach or network in its historical context is critical.

Examples of types of wood dynamics

Only a handful of field studies have the temporal and spatial scope to explicitly shed light on the different types of wood-dynamic systems. A few field examples give a sense of how patterns of wood may reveal the hypothesized dynamics in particular types of systems.

The H. J. Andrews Experimental Forest in the Cascade Range of Oregon provides examples of wood dynamics in steep, mountain streams (Swanson et al. 1998; Snyder 2000; Swanson and

Jones 2002). Throughout much of the basin, wood input to small first- through third-order streams, in the past several centuries, has been dominated by dispersed tree fall, resulting in large accumulations of wood (Harmon et al. 1986). In parts of the basin where debris flows move wood, three types of wood accumulations are observed: (1) dispersed-source control pattern in channels without debris flows for a century or more containing wood where it fell; (2) transport control pattern in barren channels recently flushed by debris flows; and (3) trapping-site control pattern in wood jams in low gradient, wide stream reaches at the ends of debris-flow tracks where wood has piled up (Swanson et al. 1976; Snyder 2000). Small channels in parts of the basin without debris flow influence primarily exhibit the dispersed-source control pattern.

An apparently cyclical pattern of shifting controls on wood conditions is represented in a series of maps spanning nearly 20 years showing wood arrangement in a fifth-order reach of lower Lookout Creek in the Andrews Experimental Forest (Nakamura and Swanson 1993; Faustini 2000; Swanson and Jones 2002). The channel at this site is about 25 m (82 ft) wide; it has a drainage area of 62 km^2 (24 mi^2) and old-growth conifer forests bordering the channel. A 1978 map (Figure 24 of Swanson and Jones 2002) shows a channel with little wood, except in small patches apparently floated into place during floods in 1964, 1965, and 1972. In the 1980s and early 1990s, several large trees toppled from the streamside area and additional wood floated in from upstream, strewing the channel with a mixture of flood-deposited and toppled wood. A major flood in 1996 flushed much of the wood out of this stream reach and formed small accumulations along the channel margins, much like those observed in 1978. Thus, this stream reach may undergo cycles of increased complexity from input of dispersed wood (dispersed-source control of pattern), punctuated by flushing events and accumulations in trapping sites (trapping-site control) during floods with recurrence intervals of about 50 years and more. Land management activities, including roadbuilding, wood removal from streams, and forest cutting, have altered the amount, transport, and arrangement of wood in parts of this basin (Swanson et al. 1976; Snyder 2000; Swanson and Jones 2002).

An example of discrete-source-patch control on wood dynamics can be found in low-gradient, alluvial rivers. In a low-gradient coastal plain river, Palik et al. (1998) and Michener et al. (1998) found that a major flood toppled 22 trees/km into Ichawaynochaway Creek, Georgia, but the wood did not move significantly downstream from the source area during this or subsequent floods. Wood input was highest in the more constrained stream reaches where current velocity was interpreted to be the greatest and as having high potential to topple trees. No trees greater than 20 cm dbh moved more than a few meters. Therefore, the pattern of wood deposits was controlled by the pattern of wood source areas because, despite flooding, transport was so limited.

Gurnell et al. (2000, 2002) presented an example of wood distribution dominated by trapping-site control. The Fiume Tagliamento, Italy, is an unconstrained river bordered by woodlands serving as the source of wood for the river. Wood along the river tended to be concentrated in areas of complex channel pattern, on exposed gravel bars, and at the heads of islands. Newly developed islands had greater amounts of trapped wood than older, established islands. Exposed gravel in multiple-thread channels had six times the wood amount as exposed gravel areas in single-thread channels. Thus, trapping sites showed variation in both the stage of island development and the context of channel form.

In each of these examples, it is appropriate to ask if the patterns observed represent long-term system behavior or are a narrow reflection of the most recent events. Some systems have stable properties of wood inputs, and others may be quite variable, which affects the strength of inferences drawn from a single sampling. Modeling and long-term field studies are needed to test and develop a more extensive and rigorous typology of wood system dynamics at multiple scales and to explore its usefulness in research and restoration. A greatly expanded set of field examples in diverse settings with longer records would be instructive.

Knowledge Gaps

A landscape perspective of wood in rivers is itself a major knowledge gap. Most rigorous research on wood in rivers is accumulating at finer scales of spatial analysis, but understanding of full reach- and basin-scales and, especially, longer time scales is quite limited. These limitations form critical gaps in our knowledge of physical patterns and dynamics of wood in rivers and responses of ecological, geomorphic, and hydrologic processes to wood. In particular, current knowledge would benefit from a framework for developing more general conceptual models of wood dynamics and their effects.

Dealing with long-term dynamics of forest sources of wood and transport capacity of rivers is particularly challenging. Some changes in wood sources are progressive (for example, the loss of wood sources from streamside areas resulting from some types of land use) and others are abrupt (for example, loss of forest cover by major floods). Broad time and space perspectives are needed to sort out the trajectories of change in these systems and to reconstruct probable scenarios of past change. Natural and human-imposed disturbances in watersheds can affect sources, transport processes, and accumulation sites of wood. A whole-system view is required because interactions among components of the system and between natural processes and management practices are complex. For example, watershed disturbance can alter peak streamflow and potentially change the size distribution of wood pieces, thereby affecting wood transport. Channel modifications, including simplification by engineering practices, such as channelization, or, by simply removing large, jam-forming pieces, alter the wood retention and transport capacities of channels. Urbanization and agriculture over entire watersheds can impose whole-system change on wood dynamics by completely eliminating forests or severely limiting the extent and role of streamside forests. Intensive plantation forestry may reduce the height of streamside forests, possibly shifting upstream zones of wood influence on streams that are determined by the scale of tree height to stream width.

Understanding of effects of changes in wood in rivers is beginning to emerge for several components of river systems. Some observations suggest that greater pool complexity can permit more species and age classes of fish to occupy habitats, which could affect distributions more broadly in a watershed (Dolloff and Warren 2003, this volume). The amount of mobile wood in a river network may influence the distribution of riparian vegetation that colonizes after a flood, such as red alder *Alnus rubra* in Cascade Range streams in Oregon. Movement of wood in the 1964–1965 floods helped open riparian habitat for alder establishment, leading to expansion of this important nitrogen-fixing species. Geographic patterns and types of wood movement in third- to fifth-order channels during the 1996 flood in this area strongly influenced disturbance patterns of aquatic and riparian areas in the river network (Swanson et al. 1998; Johnson et al. 2000; Acker et al. 2003). The extent and arrangement of wood in a river system may also affect the cumulative influence of hyporheic zone processes, which can affect water temperature and chemistry (Grimm et al. 1991; Valett et al. 1996; Edwards 1998; Wondzell and Swanson 1999). Removal of wood from some river systems has been speculated to cause channel downcutting and reduction of water storage in channel bed and floodplain aquifers, thus potentially reducing flow during dry summer months and limiting rearing habitat for certain fish species (Montgomery et al. 2003, this volume).

A practical gap in knowledge of wood dynamics in rivers is how wood structures resulting from restoration projects or other management actions differ in form and function from natural wood structures. Profound differences between the managed and wild functions of wood in rivers may have long-term ecological and geomorphic consequences.

Framework for Further Work

A landscape perspective is fundamental to examining linkages over broad time and space scales in both natural and managed systems. Such a broad perspective is an essential context for study of the contemporary state of natural and managed systems. This perspective also provides a basis for determining how various management actions individually and collectively may alter the wood regime of the river system and whether this change is stepped or gradual. Restoration practices may be most effective when predicated on understanding of natural system dynamics.

Developing a framework for future research and management begins with existing relevant frameworks. One very useful framework is presented in this volume (Benda et al. 2003), which sets forth a mathematical approach to analysis of wood routing and budgets, including sources, transport processes, and accumulation sites, and the temporal dynamics of source forest stands and wood in river networks. A second relevant framework is the channel-morphology classification scheme of Montgomery and Buffington (1997, 1998), which emphasizes the interactions of bed steepness, lateral constraint, sediment supply, and transport capacity as they affect channel morphology at a series of spatial scales within watersheds.

The sediment routing system and channel

morphology classification considered in the scheme of Montgomery and Buffington (1997, 1998) and others differs from the wood routing system in several respects important to developing a conceptual model of wood dynamics in rivers and resulting patterns of wood. I hypothesize that wood pieces have much shorter mean transport distances in rivers than inorganic sediment because of the large size and irregular shapes of wood pieces and various factors favoring retention of wood in river systems (for example, burial and deposition in stable trapping sites). Also, decomposition of wood is more rapid than the rates of physical and chemical breakdown of most forms of inorganic sediment, so most wood leaves a river system as gas to the atmosphere or in solution, and as fine fragments transported downstream, rather than as large wood. Many rivers are floored with sediment and remain so through major flow events, which involve extensive turnover of the bed. Bedforms may change little through major transport events. Wood, on the other hand, generally covers a small fraction of a channel, and wood-affected channel forms are much more likely to be profoundly modified by input and transport events; thus wood configuration is likely to more strongly reflect recent events than do sediment bedforms. These factors of limited transport distances and residence time contribute to the patchy patterns of wood in rivers. Consequently, distributions of source areas, transport distances, and deposition sites are expected to have different implications for the arrangement of wood than for inorganic sediment. The wood routing system is more conducive to a patch dynamics and a landscape analysis approach than the physics/continuum thinking applied to sediment routing and bedforms.

A typology of wood dynamics that affect the arrangement of wood pieces in channels supplements the wood budget and sediment routing/bedform conceptual frameworks for rivers. This typology would consider the dominant controls on wood patterns by source area, transport, and deposition sites. Individual types of systems may cycle between expression of different types of control. Individual types of wood dynamics would be characterized in terms of the absolute and relative amounts of different types of wood accumulations, perhaps expressed as probability density functions for amounts of wood, as proposed by Benda and Dunne (1997a, 1997b). This information could then be used to examine long-term functions of wood in ecological and other respects of river reaches and networks of different type. This perspective of characterizing the range and change of conditions maintained under different systems could form a basis for examining effects of management practices and for designing restoration projects and other management actions.

How can we advance a comprehensive theoretical framework for wood dynamics in rivers? Current trends in research funding and the broad time and space scales required for this work may preclude establishing a single, widely accepted, integrated research effort to analyze wood dynamics across a diverse range of river and forest types. Therefore, a common conceptual framework of wood dynamics within which different research groups and agencies can accumulate relevant information would be useful in refining and testing the framework and advancing understanding. An important step in understanding wood in rivers is integration of the views of forest and river ecologists, hydrologists, geomorphologists, and others in a common analytical framework. Future research and management require combinations of retrospective, long-term monitoring, and modeling approaches to address a linked set of hypotheses about controls on wood patterns and dynamics.

Acknowledgments

I thank many colleagues for discussions of wood in rivers over the years, especially when these discussions took place in and along streams, including, but not limited to, Lee Benda, Gordon Grant, Stan Gregory, George Lienkaemper, Christine May, Futoshi Nakamura, and Jim Sedell. Reviews by Lee Benda, Christine May, Stan Gregory, and Mark Meleason substantially improved the manuscript. Julia Jones helped with concept development and prepared the schematic figures This work has been supported in part by National Science Foundation grants to the H.J. Andrews Experimental Forest Long-Term Ecological Research program.

References

Abbe, T. B., and D. R. Montgomery. 1996. Large woody debris jams, channel hydraulics and habitat formation in large rivers. Regulated Rivers: Research and Management 12:201–221.

Abbe, T. B., and D. R. Montgomery. 2003. Patterns and processes of wood debris accumulation in the Queets River basin, Washington. Geomorphology 51:81–107.

Acker, S. A., S. V. Gregory, G. Lienkaemper, W. A. McKee, F. J. Swanson, and S. D. Miller. 2003. Composition, complexity, and tree mortality in riparian forests in the central western Cascades of Oregon. Forest Ecology and Management 173:293–308.

Benda, L., and T. W. Cundy. 1990. Predicting deposition of debris flows in mountain channels. Canadian Geotechnical Journal 27:409–417.

Benda, L., and T. Dunne. 1997a. Stochastic forcing of sediment supply to channel networks from landslides and debris flows. Water Resources Research 33:2849–2863.

Benda, L., and T. Dunne. 1997b. Stochastic forcing of sediment routing and storage in channel networks. Water Resources Research 33:2865–2880.

Benda, L., D. Miller, J. Sias, D. Martin, R. Bilby, C. Veldhuisen, and T. Dunne. 2003. Wood recruitment processes and wood budgeting. Pages 49–73 *in* S. V. Gregory, K. L. Boyer, A. M. Gurnell, editors. The ecology and management of wood in world rivers. American Fisheries Society, Symposium 37, Bethesda, Maryland.

Bilby, R. E., and P. A. Bisson. 1998. Function and distribution of large woody debris. Pages 324–346 *in* R. J. Naiman and R. E. Bilby, editors. River ecology and management. Springer-Verlag, New York.

Bilby, R. E., and J. W. Ward. 1989. Changes in characteristics and function of woody debris with increasing size of streams in western Washington. Transactions of the American Fisheries Society 118:368–378.

Bisson, P. A., R. E. Bilby, M. D. Bryant, C. A. Dolloff, G. B. Grette, R. A. House, M. L. Murphy, K. V. Koski, and J. R. Sedell. 1987. Large woody debris in forested streams of the Pacific Northwest: past, present, future. Pages 143–190 *in* E. O. Salo and T. W. Cundy, editors. Streamside management: forestry and fisheries interactions. Institute of Forest Resources Contribution 57, University of Washington, Seattle.

Braudrick, C. A., and G. E. Grant. 2000. When do logs move in rivers? Water Resources Research 36:571–583.

Dolloff, C. A., and M. L. Warren, Jr. 2003. Fish relationships with large wood in small streams. Pages 179–193 *in* S. V. Gregory, K. L. Boyer, and A. M. Gurnell, editors. The ecology and management of wood in world rivers. American Fisheries Society, Symposium 37, Bethesda, Maryland.

Edwards, R. T. 1998. The hyporheic zone. Pages 399–429 *in* R. J. Naiman and R. E. Bilby, editors. River ecology and management. Springer-Verlag, New York.

Faustini, J. M. 2000. Stream channel response to peak flows in a fifth-order mountain watershed. Doctoral dissertation. Oregon State University, Corvallis.

Forman, R. T. T., and M. Godron. 1986. Landscape ecology. Wiley, New York.

Frissell, C. A., W. J. Liss, C. E. Warren, and M. D. Hurley. 1986. A hierarchical framework for stream habitat classification: viewing streams in a watershed context. Environmental Management 10:199–214.

Gregory, S. V., F. J. Swanson, W. A. McKee, and K. W. Cummins. 1991. An ecosystem perspective of riparian zones. Bioscience 41:540–551.

Grant, G. E., and F. J. Swanson. 1995. Morphology and processes of valley floors in mountain streams, western Cascades, Oregon. Pages 83–101 *in* J. E. Costa, A. J. Miller, K. W. Potter, and P. Wilcock, editors. Natural and anthropogenic influences in fluvial geomorphology: the Wolman volume. Geophysical Monograph 89. American Geophysical Union, Washington, D.C.

Grimm, N. B., H. M. Valett, E. H. Stanley, and S. G. Fisher. 1991. Contribution of the hyporheic zone to stability of an arid land stream. Verhandlungen der Internationalen Vereinigung fur Theroestische und Angewandte Limnologie 20:1595–1599.

Gurnell, A. M., G. E. Petts, N. Harris, J. V. Ward, K. Tockner, P. J. Edwards, and J. Kollmann. 2000. Large wood retention in river channels: the case of the Fiume Tagliamento, Italy. Earth Surface Processes and Landforms 25:255–275.

Gurnell, A. M., H. Piégay, F. J. Swanson, and S. V. Gregory. 2002. Large wood and fluvial processes. Freshwater Biology 47:601–619.

Gurnell, A. M. 2003. Wood storage and mobility. Page 75–91 *in* S. V. Gregory, K. L. Boyer, and A. M. Gurnell, editors. The ecology and management of wood in world rivers. American Fisheries Society, Symposium 37, Bethesda, Maryland.

Harmon, M. E., J. F. Franklin, F. J. Swanson, P. Sollins, S. V. Gregory, J. D. Lattin, N. H. Anderson, S. P. Cline, N. G. Aumen, J. R. Sedell, G. W. Lienkaemper, K. Cromack, Jr., and K. W. Cummins. 1986. Ecology of coarse woody debris in temperate ecosystems. Pages 133–302 *in* A. MacFadyen and E. D. Ford, editors. Advances in ecological research. Academic Press, Orlando, Florida.

Hyatt, T. L., and R. J. Naiman. 2001. The residence time of large woody debris in the Queets River, Washington, USA. Ecological Applications 11:191–202.

Johnson, S. L., F. J. Swanson, G. E. Grant, and S. M. Wondzell. 2000. Riparian forest disturbances by a mountain flood – the influence of floated wood. Hydrological Processes 14:3031–3050.

Keller, E. A., and F. J. Swanson. 1979. Effects of large organic material on channel form and fluvial processes. Earth Surface Processes and Landforms 4:361–380.

Keller, E. A., and T. Tally. 1979. Effects of large organic debris on channel form and fluvial processes in the coastal redwood environment. Pages 169–197 *in* D. D. Rhodes and G. P. Williams, editors. Adjustments in the fluvial system. 1979 Proceedings of the Tenth Annual Geomorphology Symposium. State University of New York, Binghamton.

Lienkaemper, G. W., and F. J. Swanson. 1987. Dynamics of large woody debris in streams in old-growth Douglas-fir forests. Canadian Journal of Forest Research 17:150–156.

Martin, D. J., and L. E. Benda. 2001. Patterns of in-stream wood recruitment and transport at the watershed scale. Transactions of the American Fisheries Society 130:940–958.

Michener, W. K., E. R. Blood, J. B. Box, C. A. Couch, S. W. Golladay, D. J. Hippe, R. J. Mitchell, and B. J. Palik. 1998. Tropical storm flooding of a coastal plain landscape. BioScience 48:696–705.

Montgomery, D. R. 1999. Process domains and the river continuum. Journal of the American Water Resources Association 35:397–410.

Montgomery, D. R., and J. M. Buffington. 1997. Channel-reach morphology in mountain drainage basins. Geological Society of America Bulletin 109:596–611.

Montgomery, D. R., and J. M. Buffington. 1998. Channel processes, classification, and response. Pages 13–42 *in* R. J. Naiman and R. E. Bilby, editors. River ecology and management. Springer-Verlag, New York.

Montgomery, D. R., B. D. Collins, J. M. Buffington, and T. B. Abbe. 2003. Geomorphic effects of wood in rivers. Pages 21–47 *in* S. V. Gregory, K. L. Boyer, and A. M. Gurnell, editors. The ecology and management of wood in world rivers. American Fisheries Society, Symposium 37, Bethesda, Maryland.

Naiman, R. J., H. Decamps, M. Pollock. 1993. The role of riparian corridors in maintaining regional biodiversity. Ecological Applications 3:209–212.

Naiman, R. J., and J. R. Sedell. 1979. Benthic organic matter as a function of stream order in Oregon. Archives of Hydrobiology 87:404–422.

Nakamura, F., and F. J. Swanson. 1993. Effects of coarse woody debris on morphology and sediment storage of a mountain stream systems in western Oregon. Earth Surface Processes and Landforms 18:43–61.

Nakamura, F., and F. J. Swanson. 1994. Distribution of coarse woody debris in a mountain stream, western Cascade Range, Oregon. Canadian Journal of Forest Research 24:2395–2403.

Nakamura, F., F. J. Swanson, and S. M. Wondzell. 2000. Disturbance regimes of stream and riparian systems–a disturbance-cascade perspective. Hydrological Processes 14:2849–2860.

Nakamura, F., and F. J. Swanson. 2003. Dynamics of wood in rivers in the context of ecological disturbance. Pages 279–297 *in* S. V. Gregory, K. L. Boyer, and A. M. Gurnell, editors. The ecology and management of wood in world rivers. American Fisheries Society, Symposium 37, Bethesda, Maryland.

Palik, B., S. W. Golladay, P. C. Goebel, and B. W. Taylor. 1998. Geomorphic variation in riparian tree mortality and stream coarse woody debris recruitment from record flooding in a coastal plain stream. Ecoscience 5:551–560.

Piégay, H., and A. M. Gurnell. 1997. Large woody debris and river geomorphological pattern: examples from S. E. France and S. England. Geomorphology 19:99–116.

Piégay, H., A. Thévenet, and A. Citterio. 1999. Input, storage and distribution of large woody debris along a mountain river continuum, the Drome River, France. Catena 35:19–39.

Piégay, H. 2003. Dynamics of wood in large rivers. Pages 109–133 in S. V. Gregory, K. L. Boyer, and A. M. Gurnell, editors. The ecology and management of wood in world rivers. American Fisheries Society, Symposium 37, Bethesda, Maryland.

Reeves, G. H., L. E. Benda, K. M. Burnett, P. A. Bisson, and J. R. Sedell. 1995. A disturbance-based ecosystem approach to maintaining and restoring freshwater habitats of evolutionarily significant units of anadromous salmonids in the Pacific Northwest. Pages 334–349 *in* J. L. Nielsen, editor. Evolution and the aquatic ecosystem: defining unique units in population conservation. American Fisheries Society, Symposium 17, Bethesda, Maryland.

Risser, P. G., J. R. Karr, and R. T. T. Forman. 1984. Landscape ecology: directions and approaches. Illinois Natural History Survey, Special Publication 2, Champaign.

Snyder, K. U. 2000. Debris flows and flood disturbance in small, mountain watersheds. Master's thesis. Oregon State University, Corvallis.

Swanson, F. J., and C. T. Dyrness. 1975. Impact of clear-cutting and road construction on soil erosion by landslides in the western Cascade Range, Oregon. Geology 3:393–396.

Swanson, F. J., M. D. Bryant, G. W. Lienkaemper, and J. R. Sedell. 1984. Organic debris in small streams, Prince of Wales Island, southeast Alaska. U.S. Department of Agriculture, Forest Service, Pacific Northwest Forest and Range Experiment Station, General Technical Report PNW-166, Portland, Oregon.

Swanson, F. J., J. F. Franklin, and J. R. Sedell. 1990. Landscape patterns, disturbance, and management in the Pacific Northwest. Pages 191–213 *in* I. S. Zonneveld and R. T. T. Forman, editors. Trends in landscape ecology. Springer-Verlag, New York.

Swanson, F. J., S. V. Gregory, J. R. Sedell, and A. G. Campbell. 1982. Land-water interactions: the riparian zone. Pages 267–291 *in* R. L. Edmonds, editor. Analysis of coniferous forest ecosystems in the western United States. US/International Biological Programme Synthesis Series 14. Hutchinson Ross Publishing Co., Stroudsburg, Pennsylvania.

Swanson, F. J., S. L. Johnson, G. E. Grant, and S. M. Wondzell. 1998. Flood disturbance in a forested mountain landscape. BioScience 48:681–689.

Swanson, F. J., and J. A. Jones. 2002. Geomorphology and hydrology of the H. J. Andrews Experimental Forest, Blue River, Oregon. Pages 289–314 *in* G. W. Moore, editor. Field guide to geologic processes in Cascadia. Oregon Department of Geology and Mineral Industries, Special Paper 36, Portland, Oregon.

Swanson, F. J., G. W. Lienkaemper, and J. R. Sedell. 1976. History, physical effects, and management implications of large organic debris in western Oregon streams. U.S. Department of Agriculture, Forest Service, Pacific Northwest Forest and Range Experiment Station, General Technical Report PNW-56, Portland, Oregon.

Valett, H. M., J. A. Morrice, C. N. Dahm, and M. E. Campana. 1996. Parent lithology, surface-groundwater exchange, and nitrate retention in head water steams. Limnology and Oceanography 41:333–345.

Vannote, R. L., G. W. Minshall, K. W. Cummins, J. R. Sedell, and C. E. Cushing. 1980. The river continuum concept. Canadian Journal of Fisheries and Aquatic Sciences 37:130–137.

Wemple, B. C., F. J. Swanson, and J. A. Jones. 2001. Forest roads and geomorphic process interactions, Cascade Range, Oregon. Earth Surface Processes and Landforms 26:191–204.

Wondzell, S. M., and F. J. Swanson. 1999. Floods, channel change, and the hyporheic zone. Water Resources Research 35:555–567.

American Fisheries Society Symposium 37:315–335, 2003

Modeling the Dynamics of Wood in Streams and Rivers

STAN V. GREGORY

Department of Fisheries and Wildlife, Oregon State University, Corvallis, Oregon 97331-3803, USA

MARK A. MELEASON

National Institute of Water & Atmospheric Research, P.O. Box 11115, Hamilton, New Zealand

DANIEL J. SOBOTA

Department of Fisheries and Wildlife, Oregon State University, Corvallis, Oregon 97331-3803, USA

Abstract.—Extensive research over the last 30 years has documented the abundance and ecological functions of wood in streams and rivers. Most studies have focused on amounts and distributions of wood in streams, and a small number of studies have explored critical processes that determine quantities and patterns of wood in streams—riparian tree mortality, input, breakage, decomposition, mechanical breakdown, and transport. Empirical studies describe the outcomes of the stand dynamics, disturbance history, and human management at a site, but questions about long-term dynamics or landscape patterns and distributions are difficult to answer based on empirical observation alone. General properties of simulation models that have been developed recently to explore long-term or large-scale implications of wood dynamics are reviewed. Most existing models are not stochastic, and those that incorporate variation and unpredictable change do not incorporate interactions between processes. Models consistently indicate that forest age directly influences abundance of wood in streams, and sensitivity analysis demonstrates that most models of wood dynamics are most sensitive to estimates of decomposition rates and rates of input.

Introduction

Research since the early 1970s has documented the abundance, distribution, and ecological functions of wood in streams and rivers. Most studies have focused on amounts and distributions of wood in streams, and a small number of studies have explored critical processes that determine quantities and patterns of wood in streams—riparian tree mortality, input, breakage, decomposition, mechanical breakdown, and transport. Outcomes of stand dynamics, disturbance history, and human management at a site may be described by empirical studies, but questions about either long-term dynamics at time periods of multiple generations of trees or humans or broad landscape patterns at spatial extents of thousands of square kilometers are difficult to answer based on empirical observation.

Dynamics of wood in streams and rivers of the world reflect complex landscape processes that differ by geographic region, time interval, hydrologic regime, basin geology, channel form, network structure, forest composition, disturbance processes, and human influence. Modeling provides a major tool for integrating results of short-term empirical observations and exploring the implications of alternative management practices and long-term disturbance patterns. This chapter will describe the properties of simulation models that have been developed in recent years to explore long-term or large-scale implications of wood dynamics. We will compare approaches for modeling wood and discuss the implications of these approaches for investigating the dynamics of wood. Major contributions of modeling for understanding the dynamics of wood in world rivers are illustrated for several key characteristics of wood and ecological processes.

What is a model? Haefner (1996) defined a model simply as ". . . a description of a system. A system is any collection of interrelated objects. An object is some elemental unit upon which observations can be made, but whose internal structure either does not exist or is ignored." Such descriptions of systems are abstract representations developed by and interpreted by humans. Models can be conceptual, diagrammatic, physical, or formal mathematical models (Haefner 1996). All of the models reviewed in this chapter are formal mathematical models that were developed from conceptual descriptions of selected processes of wood dynamics. Mathematical models can be classified further as (1) mechanistic or descriptive, (2) dynamic or static, (3) continuous or discrete, (4) spatially heterogeneous or homogeneous, or (5) stochastic or deterministic (Haefner 1996).

Evaluation of models must consider the fundamental properties of model performance—realism, precision, generality (Levins 1966)—and relate these properties to the intended use of the model. Models generally are used for understanding, prediction, and control or management (Karplus 1977). Models of wood in streams and rivers have been used largely to (1) understand the processes that shape the abundance and distribution of wood at local sites or along river networks and their interactions, or (2) predict the abundance and distribution of wood that would result from different types of riparian forests or landscape dynamics as a basis for management decisions. A model developed for understanding fundamental ecological processes requires generality and realism, with less demand for precision (Haefner 1996). On the other hand, a wood model developed to predict amounts of wood in specific stream reaches requires precision and reality, but does not require generality if the model has been developed from local observations and quantitative information.

Simple mathematical relationships, such as negative exponential decay rates, directionality of tree fall, and negative exponential rates of log transport, are simple forms of models. Many studies have identified quantitative relationships between a small number of independent variables and a wood response. In this chapter, we will consider both these simple models as well as their applications in more complex, integrated models of wood dynamics.

History of Wood Models

Over the last two decades, 14 models of wood dynamics in streams and rivers have been developed for different processes, purposes, and regions (Rainville et al. 1986; Murphy and Koski 1989; Van Sickle and Gregory 1990; McDade et al. 1992; Malanson and Kupfer 1993; Minor 1997; Benda and Sias 1998, 2003; Kennard et al. 1999; Bragg 2000; Beechie et al. 2000; Downs and Simon 2001; Fleece 2002; Meleason et al. 2002, in press; Welty et al. 2002; Table 1). We will present a history of wood model development and briefly describe the general characteristics of each model.

Of the 14 models of wood dynamics in streams and rivers, 11 have been developed in the Pacific Northwest (Rainville et al. 1986; Murphy and Koski 1989; Van Sickle and Gregory 1990; McDade et al. 1992; Minor 1997; Kennard et al. 1999; Bragg 2000; Beechie et al. 2000; Fleece 2002; Meleason et al. 2002, in press; Welty et al. 2002; Benda and Sias 2003), two for the midwest region of North America (Malanson and Kupfer 1993; Downs and Simon 2001), and one for the Rocky Mountain region of North America (Bragg 2000). The strong regional bias for the Pacific Northwest, in part, reflects the longer history of wood research in the region than in other parts of the world. The lack of wood models from other countries is surprising, but some of these models are being adapted for other countries and regions. For example, the model of Meleason et al. (in press) is being parameterized for New Zealand forests through the National Institute of Water and Atmospheric Research (NIWA).

The earliest wood models were designed to simulate the delivery of wood to streams from adjacent riparian forests (Rainville et al. 1986; Van Sickle and Gregory 1990; Malanson and Kupfer 1993; Minor 1997). Murphy and Koski (1989) approximated input rates and depletion rates by measuring standing stock and age of wood (from nurse trees) and assuming that wood volume was at steady state (that is, inputs equal outputs). Recent models have attempted to describe dynamics of wood by integrating input processes, retention, decomposition, and redistribution over either long time periods and/or large portions of river networks (Kennard et al. 1999; Beechie et al. 2000; Bragg 2000; Downs and Simon 2001; Meleason et al. 2002, in press; Welty et al. 2002; Benda and Sias 2003).

Most of the existing models are deterministic models that produce single estimates of outcomes with no variance (Table 1). Disturbance

TABLE 1. A. Comparison of published simulation models of wood dynamics.

Model characteristics	Rainville et al. 1985	Murphy and Koski 1989	McDade et al. 1990	Van Sickle and Gregory 1990	Malanson and Kupfer 1993	Minor 1997
General model characteristic						
Model type	deterministic	deterministic	deterministic	deterministic	stochastic	deterministic
Purpose/goal	recruitment; harvest	depletion rate	source distance	recruitment	carbon budget	source distance
Harvest schedule	thins at 25, 75 years	pre/post	none	none	none	no thinning
Multiple reach	no	no	no	no	no	no
Both riparian sides included	no	yes	no	yes	no	no
Time interval modeled	300 years	250 years	N/A	old growth	500 years	old growth
number of iterations	1	1	1	1	10	1
Time step	10 years	1 year	N/A	10 years	1 year	N/A
Results as number or volume	number of key pieces	number	number	number by length-class and fall angle	biomass	number of key pieces
Region	Idaho	SE Alaska	PNW	PNW	Iowa River	PNW
Species	TSHE, ABGR, ABLA	TSHE/PISI	TSHE/PSME/THPL	PSMA, TSHE	Iowa floodplain spp.	PNW species
Stream width						
Riparian zone description						
Width	90 ft	>30 m	60 m	user defined	27 m wide	60 ft
Length	variable	100 m	N/A	user defined	undefined length of river	variable
Subzone definition	10 ft	N/A	N/A	user defined	27 rows, 1 m wide	2 ft for first 40 ft, 40 to 60
Stream wood definition						
Minimum diameter	N/A	10 cm	10 cm	10 cm	N/A	6 in
Minimum length	N/A	3 m	1 m	1.5 m	0.5*tree ht	3 ft

Table 1. A. Continued.

Model characteristics	Rainville et al. 1985	Murphy and Koski 1989	McDade et al. 1990	Van Sickle and Gregory 1990	Malanson and Kupfer 1993	Minor 1997
Size categories	no	4 classes; 10 cm - <90 cm		length: 5-m classes	no	2″ tree diameter classes; 12–52 in
Key pieces only	yes	no	no	no	N/A	yes
Key piece definition	10 in, 8 ft	none	none	none	N/A	24-in mean diameter, 33 ft
Riparian forest						
Dead tree size categories	6 diameter classes	N/A	N/A	height: 10-m classes	no	2″ tree diameter classes; 12–52 in
Type of forest model	growth and yield (Prognosis)	none	N/A	stand table	Gap model (FORFLO)	stand table
Sapling recruitment	no	no	N/A	no	yes	N/A
Growth included	yes	no	N/A	no	yes	N/A
Types of mortality	tree fall	tree fall; bank erosion	tree fall	tree fall	tree fall, bank erosion	tree fall
Bank undercut	first 6 ft, 20% per decade	yes	no	no	within 1 m, 70% chance	no
Tree position	center subzone	N/A	center subzone	center subzone	center subzone	center subzone
Entry						
Fall along subzone midpoint	yes	N/A	yes	yes	yes	yes
Entry Pi/360 * N	yes	N/A	yes	yes	yes	yes
Entry breakage	no	N/A	no	banks	no	banks
Fall regime	random	N/A	random	random or	random	random or
Entry mechanism	vol mort converted to cnt/dcat, then dom ht used cal	used 1/age for dcat as recruitment rate = depletion rate	same as VanSickle and Gregory	Ps = arcInt/360; vary fall angle by 5-deg interval, mean L from ht, dist cat	same as VanSickle and Gregory	same as VanSickle and Gregory, except for a function of slope in Ps

TABLE 1. A. Continued.

Model characteristics	Rainville et al. 1985	Murphy and Koski 1989	McDade et al. 1990	Van Sickle and Gregory 1990	Malanson and Kupfer 1993	Minor 1997
Instream breakage	no	N/A	no	no	no	no
Instream movement	no	in = out	no	in = out	no, but move off floodplain	no
Decomposition	no	depletion rate	no	no	terrestrial, not aquatic	no
Field data comparison	no	no	no	yes	yes	no
Sensitivity analysis	no	no	no	no	yes	no

TABLE 1. B. Comparison of published simulation models of wood dynamics. Note that Beechie et al. 2000 is fundamentally the same wood model as Kennard et al. (1999) and Welty et al. (2002).

Model characteristics	Beechie et al. 2000*	Bragg 2000	Downs and Simon 2001	Benda and Sias 1998, 2003	Meleason et al. 2003, in press	Welty et al. 2002
General model characteristic						
Model type	deterministic	stochastic	deterministic	deterministic	stochastic	deterministic
Purpose/goal	recruitment; pool formation	recruitment: individual and catastrophic mortality	recruitment from channel meandering	Recruitment; mass failure and debris flows	recruitment	recruitment, shade
Harvest schedule	thinning is user defined	clearcut	none	none	thinning is user defined	thinning is user defined
Multiple reach	no	no	no	yes	yes	no
Both riparian sides included	yes	yes	yes	yes	yes	yes
Time interval modeled	150 years	300 years	NA	800–1,800 years	500 years	240 years
number of iterations	1	20	1	1	500	1
Time step	10 years	10 years	10 years	10 years	10 years	10 years

Table 1. B. Continued.

Model characteristics	Beechie et al. 2000*	Bragg 2000	Downs and Simon 2001	Benda and Sias 2003	Meleason et al. in press	Welty et al. 2002
Results as number or volume	number, (volume), pools	number and volume	number and volume	number and volume	number and volume	number and volume
Region	PNW	Rocky Mountain	Mississippi, USA	PNW	PNW	PNW
Species	PSME/TSHE/ ALRU/ACMA	PIEN,ABLA, PICO	midwest deciduous spp.	PNW spp	PSME/THPL/ TSHE/ALRU	PSME/TSHE/ ALRU/ACMA
Stream width	5–30 m	user defined	6–20 m	user defined	user defined	user defined
Riparian zone description						
Width	ht. of tallest tree	user defined	10 m	undefined	100 m	user defined
Length	user defined	30.5 m	50 m	100 m	user defined	user defined
Subzone definition	cites Van Sickle and Gregory	9 rows, user defined	no	no	5 rows, user defined	2–4 rows
Stream wood definition						
Minimum diameter	function of BFW	10 cm	5 cm	10 cm	10 cm	10 cm
Minimum length	function of BFW	1 m		2 m	1 m	1 m
Size categories	2: small and pool forming	1–20 classes			user defined	user defined
Key pieces only	yes	both	no	no	both	both
Key piece definition	Dmin = 2.5*BFW	user defined	>0.25 m dbh		length = channel width	user defined
Riparian forest						
Dead tree size categories	N/A	N/A	5-cm intervals	N/A	N/A	N/A
Type of forest model	growth & yield; ORGANON	growth & yield	none	none	gap (modified Zelig)	growth & yield; ORGANON
Sapling recruitment	yes	yes	no	N/A	yes	yes
Growth included	yes	yes	no	N/A	yes	yes
Types of mortality	tree fall	tree fall, catastrophic	meander	tree fall, bank undercut	tree fall	tree fall
Bank undercut	no	no	yes	yes	no	no

Table 1. B. Continued.

Model characteristics	Beechie et al. 2000*	Bragg 2000	Downs and Simon 2001	Benda and Sias 2003	Meleason et al. in press	Welty et al. 2002
Tree position	center subzone	center subzone	N/A	N/A	center subzone or defined location	center subzone
Entry						
Fall along subzone midpoint	yes	yes	no	no	yes	yes
Entry Pi/360 * N	yes	no	no	yes	user defined	yes
Entry breakage	banks	yes	no	modified entry	yes	user defined, 2 pieces
Fall regime	random	directional	random	random	user defined	random or "fall bias factor"
Entry mechanism	cites Van Sickle and Gregory	tree fall	knickpoint migration	tree fall, undercutting, mass failure	same as Van Sickle and Gregory with functions for slope	tree fall
Instream breakage	no	yes	no	no	yes	no
Instream movement	no input; output is depletion	constant attrition of volume	no	yes	yes	overall depletion; user defined
Decomposition	number: depletion rate	constant attrition of volume	no	depletion rate	decay rates until piece smaller than minimum	overall depletion; user defined
Field data comparison	yes	yes	yes	no	yes	yes
Sensitivity analysis	no	yes	no	incremental	yes	yes

processes in most wood models are simulated based on fixed scenarios of long-term disturbance events. In contrast, three models are stochastic models based on probabilities of selected wood processes and rates of processes. One of these three stochastic models (Bragg 2000) is predominantly deterministic because only the processes of snag creation and fall are stochastic. The stochastic models provide both mean outcomes and the variance associated with those outcomes. These stochastic models allow analysis of both the central tendencies or means as well as the influence of stochastic factors on variance in wood dynamics. Even projections of wood dynamics and spatial and temporal variation from stochastic models are limited by the assumptions about statistical distributions of the probability functions (such as normal, lognormal, exponential, binomial) used for specific processes or rates.

All models of wood dynamics in streams and rivers must represent the adjacent riparian forests and the delivery of wood to the channel. Some models maintain a fixed riparian composition and delivery of a fixed proportion of the stand through time. These models clearly are simplistic representations of riparian forests. Stand dynamics models that represent regeneration, growth, and mortality of riparian forests are important elements of most existing wood models. Such models either operate externally or are embedded as a major component within a wood dynamics model. Stand dynamics models that have been used with existing models of wood dynamics include ORGANON (Hester et al. 1989), Forest Vegetation Simulator (Wykoff et al. 1982), ZELIG (Shugart and West 1977), and versions of forest gap models (Botkin 1993). These forest dynamics models are developed for specific tree species and ranges of stand ages, but most stand dynamics models have been developed for upland forests rather than riparian forests.

Several fundamental wood processes are represented only in one or two models. Most models are based primarily on delivery of wood from adjacent stands. Delivery from adjacent riparian forests generally is modeled as direct mortality and fall, windthrow, bank undercutting, or an overall composite mortality of all of these sources. Two models from the Midwest, USA (Malanson and Kupfer 1993; Downs and Simon 2001) simulate channel avulsion and bank erosion in rivers. Only one model (Benda and Sias 1998, 2003) from the Pacific Northwest simulates wood delivery through landslides and mass failure. Although one of the early models (Van Sickle and Gregory 1990) concluded that numbers of wood pieces are underestimated if a model does not account for breakage, only one model (STREAMWOOD, Meleason et al. 2002, in press) simulates the breakage of trees as they fall into streams and breakage of wood as it is subsequently transported. Bragg (2000) combines breakage during tree fall and breakage during storage and transport into a single breakage function. Most models also combine the processes of decomposition, breakage, and export into an overall depletion estimate, but these processes are explicitly represented only in STREAMWOOD.

Rainville model

In the mid-1980s, the first model of wood dynamics in streams was developed by Rainville and coauthors and was published in the proceedings of a conference of the Society for American Foresters (Rainville et al. 1986). The Rainville model is a deterministic model of wood input into a stream reach based on riparian forest stand conditions within 30 m of the stream. Wood enters the channel as a result of tree mortality in the stand (individual tree death) and bank undercutting, based on a table of probabilities as a function of tree size and distance from the stream. Tree fall angle is random, and trees do not break when they fall into streams. The model does not estimate or include standing stocks of wood, decomposition, or instream movement. Stand thinning and harvest intensity are evaluated as a riparian management practice. This model was the first attempt to (1) use a stand dynamics model for a forest to simulate the delivery of wood to stream channels, and (2) use modeling as a tool to evaluate the consequences of land-use practices on the delivery of wood to stream ecosystems.

Murphy and Koski model

A simple model of wood input, decomposition, and output was developed for streams in southeast Alaska (Murphy and Koski 1989). The model is based partly on empirical information and partly on assumptions of input and output. Volumes of wood were measured in Alaskan streams, and the ages of nurse trees on wood in the streams were estimated. Based on the assumption that wood volume is at steady state (inputs = outputs), the depletion rate can be estimated by dividing the standing crop by the average age of wood in the stream. This model was the first to estimate standing stock of wood in streams and incorporate a function for depletion through decay, breakage, and transport.

Van Sickle and Gregory model

Van Sickle and Gregory (1990) published a simple model of wood delivery into stream reaches from adjacent riparian forests. This model is a deterministic model based on the probability of fall and the geometric basis for the tree to intersect the stream channel as a function of tree height, distance from the channel, and fall direction. The model was evaluated by comparing simulations with data on wood characteristics in a stream in an old-growth forest.

The model concluded that simulations of wood volumes were relatively consistent with field observations, but estimates of numbers of pieces of wood delivered were underestimated if fall breakage was not included in the model. The model also identified the potential importance of directionality of fall. If tree and snag fall is not random and totally directed toward the channel, estimates of wood volume delivered are threefold greater than a model with random fall direction.

McDade model

McDade et al. (1990) measured the distance from the stream to the point on the hillslope where in-channel wood originated in streams of the Pacific Northwest. These measurements were integrated with the Van Sickle and Gregory delivery model to create a simple deterministic model of wood delivery and source distance.

This simple model was the first to identify the lateral riparian distance required to deliver specific proportions of total wood inputs to streams (for example, 85% of wood delivered from old-growth coniferous forest is located within 30 m of the channel) and to distinguish the behavior of deciduous and coniferous riparian forests (for example, 90% of total wood was delivered within 25 m for mature deciduous forests, 48 m for mature coniferous forests, and 55 m for old-growth coniferous forests).

Malanson and Kupfer model

Most models of wood dynamics have been developed for streams in the Pacific Northwest region of North America, but Malanson and Kupfer (1993) developed a stochastic model of wood delivery into large rivers of the Midwest region of North America. This also was the first model to incorporate bank erosion and channel avulsion in large rivers, which are physical processes that have received little attention in research and modeling (Piégay and Gurnell 1997). This unique application incorporates models of (1) bank erosion and channel avulsion developed for the Iowa River, and (2) dynamic floodplain forest regeneration (FORFLO). The model also represents movement of wood out of the floodplain forest and into the river. Decomposition is not modeled for wood in the river, but terrestrial wood on the floodplain decays prior to lateral transport.

This model is distinctive because it was the first stochastic model of large wood dynamics, and it was the first model of wood dynamics related to channel avulsion and floodplains in large rivers.

Minor model

A simple model of key piece delivery was developed by Minor (1997) to estimate the delivery of key pieces (>24-in diameter) from a 60-ft wide riparian zone to a single stream reach. Tree distributions and sizes are derived from fixed stand tables for composition, size, and location. Fundamentally, the model relates the number of trees delivered to the stream to the overall stand mortality rate, fall directionality, and the influence of slope on directionality.

The model illustrated the potential influence of directionality and influence of slope on wood loading into streams (empirical data on the relationships were not reported).

Riparian-in-a-Box model

A deterministic model for predicting the effects of timber harvest on wood delivery from riparian forests was developed by researchers at the University of Washington in cooperation with scientists from the Weyerhaeuser Corporation (Beechie et al. 2000; also reported in Kennard et al. 1999). Riparian-in-a-Box is a deterministic model of key piece delivery from managed riparian stands over as long as 150 years at decadal time steps and incorporates thinning and harvest. Riparian forests develop through time based on a growth and yield model. Length of wood delivered to the stream is represented in the model as a function of bank-full width, and key pieces are defined as 2.5 times bank-full width. This model also simulates pool formation in stream channels based on relationships between the relative size of wood and the dimensions of the stream channel.

Bragg model

The second major stochastic wood model was developed by Bragg (2000) for conifer stands in Wyoming. The Bragg model simulates the input, storage, and depletion of large wood from pine and fir forests of the Rocky Mountain region, expressed as means and variance of both numbers and volumes of wood. The model has the capacity to represent either scheduled timber harvest or catastrophic mortality events over 300-year periods. Wood in the channel is depleted as a result of decay, transport, and mechanical losses. This is one of the most complete representations for the ecological processes of riparian stand dynamics and instream processes related to wood dynamics.

The Bragg model was the first stochastic model to integrate stand dynamics, wood delivery into streams, and in-channel processes (transport, decay, mechanical loss). It also extended the application of wood models to forest types of the Rocky Mountain region. It included comparison of model output with field observation and sensitivity analyses. In many respects, this was the first complete model of wood dynamics in streams.

Downs and Simons model

A second model that incorporates bank erosion and channel avulsion was developed by Downs and Simon (2001) for a small river, the Yalobusha River in central Mississippi, USA. Delivery of wood to the river was modeled based on estimates of numbers, sizes, and volumes of trees in riparian forests along the banks and a civil engineering model of bank stability and knick-point migration. This channel evolution model was based on earlier models of channel formation phases or stages (Simon 1989; Hupp and Simon 1991) and included analysis of shear strength and bank stability (factor of safety analysis for current and future conditions). This deterministic model estimated future top width, calculated channel widening to stable bank angle, and used empirical knickpoint migration rates to calculate the number of trees and volume of wood recruited/m length. Field measurements of riparian forests were used to estimate the recruitment of large wood through bank erosion. Potential to trap wood was assessed on the basis of tree length to channel width and angle relative to flow.

The model results indicated that overall channel-formation processes may influence the outcome of wood delivery processes from streambank erosion. This model complements the Malanson and Kupfer model and demonstrates approaches for modeling wood that incorporate both the geomorphic processes that modify floodplains and riverbanks in large rivers and the characteristics of floodplain forests.

RAIS model

Models of wood delivery and riparian shading were combined in a deterministic model called Riparian Aquatic Interface Simulator (RAIS; Welty et al. 2002). Riparian stand dynamics are modeled with ORGANON, a growth and yield model for Douglas-fir that was adapted for western hemlock, red alder, and bigleaf maple. The model allows the user to define the thinning prescriptions and harvest rotations. Wood is delivered to the channel as a result of tree mortality based on the wood delivery model developed by Beechie et al. (2000) and Kennard et al. (1999).

This model is an extension of the Beechie model, coupled with a stand growth and yield model, for evaluating alternatives for riparian forest management. It allows the user to modify thinning and harvest approaches to assess the potential outcomes for large wood and shade in streams. This model also was the first wood model to be available to be downloaded from the Internet at http://www.weyerhaeuser.com/rais.asp.

Benda and Sias model

Landslides and mass failures are major sources of wood recruitment in steep, highly erosive landscapes. While most wood models acknowledge these sources of input, few incorporate these processes explicitly in the model. Benda and Sias (1998, 2003) developed a deterministic model of wood loading potential and transport in coastal streams of the Pacific Northwest that incorporates landslides, mass failures, and bank undercutting, as well as windthrow and adjacent stand mortality.

The Benda and Sias model was the first model of wood dynamics that explicitly modeled delivery of wood to streams through the geomorphic processes of mass failure and local bank undercutting in steep mountain streams. This model of montane streams complements the models of wood delivery through channel erosion in lowlands developed in the Midwest, USA (Malanson and Kupfer 1993; Downs and Simon 2001). The model also was the first to readily simulate wood loading for multiple stream reaches throughout a river network.

STREAMWOOD model

STREAMWOOD is a stochastic model that simulates wood dynamics in streams and riparian forests (Meleason et al. 2002, in press). Wood inputs, storage, and transport can be simulated for either single reaches or multiple reaches in a stream network. A forest gap model (modified Zelig model) is used to simulate tree regeneration, growth, and mortality and subsequently deliver wood from live trees and snags in adjacent riparian forests. Individual trees in the stand (four conifer species, alder, and a user-defined species) are modeled. A Monte Carlo procedure generates hundreds to thousands of simulations for each model scenario, estimating mean responses and variances. The model simulates delivery of trees into the channel and breakage as trees fall, based on either random or directional fall direction. Recent empirical observations of fall direction indicate that riparian trees are more likely to fall towards the stream channel (Sobota 2003); fall direction is not affected by hillslope steepness, but variance decreases with increasing slope. Individual logs in the stream are modified through time by breakage, movement, and decomposition.

STREAMWOOD is unique because it is stochastic; models both conifer and deciduous tree species; operates at single or multiple stream reaches; and includes probability functions for fall breakage, directionality of fall, in-channel breakage, transport, and decomposition. Recent research by this group also provides empirical data on directionality of fall for riparian forests in the Pacific Northwest and Intermountain regions, USA (Sobota 2003). STREAMWOOD can be downloaded from the Internet at http://www.fsl.orst.edu/lter/data/tools/models/streamwood.cfm.

Fleece model

Low-level remote sensing of riparian vegetation was used to predict wood inputs into streams within a 28-km^2 study area in the Pacific Northwest (Fleece 2002). Height and composition of riparian vegetation were estimated from Light Detection and Ranging (LIDAR) data, which detects reflectance of 0.5-m laser pulses spaced less than 6 m apart across a 4,500-m scan width. LIDAR data provide estimates of ground elevation, slope, tree height, and vegetative cover type. Wood delivery was based on the tree density, probability of falling into a stream, and the rate of tree fall (determined from mortality estimates from ORGANON growth and yield model).

Remotely sensed estimates of riparian forest density, composition, or volume can be utilized by most wood dynamics models. This simple application of tree density, fall rate, and delivery into stream channels illustrates the application of wood delivery models at larger spatial extents than single reaches and the use of remotely sensed forest cover information. The predicted rate of wood delivery into streams within this study area was within the range observed for streams in the H. J. Andrews Experimental Forest in Oregon, and the recruitment distance was similar to field observations in the region.

Major Results from Models of Wood Dynamics in Streams and Rivers

Models provide a method for exploring the implications of our current understanding of wood dynamics. The following section describes recent applications of models to explore long-term trends and spatial patterns of inputs and storage of wood in streams and rivers.

Influence of forest age on wood in streams

Models of wood dynamics have proved to be one of the most valuable tools for exploring the influence of riparian forest age on rates of delivery of wood to streams and rivers and its subsequent storage in the channels. Empirical field studies of wood abundance commonly are compromised by differences between research sites in terms of disturbance histories, human modification of the site, channel characteristics, basin characteristics, forest composition, and climate. One of the first applications of simulation models of wood dynamics was to project rates of wood input or wood storage through time. All simulation models to date have indicated that maximum rates or input or storage of wood in channels are attained 150–200 years after stand replacement by harvest or disturbance, depending on the dominant forest species. The first wood model (Rainville et al. 1986) was developed to explore the effects of riparian thinning on temporal patterns of wood recruitment to streams. In thinned stands, the mortality rate of riparian trees reached its maximum after 110–150 years, but the maximum recruitment of wood in terms of numbers of trees per distance along the stream channel was not attained until 150–180 years. Subsequent models for the Pacific Northwest (Beechie et al. 2000; Meleason et al. 2002, in press) and the Intermountain region of western North America (Bragg and Kershner 1997; Bragg 2000) also projected maximum rates of wood recruitment and storage after 150 years. These models also provide a basis for comparing temporal patterns of wood recruitment in natural riparian forests and in harvested riparian forests (Figure 1). These model results demonstrate that land use practices that decrease the age of riparian forests will ultimately lead to reduced numbers and volumes of wood in streams and rivers.

Source area for wood

One of the most important ecological questions about riparian dynamics in the late 20th century was "What distance into an adjacent riparian forest is required to deliver large wood to streams?" Field observations of down wood in riparian areas provided initial estimates (McDade 1987; Murphy and Koski 1989; Minor 1997). A simple model was developed based on empirical data to provide linear relationships of the cumulative wood recruitment as a function of lateral distance from the stream channel for different types of riparian forests (McDade et al. 1992). Because of the smaller height of deciduous trees, wood recruitment occurs within a more narrow zone adjacent to the stream in deciduous forests. For example, 90% of total wood was delivered within 25 m for mature deciduous forests, but 48 m was required in mature coniferous forests to provide an equivalent proportion of total wood loading, and 55 m was required for old-growth coniferous forests. In second-growth coniferous riparian forests in the Oregon Coast Range, 70–84% of the total instream wood was recruited from within 15 m (Minor 1997). Field studies and simulation modeling of western Washington riparian forests found that 90% of the wood loading occurred within 20 m of the stream channel (Welty et al. 2002). Empirical field studies in Alaska found that shorter distances were required to deliver equivalent proportions of wood (Murphy and Koski 1989). In these Alaskan streams, 95% of the wood in the channel was derived from trees within 20 m of the stream, and 99% of the wood loading occurred within 30 m. These studies illustrate the importance of forest composition for wood recruitment and the utility of wood models for exploring wood dynamics in different regions.

Source processes for wood

Researchers around the world have independently identified several major processes that deliver wood to streams and rivers—simple tree fall, snag fall, windthrow, bank undercutting, landslides, sediment debris flows, bank and floodplain avulsion (Harmon et al. 1986; McDade et al. 1992; Malanson and Kupfer 1993; Piégay and Gurnell 1999; Piégay et al. 1999; Benda et al. 2002, in press). Field studies in several regions have evaluated the relative magnitude of different processes that deliver wood to streams and rivers. Models provide a tool for exploring mechanisms for wood delivery and the implications for patterns of wood storage along river networks. All of the wood models described previously include a representation of recruitment of wood laterally from adjacent stands (such as tree fall, snag fall, windthrow, bank undercutting). Benda and Sias (1998, 2003) developed a model that incorporates delivery of wood from mass failures (such as landslides, sediment debris flows, hillslope slumping, and gradual mass failure). Two models simulate avulsion of floodplains and riverbanks and the subsequent delivery of wood into lowland rivers (Malanson and Kupfer 1993; Downs and Simon 2001). Malanson and Kupfer (1993) estimated that lateral bank cutting could increase wood delivery to a Midwest river by 55%. Bank failure contributed 28.3 m^3 of wood/year in the Yolabusha River, Mississippi (Downs and Simon 2001).

Influence of forest type on wood in streams

Tree species exhibit important differences in growth, stature, mortality, and decomposition. Simulation models provide a powerful tool for examining the consequences of riparian forest

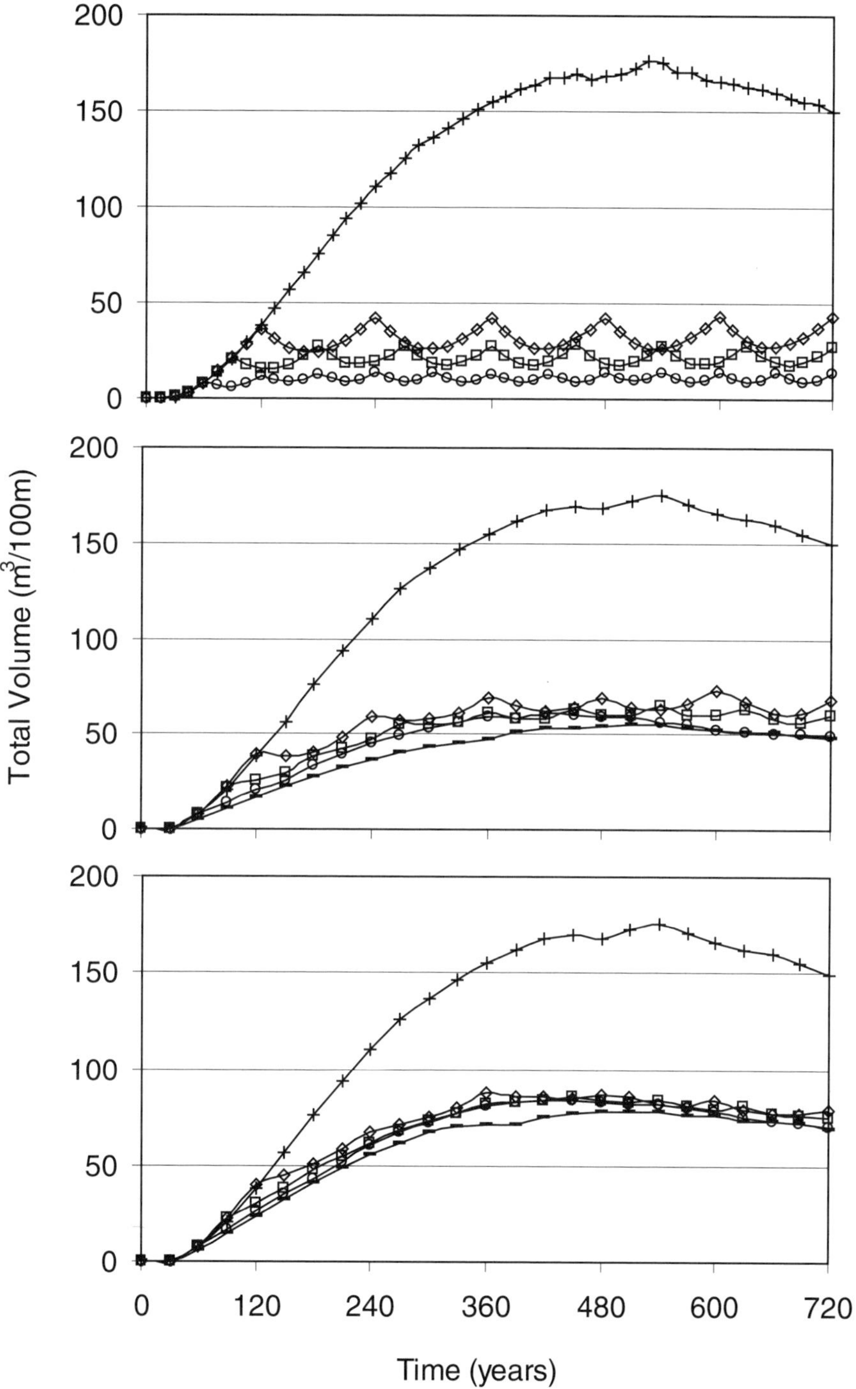

FIGURE 1. Total volume of wood stored in a stream channel through time from riparian management zone of different widths and rotation age as compared to a 75-m riparian area with no harvest over 720 years (from Meleason et al., in press). Riparian management widths are (A) 0 m, (B) 6 m, and (C) 10 m, and plantation forests are clear-cut at 60-year (○), 90-year (◇), and 120-year (□) intervals. (+ equals old-growth without harvest.) Total volume associated with the channel includes total volume of all logs intersecting at least one streambank.

composition for wood loading and storage in streams and rivers. Rainville et al. (1986) estimated that four different forest habitat types in northern Idaho produced different rates of wood loading through time. Timing of maximum wood loading ranged from 110 to 150 years after harvest and differed for the four habitat types. More commonly, simulation models have been used to compare effects of deciduous and coniferous stand composition on wood dynamics. Beechie et al. (2000) projected that the time required for riparian forests to produce enough wood to form pools was twice as long in conifer forests as in alder forests. Time required to produce large pool-forming wood was roughly half the time required to increase the abundance of wood. This essentially illustrates a time lag in wood loading dynamics, and the lag is greater in deciduous forests than in coniferous forests.

Directional fall and breakage during fall

One of the fundamental questions that quickly becomes apparent when modeling the delivery of wood to stream channels is whether riparian trees fall randomly or tend to fall toward the stream. An early model of wood delivery determined that random tree fall would contribute one-third the number of trees that would be delivered to a stream if all trees fell directly toward the stream (Van Sickle and Gregory 1990). This study examined tree fall for a 100-m section of a third-order stream and found that tree fall was not directional. A subsequent empirical study of directional tree fall in Oregon, Washington, Idaho, and Montana found that tree fall tended to be directional toward the stream (Sobota 2003). These data were incorporated into a wood dynamics model (STREAMWOOD; Meleason et al., in press) and illustrated the effect of directionality on cumulative wood loading as a function of distance from the stream (Figure 2). These model simulations estimated the directional fall observed in this field study could increase wood loading to streams from 1.6 to 2.5 times greater than random fall, depending on slope steepness.

The Van Sickle and Gregory (1990) model of wood delivery found that the model accurately predicted the orientations of pieces of large wood in a third-order Cascade Mountain stream (Oregon, USA), but it overestimated the lengths of wood pieces in the stream. The authors hypothesized that the discrepancy between model predictions and field observations of piece length was caused by the lack of a process for tree breakage during fall in the simulation model. This is a deficiency in most wood dynamics models. Sobota (2003) examined fallen trees that touch stream channels in riparian forests in the Pacific Northwest and found that approximately 40% of the trees broke during fall and produced an average of 2.7 pieces. Probability of breakage was modeled for eight tree species (Figure 3). Model results indicated that breakage increased the numbers of pieces of wood recruited from the riparian forests but decreased the number of wood pieces that could span the channel. These results have important implications for models of wood dynamics and geomorphic effects of wood.

Influence of disturbances on wood in streams

One of the most valuable applications of wood dynamics models has been the exploration of the consequences of stochastic landscape disturbances. Field studies of such events and their consequences for amounts and distributions of large wood in streams and rivers are largely descriptive, nonreplicated, and opportunistic (Lamberti et al. 1991; Nakamura and Swanson 1993; Reeves et al. 1996; Swanson et al. 1998; Benda et al. 2003; Nakamura and Swanson 2003; both this volume). Models offer the ability to simulate such stochastic events and compare the characteristics of wood under nondisturbed conditions to patterns of wood in streams that experience floods, fire, insect outbreaks, and other natural disturbances. Beechie et al. (2000) compared the effects of natural fire regimes (200-year recurrence intervals) in western hemlock forests of the Pacific Northwest to forest harvest regimes using a model of wood dynamics and channel geomorphology. The model projected the percent of riparian stands that contributed large wood capable of forming pools for small (4-m), medium (12-m), and large (15-m) streams. Under natural fire regimes, 86% of the riparian areas along small streams, 74% along medium streams, and 64% along large streams were capable of delivering pool-forming wood. Under 80-year harvest rotations, these percentages were reduced to 63%, 25%, and 0%, respectively. Under 60-year and 40-year rotations, riparian forests along medium and large streams did not contribute large pool-forming wood to the channels. In small streams, 50% of

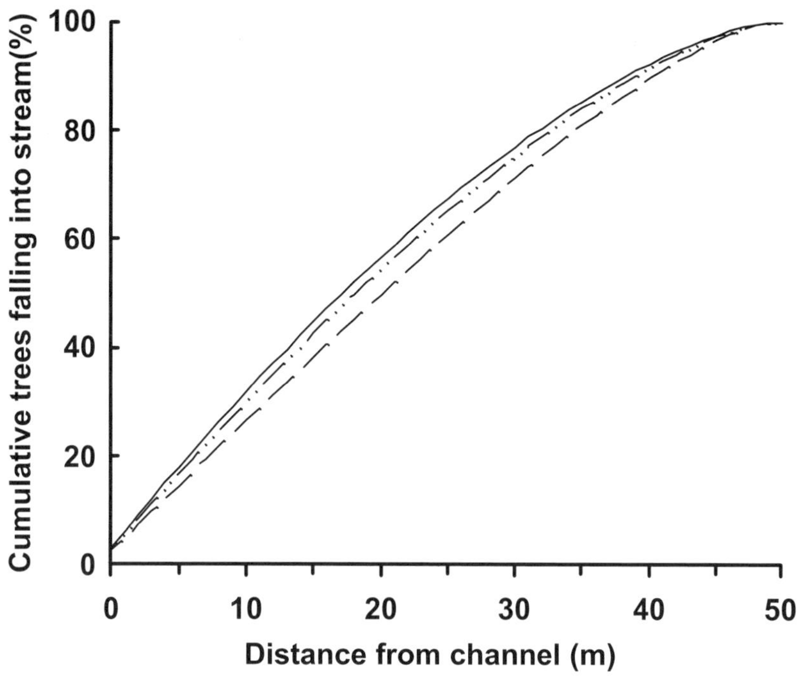

FIGURE 2. Cumulative delivery of falling trees as a function of distance from the channel for trees 50 m in height (Sobota 2003). Fall directions for each scenario were random: 1/360 chance to fall in any direction (——); side slopes 0–10%, direction toward stream, variance ± 80° (— · ·); side slopes > 90%: direction toward stream, variance ± 40° (— —).

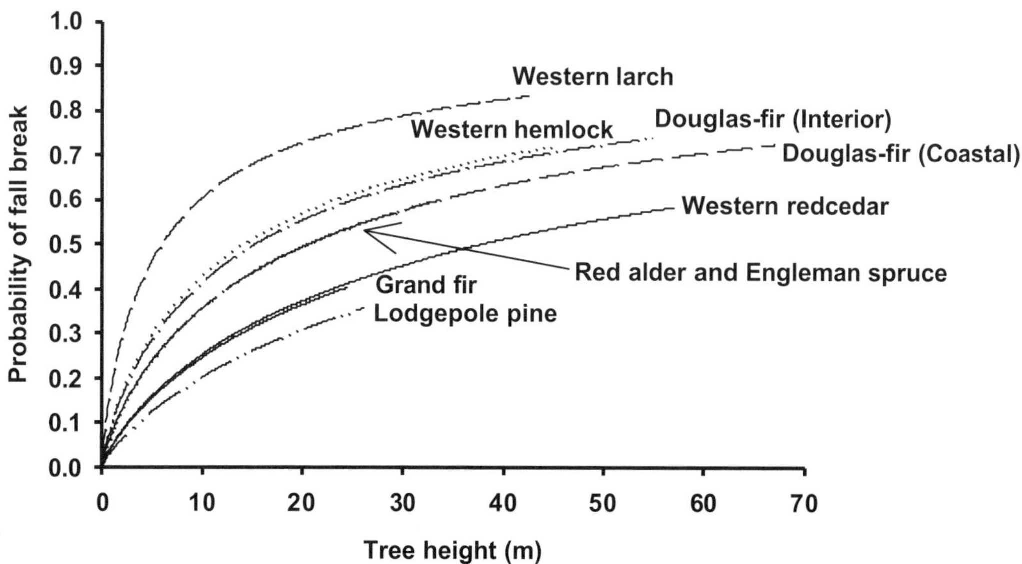

FIGURE 3. Model predictions of the probability that a riparian tree breaks on fall as a function of tree height with linear combinations of differences among species along study sites in the Pacific Northwest, USA (from Sobota 2003). Lines extend to maximum tree heights of each species observed in the study.

the stands under 60-year rotations and 33% of the stands under 40-year rotations contributed pool-forming wood.

Bragg (2000; also see Bragg and Kershner 1997) used a model of wood dynamics in the Intermountain region of western North America to compare wood recruitment in streams in forests with no disturbance to streams in forests that experienced beetle outbreaks, fire, and timber harvest. The cumulative wood loading volumes over a 300-year period for the beetle-outbreak and fire scenarios were 88% and 97% of the undisturbed scenario, respectively. Clear-cut harvest reduced the cumulative wood loading volume to 46% of the wood loading projected for the undisturbed scenario. Beetle outbreaks produced numerous spikes of wood loading and fire created bimodal peaks of wood loading related to the immediate delivery and subsequent delayed tree fall. These spikes in wood loading were greater than the maximum short-term loadings projected for the undisturbed scenario. Maximum wood loadings were observed at 30 years for clear-cutting, 80–150 years for beetle outbreaks and fire, and 250–300 years for undisturbed riparian forests.

A wood dynamics model that incorporated landslides and debris flows explored the consequences of fire cycles in coastal forests of the Pacific Northwest (Benda et al. 2002). Forests with 500-year fire recurrence intervals were projected to create greater wood recruitment, higher maximum recruitment rates, and more variable input rates than forests with 150-year fire recurrence intervals. Fire recurrence altered timing and magnitudes of wood inputs, indicating that such models could be useful for designing forest management in regions with different disturbance regimes. At low rates of mass failure, other wood delivery processes contributed the greatest proportion of wood loading, but mass failure can contribute greater amounts of wood at higher levels of disturbance.

Influence of land use practices

The first model of wood dynamics was constructed to explore the consequences of forest management on wood recruitment to streams (Rainville et al. 1986). Higher harvest rates (10% per decade) reduced tree recruitment to 20 trees/km in contrast to tree recruitment rates of 46 trees/km at lower harvest rates (3% per decade). Stand composition also affected the difference in wood loading between different harvest rates. Murphy and Koski (1989) modeled wood inputs for Alaskan streams and estimated that harvested riparian forests would require 250 years after harvest to produce 85% of the wood input of the preharvest riparian stands. This model also indicated that a 30-m riparian buffer would maintain wood input rates at preharvest levels. Bragg and Kershner (1999; also see Bragg et al. 2000) applied a wood dynamics model to compare clear-cutting, selective harvest, and no harvest for three streams in Wyoming. Both clear-cutting and selective harvest decreased wood loading substantially (77% and 49% less than no harvest, respectively) over a 300-year period in these model simulations.

Models also have been used to explore the use of thinning to create larger trees and influence wood recruitment. Rainville et al. (1986) projected a maximum riparian mortality rate at a riparian stand density of 360 trees/ha. Wood recruitment was reduced by 50% at riparian stand densities of 79 trees/ha. When the stand was thinned twice within a harvest rotation, tree recruitment to stream decreased by more than half. Kennard et al. (1999) found that small streams in thinned stands had less pool area than streams in nonthinned stands, but the opposite was true for large streams. They attributed this difference in channel response to the effect of thinning on creating large pieces of wood that were more effective in pool formation in larger streams than the smaller wood. Beechie et al. (2000) found that thinning did not increase wood loading in alder stands when channels are greater than 20 m wide, but thinning increased wood loading in coniferous riparian forests. Welty et al. (2002) found that wood loading was increased in riparian stands at tree densities greater than 200 trees/ha.

Stream clearing also impacts amounts of wood in streams and rivers. Bragg et al. (2000) used a simulation model to examine the consequences of stream clearing practices. Removal of wood created lags of several decades in initial increases in wood loading after stand removal. These streams required 80–110 years longer to attain maximum wood recruitment rates, indicating that stream alteration can create substantial lags in wood dynamics.

Sensitivity analysis

Sensitivity analysis is a valuable approach for understanding of the performance of models and the sensitivity of the model to specific relationships or interactions between functions within the model. Several researchers have explored the sensitivity of their models of wood dynamics, revealing commonalities that may be important for future research. Kennard et al. (1999) found that the Riparian-in-a-Box model was most sensitive to the quantitative representations of tree growth, mortality, and instream depletion. When depletion (that is, combined decay, breakage, and transport) was decreased from 15% to 5%, estimates of functional wood numbers increased by 100%. They also observed that relative outputs for different simulation scenarios were not as sensitive as absolute outputs. Benda and Sias (1998, 2003) found that their model was most sensitive to estimates of decay rates and bank erosion. Welty et al. (2002) observed that their model performance was most sensitive to depletions rates, stand composition, and tree densities. Meleason et al. (in press) conducted a simultaneous sensitivity analysis of multiple factors and found that the model was most sensitive to estimates of decay rates. All four models of wood dynamics that have evaluated the sensitivity of their models have concluded that the performance of the models is strongly influenced by the estimates of decay rates of wood or the estimates of overall depletion rates. If models continue to exhibit this sensitivity to estimates of instream decay or depletion, future research may provide more extensive and informative estimates of these processes and improve the performance of simulation models of wood dynamics in streams and rivers.

Development of Future Models of Wood Dynamics in Streams and Rivers

The last two decades have witnessed the development of our ability to represent complex dynamics of wood in streams and rivers over long temporal and spatial extents. Early pioneering attempts to develop models of wood processes (Rainville et al. 1986; Murphy and Koski 1989; Van Sickle and Gregory 1990; McDade et al. 1992; Malanson and Kupfer 1993) provided a basis for the development of more complex models that more fully represent the array of processes that determine the abundance and distribution of wood throughout stream and river networks. These models have allowed researchers to explore the implications of local short-term studies at specific study sites over long time frames (100–500 years) and across broad geographic extents (10–30,000 km^2). These models have provided critical information for land managers about the amounts of wood that would be expected in streams in late successional forests, the consequences of different forest composition, and the influences of geomorphic processes in streams and their basins. From the first model of wood dynamics to the present, these models have been used to evaluate the potential effects of timber harvest, thinning practices, riparian area management, stream clearing, channelization, and off-site practices, such as road development, bridge construction, and other land-use practices. The utility of models of wood dynamics in streams and rivers is clear, and several areas of research will contribute to more robust and easily applied models in the future.

As discussed earlier, models can be evaluated on their performance (realism, precision, generality) in relation to their intended uses (Levins 1966). Models of wood in streams and rivers have been used primarily to understand the processes that shape the abundance and distribution of wood and to predict the abundance and distribution of wood that would result from different management practices.

Models developed to understand fundamental ecological processes in streams and rivers will require generality and realism, with less demand for precision (Haefner 1996). Such models may require more explicit representation of basic geomorphic (channel erosion and deposition, mass failure, floodplain formation, and loss), hydrologic (flooding, drought, snowmelt, rain-on-snow events), or biological processes (tree growth and competition, tree mortality, snags, breakage, decomposition, mechanical abrasion and breakage, transport, burial). Representation of these processes in simulation models allows researchers to explore the implications of different processes, rates, and interactions that are difficult to study through field observation and manipulative experiments. Model validation or extensive comparison with field observations on the array of processes and rates may be essentially impossible; thus, detailed knowledge about the precision and accuracy of such models will be limited.

In contrast, models predicting amounts of wood in specific stream reaches or responses to a specific set of management practices require precision and reality, but do not require generality if the model has been developed from local observations and quantitative information. Such models can be calibrated to local conditions and processes by adjusting model parameters based on field observations of a limited number of processes or rates. Such calibration can improve the accuracy of the simulated amounts and distributions of wood in that local region or management application, but broader application to other systems may be limited by extensive local calibration. If the reality of the models also is simplified by combining several fundamental processes into a single overall process, prediction may be maintained or improved, but understanding of causal mechanisms will be reduced. Examples of such simplification that are common in models of wood dynamics are (1) use of overall depletion rates instead of decay, breakage, and transport; or (2) use of proportional stand delivery instead of tree regeneration, growth, competition, mortality, snag formation, and tree fall.

Development and application of future models will be guided by both the precision and accuracy of our existing data on wood dynamics and our mechanistic understanding of the processes that influence wood dynamics in streams and rivers. A conceptual diagram of the current state of knowledge in these two aspects of wood dynamics illustrates the potential implications for model development (Figure 4). Some processes are relatively well understood, and empirical data are available, such as tree mortality (though riparian studies are less common). We understand the causal mechanisms for some processes (such as wood decomposition, mass failure, snag formation) based on studies of other systems, such as upland forests. Causal mechanisms for other processes are poorly understood, such as tree fall directionality, but some empirical data are available to allow model development. Similarly, mechanisms for some processes, such as wood movement and transport, are poorly understood (Braudrick and Grant 2000), but general mathematical relationships, such as negative exponential models, provide relatively accurate representations of the overall process. And finally, some processes are poorly understood, and empirical data are almost

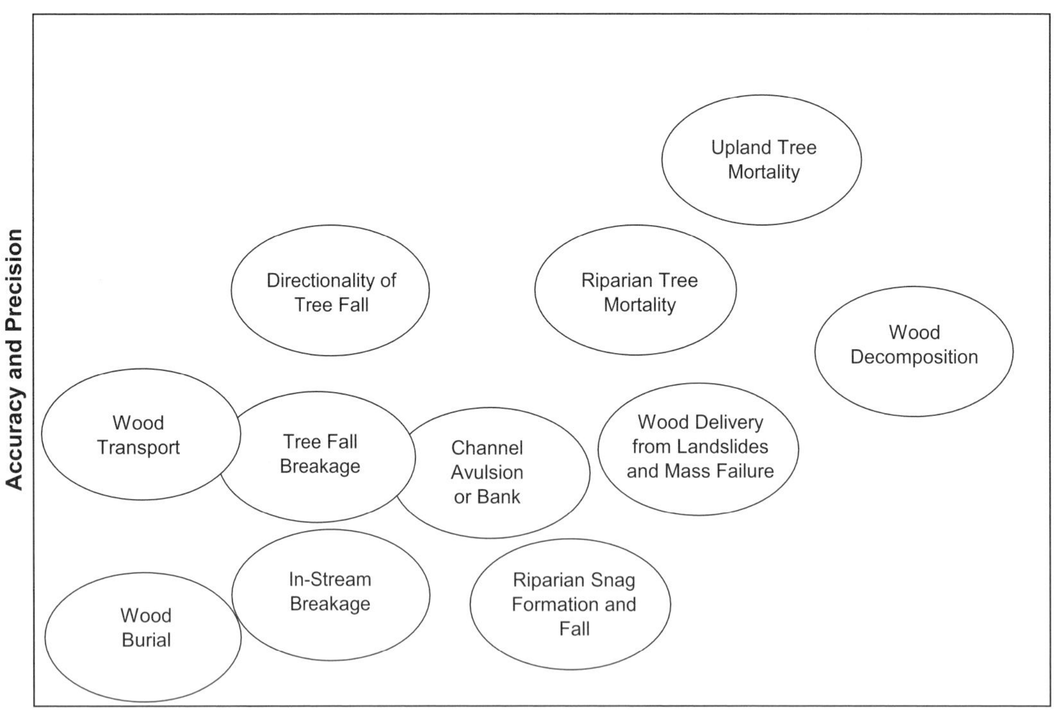

FIGURE 4. Conceptual diagram of the relative accuracy and precision of existing data on wood processes as compared to the current conceptual understanding of the mechanisms responsible for the processes. Positions of specific processes are relative and simply illustrate the potential differences in information and understanding that create challenges for modeling of wood dynamics in streams and rivers.

nonexistent, such as wood burial. Development of future models of wood dynamics can be used to guide field research to develop our understanding of the processes that shape the patterns and amounts of wood in stream and rivers.

Desirable characteristics of future models of wood dynamics

The last two decades have witnessed the application of simulation modeling for synthesizing existing information on the dynamics of wood and exploring the consequences of land and river management on wood in world rivers. Future models can build on the lessons learned from these existing models and expand the power of simulation modeling for research and management. Future models would benefit from the following attributes:

- stochastic functions that predict trends and variance in those trends
- robust stand dynamics models calibrated for riparian forests
- multiple mechanisms for delivering wood into streams and rivers
- directional fall and breakage during fall
- separate functions for decay, breakage, and transport of wood in streams
- representation of multiple reaches throughout dendritic river networks
- user-defined parameters for functions, probabilities, and probability distributions
- user-defined land-use practices (such as harvest rate, harvest rotation interval, thinning)
- sensitivity analysis of model structure and performance
- direct comparisons between model outputs and empirical observations

Additional features that would make simulation models of wood dynamics easier to use and more widely available for researchers and managers include (1) intuitive user interfaces, (2) flexible graphical output that can be defined by the user, and (3) online libraries of data for riparian forests and wood processes in streams.

Models are abstract representations of our knowledge about a given phenomenon or process and are inherently limited by the information from which they are developed. Likewise, empirical or field observations are limited by the history of the location and the portion of the landscape or river network they represent. Researchers and managers tend to champion one approach over another, sometimes rejecting models as being too abstract or field studies as being too geographically limited. The power of both simulation modeling and field investigations is greatest at the interfaces between these methods of inquiry. Research on the ecology of wood in world rivers has been strengthened by the application of simulation models for synthesizing information on the complex array of processes that result in the regeneration, growth, and mortality of riparian forests; major disturbance processes; and processes that influence wood in streams, such as decay, breakage, transport. These models can project the outcomes of these processes over long time periods and across complex networks of streams and rivers. Patterns of wood dynamics predicted by such models and the sensitivity of the models to specific processes can guide future field studies or experimental manipulations to increase our understanding of critical determinants of the dynamics of wood. Integration of sound empirical information and robust models of wood dynamics provides essential tools for resource managers to make decisions about the management of wood in the diverse streams, rivers, and riparian forests of different geographic regions of the world.

References

Beechie, T. J., G. Pess, P. Kennard, R. E. Bilby, and S. Bolton. 2000. Modeling recovery rates and pathways for woody debris recruitment in Northwestern Washington streams. North American Journal of Fisheries Management 20:436–452.

Benda, L., D. Miller, J. Sias, D. Martin, R. Bilby, C. Veldhuisen, and T. Dunne. 2003. Wood recruitment processes and wood budgeting. Pages 49–73 *in* S. V. Gregory, K. L. Boyer, and A. M. Gurnell, editors. The ecology and management of wood in world rivers. American Fisheries Society, Symposium 37, Bethesda, Maryland.

Benda, L., P. Bigalow, and T. Worsley. 2002. Processes and rates of in-stream wood recruitment in old growth and second-growth redwood forests, northern California. Canadian Journal of Forest Research 32:1460–1477.

Benda, L. E., and J. C. Sias. 1998. Landscape controls on wood abundance in streams. Earth Systems Institute, Seattle, Washington.

Benda, L. E., and J. C. Sias. 2003. A quantitative framework for evaluating the wood budget. Forest Ecology and Management 172:1–16.

Benda, L., C. Veldhuisen, and J. Black. In press. Debris flows as agents of morphological heterogeneity at low-order confluences, Olympic Mountains, Washington. Geological Society of America Bulletin.

Botkin, D. B. 1993. Forest dynamics: an ecological model. Oxford University Press, New York.

Bragg, D. C. 2000. Simulating catastrophic and individualistic large woody debris recruitment for a small riparian system. Ecology 8:1383–1394.

Bragg, D. C., and J. L. Kershner. 1997. Evaluating the long-term consequences of forest management and stream cleaning on coarse woody debris in small riparian systems of the central Rocky Mountains. Fish Habitat Relationships Technical Bulletin 21:1–9.

Bragg, D. C., and J. L. Kershner. 1999. Coarse woody debris in riparian zones. Journal of Forestry April:30–35.

Bragg, D. C., J. L. Kershner, and D. W. Roberts. 2000. Modeling large woody debris recruitment for small streams of the central Rocky Mountains. U.S. Department of Agriculture, Rocky Mountain Research Station, General Technical Report RMRS-GTR-55, Fort Collins, Colorado.

Braudrick, C. A., and G. E. Grant. 2000. When do logs move in rivers? Water Resources Research 36:571–583.

Downs, P. W., and A. Simon. 2001. Fluvial geomorphological analysis of the recruitment of large woody debris in the Yalobusha River network, central Mississippi, USA. Geomorphology 37:65–91.

Fleece, W. C. 2002. Modeling the delivery of large wood to streams with light detection and ranging (LIDAR) data. Pages 71–83 *in* W. F. Laudenslayer, Jr., P. J. Shea, B. E. Valentine, C. P. Weatherspoon, and T. E. Lisle, editors. Proceedings of the Symposium on the Ecology and Management of Dead Wood in Western Forests. USDA Forest Service General Technical Report PSW-GTR-181, Pacific Southwest Research Station.

Harmon, M. E., J. F. Franklin, F. J. Swanson, and others. 1986. Ecology of coarse woody debris in temperate ecosystems. Advances in Ecological Research 15:133–302.

Haefner, J. W. 1996. Modeling biological systems: principles and applications. Chapman and Hall, New York.

Hester, A. S., D. W. Hann, and D. R. Larson. 1989. Organon: southwest Oregon growth and yield model user manual, version 2.0. Forest Research Lab, College of Forestry, Oregon State University, Corvallis.

Hupp, C. R., and A. Simon. 1991. Bank accretion and the development of vegetated depositional surfaces along modified alluvial channels. Geomorphology 4:111–124.

Karplus, W. J. 1977. The place of systems ecology models in the spectrum of mathematical models. Pages 225–228 *in* G. S. Innis, editor. New directions in the analysis of ecological systems. Part 2. Simulation Councils Proceedings Series. Volume 5, Number 2. The Society for Computer Simulation (Simulation Councils, Inc.), La Jolla, California.

Kennard P., G. R. Pess, T. J. Beechie, B. Bilby, and D. Berg. 1998. Riparian-in-a-Box: a manager's tool to predict the impacts of riparian management on fish habitat. Pages 483–490 *in* M. K. Brewin and D. Monita, editors. Proceedings of the Forest-Fish Conference: Land Management Practices Affecting Aquatic Ecosystems. Informational Report NOR-X-356. Natural Resources Canada, Canadian Forest Service Northern Forestry Centre, Edmonton, Alberta, Canada.

Lamberti, G. A., S. V. Gregory, L. R. Ashkenas, R. C. Wildman, and K. M. S. Moore. 1991. Stream ecosystem recovery following a catastrophic debris flow. Canadian Journal of Fisheries and Aquatic Sciences 48:196–208.

Levins, R. 1966. The strategy of model building in population biology. American Scientist 54:421–431.

Malanson, G. P., and J. A. Kupfer. 1993. Simulated fate of leaf litter and woody debris at a riparian cutbank. Canadian Journal of Forest Research 23:582–590.

McDade, M. H. 1987. The source area for coarse woody debris in small streams in western Oregon and Washington. M.S. thesis. Oregon State University, Corvallis.

McDade, M. H., F. J. Swanson, W. A. McKee, J. F. Franklin, and J. Van Sickle. 1990. Source distances for coarse woody debris entering small streams in western Oregon and Washington. Canadian Journal of Forest Research 20(3):326–330.

Meleason, M. A., S. V. Gregory, and J. Bolte. 2002. Simulation of stream wood source distance for small streams in the western Cascades, Oregon. Pages 457–466 *in* W. F. Laudenslayer, Jr., P. J. Shea, B. E. Valentine, C. P. Weatherspoon, and T. E. Lisle, editors. Proceedings of the Symposium on the Ecology and Management of Dead Wood in Western Forests. USDA Forest Service, Pacific Southwest Research Station, General Technical Report PSW-GTR-181, Albany, California.

Meleason, M. A., S. V. Gregory, and J. Bolte. In press. Implications of riparian management strategies on wood in streams of the Pacific Northwest. Ecological Applications.

Minor, K. P. 1997. Estimating large woody debris recruitment from adjacent riparian areas. Master project. Oregon State University, Corvallis.

Murphy, M. L., and K. V. Koski. 1989. Input and depletion of woody debris in Alaska streams and implementation for streamside management. North American Journal of Fisheries Management 9:427–436.

Nakamura, F., and F. J. Swanson. 1993. Effects of coarse woody debris on morphology and sediment storage of a mountain stream in western Oregon. Earth Surface Processes and Landforms 18:43–61.

Nakamura, F., and F. J. Swanson. 2003. Dynamics of wood in rivers in the context of ecological disturbance. Pages 279–297 *in* S. V. Gregory, K. L. Boyer, and A. M. Gurnell, editors. The ecology and management of wood in world rivers. American Fisheries Society, Symposium 37, Bethesda, Maryland.

Piégay, H., and A. M. Gurnell. 1997. Large woody debris and river geomorphological pattern: examples from S. E. France and S. England. Geomorphology 19(1–2):99–116.

Piégay, H., A. Thevenet, and A. Citterio. 1999. Input, storage and distribution of large woody debris along a mountain river continuum, the Drome River, France. Catena 35(1):19–39.

Rainville, R. C., S. C. Rainville, and E. L. Linder. 1986. Riparian silvicultural strategies for fish habitat emphasis. Pages 186–196 *in* Forester's future: leaders or followers. Society of American Foresters National Conference Proceedings. SAF Publication 85–13, Society of American Foresters, Bethesda, Maryland.

Reeves, G., L. Benda, K. Burnett, P. Bisson, and J. Sedell. 1996 A disturbance-based ecosystem approach to maintaining and restoring freshwater habitats of evolutionarily significant units of anadromous salmonids in the Pacific Northwest. Pages 334–349 *in* J. L. Nielsen and D. A. Powers, editors. Evolution and the aquatic ecosystem: defining unique units in population conservation. American Fisheries Society, Symposium 17, Bethesda, Maryland.

Shugart, H. H., and D. C. West. 1977. Development of an Appalachian deciduous forest succession model and its application to assessment of the impact of the chestnut blight. Journal of Environmental Management 5:161–179.

Simon, A. 1989. A model of channel response in disturbed alluvial channels. Earth Surface Processes and Landforms 14(1):11–26.

Sobota, D. J. 2003. Fall directions and breakage of riparian trees along streams in the Pacific Northwest. M.S. thesis. Oregon State University, Corvallis.

Swanson, F. J., S. L. Johnson, S. V. Gregory, and S. A. Acker. 1998. Flood disturbance in a forested mountain landscape. BioScience 48:681–689.

Van Sickle, J., and S. V. Gregory. 1990. Modeling inputs of large woody debris to streams from falling trees. Canadian Journal of Forest Research 20:1593–1601.

Welty, J. W., T. Beechie, K. Sullivan, D. M. Hyink, R. E. Bilby, C. Andrus, and G. Pess. 2002. Riparian Aquatic Interaction Simulator (RAIS): a model of riparian forest dynamics for the generation of large woody debris and shade. Forest Ecology and Management 162:299–318.

Wykoff, W. R., N. L. Crookston, and A. R. Stage. 1982. User' s guide to the Stand Prognosis Model {FVS}. USDA Forest Service, Intermountain Research Station, General Technical Report INT-133, Ogden, Utah.

American Fisheries Society Symposium 37:337–353, 2003

Wood in Streams and Rivers in Developed Landscapes

Arturo Elosegi

*Department of Plant Biology and Ecology, Science Faculty
University of the Basque Country, P.O. Box 644, 48080 Bilbao, Spain*

Lucinda B. Johnson

*Natural Resources Research Institute, University of Minnesota Duluth
5013 Miller Trunk Highway, Duluth, Minnesota 55811, USA*

Abstract.—Many catchments across the world have been highly modified by human activities, including agriculture, urban development, and other land uses, that often result in a complex landscape mosaic. We define developed catchments as those dominated by activities such as agriculture or urban development, irrespective of the extent and type of riparian zone present. Far fewer papers address large wood in rivers and streams within developed catchments compared with those in more natural situations, despite the fact that residential development and agricultural activities are so pervasive worldwide. The literature highlights a clear reduction of the abundance of large wood in agricultural and urban streams and rivers, although standing stocks are highly variable depending on local conditions. As a result of its scarcity, large wood seems to play a less important physical role than in forested ones. Nevertheless, large wood still plays an important role in developed streams and rivers by providing critical habitats for invertebrates and serving as the only retention structure remaining in some channels. A lower diversity of invertebrates and/or fishes and the loss of important functions, such as retention capacity, are reported for developed rivers, compared to those in forested regions. The geomorphic role of the wood remaining in these developed systems appears to be mixed—some studies report no such role, while others report an association of wood with pools. Gaps are evident in two areas: (1) many papers fail to adequately describe the landscape in which study streams and rivers are embedded, making it impossible to discern the dominant land use or, in some cases, even the nature and extent of the riparian vegetation; and (2) studies of ecosystem properties of streams and rivers in developed landscapes are rare. We suggest that more research should be undertaken in developed systems and that addressing the role of large wood is an important component of such studies.

Introduction

Most areas of the world have been directly or indirectly modified by human activities, and rivers are affected accordingly (Hynes 1975). Subtle changes in the basin are likely to produce small shifts in the structure and function of stream and river systems, but activities directly affecting the energetic base of the ecosystem (such as sewage discharge, riparian clearing, or harvest) or its physical template (damming, channelization, water diversion) are expected to trigger major changes. These changes tend to be greatest in regions where intensive soil use (for instance, agriculture, quarries, mining) and urban development dominate or coexist in complex landscape mosaics.

Although streams and rivers across the globe differ greatly in geomorphology, hydrology, and biology, effects of agricultural development and urbanization can be generalized to a certain extent. Agricultural activities have been implicated in elevated bacterial concentration (Hunter et al. 1999), high turbidity (Cooper 1993), eutrophication from increased nutrient input (Harper 1992), and increased pesticide loading (Sweeting 1994). In many regions, agricultural practices also are associated with incised, straightened channels (Shields et al. 1998), leading to reduced wetland extent and flooding, loss of large wood from the channel and floodplain, and other major changes in stream and river ecosystem properties. The specific effects of agricultural activities, however, depend on the type and intensity of the land use (grazing pressure, fertilizer, pesticide and herbicide application rates, crop types, conservation practices; Barton 1996). Urbanization is generally associated with increased nutrient, sediment, and contaminant loading (Sweeting 1994; Paul and Meyer 2001), drastic changes in channel morphology and bank characteristics (Brookes 1994; Booth et al. 1996; Paul and Meyer 2001), and changes in hydrology (Booth 1990; Gore 1994; May et al. 1997; Paul and Meyer 2001). Effects of water extraction and wetland drainage associated with both agricultural and urban land use are observed at longer temporal and broader spatial scales as the amount of surface area drained increases (Spaling and Smith 1995; Arnold and Gibbons 1996).

Individual management practices vary by culture, economic conditions, and climate and generate different combinations of pollutants and physical changes to river channels (Watzin and McIntosh 1999). The effects of agricultural and urban activities change with spatial scale and their arrangement on the landscape. At local scales, the extent and type of riparian vegetation can influence water quality, temperature regimes, and quantity and quality of allochthonous inputs, including large wood (Peterjohn and Correll 1984; Osborne and Kovacic 1993; Johnson 1999). In some Midwestern U.S. agricultural streams, total suspended solid concentrations were best predicted by land use within 100 m of the channels, but the best predictor of nitrate concentrations was the proportion of rowcrop agriculture in a catchment (Johnson et al. 1997). Similarly, instream habitat structure was most influenced by local conditions, but hydrology and channel characteristics were most affected by regional land use patterns (Allan et al. 1997). The surficial deposits resulting from glacial action within this region strongly modify the hydrologic regime, which exerts control over many of the processes influencing nutrient dynamics, large-wood standing stocks, and biotic communities (Richards et al. 1996, 1997; Wiley et al. 1997; Johnson 1999).

Here, we will review published information on the status and role of large wood in streams and rivers running through developed landscapes. We define *developed* landscapes as areas modified by human activities to the extent that agricultural and urban uses become a major feature in a catchment, irrespective of the extent and type of riparian zone present. This definition will undoubtedly overlap that of other chapters in this book because many researchers consider a reach to be forested if riparian forests grow on both banks; the effects of agricultural and urban uses, however, often reach far beyond the local scale (Richards et al. 1996, 1997; May et al. 1997; Paul and Meyer 2001), and thus, significant effects are detected even in the presence of intact riparian corridors (Osborne and Kovacic 1993).

Because of the stark differences in land use practices across the world associated with different climatic, topographic, and economic conditions, drawing specific conclusions on the effect of catchment development on stream and river ecology is difficult. Nevertheless, fairly consistent effects of such development can be detected on large wood in streams and rivers because many human activities decrease organic matter inputs and increase log removal. Indeed, many management schemes in stream and river channels remove wood, although wood removal itself is not often considered the major impact of such management. The objectives of our review are to

- Identify literature dealing with large wood in streams and rivers draining developed catchments;
- Assess the effects of catchment development on abundance, size, distribution, and movement of large wood in stream and river channels;
- Synthesize the consequences of development on the role of large wood in physical and biological processes;
- Summarize the main effects of different types of land or channel management; and
- Detect major gaps in the knowledge of large wood in developed streams and rivers.

Literature Reviewed

An initial inspection of the literature cited in reviews of wood in streams and rivers (such as Harmon et al. 1986; Maser and Sedell 1994) shows that, despite the worldwide importance of agriculture and urbanization, most research has focused on forested systems. During our review of the literature, we found that it was often difficult to discern whether a study region fell under our definition of *developed*. Some authors described the extent of the riparian vegetation without mentioning the landscape beyond this corridor. Thus, rivers described as "seminatural," "quasi-natural," or "relatively undisturbed" may have natural riparian vegetation communities, but highly developed land beyond this zone. These descriptors also neglect both current and past management practices (such as deforestation, agricultural production, even urbanization) that could have a significant and persistent effect or land use "legacy" on current ecosystem structure and function. This legacy could lead to a skewed and misleading perception, in which streams and rivers that have been severely modified over centuries are today considered to be "natural" because they are the best preserved in a region. The problem of incomplete characterization of the study region is not unique to the large wood literature. Correll (2000) reviewed the complete literature (700 publications) on riparian zones and found that 32% of papers addressing the flux and cycling of organic matter failed to adequately describe the structure and vegetative composition of the riparian zone. Omissions such as these make meta-analyses and valid comparisons across studies difficult.

With these shortcomings in mind, we searched the literature for research on wood in developed streams and rivers. About 50 papers fit our criteria. Roughly one-half of the papers gave some insight into the abundance or distribution of wood, 7 described the size of logs, and 20 reported on movements of large wood. A dozen papers dealt with experiments on adding or removing large wood. About 10 papers reported results from multiple rivers or reaches. Several papers addressed the role of wood on different components of river ecosystems: 7 on hydraulics (patterns of current, sheer stress), 15 on geomorphology (channel width, pool surface), a dozen on wood as invertebrate habitat, and 8 on fishes. Finally, half a dozen papers dealt mainly with channel engineering (for example, stability of levees).

Landscape Development and Abundance of Large Wood

The mass of wood in streams and rivers results from the balance between inputs (riparian trees falling into the channel, plus wood coming from upstream) and outputs (scouring and breakdown). The number of trees entering the channel is greatest in unstable channels (Nakamura and Swanson 2003, this volume), the size (and stability) of individual logs is largest under mature forests (Diez et al. 2001), and the mobility of wood pieces is greatest in large rivers (Gurnell 2003, this volume). A three-dimensional model could explain most of the natural variability in wood mass between reaches (Figure 1). Human activities often result in reduced wood standing stocks due to younger riparian forests, artificially stabilized banks, and greater log mobility. Furthermore, Diez et al. (2002) showed that increased nutrient concentrations in water promote faster breakdown of wood, a fact that could affect many developed streams and rivers. The relationship between basin development and wood depletion can be obscured by several factors, including the stability of large logs in small stream channels, differing management and stewardship practices among land owners, and time elapsed since riparian zones were harvested.

Landscape development usually includes clearing vegetation for cropland or urban uses, which reduces large-wood inputs into streams and rivers. Furthermore, channel modifications associated with agricultural and urban activities (like straightening or dredging) can further reduce large-wood standing stocks through direct removal or increased mobility (Booth et al. 1996). Regardless of regional land use practices, management decisions governing the width and vegetative composition of the riparian zones are largely influenced by an individual landowner's stewardship philosophy and management guidelines such as "best management practices." Therefore, the standing stock of large wood varies widely at a reach scale, ranging from a small reduction to total depletion.

Many regions of the world provide examples of the consequences of land conversion on wood abundance in streams. Large wood was absent from an English heathland stream in the basin of the Lymington River (the Withybed), whereas forested streams in the same basin had up to 400 kg per 100 m of channel length (Table 1; Gurnell and Gregory 1995). Less marked decreases were observed in other streams in the Lymington Basin, where the mass of wood was reduced in reaches adjacent to lawns, grassland, and heathlands, compared to reaches adjacent to deciduous or coniferous forests (Gregory et al. 1993). Agriculture has also been reported to result in streams devoid of large wood in Germany (Reich 1999; Gerhard and Reich 2000). In the predominantly agricultural Minnesota River catchment, Stauffer et al. (2000) reported that large wood abundance depended significantly on the presence of riparian forests. Even in large rivers, agriculture can de-

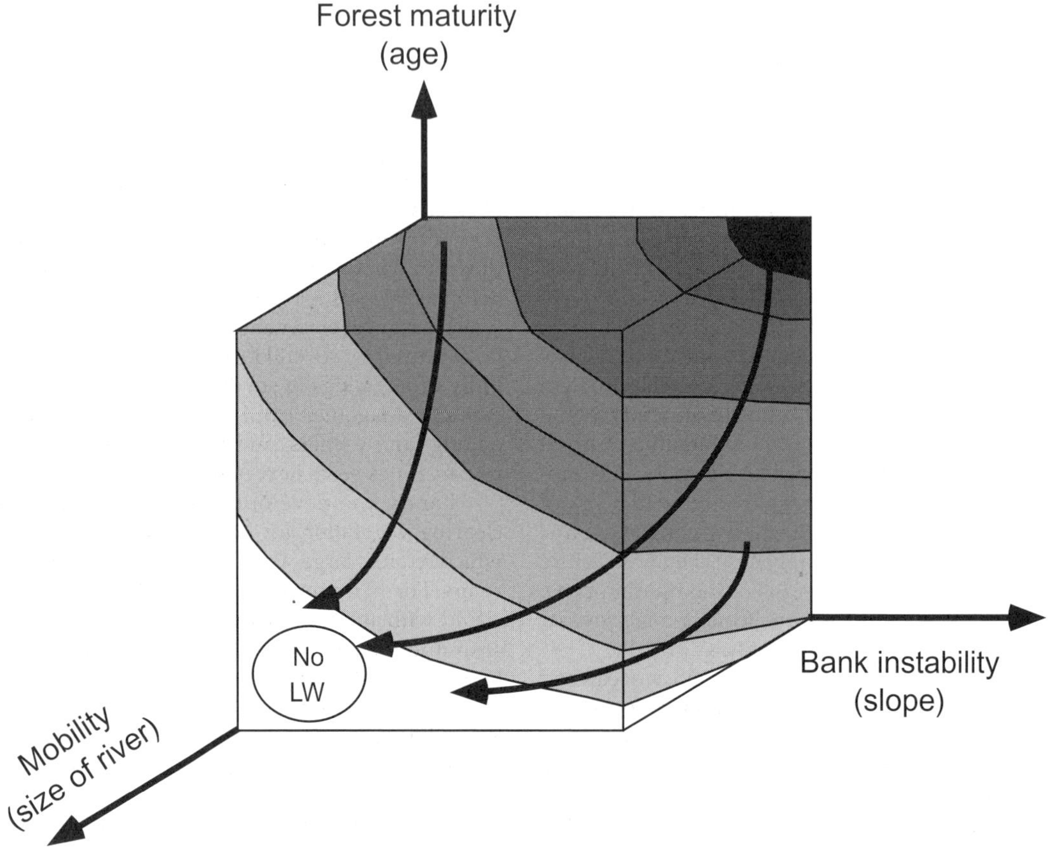

FIGURE 1. A conceptual model of the factors governing the amount of large wood in streams and rivers, and the effects of landscape development (arrows), that in the most extreme cases can lead to total wood depletion.

crease the standing stock of large wood, as shown by Piégay (1993), who found fewer wood jams adjacent to abandoned agricultural fields in the Ain, a sixth-order French meandering river with extensive riparian forests.

Low abundance of large wood in streams and rivers can reflect both past and present management practices. Wood was scarce, even in "quasi-natural" German streams (that is, reaches in which riparian forest is currently unmanaged and large wood has not been removed in the last 10 years), and completely lacking in some agricultural areas (Hering et al. 2000). Nonetheless, some reaches with meadows and forests in their floodplain had more than 5 m^3 of large wood/100 m—amounts that were very large compared to most German streams. Thus, although agriculture negatively affects the amount of wood in streams, its effects depend greatly on local conditions, such as the structure of riparian vegetation, bank stability, and channel morphology.

In 48 first- to fourth-order reaches in the Toikanbetsu and Pankenai stream systems (Japan), the number of wood pieces depended on stream type (coarse-substrate step-pool versus coarse-substrate pool-riffle, or fine-substrate pool-riffle; Table 1; Inoue and Nakano 1998). Regardless of stream type, large-wood abundance ranged from 0 to 8 m^3/100 m and was lower in grassland than in forested areas. In the Agüera stream basin (Spain), Diez (2000) compared forested and agricultural reaches, none of which were completely devoid of riparian forest. In first-order reaches running through deciduous forests, wood (diameter > 1 cm) averaged 0.014 m^3/m^2. Reaches through forest plantations had only 20% of the "reference" values, while an agricultural reach had only 1.2% (0.0002 m^3/m^2). Differences decreased downstream, and disappeared in third-order reaches. When all data were lumped, wood volume was significantly related to forest maturity and channel width (Diez et al. 2001).

In agricultural regions of the Midwestern United States, large-wood abundance was found to depend on landform type (lacustrine versus morainal) and land use (agricultural versus mixed; Johnson 1999). Mean abundances across the entire study region were extremely low, but were highest in catchments with morainal landforms and mixed land use or lacustrine landforms and agricultural land use (Table 1). The fewest large-wood accumulations were found in agricultural reaches on morainal landforms (two dams per 100 m), and reaches in mixed land use catchments on morainal landforms had the largest number (10/100 m). Although large-wood abundance was very low across this study region, the number of accumulations did not differ greatly from other studies (Table 1), suggesting that accumulations were small compared to other regions. Strong interaction effects on both large-wood abundance and number of accumulations were found be-

TABLE 1. Summary of wood and wood accumulation abundance in developed streams and rivers; units are specified in the column headings or indicated in the body of the table. Papers selected included those that described or quantified the extent of development in the catchment (see text above for a description of the criteria used).

Reference	Location; stream type	Wood density (m^3/m^2)	# of pieces wood per 100 m (range)	Wood size (range)
Zimmer and Bachman 1976[a]	Iowa, USA		0.06–3.4	
Gregory et al. 1985	Highland Water, UK; heathland.			3.7
Hauer 1989	Meyer's Creek, SC, USA; blackwater stream	0.011		
Hauer 1989	Steel Creek, SC, USA; mixed land use	0.0002		
O'Connor 1992	Pranjip-Creightons, Victoria, AUS;			≥1 cm; mean diameter 0.04 cm overall;
	grazing-affected streams	"almost no wood";		(0–40 cm)
	riparian forest intact	0.026		(0 ≥ 80 cm)
Shields and Smith 1992	S. Fk. Obion River, TN, USA;			Individual logs not measured; wood density and debris dam area estimated visually
	uncleared reach, agricultural catchment	0.043–0.094	3.5–5.8	
	cleared reach, agricultural catchment	0.009–0.033	0.6–5.8	
Diehl and Bryan 1993	12 reaches, TN, USA; agricultural	27 ± 26 m^3/km		
Evans et al. 1993	New Zealand			
	old-growth native	1.006 ± 0.107		
	120-year-old native	0.712 ± 0.364		

TABLE 1. Continued.

Reference	Location stream type	Wood density (m^3/m^2)	# of pieces wood per 100 m (range)	Wood size (range)
	10-year-old native	0.027 ± 0.008		
	10-year-old pine	0.020 ± 0.013		
Gurnell and Gregory 1995	Withybed, UK; heathland stream;	no LWD		
	nearby forested streams	up to 400 kg/ 100 m		
Gregory et al. 1993	Lymington River basin, UK; mixed land use	0.00006–0.005	0.35–2.4	>5 cm diam
Gregory and Lienkaemper[b]	TN, USA; hardwood forest	0.0126		
Richmond and Fausch 1995	Rocky Mountains, CO, USA; disturbed by forest harvest	0.0012–0.0147		
Inoue and Nakano 1998	Teshio River, Japan; some artificial grasslands in riparian zone	0–44 pieces/ 100 m 0–8 m^3/100 m		24 cm mean diameter, 3 m long; 80% of logs were <30 cm
Johnson 1999	MI, USA; 49 reaches (mean of agricultural and mixed land use combined)	0.0024 ± 0.0007 (0–0.031)	6.8 ± 0.8 (0–22)	mean 9.7 ± 0.01 cm (logs ≥ 5 cm)
Reich 1999	Germany, Habberbach, meadows;	no wood		>10 cm diameter
	Eifa, forested	3 m^3/100 m		
Diez 2000	Agüera basin, Spain; agricultural streams	first order: 0.0002 second order 0.0156		>1 cm diameter
	forested streams	first order: 0.0145 second order: 0.008		
Gerhard and Reich 2000	Germany: Lüder, third order agricultural;	no wood		>10 cm diameter
	Jossklein, plantations	7.1 m^3/100 m		
Hering et al. 2000	Germany, 69 streams, Agricultural,	no wood		>10 cm diameter
	intact riparian forest	5.0		

[a] In Shields and Smith 1992.

[b] Unpublished data, in Gurnell et al. 1995.

tween land use and landform, supporting the observations of Richards et al. (1996, 1997) that land use and surficial geology are not independent. The forested streams in most of these examples are far from pristine, and the abundance of large wood in unmanaged streams can be much greater (see Table 1; other authors, this volume).

Land use practices in urban areas also severely affect wood in streams. In Puget Sound (Washington State, USA) lowland streams, the volume of large wood decreased from 1,200 m^3/km to 0, and the number of pieces fell by 75% as the proportion of the basin covered by impervious surfaces increased from near 0% (undeveloped basins) to 60% (urban basins; May et al. 1997). In this study, substantial amounts of large wood were found only in reaches with more than 70% riparian cover and riparian zones wider than 30 m. Urban areas with intact riparian corridors still had abundant wood jams, but most were in an advanced state of decay and, thus, probably remnants from the past (May et al. 1997). In areas with a longer history of development (like most of Europe), the effect of urbanization will probably be greater. Residential development results in reduced large-wood inputs to lakes as well as rivers. In 16 lakes in Wisconsin and Michigan, large-wood abundance (diameter > 5 cm) was positively correlated with tree density and negatively with shoreline cabin density (Christensen et al. 1996).

In addition to alterations in riparian vegetation that reduce the source of large wood, streams and rivers in developed landscapes are often channelized and incised, thus promoting greater export of already scarce wood pieces. River channelization is usually practiced in low lying, flood-prone areas (Wang et al. 1997). Shields and Smith (1992) reported wood volumes of 0.043–0.094 m^3/m^2 in uncleared reaches of the South Fork Obion River (Tennessee), a channelized, agricultural river surrounded by riparian forests and 0.009–0.033 m^3/m^2 in the cleared reaches (Table 1).

Wood removal from stream and river channels is practiced by diverse government agencies and private landowners, all of whom may have different management objectives (Piégay and Gurnell 1997). In agricultural regions of the Midwestern United States, locally elected or appointed officials are responsible for maintaining drainage ditches (usually channelized headwater streams), including channelizing and clearing stream channels. Adjacent landowners periodically clear the waterways and banks themselves, both to remove potential flow obstructions and to maintain what is perceived to be an aesthetically pleasing appearance (L. Johnson, personal observations). In England, landowners favored rivers without wood versus those with wood when ranking aesthetic preferences (Gregory and Davis 1993).

The practice of clearing wood from stream channels (also known as snagging) is most common in developed areas where local flooding or clogging of bridges could be detrimental (Diehl and Bryan 1993). Thus, many channelized agricultural and urban rivers are maintained free of large wood. Some exceptions can be found, of course; in U.S. Midwestern areas undergoing land conversion from forests to low-density urban development, trees in riparian areas have presumably been retained for their aesthetic value (Johnson 1999). There, most large wood is associated with higher link numbers (that is, number of first-order tributaries; Shreve 1966) in catchments with higher percentages of low-density urban development and lower topographic relief. Johnson (1999) speculated that these areas might have wider floodplains that could support woody riparian vegetation, but which afforded limited opportunities for intensive development because of flood potential.

Other activities that can affect wood in developed channels are grazing and livestock trampling. Reaches overgrazed and eroded by sheep and cattle in the Pranjip-Creightons stream system (Victoria, Australia) had almost no wood, but downstream reaches with riparian forests had more than 0.026 m^3/m^2 (O'Connor 1992). Bank erosion caused by trampling can contribute to trees falling into channels (Trimble and Mendel 1995), but—in the Australian systems—the logs were probably buried under the high loads of sand reaching the streams, and thus made them undetectable by standard sampling techniques.

Size of Wood in Developed Streams and Rivers

Few studies of wood in developed streams or rivers explicitly report the size of logs encountered. Log size may not be reported because the wood diameter in agricultural systems seldom exceeds the 10-cm-diameter minimum applied in most studies of large wood. The paucity of studies could also arise from the perception that less abundant and smaller logs do not play a significant role in streams and rivers. Of the studies that report log sizes, logs greater than or equal to 10 cm in diameter were completely lacking in the German streams in agricultural regions (Hering et al. 2000). Similarly, Johnson (1999) noted that 27 of 49 stream reaches (55%) in an agricultural region of the Midwestern United States had no wood greater than or equal to 10 cm in diameter. Log diameter (wood ≥ 5 cm) across the 49 sites averaged 9.7 ± 0.01 cm (Table 1). O'Connor (1992) found mean log diameter to be smaller (0–40 cm) in the upstream reaches of the Pranjip Creek, subject to severe bank erosion from overgrazing, than in the lowest reaches with intact riparian vegetation (>80 cm). Over all of the reaches studied, log diameter averaged just over 40 cm.

In northern Japanese streams with some plantations and development in the riparian zone and catchment, logs averaged 24 cm in diameter and 3 m in length. Although logs were shorter in the grassland streams, diameters were not different from those of forested reaches (Inoue and Nakano 1998). In two streams in an agricultural catchment in Tennessee where Diehl and Bryan (1993) were attempting to determine the potential effect of debris jams on bridges, logs greater than 12 m long represented about 21% of the total wood volume.

The highly skewed size-distribution of logs requires some caution in evaluating statistical analyses because reaches with the highest average sizes are not always the reaches where the largest logs are found. In the Agüera stream (Spain), large logs (up to 65 cm in diameter and more than 15 m long) were present only in first-order stream reaches under mature forests (Elosegi et al. 1999; Diez 2000); they were totally absent from agricultural reaches. Analysis of variance (ANOVA) on log-transformed data showed large wood (diameter > 5 cm) to be thinner, but longer in agricultural than in other types of reaches. Clearly, the presence or absence of a few big logs is independent of the average log size, since average size is affected more by the smaller, more abundant pieces. A geometric mean is sometimes computed to address this issue, but such differences in methods make direct comparisons of mean log-sizes across studies difficult. Another issue plaguing attempts to compare large-wood abundance is that minimum log sizes vary greatly across studies, ranging from 5 to 20 cm in diameter and 1–2 m in length. A standardized definition of large wood would greatly enhance our ability to assess patterns across different stream types and regions. At the least, minimum length and diameter should be specified.

One consequence of smaller logs is that they are more mobile than larger ones. Diez (2000) estimated the turnover time of wood in 20 reaches of the Agüera stream. Turnover time at one first-order agricultural reach studied was 2 years, but in 11 forested reaches, it averaged 46.5 years. Differences decreased downstream; turnover time in the second-order agricultural reaches was 2.5 years, in contrast to 5 years in the forested reaches. In 12 agricultural and mixed land use streams in Michigan, Johnson (1999) found that, on average, more than 50% of tagged logs moved more than 5 m over an 8-month period that included a 5-year flood. Transported wood was completely replaced in most reaches over this period.

Where trees are considered by land managers to have a negative influence, either because of shading of nearby crops or because of impeded stream flow, they are frequently removed and replaced with alternative vegetation types, primarily grasses. Outright removal of riparian vegetation or harvest of riparian forests both diminish the supply of large wood to rivers and reduce the size of logs once they are introduced to the channel. Furthermore, because large logs are seldom transported far from the remnant forests in headwater reaches, only small logs can be found in the downstream developed areas.

Role of Large Wood in Developed Streams and Rivers

Large wood performs several important roles in pristine forested river systems: it influences channel morphology by controlling the distribution of pools and riffles (Montgomery et al. 2003; Mutz 2003; Gregory 2003; Dolloff and Warren 2003; Abbe et al. 2003; all this volume), it creates habitats for invertebrates and fishes and constitutes important refuges during spates (Zalewski et al. 2003; Steel et al. 2003; both this volume), it increases food availability by retaining organic matter (Bilby 2003; Wondzell and Bissonn 2003; both this volume), and it promotes richer, more productive benthic communities (Benke and Wallace 2003, this volume). The scarce wood remaining in developed streams appears to play different roles than in forested streams.

Channel morphology and hydrology

The role of large wood differs greatly between rivers of different size, power, and geomorphic pattern (Piégay and Gurnell 1997; Piégay 2003, this volume). The size of the river channel plays a major role in the rate and form of large wood retention; as a result, riparian forests exert the greatest influence on small rivers. In larger rivers running through developed landscapes, the small supply and size of wood make the formation of complete dams extremely unlikely. Hence, the role of large wood depends more on geomorphic factors, such as the braiding index, number of wooded islands, and length of eroding wooded banks upstream, than on the size of riparian trees. Meandering sections have a greater capacity to stop the migration of logs than braided sections and, thus, have higher total wood mass (Piégay and Gurnell 1997).

Centuries of management have reduced the amount and size of the wood in developed rivers, at the same time diminishing or altering its role (Hering et al. 2000). In general, the small size and unstable nature of wood and wood accumulations in developed streams and rivers reduce the potential geomorphic role of wood. Wood accumulations in developed regions are subject to washout even under moderate flow conditions and cannot increase in mass through time because accumulations usually do not completely dam the channel, even in smaller rivers and streams (Booth et al. 1996). These smaller, more mobile logs and short-lived wood accumulations in developed areas cannot divert flow over time frames sufficient to create the plunge pools and backwater habitats commonly associated with wood in forested streams. Logs did not affect channel morphology in the agricultural reaches of the Agüera Basin, but—on average—21 active pieces/100 m did affect the channel in first-order streams under deciduous forests, and 7 in second-order reaches (Diez 2000). Johnson (1999) concluded that large wood played no significant role in the control of channel morphology in Michigan streams because no correlation was observed between wood abundance and pool number or depth. Some authors, however, have found wood to be an important geomorphic agent, even in agricultural rivers with low wood abundance. Where wood was present, it influenced pool area in Nevada rangeland streams (Myers and Swanson 1997) and in urban Puget Sound streams (May et al. 1997).

Although active channel modification has not been observed in some developed streams where this phenomenon was studied, the presence of large wood in the channel undoubtedly increases hydraulic diversity at local scales and is, therefore, significant from a biotic perspective. In large, agricultural French rivers, wood jams increase hydraulic heterogeneity (Piégay 1993) as well as substrate diversity (Piégay et al. 1998). Similarly, Shields and Smith (1992) found that large wood contributed to more diverse habitats and hydrologic patterns and that sediment near the wood was finer and more organic in nature.

Channelization, wood dam removal, and riparian forest clearing also occur in urban streams and rivers. Increased impervious surfaces cause greater floods and increase bank erosion, forcing managers to reinforce banks or entire channels with rip-rap or concrete (May et al. 1997). The increased peak flows may cause the stream channel to incise to accommodate the higher discharge (Booth 1990; Cooper 1996). Bankside vegetation, instream wood, and other channel roughness elements (like rootwads and boulders) are removed, significantly decreasing Manning's n (Shields and Nunnally 1984) and exporting stored organic matter. Low channel roughness and flashy flow regimes along with generally smaller wood size result in streams lacking stable wood and wood accumulations.

Few studies have implicitly examined the relation between channelization and wood in rivers. In a comparison of channelized and unchannelized reaches in northwestern Mississippi, wood was "infrequently encountered" in the channelized streams, but the unchannelized site contained "abundant large wood of all sizes" (Shields et al. 1994). This observation was echoed in channelized and unchannelized sections of the Bunyip River, Australia (Hortle and Lake 1983). Petersen and Petersen (1991) compared the organic-matter retention capacity of channelized and unchannelized agricultural second- and third-order reaches in Sweden. They found retention to be highly reduced in channelized reaches, with spiraling distances of leaves almost sevenfold greater than unchannelized reaches. Furthermore, retention structures in channelized streams (macrophytes and alcoves) were very unstable compared to those of unchannelized streams (wedged in the substrate). Such retention processes have been observed for both particulate and dissolved substances. For example, reaches with abundant wood dams in Leicestershire (UK) were more retentive of phosphorus (Dickinson 1996; cited in Harper et al. 1999), a fact that could benefit the self-purification capacity of streams so critical in agricultural landscapes.

The effects of channelization include hydrologic as well as physical alterations to the channel that reduce wood sources and their retentive properties. This simplification of the stream channel is somewhat mimicked by large-wood clearing and snagging operations. Changes resulting from channelization and wood clearing in the Highland Water (a third-order English stream with a wide riparian corridor in a catchment dominated by heathlands) were examined by Gurnell and Sweet (1998). They observed increased erosion in channelized reaches lacking large wood because of increased slope and lack of retention structures in the channel. Sediment mobility increased in reaches where wood was removed, and pool spacing increased as a result. Similarly, wood removal in Mahogany Creek (Nevada) decreased pool stability and sometimes eliminated pools altogether (Myers and Swanson 1996).

Biota and habitat

Despite the relatively small geomorphic importance of wood in developed streams and rivers, it appears to play key roles for biota. Large wood may be the only hard substrate available in sand-bottom rivers and can thus be colonized by extremely high densities of invertebrates (Benke et al. 1985; Benke and Wallace 2003). This effect could be even more important in agricultural rivers because sediments are among the most widespread agricultural pollutants (Cooper 1993). Where habitats are more diverse, wood can function as a refugium in adverse conditions. Hax and Golladay (1998) examined recovery of macroinvertebrate communities from flow cessation in a Texas prairie stream. Sediment communities experienced greater declines and took longer to recover than those on wood.

Large wood can also affect invertebrates indirectly, through enhanced primary production. Primary production of periphyton growing on large wood can be substantial even in heavily regulated and desnagged Australian agricultural rivers (Treadwell 2000) where it can support a large proportion of the nutritional needs of herbivorous invertebrates.

Invertebrate communities can differ from woody to nonwoody habitats in agricultural streams and rivers. In Australia, O'Connor (1991) found snag invertebrates to be more abundant and diverse than streambed invertebrates because snag-dwelling species were missing in the benthos. Oligochaetes (naidids) were most common on snags, but chironomid larvae were most common in the streambed. The complex snags had richer invertebrate communities, with species groups selecting different microhabitats on the snags. Although wood was not abundant and logs were highly mobile in the Michigan and Minnesota streams studied by Johnson et al. (2003), this habitat was also found to be important for macroinvertebrates, significantly increasing taxa richness. An average of 11 additional taxa were encountered when wood habitats were present in Michigan streams. Furthermore, 85% of total taxa observed across the Michigan study region were observed on wood or in association with habitats created by wood accumulations. Twenty-seven of those taxa were unique to those habitats, and an additional 8 taxa were disproportionately observed in association with wood habitats. While similar trends were observed in Minnesota streams, fewer taxa were unique to wood habitats, but were often found on bankside vegetation as well.

In the Bunyip River, Australia (Hortle and Lake 1982) macroinvertebrate species richness, density, and standing crop did not differ between channelized reaches with little wood and unchannelized streams with abundant wood. Substrate stability was the best explanatory factor for community structure, but it did not differ consistently across the channelized and unchannelized reaches. Interestingly, fish communities differed significantly between the channelized and unchannelized sections, and the differences were explained by snag area (Hortle and Lake 1983).

Nilsen and Larimore (1973) studied invertebrate colonization of logs introduced in the Kaskaskia, an agricultural river in Illinois. Standing stocks were highest in slack, shallow water, where chironomid larvae were first to colonize, followed by rotifers and oligochaetes. In riffles, *Taeniopteryx nivalis* (a stonefly), hydropsychid larvae, and chironomid and simulid larvae were dominant. Naturally occurring logs had large populations of planarian worms, ephemeropteran and trichopteran larvae, and chironomid pupae. Fewer chironomid larvae, Taeniopterygidae, crustaceans, and oligochaetes were seen, compared to the communities on introduced logs. In a similar experiment, Spänhoff et al. (2000) found an average of 1,500 invertebrates (mainly insects) per square meter colonizing twigs in two agricultural German streams. Wood species played a minor role in colonization, suggesting that substrate was more important than the food source.

Another line of evidence on the importance of large wood for invertebrates in developed rivers comes from experimental manipulations of wood densities. For instance, Gerhard and Reich (2000) added large wood to restore two straightened German streams; remnants of nearby forest clearing were added to the Jossklein, a forested stream, and logs were planted as deflectors in the Lüder, a stream surrounded by meadows. The treatments increased habitat patchiness, invertebrate density, and species richness, but the effects were clearer at the Jossklein, where invertebrate densities increased even in microhabitats other than wood.

Management practices prevalent in many agricultural and urban areas usually result in highly simplified stream and river channels and impoverished fish communities (Hellawell 1986). Nevertheless, the few remaining logs can play a key role in providing refugia for fish or creating pool habitats, as was seen in urban streams draining into Puget Sound (May et al. 1997). In the agricultural upper Mississippi River, snags redirected current and scoured sediments downriver, creating most of the deep pools (Lehtinen et al. 1997). Higher fish biomass and abundance were found at snag sites, suggesting that fish selected snags for foraging or protection. Although species richness was similar, species composition differed between snags and other habitats. Similarly, when comparing fish communities in forested with unforested agricultural streams, Stauffer et al. (2000) found that, contrary to expectations, herbivorous fishes were more abundant in wooded reaches because the wood constituted an important substrate for periphyton and was the main physical habitat feature providing cover and hard surfaces. Siltation in the open reaches limited periphyton production. In contrast, large wood, silt, and periphyton were shown not to be important environmental factors for nursery habitat in the agricultural River Morava, Czech Republic (Jurajda 1999), where emergent vegetation was considered important.

Some researchers have experimentally tested the importance of wood abundance for fish in agricultural streams. Shields et al. (1998) added wood and stones to rehabilitate severely incised reaches in Mississippi. The treatment increased pool habitat and made streams physically more "natural." They observed only subtle changes in fish communities; fish density and species richness increased in only one of two reaches. Angermeier and Karr (1984), on the other hand, added or removed wood from several reaches in a second-order agricultural stream in Illinois. Fish and invertebrates were more abundant in the reaches with wood. Wood removal resulted in decreased water depth, benthic organic matter cover, and increased current and proportion of sandy bottom (probably associated with decreased siltation). Large fish avoided reaches without wood, whereas some smaller fish preferred them, especially in a dry year. The authors concluded that camouflage or hiding areas seemed to be more important than food in the preference of large fish for wood habitats, especially during low water levels.

At bank-full discharge in some Australian streams affected by livestock grazing, snags constituted up to 50% of the area available for macroinvertebrates (O'Connor 1992). In addition, exposed roots with diameters greater than 5 cm were common along the banks and contributed much of the surface area available for colonization at higher discharges. This habitat (rootwads) is probably important for both fish and invertebrates in developed systems but is seldom acknowledged. In the Michigan streams studied by Johnson (1999), both overhanging vegetation and root-wads trapped large amounts of organic debris, including wood. These structures, along with log dams and loose aggregations of wood, were classified as wood accumulations and enumerated. Three hundred and eighteen accumulations were found across the 49 reaches; 13 reaches had a total of 60 rootwads with trapped organic matter. A total of 54 accumulations composed of overhanging vegetation with trapped wood were encountered at 19 reaches. The role of these structures in instream primary and secondary production in channelized streams is not known.

Contrasts in Large Wood between Forested and Developed Streams

Contrasts between forested streams and those in developed landscapes are many; however, topography is a confounding factor in conducting such comparisons. In most regions, steep topography prevents or inhibits development close to a river. The most common human activities that can influence large-wood input to streams in high-gradient regions are forest harvest, mineral extraction, and road building, and specialized agricultural entities and activities such as orchards and vineyards or grazing. Large-wood input mechanisms differ between high- and low-gradient systems (Keller and Swanson 1979). Bank erosion is the primary mechanism responsible for large-wood inputs to stream channels in low-gradient streams and rivers. Storm damage (wind and ice) and disease may also cause large branches and limbs to enter the channel. In high-gradient streams, debris flows and torrents are the primary mechanisms that introduce large masses of soil and whole trees to the stream channels (Keller and Swanson 1979). If large wood becomes mobilized, it can be retained on gravel bars or islands located most commonly in the lower gradient reaches of these systems. Such retention structures are not common in either agricultural or urban streams (particularly in channelized reaches). Due to interactions between land use and landform, it is difficult to identify specific effects on the structural and functional role of large wood in streams. Sufficient evidence exists, however, to conclude that modifications to the channel and the wood supply directly and indirectly influence many ecosystem functions, including organic matter retention, nutrient cycling, primary and secondary production, and regulation of community structure. Unfortunately, few studies have rigorously documented these roles in either low-gradient or developed streams. The role of large wood in urban streams is especially poorly studied, since the major concerns of resource managers in those systems is the reduction of risk for flooding and damage to infrastructure.

Implications for Management

Wood in streams and rivers was historically viewed as an unaesthetic nuisance, an obstacle to fish migration, a hazard for navigation, a cause of bank erosion in many cases, and a risk during floods. Managers began to acknowledge the value of the wood and to adapt their approaches accordingly, first in forested areas of the United States (Bilby 1984) and later elsewhere. Although scientific evidence has shown that large wood plays a key role in the structure and function of stream and river systems (Swanson et al. 1976; Bilby and Likens 1980), many landowners and resource managers have been reluctant to change their traditional "wood-free" approach. In many developed areas, such as Europe or agricultural regions of the United States, large wood in streams and rivers is still primarily seen as a hazard. No doubt, large wood can produce both environmental benefits and societal costs, the latter especially in developed regions, where the risk of economic and human losses can be higher. When combined with restoration and/or conservation of wooded riparian buffer strips for mitigating sediment and nutrient input to streams, large wood additions could be used to also increase nutrient and sediment retention. The combined effects of a woody vegetation, providing shade, leaf material, and wood, and instream wood could enhance biodiversity at multiple trophic levels, while providing important ecosystem services.

As conservation and restoration efforts accelerate worldwide, guidelines for deploying large wood must be established that take into account both the ecological benefits and societal risks. This strategy should include managing large wood, as well as the channel and banks, in a holistic basin context, in what Winkley (1972) called "working with the river, not against it."

Needs for Future Research

The points we have discussed demonstrate that basin development severely affects the amount and role of wood in streams and rivers, thus affecting the structure and function of these ecosystems. Nevertheless, relatively few researchers have studied large wood in developed systems; thus, many information gaps remain. First, to ensure that research results can be compared more easily, we stress the need to provide a more detailed description of the channel, riparian zone and catchment, along with a description of the present and historical management practices; to use comparable units of wood mass, volume, and abundance; to use consistent terminology; and to define the size range of wood being studied.

Second, research should be expanded to different areas and physiographic settings, especially in the tropics and in large rivers. Very little is known about wood in these systems, and extrapolating conclusions from temperate areas can lead to inappropriate management decisions.

Third, we suggest taking a closer look at the way that development affects the amount, size, and role of large wood in streams and rivers. Most notable in their effects on the wood in channels are the management policies practiced in many countries that dictate removing wood from stream and river channels, followed closely by the personal actions performed by landowners to remove logs that are perceived to be unaesthetic or to threaten lives and property. Although these activities directly influence the presence and abundance of wood in streams and rivers, the indirect effect of different agricultural and urban management activities on the abundance and distribution of large wood (and, therefore, on stream and river structure and function) can differ widely. Cultural and economic factors determine the type of management practices, and the physiographic context of the river basin will, in part, determine the intensity and duration of the effects. To manage our environment in a sustainable manner, we need to understand the effects of (and interactions between) natural disturbance regimes and management activities in both the stream or river and the basin. For example, narrow riparian buffer strips maintained in harvested areas and in agricultural and urban areas are subject to blow-down during storms; the pattern of the blowdown is determined by topography and soil moisture and texture (Lee Freilich, University of Minnesota, personal communication). These "natural" disturbances contribute large wood to streams and rivers in the short term but reduce long-term sources of wood and reduce or eliminate critical ecosystem services performed by the riparian zone. Studies that emphasize the context and natural disturbance regimes of streams and rivers—including the geomorphic setting of the channel, structure and composition of riparian zones and buffer strips, and also the basin land use and hydrology—are required to make informed decisions about the costs and benefits of conservation, restoration, and management, including the introduction, preservation, or both of large wood in streams and rivers.

Finally, in developed streams and rivers, large wood can only be managed soundly through a close interaction among the general public, managers, and scientists. The way the public perceives wood in streams and rivers is a key factor that is seldom assessed in any restoration project. Thus, scientists need to assess how different groups perceive large wood and then communicate clearly why historical management practices that have depleted wood from developed streams and rivers must be abandoned. No restoration project can be successful without broad collaboration with the public.

Acknowledgments

We are grateful to the editors of this volume, Stan Gregory, Kathryn Boyer, and Angela Gurnell for their efforts and energy in coordinating the conference and leading the charge in assembling this book. Their assistant, Kelly Wildman, provided important logistical support to us during the conference and during the production of this chapter. The two anonymous reviewers provided very helpful suggestions, and their comments have vastly improved the content and flow of this chapter. Arturo Elosegi would like to acknowledge Joserra Diez, who performed much of the wood work in the Basque Country. This is contribution 341 of the Center for Water and the Environment at the Natural Resources Research Institute, University of Minnesota Duluth.

References

Abbe, T. B., A. P. Brooks, and D. R. Montgomery. 2003. Wood in river rehabilitation and management. Pages 367–389 *in* S. V. Gregory, K. L. Boyer, and A. M. Gurnell, editors. The ecology and management of wood in world rivers. American Fisheries Society, Symposium 37, Bethesda, Maryland.

Allan, J. D., D. L. Erickson, and J. Fat. 1997. The influence of catchment land use on stream integrity across multiple spatial scales. Freshwater Biology 37:149–191.

Angermeier, P. L., and J. R. Karr. 1984. Relationships between woody debris and fish habitat in a small warmwater stream. Transactions of the American Fisheries Society 113:716–726.

Arnold, C. L., and C. J. Gibbons. 1996. Impervious surface coverage. Journal of the American Planning Association 62:243–259.

Barton, D. R. 1996. The use of percent model affinity to assess the effects of agriculture on benthic invertebrate communities in headwater streams of southern Ontario, Canada. Freshwater Biology 36:397–410.

Benke, A. C., R. L. Henry, D. M. Gillespie, and R. J. Hunter. 1985. Importance of snag habitat for animal production in southeastern streams. Fisheries 10:8–13.

Benke, A. C., and J. B. Wallace. 2003. Influence of wood on invertebrate communities in streams and rivers. Pages 149–177 *in* S. V. Gregory, K. L. Boyer, and A. M. Gurnell, editors. The ecology and management of wood in world rivers. American Fisheries Society, Symposium 37, Bethesda, Maryland.

Bilby, R. E. 2003. Decomposition and nutrient dynamics of wood in streams and rivers. Pages 135–147 *in* S. V. Gregory, K. L. Boyer, and A. M. Gurnell, editors. The ecology and management of wood in world rivers. American Fisheries Society, Symposium 37, Bethesda, Maryland.

Bilby, R. E. 1984. Removal of woody debris may affect stream channel stability. Journal of Forestry 82:609–613.

Bilby, R. E., and G. E. Likens. 1980. Importance of organic debris dams in the structure and function of stream ecosystems. Ecology 61:1107–1113.

Booth, D. B. 1990. Stream channel incision following drainage basin urbanization. Water Resources Bulletin 26(3):407–417.

Booth, D. B., D. R. Montgomery, and J. Bethel. 1996. Large woody debris in urban streams of the Pacific Northwest. Pages 178–197 *in* Effects of watershed development and management on aquatic ecosystems. Proceedings of an Engineering Foundation conference. American Society of Civil Engineers, New York.

Brookes, A. 1994. River channel changes. Pages 55–75 *in* P. Calow and G. E. Petts, editors. The rivers handbook. Hydrological and ecological principles. Blackwell Scientific Publications, Oxford, UK.

Christensen, D. L., B. R. Herwig, D. E. Schindler, and S. R. Carpenter. 1996. Impacts of lakeshore residential development on coarse woody debris in north temperate lakes. Ecological Applications 64:1143–1149.

Cooper, C. M. 1993. Biological effects of agriculturally derived surface water pollutants on aquatic systems –a review. Journal of Environmental Quality 22:402–408.

Cooper, C. 1996. Hydrologic effects of urbanization in Puget Sound lowland streams. Master's thesis. University of Washington, Seattle.

Correll, D. L. 2000. The current status of our knowledge of riparian buffer water quality functions. Pages 5–10 *in* R. Beschta, editor. International conference on riparian ecology and management in multi-land use watersheds. American Water Resources Association, Middleburg, Virginia.

Dance, K. W., and H. B. N. Hynes. 1980. Some effects of agricultural land use on stream insect communities. Environmental Pollution 22:19–28.

Dickinson, S. 1996. The effect of accumulations of coarse particulate organic matter upon phosphorus retention in low order streams. Master's thesis. University of Leicester, Leicester, UK.

Diehl, T. H., and B. A. Bryan. 1993. Supply of large woody debris in a stream channel. Pages 1055–1060 *in* H. W. Shen, S. T. Su, and F. Wen, editors. Hydraulic Engineering '93 Conference, San Francisco. Proceedings: American Society of Civil Engineers.

Diez, J. R. 2000. Dinámica y función de la madera en el sistema fluvial del Agüera. Relaciones con la vegetación de la cuenca. Doctoral thesis. University of the Basque Country, Bilbao, Spain.

Diez, J. R., A. Elosegi, and J. Pozo. 2001. Woody debris in north Iberian streams: influence of geomorphology, vegetation and management. Environmental Management 28:687–698.

Diez J. R., A. Elosegi, E. Chauvet, and J. Pozo. 2002. Breakdown of wood in the Agüera stream. Freshwater Biology 47:2205–2215.

Dolloff, C. A. and M. L. Warren, Jr. 2003. Fish relationships with large wood in small streams. Pages 179–193 *in* S. V. Gregory, K. L. Boyer, and A. M. Gurnell, editors. The ecology and management of wood in world rivers. American Fisheries Society, Symposium 37, Bethesda, Maryland.

Elosegi, A., J. R. Diez, and J. Pozo. 1999. Abundance, characteristics, and movement of woody debris in four Basque streams. Archiv für Hydrobiologie 144:455–471.

Evans, B. F., C. R. Townsend, and T. A. Crowl. 1993. Distribution and abundance of coarse woody debris in some southern New Zealand streams from contrasting forest catchments. New Zealand Journal of Marine and Freshwater Research 27:227–239.

Gerhard, M., and M. Reich. 2000. Restoration of streams with large wood: effects of accumulated and built-in wood on channel morphology, habitat diversity and aquatic fauna. International Review of Hydrobiology 85:123–137.

Gore, J. A. 1994. Hydrological Changes. Pages 33–54 *in* P. Calow and G. E. Petts, editors. The rivers handbook. Hydrological and ecological principles. Blackwell Scientific Publications, Oxford, UK.

Gregory, K. J. 2003. The limits of wood in world rivers: present, past, and future. Pages 1–19 *in* S. V. Gregory, K. L. Boyer, and A. M. Gurnell, editors. The ecology and management of wood in world rivers. American Fisheries Society, Symposium 37, Bethesda, Maryland.

Gregory, K. J., and R. J. Davis. 1993. The perception of riverscape aesthetics: an example from two Hampshire rivers. Journal of Environmental Management 39:171–185.

Gregory, K. J., A. M. Gurnell, and C. T. Hill. 1985. The permanence of debris dams related to river channel processes. Hydrological Sciences Journal 30:371–381.

Gregory, K. J., R. J. Davis, and S. Tooth. 1993. Spatial distribution of coarse woody debris dams in the Lymington Basin, Hampshire, UK. Geomorphology 6:207–224.

Gurnell, A. M., and K. J. Gregory. 1995. Interactions between semi-natural vegetation and hydrogeomorphological processes. Geomorphology 13:49–69.

Gurnell, A. M., and R. Sweet. 1998. The distribution of large woody debris accumulations and pools in relation to woodland stream management in a small, low-gradient stream. Earth Surface Processes and Landforms 23:1101–1121.

Gurnell, A. M., K. J. Gregory, and G. E. Petts. 1995. The role of coarse woody debris in forest aquatic habitats: implications for management. Aquatic Conservation: Marine and Freshwater Ecosystems 5:143–166.

Gurnell, A. M. 2003. Wood storage and mobility. Pages 75–91 in S. V. Gregory, K. L. Boyer, and A. M.

Gurnell, editors. The ecology and management of wood in world rivers. American Fisheries Society, Symposium 37, Bethesda, Maryland.

Harmon, M. E., J. F. Franklin, F. J. Swanson, P. Sollins, S. V. Gregory, J. D. Lattin, N. H. Anderson, S. P. Cline, N. G. Aumen, J. R. Sedell, G. W. Lienkaemper, K. Cromack, and K. W. Cummins. 1986. Ecology of coarse woody debris in temperate ecosystems. Advances in Ecological Research 15:133–302.

Harper, D. M. 1992. Eutrophication of freshwaters. Principles, problems and restoration. Chapman and Hall, London.

Harper, D. M., M. Ebrahimnezhad, E. Taylor, S. Dickinson, O. Decamp, G. Verniers, and T. Balbi. 1999. A catchment-scale approach to the physical restoration of lowland UK rivers. Aquatic Conservation: Marine and Freshwater Ecosystems 9:141–157.

Hauer, F. R. 1989. Organic matter transport and retention in a blackwater stream recovering from flow augmentation and thermal discharge. Regulated Rivers Research and Management 4:371–380.

Hax, C. L., and S. W. Golladay. 1998. Flow disturbance of macroinvertebrates inhabiting sediments and woody debris in a prairie stream. American Midland Naturalist 139:210–223.

Hellawell, J. M. 1986. Biological indicators of freshwater pollution and environmental management. Pollution monitoring series. Elsevier, London.

Hering, D., J. Kail, S. Eckert, M. Gerhard, E. I. Meyer, M. Mutz, M. Reich, and I. Weiss. 2000. Coarse woody debris quantity and distribution in Central European streams. International Review of Hydrobiology 85:5–23.

Hortle, K. G., and P. S. Lake. 1982. Macroinvertebrate assemblages in channelized and unchannelized sections of the Bunyip River, Victoria. Australian Journal of Marine and Freshwater Research 33:1071–1082.

Hortle, K. G., and P. S. Lake. 1983. Fish of channelized and unchannelized sections of the Bunyip River, Victoria. Australian Journal of Marine and Freshwater Research 34:441–450.

Hunter, C, J. Perkins, J. Tranter, and J. Gunn. 1999. Agricultural land-use effects on the indicator bacterial quality of an upland stream in the Derbyshire Peak District in the U.K. Water Research (33):3577–3586.

Hynes, H. B. N. 1975. The stream and its valley. Verhandlungen der Internationalen Vereinigung für theoretische und angewandte Limnologie 19:1–15.

Inoue, M., and S. Nakano. 1998. Effects of woody debris on the habitat of juvenile masu salmon (*Oncorhynchus masou*) in northern Japanese streams. Freshwater Biology 40:1–16.

Johnson, L. B. 1999. Ecological implication of coarse woody debris in low gradient, Midwestern streams. Doctoral dissertation. Michigan State University, East Lansing.

Johnson, L. B., C. Richards, G. E. Host, and J. W. Arthur. 1997. Landscape influences on water chemistry in Midwestern stream ecosystems. Freshwater Biology 37:193–208.

Johnson, L. B., D. Breneman, and C. Richards. 2003. Macroinvertebrate community structure and function associated with large wood in low gradient streams. River Research and Applications 19:199–218.

Jurajda, P. 1999. Comparative nursery habitat use by 0+ fish in a modified lowland river. Regulated Rivers Research and Management 15:113–124.

Keller, E. A., and F. J. Swanson. 1979. Effects of large organic material on channel form and fluvial processes. Earth Surface Processes 4:361–380.

Larsen, P. 1996. Restoration of river corridors: German experiences. Pages 124–143 *in* G. Petts and P. Calow, editors. River restoration. Blackwell Scientific Publications, Oxford, UK.

Lehtinen, R. M., N. D. Mundahl, and J. C. Madejczyk. 1997. Autumn use of woody snags by fishes in backwater and channel border habitats of a large river. Environmental Biology of Fishes 49:7–19.

Maser, C., and J. R. Sedell. 1994. From the forest to the sea. The ecology of wood in streams, rivers, estuaries, and oceans. St Lucie Press, Boca Raton, Florida.

May, C. W., E. B. Welch, R. R. Horner, J. R. Karr, and B. W. Mar. 1997. Quality indices for urbanization effects on Puget Sound lowland streams. University of Washington, Civil Engineering Department, Water Resources Series, Technical Report 154, Seattle.

Minckley, W. L., and J. N. Rinne. 1985. Large woody debris in hot-desert streams: an historical review. Desert Plants 73:142–153.

Montgomery, D. R., B. D. Collins, J. M. Buffington, and T. B Abbe. 2003. Geomorphic effects of wood in rivers. Pages 21–47 *in* S. V. Gregory, K. L. Boyer, and A. M. Gurnell, editors. The ecology and management of wood in world rivers. American Fisheries Society, Symposium 37, Bethesda, Maryland.

Mutz, M. 2003. Hydraulic effects of wood in streams and rivers. Pages 93–107 *in* S. V. Gregory, K. L. Boyer, and A. M. Gurnell, editors. The ecology and management of wood in world rivers. American Fisheries Society, Symposium 37, Bethesda, Maryland.

Myers, T. J., and S. Swanson. 1996. Long-term aquatic habitat restoration: Mahogany Creek, Nevada, as a case study. Water Resources Bulletin 32:241–252.

Myers, T. J., and S. Swanson. 1997. Variability of pool characteristics with pool type and formative feature on small Great Basin rangeland streams. Journal of Hydrology 201:62–81.

Nakamura, F., and F. J. Swanson. 2003. Dynamics of

wood in rivers in the context of ecological disturbance. Pages 279–297 *in* S. V. Gregory, K. L. Boyer, and A. M. Gurnell, editors. The ecology and management of wood in world rivers. American Fisheries Society, Symposium 37, Bethesda, Maryland.

Nilsen, H. C., and R. W. Larimore. 1973. Establishment of invertebrate communities on log substrates in the Kaskaskia River, Illinois. Ecology 54:366–374.

Norris, R. D., and M. C. Thoms. 1999. What is river health? Freshwater Biology 41:197–209.

O'Connor, N. A. 1991. The effects of habitat complexity on the macroinvertebrates colonising wood substrates in a lowland stream. Oecologia 85:504–512.

O'Connor, N. A. 1992. Quantification of submerged wood in a lowland Australian stream system. Freshwater Biology 27:387–395.

Osborne, L. L., and D. A. Kovacic. 1993. Riparian vegetated buffer strips in water-quality restoration and stream management. Freshwater Biology 29:243–258.

Paul, M. J., and J. L. Meyer. 2001. Streams in urban landscapes. Annual Review of Ecology and Systematics 32:333–365.

Peterjohn, W. T., and D. L. Correll. 1984. Nutrient dynamics in an agricultural catchment: observations on the role of a riparian forest. Ecology 65:1466–1475.

Petersen, L. B. M., and R. C. Petersen. 1991. Short-term retention properties of channelized and natural streams. Verhandlungen der Internationalen Vereinigung für theoretische und angewandte. Limnologie 24:1756–1759.

Piégay, H. 1993. Nature, mass and preferential sites of coarse woody debris deposits in the lower Ain Valley (Mollon Reach), France. Regulated Rivers Research and Management 8:359–372.

Piégay, H., and A. M. Gurnell. 1997. Large woody debris and river geomorphological pattern: examples from S. E. France and S. England. Geomorphology 19:99–116.

Piégay, H., A. Citterio, and L. Astrade. 1998. Ligne du débris ligneux et recoupement de méandres, exemple du site de Mollon sur la Rivière d'Ain (France). Zeitschrift für Geomorphologie 42:187–208.

Piégay, H. 2003. Dynamics of wood in large rivers. Pages 109–133 *in* S. V. Gregory, K. L. Boyer, and A. M. Gurnell, editors. The ecology and management of wood in world rivers. American Fisheries Society, Symposium 37, Bethesda, Maryland.

Reich, M. 1999. Ecological, technical and economical aspects of stream restoration with large wood. Zeitschrift für Ökologie und Naturschutz 8:251–253.

Richards, C., R. J. Haro, L. B. Johnson, and G. E. Host. 1997. Catchment and reach-scale properties as indicators of macroinvertebrate species traits. Freshwater Biology 37:219–230.

Richards, C., L. B. Johnson, and G. E. Host. 1996. Landscape-scale influences on stream habitats and biota. Canadian Journal of Fisheries and Aquatic Sciences 53 (Supplement 1):295–311.

Richmond, A. D., and K. D. Fausch. 1995. Characteristics and function of large woody debris in subalpine Rocky Mountain streams in northern Colorado. Canadian Journal of Fisheries and Aquatic Sciences 52:1789–1802.

Shields, F. D., Jr., and R. H. Smith. 1992. Effects of large woody debris removal on physical characteristics of a sand-bed river. Aquatic Conservation: Marine and Freshwater Ecosystems 2:145–163.

Shields, F. D., Jr., S. S. Knight, and C. M. Cooper. 1994. Effects of channel incision on base flow stream habitats and fishes. Environmental Management 18:43–57.

Shields, F. D., Jr., S. S. Knight, and C. M. Cooper. 1998. Rehabilitation of aquatic habitats in warmwater streams damaged by channel incision in Mississippi. Hydrobiologia 382:63–86.

Shields, F. D., Jr., and N. R. Nunnally. 1984. Environmental aspects of clearing and snagging. Journal of Environmental Engineering 110:152–165.

Shreve, R. L. 1966. Statistical law of stream numbers. Journal of Geology 74:17–37.

Spaling, H., and B. Smith. 1995. A conceptual model of cumulative environmental effects of agricultural land drainage. Agriculture, Ecosystems and Environment 53:99–108.

Spänhoff, B., C. Alecke, and E. I. Meyer. 2000. Colonization of submerged twigs and branches of different wood genera by aquatic macroinvertebrates. International Review of Hydrobiology 85:49–66.

Stauffer, J. C., R. M Goldstein, and R. M. Newman. 2000. Relationship of wooded riparian zones and runoff potential to fish community composition in agricultural streams. Canadian Journal of Fisheries and Aquatic Sciences 57:307–316.

Steel, E. A., W. H. Richards, and K. A. Kelsey. 2003. Wood and wildlife: benefits of river wood to terrestrial and aquatic vertebrates. Pages 235–247 *in* S. V. Gregory, K. L. Boyer, and A. M. Gurnell, editors. The ecology and management of wood in world rivers. American Fisheries Society, Symposium 37, Bethesda, Maryland.

Swanson, F. J., G. W. Lienkaemper, and J. R. Sedell. 1976. History, physical effects, and management implications of large woody debris in western Oregon streams. U.S. Forest Service, Gen. Tech. Rep. PNW-56, Portland, Oregon.

Swanson, F. J. 2003. Wood in landscapes and river networks. Pages 299–313 *in* S. V. Gregory, K. L. Boyer, and A. M. Gurnell, editors. The ecology and management of wood in world rivers. American Fisheries Society, Symposium 37, Bethesda, Maryland.

Sweeting, R. A. 1994. River pollution. Pages 23–32 *in* P. Calow and G. E. Petts, editors. The rivers handbook. Hydrological and ecological principles. Blackwell Scientific Publications, Oxford, UK.

Treadwell, S. 2000. The importance of large woody debris surfaces for algal growth in lowland rivers. RipRap 16:6–8.

Trimble, S. W., and A. C. Mendel. 1995. The cow as a geomorphic agent: a critical review. Geomorphology 13:233–253.

Wang, L., J. Lyons, P. Kanehl, and R. Gatti. 1997. Influences of watershed land use on habitat quality and biotic integrity in Wisconsin streams. Fisheries 26:6–11.

Watzin, M. C., and A. W. McIntosh. 1999. Aquatic ecosystems in agricultural landscapes: a review of ecological indicators and achievable ecological outcomes. Journal of Soil and Water Conservation 54:636–644.

Wiley, M. J., S. L. Kohler, and P. W. Seelbach. 1997. Reconciling landscape and local views of aquatic communities: lessons from Michigan trout streams. Freshwater Biology 37:133–148.

Winkley, B. R. 1972. River regulation with the aid of nature. Pages 433–437 *in* International Commission on Irrigation and Drainage. 8th Congress Verna Transactions 29:1.

Wondzell, S. M., and P. A. Bisson. 2003. Influence of wood on aquatic biodiversity. Pages 249–263 *in* S. V. Gregory, K. L. Boyer, and A. M. Gurnell, editors. The ecology and management of wood in world rivers. American Fisheries Society, Symposium 37, Bethesda, Maryland.

Yoder, C. O., and E. T. Rankin. 1995. Biological response signatures and the area of degradation value: new tools for interpreting multimetric data. Pages 263–286 *in* W. S. Davis and T. P. Simon, editors. Biological assessment and criteria. Lewis Press, Boca Raton, Florida.

Zalewski, M., M. Lapinska, and P. B. Bayley. 2003. Fish relationships with wood in large rivers. Pages 195–211 *in* S. V. Gregory, K. L. Boyer, and A. M. Gurnell, editors. The ecology and management of wood in world rivers. American Fisheries Society, Symposium 37, Bethesda, Maryland.

Zimmer, D. W., and R. W. Bachman. 1976. A study of the effects of stream channelization and bank stabilization on warmwater sport fish in Iowa: subproject No. 4. The effects of long-reach channelization on habitat and invertebrate drift in some Iowa streams. USDI Fish and Wildlife Service, Report No. FWS/OBS-76/14, Washington, D.C.

American Fisheries Society Symposium 37:355–366, 2003

Restoring Streams with Large Wood: A Synthesis

Michael Reich

Institut für Landschaftspflege und Naturschutz, Universität Hannover, 30419 Hannover, Germany.

Jeffrey L. Kershner

USDA Forest Service, Fish and Aquatic Ecology Unit, Aquatic, Watershed and Earth Resources Utah State University, Logan, Utah 84322-5210, USA

Randall C. Wildman

Department of Fisheries and Wildlife, Nash Hall, Oregon State University Corvallis, Oregon 97331-3803, USA

Abstract.—The use of large wood in stream restoration projects has become increasingly popular in the last 20 years. We reviewed more than 30 case studies from different ecoregions and countries (Canada, Germany, Japan, United States) to evaluate the variety of approaches used and assessed their reported success. Wood inputs generally fell into two categories: fixed structural designs or placements where wood was not fixed to one location. Large wood was used in fixed designs in most studies from North America and usually built in or anchored by cables. Few projects attempted to simulate the dynamic processes of wood inputs to the floodplain. Mobile wood placements were mostly used in projects after 1990. They represented 6% of the projects in North America and 55% in Germany, where restoration projects designed with mobile wood can be found even in densely populated (200 people/km^2) rural areas, but only along second- and third-order streams. Few studies attempted to simulate historical amounts and distribution of wood in forested catchments. In most of the studies from rural areas, practical aspects like stream access or the availability of logs dominated the experimental design and placements.

Introduction

Our understanding of the role of instream large wood has undergone significant changes (Gregory 2003, this volume). Before the 1970s, large wood was generally considered a nuisance or hazard in streams throughout the world. Large wood was systematically removed from streams to benefit river navigation, prevent or decrease flooding effects, enhance log transportation, and improve fish passage (Maser and Sedell 1994). The consequences included alteration of riparian habitat, changes in nutrient cycles, the simplification of stream channels and the subsequent loss of fish habitat. These consequences have existed for hundreds of years in most European streams and for the last 150 years throughout North America. In Germany (Hering et al. 2000; Reich 2000), and probably in most parts of Europe (Gurnell et al. 1995; Elosegi et al. 1999), wood was removed from both large and small streams to "protect" them from flooding and to accelerate drainage of alluvial farmland.

The role of large wood in streams has been re-examined over the last 30 years. The understanding of that role has fundamentally shifted, as has the treatment of large wood in streams (Bisson et al. 1987). The emerging body of literature documents the role of large wood in structuring the physical template in streams (Abbe and Montgomery 1996; Bilby 2003, this volume), the importance of wood in nutrient cycles (Bisson et al. 1987), and the role of large wood in streams as fish habitat (Roni and Quinn 2001; Zalewski et al. 2003; Dolloff and Warren 2003; both this volume). The growing recognition that large wood is an important component in stream systems worldwide has caused researchers and managers to examine the potential for stream restoration or rehabilitation by adding large wood to streams.

In North America, the use of large wood in stream rehabilitation projects has a long history and probably evolved from work in England more than 100 years ago (Needham 1969). During the 1930s, large wood was used to improve habitat conditions in streams throughout the United States (Hubbs et al. 1932; Tarzwell 1937). This large wood was generally cut into small logs and then configured in various ways to create pools and holding habitat for salmonids. The use of log structures evolved to whole-tree (without rootwad) placements in the late 1970s. Logs were configured as single or multi-log structures in fixed sites by using cable and epoxy anchors tied into the streambed or large boulders (Fontaine 1987). These fixed placements have since evolved into whole-tree structures configured to mimic natural large wood frequencies and distribution throughout the channel (Gregory and Wildman 1998). These structures may be attached at one or more locations or be allowed to move freely in the channel and floodplain. In Germany, and probably in most parts of Europe, logs or trees were also used only in fixed placements to prevent bank erosion (Gunkel 1996), to protect tree plantations (soil bioengineering), or to store sediments in mountain brooks (Karl et al. 1975). The use of unattached whole trees in stream rehabilitation projects emerged during the late 1990s (Reich 2000).

During the last 20 years, interest in restoring stream habitats has been renewed throughout the world (Williams et al. 1997). As the understanding of the importance of large wood has emerged, people have recognized that this key element is missing from many streams. Consequently, a renewed interest has developed in the use of large wood in restoring and rehabilitating streams. This interest has resulted in large wood being placed in streams throughout the world. The large expenditure of time and effort in stream restoration has drawn the attention of the public, scientists, managers, and legislators. Concern over the value of restoration efforts have been raised by several authors (Sedell and Beschta 1991; Frissell and Nawa 1992; Beschta et al. 1995). Weaknesses identified in these projects include unclear objectives, design errors, and inadequate monitoring.

Although many projects have been completed, surprisingly little information is available on the success or failure of these projects worldwide. No synthesis is currently available that attempts to summarize the projects that were monitored. The original objectives of this paper were to provide a synthesis of stream restoration work worldwide using large wood and to speculate on the general usefulness of large wood as a restoration tool. Because of limited sample sizes, we will focus on the similarities and contrasts between projects in North America and Germany.

Methods

We compiled literature from a variety of sources including peer-reviewed journal articles, graduate theses, published gray literature, and various administrative reports. We developed a database of 34 papers or reports from restoration case studies that explicitly used large wood in restoration and reported results (Table 1). We attempted to generate information about project goals and objectives, land use, local human population density, year of the restoration, and the extent and type of large wood treatments. Although we tried to find as many published and unpublished reports as possible, we recognize that there may be papers that are not included in the synthesis. Not all information was available for all case studies, but the database was sufficient to draw some general conclusions and comparisons, especially between projects in North America and Germany.

Results and Discussion

We reviewed projects from the United States (17), Germany (11), and Canada (1). Projects occurred from 1976 to 2000, most of them in mountain regions. Forestry was the predominant land use in North America, while 45% of the German sites were surrounded by grassland (Figure 1).

Project goals and objectives were highly variable and fell into the general categories of restoring fish habitat, structural complexity in the channel, channel stability, and channel dynamics (Figure 2). All large-wood restoration projects in North America had goals related to restoring fish habitat, usually in combination with the restoration of structural complexity (81%). In contrast, none of the projects in Germany had fish populations as a stated goal. The primary goals for projects in Germany were to increase structural complexity in the channel (100%), usually in combination with improvements of the channel dynamics (91%). Only 13% of the projects reviewed in North America addressed the goal of improv-

TABLE 1. Stream restoration projects using large wood.

Country, state	Streams	Reference	Year of restoration
Canada:			
Quebec	NW Montreal	Burgess and Bider 1980	1976
USA:			
California	SF Salmon, EF Salmon, Elk Creek, Indian Creek	Olson and West 1989	1980+
Colorado	NF Poudre, Colorado, Walton, Jack, Beaver, St. Vrain	Riley and Fausch 1995	1988
North Carolina	Cunningham Creek	Wallace et al. 1995	1988
Oregon	Deep Creek	Grover 1996	1993
	Elk Creek	House et al. 1991; Crispin et al. 1993	1986–1989
	Grande Ronde, John Day, Tucannon, NF John Day	Taylor 2000	1990+
	Lobster, East, Moon	Solazzi et al. 2000	1990–1991
	Upper Nestucca	House et al. 1991	1986
	Quartz Creek	Gregory and Wildman 1998	1988
	Siuslaw tributaries	Armantrout 1991; Dewberry et al. 1998	1983
	Oregon coastal streams	Nickelson et al. 1992	
	Fish Creek	Reeves et al. 1997	1983
Virginia	NF Stony Creek	Hilderbrand et al. 1998	1992
Washington	NF Porter Creek	Cederholm et al. 1997	1990–1991
	SF Soleduck	Ralston et al. 1992	
	NF Stillaguamish	Abbe and White 2000	1998–1999
Wyoming	Huff Creek, Salt Creek	Binns 1982	1979
	Thomas Fork	Binns 1986	1979–1983
Germany:			
Brandenburg	Demnitzer Mühlenfließ	Kosel and Mutz 2000	2000
Hessen	Asphe	Gerhard and Reich 2001	1998
	Eifa	Reich 2000	1998
	Haberbach	Reich 2000	1998
	Ilsbach	Reich and Gerhard 2001	1998
	Jossklein	Gerhard and Reich 2000	1993
	Lehrsbach	Gerhard and Reich 2001	1998
	Lüder	Gerhard and Reich 2000	1995
	Lummersbach	Reich and Gerhard 2001	1998
	Nitzelbach	Gerhard and Reich 2001	1998
	Aabach	Reich and Gerhard 2001	1998
Japan:			
Hokkaido	Ichibangawa	Yanai, S. Hokkaido Forestry Research Institute, Bibai, Japan, personal communication	1996 + 1998
	Shakotan	Yanai, S. Hokkaido Forestry Research Institute, Bibai, Japan, personal communication	1996 + 1998

ing channel dynamics (Figure 2). In most of the German projects, the streams were straightened and severely regulated decades ago. Therefore, the major goal of the projects was to mitigate or to reverse these impacts by restoring dynamic processes. Generally, no quantifiable objectives were associated with this goal, so determining if or when the project might be successful was diffi-

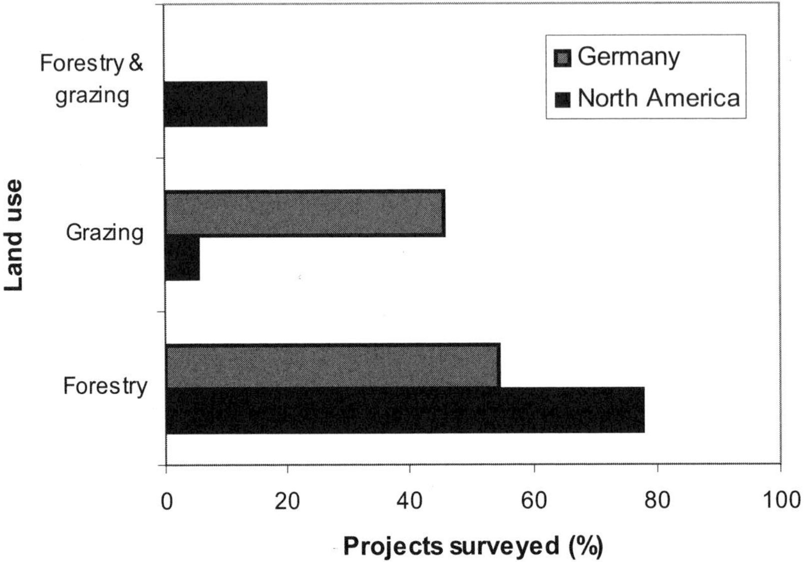

FIGURE 1. Land use in the floodplains or catchments of restoration projects in North America (n = 18) and Germany (n = 11).

cult. Note that improving fish habitat often includes improving structural complexity and changing the existing channel dynamics, however. These terms were often used interchangeably in the literature we reviewed.

The physical characteristics of the restoration project sites differed widely. The width of streams where large wood was introduced ranged from 1 to 40 m. In Germany, large wood has been predominantly used in second- and third-order streams, and occasionally in first-order streams (Hering et al. 2000; Reich 2000).

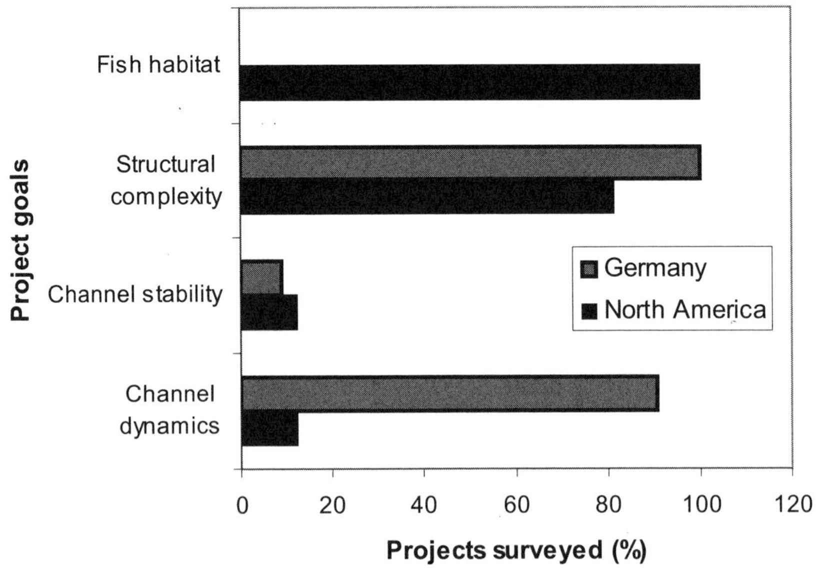

FIGURE 2. Goals of restoration projects in North America (n = 16) and Germany (n = 11).

Channel wetted width rarely exceeded 4 m (Figure 3). In contrast, streams in North America were generally wider than 4 m and ranged from 2 m to 40 m. We believe that the difference may be related to the project objectives and the potential threat of transported wood to human inhabitants and property. Because the objective of many American large wood introductions is to restore fish habitat, projects tend to be located in larger, fish-bearing streams. The primary objective for most German projects is to improve channel form, but not necessarily to improve fish habitat. There may be a number of reasons for this. Most German large wood introductions occur in streams that are situated in rural areas with 80–200 inhabitants/km^2 in the surrounding area. The risk of property damage caused by moving wood is high along all larger streams. Stream restoration with large wood started in smaller streams, where the improvement of the channel form was the dominating goal and the threat to human lives and property is minimal. Restored stream reaches in Germany were generally less than 0.5 km and contained less than five structures, but the restored reaches in North America were generally several kilometers long with numerous structures (Figure 4).

Large wood was placed in streams in one of three ways: (1) fixed structures attached to the bed, the bank, or both using cables or engineered placements; (2) whole trees or logs that were placed as either single pieces or groups of pieces and allowed to move freely at high flows; and (3) structures where large wood was attached to the bed or bank by a cable but cabling was loose enough to allow movements over several meters during high flow (partially mobile). In North America, fixed structures were used in the majority of the studies that were reviewed and were either configured (k-dams, log drops) or cabled in place (Figure 5). Mobile wood placements were mostly used in projects after 1990 and represented 6% of the projects in North America and 55% in Germany. Partially mobile wood placements represented 22% of the North American projects and 18% of the projects in Germany. In some of the North American projects, fixed and mobile structures were combined (Figure 5).

Projects that allowed large wood to move freely throughout the channel have become more common in both North America and Europe. Large wood that is freely allowed to move more closely mimics the wood input and movement processes in forested ecosystems. Whole-tree placements have been used in several areas in North America. Ashbridge (2001) used a combination of whole trees and logs in the East Fork Hood River and Clear Branch in Oregon. Large wood in the East Fork Hood River moved during two separate flood and debris flow events, but

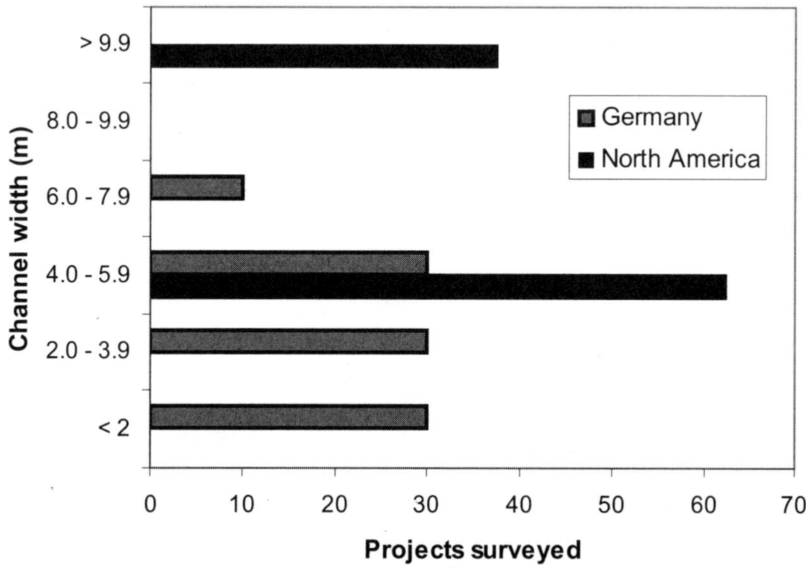

FIGURE 3. Channel width of restoration projects in North America (n = 8) and Germany (n = 10).

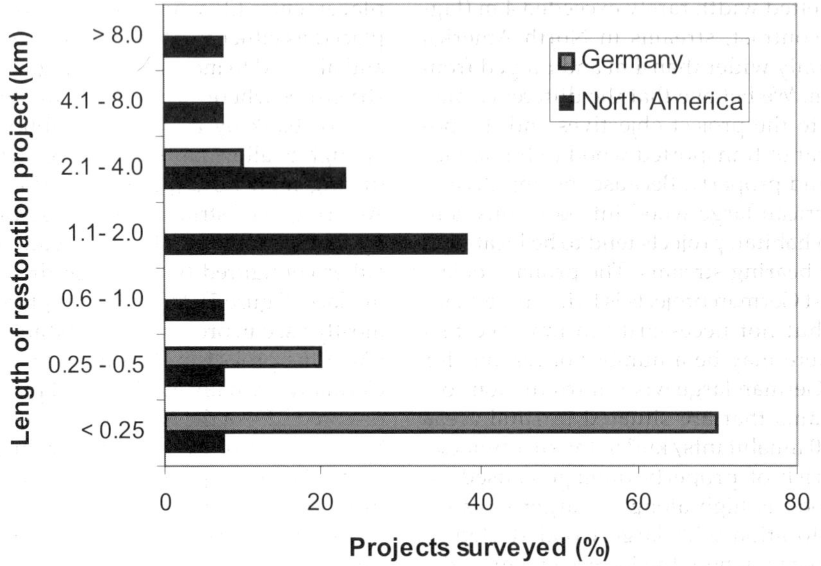

FIGURE 4. Length of restored reaches per project in North America (n = 13) and Germany (n = 10).

this movement was designed as part of the project objectives. Yager (1999) used large trees in the channel as key pieces designed to attract floating woody debris and build more complex debris jams. Similar projects have been built in British Columbia (Poulin 2001). Adding large wood to streams in Germany has primarily occurred in relatively small streams where the potential for extensive movements is low and the risk to downstream facilities is also low. Most of these projects are located along straightened channels, and re-establishing meanders is the major goal (Reich 2000). The reintroduction of mobile large wood in these channels has been shown to speed the re-

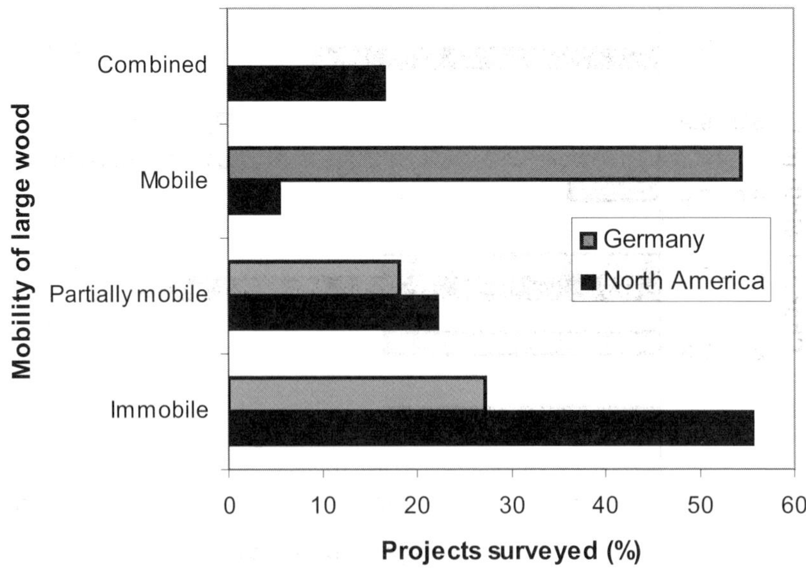

FIGURE 5. Mobility of the large wood in restoration projects in North America (n = 18) and Germany (n = 11).

establishment of channel function and structure (Gerhard and Reich 2000). The very first projects along large German rivers have been implemented with large wood configured and cabled in place (Völkl et al. 2002).

A wide variety of tree species are used in adding large wood in North America and Germany (Figure 6). In North American projects, native tree species typically represent the native riparian vegetation. Native conifers are the predominant trees used in restoration. In Germany, 82% of the tree species (poplar, spruce) used in the restoration projects were introduced to the adjacent floodplain forests decades ago. Non-native floodplain forests are now gradually being replaced by native alder, ash, and beech forests, and harvested non-native trees can be used for restoration projects during this process (Gerhard and Reich 2000).

Restoration projects using large wood generally have different goals in North America and Germany. In Germany, restoration success is based on changes in channel morphology, in substrate diversity, and in the abundance and diversity of aquatic macroinvertebrates (Figure 7). Restoration success in North America is often measured as an improvement in fish habitat and fish populations. Project monitoring results are lacking for many

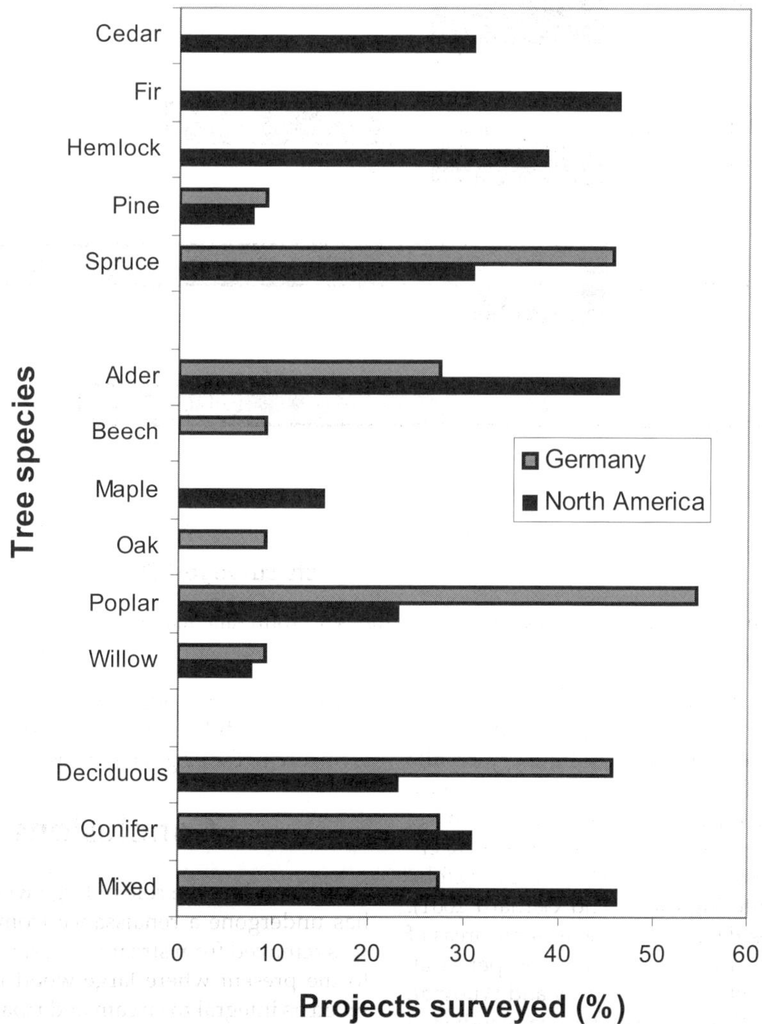

FIGURE 6. Tree species used in large-wood additions in North America ($n = 13$) and Germany ($n = 11$).

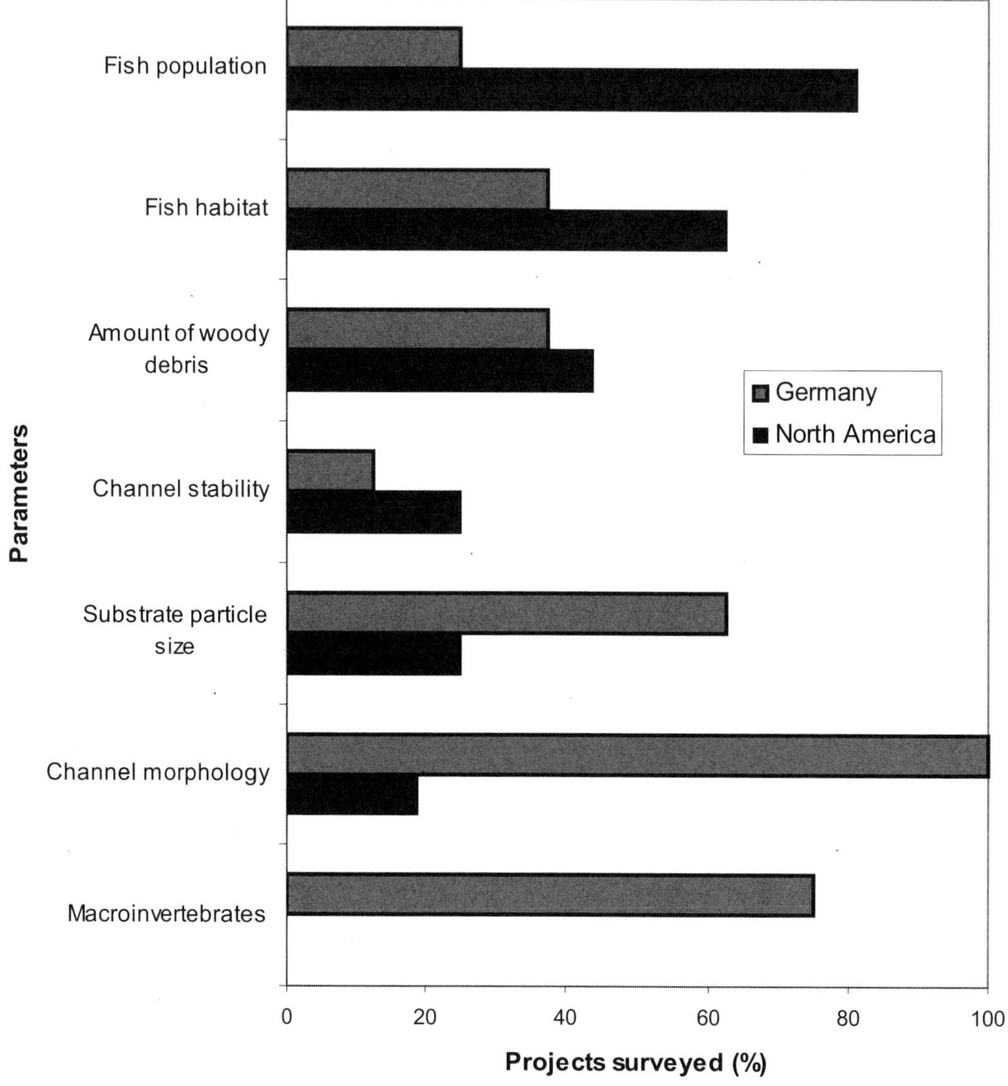

FIGURE 7. Parameters to measure success of wood additions in North America (n = 13) and Germany (n = 8).

projects in both Europe and North America. In Germany, most projects have been recently implemented, and there are only preliminary results (Table 1). It is difficult to measure success on such a short time scale, especially when channel forming floods have not yet occurred. Initial results are encouraging, but success will only be measured by long-term monitoring (Reich and Gerhard 2001). Longer-term results are available in some areas of North America (Reeves et al. 1997; Roper et al. 1997; Dewberry et al. 1998; Gregory and Wildman 1998), but few of these studies have been conducted during a time scale large enough to include multiple high flows or floods. Equally problematic is the lack of clear objectives for some of these projects. It is difficult to evaluate success when one is unsure of the original purpose.

Conclusions

Understanding the role of large wood in streams has undergone a renaissance from when wood was removed from streams as a standard practice to the present where large wood is now recognized as integral to stream and riparian function. In Germany, wood in streams was regarded primarily as a problem or a hazard before 1990. Since then, not only ecologists, but also hydraulic engineers and legislators, have begun to understand the ecological function of wood in streams. The new European water framework directive requires "good conditions" that are "not far from natural conditions" for streams throughout Europe. Clearly, large wood plays an important role to meet the intent of this policy. Although the policy recognizes the importance of large wood in streams, the problem of responsibility for damage caused by large wood is still unsolved. In North America, a similar change in philosophy has appeared over the past two decades. Before the 1970s, large wood was often viewed as an impediment to fish migration and passage, rather than a key component of streams and riparian areas. Changes in state forest practices and federal policy for large wood over the past decade reflect our growing understanding of the importance of large wood.

Large wood is an important tool in stream and riparian restoration. Probably thousands of restoration projects have been implemented that used large wood to change stream channel interactions with the floodplain. Surprisingly, the number of papers and reports documenting the results of using large wood in restoration is disproportionately small. We were able to find few papers that were peer reviewed, in gray literature, or available as graduate theses. Although many administrative reports or file reports are undoubtedly available for local use, they are often inaccessible to the public. We believe that restoration projects must have monitoring plans that document the success or failure of projects and that the results of monitoring must be available to a wide audience if we are to learn from our successes and failures. To measure success or failure, these projects will also require well-articulated, quantifiable objectives that define the parameters to be measured.

Our understanding of the linkage between riparian habitat, floodplains, and stream systems will help to refine the way large wood is used in restoration. As understanding of the role of large wood has changed, the philosophy of the role of large wood in restoration has also changed. The historical retrospective from our work suggests a switch from fixed instream placements to the use of wood in more natural, random groups of large wood free to move in the channel and floodplain. This change in philosophy most probably represents a similar emergence in the science of restoration ecology where the focus is on watershed restoration, rather than simply improving fish-habitat. Large wood additions are most often viewed in the context of stream and riparian function, rather than just fish habitat. Most often, fish-habitat objectives are developed as part of the objectives to restore stream and floodplain function rather than an endpoint. Restoration practitioners recognize that the long-term solution to the lack of large wood in streams is ultimately tied to riparian forest succession (Boyer et al. 2003, this volume). Large wood additions should generally be viewed as short-term fixes that may need to be sustained for several decades until successional processes in riparian forests can once again operate.

The literature reported here represents the results from stream restoration work in Germany and North America. Other large-wood restoration projects are ongoing in other countries like Austria (Urbanek et al. 1999), Poland (Kaczka, University of Silesia, Sosnowiec, Poland, personal communication), and Japan (Table 1). Undoubtedly, many more restoration projects have not yet been reported worldwide. As the body of literature from these projects grows, we hope that the evidence ultimately points to worldwide recovery of streams and riparian areas.

Acknowledgments

We thank Marc Gerhard, Herbert Diehl, Michael Mutz, Vollmer Consultants (Germany), and Seiji Yanai (Japan) for unpublished data. Caty Clifton, Dave Heller, Deb Konnoff, and Bob Metzger from the Pacific Northwest United States solicited unpublished papers from a variety of sources.

References

Abbe, T. B., and D. R. Montgomery. 1996. Large woody debris jams, channel hydraulics and habitat formation in large rivers. Regulated Rivers 12:201–221.

Abbe, T., and M. L. White. 2000. Wood debris tracking and ELJ performance report, North Fork Stillaguamish habitat enhancement project 1998–1999. Washington Trout Report. Duvall, Washington.

Armantrout, N. B. 1991. Restructuring streams for anadromous salmonids. Pages 136–149 *in* J. Colt, and R. J. White, editors. Fisheries Bioengineering Symposium. American Fisheries Society, Symposium 10, Bethesda, Maryland.

Ashbridge, G. 2001. Floodplain and stream channel

reconstruction: East Fork Hood River and Clear Branch Creek. Watershed Restoration Technical Bulletin 6(1):4A. Ministry of Forests and Watersheds, Vancouver.

Beschta, R. L., W. S. Platts, J. B. Kauffmann, and M. T. Hill. 1995. Artificial stream restoration – money well spent or expensive failure? Pages 76–104 *in* Proceedings of Environmental Restoration, UCOWR 1994 Annual Meeting Big Sky, MT. University Council on Water Resources, University of Illinois, Carbondale.

Bilby, R. E. 2003. Decomposition and nutrient dynamics of wood in streams and rivers. Pages 135–147 *in* S. V. Gregory, K. L. Boyer, and A. M. Gurnell, editors. The ecology and management of wood in world rivers. American Fisheries Society, Symposium 37, Bethesda, Maryland.

Bisson, P. A., R. E. Bilby, M. D. Bryant, C. A. Dollof, G. B. Grette, R. A. House, M. L. Murphy, K. V. Koski, and J. R. Sedell. 1987. Large woody debris in forest streams in the Pacific Northwest: past, present, future. Pages 143–190 *in* E. O. Salo, and T. D. Cundy, editors. Streamside management: forestry and fisheries interactions. University of Washington, College of Forest Resources, Seattle.

Binns, N. A. 1982. Improvement and evaluation of fluvial trout habitat in Wyoming. Pages 20–32 *in* Monte Seehorn, editor. Proceedings of the Trout Stream Habitat Improvement Workshop. United States Department of Agriculture – Forest Service, Southeastern Forest Experiment Station, Asheville, North Carolina.

Binns, N. A. 1986. Habitat, macroinvertebrate, and fishery response to stream improvement efforts in the Thomas Fork Bear River drainage, Wyoming. Pages 105–116 *in* J. G. Miller, J. A. Arway, and R. F. Carline, editors. Proceedings of the Fifth Trout Stream Improvement workshop. Pennsylvania Fish Commission, Harrisburg, Pennsylvania.

Boyer, K. L., D. R. Berg, and S. V. Gregory. 2003. Riparian management for wood in rivers. Pages 407–420 *in* S. V. Gregory, K. L. Boyer, and A. M. Gurnell, editors. The ecology and management of wood in world rivers. American Fisheries Society, Symposium 37, Bethesda, Maryland.

Burgess, S. A., and J. R. Bider, 1980. Effects of stream habitat improvements on invertebrates, trout populations, and mink activity. Journal of Wildlife Management 44(4):871–880.

Cederholm, C. J., R. E. Bilby, P. A. Bisson, T. W. Bumstead, B. R. Fransen, W. J. Scarlett, and J. W. Ward. 1997. Response of juvenile coho salmon and steelhead to placement of large woody debris in a coastal Washington stream. North American Journal of Fisheries Management 17(4):947–963.

Crispin, V., R. House, and D. Roberts. 1993. Changes in instream habitat, large woody debris, and salmon habitat after the restructuring of a coastal Oregon stream. North American Journal of Fisheries Management 13(1):96–102.

Dewberry, C., P. Burns, and L. Hood. 1998. After the flood. The effects of the storms of 1996 on a creek restoration project in Oregon. Restoration & Management Notes 16(2):174–182.

Dolloff, C. A., and M. L. Warren, Jr. 2003. Fish relationships with large wood in small streams. Pages 179–193 *in* S. V. Gregory, K. L. Boyer, and A. M. Gurnell, editors. The ecology and management of wood in world rivers. American Fisheries Society, Symposium 37, Bethesda, Maryland.

Elosegi, A., J. R. Diez, and J. Pozo. 1999. Abundance, characteristics, and movement of woody debris in four Basque streams. Archiv für Hydrobiologie 144(4):455–471.

Fontaine, B. L. 1987. An evaluation of the effectiveness of instream structures for steelhead trout rearing in Steamboat Creek basin. Unpublished Master's thesis. Oregon State University, Corvallis.

Frissell, C. A., and R. K. Nawa. 1992. Incidence and causes of physical failure of artificial habitat structures in streams of western Oregon and Washington. North American Journal of Fisheries Management 12:182–197.

Gerhard, M., and M. Reich. 2000. Restoration of streams with large wood: effects of accumulated and built-in wood on channel morphology, habitat diversity and aquatic fauna. International Review of Hydrobiology 85(1):123–137.

Gerhard, M., and M. Reich. 2001. Totholz in Fliessgewässern – Empfehlungen zur Gewässerentwicklung. Gemeinnützige Fortbildungsgesellschaft für Wasserwirtschaft und Landschaftsentwicklung, Mainz, Germany.

Gregory, K. J. 2003. The limits of wood in world rivers: present, past and future. Pages 1–19 *in* S. V. Gregory, K. L. Boyer, and A. M. Gurnell, editors. The ecology and management of wood in world rivers. American Fisheries Society, Symposium 37, Bethesda, Maryland.

Gregory, S. V., and R. C. Wildman. 1998. Aquatic ecosystem restoration project – Quartz Creek post-flood progress report. Oregon State University, Corvallis.

Grover, D. 1996. An evaluation of redband trout use of large woody debris in the Deep Creek drainage, Ochoco National Forest, fall 1996. Administrative Study. USDA Forest Service, Ochoco National Forest, Princeville, Oregon.

Gunkel, G., 1996. Renaturierung kleiner Fliessgewässer. Gustav Fischer Verlag, Jena, Germany.

Gurnell, A. M., K. J. Gregory, and G. E. Petts. 1995. The role of coarse woody debris in forest aquatic habitats: Implications for management. Aquatic Conservation 5:143–166.

Hering, D., J. Kail, S. Eckert, M. Gerhard, E. I. Meyer, M. Mutz, M. Reich, and I. Weiss. 2000. Coarse woody debris quantity and distribution in central European streams. International Review of Hydrobiology 85(1):5–23.

Hilderbrand, R. H., A. D. Lemly, C. A. Dolloff, and K. L. Harpster. 1998. Design considerations for large woody debris placement in stream enhancement projects. North American Journal of Fisheries Management 18(1):161–167.

House, R. A., V. Crispin, and J. M. Sutter. 1991. Habitat and channel changes after rehabilitation of two coastal streams in Oregon. Pages 150–159 in J. Colt and R. J. White, editors. Fisheries Bioengineering Symposium. American Fisheries Society, Symposium 10, Bethesda, Maryland.

Hubbs, C. L., J. R. Greeley, and C. M. Tarzwell. 1932. Methods for the improvement of Michigan trout streams. Bulletin of the Institute for Fisheries Research No. 1, University of Michigan Press, Ann Arbor.

Karl, J., J. Mangelsdorf, and K. Scheurmann. 1975. Der Geschiebehaushalt eines Wildbachsystems, dargestellt am Beispiel der Oberen Ammer. Deutsche Gewässerkundliche Mitteilungen 19:121–123.

Kosel, I., and M. Mutz. 2000. Naturgemässer Totholzeintrag als kostengünstige Methode zur Renaturierung sandgeprägter Tieflandbäche. Pages 573–578 *in* Deutsche Gesellschaft für Limnologie, Tagungsbericht 1999.

Maser, C., and J. R. Sedell. 1994. From the forest to the sea: the ecology of wood in streams, rivers, estuaries and oceans. St. Lucie Press, Delray Beach, Florida.

Needham, P. R. 1969. Trout streams: conditions that determine their productivity and suggestions for stream and lake management. Holden-Day Publishing, San Francisco.

Nickelson, T. E., M. F. Solazzi, and S. L. Johnson. 1992. Effectiveness of selected stream improvement techniques to create suitable summer and winter rearing habitat for juvenile coho salmon (*Oncorhynchus kisutch*) in Oregon coastal streams. Canadian Journal of Fisheries and Aquatic Sciences 49(4):790–794.

Olson, A. D., and J. R. West. 1989. Evaluation of instream fish habitat restoration structures in Klamath River tributaries, 1988/89. Klamath River Fisheries Resources Official Publication 89(4.25):36.

Poulin, V. A. 2001. Using thinnings from silvicultural treatments to augment or create fish habitat. Watershed Restoration Technical Bulletin 6(3):11A. Ministry of Forests and Watersheds, Vancouver.

Ralston, S. C., P. Beh, and G. L. Peterson. 1992. An example of fish habitat improvement techniques at a remote site on the Olympic Peninsula. AquaTalk: Fish Habitat Relationship Technical Bulletin: USDA Forest Service, Pacific Northwest Region, Portland, Oregon.

Reich, M. 2000. Ecological, technical and economical aspects of stream restoration with large wood. Zeitschrift für Ökologie und Naturschutz 8(4):251–253.

Reich, M., and M. Gerhard, editors. 2001. Ökologische, wasserbauliche und ökonomische Aspekte der Fliessgewässerrenaturierung mit Totholz. Technical Report, University of Hannover, Germany.

Reeves, G. H., D. B. Hohler, B. E. Hansen, F. H. Everest, J. R. Sedell, T. L. Hickman, and D. Shively. 1997. Fish habitat restoration in the Pacific Northwest: Fish Creek of Oregon. Pages 335–359 *in* J. E. Williams, C. A. Wood, and M. P. Dombeck, editors. Watershed restoration: principles and practices. American Fisheries Society, Bethesda, Maryland.

Riley, S. C., and K. D. Fausch. 1995. Trout population response to habitat enhancement in six northern Colorado streams. Canadian Journal of Fisheries and Aquatic Sciences 52(1):34–53.

Roni, P., and T. P. Quinn. 2001. Density and size of juvenile salmonids in response to placement of large woody debris in western Oregon and Washington streams. Canadian Journal of Fisheries and Aquatic Sciences 58:282–292.

Roper, B. B., K. Wieman, D. Konnoff, and D. Heller. 1997. Durability of Pacific Northwest instream structures following floods. North American Journal of Fisheries Management 18:686–693.

Sedell, J. R., and R. L. Beschta. 1991. Bringing back the "bio" in bioengineering. Pages 160–175 *in* J. Colt and R. J. White, editors. Fisheries Bioengineering Symposium. American Fisheries Society, Symposium 10, Bethesda, Maryland.

Solazzi, M. F., T. E. Nickelson, S. L. Johnson, and J. D. Rodgers. 2000. Effects of increasing winter rearing habitat on abundance of salmonids in two coastal Oregon streams. Canadian Journal of Fisheries and Aquatic Sciences 57(5):906–914.

Tarzwell, C. M. 1937. Experimental evidence on the value of trout stream improvements in Michigan. Transactions of the American Fisheries Society 66:177–187.

Taylor, C. H. 2000. Evaluation of stream habitat enhancement projects in the Umatilla National Forest, northeast Oregon and southeast Washington. Master's thesis. University of Oregon, Eugene.

Urbanek, B., M. Hinterhofer, and H. Kummer. 1999. Totholz in Fliessgewässern. Literaturrechereche und Analyse ausgewählter Aspekte. Technical report, Universität für Bodenkultur, Vienna, Austria.

Völkl, W., A. Hessberg, D. Mader, J. Metzner, P. Gerstberger, K. H. Hoffmann, H. Rebhan, and R. Krec. 2002. Natural succession in a dynamic riverine landscape and the protection of open areas. Pages 413–421 *in* B. Redecker, P. Finck, W. Härdtle, U. Reicken, and E. Schröder, editors. Pasture landscapes and nature conservation. Springer Verlag, Berlin, Heidelberg, New York.

Wallace, J. B., J. R. Webster, and J. L. Meyer. 1995. Influence of log additions on physical and biotic characteristics of a mountain stream. Canadian

Journal of Fisheries and Aquatic Sciences 52(10):2120–2137.

Williams, J. E., C. A. Wood, and M. P. Dombeck. 1997. Understanding watershed scale restoration. Pages 1–16 *in* J. E. Williams, C. A. Wood, and M. P. Dombeck, editors. Watershed restoration: principles and practices. American Fisheries Society, Bethesda, Maryland.

Yager, M. 1999. South Fork Coquille River Wood Placement Project. Watershed Restoration Technical Bulletin 4(2):10A. Ministry of Forests and Watersheds, Vancouver.

Zalewski, M., M. Lapinska, and P. B. Bayley. 2003. Fish relationships with wood in large rivers. Pages 195–211 *in* S. V. Gregory, K. L. Boyer, and A. M. Gurnell, editors. The ecology and management of wood in world rivers. American Fisheries Society, Symposium 37, Bethesda, Maryland.

American Fisheries Society Symposium 37:367–389, 2003

Wood in River Rehabilitation and Management

TIMOTHY B. ABBE

Herrera Environmental Consultants, Inc.
2200 Sixth Avenue, Suite 1100, Seattle, Washington 98121, USA

ANDREW P. BROOKS

Centre for Catchment and In-Stream Research
Griffith University, Nathan, Queensland, Australia 4111

DAVID R. MONTGOMERY

Department of Earth and Space Sciences
University of Washington, Seattle, Washington 98195, USA

Abstract.—Wood induces hydraulic, morphologic, and textural complexity into fluvial systems in forested regions around the world. Snags and logjams can create complex networks of channels and wetlands across entire river valleys and historically posed a significant obstacle to navigation. The clearing of wood from channels and riparian forest land reduced or eliminated the quantity and supply of wood into rivers in many regions of the world. Ecological restoration of fluvial environments increasingly includes the placement of wood. But few guidelines exist on appropriate methods for emulating natural wood accumulations, where and how to place wood, its longevity, the hydraulic and geomorphic consequences of wood, and how to manage systems where wood is reintroduced. Important factors to understand when placing wood in rivers include the watershed and reach-scale context of a project, the hydraulic and geomorphic effects of wood placements, possible changes in wood structures over time, and how it may impact human infrastructure and safety. Engineered logjams constructed in Washington, USA and New South Wales, Australia offer examples of how wood reintroduction can be engineered without the use of artificial anchoring to form stable instream structures as part of efforts to rehabilitate fluvial ecosystems and provide ecologically sensitive means to treat traditional problems such as bank stabilization and grade control.

Introduction

The geomorphic effects of wood on fluvial systems range in scale from controlling bed forms and influencing channel patterns to floodplain development (for example, Davis 1901; Wolff 1916; Guardia 1933; Keller and Swanson 1979; Lienkaemper and Swanson 1987; Harwood and Brown 1993; Abbe and Montgomery 1996, 2003; Buffington and Montgomery 1999; Brooks 1999; Brooks and Brierley 2002). Snags and logjams can be the principal mechanism creating habitat complexity not only within an active channel, but also by inducing localized flooding and creating and sustaining secondary channels and wetlands (Sedell and Froggatt 1984; Triska 1984; Abbe 2000; Collins and Montgomery 2002). Examples of these complex fluvial systems with numerous side channels that extend across much of a river valley are increasingly rare (Figure 1). Habitat complexity directly or indirectly related to wood clearly benefits many aquatic ecosystems (for example, Pearsons et al. 1992; Quinn and Peterson 1996; Lehtinen et al. 1997; Inoue and Nakano 1998), and reduced fish populations can reflect the extensive loss of physical complexity in fluvial systems resulting from wood removal, channelization, and floodplain development (for example, Shields and Smith 1992; Beechie et al. 2001; Collins et al. 2003; Pess et al. 2003). Rehabilitation of fluvial ecosys-

FIGURE 1. Snags and logjams contribute to the development of a complex mosaic of anastomosing channels, wetlands, and floodplain forest in river valleys such as Taiya River in southeast Alaska.

tems depends on re-establishing natural processes and conditions that create and sustain physical complexity, such as restoring instream structure and cover, mature riparian forests, floodplain connectivity, and channel migration zones (CMZs). Reintroduction of wood is an important part of river rehabilitation but must balance ecological benefits against the potential consequences to existing human development. The use of wood to restore streams and rivers should be based on sound science, engineering, and an understanding of the magnitude to which the system may change.

Although the role of wood in many parts of the world remains to be investigated, it has been found to play a significant role in the ecology and morphology of streams and rivers in a wide range of climates and physiographic regions, including Asia (for example., Inoue and Nakano 1998; Rikhari and Singh 1998), Australia and New Zealand (for example., Mosley 1981; Nanson et al. 1995; Gippel et al. 1996a; Weigelhofer and Waringer 1999; Brooks and Brierley 2002; Webb and Erskine 2003), Europe (for example, Gregory et al. 1993; Piégay 1993; Maridet et al. 1996; Piégay and Gurnell 1997; Gurnell and Sweet 1998; Hering et al. 2000a, 2000b; Diez et al. 2001; Kail 2003), northeastern North America (for example, Zimmerman et al. 1967; Thompson 1995; Beebe 1997), southeastern North America (for example, Veatch 1906; Guardia 1933; Diehl 1997; Wallerstein et al. 1997), southwestern North America (for example, Haden et al. 1999), and northwestern North America (for example, Keller and Swanson 1979; Harmon et al. 1986; Lienkaemper and Swanson 1987; Andrus et al. 1988; Robison and Beschta 1990; Nakamura and Swanson 1993; Maser and Sedell 1994). Historical management of wood in rivers has focused on the removal of snags and logjams, which were considered threats to navigation, flooding, and even fish passage (for example, Ruffner 1886; McCall 1984; Maser and Sedell 1994; Collins et al. 2002). Additional river management practices that have reduced the supply of wood include dams, channelization, bank stabilization, and removal of riparian forests. In many regions, retention of wood was further diminished by the loss of large trees capable of forming stable snags, as well as effects associated with channel straightening, levees, and navigation improvements. But it is worth noting that, through much of history, wood has been considered a viable material for instream structures and extensively utilized for infrastructure in rivers, such as bulkheads, spur dikes, bridge abutments and piers, weirs, and dams.

Large scale efforts to reintroduce wood to streams and rivers have been limited to the Pacific Northwest of North America. An increase in the passive reintroduction of wood associated with the reforestation of riparian areas, such as occurring in parts of the United States and Europe, will have physical consequences, such as increased channel width and complexity (for example, Davies 1997; Trimble 1997; Collins and Montgomery 2002), increasing flood inundation, and redirecting channels (Figure 2). Such physical effects can improve ecological conditions, but they may not be compatible with existing or future development and thus pose a significant challenge to efforts to restore streams and rivers. Channel clearing (removal of snags and logjams) continues to be a common practice in river management around the world despite international recognition of the ecological benefits of wood and studies demonstrating that significant quantities of wood can be left in some channels without adverse impacts on flow conveyance (for example, Shields and Gippel 1995; Gippel et al. 1996b). Because wood is not simply a "cosmetic" feature in streams and rivers, greater accountability for potential effects may be demanded for projects incorporating direct or passive wood reintroduction. Successful rehabilitation of fluvial ecosystems will depend on understanding the dynamics and effects of wood and identifying and resolving potential problems (for example, Maridet et al. 1996; Abbe et al. 2003). While the scientific understanding of the physical dynamics and effects of wood is still developing, substantial progress has been made regarding the patterns in which wood accumulates, its hydraulic and geomorphic effects, mechanics and longevity, and the performance of wood placements (for example, House and Boehne 1985, 1986; Murphy and Koski 1989; Gregory et al. 1993; Shields and Gippel 1995; Abbe and Montgomery 1996, 2003; Gippel et al. 1996a, 1996b; Abbe et al. 1997, 2003; Wallerstein et al. 1997, 2001, 2002; Gurnell and Sweet 1998; Syndi et al. 1998; Braudrick and Grant 2000).

Wood influences fluvial geomorphology by altering both the hydraulics and distribution of flow within a fluvial corridor and, thereby, the deposition and transport of sediment. Instream wood can provide stable flow obstructions that not only store sediment, but can also sustain and even enlarge pools in aggrading channels that have been subjected to increases in sediment supply (Lisle 1995; Lisle and Napolitano 1998). Natural wood accumulations can dam headwater channels (for example, Keller and Swanson 1979; Marston 1982) and redirect the course of large rivers (Wolff 1916; Guardia 1933; Sedell and Froggatt 1984; Triska 1984; Abbe and Montgomery 1996; Collins and Montgomery 2002). Numerous studies have described significant geomorphic effects of wood on relatively small channels (for example, Zimmerman et al. 1967; Keller and Swanson 1979). For example, instream wood can account for 8% to more than 80% of the elevation loss of montane forest channels (Tally 1980; Marston 1982; Thompson 1995; Abbe 2000). Montgomery and Buffington (1997) show that reach-scale channel morphologies associated with lower gradients can be imposed upon much steeper channels by the presence of wood. In the coast ranges of western Washington and Oregon, USA, stable log steps transform stream segments from bedrock to alluvial channels (Montgomery et al. 1996, 2003). When wood is removed from these streams, the sediment transport capacity exceeds supply and the channel segments revert to bedrock. Stable instream wood accumulations can also increase pool frequency (Andrus et al. 1988; Robison and Beschta 1990; Montgomery et al. 1995; Abbe and Montgomery 1996; Beechie and Sibley 1997), and individual snags or log jams can have both local effects on channel bed texture (Buffington and Montgomery 1999) and broader effects on stream morphology (for example, Keller and Tally 1979; Nakamura and Swanson 1993; Montgomery and Buffington 1997). In some sand-bed channels, virtually all pools can be attributed to either the direct or indirect control of wood (Brooks and Brierley 2002; Webb and Erskine 2003).

FIGURE 2. (a) The Deschutes River approximately 20 km south of Olympia, Washington on 4 February 2002. Flow is to the north (location 1). The large logjam (2) began forming in the 1990s and grew substantially in the winter of 2001–2002. As a result, river stage upstream of the logjam increased about 1.3 m during low flow conditions (~7 m^3/s versus a mean daily discharge of 14 m^3/s). Higher water elevations inundated adjacent floodplain areas (locations 3–5) and threatened low-lying homes (location 6). (b) Newly developed channel through relatively young floodplain on left side of logjam (location 3 on photo a). (c) Newly developed channel through mature cedar floodplain on right side of logjam (location 4 on photo a). All photographs were taken during low flow conditions, and arrows depict direction of flow. Despite the significant increase in the extent and complexity of aquatic habitat resulting from the logjam, it was cleared by private landowners in fall 2002.

Wood can also have reach-scale effects on channel form in large rivers (for example, Davis 1901; Russell 1909; Wolff 1916; Triska 1984; Abbe and Montgomery 1996; Collins and Montgomery 2002). The anastomosing channel pattern (that is, multiple channels separated by vegetated islands) of some large alluvial rivers has been attributed to an abundance of snags and logjams (Sedell and Froggatt 1984; Harwood and Brown 1993; Abbe 2000; Collins and Montgomery 2002; Abbe and Montgomery 2003). Effects of wood accumulations can extend beyond the confines of the bank-full channel to the development of floodplains and even terraces (Abbe 2000; Abbe and Montgomery 2003). In some circumstances, it is clear that wood accumulations were instrumental in vastly expanding the area of land subjected to inundation and fluvial processes (Ruffner 1886; Veatch 1906; Guardia 1933; Triska 1984; Collins and Montgomery 2001; Brooks et al. 2003).

Observations of geomorphic change that occurred after removal or addition of wood illustrate the magnitude to which wood can control river gradient and store immense quantities of sediment. A particularly striking example is given by clearing of wood from the Red River in Louisiana, USA, which caused portions of the river to incise more than 4 m (Veatch 1906; Guardia 1933; Harvey et al. 1988a, 1988b), increased the river gradient by an order of magnitude, and resulted in a sixfold increase in sediment transport capacity (Harvey et al. 1988a). Conversely, large wood accumulations can result in substantial aggradation of channels and floodplains. After a single year of unexceptional flows, the channel bed upstream of a logjam on Alta Creek, a tributary of the Queets River in Washington State, aggraded over 4 m (Abbe and Montgomery 2003). Channel aggradation upstream of wood accumulations results in distinctive landforms that resemble terraces but have surface gradients less than those of the valley in which they occur.

Brooks (1999) presents a dramatic example of the effect of instream wood debris and floodplain forests on the geomorphology of rivers in southeast Australia by comparing two adjacent sand-bed rivers: the relatively undisturbed Thurra River and the extensively disturbed Cann River. By early in the 20th century, the Cann River floodplain forest had been extensively cleared. The channel was then progressively cleared of wood through the 20th century. Prior to channel incision, the Cann River had not experienced any other significant types of disturbance (for example, dams, grade control), and, outside the valley bottom, the catchment remains under forest cover. Hence, channel and floodplain clearing represent the only feasible explanations for the river's present condition.

Relict channels and sedimentology of the Cann River's floodplain indicate that the river originally resembled the nearby Thurra River, which has remained undisturbed. The sand-bed Thurra River has a narrow sinuous channel with very high wood loading (0.032–0.044 m^3/m^2 within the wetted perimeter) that flows through a heavily forested floodplain (Brooks and Brierley 2002). Over the last 100 years, particularly the last 30 years, the channel capacity of the Cann River increased 700%, its morphological bank-full discharge increased 45-fold, mean annual sediment transport increased 850-fold, and peak instantaneous sediment transport capacity increased up to 40,000-fold. Bank stability parameters have also fundamentally changed, as have roughness characteristics of the channel and wood (Brooks 1999; Brooks et al. 2003). The Thurra River has exhibited no such changes, demonstrating not only the profound effect wood and riparian vegetation have on channel morphology and processes, but also on the evolution of the floodplain. A key finding from the undisturbed Thurra River was that wood comprised about 7% of the channel bed volume to a depth of 1.8 m (Brooks and Brierley 2002). The hydraulic roughness and stability imparted by this wood, coupled with macrophytes growing on the bed, prevented incision into what might otherwise be a highly unstable channel, thereby allowing the long-term aggradation of the channel and the entire floodplain (Brooks and Brierley 2002).

In gravel-bed rivers, natural logjams can form stable "hard points" within the channel migration zone, and that can limit bank erosion (Abbe and Montgomery 1996; O'Connor et al. 2003). Channel migration at these hard points is retarded, and the radius of curvature of river bends is reduced, forming tighter meanders over time (Abbe and Montgomery 2003). For example, throughout the Queets River valley in northwest Washington, USA, the radius of curvature of channel meanders associated with logjams is considerably less than for alluvial meanders with no logjams (Figure 3). This effect on channel geometry can have important consequences to hydraulic processes within the river, which, in turn, can result in additional geomorphic effects. Using a simple model for super-elevation of a water surface around a bend (Chow 1959), a reduction in the channel's radius of curvature, R_c, will

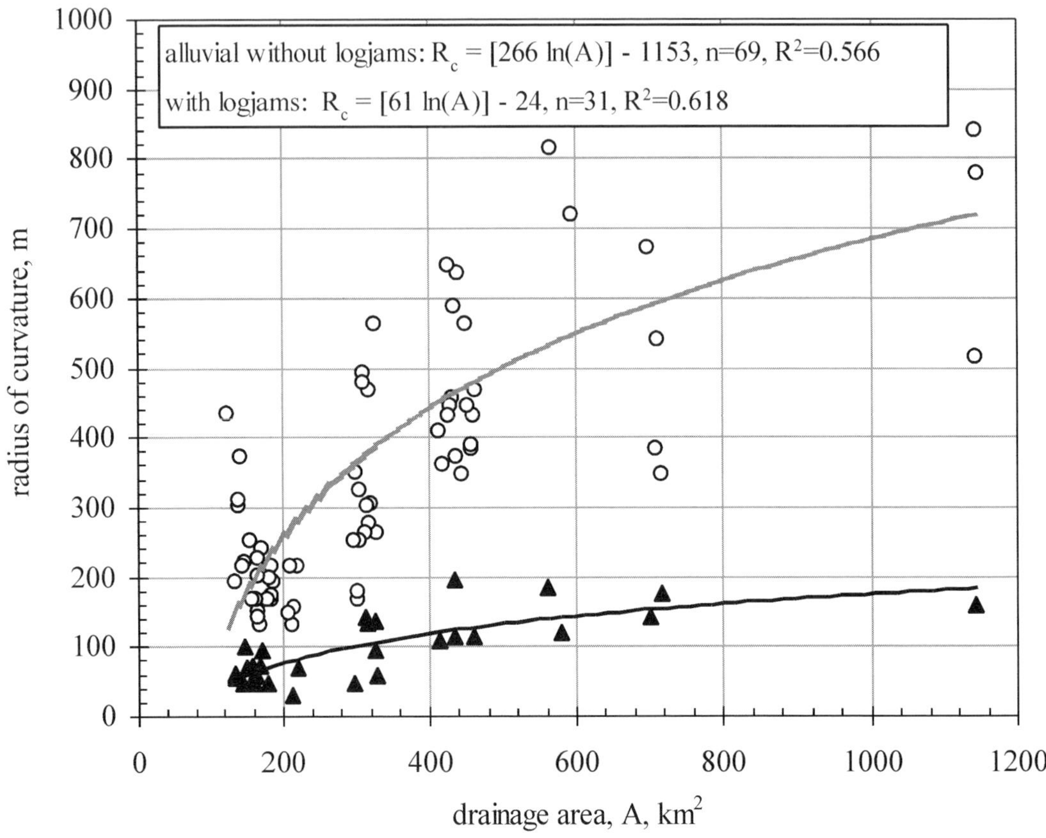

FIGURE 3. Radius of curvature of alluvial channel meanders in the Queets River, Washington. Meanders with logjams (solid triangles) have a significantly lower radius of curvature than meanders with no logjams (open circles) for the same reaches of the river system. Stable logjams form "hard points," which limit channel migration and result in the radius of curvature diminishing over time until formation of an avulsion or cutoff moves channel away from the logjam (modified from Abbe and Montgomery 2003).

increase water elevations at the outside of the bend:

$$\Delta h = \frac{W}{R_c} \times \frac{U^2}{g},$$

where Δh is the increase in water elevation around the bend (super-elevation), W is channel width, g is the acceleration of gravity, and U is the mean flow velocity around the bend. Manning's equation provides simple means of estimating velocity:

$$U = \frac{s^{1/2}\, h^{2/3}}{n},$$

where S is the hydraulic gradient, h is the mean water depth, and n is Manning's roughness coefficient. Differences in super-elevation between meanders with and without logjams were estimated using a simple model in which W, U, and R_c were predicted from empirical correlations to drainage area. The results indicate that super-elevation around bends associated with logjams is as much as 0.5 m more than meanders without logjams (Figure 4). These increased water elevations will increase the frequency of overbank flows and be more prone to initial development of avulsions and side channels.

Introducing a flow obstruction to a stream causes several indirect consequences related to the drag imposed on the flow. An increase in flow depth upstream of an obstruction can result from backwater effects. Because a flow obstruction results in a stagnation point where the horizontal velocity is essentially zero (Abbe and Montgomery 1996), the flow depth at that point is equivalent to the specific energy of the flow or $U^2/(2g)$.

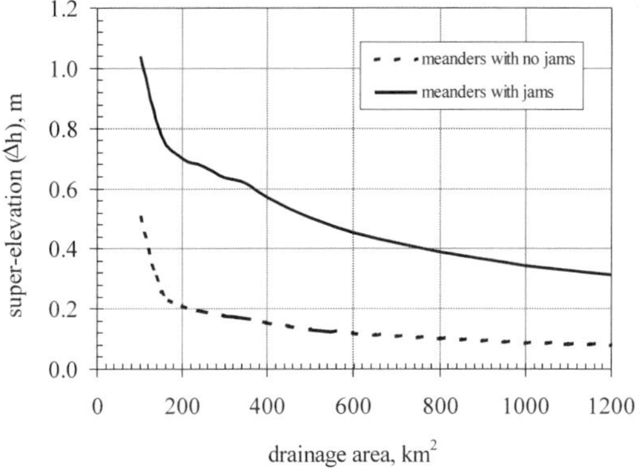

FIGURE 4. Significant reductions in a channel's radius of curvature can result in elevated water elevations around the outer bank of the meander due to super-elevation. Results of a simple model for estimating super-elevation along the Queets River, assuming only a change in radius curvature between meanders with and without logjams are plotted.

For a channel with a mean velocity of 3 m/s, the velocity head upstream of a snag or logjam would be about 46 cm, and 82 cm for a flow of 4 m/s. If the snag or logjam was located along the bank, these local increases in water elevation can deliver flow to secondary channels and the floodplain at lower discharges than it would otherwise take if the obstruction were not there. Localized increases in water elevations along riverbanks due to super-elevation and velocity head provide a mechanism to explain how logjams can contribute to the development and maintenance of side channels.

Gippel et al. (1996b) found that channel backwater effects are insignificant when wood occupies less than 5% of the bank-full cross-sectional area (a blockage coefficient of 0.05) but rapidly become more pronounced when more than 10% of the cross-sectional area is blocked. Increases in water elevation upstream of wood accumulations can be predicted given the water depth and Froude number of flow downstream of the obstructed area and the blockage and drag coefficients of the wood (Gippel et al. 1996b).

The stability of an individual piece of wood depends on whether the sum of resisting forces is greater than the sum of driving forces. The principal driving forces are buoyancy (function of displaced volume of water) and drag (function of flow velocity and cross-sectional area of wood within flow). The principal resisting forces are normal forces (function of the weight of wood and any surcharge due to overlying wood or sediment), bed friction (function of the wood area in contact with the bed or footprint), skin friction (function of the burial depth of wood into substrate), and passive earth pressures (function of the depth of sediment leeward of wood and bed material). The strength of wood can also influence its stability in small, steep channels where the wood is subjected to significant bending movements that could break a piece into smaller, more mobile pieces. Wood stability in large channels is primarily dependent on how the wood interacts with the channel bed, which is a function of the shape and size of the wood and the type of substrate (Abbe 2000). Resisting forces increase exponentially with burial of the wood due to skin friction, passive earth pressures, and surcharge.

Wood tends to be most efficiently transported through the deepest portion of a channel (thalweg), where it is subjected to the greatest buoyant and drag forces and least resistance. Just like a boat, floating wood has draft and will experience the least resistance where the water is deepest. Thus, the most stable wood accumulations are those that initiated in the thalweg where large snags become deeply embedded in the river (Figure 5). Once stable, a snag begins to redirect flow and trap other wood moving through the system (Abbe and Mongtomery 1996). As the channel continues to migrate, snags that formed in the thalweg become incorporated into bars and floodplains (Abbe and Montgomery 1996, 2003). It is important to differentiate the wood that is stable in the thalweg from the material that deposits on bars or other depositional features. There will always be smaller pieces of wood that deposit in areas of diminished depth and velocity, such as bars. But it is those pieces of sufficient size and shape to create stable flow obstructions that are crucial for altering channel morphology and retaining mobile wood or drift within the system (Keller and Swanson 1979; Nakamura and Swanson 1993; Abbe and Montgomery 1996; Abbe 2000; Abbe et al. 2003).

The draft of a tree bole increases substantially if it retains some of its root mat. The centroid of a symmetrical, homogenous tree bole will tend to be along the center line running through the middle of the bole. The centroid of a simple cylindrical log with no root mat is midway along the log's length. Situated on a smooth bed, a simple cylindrical log will distribute its mass along its entire length, resulting in relatively low normal stress and low resistance (Abbe 2000). Logs with

Figure 5. Snags embedded in thalweg of Mendenhall River, Alaska. Large trees eroded from outer bank of meander fall into river and become deeply imbedded in channel. Note crown of buried snag exposed in toe of point bar to left as river migrates to the right.

specific gravity less than unity will be buoyant at some water depth less than the log diameter. In contrast, the centroid of a bole with an attached root mat is located much closer to the basal end of the log. Weight is concentrated on a small area of the root mat perimeter in contact with the bed, thereby increasing the normal stress. The root mat also raises the centroid elevation, and the buoyant depth becomes more a function of the root mat diameter than of the bole diameter. This means that significantly greater flow depths are typically required to mobilize a log with a root mat than without. If a snag's resisting forces are still greater than the driving forces at the time bedload transport is initiated, bed deformation occurs around the snag (Abbe and Montgomery 1996, Abbe 2000). Once partially embedded, a snag can prove extremely difficult to move.

Factors that tend to increase wood stability—such as larger size, denser wood, and deposition in deep portions of the river where the wood is more likely to remain saturated—tend to also increase the wood's resistance to decay and its longevity in a system. Larger snags also have high lignin contents and lower surface area to volume that tend to further reduce decay rates (Melillo et al. 1983; Bisson et al. 1987;). Carbon dating of buried wood demonstrates that large logs can last hundreds to thousands of years in both high gradient gravel-bed and low-gradient sand-bed rivers (Murphy and Koski 1989; Nanson et al. 1995; Brooks 1999; Abbe 2000; Hyatt and Naiman 2001). Wood can remain in near pristine conditions indefinitely if it is kept either completely dry or saturated in anaerobic conditions. For example, timber piles used in the foundation of St. Mark's in Venice were found to be so well preserved they were left to support the reconstructed tower after already serving for more than 1,000 years (Jacoby and Davis 1941). Even the rapid decay rates associated with species such as black cottonwood *Populus trichocarpa* can be significantly reduced in near anaerobic conditions (Van Der Kamp and Gokhale 1979). Successful reintroduction of wood to rivers should focus on key pieces, the snags that have the most substantial geomorphic influence and longest longevity. Likewise, wood structures that incorporate key pieces will more likely persist in the system.

Integration of Wood into River Restoration

Large-scale efforts to reintroduce wood to streams in the Pacific Northwest of North America, particularly in rural forest land, were well underway in the 1980s (for example, House and Boehne 1985, 1986). While most wood reintroduction projects are well intended, the majority have been based simply on subjective decisions and have lacked rigorous scientific and engineering basis. Unfortunately, guidelines for stream restoration have little or no discussion of the mechanics and geomorphic effects of wood, nor do they provide natural analogs for the wood placements described (Rosgen and Fittante 1986; Rosgen 1996; Fischenich and Morrow 1999). The poor performance of many restoration projects is in part due to an insufficient understanding of the fluvial processes a project will be subjected to, how the project will influence these processes, and the consequences to habitat (Frissell and Nawa 1992). Preconceived perceptions that wood is inherently unstable and inadequate physical explanations of why wood was naturally stable in many streams and rivers have led to the widespread use of steel cables and artificial anchors or ballast for wood placements (for example, Fische-nich and Morrow 1999; D'Aoust and Millar 2000; Shields et al. 2000; Nichols and Sprague 2003). However, the stability of natural wood obviously never depended on such methods. For situations in which channel degradation has created conditions that are more inhospitable for wood stability than had naturally existed, such as incised channels or where large trees are no longer available, artificial means of stabilization may be necessary. A quantitative assessment of site conditions and a force balance analysis can provide the means to evaluate the stability of a proposed wood placement and help determine where artificial ballast is appropriate (Abbe et al. 1997; D'Aoust and Millar 1999, 2000; Shields et al. 2000; Castro and Sampson 2001). When cable is used in wood structures, it should only be used to secure logs tightly to one another or directly to rock ballast so that all the components act as one unified structure (D'Aoust and Millar 2000). Cable anchoring (for example, deadman or duck-billed anchors), commonly used in wood placements (Fischenich and Morrow 1999), poses significant risks that should be considered. A flexible medium, such as a cable, will not prevent wood from moving up and down or side to side with fluctuating stage or turbulence. Movement of the wood will move the cable, and an oscillating or vibrating cable will tend to cut away the material within which it is set. The cable can become exposed to create an entanglement hazard or simply fail and liberate the log that it was intended to secure. In contrast, stable wood structures designed without the use of any cable pose no such risks (Abbe et al. 1997, 2003; Brooks et al. 2001).

The observation that logjams act as a natural type of bank protection over long periods of time led to the idea that similar structures could be "engineered" to provide bank protection that better reflects the natural character of rivers than traditional engineering measures, such as rock revetments, bulkheads, and spur dikes. The term engineered logjams (ELJs) was proposed to refer to a general group of structures based on the premise of emulating natural fluvial systems using scientific observations and engineering design principles (Abbe et al. 1997). "Restoring" the natural quantities of wood loading in rivers associated with natural logjams may be unrealistic because the geomorphic consequences might be incompatible with human development and require time scales beyond the realm of planning. But there are many situations where instream wood can be reintroduced with no adverse impacts and even enhance habitat while solving traditional engineering problems. Unlike conventional river engineering solutions for flood risk reduction (for example, bank and bridge protection, grade control, snag management), ELJ technology is founded on engineering designs that emulate boundary conditions and processes of the natural fluvial system. This approach can offer the distinct advantage of contributing to the rehabilitation of riverine ecosystems, while still complying with constraints imposed by human development.

Design considerations for engineered logjams

ELJ design begins with a clear statement of the project objectives. The type of ELJ selected depends on site conditions and project objectives. The type of structure appropriate for a site is dependent on local climate, hydrology, channel and bank characteristics, bed material, objectives, and constraints. Analysis of flow conditions at the site, assessment of the conditions required for bed mobility, and a force balance of individual logs and the complete ELJ are fundamental parts of the design process. Both theoretical models and empirical relationships are utilized to evaluate factors such as buoyancy, drag, resistance, decay, and bed scour. Other design considerations include ELJ locations, size, and spacing; wood and sediment budgets; riparian vegetation; channel migration and avulsions; potential flood effects; and safety issues. Such analyses contrast with more traditional river restoration efforts that use generalized conceptual guidelines (House and Boehne 1985, 1986; Rosgen and Fittante 1986; Fischenich and Morrow 1999).

Careful consideration should be given to how and why wood is reintroduced into the fluvial environment, not only in regard to wood stabilization, but also how it is likely to respond to the system (for example, existing flux of mobile wood, channel migration rates, riparian forest conditions) and the potential effects wood placements may have (for example, increasing erosion, flood, avulsion risks, or hazards to human safety). Evaluating a project within a watershed and reach-scale context and how individual structures may alter the system or be altered (for example, changes in size) over time serves as a means of assessing project performance.

Any ELJ project begins with a geomorphic analysis of past, present, and probable future conditions of the fluvial system. Has the river changed during historical times; could these type of changes impact the project? If so, further analysis is needed to identify the factors responsible for those changes. ELJs may be used to reverse river changes previously identified or used to invoke changes to mitigate impacts elsewhere. An assessment of the potential direct and indirect consequences of predictable changes, such as significant increases in ELJ size with natural deposition of driftwood or channel changes that could affect the ELJs, can weigh the relative importance of project maintenance, which may temper the advisability of an ELJ project. An example of change commonly associated with relatively low-gradient alluvial rivers is lateral channel migration and bank erosion. The rate of channel migration and planform development of a river results from a complex interaction of numerous variables, but significant changes in certain factors, such as sediment supply, hydraulic geometry of the channel, flow regime, or bank vegetation, can accelerate or decelerate the migration rate. Assessment of the local geomorphic context, the characteristics and availability of large wood, and evidence for the type of natural wood jams is necessary to determine the appropriate type of ELJs for a particular site. The flow regime (hydraulics) of the channel and the availability of material for construction and subsequent recruitment will also influence ELJ design. Finally, infrastructure that affects hydraulic and geomorphic boundary conditions, such as bridges and the like, may also influence ELJ design.

Selecting an appropriate design for a project depends on the project objectives and constraints, the type of channel and its characteristics (gradient, size, bed material), and floodplain conditions. Several types of problems common in river engineering can be effectively addressed with different types of ELJs, such as grade control (that is., halting or retarding channel incision), bank protection, habitat rehabilitation, and creation of side channels. Geomorphic conditions influencing ELJ designs include bed and bank stability, site location in the channel network, disturbance history, and changes in channel conditions or processes.

The first step in ELJ design is to determine the appropriate type of ELJ for the project purpose and location. In some instances, such as controlling channel alignment upstream of a bridge and reducing debris accumulation on bridge piers, ELJs are selected to fit the location and constraints of the project. Another situation might prioritize establishment of pool habitat for fisheries enhancement. In this instance, the purpose and type of ELJ may be set, and the challenge is to determine the most suitable location within the stream reach of interest. The objectives of the project and the specific conditions and context of the project reach will determine how many jams may be necessary for an ELJ project.

Confidence in ELJ stability and longevity increases if the structures are evaluated for the range of flows and channel conditions expected to occur over the design life of the structure. An ELJ structure is designed to resist floatation utilizing its own mass and frictional resistance rather than employing artificial means, such as cabling. However, several additional factors reduce buoyancy and add resistance to an ELJ, so that, when submerged at some high stage, the ELJ does not become unstable. Both designed surcharge and natural sedimentation, as well as the accumulation of additional wood over time, all serve to further stabilize an ELJ. In effect, postconstruction sedimentation and wood recruitment increase the factor of safety for ELJ stability through time.

Because floatation is a key design issue, certain physical sites are well suited for ELJs. Sites with unconstrained alluvial channels and floodplains are ideally suited for wood structures that can be securely buried into the substrate. The larger shear stresses necessary to mobilize gravel make it more difficult for snags to become embedded than if they were situated in a sand-bed channel. But scour depths around a snag or logjam in a sand-bed channel are significantly greater, which can either undermine and destabilize the wood or bury it even deeper, to more firmly secure it. If buried so deep that only a small portion of the wood interacts with the flow, it may not have the desired effects. Confined channels, particularly those with bedrock beds, are more difficult sites for building stable instream wood structures. Wood structures have been applied with some success to incised channels, though typically with the addition of steel cable anchoring (House and Boehne 1985, 1986; Shields et al. 2000). In locations such as Australia, where timber with high specific gravity is available, stable wood structures without cable or artificial ballast have been built in incised, deep channels (Brooks et al. 2001).

Wood stability and longevity

Based on historical accounts, describing the difficulties encountered in removing natural snags and logjams from rivers (for example, McCall 1984; Ruffner 1886), and studies of the age of wood in a variety of river systems (for example, Nanson et al. 1995; Abbe and Montgomery 1996; Gippel 1996b; Brooks 1999; Abbe 2000), it is clear that natural wood can be remarkably stable and persist for long periods of time. Field studies of natural logjams have shown that the stability of wood accumulations is linked to the presence of one or more immobile snags or key members (Keller and Swanson 1979; Murphy and Koski 1989; Abbe and Montgomery 2003; Fox et al. 2003). Key member stability in steep channels is typically provided by interaction with pre-existing boundary conditions, such as bedrock outcroppings, boulders, or channel banks, and the strength properties of the tree, which determine its ability to withstand rupture. If the tree breaks, the smaller pieces become more susceptible to transport downstream. Log diameter, length, and strength are the principal factors influencing wood stability in steep channels (Abbe 2000).

Once the root mat becomes partially embedded in the channel bed, it becomes more difficult to move. A sediment buttress downstream of a root mat can form in situ by the accumulation of sediment in the leeward eddy or if the root mat sinks into the bed under its own weight. If an embedded rootwad begins to slide downstream, it will push up a mound of sediment, similar to a bulldozer, making further downstream movement increasingly difficult. Natural river snags are commonly observed with at least part of their root mat and bole embedded in the channel bed. Assuming a cohesionless substrate, the magnitude of resistance will depend on the depth to which the snag extends into the substrate, the surface area of the embedded portion of the root mat, the submerged weight of the alluvium, the friction angle of the alluvium, and the skin friction between the wood surface and alluvium. The importance of partial burial in the stability of wood is clearly illustrated by timber piles. A properly placed pile remains stable, even at its maximum buoyancy (fully submerged), despite having no ballast or anchoring. The stability of piles is entirely due to skin friction between the pile surface and the substrate. Frictional resistance of timber piles in various types of alluvium range from less than 1,000 N/m^2 to more than 10,000 N/m^2 (Jacoby and Davis 1941). It can take remarkably little burial to provide enough skin friction to compensate for buoyancy; for example, a completely submerged cylindrical pile 10 m long, 0.6 m in diameter, and a specific gravity of 0.5 would require a burial depth of 2.4 m for a substrate of loose sand (Figure 6). Skin friction tends to increase with increasing grain size and compaction of the substrate. Compact sand provides more than three times the skin friction of loose sand (Jacoby and Davis 1941). Since skin friction increases as a product of $2\pi R$, where R is pile

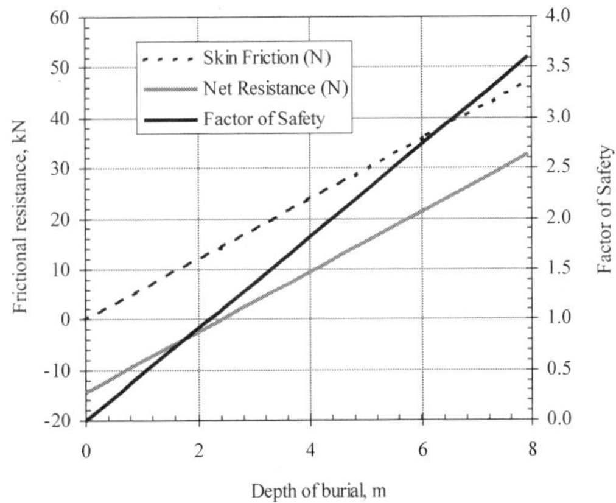

FIGURE 6. Skin friction for a timber pile as a function of burial depth in loose sand with a frictional resistance of 9.8 kN/m^2 (Jacoby and Davis 1941). Pile is 9.1 m (30 ft) in length, 0.6 m (2 ft) in diameter, and has an effective density of 500 kg/m^3. The pile is assumed to be completely submerged and thus have a constant upward buoyant force of 13.0 kN. Net resistance is equal to the skin friction—buoyant force (a negative value equals an upward force or unstable configuration). Factor of safety is net resistance divided by the buoyant force.

radius and wood volume increases as a function of $\pi^2 R$, the relative importance of skin friction resistance varies inversely with diameter.

Additional resistance to longitudinal forces is provided by the passive earth pressure of sediment downstream of the snag. This is the depth of sediment buttressing the downstream or leeward side of a buried log or root mat. The passive earth pressure per unit width of buried wood, P_p, in N/m, is a function of the depth of burial and physical characteristics of the substrate (Canadian Geotechnical Society 1985):

$$P_p = K_p \frac{\lambda_s z^2}{2} ,$$

where γ_s, = submerged bulk density of substrate in N/m^3, z is the depth of the buried wood segment, $K_p = (1 + \sin\phi)/(1 - \sin\phi)$ = coefficient of lateral passive earth pressure, and ϕ = internal friction angle of the substrate in degrees. For a buried segment of a circle, such as a partially buried root mat (Figure 7):

$$dP_p = P_p dx ,$$

where dP_p, = incremental passive earth pressure of buried segment of width $dx = (x_{i+1} - x_i)$ and the average depth of increment, z, between x_i and x_{i+1}. Summing each increment of the buried segment provides an estimate of the total resisting force imposed by passive earth pressure, F_p:

$$F_p = \Sigma dP_p dx .$$

The following example offers an illustration of how bed deformation (scour and lowering of

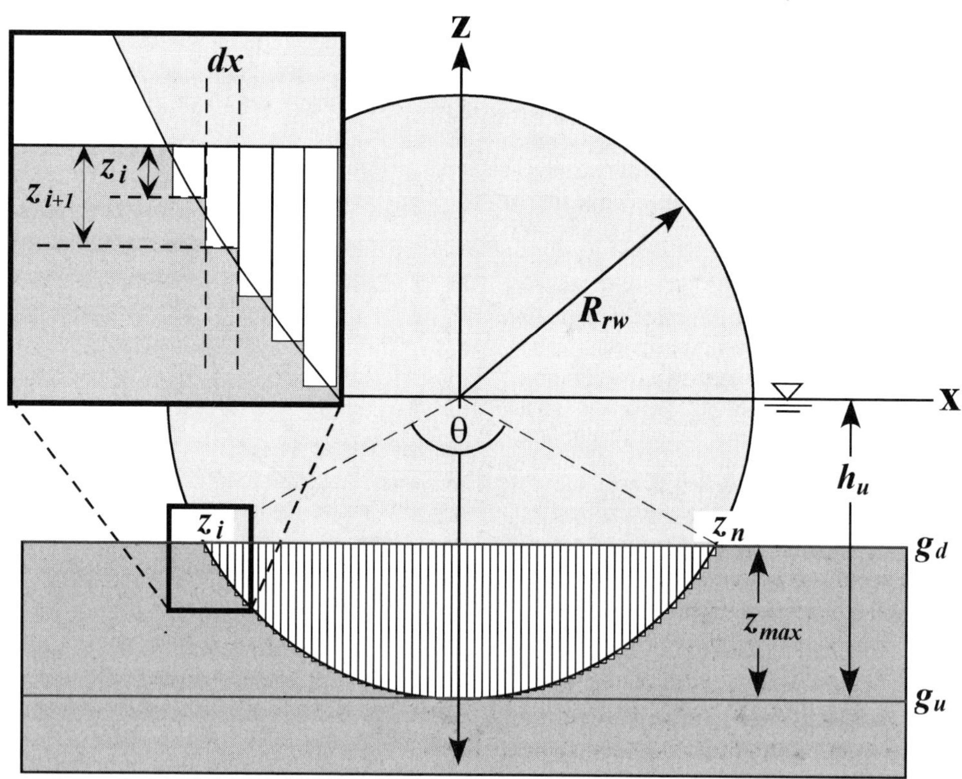

FIGURE 7. Two-dimensional representation of snag root mat with radius R_{rw} oriented normal to flow (in the downstream-facing view shown, the bole extends into the page). Water depth upstream of the root mat h_u is assumed to extend to lowest point of root mat (ground level upstream of root mat g_u), giving a conservative (high) measure of drag. Ground level downstream of root mat g_d represents the depth of sediment to the lee of the root mat for measuring passive earth pressures for each root mat increment of width dx and depth z. Angle Θ defines the arc of the buried segment of the root mat.

wood into bed and sediment deposition downstream of wood) can contribute sufficient resistance to stabilize a snag. Assuming an alluvial substrate composed of moderately rounded coarse gravel with a median grain size of 36 mm, $\phi = 35°$, then $K_p = 3.7$. The following assumptions also apply:

- substrate has a submerged bulk density of 1,200 N/m^3,
- snag has a homogeneous specific gravity = 0.5.
- snag is symmetrical about the axis of its bole
- snag bole axis is parallel to bed surface with root mat facing upstream
- snag radius = 2 m
- water depth = 2 m
- incident flow is uniformly distributed with a constant velocity = 3 m/s
- root mat is small relative to channel (no blockage effect on drag)
- drag coefficient = 1.5 (Gippel et al. 1996b)
- submerged area of root mat subjected to flow remains constant with respect to burial

Based on these assumptions, the snag is buoyant and therefore doesn't exert any normal force on the bed. The driving force exerted on the snag by flow is a function of the snag's submerged area normal to flow, the flow velocity, and the drag coefficient associated with the shape of the snag:

$$F_D = \frac{C_D A_o \rho U^2}{2},$$

where F_D is the drag imposed on the snag, C_D is the drag coefficient, U is velocity of flow, A_o is the cross-sectional area of the snag normal to flow, and ρ is the fluid density. Without any burial of the root mat, the snag has no resisting force and is unstable (Figure 8). Based on the assumptions of a constant uniform flow and that the submerged area, A_o, of the root mat remains constant, the drag force imposed on the snag remains constant (Figure 8). The resisting force rapidly increases with the depth of sediment downstream of the root mat, and a factor of safety of 1 is reached at a depth of about 1.0 m when the resisting and driving forces are equivalent at 42 kN (Figure 8). With an additional burial depth of 0.5 m (total depth of 1.5 m), the factor of safety = 2.8.

Sediment deposition on a tree bole can add substantial surcharge and thus further increase the stability of a snag or logjam. Field observations

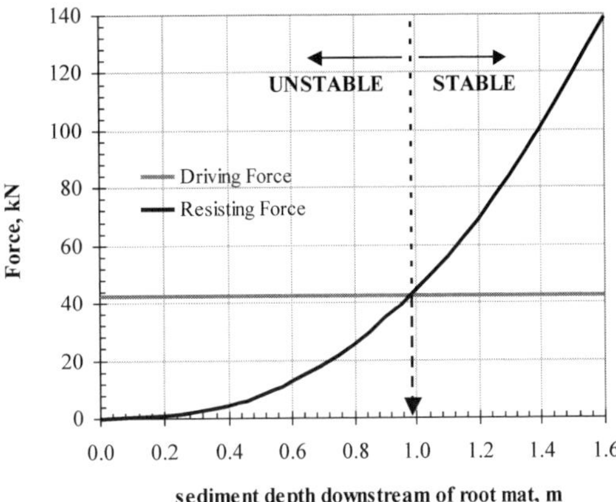

FIGURE 8. Simple example of driving and resisting forces on a snag with its root mat facing upstream and bole parallel to flow. Rootwad radius and flow depth are assumed to be 2 m, and therefore, snag is buoyant (no resisting force with no sediment leeward of root mat). Bed substrate is gravel and root mat subjected to a uniform constant flow velocity of 3 m/s.

show that sediment commonly accumulates downstream of a snag's root mat and buries part or all of the tree bole. In some cases, the entire root mat becomes buried deep within the bed. Fluvial sediments composed of coarse sand and gravel have dry bulk densities of 1,400–2,200 kg/m^3, about 3–4 times the densities of most woods, so overburden depths less than the log radius can be sufficient to negate any positive buoyancy of the wood.

If channel bed material is likely to be mobilized under flow conditions in which a tree bole remains stable, then the stability of the bole should increase due to accumulation of sediment immediately downstream. As flow moves past an obstruction (such as stable wood debris), an eddy of flow recirculation forms directly behind the obstruction, separated from the downstream flow of water by a vortex street (Abbe 2000). Sediment moving in the main flow is aggressively entrained by vortices in the shear zone and transferred into the recirculation zone where rapid deceleration leads to deposition. Such sedimentation influences debris stability in several ways: forming a buttress, which the rootwad must move up and over or plow through; adding surcharge or weight to buried portions of the tree bole; and increasing frictional resistance with burial, similar to the "embeddedness" concept by which most ship anchors work. Partial burial of a tree can increase the tree's resistance several fold and can help to explain why many snags and logjams remain stable even when completely submerged and subjected to extreme flows.

If an ELJ remains stable after bedload transport has been initiated, then accumulation of sediment in the downstream separation envelope should further stabilize the structure. In most gravel-bed channels, bedload transport begins at flows near or below bank-full stage. Hence, key members designed to be stable in flows exceeding bank-full conditions should become more stable after experiencing several bed-mobilizing events.

Another factor that can influence a force balance analysis of wood stability is the effective density of wood based on its moisture content. Most dry woods have densities less than water (1.0 kg/m^3) and cannot exceed a density of 1.5 kg/m^3, the density of the cellulose and ligin within wood (Harmon et al. 1986). As wood gets wet, its effective density can increase substantially, sometimes to values sufficient to cause it to sink, as illustrated in samples taken from wood placed in the North Fork Stillaguamish River, Washington, USA (Figure 9). Stability of wood through time can also be influenced by root cohesion and surcharge provided by vegetation growing above a snag or logjam.

Log and channel dimensions in rivers through-

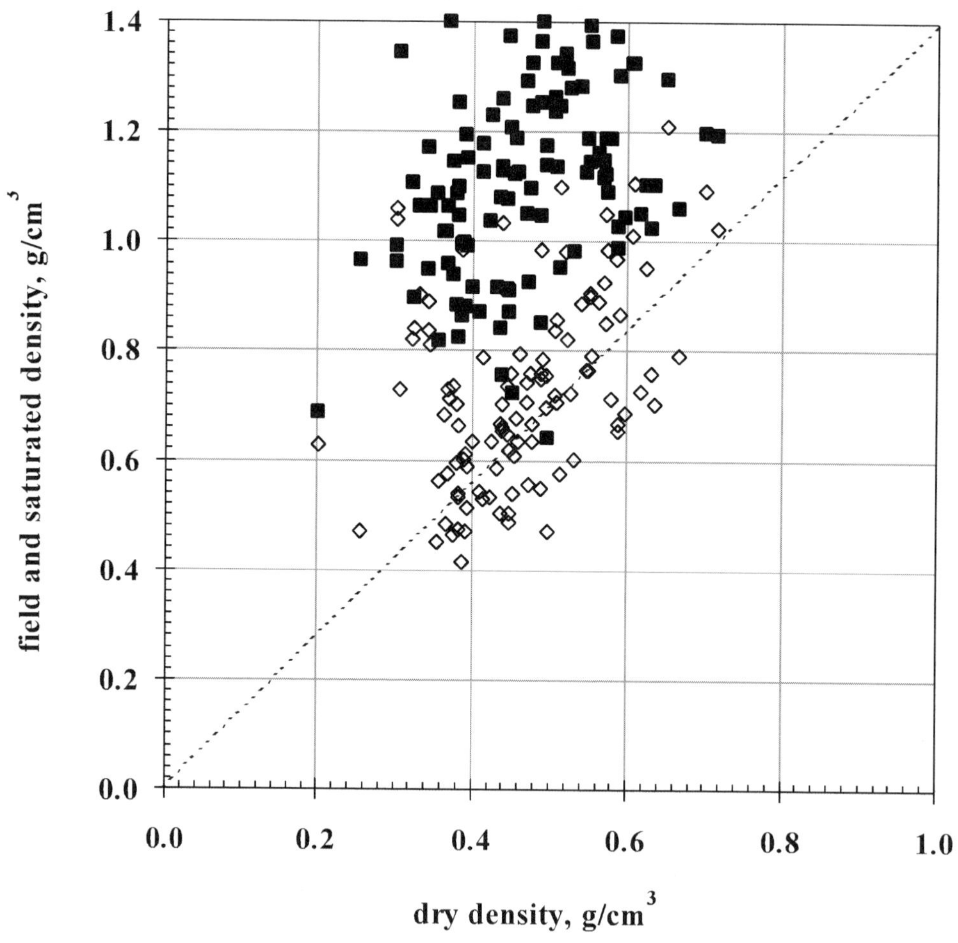

FIGURE 9. Densities of 109 core samples from trees used in engineered logjams constructed in North Fork Stillaguamish River, Snohomish County, Washington, 1998. Green (open circles) and saturated (solid squares) of same data set are plotted as a function of their oven dry densities. Tree species in data set include *Tsuga heterophylla, Thuja plicata, Picea sitchensis, Pseudotsuga menziesii,* and *Populus trichocarpa.*

out western Washington provide an empirical model of wood stability using dimensionless variables of log size: the ratio of log diameter to bank-full depth, D/h_{bf}, and the ratio of log length to channel width, L/w_{bf} (Abbe and Montgomery 2003). Stable logs, or key members, form the critical components holding wood accumulations in jams. In surveys of large alluvial channels, stable logs had minimum values of L/w_{bf} equal to 0.1 and values of D/h_{bf} rarely below 0.8. In channels exceeding 40 m in bank-full width, all key members had root mat diameters greater than the bank-full depth.

As discussed earlier, the buttressing or taper of a tree's bole concentrates a substantial portion of a tree's mass near its base, which significantly affects the log's mass and buoyant depth. The centroid elevation of a uniform cylindrical log laid upon a horizontal plane is simply equal to the log radius (as measured from the plane). A large root mat tends to concentrate mass, and the centroid moves toward that end of the bole, and the centroid elevation will approach the root mat radius. Given that bole radius is 20% or less than the root mat radius, the presence of an attached root mat can raise the centroid elevation fivefold or more. The root mat also acts similar to a plow or anchor to further increase resistance (Abbe 2000).

The availability of material to form key, stacked, and racked members is important when considering the use of ELJs. In some areas, the availability of large trees for use as key members may be a critical obstacle to the use of ELJ technology. The size of key members increases with both expected discharge and flow depth and can be related to the size of the channel, as discussed in the design section of this chapter. Costs for installation can vary greatly depending on the proximity of key members to the site location. As the size requirements for key members increase, transporting them becomes increasingly more difficult and expensive. Smaller racked members can account for 70–90% of the logs used in some types of ELJs that project into the flow, whereas some types of ELJ structures used for grade control or to harden a bank may have no racked members at all (Abbe et al. 2003).

Appropriate sizes, locations, and spacing for wood structures depend upon the project goals, the channel and its migration zone, human constraints, and acceptable risk. Bank protection projects, for example, could consist of continuous natural log revetments (for example, bank-full bench jams, flow deflection jams, chatoic cribs) and interspersed structures (for example, meander jams, flow deflection jams, crib groynes). The specific spacing of ELJ structures can be based on experimental studies of flow deflection by rock groynes and pile dikes (Klingeman et al. 1984). Our experience to date suggests that a spacing of about four times the protrusion into the main flow provides a conservative value for ELJ spacing; however, a spacing of up to seven times may prove acceptable in many situations, depending on the bend radius of curvature. A tighter bend will require closer spacing of the structures than a straight reach. Long-term project success will be better ensured by evaluating project designs under different scenarios representing the range of channel conditions, such as changes in the direction of incident flow on a structure (or set of structures), channel location, radius of curvature, bed elevations, or water surface profiles.

Wood structure dynamics over time

A key concern among many designers in regard to using wood structures is the longevity of wood as a material. How long wood lasts depends on factors such as the type of wood, the environment in which it is located, its size, and the age of tree at death. If wood remains saturated in freshwater, it can last almost indefinitely; it is not uncommon to recover premium grade timber that has been submerged in lakes for over a century. Boles of large deciduous trees found in river gravels of the ancient Rhine River in Germany have been dated at more than 18,000 years old and yet were still in excellent condition (Becker and Schirmer 1977; Becker 1980). Eucalyptus and pine logs found in floodplain sediments and across streams in southeast Australia and northwest Tasmania have yielded ^{14}C ages as old as 13,000–17,000 years b.p. (Nanson et al. 1995; Gippel et al. 1996a; Brooks and Brierley 2002). Several studies have shown that large boles comprising wood jams in relatively small streams found in coniferous forests can last for hundreds of years based on the age of trees growing on top (Keller and Tally 1979; Tally 1980; Hogan 1987; Murphy and Koski 1989). Wood from 30 buried logjams exposed in alluvial banks of the Queets River, Washington dated from modern to 1,324 ± 20 years b.p., averaging about 400 years b.p. (Abbe 2000). In contrast, logjams composed of relatively small pieces debris, such as slash commonly left behind after timber harvest, last only several years (Lisle 1986; Murphy and Koksi 1989).

Few data exist for absolute decay rates of different types of wood in fluvial environments, but there is an extensive literature on decay rates of wood on forest floors (see Mackensen and Bauhus 1999 for a recent review). Such studies of the decay rates of timber under moist aerobic conditions on a forest floor can provide a conservative estimate for timber decay used in the construction of wood structures in rivers. Accelerated laboratory tests to determine decay rates of some typical Australian eucalyptus species came up with turnover times (that is, the time taken for a block of wood 5 cm × 1.25 cm × 1.25 cm to decay to 95% of its initial mass) ranging from around 30–375 years. The authors of these experiments acknowledge that these rates are conservative for the species being tested, given that the experimental blocks have much higher ratios of surface area to volume than large logs. Species that might typically be used in river rehabilitation projects in Australia, such as river red gum *Eucalyptus camaldulensis*, forest red gum *E. tereticornis*, or iron bark *E. paniculata*, have values at the upper end of this range.

Models for wood longevity can be applied to estimate the design life of individual logs based on tree species, log diameter, and environmental setting (Abbe 2000). Decay rates can vary substantially for the same species and settings, but, in general, the longer and more continuously the wood remains submerged, the longer it will last. Assuming a constant decay rate, k, and knowing the initial mass of a log, $M(o)$, the log's mass at some time, $M(t)$, can be predicted using the simple decay model:

$$M(t) = M(o)e^{k(\Delta t)} ,$$

where Δt is the number of years from when first placed in the stream, assuming its mass was $M(o)$ when first placed. The effective diameter of a log after some time can also be predicted using the same model and assuming that decay proceeds uniformly into a cylindrical log from its perimeter (Figure 10). This type of procedure can be used to specify the size of logs that will most likely meet minimum specifications over the design life of a structure. Wood decay is a complex and highly variable process dependent on the species and physical condition of the wood, microclimate and organisms breaking down the wood, so information regarding local decay processes and rates will improve design predictions.

A substantial increase in ELJ size can occur over time if it collects debris moving through the channel. Changes in the ELJ size can subsequently obstruct a larger portion of the channel cross-sectional area and result in significantly higher water surface elevations than anticipated in the original project design. The risk will depend on changes in channel position relative to the ELJs and the flux of wood moving through the system at particular flow levels. An examination of the watershed and stream system above the site is important to get some understanding of the flux of drifting wood through the project reach and whether its accumulation at the ELJs is likely to create management problems.

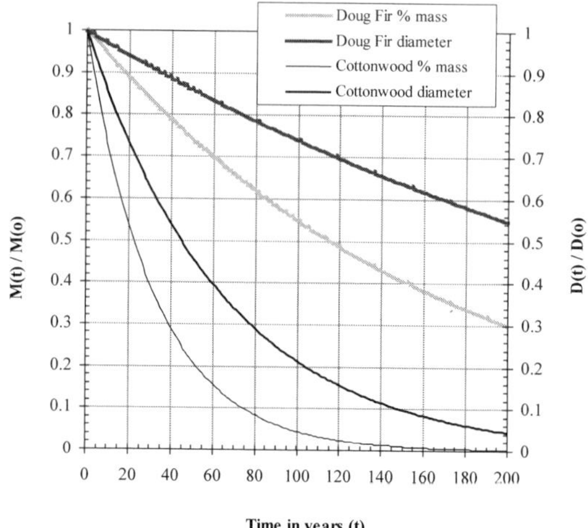

FIGURE 10. Simple decay model to estimate mass and effective diameter of a log after some time. Dimensionless mass and diameter are plotted as a function of time. $M(t)$ is its mass after time t, and $M(o)$ is the log's initial mass when placed. Diameter is derived assuming uniform decay from the perimeter of a cylindrical log to its central axis. Two different species, *Pseudotsuga menziesii* (Douglas-fir) and *Populus trichocarpa* (cottonwood), are presented using decay rates for logs lying on the ground in areas west of the Cascade Range of Oregon and Washington (Harmon et al. 1986).

Wood reintroduction projects will benefit from adaptive management plans that provide protocols for determining when additional wood accumulation may become a problem and providing guidelines for what, where, and when to remove selected pieces. These plans can also include inspection criteria for assessing the structural integrity of ELJs and what to do if maintenance becomes necessary. Failure to establish a reasonable inspection and maintenance program is likely to elevate risk of project failure or development of adverse effects due to evolution of an ELJ to conditions outside the scope of project design.

Examples of engineered logjam projects

Although ELJs remain a relatively new technology, they have been successfully used in a growing number of contexts. Demonstration projects include three ELJs constructed in the upper Cowlitz River in 1995 (Abbe et al. 1997), five ELJs in the North Fork Stillaguamish River in 1998, and seven ELJs in the Cispus River in 1999 (Abbe et al. 2003). Each of these projects have experienced multiple overtopping flows, and all remain intact in 2003.

Nineteen ELJ structures built on the Williams River, New South Wales, Australia in September–October 2000 (Brooks et al. 2001) show that ELJs may be applied beyond the Pacific Northwest. In September 2000, 19 ELJs incorporating 430 logs were constructed in a 1.1-km-long reach of the Williams River as part of an experimental reintroduction of wood into Australian Rivers. The ELJ structures included both flow deflection structures along the river's banks and channel-spanning grade control structures intended to prevent channel incision and trap additional sediment within the project reach. All of the structures were built without any artificial anchoring (that is, no cabling or imported ballast). The study site experienced six flows that overtopped most of the structures in the first 6 months after construction (Figure 11). Three of these peak flows exceeded the mean annual flood, inundating 18 of the ELJs by 2–3 m and 1 by 0.5 m. All the structures remain intact and, thus far, they are performing as intended, with a net increase of 40–60 m^3 of sediment storage per 1,000 m^2 of bed surface within the test reach and a substantial increase in channel complexity. A control reach 5 km upstream experienced a net export of 15 m^3/1,000 m over the same period.

As of 20 March 2002, the five ELJs in the North Fork Stillaguamish have been completely inundated by at least 30 flows and have successfully met objectives for increasing pool frequency and cover, bifurcating the river into two perennial channels, protecting 200 m of eroding bank, and improving a chronic problem of drift accumulation on a bridge directly downstream of the project (Abbe et al. 2003). The ELJs were designed to collect mobile wood (that is, drift or flotsam) upstream of the bridge while increasing the conveyance capacity of water and flotsam beneath the bridge by improving channel alignment. Flotsam has yet to rack up on the bridge's center piling, even though hundreds of logs have been deposited on four of five ELJs. Of the logs that moved and were subsequently found in 1999, those that were initially located upstream of at least one ELJ traveled from 200 to 600 m. Logs that moved downstream of the bridge, where there are no significant flow obstructions in the river, traveled much further, from 2 to 11 km. This data suggests that the position and frequency of logjams reduces the travel distance of wood and increases wood residence time within the fluvial system.

The successful performance of ELJ projects to date is largely attributed to the comprehensive geomorphic, engineering, and ecological analysis leading to design and implementation. Many questions remain regarding the science, engineering, risk assessment, and long-term performance of ELJs. For this reason, implementation of additional projects still should be viewed as applied research that contributes to advancing the existing body of knowledge. In situations where natural recovery of river corridors will be slow, unlikely, or unduly constrained by human development, however, ELJs offer exciting possibilities for reintroducing wood to rehabilitate fluvial environments.

Conclusions

River rehabilitation and management will benefit from emulating natural processes through a comprehensive approach to understanding current and historical conditions within a river reach and its watershed, identification of opportunities and constraints within the project reach, and application of sound engineering practice. The re-introduction of wood to rivers should follow these same principles. Projects constructed to date in the United States and Australia offer examples of engineered logjams that have experienced overtopping flows and have successfully met objectives to rehabilitate habitat, protect banks, reverse channel incision, and reduce risks to bridges. Specifically, several projects have demonstrated that ELJs (1) can remain stable in 20+ year recurrence interval floods; (2) can be effective at redirecting even large channels to control local bank erosion; (3) can dramatically enhance physical habitat, such as pool frequency and depth, and cover; and (4) need not exacerbate local flooding or erosion. The application of ELJ technology in these projects demonstrates the potential to address both river engineering and habitat concerns. As with any

FIGURE 11. View from upstream of 2 of 15 engineered logjams constructed in the Williams River, northwest New South Wales, Australia. (a) As-built structures in September 2000, (b) March 2001 flood during which peak flow overtopped structures by approximately 2 m with velocities exceeding 3 m/s, and (c) same structures in 2002 after six overtopping flows. No cables or artificial anchoring was used in the structures, and they all remain intact as of July 2003.

relatively new technology, further experimental applications are warranted before formalization of design standards. However, site-specific geomorphic and stability analyses based on general principles of ELJ design appear to provide a reasonable framework for the assessment, design, and implementation of successful projects.

The last several decades have seen an unprecedented growth of grassroots community efforts to rehabilitate urban and rural streams in both the United States and Australia. Widespread recognition by the scientific community and the public regarding the environmental and economic costs associated with human alteration of fluvial systems has probably never been greater. In many regions, dramatic changes are occurring in government policies regarding fluvial systems through increased regulation of land use within floodplains and channel migration zones. But the long-term success of efforts to rehabilitate the physical and biological attributes of fluvial systems will depend on development of management and engineering strategies that better emulate natural processes and provide valuable alternatives to traditional river engineering practice. Experience to date demonstrates that, in the overall context of these changing approaches to river management, wood can be an integral part of strategies aimed at rehabilitating and maintaining natural conditions while still attaining reliable engineered solutions to local problems.

Acknowledgments

Initial development of engineered logjam technology was made possible through a grant from the Watershed Science Institute of the Natural Resources Conservation Service, U.S. Department of Agriculture, with special thanks to Carolyn Adams. Australian research is funded by Land and Water Australia grants MQU9 and GRU29, in collaboration with the Department of Land and Water Conservation, Hunter Region, and the Hunter Catchment Management Trust. Selene Fisher was a tremendous help in editing our initial draft, and three anonymous reviewers helped to significantly improve the final manuscript. Mary Lou White and Washington Trout coordinated monitoring efforts in the North Fork Stillaguamish River. The Cispus project was made possible by Donna Ortiz de Anaya and Brenda White. And thanks to Jill Silver and Jim Turner for their steadfast commitment to protecting rivers.

References

Abbe, T. B. 2000. Patterns, mechanics and geomorphic effects of wood debris accumulations in a forest river system. Doctoral dissertation. University of Washington, Seattle.

Abbe, T. B., and D. R. Montgomery. 1996. Interaction of large woody debris, channel hydraulics and habitat formation in large rivers. Regulated Rivers Research & Management 12:201–221.

Abbe, T. B., and D. R. Montgomery. 2003. Patterns and processes of wood accumulation in the Queets River basin, Washington. Geomorphology 51:81–107.

Abbe, T. B., D. R. Montgomery, and C. Petroff. 1997. Design of stable in-stream wood debris structures for bank protection and habitat restoration: an example from the Cowlitz River, WA. Pages 809–816 *in* S. S. Y. Wang, E. J. Langendoen, and F. D. Shields, Jr., editors. Proceedings of the Conference on Management of Landscapes Disturbed by Channel Incision. University of Mississippi, University.

Abbe, T. B., G. Pess, D. R. Montgomery, and K. L. Fetherston. 2003. Integrating engineered logjam technology in river rehabilitation. Pages 443–490 *in* D. R. Montgomery, S. Bolton, D. B. Booth, and L. Wall, editors. Restoration of Puget Sound rivers. University of Washington Press, Seattle.

Andrus, C. W., B. A. Long, and F. H. Froehlich. 1988. Woody debris and its contribution to pool formation in a coastal stream 50 years after logging. Canadian Journal of Fisheries and Aquatic Sciences 45:2080–2086.

Becker, B. 1980. Progress in dating the Holocene valley development by subfossil trees. Bulletin de l'Association Francaise pour l'Etude du Quaternaire 1–2:85–86.

Becker, B., and W. Schirmer. 1977. Palaeoecological study on the Holocene valley development of the River Main, southern Germany. Boreas 4:303–321.

Beebe, J. T. 1997. Fluid patterns, sediment pathways and woody obstructions in the Pine River, Angus, Ontario. Doctoral dissertation. Wilfrid Laurier University, Waterloo, Ontario.

Beechie, T. J., and T. H. Sibley. 1997. Relationship between channel characteristics, woody debris, and fish habitat in northwestern Washington streams. Transactions of the American Fisheries Society 126:217–229.

Beechie, T. J., B. D. Collins, and G. R. Pess. 2001. Holocene and recent geomorphic processes, land use, and salmonid habitat in two north Puget Sound river basins. Pages 37–54 *in* J. B. Dorava, D. R. Montgomery, B. Palcsak, and F. Fitzpatrick, editors. Geomorphic processes and riverine habitat. American Geophysical Union, Water Science and Application 4, Washington, D.C.

Bisson, P. A., R. E. Bilby, M. D. Bryant, C. A. Dolloff, G. B. Grette, R. A. House, M. L. Murphy, K. V. Koski, and J. R. Sedell. 1987. Large woody debris in forested streams in the Pacific Northwest: past, present, and future. Pages 143 to 190 *in* E. O. Salo and T. W. Cundy, editors. Streamside management: forestry and fishery interactions. University of Washington, Institute of Forest Resources Contribution No. 57, Seattle.

Braudrick, C. A., and G. E. Grant. 2000. When do logs move in rivers? Water Resources Research 36:571–583.

Brooks, A. P. 1999. Pre- and post-European disturbance river morphodynamics in East Gippsland, Australia. Doctoral dissertation. Macquarie University, Sydney, Australia.

Brooks, A. P., T. B. Abbe, J. D. Jansen, M. Taylor, C. J. Gippel. 2001. Putting the wood back into our rivers: an experiment in river rehabilitation. Pages 73–80 *in* I. Rutherfurd, F. Sheldon, G. Brierley, and C. Kenyon, editors. Proceedings of the 3rd Australian Stream Management Conference. Brisbane, Australia.

Brooks, A. P., and G. J. Brierley. 2002. Mediated equilibrium: the influence of riparian vegetation and wood on the long term character and behaviour of a near pristine river. Earth Surface Processes and Landforms 27:343–367.

Brooks, A. P., G. J. Brierley, and R. G. Millar. 2003. The long-term control of vegetation and woody debris on channel and floodplain evolution: insights from a paired catchment study in southeastern Australia. Geomorphology 51:7–29.

Buffington, J. M., and D. R. Montgomery. 1999. Effects of hydraulic roughness on surface textures of gravel-bed rivers. Water Resources Research 35:3507–3522.

Canadian Geotechnical Society. 1985. Canadian Foundation Engineering Manual, 2nd Edition. Bitech Publishers, Vancouver.

Castro, J., and R. Sampson. 2001. Incorporation of large wood into engineering structures. U.S. Department of Agriculture, Natural Resource Conservation Service Engineering Technical Note No. 15, Boise, Idaho.

Chow, V. T. 1959. Open-channel hydraulics. McGraw-Hill Book Company, Inc., New York.

Collins, B. D., and D. R. Montgomery. 2001. Importance of archival and process studies to characterizing pre-settlement riverine geomorphic processes and habitat in the Puget Lowland. Pages 227–243 *in* J. B. Dorava, D. R. Montgomery, B. Palcsak, and F. Fitzpatrick, editors. Geomorphic processes and riverine habitat. American Geophysical Union, Water Science and Application 4, Washington, D.C.

Collins, B. D., and D. R. Montgomery. 2002. Forest development, wood jams, and restoration of floodplain rivers in the Puget Lowland. Restoration Ecology 10:237–247.

Collins, B. D., D. R. Montgomery, D. R., and A. Haas. 2002. Historic changes in the distribution and functions of large woody debris in Puget Lowland rivers. Canadian Journal of Fisheries and Aquatic Sciences 59:66–76.

Collins, B. D., D. R. Montgomery, and A. J. Sheikh. 2003. Reconstructing the historical riverine landscape of the Puget lowland. Pages 79–128 *in* D. R. Mongtomery, S. Bolton, D. B. Booth, and L. Wall, editors. Restoration of Puget Sound rivers. University of Washington Press, Seattle.

D'Aoust, S. G., and R. G. Millar. 2000. Stability of ballasted woody debris habitat structures. Journal of Hydraulic Engineering 126:810–817.

D'Aoust, S. G. D. and R. G. Millar. 1999. Large woody debris fish habitat structure performance and ballasting requirements. BC Ministry of Environment, Lands and Parks and BC Ministry of Forests, British Columbia Watershed Restoration Program, BC.

Davies, R. J. 1997. Stream channels are narrower in pasture than in forest. New Zealand Journal of Marine and Freshwater Research 31:599–608.

Davis, W. M. 1901. Physical geography. Ginn & Company, Boston.

Diehl, T. 1997. Drift in channelized streams. Pages 139–143 *in* S. S. Y. Wang, E. J. Langendoen, and F. D. Shields, Jr., editors. Proceedings of the Conference on Management of Landscapes Disturbed by Channel Incision. University of Mississippi, University.

Diez, J. R., A. Elosegi, and J. Pozo. 2001. Woody debris in North Iberian streams: influence of geomorphology, vegetation, and management. Environmental Management 28(5):687–698.

Fischenich, C., and J. Morrow, Jr. 1999. Streambank habitat enhancement with large woody debris. U.S. Army Engineer Research and Development Center, EMRRP Technical Notes Collection, ERDC TN-EMRRP-SR-13, Vicksburg, Mississippi.

Fox, M., S. Bolton, and L. Conquest. 2003. Reference conditions for instream wood in western Washington. Pages 361–393 *in* D. R. Mongtomery, S. Bolton, D. B. Booth, and L. Wall, editors. Restoration of Puget Sound rivers. University of Washington Press, Seattle.

Frissell, C. A., and R. K. Nawa. 1992. Incidence and causes of physical failure of artificial habitat structures in streams of western Oregon and Washington. North American Journal of Fisheries Management 12:182–197.

Gippel, C. J., B. L. Finlayson, and I. C. O'Neill. 1996a. Distribution and hydraulic significance of large woody debris in a lowland Australian river. Hydrobiologia 318:179–194.

Gippel, C. J., I. C. O'Neill, B. L. Finlayson, and I. Schnatz. 1996b. Hydraulic guidelines for the reintroduction and management of large woody debris in lowland rivers. Regulated Rivers Research & Management 12:223–236.

Gregory, K. J., R. J. Davis, and S. Tooth. 1993. Spatial distribution of coarse woody debris dams in the Lymington Basin, Hampshire, UK. Geomorphology 6:207–224.

Guardia, J. E. 1933. Some results of log jams in the Red River. The Bulletin of the Geographical Society of Philadelphia 31(3):103–114.

Gurnell, A. M., and R. Sweet. 1998. The distribution of large woody debris accumulations and pools in relation to woodland stream management in a small, low-gradient stream. Earth Surface Processes and Landforms 23:1101–1121.

Haden, G. A., D. W. Blinn, J. P. Shannon, and K. P. Wilson. 1999. Driftwood: an alternative habitat for macroinvertebrates in a large desert river. Hydrobiologia 397:179–186.

Harmon, M. F., J. F. Franklin, F. J. Swanson, P. Sollins, S. V. Gregory, J. D. Lattin, N. H. Anderson, S. P. Cine, N. G. Aumen, J. R. Sedell, G. W. Lienkaemper, K. Cromack, Jr., and K. W. Cummins. 1986. Ecology of coarse woody debris in temperate ecosystems. Advances in Ecological Research 15:133–302.

Harvey, M. D., D. S. Biedenharn, and P. Combs. 1988a. Adjustments of Red River following removal of the Great Raft in 1873. EOS, Transactions of the American Geophysical Union 69(18):567.

Harvey, M. D., H. S. Pranger, II, D. S. Biedenharn, and P. Combs. 1988b. Morphologic and hydraulic adjustments of Red River from Shreveport, LA to Fulton, AK, between 1886 and 1980. Pages 764–769 in S. R. Abt and J. Gessler, editors. ASCE National Conference on Hydraulic Engineering. Colorado Springs, Colorado.

Harwood, K., and A. G. Brown. 1993. Fluvial processes in a forested anastomosing river: flood partitioning and changing flow patterns. Earth Surface Processes and Landforms 18:741–748.

Hering, D., M. Mutz, and M. Reich. 2000a. Woody debris research in Germany – an introduction. International Review of Hydrobiology 85:1–3.

Hering, D., J. Kail, S. Eckert, M. Gerhard, E. I. Meyer, M. Mutz, M. Reich, and I. Weiss. 2000b. Coarse woody debris quantity and distribution in central European streams. International Review of Hydrobiology 85:5–23.

Hogan, D. L. 1987. The influence of large organic debris on channel recovery in the Queen Charlotte Islands, British Columbia, Canada. Pages 343–353 *in* R. L. Beschta, T. Blinn, G. E. Grant, F. J. Swanson, and G. G. Ice, editors. Erosion and sedimentation in the Pacific Rim. International association of Hydrological Sciences, Publication No. 165, Wallingford, UK.

House, R. A., and P. L. Boehne. 1985. Evaluation of instream enhancement structures for salmonid spawning and rearing in a coastal Oregon stream. North American Journal of Fisheries Management 5:283–295.

House, R. A., and P. L. Boehne. 1986. Effects of instream structures on salmonid habitat and populations in Tobe Creek, Oregon. North American Journal of Fisheries Management 6:38–46.

Hyatt, T. L., and R. J. Naiman. 2001. The residence time of large woody debris in the Queets River, Washington. Ecological Applications 11:191–202.

Inoue, M., and S. Nakano. 1998. Effects of woody debris on the habitat of juvenile masu salmon (*Oncorhynchus masou*) in northern Japanese streams. Freshwater Biology 40:1–16.

Jacoby, H. S., and R. P. Davis. 1941. Foundations of bridges and buildings. McGraw-Hill Book Co., New York.

Kail, J. 2003. Influence of large woody debris on morphology of six central European streams. Geomorphology 51:207–223.

Keller, E. A., and F. J. Swanson. 1979. Effects of large organic material on channel form and fluvial processes. Earth Surface Processes 4:361–380.

Keller, E. A., and T. Tally. 1979. Effects of large organic debris on channel form and fluvial processes in the coastal redwood environment. Pages 169–197 *in* D. D. Rhodes and G. P. Williams, editors. Adjustments of the fluvial system. Kendal-Hunt, Dubuque, Iowa.

Klingeman, P. C., S. M. Kehe, and Y. A. Owusu. 1984. Streambank erosion protection and channel scour manipulation using rockfill dikes and gabions. Oregon State University, Water Resources Research Institute Technical Report WRRI-98, Corvallis.

Lehtinen, R. M., N. D. Mundahl, and J. C. Madejczyk. 1997. Autumn use of woody snags by fishes in backwater and channel border habitats of a large river. Environmental Biology of Fishes 49:7–19.

Lienkaemper, G. W., and F. J. Swanson. 1987. Dynamics of large woody debris in streams in old-growth Douglas fir forests. Canadian Journal of Forest Research 17:150–156.

Lisle, T. E. 1986. Stabilization of a gravel channel by large streamside obstructions and bedrock bend, Jacoby Creek, northwestern California. Geological Society of America Bulletin 97:999–1011.

Lisle, T. E. 1995. Effects of woody debris and its removal on a channel affected by the 1980 eruption of Mount St. Helen's, Washington. Water Resources Research 31:1797–1808.

Lisle, T. E., and M. B. Napolitano. 1998. Effects of recent logging on the main channel of North Fork Caspar Creek. USDA Forest Service General Technical Report PSW-GTR-168, pp.81–85, Berkeley, California.

Mackensen, J. and J. Bauhus. 1999. The decay of coarse woody debris. National Carbon Accounting System, Technical Report No. 6. Australian Greenhouse Office, Canberra.

Maridet, L., H. Piégay, O. Gilard, and A. Thevenet. 1996. River wood jams: an ecological benefit? A factor of natural risks? La Houille Blanche 51:32–38.

Marston, R. A. 1982. The geomorphic significance of log steps in forested streams. Annals of the Association of American Geographers 72:99–108.

Maser, C., and J. R. Sedell. 1994. From the forest to the sea: the ecology of wood in streams, rivers, estuaries, and oceans. St. Lucie Press, Delray Beach, Florida.

McCall, E. 1984. Conquering the rivers: Henry Mill Shreve and the navigation of America's inland waterways. Louisiana State University Press, Baton Rouge.

Melillo, J. M., R. J. Naiman, J. D. Aber, and K. N. Eshleman. 1983. The influence of substrate quality and stream size on wood decomposition dynamics. Oecologia (Berlin) 58:281–285.

Montgomery, D. R., T. B. Abbe, N. P. Peterson, J. M. Buffington, K. Schmidt, and J. D. Stock. 1996. Distribution of bedrock and alluvial channels in forested mountain drainage basins. Nature (London) 381:587–589.

Montgomery, D. R., and J. M. Buffington. 1997. Channel reach morphology in mountain drainage basins. Geological Society of America Bulletin 109:596–611.

Montgomery, D. R., J. M. Buffington, R. Smith, K. Schmidt, and G. Pess. 1995. Pool spacing in forest channels. Water Resources Research 31:1097–1105.

Montgomery, D. R., T. M. Massong, and S. C. S. Hawley. 2003. Debris flows, log jams and the formation of pools and alluvial channel reaches in the Oregon Coast Range. Geological Society of America Bulletin 115:78–88.

Mosley, M. P. 1981. The influence of organic debris on channel morphology and bedload transport in a New Zealand forest stream. Earth Surface Processes and Landforms 6:571–579.

Murphy, M. L., and K. V. Koski. 1989. Input and depletion of woody debris in Alaska streams and implications for streamside management. North American Journal of Fisheries Management 9:427–436.

Nakamura, F., and F. J. Swanson. 1993. Effects of coarse woody debris on morphology and sediment storage of a mountain stream system in western Oregon. Earth Surface Processes and Landforms 18:43–61.

Nanson, G. C., M. Barbetti, and G. Taylor. 1995. River stabilization due to changing climate and vegetation during the late Quaternary in western Tasmania, Australia. Geomorphology 13:145–158.

Nichols, R. A., and S. G. Sprague. 2003. Use of long-line cabled logs for stream bank rehabilitation. Pages 422–442 *in* D. R. Mongtomery, S. Bolton, D. B. Booth, and L. Wall, editors. Restoration of Puget Sound rivers. University of Washington Press, Seattle.

O'Connor, J. E., M. A. Jones, and T. L. Haluska. 2003. Flood plain and channel dynamics of the Quinault and Queets rivers, Washington, USA. Geomorphology 51:31–59.

Pearsons, T. N., H. W. Li, and G. A. Lamberti. 1992. Influence of habitat complexity on resistance to flooding and resilience of stream fish assemblages. Transactions of the American Fisheries Society 121:427–436.

Pess, G., D. R. Montgomery, T. J. Beechie, and L. Holsinger. 2003. Anthropogenic alternations to the biogeography of Puget Sound salmon. Pages 129–154 *in* D. R. Mongtomery, S. Bolton, D. B. Booth, and L. Wall, editors. Restoration of Puget Sound rivers. University of Washington Press, Seattle.

Piégay, H. 1993. Nature, mass and preferential sites of coarse woody debris deposits in the lower Ain Valley (Mollon reach), France. Regulated Rivers: Research and Management 8:359–372.

Piégay, H., and A. M. Gurnell. 1997. Large woody debris and river geomorphological pattern: examples from southeast France and southern England. Geomorphology 19:99–116.

Quinn, T. P., and N. P. Peterson. 1996. The influence of habitat complexity and fish size on over-winter survival and growth of individually marked juvenile coho salmon (*Oncorhynchus kisutch*) in Big Beef Creek, Washington. Canadian Journal of Fisheries and Aquatic Sciences 53:1555–1564.

Rikhari, H. C., and S. P. Singh. 1998. Coarse woody debris in oak forested stream channels in the central Himalaya. Ecoscience 5(1):128–131.

Robison, E. G., and R. L. Beschta. 1990. Coarse woody debris and channel morphology interactions for undisturbed streams in southeast Alaska, USA. Earth Surface Processes and Landforms 15:149–156.

Rosgen, D. L. 1996. Applied river morphology. Wildland Hydrology Books, Pagosa Springs, Colorado.

Rosgen, D. L. and B. L. Fittante. 1986. Fish habitat structures - a selection guide using stream classification. Pages 163–179 *in* J. G. Miller, J. A. Arway, and R. F. Carline, editors. 5th Trout Stream Habitat Improvement Workshop. Lock Haven, Pennsylvania. Pennsylvania Fisheries Commission, Hamsburg, Pennsylvania.

Ruffner, E. H. 1886. The practice of the improvement of the non-tidal rivers of the United States, with an examination of the results thereof. Wiley, New York.

Russell, I. C. 1909. Rivers of North America. G. P. Putnam's Sons, New York.

Sedell, J. R., and J. L. Froggatt. 1984. Importance of streamside forests to large rivers: the isolation of the Willamette River, Oregon, U.S.A., from its floodplain by snagging and streamside forest removal. Verhandlungen-Internationale Vereinigung für Theoretifche und Angewandte Limnologie 22:1828–1834.

Shields, F. D., and C. J. Gippel. 1995. Prediction of effects of woody debris removal on flow resistance. Journal of Hydraulic Engineering 121(4):341–354.

Shields, F. D., and R. H. Smith. 1992. Effects of large woody debris removal on physical characteristics of a sand-bed river. Aquatic Conservation: Marine and Freshwater Ecosystems 2:145–163.

Shields, F. D., S. S. Knight, Jr., C. M. Cooper, and S. Testa, III. 2000. Large woody debris structures for incised channel rehabilitation. In R. H. Hotchkiss and M. Glade, editors. Building partnerships: proceedings of the Joint Conference on Water Resources Engineering and Water Resources Planning and Management. Environmental and Water Resources Institute of the American Society of Civil Engineers, Reston, Virginia.

Syndi, J. D., J. C. Fischenich, and S. R. Abt. 1998. Effect of woody debris entrapment on flow resistance. Journal of the American Water Resources Association 34(5):1189–1197.

Tally, T. 1980. The effects of geology and large organic debris on stream channel morphology and process for streams flowing through old-growth Redwood forests in northwestern California. Doctoral dissertation. University of California, Santa Barbara.

Thompson, D. 1995. The effects of large organic debris on sediment processes and stream morphology in Vermont. Geomorphology 11:235–244.

Trimble, S. W. 1997. Stream channel erosion and change resulting from riparian forests. Geology 25:467–469.

Triska, F. J. 1984. Role of wood debris in modifying channel morphology and riparian areas of a large lowland river under pristine conditions: a historical case study. Verhandlungen-Internationale Vereinigung für Theoretische und Angewandte Limnologie 22:1876–1892.

Van Der Kamp, B. J., and A. A. Gokhale. 1979. Decay resistance owing to near-anaerobic conditions in black cottonwood wetwood. Canadian Journal of Forest Research 9:39–44.

Veatch, A. C. 1906. Geology and underground water resources of northern Louisiana and southern Arkansas. U.S. Geological Survey, Professional Paper 46, Washington D.C.

Wallerstein, N., C. R. Thorne, and M. W. Doyle. 1997. Spatial distribution and impact of large woody debris in northern Mississippi. Pages 145–150 *in* S. S. Y. Wang, E. J. Langendoen, and F. D. Shields, Jr., editors. Proceedings of the Conference on Management of Landscapes Disturbed by Channel Incision. University of Mississippi, University.

Wallerstein, N. P., C. V. Alonso, S. J. Bennett, and C. R. Thorne. 2001. Distorted froude-scaled flume analysis of large woody debris. Earth Surface Processes and Landforms 26:1265–1283.

Wallerstein, N. P., C. V. Alonso, S. J. Bennett, and C. R. Thorne. 2002. Surface wave forces acting on submerged logs. Journal of Hydraulic Engineering 128:349–353.

Webb, A. A., and W. D. Erskine. 2003. Distribution, recruitment, and geomorphic significance of large woody debris in an alluvial forest stream: Tonghi Creek, southeastern Australia. Geomorphology 51:109–126.

Weigelhofer, G., and J. A. Waringer. 1999. Woody debris accumulations – important ecological components in a low-order forest stream (Weidlingbach, Lower Australia). International Review of Hydrobiology 84(5):427–437.

Wolff, H. H. 1916. The design of a drift barrier across the White River, near Auburn, Washington. Transactions of the American Society of Civil Engineers 16:2061–2085.

Zimmerman, R. C., J. C. Goodlett, and G. H. Comer. 1967. The influence of vegetation on channel form of small streams. Pages 225–275 *in* Symposium on River Morphology. International Association of Science, Hydrological Publication 75, Gentbrugge, Belgium.

American Fisheries Society Symposium 37:391–406, 2003

Trends in Using Wood to Restore Aquatic Habitats and Fish Communities in Western North American Rivers

PETER A. BISSON AND STEVEN M. WONDZELL

USDA Forest Service, Pacific Northwest Research Station
Forestry Sciences Laboratory, 3625 93rd Avenue SW, Olympia, Washington 98512-9193, USA

GORDON H. REEVES

USDA Forest Service, Pacific Northwest Research Station
Forestry Sciences Laboratory, 3200 Jefferson Way, Corvallis, Oregon 97331, USA

STAN V. GREGORY

Department of Fisheries and Wildlife, Oregon State University, Corvallis, Oregon 97331, USA

Abstract.—Advances in understanding wood dynamics in rivers of western North America have led to several important management trends. First, there is a trend away from using "hard" engineering approaches to anchoring wood in streams toward using "soft" placement techniques that allow some wood movement. Second, wood is being placed in locations where channel form and hydraulics favor stability and where wood is likely to accumulate. Third, there is an increased emphasis on passive recruitment of wood from natural source areas (instead of active placement) where the likelihood that it will enter streams through channel migration, windthrow, and landslides is high. Fourth, restoration targets for wood loads are incorporating landscape-scale objectives; thus, managing wood to emulate the spatial and temporal variability produced by natural disturbances is replacing fixed prescriptions for wood in individual reaches. Predicting the effects of wood restoration on individual fish populations in western North America is problematic because local biophysical conditions generate so much experimental noise that it is rarely possible to partition the effects of wood restoration from other sources of variation. Development of appropriate monitoring techniques, combined with a regional network of experimental catchments that include restored and unrestored streams, would help track changes in population status and gauge the effectiveness of wood restoration efforts.

Introduction

Why use wood? Wood has been used to improve aquatic habitat in rivers for a long time, perhaps longer than any other restoration material. Because wood is an important component of habitat and is widely available, managers have often considered it the material of choice for restoration projects (Hunt 1993). Wood provides habitat structure at relatively little cost compared to many alternatives. It floats, which aids in placement. Wood looks and functions more naturally than artificial structures, and it is an important element of the natural habitat of many fishes.

A common restoration tactic has been to add wood to aquatic ecosystems to initiate habitat recovery until natural processes recruit new material. Frissell (1997) and Frissell and Ralph (1998) provided useful discussions of aquatic restoration in western North America that include reviews of wood placement in rivers. In most cases, wood has been used to enhance habitat for one or two target fish species, not entire communities. Wood has also been used to protect property by armoring streambanks and prevent lateral channel migration, but the majority of wood placement projects in western North America have been meant to create habitat for fish.

There has been a widespread belief among managers of fish habitat that when it comes to using wood to improve habitat, more is better. The following quotes from the National Research Council's report *Restoration of Aquatic Ecosystems* (NRC 1992) is critical of the strategy of adding structure to streams solely to improve fish habitat without understanding fluvial processes.

> Efforts to improve fishing by structural means sometimes also introduce into the ecosystem undesirable, nonbiodegradable materials (e.g., rebar, wire mesh, wire rope, planks, polypropylene, hardware cloth, rubber matting, cyclone fencing, corrugated steel, or fiberglass)...Most structural efforts to enhance fish habitat rely on stone or wood dams, current deflectors and camouflaged wooden bank coverings... Some fisheries biologists believe that "water and space are going to waste" if they are not used by trout and that "...even the best streams could be made better..." by producing more trout in them. To the ecologist interested in stream or river restoration, maximizing the ecosystem for trout, or any single species, is not the same as restoring the biotic structure and function of the stream [p. 229] ...When this work is done without a profound understanding of the interactions among stream hydrology, fluvial geomorphology, and fish, the least detrimental consequence may be that mechanical structures emplaced in the stream at con- siderable expense and trouble could be of limited durability and longevity [p. 223].

Over the last 3 decades, new information on the role and dynamics of wood in aquatic ecosystems has forced a reexamination of the ways in which managers think about the value of wood for habitat and its use in stream restoration (Gregory et al. 1991; Sedell and Beschta 1991; Naiman et al. 2000; Roni et al. 2002). Wood is now generally recognized as an integral part of a restoration strategy that emphasizes natural processes (Bryant 1983; NRC 1996; Frissell and Ralph 1998; Slaney and Zaldokas 1998). The NRC (1992) report suggests broad objectives that are at the core of many habitat recovery plans:

> - Restore the natural sediment and water regime. Regime refers to at least two time scales: the daily-to-seasonal variation in water and sediment loads, and the annual to decadal patterns of floods and droughts....
> - Restore a natural channel geometry, if restoration of the water and sediment regime alone does not.
> - Restore the natural riparian plant community, which becomes a functioning part of the channel geometry and floodplain/riparian hydrology. This step is necessary only if the plant community does not restore itself upon achievement of objectives 1 and 2.
> - Restore native aquatic plants and animals, if they do not recolonize on their own. [p. 207]

These objectives, generally accepted by ecologists, have shaped emerging trends in the way wood is being used for aquatic restoration. The goal of this paper is to summarize how these trends are changing wood management in river systems of western North America. We call for increased effort in developing monitoring protocols for wood restoration projects, especially in monitoring the direct impact on targeted populations of the species of interest, to better measure the relative success of restoration efforts and to improve future restoration efforts.

Trend 1: Soft Engineering Methods Are Becoming More Common in Disturbance-Prone Systems

Until fairly recently, the approach to wood restoration has been to engineer wood structures and anchor them solidly in stream channels. The size, location, and configuration of wood structures have been engineered to provide habitat for game fishes, especially to enhance deep pools and overhanging cover. R. L. Hunt's *Trout Stream Therapy* (Hunt 1993) provides excellent illustrations of well-anchored structures designed to enhance trout habitat in North American streams in the upper Midwest, where spring-fed brooks have relatively stable flow regimes. The U.S. Forest Service's *Stream Habitat Improvement Handbook* (Seehorn 1992) discusses a similar approach in which solidly engineered wood structures have been used in streams in the southeastern United States (Figure 1). Procedures used to locate and anchor wood involve burial using heavy equipment, tethering the wood to the streambed or bank with stainless steel cable and waterproof glue, anchoring logs with metal reinforcement rods (rebar), and building clusters of logs and boulders held together by decay-resistant materials such as cable or steel mesh fence. These procedures are termed "hard" engineering because the goal is to maintain structures exactly as installed and prevent them from moving during high flow.

There are numerous examples of wood restoration projects in which individual pieces or clusters of pieces have been securely anchored to the bed, streambanks, or rock outcroppings in an attempt to keep them from moving during freshets (British Columbia Ministry of Environment 1980; Duff and Wydoski 1988). Anchoring wood structures has been based on a combination of concerns about maintaining good fish habitat in the area targeted for restoration and potential liability caused by materials floating downstream and damaging property or endangering life (Rosgen and Fittante 1986). In some instances, solidly anchored wood structures have been both durable and successful at emulating the effects of natural wood in channels. For example, the log weirs studied by Riley and Fausch (1995) and Gowan and Fausch (1996) produced a pool frequency that resembled the frequency of log-formed pools in streams flowing through old-growth forests in the Rocky Mountains of northern Colorado (Richmond and Fausch 1995).

However, there are also examples of restoration projects that were damaged or failed because structures moved, broke apart, or otherwise did not function in the way they were intended (Frissell 1997). Most projects have not been monitored, but one well-known study of habitat restoration in Oregon and Washington (Frissell and Nawa 1992) documented a relatively high loss rate of structures over a 5–10-year period that included several large floods (Figure 2). The principal causes of damage included anchor bolt and cable failures, scour underneath the structure, streambank erosion and lateral channel migration, and burial by sediment. Damage rates were highest in streams of the southern Oregon coast where intense freshets were common. The most durable structures were cabled natural large wood and logjams; the least successful structures were log weirs and log deflectors.

Concerns about project longevity and the failure of certain types of hard engineered structures to persist over time have led to new approaches in which placement of wood and methods of attachment better emulate the natural location of wood in channels (also see Reich et al. 2003, this volume). New methods are less reliant on cables, bolts, and rebar, and greater attention is given to utilizing wood of the appropriate size and species for the site. There is evidence that this "soft" engineering approach has led to project improvements. In one recent survey (Roper et al. 1998), persistence of habitat restoration structures, many of which were emplaced using soft engineering methods, during floods increased. Of 3,946 structures surveyed in 94 streams in Oregon and Washington, they found that fewer than 20% had been removed from sites experiencing floods ranging from 5- to 150-year return intervals (Table 1). Habitat structures were installed with heavy equipment and usually involved large logs greater than 30-cm diameter. Many structures were secured to the stream by cables or partial burial, so in effect, they combined soft and hard engineering techniques. However, logs were placed in such a way as to resemble natural wood location and to withstand heavy flooding characteristic of the Pacific coastal ecoregion. Most did not employ design specifications typical of wood restoration projects in eastern North America. Habitat structures made of partially buried logs or boulders proved to be more durable than structures made of clusters of logs and boulders cabled together. Logs with one or both ends anchored to or buried in streambanks were more durable than logs tethered to the middle of the channel. There was a decrease in durability as stream size increased, and structures in drainages with high landslide frequency were less durable than those in areas with stable valley walls.

The question of whether to cable or otherwise anchor logs to the streambed or streambank is addressed by several authors elsewhere in this book (for example, Gregory 2003; Abbe et al. 2003; Montgomery et al. 2003; and Reich et al. 2003; all this volume). Proponents of cabling argue that preventing movement of wood placed in the channel is needed to safeguard lives and property downstream and that habitat may be lost on site if structures are displaced. Opponents of cabling argue that wood movements are a natural feature of channel dynamics and that floated wood will assume locations within the stream that are fa-

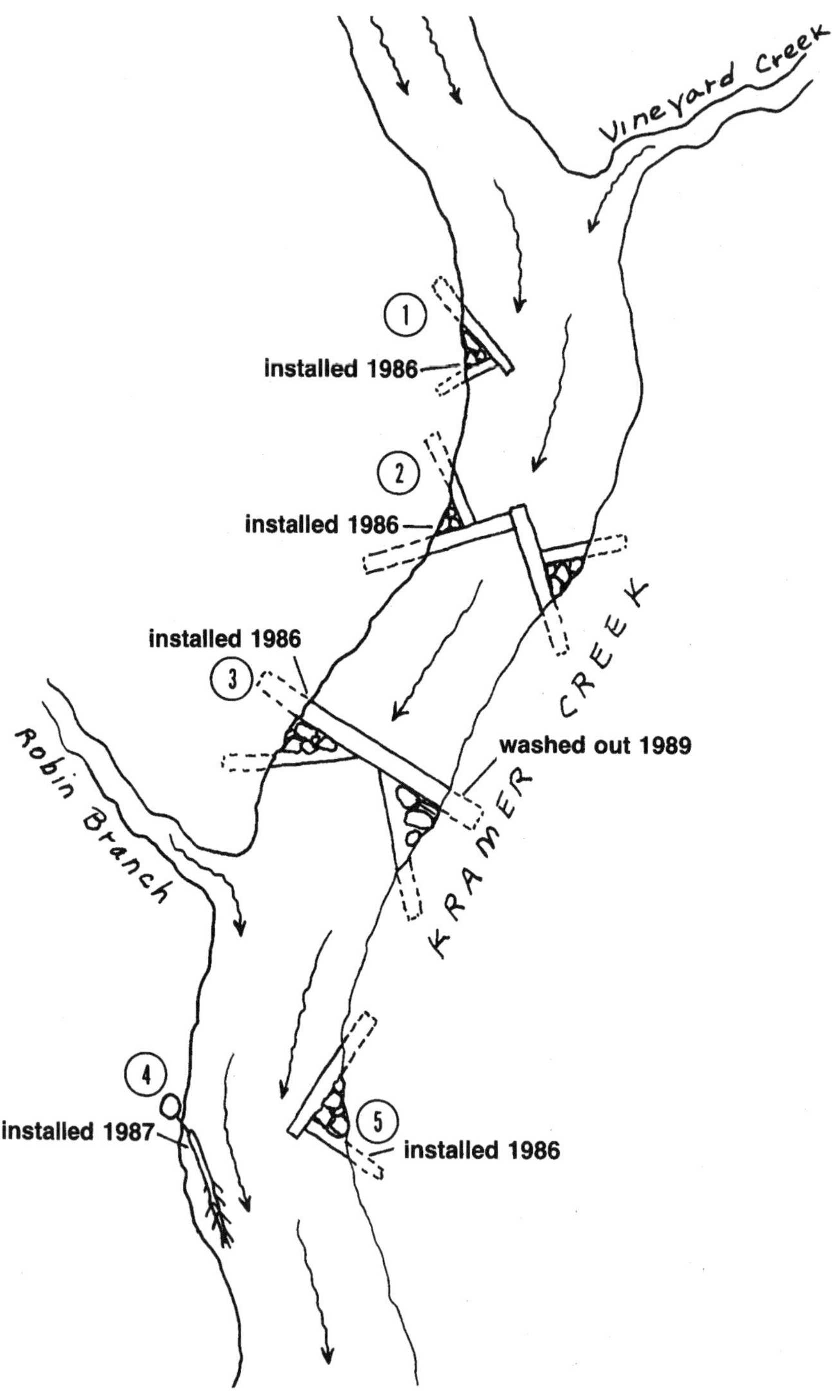

FIGURE 1. Examples of stream habitat restoration structures utilizing hard engineering techniques and relatively uniform spacing (from USDA 1992).

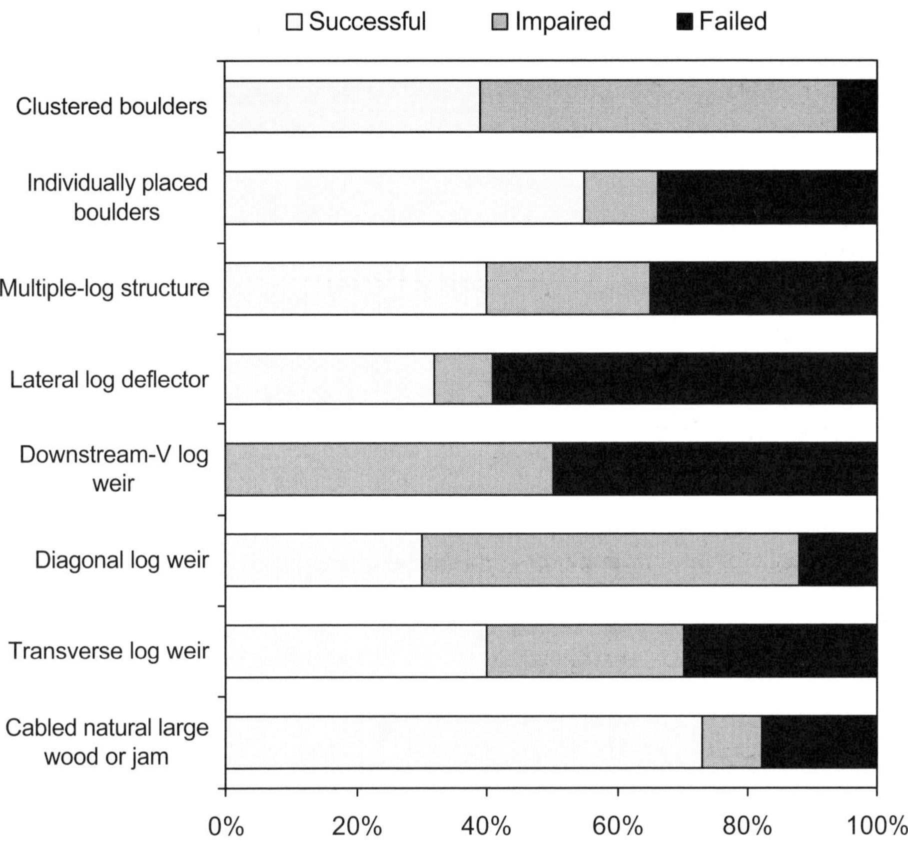

FIGURE 2. Rates of impairment and failure for 161 structures in western Oregon and Washington streams 1–5 years after emplacement (after Frissell and Nawa 1992).

TABLE 1. The persistence of 3,839 habitat restoration structures in streams of Oregon and Washington following severe flooding in 1995 and 1996, based on Roper et al. (1998). Sample size refers to the number of individual structures surveyed. "In place," "shifted on site," and "removed" structure categories approximate the "successful," "impaired," and "failed" categories, respectively, in Frissell and Nawa (1992).

Stream order	Sample size	Flood frequency (years)	Structure movement category: In place (%)	Shifted on site (%)	Removed (%)
2	176	<40	83	8	9
2	157	≥40	70	20	10
3	794	<40	70	21	9
3	702	≥40	64	19	17
4	711	<40	76	15	9
4	847	≥40	51	31	18
5	118	<40	60	20	20
5	334	≥40	41	17	42
		Average[a]	64	21	16

[a] Averages were weighted for sample size.

vorable for the full spectrum of ecological functions it provides. Because the question involves both ecological and social considerations, there is no categorically correct or incorrect answer. However, there is general agreement that engineered structures have a higher probability of remaining in place if they are positioned in such a way that they can withstand high flows, and this may or may not require the use of artificial materials to increase stability. The vast majority of recent improvements in project durability have been made possible by placing wood in locations where channel form and hydraulics naturally favor stability.

Trend 2: Wood Placement Is Guided by Fluvial Patterns

Attempts to use wood to improve fish habitat have sometimes involved placing structures at more or less fixed intervals with little regard for natural patterns or even the appropriateness of placing wood in a particular stream reach (Heede and Rinne 1990; NRC 1992). More attention is now being given to placing wood in areas of the stream and its floodplain where it is likely to occur naturally and where it can provide for a more complete range of ecological functions (Sedell and Beschta 1991). An illustration is seen in the design of a large river restoration project in western Washington (Abbe et al. 1997). The location and structure of wood accumulations in a comparatively pristine system, the Queets River on Washington's Olympic Peninsula, were thoroughly studied and stable logjams mapped relative to channel and floodplain features (Figure 3). Insights into naturally stable accumulations of wood (Abbe and Montgomery 1996) were used to design logjams in the Cowlitz River (Figure 4) that created deep pools, provided protection against continued streambank erosion, and anchored the formation of vegetated gravel bars. The cost of the project was less than 2% of the cost of comparable streambank protection involving rock groins and revetments (Abbe et al. 1997).

Mobility of wood pieces and the degree of clumping along streambanks is strongly influenced by stream size and the intensity of high flows (Bilby and Ward 1989, 1991; Gurnell 2003, this volume). In small streams, large wood moves only short distances before it is trapped by channel or bank roughness elements (other logs, large boulders, confined valley walls). In larger streams, fluvial transport of wood typically moves pieces longer distances and results in jams along the channel edge (Richmond and Fausch 1995; Bilby and Bisson 1998). Understanding the transport characteristics of wood as a function of stream size aids resource managers in placing wood to mimic natural processes.

Many projects enhance both durability and ecological function by using very large logs and rootwads, either individually or in combination with smaller logs, to provide a secure foundation for the structure. These logs, termed "key pieces," are often larger than needed simply to resist movement in the channel. They act as a nucleus of accumulation, trapping smaller logs and branches, coarse sediment, and particulate organic matter. Habitat structures anchored by large key pieces better resemble those found in natural accumulations than those found in single small to medium-sized logs or simple clusters. In addition to providing pool habitat, complex cover exists within the jam. A low-gradient riffle is often created where sediment accumulates above the jam, and the entire wood-sediment complex serves as a site for organic matter processing and nutrient regeneration (Bilby and Bisson 1998; Bilby 2003; Wondzell and Bisson 2003; both this volume). Wood-sediment complexes also promote development of complex hyporheic water pathways that serve a variety of ecological functions (Triska et al. 1989; Duff and Triska 1990; Haggerty et al. 2002). Use of key pieces is central to many wood restoration project designs (Doppelt et al. 1993), and project managers are finding that the increased cost of large logs is offset by their long-term durability and improved ecological function. Managers are recognizing the detrimental outcomes when wood is installed in channels where it would not normally be abundant (Dominguez and Cederholm 2000). Certain types of channels (for example, those in low gradient meadows, high gradient cascades, and narrowly constrained canyons) are not appropriate candidates for wood restoration projects.

It can be very difficult to restore wood to levels that approximate the range of conditions occurring before human disturbance at the scale of an entire catchment. Reeves et al. (1997) describe a case study in western Oregon in which habitat restoration was attempted throughout a 171-km^2 drainage system that had been extensively logged and had experienced a major flood in 1964. Fish Creek, a fifth-order tributary of the Clackamas River, had abundant wood and pool habitat prior to timber harvest and a 100-year flood (pools were estimated at 45% of the stream area in 1959). Im-

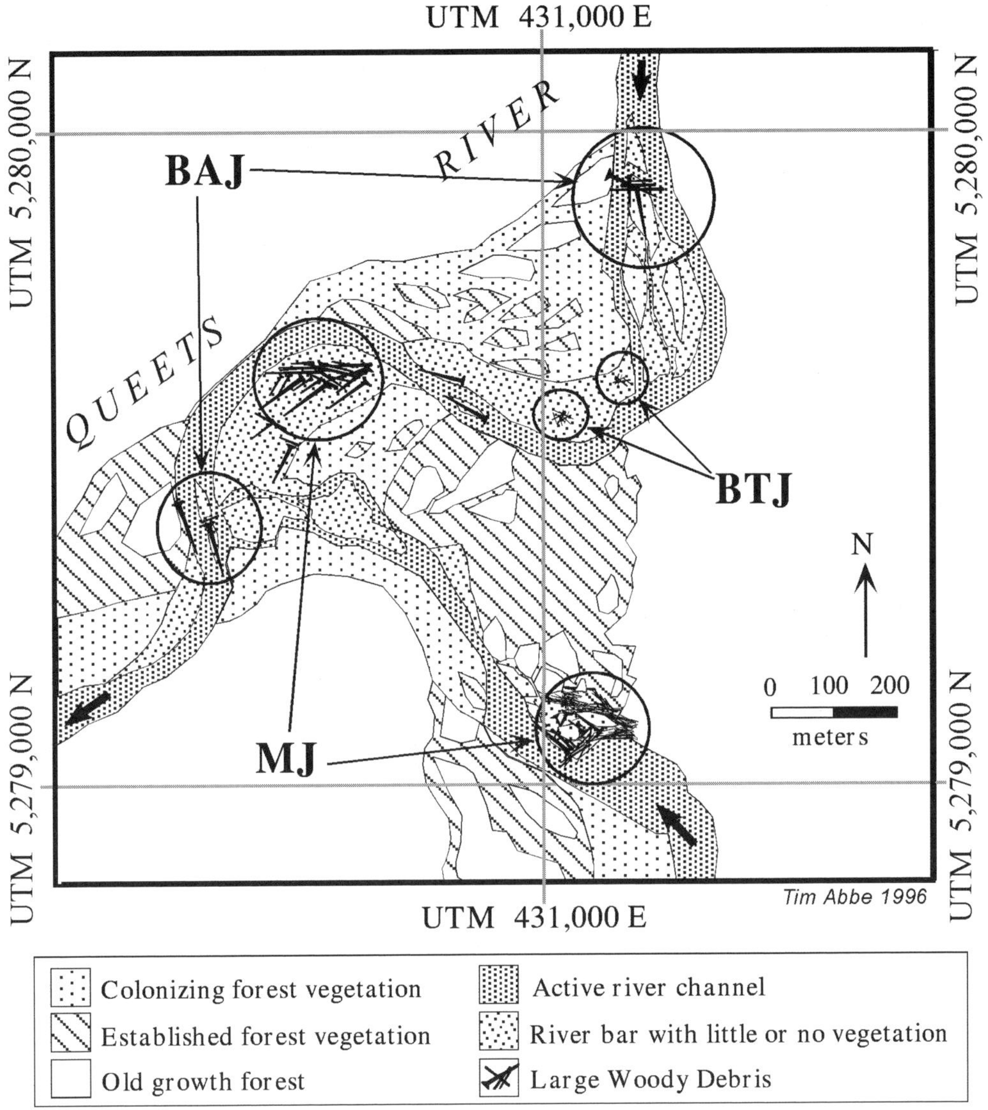

FIGURE 3. Distribution of natural wood accumulations in or near the channel of a large, relatively pristine river (Queets River) in western Washington. BAJ refers to bar apex jam, BTJ refers to bar transverse jam, and MJ refers to meander jam. From Abbe and Montgomery (1996) with permission from Regulated Rivers: Research and Management and John Wiley and Sons Limited.

mediately after the flood, pools comprised only 27% of the channel area, and by 1982, they constituted only 11% of the stream area as a consequence of wood removed from the channel during salvage logging operations. From 1982 to 1988, approximately 1,400 large wood and boulder structures were placed in Fish Creek at a cost of several hundred thousand dollars to create rearing habitat for salmon and trout. Many of the structures were solidly cabled to the streambed to prevent movement during high flow, and most were designed to withstand the hydrologic rigors of the region. During the period when structures were installed (1982–1988), pools ranged from 8 to 21% of the channel area; from 1989 to 1995, after completion of the restoration project, pools increased to 19–39%. Large storms in November 1995 and February 1996 resulted in major flooding and 236 landslides in the

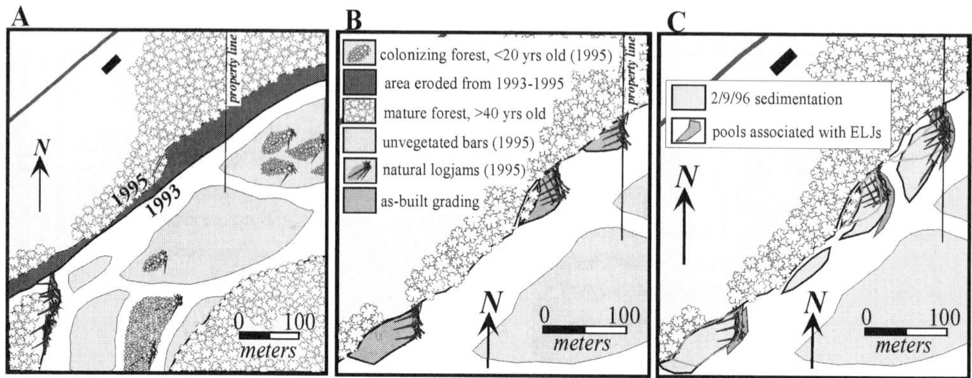

FIGURE 4. Plan view of a wood restoration project modeled after the location and structure of wood accumulations in a natural river (see Figure 3). Panel A shows the reach between 1993 and 1995; panel B shows the channel after the installation of three engineered log jams; panel C shows the channel after a major flood in February 1996. ELJs refer to engineered log jams. From Abbe et al. (1997) with permission.

watershed. Reeves et al. (1997) estimated that 49% of the structures were removed from the drainage by high discharge and debris flows. Following these two large storms, pool area declined to 33%, a percentage similar to that observed before the restoration period.

The high cost of restoration and the limited durability of wood structures in Fish Creek illustrate the difficulty of re-establishing wood in a relatively high-energy catchment. Much of Fish Creek consists of geomorphically constrained, high gradient (>4%) channels, and the large boulder and bedrock dominated substrate is typical of streams draining the Cascade Mountains. Previously wood was abundant but dominated by pieces from old-growth forests with partially intact rootwads and branches—characteristics needed for durability in high-energy systems but very difficult to replace. The study shows that even securely anchored structures have high failure rates during large floods and debris flows. Faced with this dilemma, managers are placing a higher priority on protecting and restoring natural source areas for future wood recruitment rather than on expensive, high maintenance, in-channel structures. This has resulted in the third general trend in wood management.

Trend 3: Emphasis Is Shifting from Active Intervention to Passive Restoration

Except for extensive restoration efforts such as in Fish Creek, projects involving deliberate placement of wood usually affect only a small percentage of the channel network. Provisions for long-term wood recruitment are likely to be more important than short-term wood placement (Sedell and Beschta 1991). Large natural disturbances (fires, floods, and insect and disease outbreaks) create a landscape mosaic of different forest ages and species. Individual streams cycle from low to high wood loads as channels fill with, and flush, accumulated sediment and wood over periods that may span centuries (Benda et al. 1998). At any point in time, some streams will be highly loaded with wood, while in others, large wood will be relatively scarce. In montane landscapes, aquatic productivity changes from low to high to low levels again as streams accumulate wood and coarse sediment gradually through windthrow or channel meandering or rapidly through landslides and then lose this material gradually through decomposition or rapidly during brief but intense disturbances such as debris flows (Benda and Dunne 1997). From a management standpoint, the outcome of natural disturbance and recovery cycles is that some areas are highly productive for fishes and other aquatic species, while others remain relatively unproductive until erosion and fluvial transport restores structure and complexity to stream channels (Reeves et al. 1995). Viewed in this way, protection of wood source areas—riparian zones and landslide-prone hillslopes—and the processes that deliver wood to streams—windthrow, lateral channel movements, mass wasting—become important landscape goals (Reeves et al. 1995; Gregory and

Bisson 1997; Connolly and Hall 1999; Beechie et al. 2000; Benda et al. 2002; Boyer et al. 2003, this volume).

Passive restoration approaches that rely on natural processes to improve streams are recommended by the National Research Council to recover aquatic habitats where anthropogenic damage has not caused irreversible loss (NRC 1996). This is not to say that deliberate addition of wood has no place in habitat restoration, but rather that decisions about wood placement should be grounded in the context of the overall condition of the watershed and focused primarily on streams in which recruitment of new wood from adjacent riparian areas and hillslopes, if it occurred historically, is no longer possible. Wood restoration may be desirable to rapidly improve habitat while trees grow in source areas and are recruited to the stream, but with the understanding that the longevity of wood structures in river basins prone to flow extremes and high rate of erosion is limited.

Some landscape management plans allow future disturbances to deliver wood to channels by leaving protected riparian zones and headwater areas in locations likely to deliver wood to the stream network over time (Sedell et al. 1994; Cissel et al. 1998). Concerns are often raised that extreme natural disturbances, such as very large floods or wildfires, cause such damage to aquatic habitats that ecosystem recovery is unacceptably delayed or even rendered impossible without aggressive restoration. Certainly, western North America has experienced large natural disturbances over the last century, including floods, extensive wildfires, prolonged droughts, and volcanic eruptions. Yet the resiliency of native flora and fauna has been widely documented (for example, Franklin et al. 1985; Minshall et al. 1989; Swanson et al. 1998), and studies of ecological recovery after disturbances affecting wood abundance suggest that long-term restoration effectiveness is strongly influenced by the re-establishment of wood source areas that may include areas immediately adjacent to the stream as well as upslope sources (Benda and Cundy 1990; Reeves et al., in press). Depending on the extent of human disturbance and the potential for recovery, passive restoration may not require commitment to extensive instream structure addition as long as natural wood recruitment processes are preserved. (Boyer et al. 2003) The need to balance active and passive wood restoration approaches over entire drainage networks has resulted in a fourth trend.

Trend 4: Wood Restoration Goals Are Being Defined at Larger Scales

Fixed restoration targets for wood abundance, usually expressed as numbers of pieces of large wood per unit length of channel, tend to be based on expected reference conditions, usually counts of wood within a relatively pristine river reach. However, fixed wood abundance targets do not easily account for natural variation among reaches or the habitat requirements of different aquatic species. When management plans set fixed standards, spatial and temporal variation in wood abundance is likely to be considered a "sampling problem." One consequence is that the differences in habitat requirements among species are often ignored. A more recent trend is for management plans to recognize that variation in wood abundance is an important part of the spatial and temporal variation present within landscapes and to include that variability in restoration planning.

Natural disturbances generating habitat variation are part of the environmental template for the evolution of aquatic species. The maintenance of genetic and phenotypic variation needed to cope with environmental change allows populations to persist in the face of habitat disruption (Scudder 1989; Poff and Ward 1990; Reice et al. 1990). Complex mosaics of aquatic and floodplain habitats are necessary to fulfill different fish life history requirements and for a full range of environmental conditions needed to maintain regional biodiversity. Fixed reach-level wood targets have the effect of homogenizing wood loads across a drainage network. Attempting to maintain constant levels of wood in all streams at all times may be unrealistic and contrary to natural patterns and processes.

Alternative approaches to restoring wood in river systems consider how disturbance processes affect wood distribution and abundance throughout the fluvial network. Roni et al. (2002) present a hierarchical strategy for prioritizing habitat restoration for salmon in the Pacific Northwest. The strategy begins with identification of those parts of a watershed that remain ecologically healthy or nearly so (that is, that contain high quality habitats and support the range of ecological functions needed to maintain them). After such areas have been located, corridors of good habitat are established between them that facilitate movements of fishes and other organisms in response to life cycle requirements. Often, re-establishing connectivity can be as simple as removing artificial obstructions to migration. Once these essential connections are restored, emphasis is placed on recovering aquatic ecosystems by eliminating or reducing anthropogenic stresses and by facilitating the beneficial effects of natural disturbances. Only after the preceding steps occur are deliberate short-term habitat improvements such as wood additions considered. Not all stream reaches will be productive for target species such as salmon; some settings will be far more productive than others (Reeves et al. 1995; Pess et al. 2002). The strategy proposed by Roni et al. (2002) forces a view of habitat improvement that goes beyond restoring wood to individual reaches and instead considers conditions and processes at much larger scales. Landscape management plans based on natural disturbance regimes (for example, Cissel et al. 1998) are likely to be successful over time because they accommodate watershed-scale features and broaden our perspective beyond the simple restoration of large wood in channels.

Monitoring

Accountability for restoration effectiveness requires that wood restoration projects be monitored. In many cases, monitoring has focused on measuring the durability of structures and how the structures influenced the development of aquatic habitats (Crispin et al. 1993). Far fewer studies have examined the effects of stream restoration on target species (Botkin et al. 2000), and results from these investigations have been somewhat equivocal. Monitoring efforts have usually demonstrated an increase in fish utilization of the stream reach to which wood has been added (House and Boehne 1986), but very few studies have documented a sustained increase in fish populations over an entire drainage system.

A number of monitoring efforts have documented that fish do indeed utilize wood structures (House and Boehne 1985; Crispin et al. 1993; Cederholm et al. 1997; Roni and Quinn 2001). Of those studies where attempts to measure the response of entire populations to wood additions have taken place (for example, Gowan and Fausch 1996; Solazzi et al. 2000), monitoring may be limited to less than 10 years. This time period may be insufficient to detect treatment-related changes at the population level, especially for long-lived anadromous species (Hall and Knight 1981; Hilborn and Winton 1993). Power analyses based on estimates of interannual population variability of Pacific salmon have shown that decades of monitoring will be needed to detect treatment effects on some populations, even at relatively coarse statistical confidence levels (Bisson et al. 1997; Ham and Pearsons 2000). For resident freshwater fishes that are not subject to the environmental extremes faced by Pacific salmon, monitoring requirements may be less daunting. Gowan and Fausch (1996) found that an 8-year monitoring period was sufficient to detect a 44% increase in cutthroat trout *Oncorhyn-chus clarki* abundance after wood enhancement of several northern Colorado streams.

Other environmental factors often obscure the effects of wood restoration. For example, abrupt changes in freshwater and marine survival of anadromous salmonids may accompany climate changes that persist for decades (Ward 2000). If such a change occurs during a post-treatment monitoring period, results are easily misinterpreted. The Keogh River study in British Columbia provides a case in point. The Keogh River is a medium size drainage system on Vancouver Island draining a catchment of 130 km^2 (Ward and Slaney 1979). Populations of steelhead *O. mykiss*, coho salmon *O. kisutch*, and char have been monitored there since 1976, making the Keogh River system one of the longest continuously monitored catchments in western North America. Attempts to improve steelhead and coho salmon production have involved physical habitat improvements with logs and boulders, and during the 1980s, inorganic nutrients were added to the river to stimulate primary and secondary production. An increase in adult steelhead was observed in the Keogh River after these measures had been implemented. Analyses of long-term monitoring data have suggested that physical habitat manipulations have not improved smolt production as much as nutrient additions, but the overall productivity of anadromous salmonids in the system has been driven primarily by climate shifts influencing freshwater and marine survival (Ward 2000). Furthermore, a sharp decline after 1990 in returning adult steelhead, the dominant anadromous salmonid in the Keogh River, suggests that unfavorable climate shifts have masked restoration efforts and placed fish populations at risk of extirpation. Such a conclusion would not have been possible without decades of careful, time-consuming monitoring.

Long-term restoration monitoring studies,

such as those at Keogh River, BC and Fish Creek, Oregon, are rare in western North America but have contributed greatly to our understanding the influence of climate shifts, floods, and other natural disturbances on fish populations. Because these studies have spanned decades, they have involved extraordinary commitment from their funding organizations and participating scientists. Likewise, other long-term studies of forestry operations on fish populations (for example, Alsea watershed study and Carnation Creek study) have produced invaluable information on the effects of logging on aquatic resources (Hartman and Scrivener 1990; Stednick and Hall, in press). Elsewhere, we have argued for the establishment of a regional network of experimental catchments in which habitat restoration and other management trials can be carried out at a landscape level appropriate to fish populations and with a commitment to long-term monitoring (Bisson et al. 1992). We repeat that call here.

Integrative indices of environmental quality (for example, community-based measures such as IBI, the Index of Biotic Integrity [Karr 1998]) usually combine data on physical habitat, water quality, and the abundance of certain indicator organisms. They are of limited value in determining the *specific* effects of wood restoration on fishes, although they may provide information on overall aquatic ecosystem condition relative to more pristine sites (Karr and Chu 1998). Wallace et al. (1996) successfully used the North Carolina Biotic Index (NCBI) and the Ephemeroptera + Plecoptera + Trichoptera (EPT) index to track the invertebrate community and ecosystem-level responses to experimental insecticide application in a headwater stream at the Coweeta Hydrologic Laboratory in North Carolina. Wallace et al. (1995) examined invertebrate community responses to wood addition and documented immediate changes in composition: abundances and biomass of scrapers and filterers decreased; collectors and predators increased; overall shredder biomass was unchanged, but biomass of trichopteran and dipteran shredders increased while most plecopteran shredders decreased; and plecopteran predators also decreased despite greater abundances of potential prey. These authors found shifts in functional group abundances, biomass, and production between control and wood-enriched sites, which emphasized the importance of wood in structuring invertebrate communities within mountain streams (see also Benke and Wallace 2003, this volume).

Additional biological performance measures of wood restoration effectiveness that incorporate natural variability are needed, particularly as related to the response of fishes. Promising areas for investigation include changes in the guild structure of fish communities, physiological responses to wood restoration such as changes in growth, and changes in the availability of prey related specifically to wood addition. One measure of performance that was formerly popular among aquatic scientists, but is now uncommon in most American investigations, is production—a measure that integrates population density and average individual growth rates (Chapman 1978). Estimates of fish production may be especially valuable in studies involving a before-after control-impact (BACI) experimental design. There have not been any studies of fish populations in western North America in which annual production has been estimated over long periods of time, so it is not known whether production is less variable than simple population censuses. Salmonid fishes often exhibit density-dependent growth in streams (Warren 1971), which may serve to dampen variation in annual production estimates. If so, production may be a less variable indicator of the habitat capacity of the stream than annual density estimates, and detection of population responses to wood restoration could be achieved with shorter monitoring intervals. But production estimates are costly and can be confounded by undetected movements of fish between streams (Fausch and Young 1995). Thus, a suite of performance measures, including growth, age and size structure, abundance, movement, and production would be helpful in obtaining an accurate picture of the effects of a wood restoration project.

Finally, it is useful to include some measures of the ecological response of a stream and its riparian zone to wood restoration that may not directly impact fish survival and growth but may influence fish communities through indirect pathways. For example, it is well known that large wood traps salmon carcasses and facilitates the entry of marine-derived nutrients into aquatic food webs (Cederholm et al. 1989; Bilby and Bisson 1998). Other questions remain. For example, how does wood restoration affect the abundance of aquatic invertebrates (potential prey)? Also, how does wood restoration affect riparian vegetation, either by trapping carcasses and other organic matter or by changing the patterns of vegetated gravel bars? Additionally, the significance of wood restoration to the pattern and extent of hyporheic water and nutrient storage in small streams (Haggerty et al. 2002) deserves additional investigation because hyporheic flows influence stream temperature and nutrient availability.

Summary

Today, more managers are viewing wood restoration as a component of ecosystem management rather than a simple, stand-alone mitigation tool. Greater efforts are being made to manage landscapes in a manner promoting or emulating natural processes that distribute wood throughout the channel network. Although most restoration projects focus on one or two fish species of commercial or recreational importance, the role of wood in creating complex habitats is better understood and the link between habitat complexity and fish populations is more widely appreciated.

As wood restoration projects increasingly resemble natural wood distribution patterns, aquatic communities as a whole will benefit. Landscape perspectives that recognize and allow for disturbance and recovery processes over space and time are gradually replacing prescriptions that set fixed targets for wood in rivers. Broader landscape perspectives draw our attention back from reach scales to entire drainage systems with the understanding that wood loads will naturally increase and decrease in different parts of the river basin over time. Provisions for long-term recruitment of wood are starting to receive higher priority than placement of wood structures in streams.

Monitoring the effects of wood restoration on fish populations presents many challenges, and refinement of alternative performance measures deserves greater attention. Highly variable fish populations and application of wood restoration to limited portions of the drainage system often prevent statistical detection of possible improvements at the population level. Several biological measures may be needed to detect the effects of stream habitat changes brought about by wood additions, including community structure, biomass and abundance, age and size structure, physiological performance, and production. Development of appropriate biological monitoring techniques remains a significant challenge to assessing wood restoration effects. A regional network of experimental catchments or network segments in which wood restoration can be studied would help scientists and managers address questions about restoration effectiveness that currently remain unanswered.

Acknowledgments

We thank the USDA Forest Service Pacific Northwest Research Station and Oregon State University for financial support and the many attendees of the Wood in World Rivers Conference who made helpful comments. Bob Bilby, Deanna Stouder, and two anonymous reviewers provided constructive suggestions for the manuscript.

References

Abbe, T. B., A. P. Brooks, and D. R. Montgomery. 2003. Wood in river rehabilitation and management. Pages 367–389 *in* S. V. Gregory, K. L. Boyer, and A. M. Gurnell, editors. The ecology and management of wood in world rivers. American Fisheries Society, Symposium 37, Bethesda, Maryland.

Abbe, T. B., and D. R. Montgomery. 1996. Large woody debris jams, channel hydraulics, and habitat formation in large rivers. Regulated Rivers Research and Management 12:201–221.

Abbe, T. B., D. R. Montgomery, and C. Petroff. 1997. Design of stable in-channel wood debris structures for bank protection and habitat restoration: an example from the Cowlitz River, WA. Pages 809–814 *in* S. S. Y. Wang, E. J. Langendoen, and F. D. Shields, Jr., editors. Proceedings of the conference on management of landscapes disturbed by channel incision. Center for Computational Hydroscience and Engineering, The University of Mississippi, Oxford.

Beechie, T. J., G. Pess, P. Kennard, R. E. Bilby, and S. Bolton. 2000. Modeling recovery rates and pathways for woody debris recruitment in northwestern Washington streams. North American Journal of Fisheries Management 20:436–452.

Benda, T. W., P. Bigelow, and T. M. Worsley. 2002. Recruitment of wood to streams in old-growth and second-growth redwood forests, northern California, USA. Canadian Journal of Forest Research 32:1460–1477.

Benda, L. E., and T. W. Cundy. 1990. Predicting deposition of debris flows in mountain channels. Canadian Geotechnology Journal 27:409–417.

Benda, L. E., and T. Dunne. 1997. Stochastic forcing of sediment supply to channel networks from landsliding and debris flow. Water Resources Research 33:2849–2863.

Benda, L. E., D. J. Miller, T. Dunne, G. H. Reeves, and J. K. Agee. 1998. Dynamic landscape systems. Pages 261–288 *in* R. J. Naiman and R. E. Bilby, editors. River ecology and management. Springer, New York.

Benke, A. C., and J. B. Wallace. 2003. Influence of wood on invertebrate communities in streams and rivers. Pages 149–177 *in* S. V. Gregory, K. L. Boyer, and A. M. Gurnell, editors. The ecology and management of wood in world rivers. American Fisheries Society, Symposium 37, Bethesda, Maryland.

Bilby, R. E. 2003. Decomposition and nutrient dynamics of wood in streams and rivers. Pages 135–147 *in* S. V. Gregory, K. L. Boyer, and A. M. Gurnell, editors. The ecology and management of wood in world rivers. American Fisheries Society, Symposium 37, Bethesda, Maryland.

Bilby, R. E., and P. A. Bisson. 1998. Function and distribution of large woody debris. Pages 324–346 *in* R. J. Naiman and R. E. Bilby, editors. River ecology and management. Springer, New York.

Bilby, R. E., and J. W. Ward. 1989. Changes in characteristics and function of woody debris with increasing stream size in western Washington. Transactions of the American Fisheries Society 118:368–378.

Bilby, R. E., and J. W. Ward. 1991. Characteristics and function of large woody debris in streams draining old-growth, clear-cut, and second-growth forests in southwestern Washington. Canadian Journal of Fisheries and Aquatic Sciences 48:2499–2508.

Bisson, P. A., T. P. Quinn, G. H. Reeves, and S. V. Gregory. 1992. Best management practices, cumulative effects, and long-term trends in fish abundance in Pacific Northwest river systems. Pages 189–232 *in* R. J. Naiman, editor. Watershed management: balancing sustainability and environmental change. Springer-Verlag, New York.

Bisson, P. A., G. H. Reeves, R. E. Bilby, and R. J. Naiman. 1997. Watershed management and Pacific salmon: desired future conditions. Pages 447–474 *in* D. J. Stouder, P. A. Bisson, and R. J. Naiman, editors. Pacific salmon and their ecosystems: status and future options. Chapman and Hall, New York.

Botkin, D. B., D. L. Peterson, and J. M. Calhoun, technical editors. 2000. The scientific basis for validation monitoring of salmon for conservation and restoration plans. Olympic Natural Resources Center Technical Report, University of Washington, Forks.

Boyer, K. L., D. R. Berg, and S. V. Gregory. 2003. Riparian management for wood in rivers. Pages 407–420 *in* S. V. Gregory, K. L. Boyer, and A. M. Gurnell, editors. The ecology and management of wood in world rivers. American Fisheries Society, Symposium 37, Bethesda, Maryland.

British Columbia Ministry of Environment. 1980. Stream enhancement guide. Ministry of Environment, Fish and Wildlife Branch, Vancouver.

Bryant, M. D. 1983. The role and management of woody debris in west coast salmonid nursery streams. North American Journal of Fisheries Management 3:322–330.

Cederholm, C. J., R. E. Bilby, P. A. Bisson, T. W. Bumstead, B. R. Fransen, W. J. Scarlett, and J. W. Ward. 1997. Response of juvenile coho salmon and steelhead to placement of large woody debris in a coastal Washington stream. North American Journal of Fisheries Management 17:947–963.

Cederholm, C. J., D. B. Houston, D. L. Cole, and W. J. Scarlett. 1989. Fate of coho salmon (*Oncorhynchus kisutch*) carcasses in spawning streams. Canadian Journal of Fisheries and Aquatic Sciences 46:1347–1355.

Chapman, D. W. 1978. Production in fish populations. Pages 5–25 *in* S. D. Gerking, editor. Ecology of freshwater fish production. Blackwell Scientific Publications, Oxford.

Cissel, J. H., F. J. Swanson, G. E. Grant, D. H. Olson, S. V. Gregory, S. L. Garman, L. R. Ashkenas, M. G. Hunter, J. A. Kertis, J. H. Mayo, M. D. McSwain, S. G. Swetland, K. A. Swindle, and D. O. Wallin. 1998. A landscape plan based on historical fire regimes for a managed forest ecosystem: the Augusta Creek study. U.S. Department of Agriculture, Forest Service, Pacific Northwest Research Station, General Technical Report PNW-GTR-422, Portland, Oregon.

Connolly, P. J., and J. D. Hall. 1999. Biomass of coastal cutthroat trout in unlogged and previously clear-cut basins in the central coast range of Oregon. Transactions of the American Fisheries Society 128:890–899.

Crispin, V., R. House, and D. Roberts. 1993. Changes in instream habitat, large woody debris, and salmon habitat after the restructuring of a coastal Oregon stream. North American Journal of Fisheries Management 13:96–102.

Dominguez, L. G., and C. J. Cederholm. 2000. Rehabilitating stream channels using large woody debris with considerations for salmonid life history and fluvial geomorphic processes. Pages 545–563 *in* E. E. Knudsen, C. R. Steward, D. D. MacDonald, J. E. Williams, and D. W. Reiser, editors. Sustainable fisheries management: Pacific salmon. CRC Press LLC, Boca Raton, Florida.

Doppelt, B., M. Scurlock, C. Frissell, and J. Karr. 1993. Entering the watershed: a new approach to save America's river ecosystems. Island Press, Washington, D.C.

Duff, D. A., and R. S. Wydoski. 1988. Indexed bibliography on stream habitat improvement. Technical Report, U.S. Department of Agriculture, Forest Service, Intermountain Research Station, Ogden, Utah.

Duff, J. H., and F. J. Triska. 1990. Denitrification in sediments from the hyporheic zone adjacent to a small forested stream. Canadian Journal of Fisheries and Aquatic Sciences 47:1140–1147.

Fausch, K. D., and M. K. Young. 1995. Evolutionarily significant units and movement of resident stream fishes: a cautionary tale. Pages 360–370 *in* J. L. Nielsen, editor. Evolution and the aquatic ecosystem: defining unique units in population conservation. American Fisheries Society, Symposium 17, Bethesda, Maryland.

Franklin, J. F., J. A. MacMahon, F. J. Swanson, and J. R. Sedell. 1985. Ecosystem responses to the eruption of Mount St. Helens. National Geographic Research 1:198–216.

Frissell, C. A. 1997. Ecological principles. Pages 96–115 *in* J. E. Williams, C. A. Wood, and M. P. Dombeck, editors. Watershed restoration: principles and practices. American Fisheries Society, Bethesda, Maryland.

Frissell, C. A., and R. K. Nawa. 1992. Incidence and causes of physical failure of artificial habitat structures in streams of western Oregon and Washington. North American Journal of Fisheries Management 12:182–197.

Frissell, C. A., and S. C. Ralph. 1998. Stream and watershed restoration. Pages 599–624 *in* R. J. Naiman and R. E. Bilby, editors. River ecology and management: lessons from the Pacific Coastal Ecoregion. Springer-Verlag, New York.

Gowan, C., and K. D. Fausch. 1996. Long-term demographic responses of trout populations to habitat manipulation in six Colorado streams. Ecological Applications 6:931–946.

Gregory, S. V. 2003. Modeling the dynamics of wood in streams and rivers. Pages 315–335 *in* S. V. Gregory, K. L. Boyer, and A. M. Gurnell, editors. The ecology and management of wood in world rivers. American Fisheries Society, Symposium 37, Bethesda, Maryland.

Gregory, S. V., and P. A. Bisson. 1997. Degradation and loss of anadromous salmonid habitat in the Pacific Northwest. Pages 277–314 *in* D. J. Stouder, P. A. Bisson, and R. J. Naiman, editors. Pacific salmon and their ecosystems: status and future options. Chapman and Hall, New York.

Gregory, S. V., F. J. Swanson, and W. A. McKee. 1991. An ecosystem perspective of riparian zones. BioScience 40:540–551.

Gurnell, A. M. 2003. Wood storage and mobility. Pages 75–91 *in* S. V. Gregory, K. L. Boyer, and A. M. Gurnell, editors. The ecology and management of wood in world rivers. American Fisheries Society, Symposium 37, Bethesda, Maryland.

Haggerty, R., Wondzell, S. M., and Johnson, M. A. 2002. Fractal scaling of residence time distribution in the hyporheic zone of a 2nd-order mountain stream. Geophysical Research Letters 29:18-1–18-4.

Hall, J. D., and N. J. Knight. 1981. Natural variation in abundance of salmonid populations in streams and its implications for design of impact studies. Environmental Protection Agency, Report EPA-600/S3–81-021, Corvallis, Oregon.

Ham, K. D., and T. N. Pearsons. 2000. Can reduced salmonid population abundance be detected in time to limit management impacts? Canadian Journal of Fisheries and Aquatic Sciences 7:17–24.

Hartman, G. F., and J. C. Scrivener. 1990. Impacts of forestry practices on a coastal stream ecosystem, Carnation Creek, British Columbia. Canadian Bulletin of Fisheries and Aquatic Sciences 223, Ottawa.

Heede, B. H., and J. N. Rinne. 1990. Hydrodynamic and fluvial morphologic processes: implications for fisheries management and research. North American Journal of Fisheries Management 10(3):249-268.

Hilborn, R., and J. Winton. 1993. Learning to enhance salmon production: lessons from the Salmonid Enhancement Program. Canadian Journal of Fisheries and Aquatic Sciences 50:2043–2056.

House, R. A., and P. L. Boehne. 1985. Evaluation of instream enhancement structures for salmonid spawning and rearing in a coastal Oregon stream. North American Journal of Fisheries Management 5:283–295.

House, R. A., and P. L. Boehne. 1986. Effects of instream structures on salmonid habitat and populations in Tobe Creek, Oregon. North American Journal of Fisheries Management 6:38–46.

Hunt, R. L. 1993. Trout stream therapy. University of Wisconsin Press, Madison.

Karr, J. R. 1998. Rivers as sentinels: using the biology of rivers to guide landscape management. Pages 502–528 *in* R. J. Naiman and R. E. Bilby, editors. River ecology and management: lessons from the Pacific coastal ecoregion. Springer-Verlag, New York.

Karr, J. R., and E. W. Chu. 1998. Restoring life in running waters: better biological monitoring. Island Press, New York.

Minshall, G. W., J. T. Brock, and J. D. Varley. 1989. Wildfire and Yellowstone's stream ecosystems. BioScience 39:707–715.

Montgomery, D. R., B. D. Collins, J. M. Buffington, and T. B. Abbe. 2003. Geomorphic effects of wood in rivers. Pages 21–47 *in* S. V. Gregory, K. L. Boyer, and A. Gurnell, editors. The ecology and management of wood in world rivers. American Fisheries Society, Symposium 37, Bethesda, Maryland.

Naiman, R. J., R. E. Bilby, and P. A. Bisson. 2000. Riparian ecology and management in the Pacific coastal rain forest. BioScience 50:996–1011.

NRC (National Research Council). 1992. Restoration of aquatic ecosystems. National Academy Press, Washington, D.C.

NRC (National Research Council). 1996. Upstream: salmon and society in the Pacific Northwest. National Academy Press, Washington, D.C.

Pess, G. R., D. R. Montgomery, E. A. Steel, R. E. Bilby, B. E. Feist, and H. M. Greenberg. 2002. Landscape characteristics, land use, coho salmon (*Onco-*

rhynchus kisutch) abundance, Snohomish River, Washington State, U.S.A. Canadian Journal of Fisheries and Aquatic Sciences 59:613–623.

Poff, N. L., and J. V. Ward. 1990. The physical habitat template of lotic systems: recovery in the context of historical pattern of spatio-temporal heterogeneity. Environmental Management 14:629–646.

Reeves, G. H., L. E. Benda, K. M. Burnett, P. A. Bisson, and J. R. Sedell. 1995. A disturbance-based ecosystem approach to maintaining and restoring freshwater habitats of evolutionarily significant units of anadromous salmonids in the Pacific Northwest. Pages 334–349 *in* J. L. Nielsen, editor. Evolution and the aquatic ecosystem. American Fisheries Society, Symposium 17, Bethesda, Maryland.

Reeves, G. H., D. B. Hohler, B. E. Hansen, F. H. Everest, J. R. Sedell, T. L. Hickman, and D. Shively. 1997. Fish habitat restoration in the Pacific Northwest: Fish Creek of Oregon. Pages 335–359 *in* J. E. Williams, C. A. Wood, and M. P. Dombeck, editors. Watershed restoration: principles and practices. American Fisheries Society, Bethesda, Maryland.

Reeves, G. H., K. M. Burnett, and E. V. McGarry. In press. Sources of wood in a pristine watershed in Coastal Oregon. Canadian Journal of Forestry.

Reice, S. R., R. C. Wissmar, and R. J. Naiman. 1990. Influence of spatial-temporal heterogeneity and background disturbance regime on the recovery of lotic ecosystems. Environmental Management 14:647–659.

Reich, M., J. L. Kershner, and R. C. Wildman. 2003. Restoring streams with large wood: a synthesis. Pages 355–366 *in* S. V. Gregory, K. L. Boyer, and A. M. Gurnell, editors. The ecology and management of wood in world rivers. American Fisheries Society, Symposium 37, Bethesda, Maryland.

Richmond, A. D., and K. D. Fausch. 1995. Characteristics and function of large woody debris in mountain streams of northern Colorado. Canadian Journal of Fisheries and Aquatic Sciences 52:1789–1802.

Riley, S. C., and K. D. Fausch. 1995. Trout population response to habitat enhancement in six northern Colorado streams. Canadian Journal of Fisheries and Aquatic Sciences 52:34–53.

Roni, P., and T. P. Quinn. 2001. Density and size of juvenile salmonids in response to placement of large woody debris in western Oregon and Washington streams. Canadian Journal of Fisheries and Aquatic Sciences 58:1–11.

Roni, P., T. J. Beechie, R. E. Bilby, F. E. Leonetti, M. M. Pollock, and G. R. Pess. 2002. A review of stream restoration techniques and a hierarchical strategy for prioritizing restoration in Pacific Northwest watersheds. North American Journal of Fisheries Management 22:1–20.

Roper, B., D. Konnoff, D. Heller, and K. Wieman. 1998. Durability of Pacific Northwest instream structures following floods. North American Journal of Fisheries Management 18:686–693.

Rosgen, D., and B. L. Fittante. 1986. Fish habitat structures: a selection guide using stream classification. Proceedings of the Fifth Trout Stream Habitat Improvement Workshop, Lock Haven, Pennsylvania.

Scudder, G. G. E. 1989. The adaptive significance of marginal populations: a general perspective. Pages 180–185 *in* C. D. Levings, L. B. Holtby, and M. A. Henderson, editors. Proceedings of the National Workshop on Effects of Habitat Alteration on Salmonid Stocks. Canadian Special Publication of Fisheries and Aquatic Sciences 105, Ottawa.

Sedell, J. R., and R. L. Beschta. 1991. Bringing back the "bio" in bioengineering. Pages 160–175 *in* J. Colt and R. J. White, editors. Fisheries Bioengineering Symposium. American Fisheries Society, Symposium 10, Bethesda, Maryland.

Sedell, J. R., G. H. Reeves, and K. M. Burnett. 1994. Development and evaluation of aquatic conservation strategies. Journal of Forestry 92:28–31.

Seehorn, M. E. 1992. Stream habitat improvement handbook. U.S. Department of Agriculture, Forest Service, Southern Region, Technical Publication R8-TP 16, Atlanta, Georgia.

Slaney, P. A., and D. Zaldokas, editors. 1998. Fish habitat rehabilitation procedures. Ministry of Environment, Lands, and Parks, Watershed Restoration Technical Circular Number 9, Vancouver.

Solazzi, M. F., T. E. Nickelson, S. L. Johnson, and J. D. Rogers. 2000. Effects of increasing winter rearing habitat on abundance of salmonids in two coastal Oregon streams. Canadian Journal of Fisheries and Aquatic Sciences 57:906–914.

Stednick J., and J. D. Hall, editors. In press. The Alsea watershed: hydrological and biological responses to temperate coniferous forest practices. Springer-Verlag, New York.

Swanson, F. J., S. L. Johnson, S. V. Gregory, and S. A. Acker. 1998. Flood disturbance in a forested mountain landscape. BioScience 48:681–689.

Triska, F. J., V. C. Kennedy, R. J. Avazino, G. W. Zellweger, and K. E. Bencala. 1989. Retention and transport of nutrients in a third order stream: hyporheic processes. Ecology 70:1893–1905.

USDA (United States Department of Agriculture), United States Forest Service. 1992. Stream habitat improvement handbook. USDA Forest Service, Southern Region, Technical Publication R8-TP 16, Atlanta.

Wallace, J. B., J. W. Grubaugh, and M. R. Whiles. 1996. Influences of coarse woody debris on stream habitats and invertebrate biodiversity. Pages 119–129 *in* J. E. McMinn and D. A. Crossley, Jr., editors. Biodiversity and coarse woody debris in southern forests. Proceedings of the Workshop on Coarse Woody Debris in Southern Forests: Effects on Biodiversity, October 18–20, 1993. U.S. Department of Agriculture, Forest Service, Southern Research Station, General Technical Report SE-94, Athens, Georgia.

Wallace, J. B., J. R. Webster, and J. L. Meyer. 1995. Influence of log additions on physical and biotic characteristics of a mountain stream. Canadian Journal of Fisheries and Aquatic Sciences 52:2120–2137.

Ward, B. R. 2000. Declivity in steelhead (*Oncorhynchus mykiss*) recruitment at the Keogh River over the past decade. Canadian Journal of Fisheries and Aquatic Sciences 57:298–306.

Ward, B. R., and P. A. Slaney. 1979. Evaluation of in-stream enhancement structures for the production of juvenile steelhead trout and coho salmon in the Keogh River: progress 1977 and 1978. British Columbia Ministry of Environment, Fisheries Technical Circular 45, Victoria, BC.

Warren, C. E. 1971. Biology and water pollution control. Saunders, Philadelphia.

Wondzell, S. M., and P. A. Bisson. 2003. Influence of wood on aquatic biodiversity. Pages 249–263 *in* S. V. Gregory, K. L. Boyer, and A. M. Gurnell, editors. The ecology and management of wood in world rivers. American Fisheries Society, Symposium 37, Bethesda, Maryland.

American Fisheries Society Symposium 37:407–420, 2003

Riparian Management for Wood in Rivers

KATHRYN L. BOYER

USDA Natural Resources Conservation Service, Wildlife Habitat Management Institute Department of Fisheries and Wildlife, Oregon State University, Corvallis, Oregon 97331, USA

DEAN RAE BERG

Sivicultural Engineering, 15806 60th Avenue W., Edmonds, Washington 98026, USA

STAN V. GREGORY

Department of Fisheries and Wildlife, Oregon State University, Corvallis, Oregon 97331, USA

Abstract.—Riparian and floodplain forests are vital components of landscapes. They are transitional zones (ecotones) between river and upland ecosystems where ecological processes occurring in riparian areas and floodplains connect and interact with those of rivers and streams. These forests are the major source of large wood for streams and rivers. Extensive loss of riparian and floodplain forests around the globe is evident from the dramatically reduced supply of large wood in rivers. Clearly, it is necessary to conserve and restore riparian forests to sustain a supply of wood for rivers. This chapter discusses river and land management practices that are designed to provide a continuous source of large wood for rivers and retain wood once it has entered the channel or floodplain. These management practices include conservation of intact riparian and floodplain forests, restoration of ecological processes necessary to sustain riparian forests in the long term, and management of riparian forests specifically to accelerate recruitment of large wood to rivers and streams.

Ecological Functions of Riparian Forests

Large wood is a critical component of rivers and streams of forested ecosystems throughout the world. It provides structure and organic matter that create and enhance habitat diversity and food sources for many riparian and aquatic organisms (Benke and Wallace 2003; Bilby 2003; Dolloff and Warren 2003; Pollock et al. 2003; Steel et al. 2003; Wondzell and Bisson 2003; Zalewski et al. 2003; all this volume). Large wood in rivers and streams originates from both riparian and upland forests (Gurnell 2003; Swanson 2003; both this volume). Upland contributions of wood to streams are highly variable and generally a result of landslides from adjacent hillslopes in headwaters and steeper portions of watersheds (Keller and Swanson 1979; Benda et al. 2003; Nakamura and Swanson 2003; both this volume). The proportion of wood in streams and rivers from landslides historically is relatively small in relation to that recruited cumulatively along the longitudinal miles of intact riparian forests from headwaters to the mainstems of large river systems (Keller and Swanson 1979; Sedell and Froggatt 1984; Townsend 1996; Piégay et al. 1999; Piégay 2003, this volume).

Native riparian forests develop and function in complex and cyclical ways, primarily through disturbances such as floods, fires, pest and disease outbreaks, and hurricanes. These natural disturbances result in regeneration and succession of floodplain and riparian forests (Junk et al. 1989; Stromberg et al. 1993; Piégay and Bravard 1997; Scott et al. 1997; Middleton 2002), channel avulsion and lateral migration, "pulses" of wood transport and relocation, and floodplain development (Agee 1988). Riparian and floodplain forest composition, structure, and successional attributes are affected both by local conditions (such as land management of individual parcels of land and small-scale disturbance regimes) and large-scale changes in climate and human alterations of hydrologic regimes, channel morphology, and land use. Over the past 150–300 years in North America, the past 1,000+ years in Europe, and perhaps even longer in Africa and Asia, riparian forests and their important ecological functions have been chronically and cyclically compromised by human actions at multiple scales (Décamps et al. 1988; Tabacchi et al. 1996; Elosegi and Johnson 2003; Montgomery et al. 2003; both this volume). Throughout the world, riparian forests have high ecological, economic, and intrinsic values (such as natural beauty). These values subject riparian forests to controversy and conflict in an increasingly populous human landscape of multiple jurisdictions and conflicting points of view.

Ecological and physical linkages among riparian forests, rivers and their floodplains are critical to the processes that maintain their many functions (Gregory et al. 1991; Nilsson 1991). Riparian processes occur in three spatial dimensions (longitudinal, lateral, and vertical) and over time within a basin. The conservation, enhancement and restoration of linkages among these processes—necessarily at multiple spatial and temporal scales—is likely one of the most complex land management challenges of the 21st century. Watershed management strategies that recognize relationships among processes acting on riparian forests in space and time are now being incorporated into land management actions and long-term planning (Gregory et al. 1998; Wissmar and Beschta 1998). Land managers and planners are beginning to identify mechanisms that can restore ecological processes of watersheds and should sustain a more continuous, albeit patchy, supply of large wood to rivers (Oliver et al. 1992; Beechie and Bolton 1999; Zalewski et al. 2003). Conservation or management of uplands, rivers, streams, and their floodplains for the purposes of maintaining or restoring riparian function demands technical acuity and interdisciplinary cooperation in forest and riparian ecology, fisheries biology, fluvial geomorphology, hydrology, soil science, silviculture and forest stand dynamics, forest and civil engineering, resource economics, sociology, and other disciplines (Mitsch and Jorgensen 1989; Berg 1995; Montgomery 1997; Zalewski et al. 2003).

Riparian forests affect, and are affected by, their streams and rivers (Junk et al. 1989; Gregory et al. 1991; Bren 1993; Brookes et al. 1996; Hupp and Osterkamp 1996; Huggenberger et al. 1998). Interactions between intact riparian and floodplain forests and their rivers are reciprocal for numerous riparian and riverine processes and functions, especially the exchange of organic matter, including large wood. Land management practices that maintain these functional linkages should be considered when formulating long-term land/river management goals.

Geomorphological linkages of riparian forests: considerations for riparian management

The dynamic nature of riparian forest composition and structure reflect the complex linkages among geomorphic processes acting on rivers and adjacent floodplains (Malanson 1993; Hupp and Osterkamp 1996; Hughes 1997). The geomorphology of the channel and its floodplain will affect rates, amounts, and spatial distribution of wood recruitment resulting from bank erosion and lateral channel migration. Recent studies in the Pacific Northwest of North America indicate that recruitment of wood from forest stands of mountainous streams is greater along unconstrained stream reaches compared to constrained reaches (Acker et al. 2003). Valley form, as well as soil type and quality, also influences the fall direction of a tree potentially available as river wood (Sobota 2003). Studies in montane streams of Oregon draining forests of different management regimes demonstrated that, during a flood, floating large wood and the amount mobilized (congested versus uncongested wood transport) influenced the degree to which floods affect riparian tree toppling. In addition, this waterborne wood significantly influenced the consequent amount of large wood that was recruited to the system during a flood event, with longer-term affects likely influenced by forest management practices that occurred over the previous decades (Johnson et al. 2000). In essence, congested wood transport was a function of recent and historical land-use practices and toppled more riparian trees and deposited more jams in the channel after a flood.

The interaction of in-channel wood, floodborne deposits, and riparian forest development is essentially a positive feedback loop for wood recruitment. Instream wood accumulations are roughness elements that strongly influence fine sediment and gravel deposition in rivers and floodplains. In turn, these deposits provide substrate suitable for riparian tree seedlings (Swanson and Lienkaemper 1982; Fetherston et al. 1995; Gerhard and Reich 2000). Floodplain and island trees that are able to withstand floods and grow to large size are future sources of large wood for the channel. For example, in a recent study on the Tagliamento River in northeastern Italy, van der Nat et al. (2003) demonstrated that large wood mass in vegetated island-braided reaches was higher (100–150 t/ha) than in a bar-braided reaches (15–70 t/ha).

Seral stage of the riparian forest stand influences amount, type, and size of wood that is recruited through processes of bank failure, windthrow, and mortality from disease (Lienkaemper and Swanson 1987; Hedman et al. 1996; Benda et al. 2003). In large rivers of France, the successional stage of riparian forests coupled with bank erosion rates differ between meandering channels of the Ain River and braided channels of the Drôme River, with consequent differences in wood input volumes (Citterio 1996; Piégay 2003): wood inputs in the Ain River study reach were twice that of the Drôme over the same 8-year period.

Hydrological linkages with riparian forests: considerations for riparian management

Annual hydrological patterns, including flood frequency and magnitude, affect active recruitment of riparian trees to the channel and regeneration success of newly established seedlings on the floodplain, especially for flood-dependent species like cottonwood (*Populus* spp; Howe and Knopf 1991; Décamps 1996; Scott et al. 1997; Shafroth et al. 1998; Stromberg 2001). Natural and altered hydrologic regimes also affect regeneration rates as well as species composition of riparian forests (Johnson 1992; Tabacchi et al. 1996). For example, in arid watersheds of the western United States, exotic species such as saltcedar *Tamarix chinensis* thrive under altered flow regimes that inhibit success of native cottonwood *Populus fremontii* recruitment (Busch and Smith 1995). Hydrological regimes thus influence the susceptibility of a particular river system to invasion by exotic plant species, which, in turn, determines plant community dynamics for long time periods (decades and centuries). This is very evident in the riparian forests along the Garrone River of France, which has a high diversity of tree species (about 200), but greater than 50% of these are exotic species (E. Tabacchi, University of Toulouse, personal communication).

Riparian forest composition and nutrient dynamics: considerations for riparian management

Riparian forest composition affects both spatial and temporal dynamics of wood in streams and rivers, including arrangement and stability of individual pieces and accumulations of wood and decomposition rates. Decomposition rates of large wood in streams and rivers vary widely and depend on tree species, piece size, wood quality and condition, and location within the riparian/aquatic system (Harmon et al. 1986; Beechie and Bolton 1999; Bilby 2003). In forested regions of North America, conifers provide the most desirable structural elements in streams and rivers because they are resistant to movement and decompose slowly (Bisson et al. 2003; Dolloff and Warren 2003). Though deciduous trees generally decompose more quickly than conifers, submerged hardwood species in beaver ponds decompose very slowly, in several reported cases as low as 1.1% per year (Hodkinson 1975).

The contiguity of the riparian forest in a watershed and connectivity of the river with its floodplain affect flow discharge rates, channel migration rates, exchange of nutrients between surface and subsurface flows, all of which in turn affect dynamics of riparian plant communities. Intricacies of these linkages are well illustrated in nutrient cycling processes of temperate forests of the northwest Pacific rim. In this region, returning anadromous salmon and steelhead *Oncorhynchus* spp. are significant nutrient sources to rivers, streams, and adjacent riparian forests. Adult fish, both through excretion and releasing gametes in the month or so following return to freshwater and prior to death, contribute substantial amounts of marine-derived nitrogen (approximately 30% of the total) to a stream (Cederholm et al. 1999). Spawned-out carcasses are the other main source of returning nutrients and, under natural conditions, are distributed extensively throughout the aquatic system where salmon are able to spawn (Cederholm and Peterson 1985; Michael 1995; Bilby et al. 1996; Larkin and Slaney 1997). Recent studies in Alaska compared growth rates of trees in riparian forests adjacent to salmon spawning sites to growth rates of trees in riparian areas adjacent to streams where no spawning historically occurred (Helfield and Naiman 2003). Total annual growth per unit forest area was more than three times higher along spawning reaches. In the same study, tree ring data indicated that trees reached large enough size to fall into and persist in the spawning streams 200 years earlier than in nonspawning reaches. Thus marine-derived nitrogen from spawning salmon appears to fertilize riparian trees that eventually enter the stream and, in turn, may enhance habitat for future generations of salmon.

In the Pacific Northwest of the United States, coastal riparian forests in early seral stages following logging have a large red alder *Alnus rubra* component. Recent studies in Washington State demonstrated that red alder is also a significant source of nitrogen to streams and, thus, may be an important component of riparian forest nutrient cycles, particularly since its nitrogen fixation is directly involved in riparian hardwood production (Volk et al. 2003). Alder-derived nitrogen may offset the lower marine-nitrogen contributions that have occurred with drastic declines in returning adult salmon.

Management Applications for Maintaining a Source of Large Wood for Rivers

The complexity of interactions among the forest and river, its physical setting, its ecological role, and disturbance history poses an intellectual challenge to natural resource scientists and land managers trying to design strategies to restore ecological functions, including those provided by wood in rivers. Add to this inherent complexity the mix of jurisdictional boundaries and socioeconomic concerns of multiple landowners within a watershed, and the land management challenge becomes even more daunting. Wood budgets and models are effective tools in light of this complexity (Benda et al. 2003; Gregory 2003; this volume). Several wood models have been linked to forest models and used to explore the long-term consequences of various riparian management regimes on wood dynamics in streams (Prognosis, Rainville et al. 1985; ORGANON, Berg 1995; FVS, Bragg et al. 2000; RAIS, Welty et al. 2002; Meleason et al., in press) and the fate of carbon from riparian forests (Malanson and Kupfer 1993). Other wood models have been linked to forest stand data to explore wood recruitment characteristics, such as source distance (McDade et al. 1990), frequency distributions of piece volume, length and orientation (Van Sickle and Gregory 1990), and influence of valley wall slope on tree fall angle (Minor 1997; Sobota 2003). A wood budget approach has also been used to explore wood standing stock at the reach- (Murphy and Koski 1989) and network-scale (Benda et al. 2003). Collectively, a range of insights into forest–stream interactions has resulted from wood simulation studies that would otherwise be difficult to obtain through field research alone (Gregory 2003).

Interim- or smaller-scale approaches to improve river and stream fish habitats and to alter fluvial dynamics have focused on the intentional placement of wood in channels (Abbe et al. 2003; Bisson et al. 2003; Reich et al. 2003; all this volume). These short-term approaches can expedite improvement of degraded rivers and streams. However, these approaches are likely to prove even more effective when designed to complement long-term riparian forest management objectives that focus on recovery of a sustainable source of wood for rivers as a central goal. Long-term riparian forest management strategies that highlight wood recruitment as an objective include (1) conservation of intact riparian and floodplain forests, (2) restoration of degraded riparian forests and their ecological processes and functions, and (3) active forestry management that prescribes silvicultural treatments specifically for wood recruitment.

Conservation of intact functional riparian and floodplain forests

Numerous estimates of the dramatic loss of riparian and floodplain forests around the world are alarming in their implications for water quality and aquatic habitats (Welcomme 1979; Tabacchi et al. 1990; NRC 1992, 2002a, 2002b; Malanson 1993). Because of the rapid loss of functional wetlands and riparian areas around the world, the scientific community is calling for aggressive protection of these limited resources. Conservation of intact riparian areas may prove to be the most cost-effective management approach for initial restoration of ecological functions to watersheds, including delivery of wood to channels (Frissell and Nawa 1992; Naiman et al. 1992; Gregory 1997). For example, conservation of soil and soil quality on managed riparian areas can be accomplished indirectly with sound land use practices that protect riparian soils and woody vegetation, such as fencing to exclude livestock from sensitive riparian areas. On private lands, incentive programs that compensate landowners for the purchase of conservation easements or provide tax incentives for riparian conservation are promising approaches to protect intact riparian and floodplain forests as a source for large wood in streams and rivers. However, determining the appropriate value of these lands is not without controversy in market-based economies where ecological goods and services are not customarily considered in real estate appraisals (Daily et al. 1997).

Restoration of degraded riparian and floodplain forests

Management actions that restore long-term dynamics of riparian forests should eventually provide a renewable source of wood for rivers. Restoration of ecological functions may take decades, and in some cases centuries, of concerted management focus to achieve recovery of riparian forests and complex river and stream habitats. This requires patience and perseverance by the people and governments that pay for and implement restoration of impaired ecosystems. In the United States, federal programs such as those of the Farm Bill provide funding for landowners interested in improving fish and wildlife habitat, water quality, and soil erosion, all of which can influence riparian conditions on private lands. However, inadequately trained program managers responsible for developing incentive program objectives may not address the ecological complexity of river-riparian systems and underestimate the time required to restore desired ecological functions. For example, the United State Department of Agriculture's (USDA) Conservation Reserve Enhancement Program provides incentive payments to agricultural landowners in more than 27 states to limit agricultural production in converted riparian areas and replant them with native riparian species. However, out of the 1.4 million acres available for riparian improvement as of 2002, less than 14% have as their primary objective improving riparian areas as a source of large wood for aquatic habitat (USDA, unpublished data). In addition, many of the agreements made between USDA and the landowner require only a 15-year contract period. Longer-term (30–50 years) leases are needed to restore ecological functions to riparian areas, such as wood recruitment to adjacent waters.

Riparian forest restoration goals need not be old-growth forests per se. In most scenarios around the world, restoring riparian forest to late-successional seral stages may not be feasible because managers must work within shorter time frames, infrastructure constraints, and political and economic mandates. In some cases, mimicking or restoring natural flow regimes are promising approaches for riparian restoration (Hughes 1997; Molles et al. 1998; NRC 2002b). Reconnecting floodplains and rivers with connected riparian forest patches of various structure and widths also deserves consideration (Gore and Bryant 1988; Wissmar and Beschta 1998; Ward et al. 1999; Mutz 2000). This approach has potential for restoring large-scale ecological functions to river systems as a whole. Planning and implementing restoration actions as controlled experiments, collecting baseline data, and comparing results to relevant reference sites encourages evaluation and learning about the effectiveness of these innovative approaches and their efficacy for long-term adaptive watershed management (Berg et al. 1998; Zalewski et al. 2003).

Riparian forest management: silvicultural treatments

Stochastic natural disturbances, such as floods, windstorms, fires, landslides, and flood-induced bank erosion, can be expected to deliver most of the potential large wood to rivers and streams, but may be ineffective in reaching supplementation goals for today's wood-deprived rivers. Therefore, a key component of any ambitious watershed restoration strategy should be riparian silviculture. Silvicultural designs that consider principles of stream ecology and forest stand dynamics when developing management prescriptions can accelerate riparian forest succession and wood recruitment. In the Pacific Northwest, goals for stand structure are based on reference conditions of the few remaining old-growth stands, as well as habitat needs of aquatic species (such as complex structure and water quality), especially fish and amphibians. Appropriate riparian forest silvicultural approaches must consider the interactions of aquatic and terrestrial ecosystem processes (Gregory et al. 1991) coupled with life history requirements for a variety of salmon and other aquatic species that live in adjacent streams and rivers. The declines of economically and culturally important salmon have led to numerous studies to determine important freshwater habitat elements, including the size and amounts of instream large wood (Bryant 1983; Bilby and Bisson 1991; Bjornn and Reiser 1991; Bisson et al. 1992; Fausch and Northcote 1992). Scientists and land managers have applied these studies to determine appropriate wood loading standards for maintaining complex salmon and trout habitats, especially on federal lands. As described in previous sections, wood dynamics are complex, and thus, prescribing uniform standards makes less sense than establishing a range of desired conditions tailored to the local area. It is possible to determine the size and amount of wood a riparian forest is expected to provide for specific reaches of stream and river habitat and then determine if this amount lies within a range considered acceptable by aquatic habitat specialists. Tree growth models (for example, Wykoff et al. 1982) have been developed that link site-specific stream wood objectives to the size and density of potential large wood in the riparian forest. Silvicultural systems can be designed to focus on wood recruitment to streams and rivers and incorporate forest growth models for stand projection, site preparation and maintenance regimes, planting density and stand development plans, thinning prescriptions, and monitoring protocols. Each of these aspects is described below:

Stand projections of riparian forests.—The USDA Forest Service Forest Vegetation Simulator (FVS, Wykoff et al. 1982) is an example of a public-domain growth and yield model that can be used to compare growth of various stands once they are established under a silvicultural system. In addition to modeling growth of multispecies stands, FVS can also predict yields in stands of mixed composition (for example, 50% hardwood and 50% conifer species). While there are numerous other public-domain models (for example, DFSIM, Curtis et al. 1981; TASS, Mitchell and Cameron 1985; ORGANON, Hester et al. 1989), FVS has many versions adaptable to most regions of North America.

While most simulation models are easily accessible, several other methods can estimate riparian forest stand structure through time. Empirical yield tables (McArdle et al. 1949; Minore 1983; Nystrom et al. 1984) predict the growth of a particular species at specific sites. Density management diagrams (DMD) show the relationship between stocking and various growth parameters of a species (for example, Hibbs 1987; Smith 1987; Long et al. 1988). Length of time to desired conditions under different stocking levels can be estimated using these methods.

Site preparation and early stand maintenance.—Riparian forests may be difficult to re-establish in areas where trees have not existed for long periods of time, as in lowland agricultural areas. Often, disturbed sites are prone to invasion by exotic weeds and shrubs, especially immediately following discontinued use of herbicides and pesticides for crop production. Early, intensive site preparation is one key to suppressing weed competition. Nonnative plants (weeds) will likely sprout and grow among planted stock regardless of soil condition. Generally, this is a temporary nuisance because these species are often shade intolerant and do not survive once sapling canopies begin to close (usually within 2–5 years). Some more aggressive exotic species, such as Himalayan blackberry, will competitively exclude desired riparian shrub species for decades and should be controlled early on.

Native riparian understory species also provide important ecological functions of riparian communities, including structural diversity, food sources, microclimate, nutrient cycling, and habitat for riparian species that may influence reciprocal habitat subsidies between riparian areas and streams (Hilderbrand et al. 1999; Steel et al. 2003). These understory plant species should also be considered when designing silvicultural systems for the re-supply of large wood to streams and rivers.

Planting density and stand development.—Initial planting density significantly affects forest stand development. High-density plantations have smaller trees compared to low-density plantations over similar periods of growth. Under high-density scenarios, larger numbers of stems become available for use in stream habitat improvement projects. Actively selecting trees to be thinned, based on the current size or species, can optimize growth. Mixed stock plantations have advantages. Diseases associated with mono-cultures are minimized because pathogen migration is interrupted by plants of different genera, which are resistant to each other's diseases.

Planting pattern.—Once the stand is established and undesirable species are controlled, silvicultural systems can be modified to create desired ecological structure and function (Berg et al. 1998). Numerous species and spatial patterns are possible (Franklin et al. 1996). Two important functions should be considered when establishing riparian forests: (1) adequate trees of the proper size and species should be grown to supply streams or rivers with sufficient volumes of wood, and (2) stands should be managed to maintain adequate shade to moderate stream temperatures.

Competition between planted trees and invasive, exotic vegetation can be reduced by mechanically controlling weeds as frequently as possible. Conventional herbicides are an alternative to mechanical removal, but their use near streams may contaminate surface waters or harm organisms.

Thinning impacts on stand development.—Thinning riparian forest trees can accelerate the time required to reach desired stand conditions by concentrating growth on fewer stems (Daniel et al. 1979; Curtis et al. 1997; Beach and Halpern 2001; Zeide 2001). Competition for nutrients and sunlight among plants is well understood and has been applied to forest stands (Beach and Halpern 2001). As forest stands mature, competition for limited light and nutrients increases. Thinning less desirable species is intended to establish or accelerate development of species of greater economic and/or ecological value, such as providing large, slowly decomposing wood for rivers. Rainville and others (1985) suggest that thinning, if not done properly, will reduce the amount of large wood available for recruitment. While thinning has potential benefits, these methods are largely untested in riparian ecological applications (Berg 1995, 1997; Beechie et al. 2000). Forest managers who invest time and resources must recognize some level of risk, such as tree loss, because of the physically active and dynamic nature of floodplains and riparian areas. Simulations from FVS models predict that thinning produces larger numbers of trees that meet the desired size criteria within the given time for both Douglas-fir *Pseudotsuga menzesii* and western redcedar *Thuja plicata* (Table 1). Though average diameters of two stands may be comparable, small trees are far more numerous in an un-thinned stand (Berg, author's unpublished data). Thinning concentrates growth in fewer individuals so that the sizes of trees are larger though density is lower. This equates to a potentially higher value from both financial and ecological standpoints. If the objective is to maximize the number of trees of a desired size, stands could be planted at high densities along all streams up to 20 m wide, since by year 100 (earlier on the smaller streams), enough wood could be recruited to the stream to meet the desired outcomes. Other stocking levels produce larger diameter trees but at lower overall density, limiting the available large wood for recruitment to streams and rivers. The former might be necessary where streams are devoid of large wood, while the latter may be useful on streams with desired amounts of large wood, and regenerating standing stock in the riparian forest would be available as a long-term source of wood for the channel.

Thinning can be done in patches or applied uniformly across a forest stand (Franklin et al. 1996). Various patterns of thinning can affect the type and amount of natural regeneration of riparian forest. For example, openings of one-quarter to one-half acre have been found to provide sufficient growing space for Douglas-fir regeneration (Isaac 1943; Curtis et al. 1997). Because of the proximity to edges in many riparian stands (streamside and fieldside), light availability may be greater than in forest interiors, resulting in greater natural regeneration.

Monitoring.—Monitoring managed riparian forests provides a framework for systematic and quantitative measurements of change over time. Monitoring silvicultural systems that are designed and implemented specifically as a source for wood in rivers can generate timely information about the progress of riparian forest development and the effectiveness of silvicultural treatments to meet wood recruitment goals. Through monitoring, problems can be identified and silvicultural practices can be modified to better achieve goals and objectives of the landowner (Berg 1997).

Vegetation monitoring generally includes estimating seedling survival and density, measuring annual growth rates, estimating cover and/or biomass production, and quantifying species diversity. Performance standards for vegetation are based on the initial planting density, which is site-specific, and a function of the managed stem density. High survival is desired because high density will block invasive exotic plants and provide the broadest possible selection for thinning. Data collected when monitoring a stand can be used to develop contingency plans for corrective measures to take in the event that performance standards are not met.

Monitoring channel conditions to assess wood loading and habitat changes provides a way to evaluate effectiveness of forest management

TABLE 1. FVS modeled stands for Douglas-fir and western redcedar, demonstrating the effect of thinning on tree diameter (DBH) and height. Stand initiation conditions for these silvicultural systems were 4,064 total stems per hectare (SPH); thinning was from below (removing the smallest stems first) and removed 50% of the stems at each entry.

	Douglas-fir			Western redcedar			
Stand treatment	Mean DBH (cm)	Mean tree height (m)	SPH	Mean DBH (cm)	Mean tree height (m)	SPH	Total SPH
50 years							
No thin	29.4	26	540	29.4	26	529	1069
Thin twice at age 20 and 40	35.6	28	716	35.4	28	54	303
Thin once at age 25	33.9	28	635	33.8	28	214	849
100 years							
No thin	51.6	39	179	51.6	39	175	354
Thin twice at age 20 and 40	56.0	41	287	56.0	41	20	307
Thin once at age 25	55.7	41	235	55.7	41	76	312

and the rate at which aquatic habitat is improving as a result of riparian silvicultural practices. Where target numbers of large wood pieces have been achieved by placing logs from the adjacent stand after a thinning or from off-site sources, periodic monitoring (annually for 5–10 years) is necessary to evaluate the long-term effectiveness of the project. This includes surveys to determine if large wood remains within the project reach or if replacement from upstream or the riparian forest has kept the reach in desired condition. Amounts of wood can change if the export of wood exceeds the import or if the wood traps other pieces that have moved from upstream. If large wood is exported faster than it is delivered from the adjacent riparian area or from upstream, wood supplements may need to be added. This additional wood should only come from the adjacent riparian forest if monitoring data suggest that trees of sufficient size exist in adequate numbers to sustain a long-term source of wood into the future.

In addition, characteristics of the riparian forest (density of trees, average and maximum sizes, crown structure, rate of growth and regeneration, understory vegetation, and overall community structure) should be monitored to guide ongoing riparian management. These attributes of riparian forests are related to the effectiveness and validation of the proposed or implemented silvicultural system.

A multifunctional silvicultural plan designed to improve riparian forest conditions for long-term recruitment of wood to rivers should be designed to evaluate the effectiveness of different riparian management strategies in meeting ecological objectives. A network of well-documented and monitored riparian silvicultural systems then becomes a robust set of treatments to test various prescriptions and ecological responses in the watershed. Riparian forest management does come at some price. The costs of tending the stand may not be recovered from harvest revenue and are thus an unrewarded investment in ecological services for the benefit of fish, water quality, and riparian forests. If designed, implemented, and monitored carefully, these systems can also serve as demonstration sites of successful forestry applications for agencies or landowners seeking to justify the costs of providing ecological services to society.

Conclusion

Riparian forest management applications presented here integrate what we know about riparian processes with specific techniques to restore or improve riparian functions. They are based on current knowledge of natural ecosystem functions and practical considerations (Berg 1995; Gregory 1997; Bilby and Bisson 1998; Naiman et al. 1999). Hypotheses about riparian forest improvement and subsequent wood recruitment to rivers can be tested using an experimental, science-based approach with cooperation from stakeholders who are applying innovative techniques on their land (Franklin 1997; Hulse et al. 2002). Concurrently, managers and stakeholders must acknowledge the dynamic nature of riparian and floodplain forests and the amount of time required to restore ecological functions in a constantly changing landscape. The spatial significance of the effects of riparian forests on river landscapes and the globally significant conditions that are likely to impact their conservation must also be recognized. Climate and land-use changes coupled with their impacts on hydrological regimes have clear consequences on the quality, quantity, and dynamics of wood in rivers. Restoring function to altered landscapes may be daunting, and a return to what we think of as historical presettlement conditions is not feasible. Restoring even a fraction of a managed riparian landscape is warranted if, in doing so, key processes important for clean water, fish and wildlife habitat, and intrinsic values are sustained for future generations.

Acknowledgments

We are grateful to the following colleagues for technical assistance with this chapter: Mark Meleason assisted with wood modeling text and content. Art Mckee, Mike Maki, and Dan Sobota provided timely and thorough editorial review and suggestions for improvement.

References

Abbe, T. A., A. P. Brooks, and D. R. Montgomery. 2003. Wood in river rehabilitation and management. Pages 367–389 *in* S. V. Gregory, K. L. Boyer, and A. M. Gurnell, editors. The ecology and management of wood in world rivers. American Fisheries Society, Symposium 37, Bethesda, Maryland.

Acker, S. A., S. V. Gregory, G. Lienkaemper, W. A. McKee, F. J. Swanson, and S. D. Miller. 2003. Composition, complexity, and tree mortality in riparian forests in the central Western Cascades of Oregon. Forest Ecology and Management 173:293–308.

Agee, J. K. 1988. Successional dynamics in forest riparian zones. Pages 31–43 *in* K. J. Raedeke, editor. Streamside management: forestry and fishery interactions. Institute of Forest Resources, University of Washington, Seattle.

Beach, E. W., and C. B. Halpern. 2001. Controls on conifer regeneration in managed riparian forests: effects of seed source, substrate, and vegetation. Canadian Journal of Forest Research 31:471–482.

Beechie, T., and S. Bolton. 1999. An approach to restoring salmonid habitat-forming processes in Pacific Northwest watersheds. Fisheries 24(4):6–15.

Beechie, T., G. Pess, P. Kennard, R. E. Bilby, and S. Bolton. 2000. Modeling recovery rates and pathways for woody debris recruitment in Northwestern Washington streams. North American Journal Fisheries Management 20:436–452.

Benda, L., D. Miller, J. Sias, D. Martin, R. Bilby, C. Veldhuisen, and T. Dunne. 2003. Pages 49–73 *in* S. V. Gregory, K. L. Boyer, and A. M. Gurnell, editors. The ecology and management of wood in world rivers. American Fisheries Society, Symposium 37, Bethesda, Maryland.

Benke, A. C., and J. B. Wallace. 2003. Influence of wood on invertebrate communities in streams and rivers. Pages 149–177 *in* S. V. Gregory, K. L. Boyer, and A. M. Gurnell, editors. The ecology and management of wood in world rivers. American Fisheries Society, Symposium 37, Bethesda, Maryland.

Berg, D. R. 1995. Riparian silvicultural system design and assessment in the Pacific Northwest Cascade Mountains. Ecological Applications 5:87–96.

Berg, D. R. 1997. Active management of riparian habitats. Pages 50–61 *in* K. B. McDonald and F. Weinmann, editors. Wetland and riparian restoration: taking a broader view. US EPA, Region 10, Publication EPA 910-R-97–007, Seattle.

Berg, D. R., P. Stevenson, and S. Hashisaki. 1998. Riparian silvicultural trials in Washington State. Pages 23–25 *in* Ecosystem restoration: turning the tide; proceedings 1998 Society of Ecological Restoration Northwest Chapter Symposium. Society of Ecological Restoration and University of Washington Center for Streamside Studies.

Bilby, R E., and P. A. Bisson. 1991. Enhancing fisheries resources through active management of riparian areas. Pages 201–209 *in* B. White and I. Guthrie, editors. Proceedings of the 15th Northeast Pacific Pink and Chum Salmon Workshop. Pacific Salmon Commission, Vancouver.

Bilby, R. E., B. R. Fransen, and P. A. Bisson. 1996. Incorporation of nitrogen and carbon from spawning coho salmon into the trophic system of small streams: evidence from stable isotopes. Canadian Journal of Fisheries and Aquatic Sciences 53:164–173.

Bilby, R. E., and P. A. Bisson. 1998. Function and distribution of large woody debris. Pages 324–346 *in* R. J. Naiman and R. E. Bilby, editors. River ecology and management lessons from the Pacific coastal ecoregion. Springer-Verlag, New York.

Bilby, R. E. 2003. Decomposition and nutrient dynamics of wood in streams and rivers. Pages 135–147 *in* S. V. Gregory, K. L. Boyer, and A. M. Gurnell, editors. The ecology and management of wood in world rivers. American Fisheries Society, Symposium 37, Bethesda, Maryland.

Bisson, P. A., T. P. Quinn, G. H. Reeves, and S. V. Gregory. 1992. Best management practices, cumulative effects, long-term trends in fish abundance in Pacific Northwest river systems. Pages 189–232 *in* R. J. Naiman, editor. Watershed management: balancing sustainability and environmental change. Springer-Verlag, New York.

Bisson, P. A., G. H. Reeves, and S. V. Gregory. 2003. Trends in using wood to restore aquatic habitats and fish communities in western North American rivers. Pages 391–406 *in* S. V. Gregory, K. L. Boyer, and A. M. Gurnell, editors. The ecology and management of wood in world rivers. American Fisheries Society, Symposium 37, Bethesda, Maryland.

Bjornn, T. C., and D. W. Reiser. 1991. Habitat requirements of salmonids in streams. Pages 83–138 *in* W. R. Meehan, editor. Influences of forest and rangeland management on salmonid fishes and their habitats. American Fisheries Society, Special Publication 19, Bethesda, Maryland.

Bragg, D. C., J. L. Kershner, and D. W. Roberts. 2000. Modeling large woody debris recruitment for small streams of the Central Rocky Mountains. United States Department of Agriculture Forest Service, Rocky Mountain Research Station General Technical Report RMRS-GTR-55, Fort Collins, Colorado.

Bren, L. J. 1993. Riparian zone, stream, and floodplain issues: a review. Journal of Hydrology 150:277–299.

Brookes, A., J. Baker, and C. Redmond. 1996. Floodplain restoration and riparian zone management. Pages 201–228 *in* A. Brookes and J. F. Shields, editors. River channel restoration: guiding principles for sustainable projects. Wiley, Chicester, UK.

Bryant, M. D. 1983. The role and management of woody debris in West Coast salmonid nursery streams. North American Journal of Fisheries Management 3:322–330.

Busch, D. E., and S. D. Smith. 1995. Mechanisms associated with decline of woody species in riparian ecosystems of the southwestern U.S. Ecological Monographs 65:347–370.

Cederholm, C. J., and N. P. Peterson. 1985. The retention of coho salmon (*Oncorhynchus kisutch*) carcasses by organic debris in small streams. Canadian Journal of Fisheries and Aquatic Sciences 42:1222–1225.

Cederholm, C. J., M. D. Kunze, T. Murota, and A. Sibatani. 1999. Pacific salmon carcasses: essential contributions of nutrients and energy for aquatic and terrestrial ecosystems. Fisheries 24(10):6–15.

Citterio, A. 1996. Dynamique de dépôts et de prise en charge du bois mort sur deux hydrosystèmes l'Ain et la Drôme (The dynamics of deposition and mobilisation of dead wood in two hydrosystems: the River Ain and the River Drôme). Master's Thesis. Université Lyon, Lyon, France.

Curtis, R. O., G. W. Clendenen, and D. J. DeMars. 1981. A new stand simulator for coastal Douglas-fir: DFSIM users guide. USFS Pacific Northwest Forest and Range Experiment Station, General Technical Report PNW-128, Portland, Oregon.

Curtis, R. O., D. D. Marshall, and J. F. Bell. 1997. LOGS: A pioneering example of silvicultural research in coast Douglas-fir. Journal of Forestry 95:19–25.

Daily, G. C., S. Alexander, P. R. Ehrlich, L. Goulder, J. Lubchenco, P. A. Matson, H. A. Mooney, S. Postel, S. H. Schneider, D. Tilman, and G. M. Woodwell. 1997. Ecosystem services: benefits supplied to human societies by natural ecosystems. Issues in Ecology 2:1–16.

Daniel, T. W., J. A. Helms, and F. S. Baker. 1979. Principles of silviculture. 2nd edition. McGraw-Hill, New York.

Décamps, H., M. Fortune, F. Gazelle, and G. Pautou. 1988. Historic influence of man on the riparian dynamics of a fluvial landscape. Landscape Ecology 1:163–173.

Décamps, H. 1996. The renewal of floodplain forests along rivers: a landscape perspective. Verhandlungen Internationale Vereinigung für Theoretische und Angewandte Limnologie 26:35–59.

Dolloff, C. A., and M. L. Warren, Jr. 2003. Fish relationships with large wood in small streams. Pages 179–193 *in* S. V. Gregory, K. L. Boyer, and A. M. Gurnell, editors. The ecology and management of wood in world rivers. American Fisheries Society, Symposium 37, Bethesda, Maryland.

Elosegi, A., and L. B. Johnson. 2003. Wood in streams and rivers in developed landscapes. Pages 337–353 *in* S. V. Gregory, K. L. Boyer, and A. M. Gurnell, editors. The ecology and management of wood in world rivers. American Fisheries Society, Symposium 37, Bethesda, Maryland.

Fausch, K. D., and T. G. Northcote. 1992. Large woody debris and salmonid habitat in a small coastal British Columbia stream. Canadian Journal of Fisheries and Aquatic Sciences 49:682–693.

Fetherston, K. L., R. J. Naiman, and R. E. Bilby. 1995.

Large woody debris, physical process, and riparian forest succession in montane river networks of the Pacific Northwest. Geomorphology 13:133–144.

Franklin, J. F., D. R. Berg, D. A. Thornburgh, and J. C. Tappener. 1996. Alternative silvicultural approaches to timber harvesting: variable retention harvest systems. Pages 111–130 *in* K. A. Kohm and J. F. Franklin., editors. Creating a forestry for the 21st century: the science of ecosystem management. Island Press, Washington, D.C.

Franklin, J. F. 1997. Ecosystem management: an overview. Pages 21–53 *in* A. W. Haney and Mark S. Boyce, editors. Ecosystem management: applications for sustainable forest and wildlife resources. Yale University Press, New Haven, Connecticut.

Frissell, C. A., and R. K. Nawa. 1992. Incidence and causes of physical failure of artificial habitat structures in streams of western Oregon and Washington. North American Journal of Fisheries Management 12:182–197.

Gerhard, M., and M. Reich. 2000. Restoration of streams with large wood: effects of accumulated and built-in wood on channel morphology, habitat diversity and aquatic fauna. International Review of Hydrobiology 85:123–137.

Gore, J. A., and F. L. Bryant. 1988. River and stream restoration. Pages 23–38 *in* Rehabilitating damaged ecosystems. CRC Press Inc., Boca Raton, Florida.

Gregory, S. V., F. J. Swanson, W. A. McKee, and K. W. Cummins. 1991. An ecosystem perspective of riparian zones: focus on links between land and water. Bioscience 41:540–551.

Gregory, S. V. 1997. Riparian management in the 21st century. Pages 69–85 *in* K. A. Kohm and J. F. Franklin, editors. Creating a forestry for the 21st century. Island Press, Washington, D.C.

Gregory, S. V., D. W. Hulse, D. H. Landers, and E. Whitelaw. 1998. Integration of biophysical and socioeconomic patterns in riparian restoration of large rivers. Pages 231–247 *in* H. Wheater and C. Kirby, editors. Hydrology in a changing environment. Wiley, Exeter, UK.

Gregory, S. V. 2003. Modeling the dynamics of wood in streams and rivers. Pages 315–335 *in* S. V. Gregory, K. L. Boyer, and A. M. Gurnell, editors. The ecology and management of wood in world rivers. American Fisheries Society, Symposium 37, Bethesda, Maryland.

Gurnell, A. M. 2003. Wood storage and mobility. Pages 75–91 *in* S. V. Gregory, K. L. Boyer, and A. M. Gurnell, editors. The ecology and management of wood in world rivers. American Fisheries Society, Symposium 37, Bethesda, Maryland.

Harmon M. E., J. F. Franklin, F. J. Swanson, P. Sollins, S. P. Cline, N. G. Aumen, J. R. Sedell, G. W. Lienkaemper, K. Cromack, Jr., and K. W. Cummins. 1986. The ecology of coarse woody debris in temperate ecosystems. Pages 133–302 *in* A. MacFadyen and E. D. Ford, editors. Advances in ecological research. Volume 15. Academic Press, New York.

Hedman, C. W., D. H. Van Lear, and W. T. Swank. 1996. In-stream large woody debris loading and riparian forest seral stage associations in southern Appalachian Mountains. Canadian Journal of Forestry Research 26:1218–1227.

Helfield, J. M., and R. J. Naiman. 2003. Effects of salmon-derived nitrogen on riparian forest growth and implications for stream productivity. Ecology 82:2403–2409.

Hester, A. S., D. W. Hann, and D. R. Larson. 1989. Organon: southwest Oregon growth and yield model user manual, version 2.0. Forest Research Lab, College of Forestry, Oregon State University, Corvallis.

Hibbs, D. E. 1987. The self-thinning rule and red alder management. Forest Ecology and Management 18:273–281.

Hilderbrand, G. V., T. A. Hanley, C. T. Robbins, and C. C. Schwartz. 1999. Role of brown bears (*Ursus arctos*) in the flow of marine nitrogen into a terrestrial ecosystem. Oecologia 121:546–550.

Hodkinson, I. D. 1975. Dry weight loss and chemical changes in vascular plant litter of terrestrial origin, occurring in a beaver pond ecosystem. Journal of Ecology 63:131–142.

Howe, W. H., and F. L. Knopf. 1991. On the imminent decline of the Rio Grande cottonwoods in central New Mexico. Southwestern Naturalist 36:218–224.

Huggenberger, P., E. Hoehn, R. Beschta, and W. Woessner. 1998. Abiotic aspects of channels and floodplains in riparian ecology. Freshwater Biology 40:407–425.

Hughes, F. M. R. 1997. Floodplain biogeomorphology. Progress in Physical Geography 21:501–529.

Hulse, D., S. Gregory, and J. Baker. 2002. Willamette River basin planning atlas: trajectories of environmental and ecological change. Oregon State University Press, Corvallis.

Hupp, C. R., and W. R. Osterkamp. 1996. Riparian vegetation and fluvial geomorphic processes. Geomorphology 14:277–295.

Isaac, L. A. 1943. Reproductive habits of Douglas-fir. C. L. Pack Forestry Foundation, Washington D.C.

Johnson, W. C. 1992. Dams and riparian forests: case study from the upper Missouri River. Rivers 3:229–242.

Johnson, S. L., F. J. Swanson, G. E. Grant, and S. M. Wondzell. 2000. Riparian forest disturbances by a mountain flood—the influence of floated wood. Hydrological Processes 14:3031–3050.

Junk, W. J., P. B. Bayley, and R. E. Sparks. 1989. The flood pulse concept in river-floodplain systems.

Canadian Special Publications in Fisheries and Aquatic Sciences 106:110–127.

Keller, E. A., and F. J. Swanson. 1979. Effects of large organic material on channel form and fluvial processes. Earth Surface Processes 4:361–380.

Larkin, G. A., and P. A. Slaney. 1997. Implications of trends in marine-derived nutrient influx to south coastal British Columbia salmonid production. Fisheries 22(11):16–24.

Lienkaemper, G. W., and F. J. Swanson. 1987. Dynamics of large woody debris in streams in old-growth Douglas-fir forests. Canadian Journal of Forest Research 17:150–156.

Long, J. N., J. B. McCarter, and S. B. Jack. 1988. A modified density management diagram for coastal Douglas-fir. Western Journal Applied Forestry 3(3):88–89.

Malanson, G. P. 1993. Riparian landscapes. Cambridge University Press, Cambridge, UK.

Malanson, G. P., and J. A. Kupfer. 1993. Simulated fate of leaf litter and woody debris at a riparian cutbank. Canadian Journal of Forest Research 23:582–590.

Meleason, M. A., S. V. Gregory, and J. P. Bolte. In press. Implications of riparian management strategies on wood in streams of the Pacific Northwest. Ecological Applications.

McArdle, R. E., W. H. Meyer, and D. Bruce. 1949. Yield of Douglas-fir in the Pacific Northwest, Revised. USDA Technical Bulletin 201, Washington, D.C.

McDade, M. H., F. J. Swanson, W. A. McKee, J. F. Franklin, and J. V. Sickle. 1990. Source distances for coarse woody debris entering small streams in western Oregon and Washington. Canadian Journal of Forest Research 20:326–330.

Michael, J. H. 1995. Enhancement effects of spawning pink salmon on stream rearing juvenile coho salmon: managing one resource to benefit another. Northwest Science 69(3):228–233.

Middleton, B. 2002. Flood pulsing in wetlands: restoring the natural hydrological balance. Wiley, New York.

Minor, K. P. 1997. Estimating large woody debris recruitment from adjacent riparian areas. Master's thesis. Oregon State University, Corvallis.

Minore, D. 1983. Western redcedar: a literature review. USDA USFS General Technical Report PNW-150, Pacific Northwest Forest and Range Experiment Station, Portland, Oregon.

Mitchell, K. J., and I. R. Cameron. 1985. Managed stand yield table for coastal Douglas-fir: initial density and precommercial thinning. Ministry of Forests, Victoria, BC.

Mitsch, W. J., and S. E. Jorgensen. 1989. Ecological engineering. Wiley and Sons, Inc., New York.

Molles, M. C. Jr., C. S. Crawford, L. M. Ellis, H. M. Vallett, and C. N. Dahm. 1998. Managed flooding for riparian ecosystem restoration. BioScience 48:749–756.

Montgomery, D. R. 1997. What's best on the banks? Nature 388:328–329.

Montgomery, D. R., B. D. Collins, J. M. Buffington, and T. B. Abbe. 2003. Pages 21–47 *in* S. V. Gregory, K. L. Boyer, and A. M. Gurnell, editors. The ecology and management of wood in world rivers. American Fisheries Society, Symposium 37, Bethesda, Maryland.

Murphy, M. L., and K. V. Koski. 1989. Input and depletion of woody debris in Alaska streams and implications for streamside management. North American Journal of Fisheries Management 9:427–436.

Mutz, M. 2000. Influences of woody debris on flow patterns and channel morphology in a low-energy, sand-bed stream reach. International Review of Hydrobiology 85:107–121.

Naiman, R. J., T. J. Beechie, L. E. Benda, D. R. Berg, P. A. Bisson, L. H. MacDonald, M. D. O'Conner, P. L. Olsen, and E. A. Steele. 1992. Fundamentals of ecologically healthy watersheds in the Pacific Northwest coastal ecoregion. Pages 127–188 *in* R. J. Naiman, editor. Watershed management: balancing sustainability and environmental change. Springer-Verlag. New York.

Naiman, R. J., S. R. Elliott, J. M. Helfield, and T. C. O'Keefe. 1999. Biophysical interactions and the structure and dynamics of riverine ecosystems: the importance of biotic feedbacks. Hydrobiologia 410:79–86.

Nakamura, F., and F. J. Swanson. 2003. Dynamics of wood in rivers in the context of ecological disturbance. Pages 279–297 *in* S. V. Gregory, K. L. Boyer, and A. M. Gurnell, editors. The ecology and management of wood in world rivers. American Fisheries Society, Symposium 37, Bethesda, Maryland.

Nilsson, C. 1991. Conservation management of riparian communities. Pages 352–372 *in* L. Hansson, editor. Ecological principles of nature conservation. Elsevier Press, Amsterdam.

NRC (National Research Council). 1992. Restoration of aquatic ecosystems: science, technology, and public policy. National Academy Press, Washington, D.C.

NRC (National Research Council). 2002a. The Missouri River ecosystem: exploring the prospects for recovery. National Academy Press, Washington, D.C.

NRC (National Research Council). 2002b. Riparian areas: functions and strategies for management. National Academy Press, Washington, D.C.

Nystrom, M. N., D. S. DeBell, and C. D. Oliver. 1984. Development of young growth Western redcedar stands. USFS Research paper. PNW 324. Pacific Northwest Forest and Range Experiment Station, Portland, Oregon.

Oliver, C. D., D. R. Berg, D. R. Larsen, and K. L. O'Hara. 1992. Integrating management tools, ecological knowledge, and silviculture. Pages 361–382 *in* R. J. Naiman, editor. Watershed management: balancing sustainability and environmental change. Springer-Verlag, New York.

Piégay, H., and J.-P. Bravard. 1997. Response of a Mediterranean riparian forest to a l in 400 year flood. Ouveze River, Drome-Vaucluse, France. Earth Surface Processes and Landforms 22:31–43.

Piégay, H., A. Thevenet, and A. Citterio. 1999. Input, storage and distribution of large woody debris along a mountain river continuum, the Drôme River, France. Catena 35:19–39.

Piégay, H. 2003. Dynamics of wood in large rivers. Pages 109–133 *in* S. V. Gregory, K. L. Boyer, and A. M. Gurnell, editors. The ecology and management of wood in world rivers. American Fisheries Society, Symposium 37, Bethesda, Maryland

Pollock, M. M., M. Heim, and D. Werner. 2003. Hydrologic and geomorphic effects of beaver dams and their influence on fishes. Pages 213–233 *in* S. V. Gregory, K. L. Boyer, and A. M. Gurnell, editors. The ecology and management of wood in world rivers. American Fisheries Society, Symposium 37, Bethesda, Maryland.

Rainville, R. P., S. C. Rainville, and E. L. Lider. 1985. Riparian silvicultural strategies for fish habitat emphasis. Pages 186–196 *in* Proceedings of 1985 Society of American Foresters National Convention. Fort Collins, Colorado.

Reich, M., R. Wildman, and J. L. Kershner. 2003. Restoring streams with large wood: a synthesis. Pages 355–366 *in* S. V. Gregory, K. L. Boyer, and A. M. Gurnell, editors. The ecology and management of wood in world rivers. American Fisheries Society, Symposium 37, Bethesda, Maryland.

Sedell, J. R., and J. L. Froggatt. 1984. Importance of streamside forests to large rivers: The isolation of the Willamette River, Oregon, U.S.A., from its floodplain by snagging and streamside forest removal. Verhandlungen Internationale Vereinigung für Theoretische und Angewandte Limnologie 22:1828–1834.

Scott, M. L., G. T. Auble, and J. M. Friedman. 1997. Flood dependency of cottonwood establishment along the Missouri River, Montana, USA. Ecological Applications 7:677–690.

Shafroth, P. B., G. T. Auble, J. C. Stromberg, and D. T. Patten. 1998. Establishment of woody riparian vegetation in relation to annual patterns of streamflow, Bill Williams River, Arizona. Wetlands 18:577–590.

Smith, N. J. 1987. Stand density control diagram for Western Redcedar, *Thuja plicata*. Forest Ecology and Management 27:235–244.

Sobota, D. J. 2003. Fall directions and breakage of riparian trees along streams of the Pacific Northwest. M. S. Thesis. Oregon State University. Corvallis.

Steel, E. A., W. H. Richards, and K. A. Kelsey. 2003. Wood and wildlife: benefits of river wood to terrestrial and aquatic vertebrates. Pages 235–247 in S. V. Gregory, K. L. Boyer, and A. M. Gurnell, editors. The ecology and management of wood in world rivers. American Fisheries Society, Symposium 37, Bethesda, Maryland.

Stromberg, J. C., S. D. Wilkens, and J. A. Tress. 1993. Vegetation-hydrology models as management tools for velvet mesquite (*Prosopis velutina*) riparian ecosystems. Ecological Applications 3:307–314.

Stromberg, J. 2001. Restoration of riparian vegetation in the south-western United States: importance of flow regimes and fluvial dynamism. Journal of Arid Environments 49:17–34.

Swanson, F. J. 2003. Wood in rivers: a landscape perspective. Pages 299–313 *in* S. V. Gregory, K. L. Boyer, and A. M. Gurnell, editors. The ecology and management of wood in world rivers. American Fisheries Society, Symposium 37, Bethesda, Maryland.

Swanson, F. J., and G. W. Lienkaemper. 1982. Interactions among fluvial processes, forest vegetation, and aquatic ecosystems, South Fork Hoh River, Olympic National Park. Pages 30–34 *in* E. E. Starkey, J. F. Franklin and J. W. Matthews, editors. Ecological research in national parks of the Pacific Northwest. Oregon State University Press, Corvallis.

Tabacchi, E., A. M. Planty-Tabacchi, and H. Décamps. 1990. Continuity and discontinuity of the riparian vegetation along a fluvial corridor. Landscape Ecology 5:9–20.

Tabacchi, E., A. Planty-Tabacchi, M. J. Salinas, and H. Décamps. 1996. Landscape structure and diversity in riparian plant communities: a longitudinal comparative study. Regulated Rivers: Research and Management 12:367–390.

Townsend, C. R. 1996. Concepts in river ecology: pattern and process in the catchment hierarchy. Archiv Fur Hydrobiologie Supplement 113:4–21.

van der Nat, D., K. Tockner, P. J. Edwards, and J. V. Ward. 2003. Large wood dynamics of complex Alpine river floodplains. Journal of the North American Benthological Society 22:35–50.

Van Sickle, J., and S. V. Gregory. 1990. Modeling inputs of large woody debris to streams from falling trees. Journal of Forest Research 20:1593–1601.

Volk, C. J., P. M. Kiffney, and R. L. Edmonds. 2003. Role of riparian red alder in the nutrient dynamics of coastal streams of the Olympic Peninsula, Washington, USA. Pages 213–225 *in* J. Stockner, editor. Nutrients in salmonid ecosystems: sustaining production and biodiversity. American Fisheries Society, Symposium 34, Bethesda, Maryland.

Ward, J. V., K. Tockner, and F. Schiemer. 1999. Biodiversity of floodplain river ecosystems: ecotones and connectivity. Regulated Rivers: Research and Management 15:125–139.

Welcomme, R. 1979. Fisheries ecology of floodplain rivers. Longman, New York.

Welty, J. J., T. Beechie, K. Sullivan, D. M. Hyink, R. E. Bilby, C. Andrus, and G. Pess. 2002. Riparian Aquatic Interaction Simulator (RAIS): a model of riparian forest dynamics for the generation of large woody debris and shade. Forest Ecology and Management 162:299–318.

Wissmar, R. C., and R. L. Beschta. 1998. Restoration and management of riparian ecosystems: a catchment perspective. Freshwater Biology 40:571–585.

Wondzell, S. M., and P. A. Bisson. 2003. Influence of wood on aquatic biodiversity. Pages 249–263 *in* S. V. Gregory, K. L. Boyer, and A. M. Gurnell, editors. The ecology and management of wood in world rivers. American Fisheries Society, Symposium 37, Bethesda, Maryland.

Wykoff, W. R., N. L. Crookston, and A. R. Stage. 1982. User's guide to the Stand Prognosis Model {FVS}. USDA, Forest Service, Intermountain Research Station, General Technical Report INT-133, Ogden, Utah.

Zalewski, M., M. Lapinska, and P. B. Bayley. 2003. Fish relationships with wood in large rivers. Pages 195–211 *in* S. V. Gregory, K. L. Boyer, and A. M. Gurnell, editors. The ecology and management of wood in world rivers. American Fisheries Society, Symposium 37, Bethesda, Maryland.

Zeide, B. 2001. Thinning and growth. Journal of Forestry 99:20–25.

American Fisheries Society Symposium 37:421–431, 2003

Rivers and Wood: A Human Dimension

GEOFF PETTS

University of Birmingham, Edgbaston, Birmingham B15 2TT, UK

ROBIN WELCOMME

Department of Environmental Science and Technology, Imperial College London, London SW7 2AZ, UK

Abstract.—Water and wood management are considered as the bases for early civilizations. Rivers were—and are—vital for civilizations, especially for irrigation but increasingly for urban and industrial water supplies, navigation, and power. Wood presents both benefits and hazards for human societies today as in the past. But the literature provides only sparse evidence to confirm human responses to wood in rivers. It is postulated that the removal of large wood from rivers dramatically affected fluvial systems and early societies; yet researchers of human anthropology and environmental change have generally overlooked these effects. From the Neolithic period onwards, the fear of dense, dark woodlands, the need for fuel, the hazards of flood and fire, and the need for rivers to be developed as routeways may explain the probable loss of wood from rivers. In Europe, this may have happened as early as 4,000 years ago. Today, except where a wilderness culture can be advanced, modern societies are still motivated by the same drive for a clean, tidy, and safe environment. Major institutions have shown little interest in the roles of wood in rivers, but opportunities for advancing a new understanding of these roles may arise through developing biodiversity protocols and restoration programs. If the issue of wood in streams and rivers is to expand from its present heartland in the northwestern United States, those responsible for resource programs in the various agencies need to be made more aware of the ecological significance in the management of their rivers and streams.

Introduction

In an age of general preoccupation with the global environment, the rapid expansion in international efforts to tackle the ecological problems is not surprising. Such efforts range from high-level consultations leading to international agreements to small-scale projects on particular aspects of the problem. The effects are as much conditioned by political perceptions of the global ecosystems as by any focused scientific concerns. They also represent a network of activities that have great potential to improve the effectiveness of conservation and restoration by raising awareness, giving an intergovernmental sanction to national activities and providing legal frameworks that constrain countries' individual actions.

The inland water ecosystems of the world generally are recognized as particularly at risk (Boon et al. 2000), and some agencies have developed programs to conserve and restore ecosystems. Such programs are usually extremely complex because of the growing demand for water for many human activities: irrigation for agriculture, power generation, waste disposal, domestic drinking water supply, navigation, fisheries, and flood control. All these activities affect the quality, quantity, and timing of water available in the system, the structure of the ecosystems, and the composition and abundance of aquatic organisms.

A growing body of evidence suggests that wood was once a common and perhaps dominant feature of streams and ecosystems throughout the forest regions of the planet (Gurnell and Petts 2002). Investigations from the Pacific Northwest, USA emphasize the abundance and ecological importance of wood in rivers in the 19th century, before the era of major human effects (Sedell and Froggatt 1984; Triska 1984, Maser and Sedell 1994). The effects of clearing wood from streams has received little attention. Today, watercourses in all developed regions are almost devoid of large wood (or driftwood), and modern societies often demand that obstructions be removed from all river channels. Much has been written about the development of modern societies, the taming of rivers, and the widespread effects of deforestation.

This paper reviews the relationship among society, rivers, and wood to evaluate the cultural basis for restoring large wood to streams and rivers. Our focus is the rest of the world beyond America's Pacific Northwest. It then examines briefly the roles of the various types of agencies that manage and conserve aquatic ecosystems and explores their present and possible future interest in wood in rivers. To evaluate the impact of wood on the natural history and human use of rivers, information was drawn from (1) a literature review of past relationships of rivers, wood, and society, and (2) a review of the activities of major international river-management agencies.

Rivers, Wood, and Society in History

A literature search was undertaken to seek information on the use, or perception, of wood in streams and rivers by indigenous peoples. The search used three resources: GEOREF database 1990–1998, University of Birmingham Library search, and Online Scientific Journal search. The aim was to find not only thematic works but also records of eyewitness accounts of aboriginal peoples and wood by early explorers. Thirty keywords were entered to the search criteria, including a variety of terms for natural wood (wood, driftwood, fire wood, woody debris, logs) and for artificial wood (fish traps, weirs) in rivers; archaeological and recent timescales; and references to indigenous populations in various places.

The search failed to find any substantive examples of interactions between people and wood in streams and few references to the value of wood along rivers to indigenous cultures. None of the texts on "environmental change" and "human impacts" (such as Goudie 1986, 1997; Chambers 1993; Butlin and Roberts 1995; and Roberts 1998) recognized the term "wood" or "woody debris" or its relevance to geomorphology, ecology, or anthropology. All of the scientific articles dealing with wood in rivers (such as Abbe and Montgomery 1996; Gippel et al. 1996; Braudrick et al. 1997; Piégay and Gurnell 1997; Thevenet et al. 1998; Hagan and Grove 1999) were modern examinations of measuring and managing wood or studies about removing woody debris by postindustrial societies (Warner 1991, 1995; Byers 1996; Gale et al. 1997).

The literature on the relationships between people (especially indigenous settlements) and streams or rivers is significant. Bishop et al. (1994) in describing the medieval Thai city on the banks of the Yom River do not mention wood accumulations nor wood buried in the thick floodplain deposits. Van Andel and Runnels (1995), in their discussion of ancient farmers in Europe, do not make reference to large wood in streams, nor does Denevan (1996) in describing prehistoric settlements in Amazonia. Some studies of sedimentary deposits have shown that buried logs in some environments often remain in situ for hundreds (Featherston et al. 1995) or even thousands (Salisbury et al. 1984) of years. Nevertheless, it is possible to piece together an evolutionary scenario despite the lack of direct information on relationships between human societies and large wood.

The hydraulic basis for civilizations

The foundation for the earliest civilizations has been proposed to be the efficient and equitable distribution of water, through advancing both technology and management systems. These highly structured "hydraulic civilizations" (Wittfogel 1956) developed along the Tigris, Euphrates, Nile, and Indus rivers and in China. By 4,000 years B.P., irrigation for agriculture, long-distance potable water supplies to urban centers, and elementary flood control were widespread. Even in more humid areas, water-engineering technology and advanced administration systems were introduced by the Romans (Binnie 1987). Water projects were seen as symbols of social advancement and technological prowess, and this was as much a feature of ancient Egypt as it was during the 20th century. Thus, the Sumerian civilization, depended for its survival not only on effective control but also on effective administration of the waters of the lower Tigris-Euphrates (Toynbee 1976). The first revolution in water technology followed the introduction of an indispensable new tool—script. Writing became necessary for organizing people, water, and soil in quantities too vast to be handled efficiently by the unrecorded memorizing of oral arrangements and instructions (Toynbee 1976, p. 51). Later, before the modern age of dam building, improved coordination and administration schemes for managing water and land had to be developed (Cosgrove and Petts 1990). Modern river management is founded in the observations of Leonardo da Vinci, who in his study of the Arno (1502–1503) provided the basis for linking hydrology and hydraulics (Newson 1992). The roots of Modernism are found in Europe in the late 16th and early 17th centuries, a period characterized by commercial expansion, rapid advances in hydraulic engineering, and a change from feudal to early capitalist systems of land evaluation. Thus, from about 1800, rivers across Europe were controlled (Petts et al. 1989).

The use of wood

An alternative view suggests that wood, as well as water, played important roles in developing civilizations. Perlin (1989) considers (p. 25) that "wood is the unsung hero of the technological revolution that has brought us from a stone and bone culture to our present age." Wood provided two needs for aboriginal peoples: building materials and fuel. Tools, houses, boats, wheels, and wagons were all made of wood. Wood as a fuel enabled peoples to inhabit cool climates, to cook grain making it edible, to extract metal from stone, to make salt from brine, to bake bread, to boil mixtures into useful products, and to make glass from potash and sand.

The history of deforestation to provide fuel, timber, and agricultural land has been well documented. Rackham (1986) describes the situation in England where Neolithic peoples cleared much land; by the early Iron Age, about 50% of the natural forests had been converted to farmland and heath. The decline in forest cover in England continued for centuries; in 1086, only 15% of the country was wooded. By 1350, it was reduced to 10%—a loss of 17.5 acres per day (Rackham 1986). The decline in forest cover in England was stopped by the Black Death of 1349 and the rise in value of wood from about 1550 with rising population, rising standards of domestic heating with the introduction of chimneys, and the onset of the Little Ice Age.

Dead trees with no felling cuts, with evidence of fluvial abrasion, and aligned to the paleaoflow direction have been reported from gravel excavations along the River Trent, UK (Salisbury et al. 1984). These include bog oaks found at five sites and elm and ash at another. Dendrochronology has dated the bog oaks at 3420–5335 years B.P.

Early peoples had quickly discovered the self-renewing power of trees and that regrowth shoots are more use than the original tree. Thus, from the early Neolithic period, underwood—consisting of coppice poles, young suckers, or pollard poles—was recognized as a more valuable product than timber. It is tempting to suggest that observations on the regrowth of pieces of wood along rivers, and the different forms of regrowth of different species (such as willow), might have inspired this discovery.

Wood resources and tamed rivers were key symbols of political power (Perlin 1989). The achievements of early civilizations were based on little scientific understanding but good organization. The scarcity of wood has triggered major technological revolutions, including transition from the Bronze Age to the Iron Age and the fossil-fuel era, substituting coal for wood as the principal fuel. Forest conquest was one motivating force behind colonialism and international agreements (Rackham 1986; Perlin 1989). Thus, the rich forests of Crete and Macedonia enabled these peoples to rise to prominence in the 2nd millennium B.C. and the 4th century B.C., respectively, after the Greeks had exhausted their own supplies of wood. The Roman Empire expanded west to Spain and Gaul to take advantage of the abundant forests to fuel their mining operations and industry.

Towards a civilized or managed environment

Although recent research has demonstrated the benefits of wood in rivers, illustrated not least by the papers in this volume, the view of wood in rivers as a hazard appears to have dominated in the past and remains so in the present. Clearing wood from streams was likely one of the first actions taken by early and aboriginal societies. The use of fire as a tool to exercise stewardship over country in northern Australia led Head (1996) to consider that the human desire to clean up the country is a universal ethic. Many societies saw the natural environment as a constraint on development, yield, or use. Early settlements developed along rivers and at the forest edge. Rivers provided routeways into the interior, but riparian and floodplain forests were major obstacles to passage.

Early stories from Mesopotamia, around 5,000 years B.P., describe native forests as dark, damp, and dangerous (Perlin 1989). Romans cleared routeways to counteract guerilla warfare, and in the UK, the Barnwell Chronicles of 1285 commanded clearance of underwood and bushes along roadways where men "may lurk to do evil," and about the same time, coppicing was encouraged to deny cover to wolves and criminals.

Wood in rivers could have been collected for fuel, and Australian aboriginals use flood debris from dry ephemeral streams for firewood (D. M. J. S. Bowman, personal communication). Similarly, Sauer (1972), reporting the original observations of Joseph Billings between 1785 and 1794 on an expedition to Siberia, refers to driftwood on the floodplain being used by natives for firewood. A narrative of the search for Lieutenant-Commander DeLong in the Lena Delta by a chief engineer in the U.S. navy, published in 1885, referred to driftwood collected near huts; driftwood littering the beach; and the use of driftwood for fires. Further (p. 325), Melville describes the Yakut mode of building camp fires: "a dry piece of wood is procured from the high riverbanks, many sticks being cut with the axe and rejected, until one entirely free from moisture and fit for kindling is found; which is then carefully split and kept dry. The best of the driftwood is next selected and also split up and chopped into proper lengths." Wood is likely to have been cleared to prevent natural fires as in drought years; wood fragments and wastewood can become so dry that it can act as kindling for the smallest spark (Perlin 1989). Wood also would be cleared to ease passage along the riverbank and for hauling boats. Sauer (1972) reported that some Siberian rivers were not passable by canoe because of fallen trees. Log floating would also require channel clearance, and this practice was certainly common by medieval times in Europe (Thirgood 1981).

Clearance of wood from rivers and streams to relieve hazards associated with driftwood is likely to have been an early activity to facilitate passage and to prevent damage to hydraulic structures. The bridge has been described as a symbol of human conquest of nature, and the river crossing is prominent in social, economic, and military history (de Mare 1954). Worldwide, folklore indicates that bridges had an important place in the lives of early communities and a strong ritualistic character. River gods were angered by the building of bridges that deprived them of their regular toll of drowning people and animals, and sacrifices were deliberately made during construction to appease the angry spirits (de Mare 1954). The earliest bridges had narrow spans between stone or wooden piers and would have been vulnerable to flood damage caused by the piling of large wood. The arch was invented by the Sumerians around 6,000 years B.P., and the Persians, around 2,500 B.P., were expert bridge builders (de Mare 1954). But the Romans were the greatest bridge builders in ancient history. The "control" of floating wood must have been a major concern for these early communities.

Break-up of wood jams during rare floods could have caused catastrophic changes to channel form through avulsion and erosion. Wood jams retain large amounts of sediments, especially in headwater streams, and jam break-up is likely to flush large volumes of sediment and debris along a river. Jams also pond flows, and the drainage of these ponds and their associated wetlands can sustain river flows during periods of drought. These lateral habitats can be valuable for a diversity of species, including fish, waterfowl, and mammals (Maser and Sedell 1994), providing important food resources for early human populations.

It is well known that aboriginal fishing often uses wooden weirs and fish traps, not least to catch eels (Williams 1988), and this practice was common in the UK until the modern era. Again, fish drives may have been one of the earliest uses of large wood jams. Certainly there is much evidence of indigenous peoples in groups driving fish into brush barricades and shallows where they were speared (Dortch 1997). Yost and Kelley (1983) refer to the wooden weirs used by the Waorani tribe in Amazonia. In Australia, aboriginals may spare flood debris in some creeks to preserve food resources within the streams (C.Doyle, Wollongong University, personal communication). Today, recreational fishermen also value large wood. For example, a recent article in the Australian fishing magazine *Modern Fishing* (December 2001) included a statement that clearly recognizes the value of wood accumulations: "The autumn floods have left us with some great new snags that, like good wine, are aging rather nicely."

Unfortunately, in most river basins, especially in the developed economies, wood is still seen as a hazard not only from flood management and navigation (including recreational boating) perspectives, but also from the viewpoint of public amenity. The land-water interface is highly regarded for its visual amenity, especially when characterized by smoothness, tidiness, and passiveness (Petts 1990). "Naturalness" is a key criterion for nature conservation in developed economies (Hellawell 1988; Boon et al. 2000). However, "naturalness" is often interpreted as visible attributes that indicate environmental conditions favoring survival of humans (Petts 1990). Dense, wooded riverbanks, with refuge and concealment from view and numerous entry and exit points, are often seen as dark, difficult, and dangerous and as unhealthy and untidy waste areas. In contrast, the "panoramic" designs of Capability Brown and Andre Le Notre in the 18th century are seen as highly desirable, prospect-orientated landscapes. Today, parkland landscapes with neat, well-organized patches, including open woodland and dominated by a high measure of prospect, are most valued. If rivers are to be restored, then the ecological benefits of dense wooded riverbanks, driftwood, and wood jams must be recognized. One way this might be achieved is through the institutions that promote river management and nature conservation.

Wood and Institutions in the 21st Century

Several types of international agencies are involved with inland water environments (Table 1). These include agencies involved with negotiating multilateral agreements and the collection, interpretation, and dissemination of international information. Intergovernmental agencies assemble representatives of national governments. They, therefore, have a mandate to reach international treaties and conventions. They are mostly grouped within the United Nations family of organizations, each of which is independent with its own charter and functions.

The United Nations family

The central, major coordinating mechanism within the UN system, the United Nations Secretariat, has contributed to international development efforts as a funding agency through the United Nations Development Program (UNDP), working mainly with specialized agencies such as the Food and Agriculture Organization (FAO) and UNESCO as executing agencies. More recently, the recipient countries themselves execute the UNDP programs. These are usually developmental in nature, concerned with increasing food supply and with building the capacity of member nations to gradually assume responsibility for their own development. As a major formulator of international policy, the United Nations strongly advocates management of rivers at the basin level and therefore has an interest in all mechanisms leading to the conservation and sustainable use of running waters.

The FAO has several concerns germane to the issues of wood in rivers. First, it is concerned with fish production systems and, as such, has been involved in the past with work on brush parks (Welcomme 2002). In particular, basic studies on fishing gear were carried out in the Benin lagoons in the 1960s (Welcomme 1979). Second, it is concerned about rehabilitating stream and riverine ecosystems, and it has incorporated methods for using wood in rehabilitation structures (Cowx and Welcomme 1998). This work has been pursued principally with one of its subsidiary bodies, the European Inland Fisheries Advisory Commission (EIFAC), which is continuing to explore methods for rehabilitating river and lake ecosystems. Third, during the earlier days of dam construction in Africa, a series of field projects responsible for pre- and post-impoundment studies on new reservoirs were much occupied with the need for and impact of deforestation of the area to be flooded.

The concern of UNESCO for inland waters is expressed partly through its Man and the Biosphere (MAB) program, which is concerned with the setting up of biosphere reserves, partly through the International Hydrological Program (IHP). The IHP aims principally at the rational utilization of water resources, including the protection of the environment. This program includes aspects of river management within its remit that find expression mainly within its project on ecohydrology (see Zalewski et al. 2003, this volume). The ecohydrology project is largely concerned with the relationships between the hydrological cycle and living aquatic resources. These are mainly expressed through changes in the ecotonal structure of floodplains and riparian zones. As such, the project has a legitimate interest in the role of large wood as part of ecosystem function in the river channels and floodplain water bodies. Some activities within the project have examined environmental aspects of large wood (Zalewski et al. 1994)

The UN Environment Program is explicitly concerned with all aspects of protecting and sustainably managing the environment. It does not execute projects itself but catalyzes activities by other UN agencies, and some non-UN organizations, on areas of environmental interest. UNEP has concentrated mainly on water quality issues in inland waters and has had no concern with large wood in rivers and streams.

TABLE 1. World institutions concerned with river management and their function.

	Role
UN Food and Agricultural Organization (FAO)	Concerned with sustainable food production: includes Lands and Waters Division of the Agriculture Department, occupied with the supply of irrigation water, and Fishery Department., concerned with all issues to do with fisheries including aquaculture
UN Educational, Scientific, and Cultural Organization (UNESCO)	More oriented to pure science than the more technologically oriented FAO; concerned with preserving the world's natural, social, and cultural heritage
The World Bank	A funding agency that finances major projects throughout the world; one of its highest profile sectors is its capacity to fund large dam and river training-projects with the need to carry out impact assessments as an integral part of project formulation
Economic-integrating organizations	Free associations of nations subscribing to a common market and agreeing to standardized economic policies to ensure that the market remains fair and functional; they can also adopt legislation that affects other aspects of national life, including the environment
River basin commissions	A totally different type of intergovernmental agency; commissions have been formed on many of the world's larger rivers and exist expressly to ensure administrative coordination within the basin between different riparian political entities, which may be at the scale of local administrative units in one country, federal states in a federated nation, or between states as in large international river systems
Nongovernmental organizations	Agencies operating at international scales but without the formal adhesion of governments; they often work more directly with academic institutions and nongovernmental organizations (NGOs), so their direct effects on legislation are therefore somewhat less, although they often have powerful lobbies whereby they influence policy

Economic-integrating organizations

The European Community has grown in size and complexity since it was first conceived as a vehicle for a common agricultural policy for western Europe. It is now heavily involved with conserving the European environment through a series of directives that are binding on the member states and have to be incorporated into national legislation. One of the most recent of these is the EC Water Framework Directive (2000/60/EC). This establishes a strategic framework for managing the aquatic environment, including chemical quality objectives and, for the first time, ecological quality objectives. It requires member countries to produce strategic management plans for each river basin, setting out how the objectives set for the water bodies within the basin are to be achieved. The EC has no explicit interest in wood in rivers, and in general, European countries have a statutory obligation to maintain their watercourses clear of obstructions.

River basin commissions

Normally, basin commissions concentrate on issues of major economic importance, such as allocation of water resources, water quality, and navigation. They rarely concern themselves effectively with living aquatic resources, including fisheries, nor do they consider the smaller tributaries, concentrating rather on the main stem and major tributaries of the river. The recent emphasis on conserving biological diversity and sustaining food supply at the basin level is forcing increased attention to the wider dimensions of the basin and to environmental issues. Thus, although, at present, no basin commission to date would address the question of wood in streams, other than to remove it to keep the navigation channel clear; future developments may well consider the structure and function of the smaller headwater streams in the system.

Nongovernmental agencies

Several agencies have a direct interest in the aquatic environment. As a leading independent environmental agency, the World Conservation Union (IUCN) frequently works closely with UNEP to assist countries in conserving the diversity of their landscapes and the species living in them. The IUCN has a Wetlands and Water Resources program that aims at ensuring the sustainability of biological diversity associated with aquatic systems. It is particularly concerned with the impacts of dams but is also interested in the broader issues of river basin conservation. To this end, it has formed partnerships with several academic institutions to develop the scientific basis for protecting aquatic ecosystems in the face of a variety of human impacts. It also has larger projects on major river systems, such as the Mekong (in association with the Mekong River Commission), the Rufigi, and the Zambezi.

The World Wide Fund for Nature (WWF) is another conservation agency that is emphasizing river systems through its living-waters campaign. This campaign aims at managing rivers better by recognizing the vital interdependence of land and water ecosystems. It also aims to increase wetland conservation areas and improve their management and use.

Ramsar is an agency formed as a secretariat for the 1971 Ramsar Convention. The convention aims at protecting wetlands and collaborates with participating countries to establish protected wetlands as Ramsar designated sites. There are now 124 contracting parties with 1,073 protected wetlands of international importance covering some 81 million ha. Originally, the scope of the designated sites was to protect areas valuable to water birds, but subsequently, other categories of aquatic organisms, especially fish, can be used to justify the designation of a particular wetland. Ramsar acknowledges the need for planning at a river basin scale, which involves integration of water management and wetland conservation. The seventh conference of the parties to Ramsar adopted guidelines for integrating wetland conservation and wise use into river basin management. These do not mention wood or woody debris.

The review of the major international players in the field of river and wetland conservation showed that few have any expressed interest in wood in the aquatic environment. Some, such as FAO, have peripheral interest and experience where some aspect of wood or woody debris affects their regular work. The lack of interest arises mainly because there is little awareness of the issue of wood in rivers and streams, as expressed in the north-temperate literature, because priorities for research have concentrated on the major food fisheries of larger rivers and lakes. In addition, the problems of the north-temperate salmon are not seen as relevant to the hundreds of species found in tropical systems, and considerable research is needed to extend knowledge of the responses of tropical fish communities to wood in systems.

River Management, Nature Conservation and Wood

The goal of most international agencies is to improve human existence by assuring sustainable food supplies. This creates a fundamental difference between the goals and strategies of the tropical world, which places great emphasis on food production and economic improvement and those of the temperate world, which is increasingly preoccupied with conserving and rehabilitating wetland habitats. The growing pressure on the water resources of the world that is leading to growing scarcity in water supply amplifies these concerns. Even though many countries in the developing tropical world have acquired obligations to conserve and protect their biological and ecosystem diversity, these obligations are, in most cases, secondary to the expansion of the food producing, urban, and industrial sectors.

The preoccupation with increasing food production concentrates research and development on large ecosystems where significant quantities of fish may be produced in any one area. Generally, smaller streams have remained uninvestigated despite the fact that cumulatively they probably contribute as much fish to the subsistence economy as the major fisheries do to the market. (Welcomme 1976)

Because of the low research input and the almost negligible individual contribution of small water bodies to fisheries, there is little understanding of the ecology and functioning of such systems, although it is precisely here that much of the wood is found. However, attitudes of national authorities and international agencies that serve them are changing rapidly. The deepening concern about an approaching crisis in the availability of freshwater in many parts of the world and a growing concern over the fragility of the existing wetland areas and their biodiversity are leading to a certain consensus among some policy makers that aquatic ecosystems need to be addressed holistically. Commonly, those who wish for economic development, especially in the poorer countries, continue to offset this lobby to a major degree.

The message of the 1992 UN Conference on the Environment has now been assimilated by most countries in the world, and their general adherence to the Convention on Biological Diversity provides a framework for individual and collective national efforts for conservation of species and ecosystems. Unfortunately, the knowledge needed to adequately conserve some aquatic ecosystems is lacking, particularly for the small streams in the tropics. Basin planning has emerged as the primary approach to conserving aquatic environments. This is now being advocated by the UN system as a whole, supported by the conservation agencies, as a framework for all countries. The EC has taken a step further in making this a legal obligation for its members, and, by association, eastern European aspirants to join the Community are adopting the same approach. Whether planning efforts within the river basin framework will concentrate on the higher order rivers of immediate economic importance, or will extend their plans to the small headwater streams, remains to be seen. If so, then the issues of large wood in streams and rivers may well enter into the arguments. In any case, the ecological value of wood in streams and rivers should become one element of decisions concerning forest restoration and management.

Opportunities for change

With a change of focus from development of natural resources to issues of governance for their sustainable management, organizations may become more open to issues of functional-ecosystem management. In this context, large wood may come to be seen by organizations such as FAO as a tool for river restoration and fisheries management. Economic-integrating organizations such as the EEC may make provisions for conserving riparian woods and wood in some water courses. Possible interests in wood in rivers could arise under the headings of biodiversity loss and degradation of international waters. One opportunity may be through the Global Environment Facility (GEF), a partnership between the UNDP, UNEP, and the World Bank, which aims to assist member countries of the UN system to improve aspects of their environment. The Bank has funded several large-scale rehabilitation projects on major rivers, the Danube for example. It is, however, not concerned with smaller streams where wood is likely to be an issue.

The Convention on Biological Diversity (CBD) was formulated as a co-lateral activity to the Rio Conference and open for signature at the summit in 1992. It is the only international convention on the environment that is legally binding on its members. It has been signed and ratified by 183 countries. For the purposes of the convention, biological diversity means the variability among living organisms from all sources including, inter alia, terrestrial, marine, and other aquatic ecosystems and the ecological complexes of which they are part, including diversity within species, between species, and of ecosystems. Inland-waters biodiversity was adopted as a CBD thematic area at the fourth meeting of the Conference of the Parties held in Bratislava. It now forms a program of the convention that works on biological diversity of inland-water ecosystems: it promotes watershed management; appropriate technologies; research, monitoring and assessment; and cooperation with other conventions and organizations through Joint Work Plans.

While the convention has no explicit interest in wood in streams, it is concerned with the structure and diversity of ecosystems and thus could well consider that maintaining structural diversity, such as that created by wood in rivers and streams, is a worthwhile conservation target. If it can be established that certain species, particularly in tropical waters, are totally dependent on large wood in the water, it should be a national goal to preserve such habitats within the general structure of conservation reserves.

Apart from these more general views of the treatment of wood in inland aquatic systems five particular themes appear to be important.

(1) Conservation. The role of wood in conserving aquatic biodiversity needs to be clarified.

The relationship between wood in streams and salmonids has been widely investigated in North America and, to a lesser extent, in Europe. The situation with the coarse fish populations of lowland rivers is less clear, as a couple of millennia of clearance of woody obstructions for rivers in Europe and more recent concentrated efforts in North America provide little idea how the original ecosystems functioned. The situation is even more difficult in the tropics, although there are indications from observations, such as those of Lowe-McConnell (1977), to suggest that certain species depend heavily on wood for microhabitat and refuge. It is also well established that many species rely heavily on flooded forests in equatorial and tropical rivers for refuge, breeding, and feeding areas during the floods. More studies are needed to clarify these issues in support of biodiversity protection programs.

(2) Restoration and rehabilitation. The use of wood and wood accumulations in the restoration and partial rehabilitation of damaged riverine and wetland ecosystems arises from its possible role in species conservation. It is a widely expressed principle when building structures, such as weirs, dams, and deflectors, and in the protection and reconstruction of riparian areas that natural materials such as wood be used wherever possible. However, it is difficult under the present obligations to keep channels clear, for authorities undertaking rehabilitation projects to create artificial log dams. Here, it would be essential to establish that there are good scientific grounds for conservation and ecosystem functioning that would justify a change in regulations, at least in individual, local cases.

(3) Clearing of wood from reservoir sites. In the early days of reservoir construction in the 1960s and 1970s, there was much debate as to whether or not the vegetation should be removed from areas to be flooded by new dams (see, for example, Obeng 1968) In some cases, limited areas were deforested in the area to be flooded by major dams such as Lake Volta, and some of the smaller dams were cleared. The practice was generally abandoned in all but the smallest dams for economic reasons. However, there have been many repercussions of this practice for fisheries, water quality, and human health in dams in the equatorial forest of Suriname (Lake Brakopondo) and the Amazon (Balbinas and Tucurui). The dynamics of wood in the flooded reservoirs is little known, so the effectiveness of deforestation in new reservoirs is still difficult to evaluate.

(4) Brush park fisheries. The use of wood in brush parks for trapping and culturing fish is widespread in the tropics (Welcomme 2002). This method of fishing gives high yields of fish in exchange for large quantities of branches. Areas where the method is common have become deforested, and it has been necessary, in some cases, to create plantations of the right kinds of wood to support the fisheries. The method is still poorly studied, particularly with regard to the most efficient use of wood. Continuing interest in brush-park fisheries in many parts of the tropics justifies further study.

(5) Harvest reserves. As an extension of the brush-park principle, some countries are now creating reserves within rivers, floodplain wetlands, and lakes as sites to protect part of the stock of fish. In many tropical systems, the most appropriate type of environment is the wooded floodplain or riparian margin. At present, there have been only a few theoretical studies to try to establish the appropriate size of a harvest reserve relative to the main water body.

Conclusion

Today, even in the developed economies where ecosystem management and collaborative decision making is emerging to underpin environmental restoration and rehabilitation policies, wood in rivers is rarely considered as a component of such schemes outside wilderness areas. It is apparent that there is a range of topics related to wood in streams and rivers that could benefit by better understanding of the processes involved. If the issue of wood is to expand from its present heartland in the northwestern United States, it would appear essential that those responsible for programming in the various agencies are made more aware of the possible significance of this topic to their work. Research programs on ecosystems that better define the role of wood in these systems should be developed as a result of this improved understanding. In this way, improved understanding of the role of wood in aquatic ecosystems could strengthen efforts of conservation and river basin planning especially in lower-order streams.

Acknowledgments

We are grateful to Chris Doyle for his efforts in undertaking the literature searches.

References

Abbe, D. B., and D. R. Montgomery. 1996. Large woody debris jams, channel hydraulics and habitat formation in large rivers. Regulated Rivers: Research and Management 12:201–221.

Binnie, G. M. 1987. Early dam builders in Britain. Thomas Telford, London.

Bishop, P., D. Hein, M. Barbetti, and T. Sutthinet. 1994. Twelve centuries of occupation of a riverbank setting: Old Sisatchanalai, northern Thailand. Antiquity 68(261):745–757.

Boon, P. J., B. R. Davies, and G. E. Petts, editors. 2000. Global perspectives on river conservation. Wiley, Chichester, UK.

Braudrick, C. A., G. E. Grant, Y. Ishikawa, and H. Ikeda. 1997. Dynamics of wood transport in streams: a flume experiment. Earth Surface Processes and Landforms 22:669–683.

Butlin, R. A., and N. Roberts, editors. 1995. Ecological relations in historic times: human impact and adaptations. Blackwell Scientific Publications, Oxford, UK.

Byers, A. C. 1996. Historical and contemporary human disturbance in the upper Barun Valley, Makalu-Barun National Park and Conservation Area, East Nepal. Mountain Research and Development 16(3):235–247.

Chambers, F. M., editor. 1993. Climate change and human impact on the landscape: studies in palaeoecology and archaeology, Chapman and Hall, London.

Cosgrove, D. E., and G. E. Petts, editors. 1990. Water, engineering and landscape. Belhaven, London.

Cowx, I. G., and R. Welcomme, editors. 1998. Rehabilitation of rivers for fish. Food and Agricultural Organisation of the United Nations, Rome.

de Mare, E. 1954. The bridges of Britain. B. T. Batsford, London.

Denevan, W. M. 1996. A bluff model of riverine settlement in prehistoric Amazonia. Annals-Association of American Geographers 86(4):654–681.

Dortch, C. E. 1997. New perceptions of the chronology and development of aboriginal fishing in south-western Australia. World Archaeology 29(1):15–35.

Featherston, K. L., R. J. Naiman, and R. E. Bilby. 1995. Large woody debris, physical processes and riparian forests development in montane river networks of the Pacific Northwest: Proceedings of the 26th Binghampton Symposium in Geomorphology. Geomorphology 13:133–144.

Gale, S. J., R. J. Howarth, and P. C. Pisanau. 1997. Human impact on the natural environment in early colonial Australia. Australian Geographer 28:23–35.

Gippel, C. J., I. C. O'Neil, B. L. Finlayson, and I. Schnatz. 1996. Hydraulic guidelines for the reintroduction and management of large woody debris in lowland rivers. Regulated Rivers: Research and Management 12:223–236.

Goudie, A. 1986. Human impact on the natural environment. Blackwell Scientific Publications Publishers, London.

Goudie, A., editor, 1997. The human impact reader. Blackwell Scientific Publications Publishers, London.

Gurnell, A. M., and G. E. Petts. 2002. Island-dominated landscapes of large floodplain rivers, a European perspective. Freshwater Biology 47(4):581–600.

Hagan, J. M., and S. L. Grove. 1999. Coarse woody debris. Journal of Forestry 97(1):6–11.

Head, L. 1996. Rethinking the prehistory of hunter-gatherers, fire and vegetation change in northern Australia. The Holocene 6:481–487.

Hellawell, J. M. 1988. River regulation and nature conservation. Regulated Rivers 2:425–444.

Lowe-McConnell, R. H. 1977. Ecology of fishes in tropical waters. Institute of Biology Studies in Biology 76.

Maser, C., and J. R. Sedell. 1994. From the forest to the sea: the ecology of wood in streams, rivers, estuaries, and oceans. St. Lucie Press, Florida.

Melville, G. W. 1885. In the Lena Delta. Longmans, Green and Co., London.

Newson, M. D. 1992. Land, water and development. Sustainable management of river basin systems. Routledge, London.

Obeng, L. 1968. Man-made lakes: the Accra symposium. University Press, Accra, Ghana.

Perlin, J. 1989. A forest journey. Harvard University Press, Cambridge, Massachusetts.

Petts, G. E., H. Moller, and A. L. Roux, editors. 1989. Historical change of large alluvial rivers: western Europe. Wiley, Chichester, UK.

Petts, G. E. 1990. The role of ecotones in aquatic landscape management. Pages 227–262 *in* R. J. Naiman and H. Decamps, editors. The ecology and management of aquatic-terrestrial ecotones. UNESCO, Paris and Parthenon, Carnforth, UK

Piégay, H., and A. M. Gurnell. 1997. Large woody debris and river geomorphological pattern; examples from S. E. France and S. England. Geomorphology 19(1–2):96–116.

Rackham, O. 1986. The history of the countryside. Phoenix Giant, London.

Roberts, N. 1998. The holocene; an environmental history. Blackwell Scientific Publications Publishers, Oxford, UK.

Salisbury, C. R., P. J. Whitley, C. D. Litton, and J. L.

Fox. 1984. Flandrian courses of the River Trent at Colwick, Nottingham. Mercian Geologist 9(4):189–207.

Sauer, M. 1972. Expedition to the northern parts of Russia. The Richmond Publishing Co. Ltd, Richmond, UK.

Sedell, J. R., and J. L. Froggatt. 1984. Importance of streamside forests to large rivers: the isolation of the Willamette River, Oregon, U.S.A., from its floodplain by snagging and streamside forest removal. Verhandlungen Internationale Vereinigung für Theoretische und Angewandte Limnologie 22:1828–1834.

Thevenet, A., A. Citterio, and H. Piégay. 1998. A new methodology of large woody debris accumulations on highly modified rivers. (Example of two French piedmont rivers). Regulated Rivers: Research and Management 14:467–483.

Thirgood, J. V. 1981. Man and the Mediterranean Forest: a history of resource depletion, Academic Press, London.

Toynbee, A. 1976. Mankind and Mother Earth. Oxford University Press, Oxford, UK.

Triska, F. J. 1984. Role of woody debris in modifying channel geometry and riparian areas of a large lowland river under pristine conditions: a historical case study. Verhandlungen Internationale vereiningung für Theorestische und Angewandte Limnologie 22:1876–1892.

Van Andel, T. H., and C. N. Runnels. 1995. The earliest farmers in Europe. Antiquity 69(264):481–500.

Warner, R. F. 1991. Impacts of environmental degradation on rivers, with some examples from the Hawkesbury-Nepean system. Australian Geographer 22:1–13.

Warner, R. F. 1995. Human impacts on Australian rivers. Australian Geographical Studies 33:3–5.

Welcomme, R. L. 1976. Some general and theoretical considerations on the fish yield of African rivers. Journal of Fisheries Biology 8:351–364.

Welcomme, R. L. 1979. Fisheries ecology of floodplain rivers. Longman, London.

Welcomme, R. L. 2002. An evaluation of tropical brush and vegetation park fisheries. Fisheries Management and Ecology 9:175–188.

Williams, E. 1988. Complex hunter-gatherers: late Holocene example from temperate Australia. BAR International Series 423.

Wittfogel, K. A. 1956. The hydraulic civilisations. Pages 152–164 *in* W. L. Thomas, editor. Man's role in changing the face of the earth. University of Chicago Press, Chicago, Illinois.

Yost, J. A., and M. Kelley. 1983. 'Shotguns, blowguns, and spears: the analysis of technological efficiency'. Pages 189–224 *in* R. B. Hames and W. T. Vickers, editors. Adaptive responses of native Amazonians. Academic Press, New York.

Zalewski, M., W. Puchelski, P. Frankiewiez, and B. Bis. 1994. Riparian ecotones and fish commentaries in rivers – intermediate complexity hypothesis. Pages 152–160 *in* I. F. Curse, editors. Rehabilitation of freshwater fisheries. Fishing News Book, Oxford, UK.

Zalewski, M., M. Lapinska and P. B. Bayley. 2003. Fish relationships with wood in large rivers. Pages 195–211 *in* S. V. Gregory, K. L. Boyer, and A. M. Gurnell, editors. The ecology and management of wood in world rivers. American Fisheries Society, Symposium 37, Bethesda, Maryland.